Communications in Computer and Information Science 2780

Series Editors

Rationale

The CCIS series is devoted to the publication of proceedings of computer science conferences. Its aim is to efficiently disseminate original research results in informatics in printed and electronic form. While the focus is on publication of peer-reviewed full papers presenting mature work, inclusion of reviewed short papers reporting on work in progress is welcome, too. Besides globally relevant meetings with internationally representative program committees guaranteeing a strict peer-reviewing and paper selection process, conferences run by societies or of high regional or national relevance are also considered for publication.

Topics

The topical scope of CCIS spans the entire spectrum of informatics ranging from foundational topics in the theory of computing to information and communications science and technology and a broad variety of interdisciplinary application fields.

Information for Volume Editors and Authors

Publication in CCIS is free of charge. No royalties are paid, however, we offer registered conference participants temporary free access to the online version of the conference proceedings on SpringerLink (http://link.springer.com) by means of an http referrer from the conference website and/or a number of complimentary printed copies, as specified in the official acceptance email of the event.

CCIS proceedings can be published in time for distribution at conferences or as post-proceedings, and delivered in the form of printed books and/or electronically as USBs and/or e-content licenses for accessing proceedings at SpringerLink. Furthermore, CCIS proceedings are included in the CCIS electronic book series hosted in the SpringerLink digital library at http://link.springer.com/bookseries/7899. Conferences publishing in CCIS are allowed to use Online Conference Service (OCS) for managing the whole proceedings lifecycle (from submission and reviewing to preparing for publication) free of charge.

Publication process

The language of publication is exclusively English. Authors publishing in CCIS have to sign the Springer CCIS copyright transfer form, however, they are free to use their material published in CCIS for substantially changed, more elaborate subsequent publications elsewhere. For the preparation of the camera-ready papers/files, authors have to strictly adhere to the Springer CCIS Authors' Instructions and are strongly encouraged to use the CCIS LaTeX style files or templates.

Abstracting/Indexing

CCIS is abstracted/indexed in DBLP, Google Scholar, EI-Compendex, Mathematical Reviews, SCImago, Scopus. CCIS volumes are also submitted for the inclusion in ISI Proceedings.

How to start

To start the evaluation of your proposal for inclusion in the CCIS series, please send an e-mail to ccis@springer.com

Hongbin Ma · Jinhua She · Bin Xin · Yaping Dai
Editors

Advanced Computational Intelligence and Intelligent Informatics

9th International Workshop, IWACIII 2025
Zhuhai, China, October 31 – November 4, 2025
Proceedings, Part I

Springer

Editors
Hongbin Ma
Beijing Institute of Technology
Beijing, China

Jinhua She
Tokyo University of Technology
Tokyo, Japan

Bin Xin
Beijing Institute of Technology
Beijing, China

Yaping Dai
Beijing Institute of Technology (Zhuhai)
Zhuhai, China

ISSN 1865-0929 ISSN 1865-0937 (electronic)
Communications in Computer and Information Science
ISBN 978-981-95-6729-4 ISBN 978-981-95-6730-0 (eBook)
https://doi.org/10.1007/978-981-95-6730-0

This Springer imprint is published by the registered company Springer Nature Singapore Pte Ltd.
The registered company address is: 152 Beach Road, #21-01/04 Gateway East, Singapore 189721, Singapore

Preface

It is our great pleasure to present the Proceedings of the 9th International Workshop on Advanced Computational Intelligence and Intelligent Informatics (IWACIII 2025), a landmark academic gathering held in Zhuhai, China, from October 31 to November 3, 2025, with a pre-conference Sino-Canadian Bilateral Workshop organized in Beijing on October 29, 2025.

IWACIII 2025 brought together over 300 participants from academia, industry, and research institutions worldwide, serving as a dynamic platform for the exchange of ideas and collaboration in the rapidly evolving fields of computational intelligence and intelligent informatics. The conference agenda encompassed both theoretical innovations and practical applications, aiming to advance intelligent technologies that address real-world challenges and societal needs.

The conference featured 14 inspiring keynote speeches delivered by internationally renowned experts—including IEEE Fellows and editors of leading journals—who provided profound insights into diverse topics such as learning and feedback in uncertain systems, the evolution of machine learning paradigms, intelligent sensing, and next-generation intelligent control. In addition to the plenary sessions, IWACIII 2025 hosted 18 organized sessions (OS) and 15 general sessions (GS), offering a comprehensive view of emerging research directions and interdisciplinary developments. The AI Contest and AI Innovation Roadshow also provided a vibrant stage for young researchers and students to demonstrate creativity through innovative applications.

To reflect the depth and breadth of the research presented, the proceedings have been divided into four thematic volumes, each representing a key frontier in computational intelligence and intelligent informatics:

Volume 2780: Intelligent Algorithmic Theories and Optimization Methods
Volume 2781: Smart Industry and Power Systems
Volume 2782: Perception, Vision, and Intelligent Detection
Volume 2783: Unmanned Systems, Robotics, and Control

Each volume comprises carefully selected, peer-reviewed papers that capture the latest theoretical advancements, methodological innovations, and application-oriented breakthroughs. All submissions underwent a rigorous double-blind peer-review process managed by the IWACIII 2025 Technical Program Committee, with at least two independent reviewers assessing each paper for originality, technical quality, clarity, and relevance. Out of 393 submissions, 173 papers were accepted across the four volumes.

These four volumes collectively showcase the state of the art in computational intelligence—spanning intelligent optimization algorithms, industrial automation, perception and vision systems, and robotic control—demonstrating IWACIII's continuing commitment to fostering innovation, interdisciplinarity, and global collaboration in intelligent systems research.

We would like to express our deepest gratitude to all who contributed to the success of IWACIII 2025. Special thanks go to the keynote speakers, reviewers, session organizers, and members of the organizing committee, including representatives from the International Society of Computational Intelligence, the Beijing Society of Automation, and the Beijing Institute of Technology (Zhuhai). We also gratefully acknowledge the strong support of the Zhuhai Science and Technology Association and the Zhuhai Bureau of Science and Technology Innovation, whose assistance ensured the smooth execution of this event.

Finally, we extend our appreciation to all authors and participants, whose creativity and dedication formed the foundation of IWACIII 2025. We hope that these four volumes will serve as a valuable reference for researchers, practitioners, and students, inspiring further advancements and collaborations in the exciting field of computational intelligence and intelligent informatics.

October 2025

Hongbin Ma
Bin Xin
Jinhua She
Yaping Dai

Organization

Honorary Chairs

Kaoru Hirota	Beijing Institute of Technology, China Tokyo Institute of Technology, Japan
Meiling Wang	Beijing Institute of Technology (Zhuhai), China
Liqun Han	Beijing Technology and Business University, China

General Chairs

Hongbin Ma	Beijing Institute of Technology, China
Shinichi Yoshida	Kochi University of Technology, Japan
Xingguang Duan	Beijing Institute of Technology (Zhuhai), China

Advisory Board

Fuchun Sun	Tsinghua University, China
Weidong Hu	Beijing Institute of Technology, China
Toshio Fukuda	Meijo University, Japan
Naoyuki Kubota	Tokyo Metropolitan University, Japan

Program Committee Chairs

Bin Xin	Beijing Institute of Technology, China
Jinhua She	Tokyo University of Technology, Japan

Organized Session Chairs

Xiaoyan Zhao	University of Science and Technology Beijing, China
Takenori Obo	Tokyo Metropolitan University, Japan
Kewei Chen	Ningbo University, China
Fangyan Dong	Ningbo University, China

Publication Chairs

Yuan Xu	Beijing University of Chemical Technology, China
Sijie Yin	Beijing Institute of Technology, China

Local Organizing Chairs

Yaping Dai	Beijing Institute of Technology (Zhuhai), China
Yuan Li	Beijing Institute of Technology, China

Publicity Chairs

Eri Sato-Shimokawara	Tokyo Metropolitan University, Japan
Qing Wang	Beijing Institute of Technology, China

Registration Chairs

Weidong Zou	Beijing Institute of Technology (Zhuhai), China
Zhiyang Jia	Beijing Institute of Technology, China

Competition Chairs

Guiping Lu	Beijing Institute of Technology (Zhuhai), China
Minling Zhu	Beijing Information Science and Technology University, China

General Affairs

Fengmiao Chen	Beijing Institute of Technology (Zhuhai), China
Rongli Li	Beijing Institute of Technology, China
Genjing Chang	CSEDS-AIRE, China

Program Committee

A. Kecskemethy	University of Duisburg-Essen, Germany
A.M. Iliyasu	Salman Bin Abdulaziz University, Saudi Arabia
A. Ralescu	University of Cincinnati, USA
A. Shibata	Advanced Institute of Industrial Technology, Japan
Bin Xin	Beijing Institute of Technology, China
Bin Xu	Northwestern Polytechnical University, China
C.N. Nyirenda	University of Namibia, Namibia
Changhe Li	China University of Geosciences, Wuhan, China
Cheng Tan	Qufu Normal University, China
Chuan-Ke Zhang	China University of Geosciences, Wuhan, China
D. Kim	Mokpo National Maritime University, South Korea
D. Ralescu	University of Cincinnati, USA
Dawei Shi	China University of Geosciences, China
E.F. Fukushima	Tokyo University of Technology, Japan
E. Shimokawara	Tokyo Metropolitan University, Japan
Elmer Dadios	De La Salle University, Philippines
Fangyan Dong	Ningbo University, China
Fei Yan	Beijing Academy of Quantum Information Sciences, China
Feng Liu	Beijing Jiaotong University, China
Feng Pan	Beijing Institute of Technology, China
Gyunghoon Park	University of Seoul, South Korea
G. Vachkov	University of the South Pacific, Fiji
Gao Wang	Jinan University, China
Guohun Zhu	University of Queensland, Australia
H. Ishibuchi	Southern University of Science and Technology, China
H. Iyatomi	Hosei University, Japan
H. Masuta	Toyama Prefectural University, Japan
H. Nobuhara	University of Tsukuba, Japan
H. Ohtake	Kyushu Institute of Technology, Japan
H. Sakaniwa	Hitachi Central Research Lab, Japan
H. Seki	Osaka University, Japan
H. Takahashi	Shonan Institute of Technology, Japan
Haitao Li	Shandong Normal University, China
Hao Fang	Chinese Academy of Sciences, China
Haobin Dong	China University of Geosciences, Wuhan, China
Hongbin Ma	Beijing Institute of Technology, China

Hongsheng Qi	Beijing Institute of Technology, China
Hongtao Sun	Qufu Normal University, China
Hua Yu	Zhejiang University, China
Hui Chen	Beijing Institute of Technology, China
Ichiro Kobayashi	Ochanomizu University, Japan
I. Ono	Tokyo Institute of Technology, Japan
J. Kacprzyk	Polish Academy of Sciences, Poland
Joseph W. Spencer	University of Liverpool, UK
Jian Sun	Beijing Institute of Technology, China
Jianqiang Yi	Chinese Academy of Sciences, China
Jie Chen	Harbin Institute of Technology, China
Jing Wang	North China University of Technology, China
Jinhua She	Tokyo University of Technology, Japan
Jun-e Feng	Shandong University, China
Junming Wang	Beijing Institute of Technology, China
Junzheng Wang	Beijing Institute of Technology, China
Krzysztof Galkowski	University of Zielona Góra, Poland
K. Kawamoto	Chiba University, Japan
K. Kurashige	Muroran Institute of Technology, Japan
K. Nitta	Tokyo Institute of Technology, Japan
K. Okamoto	University of Electro-Communications, Japan
K. Tachibana	Kogakuin University, Japan
K. Uehara	Ibaraki University, Japan
K. Yoshida	Kyushu Institute of Technology, Japan
Katsushige	FujimotoFukushima University, Japan
Kevin Wong	Murdoch University, Australia
L.T. Kóczy	Szent István University, Hungary
Li Xu	Tongji University, China
Liqun Han	Beijing Technology and Business University, China
Luefen Chen	China University of Geosciences, Wuhan, China
M. Inuiguchi	Osaka University, Japan
Masahiro Moniwa	Tokyo University of Technology, Japan
Min Wu	China University of Geosciences, Wuhan, China
Minggang Gan	Beijing Institute of Technology, China
N.H. Phuong	Thang Long University,Vietnam
N. Ikoma	Nippon Institute of Technology, Japan
Ning Liu	Jinan University, China
Pingli Lu	Beijing Institute of Technology, China
Qing Wang	Beijing Institute of Technology, China
Qinghe Wu	Beijing Institute of Technology, China
Quan Pan	Northwestern Polytechnical University, China

Yifen Mu	University of Chinese Academy of Sciences, China
Yimin Sun	Zhongshan University, China
Yong He	China University of Geosciences,Wuhan, China
Yonghua Xiong	China University of Geosciences, Wuhan, China
Yongqiang Bai	Beijing Institute of Technology, China
Yuan Li	Beijing Institute of Technology, China
Yuan Xu	Beijing University of Chemical Technology, China
Yuanqing Xia	Beijing Institute of Technology, China
Yuqiang Wu	Xi'an Jiaotong University, China
Zhi Geng	Beijing Technology and Business University, China
Zhen Chen	Beijing Institute of Technology, China
Zhengqiang Zhang	Qufu Normal University, China
Zhentao Liu	China University of Geosciences, Wuhan, China
Zhijia Zhao	Guangzhou University, China
Zhiqiang Zuo	Tianjin University, China
Zhiwen Li	Southeast University, China
Zhiyang Jia	Beijing Institute of Technology, China
Zhuoyue Song	Beijing Institute of Technology, China

Additional Reviewers

Xianrui Bao
Ziyi Chai
Dan Chen
Gaohua Chen
Ge Chen
Hui Chen
Jingchuan Chen
Taoying Chen
Zhuxiang Chen
Junzhe Cheng
Lan Cheng
Zhang Chuang
Anjie Dai
Qiyuan Deng
Ao Ding
Junyi Ding
Long Ding
Yanchun Du
Yang Du
Huahao Fan
Ruize Fan
Yehao Fang
Jinfeng Gao
Zhipeng Gou
Hongge Guo
Xinxiang Guo
Zhentao Guo
Minglei Han
Jiang Hanlin
Fanzhu Hao
Yukinobu Hoshino
Yongjie Hou
Bin Hu
Hongjian Hu
Chi Huang
Jingtao Huang

Siying Huang
Zhichao Huang
Chuan Ji
Xianrui Ji
Xin Ji
Lingyang Jiang
Meng Jiang
Zhitao Jiang
Chengzhu Li
Guangxing Li
Jianzhen Li
Jinjin Li
Tong Li
Xiaoli Li
Xinheng Li
Yehui Li
Yisong Li
Rufu Lin
Bo Liu
Chenling Liu
Haoting Liu
Mingwei Liu
Xiaobo Liu
Yijian Liu
Yin Liu
Yitong Liu
Fei Long
Hangyu Lou
Youcheng Lou
Jianquan Lu
Kaineng Lu
Yulin Luo
Haijie Mao
Qicheng Mei
Yifen Mu
Ning Nan
Takenori Obo
Leyun Peng
Xiaolei Peng
Rongbin Qi
Zixi Qi
Jianing Qin
Chengzhi Ren
Xianyong Ruan
Eri Sato-Shimokawara
Fanglue Shan
Weiguo Shi
Yuyuan Shi
Fan Song
Ying Song
Pengpeng Su
Licheng Sun
Wei Sun
Tamanna Masuma Rifah
Huaibin Tang
Jiaxin Tang
Lili Tang
Weijie Tang
Yifan Tang
Chenxiao Wang
Haipeng Wang
Hanyuan Wang
Haoran Wang
Haozhu Wang
Miaomiao Wang
Tianhao Wang
Xiaoling Wang
Xinyang Wang
Yuyang Wang
Jiangxia Wei
Yanran Wei
Yi Wei
Xin Wen
Jian Wu
Keyu Wu
Qiong Wu
Ruoyan Wu
Xiangdong Wu
Xiaolan Wu
Xinwei Wu
Yaxi Wu
Zhouming Wu
Zongyang Wu
Benlong Xiang
Liping Xie
Xingwang Liu
Xiangdong Xu
Yingqi Xu
Yun Xu
Yumeng Yan

Wei Yang
Xiaohui Yang
Xiaoli Yang
Jiale Yi
Zhihao Yu
Hengyang Yuan
Xiangyu Zeng
Zhiyuan Zeng
Chuan Zhang
Chunmei Zhang
Han Zhang
Jingchuan Zhang
Junyue Zhang
Leyuan Zhang
Mingqing Zhang
Ruiyou Zhang
Shitong Zhang
Weili Zhang
Weiqiang Zhang
Wen Zhang
Wenjie Zhang
Xuanming Zhang
Ying Zhang
Yuyan Zhang
Zhaotong Zhang
Zhe Zhang
Zhiqiang Zhang
Yuan Zhao
Zhixin Zhao
Yifan Zheng
Zhi Zheng
Ding Zhong
Xiurong Zhong
Jin Zhou
Qingchun Zhou
Zhikai Zhou
Wei Zhu

Contents

Intelligent Decision-Making and Optimization

A Multi-operator Evolutionary Algorithm for the Multi-objective WTA Problem

Chengxin Wen[1,3], Fuhao Liu[1], Ting Wang[1(✉)], Hongbin Ma[2,3], and Yanhuan Jiang[4]

[1] Beijing Institute of Astronautical Systems Engineering, Beijing 100079, China
wangtingood@163.com

[2] School of Automation, Beijing Institute of Technology, Beijing 100081, China

[3] National Key Lab of Autonomous Intelligent Unmanned Systems, Beijing 100081, China

[4] Changcheng Institute of Metrology and Measurement, Beijing 100091, China

Abstract. The Weapon Target Assignment (WTA) problem is a fundamental optimization problem that seeks to determine the optimal allocation of weapons to targets to maximize specific performance indicators. This paper addresses the WTA problem from a multi-objective optimization perspective, incorporating three key objectives: target damage probability, strike balance, and strike benefit. To effectively tackle this problem, a multi-operator evolutionary algorithm is proposed, integrating various evolutionary operators to enhance both exploration and exploitation capabilities within the solution space. Furthermore, a dynamic adaptive strategy is introduced to adjust the selection of operators based on their performance during the optimization process. In the experimental evaluation, nine multi-objective WTA instances of varying scales are generated to assess the performance of the proposed algorithm. The results are compared against five state-of-the-art multi-objective optimization algorithms. Experimental findings demonstrate that the proposed algorithm exhibits strong competitiveness in solving the multi-objective WTA problem, outperforming or matching the performance of existing approaches.

Keywords: Weapon-Target Assignment · Multi-Objective Optimization · Evolutionary Algorithms

1 Introduction

The WTA problem is a classical optimization problem with broad applications in economics, game theory, military strategy, enterprise management, and other domains [1,2,6]. It focuses on the optimal allocation of limited weapon resources to multiple targets to maximize specific performance metrics or achieve predefined strategic objectives. Solving this problem involves addressing various constraints, including the heterogeneous characteristics of weapon systems and targets, differing target priorities, and strict operational constraints [14]. These complexities impose significant challenges on solution methodologies, necessitating the development of sophisticated optimization techniques [17].

H. Ma et al. (Eds.): IWACIII 2025, CCIS 2780, pp. 3–14, 2026.
https://doi.org/10.1007/978-981-95-6730-0_1

The simplest form of the WTA problem can be formulated as a combinatorial optimization problem, where the objective is to maximize the strategic benefits of a single allocation. However, in large-scale allocation scenarios, the problem becomes significantly more complex due to the vast number of weapon resources and targets involved, the interdependencies among targets, and the time-dependent nature of evaluation metrics. This complexity firmly classifies the WTA problem as an NP-hard problem [12], necessitating the development of more efficient algorithms to achieve effective and computationally feasible solutions.

Existing approaches for solving the WTA problem primarily employ algorithms designed for optimal allocation, which can be broadly categorized into exact methods and heuristic algorithms [13]. While exact methods guarantee optimal solutions for small- and medium-sized problem instances, their computational complexity grows exponentially with problem size, rendering them impractical for large-scale scenarios [5]. In contrast, heuristic methods offer more scalable alternatives but often sacrifice solution optimality. Moreover, as the number of targets increases significantly, the search space of the WTA problem expands exponentially [8], leading to a substantial decrease in the convergence speed of heuristic algorithms [7].

Furthermore, in real-world applications, multiple optimization criteria must be considered [18]. Solutions derived solely based on target damage values may not be applicable in complex battlefield environments [4]. To address these challenges, this paper proposes a multi-operator evolutionary algorithm for solving large-scale WTA problems. The proposed method leverages state-of-the-art optimization techniques and advanced computational strategies to achieve high-quality solutions within a feasible time frame. Specifically, the algorithm integrates multiple evolutionary operators to enhance both exploration and exploitation capabilities within the solution space. Additionally, a dynamic adaptive strategy is introduced to adjust operator selection based on their performance during the optimization process, ensuring a balanced and efficient search.

The rest of this paper is organized as follows: Sect. 2 introduces the mathematical model of the WTA problem studied in this paper. Section 3 presents the designed approach for solving the multi-target weapon target assignment problem. Section 4 evaluates the proposed algorithm through extensive experiments using realistic large-scale scenarios and compares its performance with contrasting algorithmic approaches. Finally, Sect. 5 concludes this paper and discusses potential directions for future research.

2 Problem Formulation

Consider a set $J = \{j_1, j_2, ...j_m\}$ consisting of m indivisible weapon resources and a set $T = \{t_1, t_2, ..., t_n\}$ comprising n targets to be assigned. Each available weapon resource can be allocated to only one target, whereas each target can receive multiple weapon resources. Both weapon resources and targets belong to multiple categories, and the effectiveness of assigning a specific weapon resource to a target varies depending on their respective categories. Let γ_{ij} denote the benefit obtained when weapon resource j_i is assigned to target t_j.

Define x_{ij} as the weapon resource assignment variable, where $x_{ij} = 1$ if weapon resource j_i is assigned to target t_j, and $x_{ij} = 0$ otherwise. For any target t_j, its associated value is denoted as k_j, while for any weapon resource j_i, the cost of its utilization is represented by m_i.

To provide a more comprehensive representation of the problem, a damage probability model is introduced . Given the assignment variable x_{ij} for each weapon resource, the possession probability p_j can be uniquely determined. The relationship between the possession probability and the weapon resource assignment variable is expressed in Eq 1, Where h_0 is the critical damage value of each target.

$$p_j = \max\{0, 1 - \frac{1}{\exp(\sum_{i=1}^{m} r_{ij}x_{ij} - h_0)}\} \tag{1}$$

The problem examined in this study is formulated as a multi-objective weapon target allocation problem, aiming to simultaneously optimize three minimization objectives. The detailed mathematical model is presented as follows:

$$\min F = \{f_1, f_2, f_3\} \tag{2}$$

$$f_1 = \sum_{j=1}^{n} \frac{1}{p_j} \tag{3}$$

$$f_2 = \sum_{i=1}^{m}\sum_{j=1}^{n} x_{ij} \tag{4}$$

$$f_3 = -\sum_{j=1}^{n} p_j k_j + \sum_{i=1}^{m} m_i x_{ij} \tag{5}$$

$$s.t. \quad \sum_{j=1}^{n} x_{ij} \leq 1, \forall i \in \{1, 2, ..., m\} \tag{6}$$

$$x_{ij} \in 0, 1, \forall i \in \{1, 2, ...m\}, \forall j \in \{1, 2, ..., n\} \tag{7}$$

Equation 2 defines the objective function, while Eq. 6 and Eq. 7 represents the set of constraints. Specifically, the Eq. 3 aims to maximize the total probability of target damage, the Eq. 4 seeks to minimize the consumption of weapon resources, and the Eq. 5 maximizes the operational benefits. Meanwhile, the Eq. 6 ensures that each weapon resource can be assigned to only one target, while the Eq. 7 defines the domain of the decision variable.

3 The Proposed Method

To effectively address the multi-objective WTA problem, this paper proposes a Multi-Objective Multi-Operator Evolutionary Algorithm (MOMOEA). The proposed algorithm adaptively adjusts the weights of multiple crossover operators during the search process, thereby enhancing search efficiency and improving solution quality.

3.1 General Framework

The pseudo-code for the MOMOEA algorithm framework is presented in Algorithm 1. Following the general structure of evolutionary algorithms, the MOMOEA algorithm begins by initializing the population and the weight matrix E associated with each operator. During each generation of evolution, the number of offspring produced by the three operators is determined based on the values in E. Subsequently, the parent and offspring populations are merged, and an environmental selection process is applied to retain high-quality solutions. Finally, the weights of the operators are adjusted dynamically based on the quality of the offspring generated by each operator, ensuring adaptive search optimization.

Algorithm 1: The framework of MOMOEA

Data: maximum number of function evaluations, FEmax, population size, N.
Result: final population, P

Initialize population P;
E = [1 1 1];
FE = N;
while $FE \leq FEmax$ **do**
  Q = OffspringGenerate(P,N,E);
  P = EnvironmentSelction($P \bigcup Q$);
  E = FactorAdjust(E);
  FE = FE + Q;
end

3.2 Solution Encoding and Offspring Generation

The allocation scheme in this paper employs a single-part gene expression method, wherein the allocation is encoded as a vector of length n, with each element representing the target ID to which the corresponding weapon resource is assigned. For instance, if the fifth element in the vector has a value of 12, it indicates that the weapon resource indexed as 5 is allocated to target 12.

For offspring generation, MOMOEA utilizes three crossover operators: single-point crossover, two-point crossover, and multi-point crossover. The detailed process of these crossover operations is illustrated in Fig. 1.

3.3 Adaptive Operator Selection Mechanism

MOMOEA incorporates an adaptive mechanism to dynamically adjust the number of offspring generated by different crossover operators. This mechanism consists of the following four steps:

1. **Evaluation of Offspring Proportions:** Over a span of ten generations, the proportion of offspring produced by each of the three crossover operators is computed based on environmental selection.

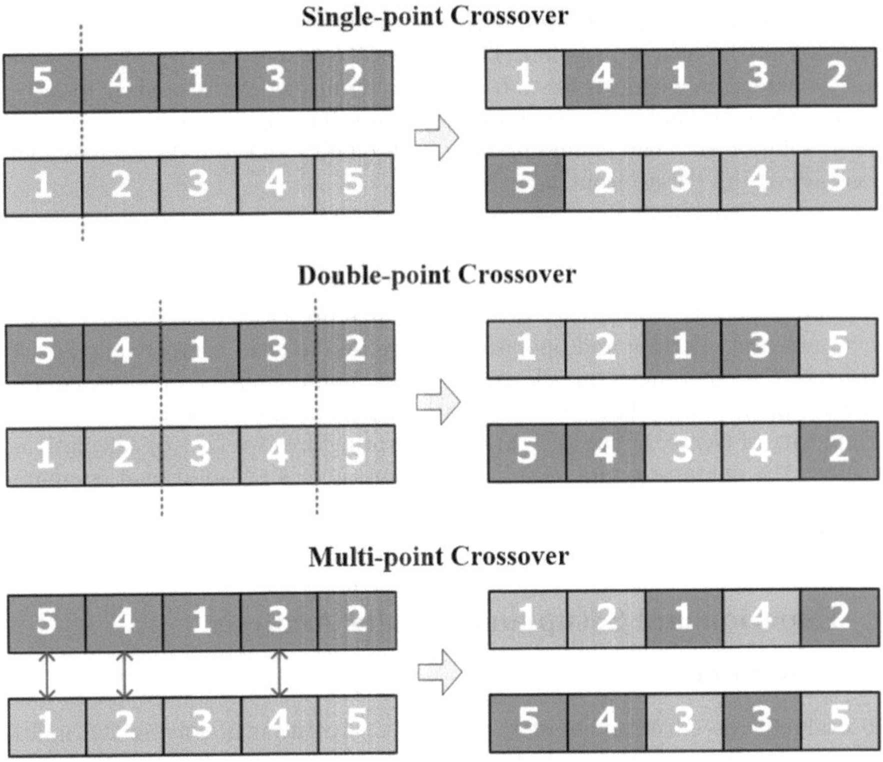

Fig. 1. Illustration of three crossover operators.

2. **Adjustment of Operator Factors**: The three operators are ranked based on their offspring proportions. The factor corresponding to the highest-proportion operator is increased by 0.1, while the factor of the lowest-proportion operator is decreased by 0.1.
3. **Boundary Constraints**: The adjusted factors are checked to ensure they remain within predefined bounds. If an operator factor exceeds 0.8, it is capped at 0.8; if it falls below 0.1, it is raised to 0.1.
4. **Normalization**: All operator factors are normalized to maintain a valid probability distribution.

After determining the final proportion factors for the three crossover operators, the number of offspring generated by each crossover operation is allocated based on these proportions, considering the overall population size.

3.4 Environmental Selection

The environmental selection mechanism in MOMOEA follows the same procedure as that in NSGA-II. Specifically, the selection process begins with a non-dominated sorting of the candidate population. Individuals are ranked based on their dominance relationships, with lower-rank solutions being prioritized for inclusion in the next generation. This process continues until the last feasible rank, where a crowding distance-based truncation method is applied to select the most diverse and well-distributed solutions. The final outcome is an optimized solution set that maintains both high-quality and well-distributed solutions across the Pareto front.

3.5 Complexity Analysis

The computational complexity of MOMOEA is primarily concentrated in the environmental selection and operator adaptive adjustment components. Specifically, in the environmental selection process, non-dominated sorting requires $O(mn^2)$ operations, while the computation of crowding distance also incurs a complexity of $O(mn^2)$. The generation of offspring involves $O(mn)$ calculations, whereas the operator adaptive adjustment introduces an additional computational cost of $O(mn^3)$. Consequently, the overall computational complexity of the MOMOEA algorithm is $O(mn^3)$.

4 Experimental Setup and Results Analysis

4.1 Test Instances

To evaluate the effectiveness of the proposed algorithm in solving the multi-objective WTA problem, nine test cases are designed with progressively increasing problem scales. The detailed parameters of these test cases are presented in Table 1.

Table 1. Instance parameter settings.

Problem	$\|J\|$	$\|T\|$	kw	kt	FEmax
Instance-1	50	20	[15,5,10,13,7]	[5 7 8]	20000
Instance-2	100	40	[19,25,25,15,16]	[11,13,16]	50000
Instance-3	200	80	[34,41,41,39,45]	[23,31,26]	80000
Instance-4	250	100	[43,49,59,61,38]	[26,44,30]	120000
Instance-5	350	150	[81,74,58,60,77]	[49,54,47]	150000
Instance-6	450	200	[91,83,84,110,82]	[70,71,59]	200000
Instance-7	600	250	[108,117,124,123,128]	[74,90,86]	250000
Instance-8	1000	300	[208,183,192,205,212]	[110,91,99]	300000
Instance-9	1500	400	[308,313,279,309,291]	[160,118,122]	400000

4.2 Evaluation Metrics

To comprehensively evaluate the performance of multi-objective optimization algorithms applied to the WTA problem, this study adopts hypervolume (HV) and inverted generational distance (IGD) as the primary evaluation metrics. These indicators quantitatively assess the convergence, distribution, and proximity of the obtained solution set to the true Pareto front. HV emphasizes a global assessment of solution quality, whereas IGD captures local convergence characteristics. Given the multi-modal nature of the WTA problem, the combined use of these two indicators mitigates the limitations inherent in each individual metric—such as the sensitivity of HV to reference point selection and the dependence of IGD on the density of sampled points along the Pareto front.

Table 2. Hypervolume normalization parameter settings for each instance.

Problem	Ideal point			Nadir point		
	f_1	f_2	f_3	f_1	f_2	f_3
Instance-1	3.52E-01	2.20E+01	6.57E-01	6.76E-01	5.00E+01	1.44E+00
Instance-2	3.43E-01	6.60E+01	9.45E-01	5.05E-01	1.00E+02	1.39E+00
Instance-3	2.92E-01	1.67E+02	1.06E+00	3.77E-01	2.00E+02	1.23E+00
Instance-4	2.80E-01	2.12E+02	9.88E-01	3.43E-01	2.50E+02	1.12E+00
Instance-5	3.12E-01	3.11E+02	1.18E+00	3.72E-01	3.40E+02	1.23E+00
Instance-6	3.07E-01	4.15E+02	1.06E+00	3.86E-01	4.42E+02	1.15E+00
Instance-7	3.18E-01	5.64E+02	1.18E+00	3.56E-01	5.92E+02	1.21E+00
Instance-8	1.91E-01	9.56E+02	1.64E+00	3.40E-01	1.00E+03	1.89E+00
Instance-9	1.54E-01	1.45E+03	1.80E+00	2.72E-01	1.48E+03	1.96E+00

For the calculation of HV, the objective values were normalized. The nadir point and ideal point employed in this normalization process are presented in Table 2. For the computation of IGD, reference points are uniformly sampled in the objective space—such as via Latin Hypercube Sampling—to ensure a representative and evenly distributed approximation of the true Pareto front, while retaining only non-dominated solutions for accuracy in evaluation.

4.3 Comparison Algorithms

For comparative analysis, five state-of-the-art multi-objective optimization algorithms were selected: AdaW [10], BiCo [11], NSGA-II [3], PeEA [9], and VaEA [16]. All algorithm implementations were sourced from the PlatEMO [15] platform. To ensure fairness in comparison, all algorithms employed the same crossover and mutation operators, and the termination criterion was uniformly defined as a fixed number of function evaluations. Parameter settings for each algorithm followed their respective default configurations.

4.4 Experimental Platform and Parameter Configuration

All experiments were conducted on a PC equipped with an Intel i7-12700 CPU (2.11 GHz) and 12 GB of RAM. To reduce the impact of stochastic variability, each algorithm was independently executed 30 times for each test instance. Additionally, the Wilcoxon rank-sum test at a 95% confidence level was employed to perform pairwise comparisons between algorithms. In the result tables, the symbols "+", "−", and "≈" indicate that the compared algorithm performs significantly better than, significantly worse than, or shows no significant difference from the proposed algorithm, respectively.

4.5 Analysis of Experimental Results

A statistical comparison of the HV values achieved by MOMOEA and the five baseline algorithms on nine benchmark test instances is presented in Table 3. The best HV result for each instance is highlighted to facilitate direct performance comparison. The experimental results demonstrate that the proposed MOMOEA algorithm consistently outperforms the five comparison algorithms—AdaW, BiCo, NSGA-II, PeEA, and VaEA—in terms of HV across all test cases. This indicates that MOMOEA is more effective in generating well-converged and widely distributed Pareto-optimal solutions. The superior HV performance reflects the algorithm's strong global search capability and its ability to maintain diversity in the solution set. These results validate the effectiveness of the multi-operator adaptive mechanism in enhancing overall optimization performance.

Table 3. Statistical results of HV values obtained with AdaW, BiCo, NSGA-II, PeEA, VaEA and MOMOEA on Instance 1–9 problems.

Problem	AdaW	BiCo	NSGA-II	PeEA	VaEA	MOMOEA
Instance-1	2.45e-01(6.77e-02)−	3.35e-01(6.25e-02)−	2.87e-01(7.57e-02)−	2.38e-02(3.29e-02)−	3.45e-01(6.82e-02)−	4.41e-01(9.88e-02)
Instance-2	9.47e-02(5.26e-02)−	1.94e-01(7.25e-02)−	2.03e-01(5.66e-02)−	8.66e-04(3.77e-03)−	1.82e-01(5.47e-02)−	3.78e-01(1.08e-01)
Instance-3	2.06e-02(3.23e-02)−	1.51e-01(8.34e-02)−	2.08e-01(6.79e-02)−	0.0e+00(0.0e+00)−	1.35e-01(8.56e-02)−	3.42e-01(1.17e-01)
Instance-4	1.23e-03(4.72e-03)−	6.09e-02(6.01e-02)−	1.49e-01(6.94e-02)≈	0.0e+00(0.0e+00)−	5.96e-02(7.03e-02)−	1.70e-01(1.20e-01)
Instance-5	0.0e+00(0.0e+00)−	0.0e+00(0.0e+00)−	8.13e-03(2.60e-02)−	0.0e+00(0.0e+00)−	0.0e+00(0.0e+00)−	3.08e-02(9.72e-02)
Instance-6	0.0e+00(0.0e+00)−	8.33e-03(2.56e-02)−	1.73e-01(1.04e-01)+	0.0e+00(0.0e+00)−	2.14e-02(3.49e-02)−	9.94e-02(1.26e-01)
Instance-7	0.0e+00(0.0e+00)−	0.0e+00(0.0e+00)−	2.78e-02(5.16e-02)+	0.0e+00(0.0e+00)−	0.0e+00(0.0e+00)−	1.30e-02(4.74e-02)
Instance-8	1.91e-01(4.80e-02)−	3.33e-01(6.60e-02)−	4.51e-01(6.69e-02)−	5.51e-03(8.35e-03)−	3.07e-01(7.47e-02)−	5.90e-01(9.23e-02)
Instance-9	5.27e-02(4.64e-02)−	1.01e-01(5.61e-02)−	1.24e-01(9.03e-02)−	1.18e-03(3.10e-03)−	9.92e-02(6.29e-02)−	4.43e-01(1.64e-01)
+/−/≈	0/9/0	0/9/0	2/6/1	0/9/0	0/9/0	

Table 4 summarizes the numerical statistics of the IGD values obtained from the experimental runs of MOMOEA and the five comparison algorithms on nine benchmark test instances. In terms of the inverted generational distance metric, the proposed MOMOEA algorithm also demonstrates superior performance compared to the five benchmark algorithms—AdaW, BiCo, NSGA-II, PeEA, and VaEA—on the majority of test cases. The lower IGD values achieved by MOMOEA indicate its strong ability to generate solutions that are both close to the true Pareto front and evenly distributed along it. This highlights the effectiveness of the adaptive operator selection strategy in guiding the algorithm toward high-quality solutions with good convergence and diversity. The results further confirm that MOMOEA provides a more precise approximation of the Pareto front in the context of the WTA problem.

Table 4. Statistical results of IGD values obtained with AdaW, BiCo, NSGA-II, PeEA, VaEA and MOMOEA on Instance 1–9 problems.

Problem	AdaW	BiCo	NSGA-II	PeEA	VaEA	MOMOEA
Instance-1	3.42e-01(7.69e-02)−	2.33e-01(5.07e-02)−	2.60e-01(7.15e-02)−	8.12e-01(1.26e-01)−	2.34e-01(5.31e-02)−	1.57e-01(6.39e-02)
Instance-2	5.71e-01(1.27e-01)−	3.46e-01(9.83e-02)−	2.97e-01(7.25e-02)−	1.19e+00(2.19e-01)−	3.56e-01(7.04e-02)−	2.01e-01(8.08e-02)
Instance-3	9.47e-01(2.66e-01)−	4.66e-01(1.59e-01)−	3.65e-01(9.59e-02)−	2.49e+00(3.19e-01)−	5.03e-01(1.81e-01)−	2.55e-01(9.89e-02)
Instance-4	1.22e+00(3.03e-01)−	6.16e-01(1.99e-01)−	3.62e-01(1.01e-01)+	2.52e+00(3.60e-01)−	6.42e-01(2.15e-01)−	3.84e-01(1.66e-01)
Instance-5	3.77e+00(8.03e-01)−	2.52e+00(6.46e-01)−	1.36e+00(3.87e-01)≈	1.00e+01(9.31e-01)−	2.37e+00(5.33e-01)−	1.28e+00(5.13e-01)
Instance-6	1.51e+00(2.50e-01)−	1.02e+00(2.28e-01)−	5.78e-01(1.30e-01)+	4.31e+00(6.14e-01)−	9.24e-01(2.09e-01)−	6.78e-01(1.98e-01)
Instance-7	4.10e+00(8.86e-01)−	2.66e+00(5.81e-01)−	9.05e-01(3.36e-01)+	1.22e+01(1.39e+00)−	2.69e+00(6.99e-01)−	1.42e+00(5.68e-01)
Instance-8	5.24e-01(8.09e-02)−	3.51e-01(5.42e-02)−	2.39e-01(2.28e-02)−	1.25e+00(9.82e-02)−	3.77e-01(8.52e-02)−	1.51e-01(6.28e-02)
Instance-9	7.07e-01(1.47e-01)−	5.38e-01(8.74e-02)−	3.69e-01(8.87e-02)≈	1.40e+00(1.16e-01)−	5.57e-01(1.19e-01)−	3.40e-01(7.14e-02)
+/−/≈	0/9/0	0/9/0	3/4/2	0/9/0	0/9/0	

The runtime comparison results, measured in seconds, are illustrated in Fig. 2. As shown, the execution time of all algorithms increases approximately linearly with the number of function evaluations, aligning with the expected computational behavior of evolutionary algorithms. Among the six algorithms, PeEA consistently exhibits the highest runtime across all test instances, which can be attributed to its relatively complex environmental selection and update mechanisms. In contrast, the proposed MOMOEA maintains one of the shortest runtimes throughout, demonstrating superior computational efficiency while still achieving competitive optimization performance.

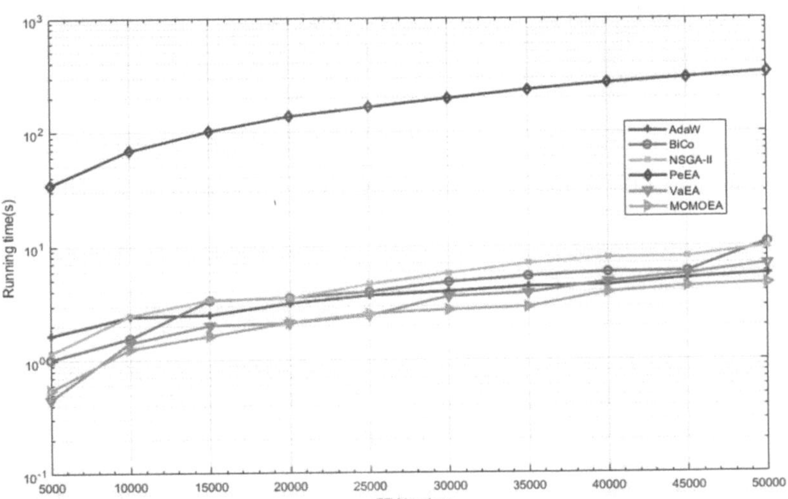

Fig. 2. Runtime comparison among AdaW, BiCo, NSGA-II, PeEA, VaEA and MOMOEA on Instance-2.

Figure 3 illustrates the dynamic proportions of the three crossover operators (single-point, double-point, and multi-point crossover) utilized by MOMOEA throughout the optimization process. For the convenience of representation, the three operators are abbreviated as SPC, DPC and MPC respectively. The results show that the usage of single-point crossover increases significantly during the early stages of the search, indicating its effectiveness in promoting rapid convergence at the initial phase. However, its proportion gradually decreases as the search progresses. In contrast, the proportion of multi-point crossover steadily increases in the later stages, suggesting its role in enhancing solution diversity and preventing premature convergence. The two-point crossover maintains a relatively stable usage throughout the process, providing a balance between exploration and exploitation. This adaptive operator allocation reflects the effectiveness of MOMOEA's multi-operator mechanism in dynamically guiding the search toward high-quality solutions.

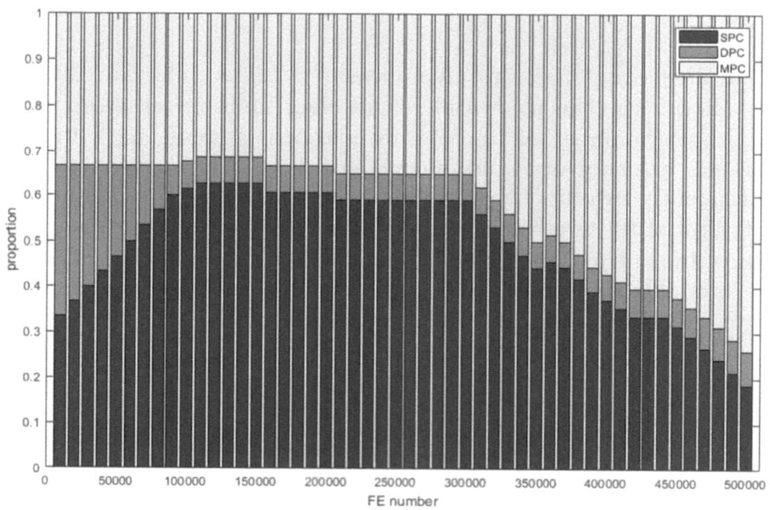

Fig. 3. Evolution of the proportional usage of the three operators in MMOEA throughout the optimization process.

5 Conclusion

This paper proposes an evolutionary algorithm incorporating a multi-operator adaptive mechanism for addressing the multi-objective WTA problem. During the evolutionary process, the algorithm dynamically adjusts the weights of different operators to enable efficient exploration of the decision space. In the experimental evaluation, the proposed method was compared against five state-of-the-art multi-objective optimization algorithms across nine benchmark test cases.

The results demonstrate that the proposed algorithm achieves competitive or superior performance in the majority of cases.

As part of future work, we plan to extend the adaptive mechanism by designing two separate operator sets—one for crossover and the other for mutation. These operator sets will be adaptively mapped and selected during the evolutionary process to further enhance search efficiency and solution quality.

Acknowledgements. This work was partially supported by the National Natural Science Foundation of China (General Program, Grant No. 62473052), and the Fund of Science and Technology on Metrology and Calibration Laboratory under Grant JLKG2024001B004.

References

1. Ahuja, R.K., Kumar, A., Jha, K.C., Orlin, J.B.: Exact and heuristic algorithms for the weapon-target assignment problem. Oper. Res. **55**(6), 1136–1146 (2007)
2. Cai, H., Liu, J., Chen, Y., Wang, H.: Survey of the research on dynamic weapon-target assignment problem. J. Syst. Eng. Electron. **17**(3), 559–565 (2006)
3. Deb, K., Pratap, A., Agarwal, S., Meyarivan, T.: A fast and elitist multiobjective genetic algorithm: NSGA-II. IEEE Trans. Evol. Comput. **6**(2), 182–197 (2002)
4. Gao, C., Kou, Y., Li, Y., Li, Z., Xu, A.: Multi-objective weapon target assignment based on d-nsga-iii-a. IEEE Access **7**, 50240–50254 (2019)
5. Johansson, F., Falkman, G.: An empirical investigation of the static weapon-target allocation problem. In: Proceedings of the 3rd Skövde Workshop on Information Fusion Topics, pp. 63–67 (2009)
6. Kline, A., Ahner, D., Hill, R.: The weapon-target assignment problem. Comput. Oper. Res. **105**, 226–236 (2019)
7. Kline, A.G., Ahner, D.K., Lunday, B.J.: Real-time heuristic algorithms for the static weapon target assignment problem. J. Heurist. **25**, 377–397 (2019)
8. Li, J., Chen, J., Xin, B., Dou, L.: Solving multi-objective multi-stage weapon target assignment problem via adaptive nsga-ii and adaptive moea/d: a comparison study. In: 2015 IEEE Congress on Evolutionary Computation (CEC), pp. 3132–3139. IEEE (2015)
9. Li, L., Yen, G.G., Sahoo, A., Chang, L., Gu, T.: On the estimation of pareto front and dimensional similarity in many-objective evolutionary algorithm. Inf. Sci. **563**, 375–400 (2021)
10. Li, M., Yao, X.: What weights work for you? adapting weights for any pareto front shape in decomposition-based evolutionary multiobjective optimisation. Evol. Comput. **28**(2), 227–253 (2020)
11. Liu, Z.Z., Wang, B.C., Tang, K.: Handling constrained multiobjective optimization problems via bidirectional coevolution. IEEE Trans. Cybern. **52**(10), 10163–10176 (2021)
12. Lloyd, S.P., Witsenhausen, H.S.: Weapons allocation is np-complete. In: 1986 Summer Computer Simulation Conference, pp. 1054–1058 (1986)
13. Malcolm, W.P.: On the character and complexity of certain defensive resource allocation problems. Technical report (2004)

14. Rosenberger, J.M., Hwang, H.S., Pallerla, R.P., Yucel, A., Wilson, R.L., Brungardt, E.G.: The generalized weapon target assignment problem. In: 10th International Command and Control Research and Technology Symposium, pp. 1–12. Citeseer (2005)
15. Tian, Y., Cheng, R., Zhang, X., Jin, Y.: Platemo: a matlab platform for evolutionary multi-objective optimization [educational forum]. IEEE Comput. Intell. Mag. **12**(4), 73–87 (2017)
16. Xiang, Y., Zhou, Y., Li, M., Chen, Z.: A vector angle-based evolutionary algorithm for unconstrained many-objective optimization. IEEE Trans. Evol. Comput. **21**(1), 131–152 (2016)
17. Xin, B., Chen, J., Peng, Z., Dou, L., Zhang, J.: An efficient rule-based constructive heuristic to solve dynamic weapon-target assignment problem. IEEE Trans. Syst. Man Cybern.-Part A: Syst. Humans **41**(3), 598–606 (2010)
18. Yi, X., Yu, H., Xu, T.: Solving multi-objective weapon-target assignment considering reliability by improved moea/d-am2m. Neurocomputing **563**, 126906 (2024)

An Adaptive Multi-granularity Dynamic Task Reallocation Method Based on CBBA

Binhua Guo[1(✉)], Bin Xin[1,2], and Yixi Yang[1]

[1] School of Automation, Beijing Institute of Technology, Beijing 100081, China
3220231013@bit.edu.cn

[2] National Key Laboratory of Autonomous Intelligent Unmanned Systems, Beijing 100081, China

Abstract. Aiming at the bottleneck of collaborative efficiency caused by multi-source heterogeneous dynamic disturbances (such as task addition, deletion/change, agent failure/online) in post-disaster rescue scenarios, this paper proposes an improved consensus bundling algorithm framework (CBBA-AMG) that integrates adaptive multi-granularity response and load balancing optimization. By constructing a time-resource composite load model and introducing a load balancing penalty term based on Jensen-Shannon divergence, the problem of uneven task distribution in the traditional CBBA algorithm in heterogeneous multi-agent systems is effectively alleviated. At the same time, an event-driven dynamic reallocation mechanism is designed to dynamically trigger differentiated response granularities (global reset/cluster collaboration/individual adjustment) based on the disturbance impact metric function $I(\epsilon)$ to minimize the impact of replanning on system stability. Simulation results show that the proposed load balancing mechanism reduces the Gini coefficient of task allocation by 3.9%, and the maximum-minimum load difference is reduced by 16%; the adaptive multi-granularity strategy is 6.9 times faster than the global reallocation response speed, and the cumulative task revenue is increased by 12.7%, significantly improving the task completion rate and robustness of the system in a dynamic environment.

Keywords: multi-agent system · task redistribution · uncertainty · CBBA

1 Introduction

Disaster rescue is a core part of emergency management, and its timeliness is directly related to the safety of people's lives and property and the post-disaster recovery process. In the complex and changeable post-disaster harsh environment with limited information, it is crucial to efficiently coordinate multiple forces for search and rescue and disposal [1]. In recent years, Multi-Agent Systems (MAS) have shown a wide range of application scenarios in the field of post-disaster

H. Ma et al. (Eds.): IWACIII 2025, CCIS 2780, pp. 15–30, 2026.
https://doi.org/10.1007/978-981-95-6730-0_2

collaborative search and rescue with their advantages of distribution, flexibility and robustness [2,3].

However, actual disaster scenarios are highly dynamic and uncertain, mainly reflected in the frequent disturbances at the task level (emergence of new tasks, information changes/cancellations), agent level (sudden failures, energy exhaustion and offline, new units joining), and environment level (random obstacles appear, road clearance changes accessibility and task execution efficiency) [4,5]. Static task allocation is difficult to ensure stability and adaptability, and a dynamic allocation mechanism is urgently needed to respond. To address this problem, existing studies have proposed a variety of strategies. Huang et al. [6] studied the loss detection and automatic replacement scheme of distributed drones, and proposed a recovery algorithm for the recovery of newly inserted drones or lost drones from failures. Bi et al. [7] introduced a timeliness parameter to adjust the bidding strategy, and optimized the task selection strategy based on the time-order priority insertion principle to quickly generate a task reallocation plan. Yang [8] et al. established a distributed framework for dynamic event type determination processing strategy for multiple UAVs performing search and rescue (SAR) missions, and proposed a partial reallocation algorithm (TRA), considering five dynamic events: finding new tasks, canceling original tasks, updating task coordinates, updating task deadlines, and updating task duration. Wang et al. [9] established a multi-UAV collaborative reconnaissance task reallocation model (DE-MCRTR) [10] in a dynamic environment. The DE-MCRTR model solves the task reallocation problem in dynamic events, including UAV damage, new target appearance, target position change, and target time window change, while comprehensively considering various factors such as UAV performance differences, target area size, and time window constraints. However, existing methods mostly focus on a single type of disturbance (such as only task changes or only agent failures), lacking a systematic design for collaborative response to multi-source heterogeneous disturbances; at the same time, the reallocation strategy often adopts a fixed response granularity (such as global reset or local fine-tuning), which is difficult to adaptively match the spatiotemporal impact range of different disturbance events.

For the task reallocation problem in dynamic environments, distributed task allocation algorithms have shown significant advantages due to their decentralized characteristics. CBBA (Consensus-Based Bundle Algorithm) [11] is widely used due to its good scalability and real-time adaptability to dynamic environments. Buckman et al. [12] proposed CBBA with partial replanning (CBBA-PR), which maximizes time-discounted rewards by allowing a group of drones to reallocate low-priority tasks together with newly discovered tasks to reduce the number of communication rounds and running time. The CBBA-PR method minimizes interference with ongoing tasks and balances fast response time and task coordination in dynamic environments. Chen et al. [13] introduced CBBA with local replanning (CBBA-LR), focusing on maximizing time-discounted task rewards and minimizing the distance between drones and task locations. The algorithm allows drones with available resources to replan tasks that overlap in time windows and are closest to new tasks. However, in large-scale heterogeneous task scenarios, the reward maximization orientation of traditional CBBA easily leads to a large disparity in the amount of tasks of similar agents (partial overload, partial restriction), which restricts the overall rescue efficiency and robustness.

In view of the timeliness and high uncertainty of the post-disaster environment, this paper establishes a multi-unmanned vehicle task allocation and redistribution model in a dynamic environment, which simultaneously considers time constraints, agent capability matching, and load balancing. An adaptive multi-granularity dynamic redistribution strategy is proposed to expand the CBBA framework. According to the event type, impact scope and task urgency, it dynamically triggers differentiated redistribution granularity (global reset/cluster collaboration/individual local adjustment), and comprehensively considers task addition and deletion, agent damage and online. Task priority is updated in real time to trigger the corresponding level of redistribution. Introduce load balancing factors in the CBBA bidding stage to constrain the task load of agents and eliminate the "uneven busy and idle" phenomenon.

2 Problem Description and Modeling

2.1 Problem Description

The problem studied in this paper belongs to the XD[ST-SR-TA] category in the iTax taxonomy [14], namely, the single-task robot, single-robot task, and time-extended allocation problem. In the post-disaster rescue scenario, an unmanned vehicle is set to perform only one task at a time, and each task must be assigned to a unique drone. The drone must perform the task within its specified deadline. Due to the limitation of carrying capacity, each unmanned vehicle can only perform at most L tasks. The goal of task allocation is to seek the optimal conflict-free matching between tasks and robots and determine the task execution sequence of each agent under the premise of a given rescue task set $\mathcal{T} = \{t_1, \ldots, t_m\}$ and a rescue agent set $\mathcal{A} = \{a_1, \ldots, a_n\}$ to maximize the global benefit while satisfying the time window constraint and capability matching constraint.

(1) The set of heterogeneous unmanned vehicle groups is $\mathcal{A} = \{a_1, ..., a_n\}$, n is the total number of unmanned vehicles. Each agent is a four-tuple: $a_i = (ax_i, ap_i, ac_i, as_i, v_i)$, which are state, position, ability, resource capacity vector, and cruising speed respectively. Among them, the initial resource capacity vector $as_i = \{s_{il}\}\,(l = 1, 2, ..., S)$. s_{il} is the value of the lth resource capacity of a_i, which generally includes rescue resources, the amount of materials carried, and endurance. S is the number of resource types, and different types of rescue robots carry different types of resources;

(2) Tasks $\mathcal{T} = \{t_1, ..., t_m\}$, m is the total number of tasks. Each task is a seven-tuple $t_j = (tx_j, xp_j, tc_j, ts_j, ty_j, tw_j)$, which are state, location, task type, resource demand vector, task value, task time window constraint, and task

execution time. Among them, $ts_j = \{s_{jl}\}\,(l = 1, 2, ..., S)$, s_{jl} is the value of the lth resource capacity required to execute task t_j; $tw_j = \left(t_j^{start}, t_j^{end}\right)$, t_j^{start} represents the earliest start time set for task t_j, t_j^{end} represents the latest completion time set for task t_j, and the time window must satisfy $t_j^{end} - t_j^{start} > t_j^{\Delta}$; t_j^{Δ} is the task execution time, and the task execution time is not $t_j^{execute}$.

(3) Assign variables φ_{ij}: If task t_j is assigned to a_i, then $\varphi_{ij} = 1$; otherwise $\varphi_{ij} = 0$

2.2 Mathematical Model

Taking the maximization of total task benefits as the goal, considering task resource constraints and time window constraints, a dynamic optimization model for multi-agent task reallocation is established as follows.

$$\max J = \sum_{i=1}^{n}\sum_{j=1}^{m}\left[e^{-\lambda\tau_{ij}} * ty_j \cdot \varphi ij\right] \tag{1}$$

$$\sum_{i=1}^{n} x_{ij} \leq 1, \quad \forall j \in \mathcal{T} \tag{2}$$

$$\sum_{j=1}^{m} x_{ij} \leq L, \quad \forall i \in \mathcal{A} \tag{3}$$

$$t_j^{start} \leq t_j^{execute} \leq t_j^{end} - t_j^{\Delta} \tag{4}$$

$$\sum_{j} \varphi_{ij} \cdot ts_{jl} \leq a_{sil}, \quad \forall i \in \mathcal{A}, l = 1, \ldots, S \tag{5}$$

$$\varphi_{ij} \leq \mathbb{I}(ax_i = 1), \forall i \in \mathcal{A} \tag{6}$$

where $e^{-\lambda\tau_{ij}} * ty_j \cdot \varphi ij$is the value of task t_j at completion time c_j, $\tau_{ij} = \frac{\|ap_i - tp_j\|}{v_i}$: the travel time for agent a_i to reach task t_j, and $\lambda$is the time cost weight coefficient. Equation (2) is the task constraint, a task can only be assigned to one agent. Equation (3) is the agent constraint, an agent can only complete one task at a time. Equation (4) indicates that the task execution time must be greater than the earliest start time, and the latest completion time must be greater than or equal to the task execution time. (5) is the resource constraint of different types for each agent. Equation (6) is the resource capacity matching constraint, that is, the allocation is allowed only when the agent meets the capacity and resource requirements of the task and the agent is available.

2.3 Load Balancing

When there are many agents of the same type, the task types are unevenly distributed, or the task space is unevenly distributed, it is easy to have uneven load (some agents are assigned many tasks, some are assigned few or even none). Therefore, a load balancing mechanism is designed to make the number of tasks assigned to agents of the same type as balanced as possible.

To solve the problem of uneven task distribution in the traditional CBBA algorithm in heterogeneous unmanned vehicle rescue scenarios, this paper proposes a multi-dimensional load balancing mechanism. The load L_i of agent a_i is defined as a time-resource composite indicator. The time load (accumulated travel time), resource load rate (resource consumption ratio) and task saturation (task quantity ratio) of agent a_i are comprehensively considered.

$$L_i = \omega_t \cdot \frac{\sum_{j\in\mathbf{b}_i} \tau_{ij}}{\tau_{\max}} + \omega_r \cdot \frac{1}{L}\sum_{l=1}^{L} \frac{\sum_{j\in\mathbf{b}_i} ts_{jl}}{as_{il}} + \omega_c \cdot \min\left(1, \frac{|T_i|}{L}\right) \tag{7}$$

where $\mathbf{b}_i$ is the set of tasks assigned to a_i, and $\omega_t + \omega_r + \omega_c = 1$ is the weight coefficient. $\tau_{\max} = \max_i(\sum_j \tau_{ij})$ realizes cross-platform time load normalization. The resource item adopts the arithmetic average of the L class resource consumption ratio to avoid single resource dominance.

Based on this definition, the load balancing penalty item is quantified by the improved Jensen-Shannon divergence, and the quantified result of JS divergence is used as a load imbalance penalty item, so that the greater the load imbalance, the greater the penalty item, thereby affecting the task allocation decision.

$$\Phi(\mathbf{L}) = \frac{1}{2}\left[D_{\mathrm{KL}}(P\|M) + D_{\mathrm{KL}}(Q\|M)\right], \quad M = \frac{P+Q}{2} \tag{8}$$

P is the actual load distribution $[L_1, \ldots, L_n]$, Q is the ideal uniform distribution $\mathcal{U}(\bar{L} - \delta, \bar{L} + \delta)$, $\mathbf{M}$ is the mean distribution of these two distributions, indicating the "average" state of the two loads. $D_{\mathrm{KL}}(P\|M)$ and $D_{\mathrm{KL}}(Q\|M)$ are Kullback-Leibler divergences, which measure how much two distributions (P and Q) deviate from the middle distribution M. Kullback-Leibler divergence (KL divergence) is defined as:

$$D_{\mathrm{KL}}(P \parallel M) = \sum_i P(i) \log\left(\frac{P(i)}{M(i)}\right)$$

This measure is more sensitive to load outliers and has a bounded range of $[0, 1]$. The penalty term is embedded in the objective function in the form of negative weights $-\mu \cdot \Phi(\mathbf{L})$.

$$\max, J_{\text{enhanced}} = \sum_{i=1}^{n}\sum_{j=1}^{m}\left[e^{-\lambda\tau_{ij}} \cdot ty_j \cdot \varphi_{ij}\right] - \mu \cdot \Phi(\mathbf{L}) \tag{9}$$

By introducing JS divergence as a penalty term for load balancing, the objective function will allocate tasks according to the maximum value of the tasks. When the load imbalance is large, JS divergence will give a higher penalty, which will encourage more consideration of load balancing in the task allocation process to avoid overloading some agents while other agents are idle. It can effectively avoid excessive load on a single agent and improve the overall task completion efficiency and reliability of the system.

3 CBBA-AMR

Consensus-based bundle algorithm (CBBA), which uses a two-stage structure, iterating between bundle construction and conflict resolution. In the auction stage, tasks are selected to be added to their own task sequence, and in the consensus stage, conflicting tasks are removed, and finally a solution without conflicts between all robots is obtained. The algorithm is highly adaptable, handles multiple objectives, and ensures global consistency, robustness and scalability.

3.1 Information Structure

During the task allocation process, the information structure that agent i needs to store and update is:

(1) Task bundle: Task bundle $\mathbf{b}_i$ is a sequence of tasks that agent i has selected but not yet executed, recorded in order of addition time. Each agent can hold at most L tasks (whether assigned or planned), $\mathbf{b}_i \in (\mathcal{T} \cup \{\emptyset\})^L$

$$\boldsymbol{b}_i = \boldsymbol{b}_i^1 \oplus_{\text{end}} \boldsymbol{b}_i^k, \; |\boldsymbol{b}_i| \leqslant L \tag{10}$$

where $\oplus_{\text{n}}$ represents the operation of inserting the list after the symbol into the nth element of the list after the symbol, and $\oplus_{\text{end}}$ is the operation of inserting to the end of the task list. $\boldsymbol{b}_i^k$ represents the task at the kth layer among all the tasks performed by agent i. Tasks are arranged in the order in which they are added to the package.

(2) Task path: The task path $\mathbf{p}_i$ records the order in which tasks are executed according to the insertion order in the task execution. $\mathbf{p}_i \in (\mathcal{T} \cup \{\emptyset\})^{L_t}$.

$$\boldsymbol{p}_i = \boldsymbol{p}_i^1 \oplus_{\text{optimal}} \boldsymbol{p}_i^k, \; |\boldsymbol{p}_i| \leqslant L_t \tag{11}$$

where $\boldsymbol{b}_i^k$ represents the task at the kth layer among all the tasks performed by agent i, which is the insertion point that minimizes the cost or maximizes the benefit.

(3) Winning bid list: The winning bid list is represented by the matrix $\boldsymbol{y}_i$,$y_i \in \mathbb{R}_+^m$. The rows of the matrix represent the target task layer and the columns represent the target served by the formation. The winning bid list of agent i can be represented as $\boldsymbol{y}_i \in \mathbb{R}_+^{m \times n}$. Among them, y_{ij}^k represents the winning bid of the kth task of target j.

(4) List of winning agents: The list of winning agents is represented by the matrix $\boldsymbol{z}_i$,$z_i \in \mathcal{I}^m$, which stores the information of which agent i currently has which agent has won which tasks, which can be expressed as $\boldsymbol{z}_i \in \mathcal{I}^{m \times n}$, z_{ij}^k represents the winning agent of the task k of target j.

(5) Timestamp: The timestamp $\boldsymbol{s}_i \in \mathbb{R}^n$ represents the update timestamp of the information obtained by agent i from other members of the formation. This vector is an important indicator in the conflict elimination phase, which is used to indicate the newness of the information obtained by the agent from other agents.

3.2 Bundle Construction

The first stage of the CBBA algorithm is the task bundle construction process. Each agent continues to add tasks to its task bundle until no more tasks can be added. The task list is iteratively updated as follows.

Each task has a time window requirement (must be completed in a certain time period). Add a time-driven task allocation mechanism to optimize the order of task allocation according to the timeliness requirements of each task. Let $S_i^{\mathbf{p}_i}$ be defined as the total reward value of the task on the execution path $\mathbf{p}_i$ of agent i.

$$S_i^{\mathbf{p}_i} = \sum_j e^{-\lambda \tau_i^j(\mathbf{p}_i)} * ty_j - \mu \cdot \Phi(\mathbf{L}) \tag{12}$$

where λ is the discount factor of the task, $\tau_i^j(\mathbf{p}_i)$ is the estimated time for agent i to reach the location of task j along path $\mathbf{p}_i$, and ty_j is the static score related to executing task j.

$$\tau_i^j(\mathbf{p}_i) = \begin{cases} \max\left(t_j^{start}, \frac{\sqrt{(x_{a_i} - x_{t_j})^2 + (y_{a_i} - y_{t_j})^2}}{v_i}\right), & \text{Task} j \text{ is a task in the task package,} \\ \max\left(t_j^{start}, t_{prev}^{end} + \frac{\sqrt{(x_{t_{\text{prev}}} - x_{t_j})^2 + (y_{t_{\text{prev}}} - y_{t_j})^2}}{v_i}\right), & \text{Task} j \text{ is not the first task in the task bundle.} \end{cases} \tag{13}$$

If task j is added to the task bundle $\mathbf{b}_i$, the marginal score improvement will be calculated as

$$c_{ij}[\mathbf{b}_i] = \begin{cases} 0, & \text{if } j \in \mathbf{b}_i \\ \max\limits_{n \le |\mathbf{p}_i|} \left\{ S_i^{\mathbf{p}_i \oplus_n \{j\}} - S_i^{\mathbf{p}_i} \right\}, & \text{otherwise} \end{cases} \tag{14}$$

where $|\cdot|$ The CBBA scoring mechanism inserts new tasks at the position that produces the largest score improvement, which is the marginal score associated with the task given the current path. Therefore, if the task is already included in the path, it will not bring additional score improvement. In addition, it is assumed that adding any new task will bring a non-trivial reward, that is, $c_{ij}[\mathbf{b}_i] \ge 0$, and the equality holds only when $j \in \mathbf{b}_i$.

The scoring function is initialized to $S_i^{\{\emptyset\}} = 0$, and the paths and task bundles are recursively updated as follows:

$$\mathbf{b}_i = \mathbf{b}_i \oplus_{\text{end}} \{J_i\}, \quad \mathbf{p}_i = \mathbf{p}_i \oplus_{n_{i,J_i}} \{J_i\} \tag{15}$$

where $J_i = \arg\max_j \left(c_{ij}[\mathbf{b}_i] \times h_{ij}\right)$, $n_{i,J_i} = \arg\max_n \left\{ S_i^{\mathbf{p}_i \oplus_n \{J_i\}} \right\}$, and $h_{ij} = \mathbb{I}(c_{ij} > y_{ij})$ ($\mathbb{I}$ is an indicative function, which takes 1 when the condition is met, otherwise takes 0). The above recursive process continues until $|\mathbf{b}_i| = L_t$ or $h_{ij} = 0$.

Algorithm 1. Task bundle construction process for agent i

Require: Task bundles of each task layer in the last iteration $\boldsymbol{b}_i^k$
Task path $\boldsymbol{p}_i^k$
Winning bid list $\boldsymbol{y}_i$
Winning agent list $\boldsymbol{z}_i$
for $k = 1$ **to** m **do**
  Construct task bundle and path: $\boldsymbol{b}_i$ $\boldsymbol{p}_i$
  Calculate task benefit: $S_i^{\boldsymbol{p}_i}$
  if $|\boldsymbol{b}_i| < L_m$ **and** $L_m = \min(mL, n)$ **then**
    Calculate marginal gain: $c_{ij}^k, \forall j \in \mathcal{T}$
  else
    return $\boldsymbol{b}_i^k, \boldsymbol{p}_i^k, \boldsymbol{y}_i, \boldsymbol{z}_i$
  end if
  Update indicator variables: $h_{ij}^k \leftarrow \mathbb{I}(c_{ij}^k > y_{ij}^k), \forall j \in \mathcal{J}$
  if $\boldsymbol{h}_i^k \neq \boldsymbol{0}$ **then**
    Select the best task: $J_i^k \leftarrow \arg\max_j \left(c_{ij}^k h_{ij}^k\right)$
    Find the insertion position: $n_{i,J_i^k}^k \leftarrow \arg\max_{n_k} S_i^{\boldsymbol{p}_i^k \oplus_{n_k} \{J_i^k\}}$
    Update task package and path: $\boldsymbol{b}_i^k \leftarrow \boldsymbol{b}_i^k \oplus_{\mathbf{end}} \{J_i^k\}$ $\boldsymbol{p}_i^k \leftarrow \boldsymbol{p}_i^k \oplus_{n_{i,J_i^k}^k}$ $\{J_i^k\}$
    Update winning information: $y_{i,J_i^k}^k \leftarrow c_{i,J_i^k}^k$ $z_{i,J_i^k}^k \leftarrow i$
  end if
end for
return $\boldsymbol{b}_i^k, \boldsymbol{p}_i^k, \boldsymbol{y}_i, \boldsymbol{z}_i$

3.3 Conflict Resolution Phase

After the task package construction process is completed, the agent shares the task list, task path list and overall benefit information with the adjacent agents. According to certain action rules, the corresponding task information is updated to obtain conflict-free task allocation. The conflict resolution process uses the timestamp vector to determine the latest allocation information. The task urgency is introduced into the conflict resolution process, so as to give priority to the allocation of urgent tasks in the distributed task allocation. Each time a message is transmitted, the time vector is filled as follows:

$$s_{ik}^{\text{new}} = \begin{cases} \max(T_r, s_{ik} + \Delta_p) & \text{if } i \in \mathcal{A}_c \\ T_r & \text{if } z_{ik} \neq z_{ik}^{\text{prev}} \\ s_{ik} & \text{otherwise} \end{cases} \tag{16}$$

where T_r is the message receiving time. The priority compensation term Δ_p is to make the urgent task gain an advantage in timestamp, but not completely abandon the time factor.

$$\Delta_p = \beta \cdot \sum_{j \in \mathbf{b}_i} \eta_j, \quad \eta_j = e^{-\lambda(t - t_j^{\text{start}})} \tag{17}$$

When agent i receives a message from agent k, z_i and s_i are used to determine which agent has the latest information for each task. Agent i can take three actions for task j:

1. Update: $y_{ij} = y_{kj}, z_{ij} = z_{kj}$;
2. Reset: $y_{ij} = 0, z_{ij} = \emptyset$;
3. Keep: $y_{ij} = y_{ij}, z_{ij} = z_{ij}$.

3.4 Task Reassignment Mechanism

This paper adopts event-driven, minimum disturbance and multi-granularity strategies to optimize the task allocation process. The event-driven mechanism means that the recalculation of task allocation will be triggered only when a "dynamic event" occurs in the system, thereby avoiding unnecessary frequent global allocation. In view of the actual situation, we consider the uncertainty of tasks and robots, and define the dynamic event set $\mathcal{E} = \epsilon^{\text{new}}, \epsilon^{\text{cancel}}, \epsilon^{\text{fail}}, \epsilon^{\text{update}}$ corresponding to: new task arrival, task cancellation, agent failure, and agent online. When task uncertainty occurs, considering the completed tasks, it is necessary to filter out the "current valid task set" $\mathcal{T}_{\text{valid}}(t_{\text{cur}}) = \{t_j \in \mathcal{T}(t_{\text{cur}}) \mid e_j > t_{\text{cur}} \text{ and } t_j \text{ not completed}\}$ before each reallocation. Determine the affected agent A_{aff}. Freeze the allocation of unaffected agents to minimize the difference with the initial allocation. Under the premise of keeping the allocation of unaffected agents unchanged, only optimize the affected agent A_{aff} and the current valid task set $\mathcal{T}_{\text{valid}}(t_{\text{cur}})$. The optimization domain constraint is $\mathcal{A}_{\text{replan}} \subseteq \mathcal{A}, \quad \mathcal{T}_{\text{replan}} \subseteq \mathcal{T}_{\text{valid}}(t_{\text{cur}})$. The agent set changes dynamically as $\mathcal{A}(t) = \mathcal{A}0 \cup \mathcal{A}\text{new}(t) \setminus \mathcal{A}_{\text{fail}}(t)$, The task set changes dynamically as $\mathcal{T}(t) = \mathcal{T}0 \cup \mathcal{T}\text{new}(t) \setminus \mathcal{T}_{\text{cancel}}(t)$

The multi-granularity strategy dynamically selects the redistribution granularity (global redistribution, local redistribution or individual adjustment) based on the type of event, the scope of impact (how many agents and tasks are affected by this event) and the urgency of the task to optimize the efficiency and stability of the system, ensuring that the unaffected agents and task allocations remain unchanged as much as possible, and only the affected parts are redistributed. Event impact measurement function:

$$I(\epsilon) = \alpha \cdot \Delta J_{\text{norm}} + \beta \cdot \left(\frac{|\mathcal{A}_{\text{aff}}|}{|\mathcal{A}|}\right) + \gamma \cdot \max_{j \in \mathcal{T}_\epsilon} \left(\frac{ty_j \cdot w_j}{\max(\delta, t_j^{\text{end}} - t_{\text{cur}})}\right) + \eta \cdot d_{\min} \tag{18}$$

$$\Delta J_{\text{norm}} = \frac{|J_{\text{new}} - J_{\text{old}}|}{J_{\max}} \in [0, 1] \tag{19}$$

$$d_{\min} = \min_{i \in \mathcal{A}} \frac{\|\mathbf{x}_i - \mathbf{x}_\epsilon\|}{R_{\text{com}}} \in [0, 1] \tag{20}$$

where ΔJ_{norm} is the normalized return shock, which is used to quantify the impact of the event on the total return of the system, J_{old}: the total return of the system before the event, J_{new}: the total return of the system after the event, $J_{\max}$: the theoretical maximum return of the system (normalized benchmark). w_j is the event type weight factor, which is used to amplify the urgency of a specific event; δ is the minimum time protection to avoid zero division errors.

$d_{\min}$ normalizes the minimum spatial distance and quantifies the spatial compactness of events; $\mathbf{ax}_i$ is the current position coordinate of agent i, $\mathbf{ax}_\epsilon$: the position coordinate of event ϵ, R_{com}: the communication radius of the agent (distance normalization benchmark). $\alpha, \beta, \gamma, \eta$ are weight coefficients used to balance the contribution of each dimension.

Dynamically select the granularity of redistribution according to the event impact measurement function, and select the adaptive response strategy: Global redistribution (global): The impact range is large or there are extremely urgent tasks, and all agents and tasks participate in the redistribution. Local redistribution (local/cluster): The impact range is medium, and only the affected agents and tasks participate in the redistribution. Individual adjustment: The impact range is extremely small and the task is not urgent, so only a very small range of adjustment is made. The unaffected allocation is frozen and restored through the dictionary mapping $(agent_{id}isthekey)$ to ensure allocation consistency and robustness.

$$\text{Response}(\epsilon) = \begin{cases} \text{Global} & \text{if } I(\epsilon) \geq \theta_g \\ \text{Cluster} & \text{if } \theta_c \leq I(\epsilon) < \theta_g \\ \text{Local} & \text{otherwise} \end{cases} \tag{21}$$

where $\theta_g > \theta_c > 0$ is the response threshold. $d(a_i, \epsilon)$ is the spatial/logical distance between the agent and the event Dynamically adjust the conflict handling strategy according to the current redistribution level of the system (global/local/individual). Its core innovation is to align the conflict resolution granularity with the task reallocation granularity, and balance the system consistency and efficiency through three-level differentiation processing (Table 1).

Table 1. Replanning strategy

Strategy	List of affected agents and tasks
Global	$\mathcal{A}_{\text{replan}} = \mathcal{A}, \quad \mathcal{T}_{\text{replan}} = \mathcal{T}$
Cluster	$\mathcal{A}_{\text{replan}} = \{i \in \mathcal{A} \mid d(a_i, \epsilon) \leq R_c\}, \quad \mathcal{T}_{\text{replan}} = \mathcal{T} \cap \mathcal{T}_\epsilon$
Local	$\mathcal{A}_{\text{replan}} = i_{\text{most\_affected}}, \quad \mathcal{T}_{\text{replan}} = \mathcal{T}_{\text{new}} \cup \mathcal{T}_{\text{related}}$

Algorithm 2. Adaptive multi-granularity CBBA dynamic reallocation (AMG-CBBA)

Require: current state $\{\mathbf{b}_i, \mathbf{p}_i, \mathbf{y}_i, \mathbf{z}_i\}$, event $\mathcal{E}$
Ensure: updated allocation scheme

detect dynamic events $\mathcal{E} \leftarrow$ EventDetector()
if $\Psi(\mathcal{E}) =$ True **then**
  calculate event impact $I(\mathcal{E}) \leftarrow \alpha \Delta J_{\text{norm}} + \beta |\mathcal{A}_{\text{aff}}|/|\mathcal{A}| + \gamma \cdot \text{Urgency}(\mathcal{E}) + \eta d_{\min}$
  Determine the reallocation granularity $\mathcal{G} \leftarrow \text{ResponseSelector}(I(\mathcal{E}), \theta_g, \theta_c)$
  $\mathcal{G}$ Global reallocation
  $\{\mathbf{b}_i, \mathbf{p}_i\} \leftarrow \text{CBBA\_Full}(\mathcal{T}_{\text{valid}}, \mathcal{A})$
  local redistribution
  $(\mathcal{A}_{\text{aff}}, \mathcal{T}_{\text{aff}}) \leftarrow \text{ComputeAffectedDomain}(\mathcal{E})$
  Unbind task: $\forall i \in \mathcal{A}_{\text{aff}}, \mathbf{b}_i \leftarrow \mathbf{b}_i \setminus \mathcal{T}_{\text{aff}}$
  Reset winning bid: $\forall j \in \mathcal{T}_{\text{aff}}, y_{\cdot j} \leftarrow 0, z_{\cdot j} \leftarrow \emptyset$
  // CBBA phase 1: Local package build
  for $i \in \mathcal{A}_{\text{aff}}$ **do**
    $\mathbf{b}_i^{\text{new}} \leftarrow \text{CBBA\_BundleBuild}(i, \mathcal{T}_{\text{aff}})$
  end for
  // CBBA phase 2: load-aware conflict resolution
  $\{\mathbf{y}_i, \mathbf{z}_i\}_{i \in \mathcal{A}_{\text{aff}}} \leftarrow \text{LoadAwareConsensus}(\mathcal{A}_{\text{aff}})$
  Update path: $\forall i \in \mathcal{A}_{\text{aff}}, \mathbf{p}_i \leftarrow \text{PathFromBundle}(\mathbf{b}_i^{\text{new}})$ Individual adjustment {Lightweight adjustment}
  $\mathcal{A}_{\text{adj}} \leftarrow \{i \mid d(a_i, \mathcal{E}) \leq R_{\text{adj}}\}$
  for $i \in \mathcal{A}_{\text{adj}}$ **do**
    $\mathbf{p}_i^{\text{new}} \leftarrow \text{TaskReschedule}(\mathbf{p}_i, \mathcal{E})$
    $\mathbf{b}_i^{\text{new}} \leftarrow \text{BundleUpdate}(\mathbf{b}_i, \mathbf{p}_i^{\text{new}})$
  end for
  Update global consensus: $\{\mathbf{y}_i, \mathbf{z}_i\} \leftarrow$ GlobalConsensusUpdate()
end if

4 Simulation Results

Suppose that after a disaster in a certain area, some rescue tasks are searched and a group of heterogeneous rescue vehicles are required to carry out the rescue. Our goal is to use heterogeneous vehicles to provide emergency rescue support as quickly as possible and obtain the maximum value. According to the tasks that may appear after the disaster, we consider four typical post-disaster task types: demolition, simple first aid, advanced first aid, and transportation. In the simulation, unmanned vehicles and rescue tasks are randomly generated. Assume that 8 unmanned vehicles from the same base are assigned to perform 20 tasks in a 1 km × 1 km two-dimensional task area.

4.1 Load Balancing

This experiment constructs an extremely unbalanced scenario to verify the effectiveness of the load balancing mechanism. Set Agent_0 as a high-speed node

($v_0 = 10.0$ m/s), and the rest of the agents as low-speed nodes ($v_i = 0.1$ m/s, $i \geq 1$), with a speed difference of 100 times. By comparing the task allocation results when the load balancing mechanism is turned off ($\mu = 0$) and turned on ($\mu = 100$), three quantitative indicators are used to evaluate the system load distribution characteristics:

- **JS divergence**: Characterizes the difference between the actual load distribution P and the ideal uniform distribution Q, calculated as
- **Maximum-minimum difference**$\Delta L = \max_i(L_i) - \min_i(L_i)$, which is used to reflect the extreme difference in load, L_i represents the load of the unmanned vehicle a_i.
- **Gini coefficient**: quantifies the degree of inequality in load distribution.

$$G = \frac{\sum_{i=1}^{n}\sum_{j=1}^{n}|L_i - L_j|}{2n^2\bar{L}} \tag{22}$$

where $\bar{L}$ is the average load, n is the number of Agents, and the value range is $[0, 1]$, where 0 represents complete balance.

Table 2 presents the comparison results of indicators before and after the load balancing mechanism is enabled:

Table 2. Load balancing mechanism performance indicator comparison

Evaluation indicators	not enabled	enabled	Change rate
JS divergence	0.0961	0.0957	−0.42%
Maximum-minimum difference	0.5000	0.4200	−16.00%
Gini coefficient	0.2692	0.2587	−3.90%

As can be seen from the Table 2, the uniformity of load distribution has improved: JS divergence has decreased by 0.42% (from 0.0961 to 0.0957), and the Gini coefficient has decreased by 3.90% (from 0.2692 to 0.2587). This change is mainly due to the load balancing mechanism's bidding penalty on high-load agents, forcing tasks to migrate to low-speed nodes. The designed load balancing mechanism performs well in effectively suppressing extreme differences: the maximum-minimum difference has significantly decreased by 16.00% (from 0.5000 to 0.4200), proving that the system has successfully alleviated the "busy and idle polarization" phenomenon. As shown in the Fig. 1, after the mechanism is enabled, the low-speed agent shares more tasks. Experimental results show that the load balancing mechanism achieves global fairness by trading local efficiency. Specifically, bidding suppression increases the task cost of high-load agents, prompts task migration, ensures that slow agents can obtain new tasks, thereby achieving global optimization and reducing the system's Gini coefficient. Although the execution efficiency of slow agents is low, tasks are still actively assigned to these nodes to avoid overloading fast agents. This strategy remains robust when the performance difference reaches 100 times.

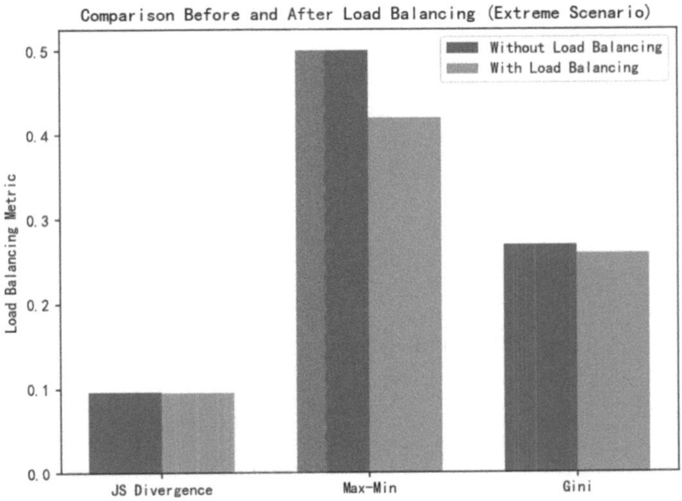

Fig. 1. Load balancing comparison results.

4.2 Reassignment Results

During the simulation, dynamic events mainly include: new task arrival (such as task 999 added at 5.0 s), task cancellation (such as task deleted at 12.0 s), task attribute change (such as task value change at 18.0 s), agent online/offline (agent 888 added at 22.0 s, agent 0 deleted at 32.0 s). These events are automatically detected and triggered during the simulation, and the system will dynamically adjust the task allocation plan according to the event type and impact range.

Figure 2 is the initial allocation plan. During the simulation, four types of dynamic events (task addition, task deletion, agent addition, agent deletion) are set at different time points. In the event of the addition of new task 999 at $t = 5s$, the adaptive mechanism triggers the local reallocation strategy, and only adjusts the task sequence of the affected agent 0. Figure 3 shows that only the task sequence of Agent 0 is adjusted, from the original task [8,16,4] to the new task [999,8,16,4], which does not affect other agents. When $t = 12s$ high-value task 0 is deleted, the system performs cluster-level reallocation, and the optimization range covers the adjacent agents. Tasks 3, 16, and 19 have been completed, and only the affected agents are reallocated. At $t = 22s$, agent 888 is added. The new agent and related agents trigger cluster-level reallocation. The new agent immediately takes over task 12, and the system structure changes. $t = 32s$ core agent 0 fails and is automatically upgraded to global reallocation, and the tasks of the original agent 0 are reallocated to other agents. Compare the allocation time and system cumulative benefits of CBBA-AMR and global allocation based on CBBA when dynamic events occur. The experimental results Figs. 7, 8 show that the allocation time of adaptive multi-granularity reallocation in each event response is significantly lower than that of global reallocation, usually only 1 to

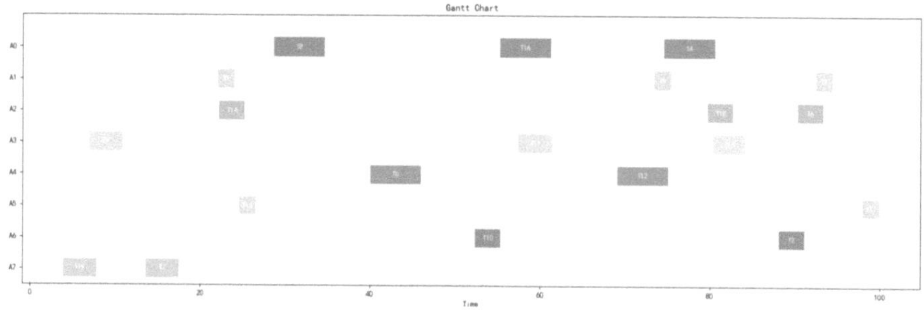

Fig. 2. Initial allocation result Gantt chart.

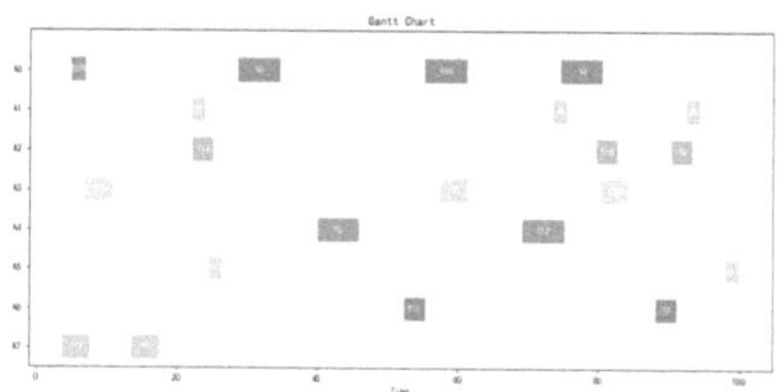

Fig. 3. Gantt chart of the result of the reallocation of the newly added task 999.

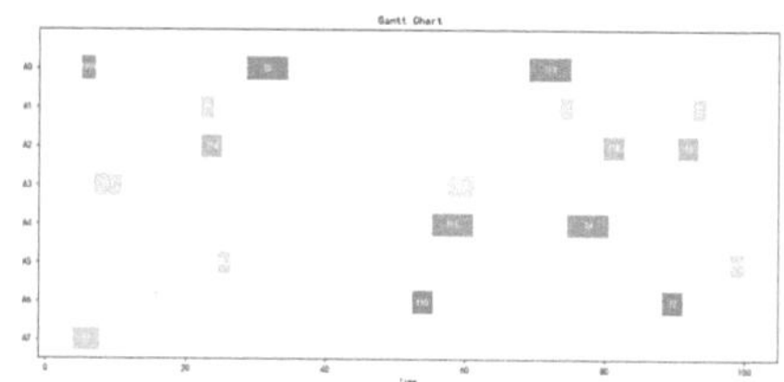

Fig. 4. Delete Task 0 and reassign the result Gantt chart.

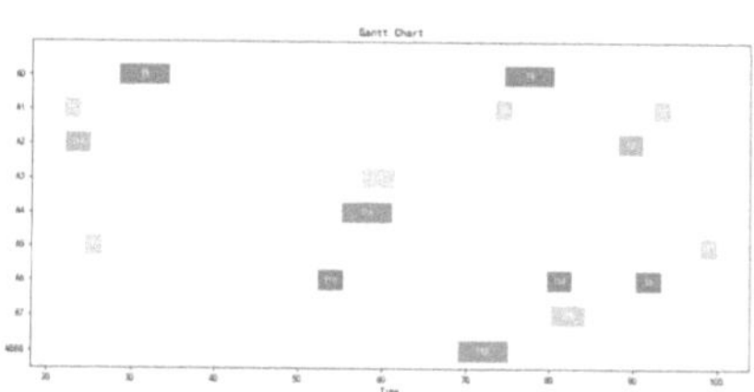

Fig. 5. Added Gantt chart of agent 888 reallocation results.

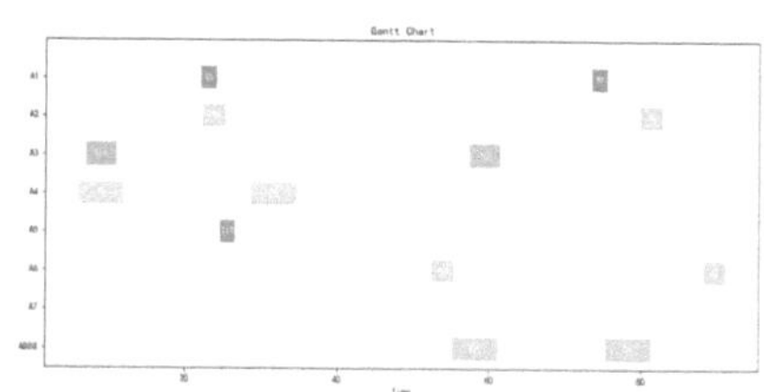

Fig. 6. Delete the Smart Weight Distribution Result Gantt Chart.

70 milliseconds, while global reallocation requires about 100 milliseconds. The average response time of the adaptive mechanism under each event is 0.015 s, which is 6.9 times faster than global reallocation. Especially in lightweight events such as task attribute changes, the response speed is improved by more than 80 times. Thanks to the adaptive mechanism that only reallocates the affected local agents and tasks, the real-time and scalability of the system are greatly improved. At the same time, under various dynamic events, the cumulative benefits of the adaptive scheme are higher than those of global reallocation (Figs. 4, 5 and 6).

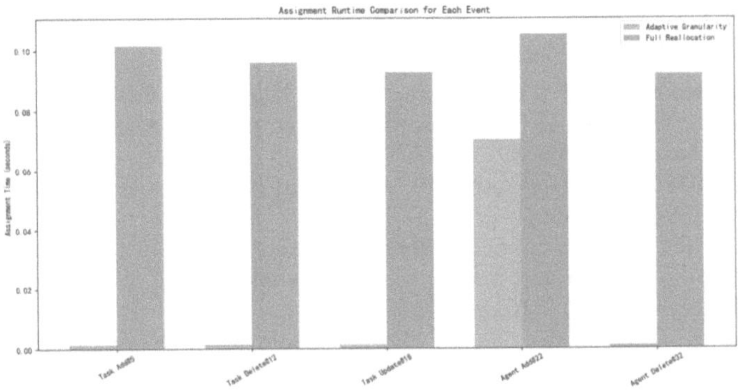

Fig. 7. Adaptive granular redistribution and global redistribution time histogram.

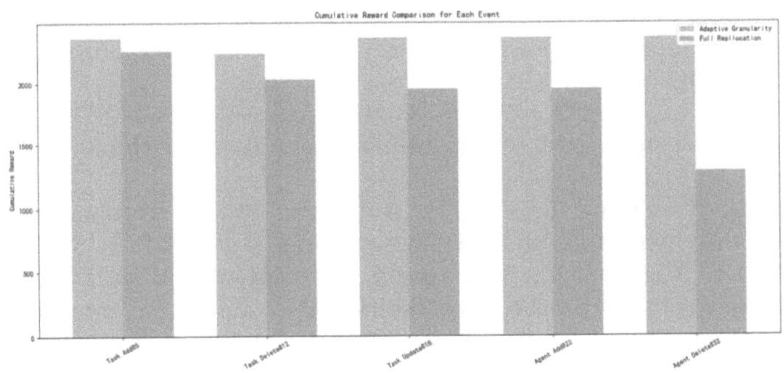

Fig. 8. Adaptive granular redistribution and global redistribution benefit bar chart.

5 Conclusion

In this paper, we constructed a mathematical model for post-disaster rescue task allocation and designed a load balancing optimization mechanism to solve the problem of "uneven busyness" of traditional CBBA agents. We constructed an event response measurement function, designed a multi-level response strategy, and proposed an adaptive multi-granularity dynamic reallocation mechanism CBBA-AMR to provide a lightweight response solution for post-disaster dynamic environments. Simulation results show that in extremely heterogeneous scenarios, the load balancing mechanism significantly improves the task distribution, and CBBA-AMR is effectively improved in both allocation time and cumulative task benefits compared to the global reallocation scheme. In future research, we will further explore optimizing this mechanism to cope with more complex environmental changes and improve the efficiency of collaboration between agents.

References

1. Luo, C., Espinosa, A.P., Pranantha, D., De Gloria, A.: Multi-robot search and rescue team. In: 2011 IEEE International Symposium on Safety, Security, and Rescue Robotics, pp. 296–301. IEEE (2011)
2. Drew, D.S.: Multi-agent systems for search and rescue applications. Curr. Robot. Rep. **2**(2), 189–200 (2021)
3. Queralta, J.P., Taipalmaa, J., Pullinen, B.C., Sarker, V.K., Gia, T.N.: Collaborative multi-robot search and rescue: planning, coordination, perception, and active vision. IEEE Access **8**, 191617–191643 (2020)
4. Ghauri, S.A., Sarfraz, M., Qamar, R.A., Sohail, M.F., Khan, S.A.: A review of multi-UAV task allocation algorithms for a search and rescue scenario. J. Sensor Actuator Netw. **13**(5), 47 (2024)
5. Alqefari, S., Menai, M.E.B.: Multi-UAV task assignment in dynamic environments: current trends and future directions. Drones **9**(1), 75 (2025)
6. Huang, S., Teo, R.S., Kwan, J.L., Liu, W., Dymkou, S.M.: Distributed UAV loss detection and auto-replacement protocol with guaranteed properties. J. Intell. Robot. Syst. **93**(1–2), 303–316 (2019)
7. Bi, W., Shen, J., Zhou, J., Zhang, A.: Heterogeneous multi-UAV mission reallocation based on improved consensus-based bundle algorithm. Drones **8**(8), 345 (2024)
8. Yang, M., Bi, W., Zhang, A., Gao, F.: A distributed task reassignment method in dynamic environment for multi-UAV system. Appl. Intell. **52**(2), 1582–1601 (2022)
9. Wang, G., Lv, X., Yan, X.: A two-stage distributed task assignment algorithm based on contract net protocol for multi-UAV cooperative reconnaissance task reassignment in dynamic environments. Sensors **23**(18), 7980 (2023)
10. Lv, X., Wang, G., Chen, J.: Multi-UAV cooperative reconnaissance task allocation based on ieppso algorithm. In: Sun, Y., Lu, T. (eds.) Computer Supported Cooperative Work and Social Computing, pp. 119–129. Springer, Singapore (2024)
11. Choi, H.-L., Brunet, L., How, J.P.: Consensus-based decentralized auctions for robust task allocation. IEEE Trans. Robot. **25**(4), 912–926 (2009)
12. Buckman, N., Choi, H.L., How, J.P.: Partial replanning for decentralized dynamic task allocation. In: AIAA Scitech 2019 Forum. American Institute of Aeronautics and Astronautics (2019)
13. Chen, J., Qing, X., Ye, F., Xiao, K., You, K., Sun, Q.: Consensus-based bundle algorithm with local replanning for heterogeneous multi-UAV system in the time-sensitive and dynamic environment. J. Supercomput. **78**(2), 1712–1740 (2022)
14. Korsah, G.A., Stentz, A., Dias, M.B.: A comprehensive taxonomy for multi-robot task allocation. Int. J. Robot. Res. **32**(12), 1495–1512 (2013)

Complete Coverage Path Planning Based on Comprehensive Improved Particle Swarm Optimization Algorithm

Jian Fu(✉) and Xinlei Yin

Wuhan University of Technology, 430070 Wuhan, China
fujian@whut.edu.cn

Abstract. Complete Coverage Path Planning (CCPP) in unstructured environments presents a critical challenge due to the necessity of balancing high coverage efficiency and path accuracy. Traditional Particle Swarm Optimization (PSO) is effective but tends to converge prematurely to local optima in complex scenarios. The paper proposes a Comprehensive Improved Particle Swarm Optimization (CIPSO) algorithm, which enhances conventional PSO through three targeted modifications to tackle the challenge in scenarios involving randomly distributed stains: (1) Dynamic merging of redundant path points reduces excessive revisitation; (2) A particle evaluation parameter mechanism is presented to adjusts learning factors to alleviate premature convergence; and (3) Hybridization with genetic algorithm strategies enhances global exploration capabilities. Experimental evaluations were performed in an 80 cm × 100 cm blackboard with randomly distributed stains, comparing the proposed CIPSO against standard PSO and dynamic-weight PSO variants. Results indicate that CIPSO achieves a 23% reduction in comprehensive costs compared to standard PSO and outperforms dynamic-weight IPSO by 15%. Furthermore, CIPSO shows strong stability across randomized trials, effectively balancing efficiency and accuracy in unstructured path planning, offering a promising CCPP optimization framework.

Keywords: Unstructured Environment · Complete Coverage Path Planning · Particle Swarm Optimization Algorithm · Genetic Algorithm · Gaussian Mutation

1 Introduction

With the gradual penetration of robots into precision manufacturing, medical surgery, disaster rescue and other fields with high requirements for operation accuracy and environmental adaptability, the limitations of traditional robot technology in path planning efficiency have become increasingly prominent. When facing unstructured scenarios such as fieldwork and complex terrain environments, as well as complex coverage areas, the accuracy and computational efficiency have decreased. To solve such complex problems, researchers have conducted extensive research and innovation and proposed many methods to solve the problem of full coverage path planning in unstructured environments.

H. Ma et al. (Eds.): IWACIII 2025, CCIS 2780, pp. 31–47, 2026.
https://doi.org/10.1007/978-981-95-6730-0_3

Zhang et al. [1] proposed the spiral A* algorithm based on raster maps. When dealing with regular planes like the floor of a library, it can effectively solve the problem that traditional raster maps fall to local optimum. However, when dealing with complex edges with irregular shapes, the full coverage rate will be significantly reduced. Meanwhile, Zhao et al. [2] reduced the search space and time of the path planning algorithm through the map simplification strategy and were able to calculate high-quality paths more quickly when facing unstructured environments. However, for the path planning algorithm with more concentrated path points in the full coverage path problem, the optimization performance of the algorithm was significantly reduced. Wu et al. [3] proposed an improved genetic algorithm based on the Boustrophedon decomposition method, which effectively reduced the repetition of the total path with full coverage and improved the efficiency and quality of mechanical operations.

The Travelling Salesman Problem (TSP) and the Coverage Path Planning (CPP) problem, which are the problems of searching for the shortest path at fixed points, have been studied relatively maturely. Jha et al. [4] expressed the TSP problem as an NP problem of graph optimization. By using geometric methods such as computational geometry and convex optimization, they effectively found the route with a length close to the optimal one. Xie et al. [5] combined the TSP problem with the CPP problem, introduced a CPP method covering a single polygonal region, and provided a heuristic method for dynamic programming optimization and heuristic efficient path generation, effectively ensuring the optimality and efficiency of path planning. This path optimization method, which requires pre-determining the path points and then sorting them in the order of the shortest connection lines, has high requirements for the performance of the model algorithm for initially extracting the path points. And in an unstructured environment, the emerging area is not necessarily a fixed large polygonal area. It could also be a Complete Coverage Path Planning (CCPP) problem with small block partitions, which required block processing. Or dynamically adjust the path points to adapt to this unstructured problem.

In contrast, the Particle Swarm Optimization (PSO) algorithm, which uses individual and social optimal solutions to find new optimal solutions [6], does not rely entirely on the problem itself but approximates the temporary optimal solution, and is more suitable for solving problems in unstructured environments. The PSO has the advantages of strong global search ability, good dynamic adaptability and fast convergence velocity. However, at the same time, it also has problems such as being prone to fall into local optimum [7]. Many researchers have also conducted corresponding optimization studies. Zhang et al. [8] proposed an optimization method combining domain heuristic search and Gaussian cloud learning, which effectively improved the search velocity of the algorithm, balanced the comprehensive ability of global exploration and local optimization of path optimization, and verified the robustness and accuracy of the algorithm through experiments. Huang et al. [9] faced complex function problems such as nonlinearity and non-convex. Through strategies such as adaptive nonlinear weights, they effectively improved the convergence velocity of the algorithm in the later stage. They successfully verified the effectiveness of the algorithm through multiple experiments using multiple sets of basic functions. Such optimization measures have certain guiding significance for discrete point actual path planning problems like CCPP. Wang et al. [10] utilized the improved particle swarm optimization algorithm to adaptively adjust the proportional, integral and differential parameters of the PID control system for electric vehicle charging piles, thereby enhancing the adaptability and anti-interference ability of the system.

In view of the problems that the existing algorithms must rely on the accuracy of path point extraction when facing unstructured CCPP tasks, and are prone to fall into local optimum during the optimization process from the traditional dynamic weight PSO algorithm [11], we propose a comprehensive improved CIPSO algorithm to address the unstructured stain CCPP problem on the rule board, which adds particle evaluation parameters and genetic hybridization algorithm on the basis of the standard PSO algorithm. The state of the particles during optimization is predicted, and through this evaluation parameter, each particle is iterated using an appropriate strategy in the calculation process, to achieve the purpose of improving the accuracy and stability of the algorithm.

To optimize the target algorithm and conduct relevant validation, this paper proposes to introduce innovations in the following aspects:

- The dynamic adjustment of the intermediate path points is introduced in this algorithm. In practice, when the robot moves along the path, there must be a corresponding width. The distance between each path is regarded as the optimization target. When the distance between the paths is lower than the preset threshold, it will be considered as repeated coverage. Merging will be carried out without affecting the final coverage effect to achieve the purpose of reducing redundant path points, accelerating the convergence of the algorithm and reducing the repeated coverage rate.
- The particle evaluation parameters combined with a genetic algorithm hybridization technique enhanced by Gaussian variation is presented. Each particle determines whether to conduct local optimization or global exploration based on the state of the corresponding particle at this time. The individual learning factor c_1 and the social learning factor c_2 are dynamically adjusted. When the change value of the cost function of the path is less than a certain range, the hybridization and mutation of the genetic algorithm are triggered to improve the ability of the model to find the global optimal solution.
- Experimental verifications are conducted. Under the randomly generated area to be covered, comparative experiments were conducted using the traditional PSO algorithm and the optimized IPSO algorithm to verify the global optimization of the improved algorithm and the robustness of the algorithm.

The rest of this paper is organized as follows. In the second section, we introduced the basic formulas and algorithm flow of the particle swarm optimization algorithm. Then, in the third section, we describe the methodological improvements based on the basic architecture of the particle swarm to address the shortcomings and defects of the traditional algorithm. We proposed CIPSO and provided the corresponding formulas and improvement flowcharts. In the fourth section of the article, we built the corresponding model framework and designed a series of experiments to verify the performance of the model we proposed. Finally, through the comprehensive analysis of the experimental data, we drew a conclusion and looked forward to the future research direction.

2 Theoretical Basis

2.1 Particle Swarm Optimization

PSO is a meta-heuristic optimization algorithm inspired by the foraging behavior of bird flocks. The core idea is through collaboration and information sharing between individuals in the group, to find the optimal solution in the search space. In PSO, each particle has the position $X_i = (x_{i1}, x_{i2}, \ldots, x_{iD})$ and velocity $V_i = (v_{i1}, v_{i2}, \ldots, v_{iD})$ in the D-dimensional space [12]. The update formulas for its velocity and position are as follows:

$$v_{id}(t+1) = w * v_{id}(t) + c_1 * r_1 * (p_{id} - x_{id}(t)) + c_2 * r_2 * \left(p_{gd} - x_{id}(t)\right) \quad (1)$$

$$x_{id}(t+1) = x_{id}(t) + v_{id}(t+1) \quad (2)$$

w is the inertia weight, which controls the degree to which the particle inherits the historical velocity. c_1 and c_2 are acceleration constants, which are used to adjust the step sizes for particles to learn from the individual optimal and global optimal positions, respectively. r_1 and r_2 are random numbers within the interval of [0, 1]. p_{id} and p_{gd} respectively represent the current individual optimal path and the global optimal path.

Meanwhile, in order balance the global search and local development capabilities of the algorithm, inertial linearly decreasing weights are often adopted [13]:

$$w(t) = w_{max} - \frac{w_{max} - w_{min}}{T_{max}} * t\# \quad (3)$$

w_{max} and w_{min} are the maximum and minimum inertial weights. T_{max} is the maximum number of iterations.

The basic algorithm process of PSO is: Firstly, initialize the position and velocity of the particle swarm, and calculate the fitness value of each particle; Then in each iteration, the velocity and position of the particles are updated according to the above formula, and the individual optimum and global optimum are updated. Finally, when the convergence conditions such as the number of iterations or the fitness change threshold are met, the global optimal solution is output.

2.2 Theoretical Issues

The core features of an unstructured environment are irregular coverage areas, the need for dynamic adjustment of path points, and the coexistence of requirements for coverage integrity and path efficiency. However, the inherent design of traditional PSO makes it difficult to adapt to these requirements.

Firstly, the traditional PSO algorithm ignores the physical coverage width of the robot and ensures the efficiency of the algorithm search by selecting the initial path points. However, in an unstructured environment, it lacks stability of algorithm optimization.

Secondly, the static adaptation of parameters is insufficient. w only decreases linearly with the number of iterations. c_1 and c_2 are fixed and cannot dynamically switch between global exploration and local optimization states, which may lead to incomplete coverage or path redundancy.

Finally, it is easy to fall into local optimum. Each iteration of optimization relies on the guidance of p_{id} and p_{gd}, resulting in the convergence of particle velocities and the contraction of the search space. Without a strategy that breaks away from the local optimum, it is prone to premature convergence in the regular areas of the unstructured environment and cannot guarantee complete coverage.

2.3 Optimization Strategy

The optimization efficiency of traditional Particle Swarm Optimization (PSO) heavily relies on the selection of initial path points. To address this limitation, a dynamic merging strategy has been developed. This strategy identifies and merges redundant path points using the physical coverage width of the robot as a criterion. By doing so, it eliminates duplicate coverage areas, shortens the total path length, accelerates convergence, and effectively resolves the issue of path redundancy. Meanwhile, to ensure the adaptive change of the iterative update strategy, we have introduced the dynamic adjustment strategies of w, c_1, and c_2, and introduced hybridization and mutation operations to break the local optimum constraints, addressing the defects of static parameters and the tendency to fall into local optimum.

3 Improved Algorithm

The CIPSO algorithm in this paper mainly focuses on full-coverage path planning for randomly generated paths with widths. Next, we will introduce the problems existing in the traditional PSO algorithms mentioned above, including dynamic merge, dynamic weight and genetic update:

3.1 Dynamic Merge

Since the research background of this paper is an unstructured environment, the grid uniform division extraction path algorithm of the initial extraction path cannot guarantee the optimum. Therefore, the path points need to be dynamically adjusted. After the optimal path is found through optimization, the path points need to be adjusted before updating the individual optimal path and the global optimal path. For the path points that are close enough, they will be merged. Meanwhile, the regional coverage degree and cost function of the new path are re-evaluated to achieve the goal of minimizing the repetition of the path and reducing the intermediate path points under the condition of meeting the requirements.

Taking the grid structure as the background, when optimizing a path, the path points will first be placed in the corresponding squares in sequence, and at the same time, the squares with the specified radius will be filled. When a certain point is found to have been filled during the traversal, the dynamic merging of path points strategy will be triggered.

For points p_i and p_j that meet the conditions, point p_j will be directly merged with p_i and will not be added to the optimized path:

$$p'_{ix} = 0.5 * (p_{ix} + p_{jx})\# \tag{4}$$

$$p'_{iy} = 0.5 * \left(p_{iy} + p_{jy}\right)\# \tag{5}$$

And to avoid large-scale gaps as much as possible after the path points are merged, the corresponding path starting points of the two points p_{i-1} and p_{j-1} will be corrected to be brought closer together:

$$p'_{(i-1)x} = 0.25 * p_{(j-1)x} + 0.75 * p_{(i-1)x}\# \tag{6}$$

$$p'_{(i-1)y} = 0.25 * p_{(j-1)y} + 0.75 * p_{(i-1)y}\# \tag{7}$$

$$p'_{(j-1)x} = 0.75 * p_{(j-1)x} + 0.25 * p_{(i-1)x}\# \tag{8}$$

$$p'_{(j-1)y} = 0.75 * p_{(j-1)y} + 0.25 * p_{(i-1)y}\# \tag{9}$$

After traversing all the points, a new path that has been dynamically merged will be obtained. To ensure that the new path can fully cover all the points along the path, the coverage rate of this new path needs to be calculated once. If the coverage rate is lower than 0.95, this merge will be abandoned.

The algorithm flowchart is shown in Fig. 1.

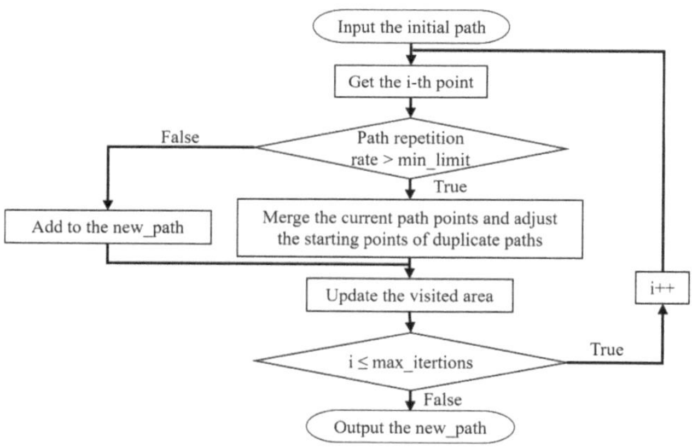

Fig. 1. Dynamic merging path point judgment process.

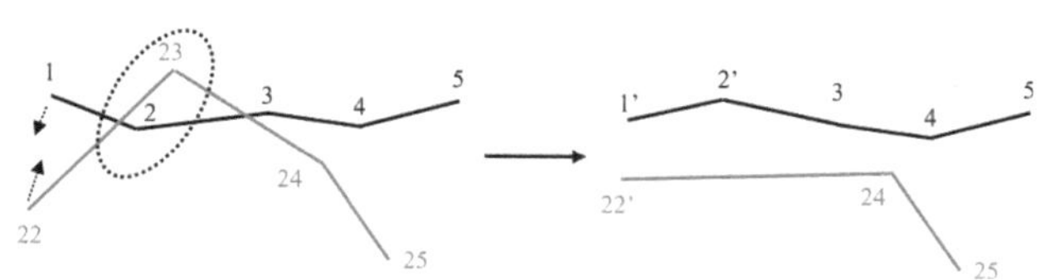

Fig. 2. Schematic diagram of merging path points.

When two fragment paths of a particle intersect during the iteration process, their path endpoints will first be merged, and then the starting points of the two paths will be brought closer to each other. When the coverage rate of the merged paths meets the requirements, one iteration is completed, ensuring the overall coverage effect while reducing the repetition of the paths. The schematic diagram is shown in Fig. 2.

3.2 Dynamic Weight

To ensure an appropriate strategy is consistently applied during particle updates, we maintain records of historical particle states to predict their current state and dynamically adjust the weight parameters (w,c_1,c_2) of the iterative algorithm.

w represents the inertia weight, mainly indicating the degree of influence of the historical velocity of the particle. The larger its value is, the greater the particle's movement velocity, the wider the movement range, and the stronger the global search ability. Conversely, it tends to be local optimization. Dynamic weights are adopted in this paper as:

$$\begin{cases} w(t) = w_{min} + (w_{max} - w_{min}) * sigmod(s_{id}(t))\# \\ s_{id}(t) = \sum_{x=1}^{\min(count,\ t)} \frac{v_{id}(t-x)}{count} \qquad \#\# \end{cases} \tag{10}$$

where w_{min} and w_{max} represent the minimum and maximum weights respectively, while s_{id} represents the historical evaluation state of the current particle, which is used to represent the changing trend of the particle. i represents the index of the current particle, and d represents the dimension of the particle.

Here, we use s_{id} instead of the number of iterations t to determine the current corresponding strategy pattern in real time based on s_{id} instead of the fixed strategy of global search first and then local search in the traditional PSO algorithm. Meanwhile, to ensure that the inertia weight of each iteration does not increase infinitely due to the recent particle movement velocity evaluation state s_{id}, we introduce the sigmoid function to wrap s_{id}, ensuring that the inertia weight remains within a limited range and guaranteeing that the algorithm can converge effectively.

c_1 is the individual learning factor, mainly controlling the tendency of the particle to move towards its own historical optimal position p_i. It can be understood as the self-learning intensity of the particle. The larger its value is, the more inclined it is to search near the historical optimal, and the local optimization effect of the model is obvious. c_2 is the social learning factor, which controls the tendency of particles to move towards the global optimal p_g of the population. It can be regarded as the magnitude of the social influence received by the particles. The larger this value is, the stronger the global influence of the particle is, and the global search effect of the model is obvious.

For c_1 and c_2, these are a pair of contradictory parameters. It is impossible for a particle to be in both the state of global exploration and local optimization simultaneously [16]. Meanwhile, ensure that the velocity state of the particle does not increase infinitely, causing the model to be difficult to converge, the following strategy is adopted for parameter update:

$$c_2(t) = C * sigmod(s_{id}(t)) + C_1\# \tag{11}$$

$$c_1(t) = C - c_2(t)\# \quad (12)$$

C represents a fixed value, which is used to limit the maximum values of c_1 and c_2 to prevent the parameters from being adjusted too high, resulting in abnormal particle motion that is difficult to converge.

C_1 represents a correction value to ensure that the values of c_1 and c_2 are within a reasonable range.

Thus, the following improved and updated recursive formula is given:

$$v_{id}(t+1) = w(t) * v_{id}(t) + c_1(\mathrm{t}) * r_1 * (p_{id} - x_{id}(t)) + c_2(\mathrm{t}) * r_2 * \left(p_{gd} - x_{id}(t)\right) \quad (13)$$

$w(t)$ is the dynamic inertia weight, which changes dynamically according to the historical velocity of the particle to achieve the purpose of predicting the state of the particle. $c_1(t)$ and $c_2(t)$ are dynamic acceleration constants, which are used to dynamically adjust the step size for particles to learn from the individual optimal and global optimal positions. p_{id} and p_{gd} respectively represent the current individual optimal path and the global optimal path.

When the movement velocity of the particles is lower than a certain threshold, in order to prevent falling into local optimum, the hybridization and mutation of the genetic algorithm will be used to re-stitch and exchange the path, ultimately achieving the purpose of getting out of local optimum and re-exploring the global situation, and improving the adaptability of the algorithm.

3.3 Genetic Update

At the particle hybridization stage, particles will be recombined and connected with their own individual optima and global optima [14]. The pseudo-code is shown in Fig. 3.

Particle hybridization generates a new path by recombining fragments of the current particle path and the optimal path, avoiding particles getting stuck in local optimal stagnation. The specific steps are as follows:

Trigger condition judgment: When the cost function change value of the particle path is less than the preset range, it is determined that the particle may have fallen into local optimum, triggering the hybridization operation.

Extract continuous sub-paths: Randomly select a continuous path point from the individual optimal path and the global optimal path respectively as the optimized sub-path fragment.

Path fragment concatenation: Based on the current particle path, the continuous sub-path fragments of the individual optimal and global optimal paths respectively replace the random continuous fragments of the same length in the current path. At the same time, the mean value of the transition path points at the splicing points is calculated to ensure the smoothness of the new path.

Screen the optimal path: Calculate the comprehensive cost function value of the two candidate paths generated after concatenation and the original current path, and select the path with the lowest cost as the new path after particle update; If the costs of the

Algorithm: **Single Crossover Process**

Input: CP (current iteration path), **GBP** (global optimal path), **IBP** (individual optimal path), **calcF** (Path) (fitness/cost calculation function)

Output: FinalPath (minimum cost path of this iteration)

```
CrossoverP1=cross(CP,GBP) // Crossover with global optimal path
CrossoverP2=cross(CP,IBP) // Crossover with individual optimal path
Cost_Current= calcF(CP)
Cost_Cross1= calcF(CrossoverP1)
Cost_Cross2= calcF(CrossoverP2)
MinCost = min(Cost_Current, Cost_Cross1, Cost_Cross2)
if MinCost = = Cost_Current:
   FinalPath = CurrentPath
elif MinCost = = Cost_Cross1:
   FinalPath = CrossoverPath1
else:
   FinalPath = CrossoverPath2**

return FinalPath
```

Fig. 3. Particle hybridization algorithm.

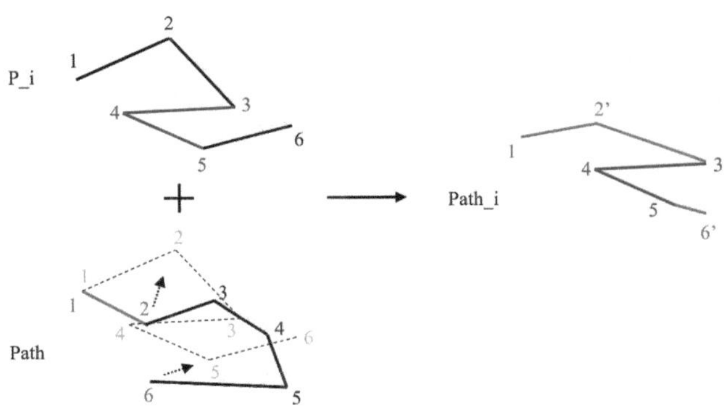

Fig. 4. Schematic diagram of single path hybridization.

candidate paths are all higher than those of the original path, the original path will be retained.

During the hybridization process between two paths (as illustrated in Fig. 4), we begin by selecting a random continuous segment of length 3–5 from an individual's optimal path. Subsequently, we identify the remaining points (excluding those within the chosen segment) from the current path, specifically points 1, 2, and 6, and splice these into the new path configuration. To ensure smooth transitions and comprehensive coverage, intermediate points (e.g., points 2 and 6 in the revised path) are recalculated by averaging their coordinates with corresponding points from the original optimal path. This method generates a novel path structure that enhances the model's ability to escape local optima.

Gaussian Mutation perturbs individual by using Gaussian distribution to make particles change randomly, so that it can search more widely in the solution space and avoid the algorithm falling into local optimum prematurely [15]:

$$x'_{id}(t) = x_{id}(t) + sigmod(s_{id}(t)) * \mathbf{N}(0, 1)\# \tag{14}$$

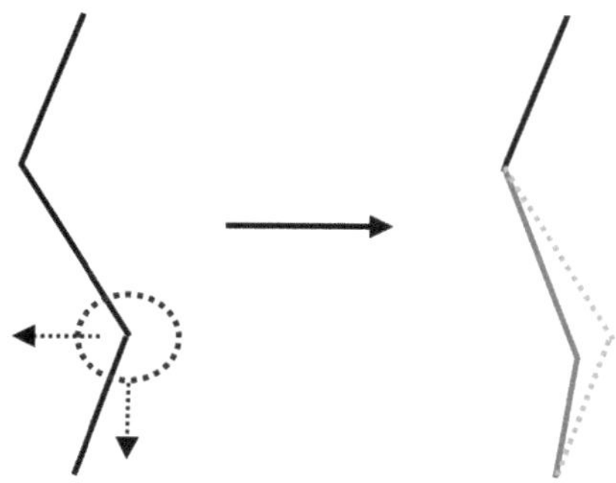

Fig. 5. Single path point Gaussian Mutation.

The specific operation diagram is shown in Fig. 5. When the model triggers Gaussian Mutation, we select several points for random offsets. Since our experiment is conducted on a two-dimensional plane, there are offsets in both the horizontal and vertical directions in the figure.

The overall process of the above-mentioned optimization strategy in the entire algorithm is shown in Fig. 6.

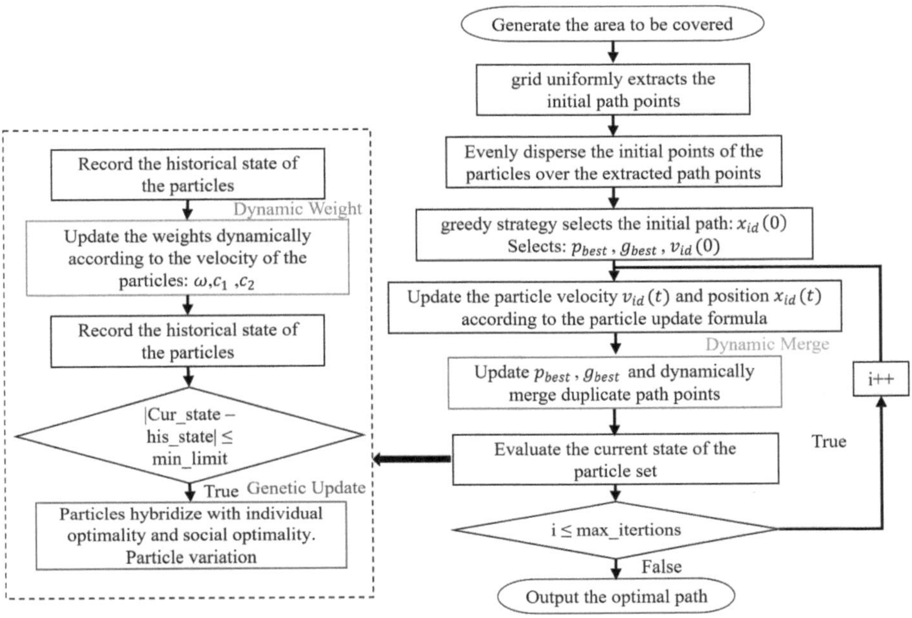

Fig. 6. CIPSO algorithm flowchart.

4 Experiment

In order verify the comprehensive improved CIPSO algorithm proposed in this paper, a comparative experiment was conducted with the standard PSO algorithm and the IPSO algorithm with only dynamic weights. We use a two-dimensional array to simulate the plane to be cleaned. In this array, the target area is randomly generated to simulate the area to be covered. In the same area, three algorithms are used for path planning to analyze the advantages and disadvantages of the algorithms. Each algorithm explores the optimal path in the same randomly generated area to be covered. The cost function values that integrate the motion path, the completion degree of cleaning, and the cumulative turning Angle are used to characterize the quality of the path.

$$f_{cost}(t) = \alpha * A(t) * L(t) + \beta * \theta(t)\# \tag{15}$$

$$A(t) = \begin{cases} \infty & c(t) < 0.95 \\ \frac{1}{c(t)} & c(t) > 0.95 \end{cases} \tag{16}$$

This article mainly focuses on optimizing the coverage time in the full coverage path planning, aiming to complete the full coverage of the designated area in a shorter time. Therefore, the path length should be considered first. However, given that too many large angle turns can also lead to a decrease in movement velocity, we prioritize the coverage path and then consider the cumulative corner size. Here, we make use of $s = \theta(t) * R$

$(R = \beta/\alpha)$to make an approximate transformation in order to reconcile two criteria length and angle.

α represents the weight of the covered path length, which is taken as 1 in this paper; $A(t)$ represents the generation value coefficient corresponding to the path coverage rate; $c(t)$ represents the path coverage rate; $L(t)$ represents the total path length; β represents the weight coefficient of the cumulative turning Angle; $\theta(t)$ represents the cumulative Angle of the turn.

The number of iterations is uniformly selected as 500 times. The experiment was run on python 3.9.13.

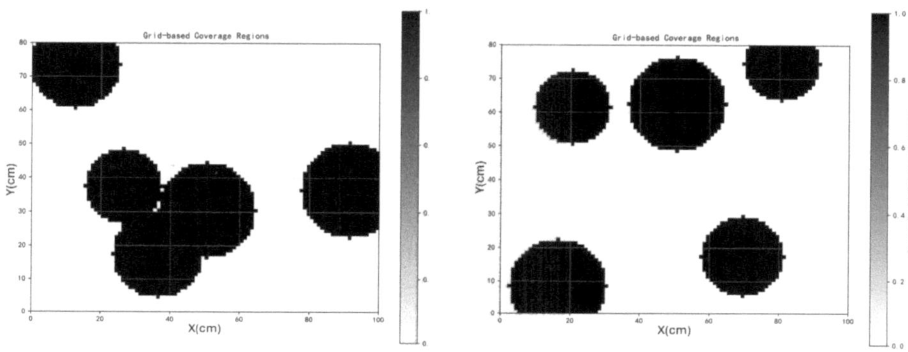

Fig. 7. Randomly generated areas to be covered.

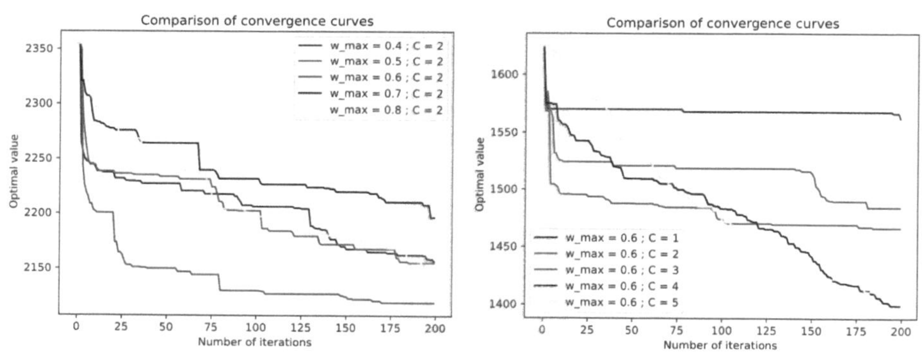

Fig. 8. Comparison of Parameters in different groups. The comparative experiment on $\boldsymbol{w_{max}}$ (left); A comparative experiment on $\boldsymbol{C}$ (right)

4.1 Modeling

In an unstructured context, the area to be covered cannot be a completely regular figure. It is very likely to be a complex area composed of multiple irregular figures and multiple blocks. For this purpose, this paper adopts randomly generated regions for experimental verification. On an 80*100 background board, circular regions with radii ranging from 10 to 20 units of length are randomly formed at certain points to simulate the areas to be covered by multiple irregular contour graphics, as shown in Fig. 7. Meanwhile, we set the end of the robot used to cover the target area as a rhombus with a diagonal length of 5 to simulate the coverage situation under the real motion trajectory.

4.2 Parameter Setting

The parameter setting of the CIPSO algorithm aims to determine the optimal initial parameter combination that adapts to the unstructured environment CCPP task, ensuring the algorithm balances global exploration, local optimization, and convergence stability. This section conducts targeted optimization experiments based on the paper's experimental framework (80*100 randomly generated coverage area, robot end as a rhombus with diagonal length 5, 500 iterations, Python 3.9.13 environment), focusing on the two core parameters of inertia weight maximum w_{max} and fixed value C, and clarifies the setting logic and basis.

The selection of w_{max} and C is closely linked to the algorithm's performance in unstructured environments: w_{max} is the upper limit of the dynamic inertia weight (Formula 10). Its value directly affects the global exploration ability of particles, too large value leads to particle motion oscillation and increased path redundancy, while too small a value results in insufficient exploration and difficulty in covering scattered areas.

C limits the maximum sum of the dynamic learning factors c_1 and c_2 (Formulas 11, 12). It prevents excessive adjustment of c_1 and c_2 from causing particle velocity loss control, which is crucial for maintaining the stability of path optimization.

The optimization experiment results are shown in Fig. 8.

First, we fix the values of the individual and social factors and C as the empirical value 2 [12]. The value of w_{max} changes from 0.4 to 0.8. The optimal path effect is optimal when w_{max} is set to 0.6, and the iteration efficiency is the highest currently. So, we take the value of w_{max} as 0.6. After fixing the value of w_{max} to the local optimal result of 0.6, the value of C is optimized. The model achieves optimal performance in terms of optimization when C is set to 4 or 5, as illustrated in the figure. At these values, the model demonstrates the most effective ability to escape local optima, balancing efficiency and accuracy in its convergence behavior. Here, we select the smaller one, which is 4.

Finally, we choose the initial weight parameters w_{max} of the model as 0.6 and C as 4.

4.3 Comparative Experiment

Under the condition of selected parameters, the standard PSO algorithm, the IPSO algorithm with dynamic weights and the CIPSO algorithm proposed in this paper were used for comparative experiments. The random area to be covered in the experiment is shown in Fig. 9. The experimental results are illustrated in Fig. 10, with data displayed sequentially from top to bottom for PSO, IPSO, and CIPSO respectively.

Compared with the PSO algorithm, the IPSO has a smaller optimal value. This is because the IPSO introduces dynamic weights related to the current state of the particles.

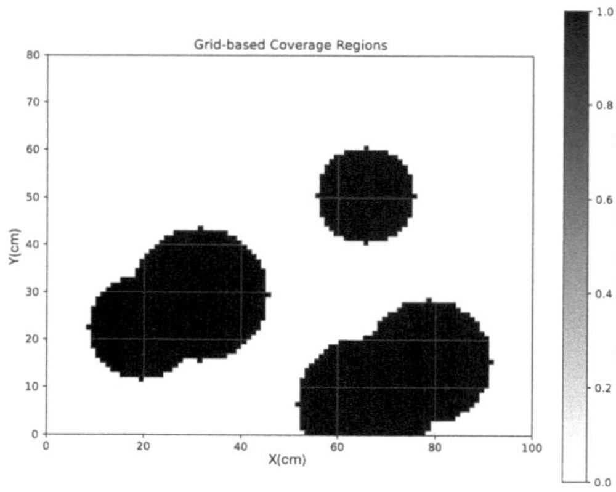

Fig. 9. Comparative experimental area.

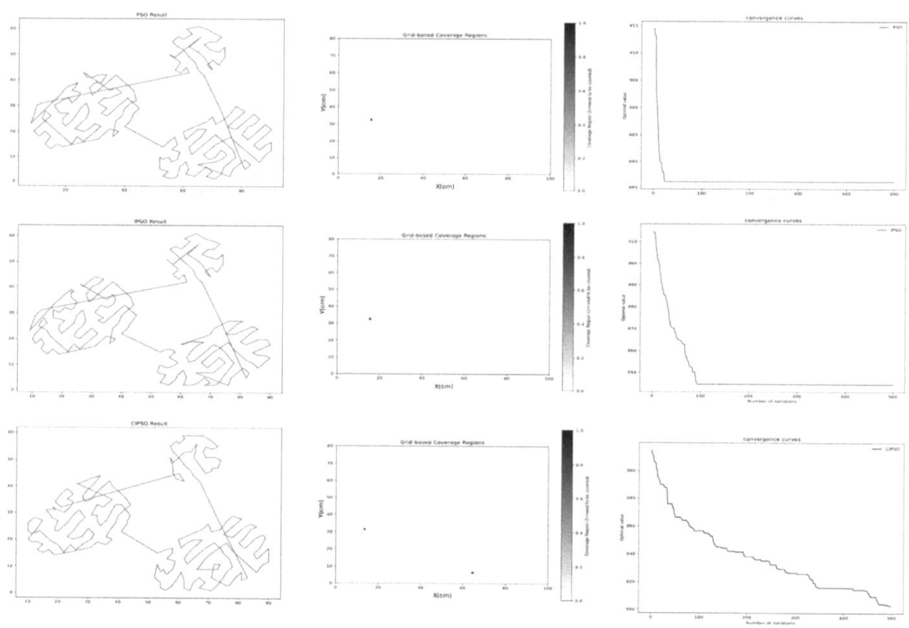

Fig. 10. The optimization results in an unstructured environment. Optimal paths of different algorithms (left column); The final remaining uncovered area (middle column); The iterative evolution of loss values (right column)

Such parameters enable the model to adjust the parameters for global search or local optimization at different stages, allowing the model to find a better convergence value.

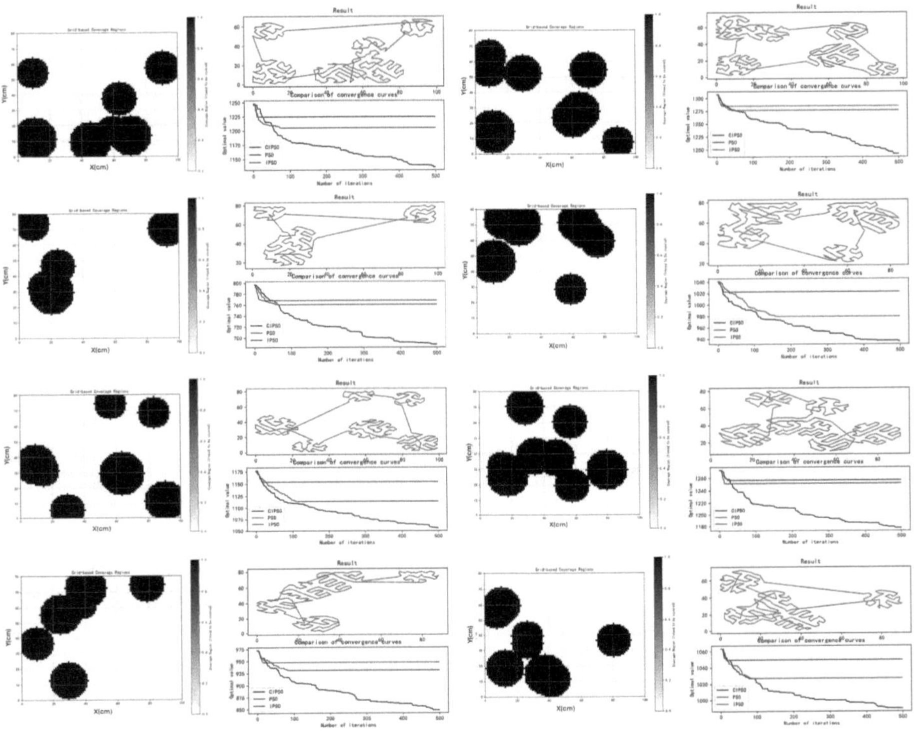

Fig. 11. Multiple sets of comparative experiments. Each image (total 8 images) include: the area to be covered (left); The optimal path of the CIPSO algorithm (upper right); The iterative evolution of loss values of different algorithms (lower right).

The CIPSO algorithm combines the hybridization and mutation of particle state prediction and genetic algorithm. When the model approaches convergence, it can trigger large-scale reconstruction of particle paths, enabling global exploration to be reconducted and moving closer to the current optimal path. It can be seen from the figure that after the CIPSO model algorithm enters the local optimum, the generational value still shows a downward trend. This is because the model undergoes genetic variation of particles, reconstructs the particle paths and conducts global exploration again. Theoretically speaking, this can always find a relatively better solution and will not fall to the local optimum due to the interference of the complex environment. Moreover, the comprehensive cost function values (path length, coverage completion degree, and turning Angle) are 23% lower than those of the standard PSO and 15% lower than those of the IPSO.

Ensuring the stability of the model proposed in this paper, we also conducted multiple sets of randomized experiments. The corresponding experimental results are shown in Fig. 11. It can be seen from the figure that the randomly generated model causes significant fluctuations in the optimization capabilities of the standard PSO algorithm and the IPSO algorithm. The IPSO algorithm with only conventional dynamic weight coefficients cannot guarantee that it can get out of the local optimum every time. It has already fallen to the local optimum within about 200 or even 100 iterations. In contrast, we can see that the CIPSO algorithm can always be in a stable descending optimization state and can find a better path. Under the condition of ensuring the excellence of the path, sometimes the optimization efficiency even exceeds that of the other two algorithms. This proves that the comprehensive improvement algorithm proposed in this paper has good stability.

5 Conclusion

This paper introduces the CIPSO algorithm, designed to address the limitations of conventional algorithms in optimizing within unstructured environments. It specifically tackles the tendency of traditional Particle Swarm Optimization (PSO) methods to prematurely converge on suboptimal solutions. Through three key innovations—the integration of adaptive strategies for consolidating redundant path nodes, dynamic weight adjustments based on real-time particle states, and hybridization with genetic algorithm mechanisms—the proposed framework significantly enhances optimization efficacy. Theoretical analysis and simulation experiment results show that the CIPSO algorithm has better optimization efficiency and better robustness when dealing with the full-coverage path optimization problem in unstructured environments.

There are a few interesting future research directions along this topic. First, in this paper, we only compare CIPSO with PSO and IPSO. We plan to expand experimental comparisons to include additional advanced optimization algorithms. Second, we aim to investigate the integration of a curve-fitting model to ensure optimized paths exhibit improved smoothness.

References

1. Zhang, H., Hong, W., Chen, M.: A path planning strategy for intelligent sweeping robots [C]. In: IEEE International Conference on Mechatronics and Automation (ICMA), vol. 2019, pp. 11–15 (2019)
2. Zhao, J., Deng, C., Yu, H., et al.: Path planning of unmanned vehicles based on adaptive particle swarm optimization algorithm [J]. Comput. Commun. **216**, 112–129 (2024)
3. Wu, X., Bai, J., Hao, F., et al.: Field complete coverage path planning based on improved genetic algorithm for transplanting robot [J]. Machines. **11**(6), 659 (2023)
4. Jha, R., Mangal, S.: Geometric approaches to solving the traveling salesman problem [J]. Modern Dynamics: Mathematical Progressions. **1**(1), 22–25 (2024)
5. Xie, J., Garcia Carrillo, L.R., Jin, L.: Path planning for UAV to cover multiple separated convex polygonal regions [J]. IEEE Access. **8**, 51770–51785 (2020)
6. Clerc, M., Kennedy, J.: The particle swarm - explosion, stability, and convergence in a multidimensional complex space [J]. IEEE Trans. Evol. Comput. **6**(1), 58–73 (2002)
7. Tianshu, H., Shaoyong, W.: Enhanced Particle Swarm Optimization Algorithm Based on Kalman Filter Principle [Z](2005–01–01)
8. Zhan, D., Lu, H., Hao, W., et al.: Improving particle swarm optimization: using neighbor heuristic and Gaussian cloud learning [J]. Intell. Data Anal. **20**(1), 167–182 (2016)

9. Huang, J., Guo, D., Zhang, J.: The particle swarm optimization algorithm improved based on multiple strategies including adaptive nonlinear weights [C]. In: Proceedings of the 4th International Conference on Artificial Intelligence and Computer Engineering, pp. 176–181 (2023)
10. Wang, X., Tang, Y., Li, Z., et al.: Research on charging strategy based on improved particle swarm optimization PID algorithm [J]. Complex Intell. Syst. **10**(5), 6421–6433 (2024)
11. Hu, G., Guo, Y., Zhao, W., et al.: An adaptive snow ablation-inspired particle swarm optimization with its application in geometric optimization [J]. Artif. Intell. Rev. **57**(12), 332 (2024)
12. He, S.: Mathematical basis of particle swarm optimization algorithm [M]//. In: Lecture Notes on Data Engineering and Communications Technologies, pp. 903–908. Springer Singapore (2022)
13. Choudhary, S., Sugumaran, S., Belazi, A., et al.: Linearly decreasing inertia weight PSO and improved weight factor-based clustering algorithm for wireless sensor networks [J]. J. Ambient. Intell. Humaniz. Comput. **14**(6), 6661–6679 (2021)
14. Poohoi, R., Puntusavase, K., Ohmori, S.: A novel crossover operator for genetic algorithm: stas crossover [J]. Decis. Sci. Lett. **12**(3), 515–524 (2023)
15. Wu, S., Heidari, A.A., Zhang, S., et al.: Gaussian bare-bone slime mould algorithm: performance optimization and case studies on truss structures [J]. Artif. Intell. Rev. **56**(9), 9051–9087 (2023)
16. Lin, Z., Liu, G., Sun, L.: Parameters setting of particle swarm optimization based on orthogonal experiment analysis [J]. sJ. Fujian Univ. Technol. **14**(1), 55–61 (2016)

Multi-UGV Task Allocation Based on Improved NSGA-II Algorithm for Forest Firefighting

Zijian Zhou[1,2], Jia Zhang[1,2](✉), and Mingchang Chen[1,2]

[1] Beijing Institute of Technology, Beijing 10081, China
[2] National Key Laboratory of Autonomous Intelligent Unmanned Systems, Beijing Institute of Technology, Beijing 100081, China
zhangjia@bit.edu.cn

Abstract. Addressing the time-sensitive and multi-constrained characteristics of multi-Unmanned Ground Vehicle (UGV) firefighting task allocation problem in forest fires, this paper constructs a multi-objective optimization model with time-varying task states. To tackle challenges such as large discrete decision space, complex constraints, and dynamic fire changes inherent in this problem model, an improved Non-dominated Sorting Genetic Algorithm II (NSGA-II) is proposed. The algorithm employs integer encoding to match tasks with vehicles, and introduces dynamic crowding distance sorting and an adaptive mutation strategy to enhance solution convergence and distribution diversity. Simulation experiments in various scale scenarios demonstrate that the proposed algorithm outperforms comparative algorithms, in terms of superiority, diversity, and stability. It can significantly shorten the firefighting cycle and reduce resource consumption, providing an intelligent multi-UGV task allocation solution for mitigating forest fire losses, thus validating its practical value in forest fire suppression scenarios.

Keywords: Forest firefighting · Multi-UGV · Task allocation · Multi-objective optimization · NSGA-II algorithm

1 Introduction

In recent years, frequent extreme climates have led to a continuous increase in the frequency of global forest fires, and the average number of casualties caused by forest fires has shown a fluctuating upward trend [1]. Traditional manual firefighting mode faces severe challenges in terms of response speed and safety. With the rapid development of intelligent technologies, multi-agent systems, represented by Unmanned Ground Vehicles (UGVs), have become a crucial component of intelligent emergency systems due to their rapid response, flexible deployment, and environmental adaptability. However, achieving efficient collaborative multi-UGV task allocation in dynamic fire environments still presents complex challenges such as task coupling, resource constraints, and path constraints.

H. Ma et al. (Eds.): IWACIII 2025, CCIS 2780, pp. 48–60, 2026.
https://doi.org/10.1007/978-981-95-6730-0_4

According to [2], task allocation methods are mainly divided into three categories: centralized methods, distributed methods, and learning-based methods.

Centralized methods rely on a central control unit for unified planning, which can use global information to obtain high-precision global optimal or suboptimal solutions, and is relatively simple to implement and maintain [3]. Representative algorithms include deterministic approaches such as integer programming [4] and branch-and-bound [5], as well as meta-heuristics like genetic algorithms [6] and particle swarm optimization [7]. Among them, the NSGA-II algorithm [8] is one of the most popular research methods for multi-objective optimization problems due to its good convergence and diversity of solutions [9]. Ma et al. [10] effectively solved the problem of multi-UAV collaborative search and rescue path planning under severe sea conditions by improving the population initialization mechanism of the NSGA-II algorithm and introducing a multi-task collaborative optimization framework. Centralized algorithms are suitable for problems where the environment is known and there is a high demand for optimal allocation schemes.

Distributed methods endow agents with a high degree of autonomy, leveraging local interactions and decentralized decisions to achieve stronger robustness and real-time performance [11]. Duan et al. [12] introduced a hybrid two-stage auction algorithm that effectively handles the dynamic task-allocation problem for heterogeneous UAV swarms in complex and evolving settings. Qi et al. [13] further presented a market-mechanism-based strategy that jointly accounts for distance, fire intensity, and temporal factors to realise distributed, dynamic optimisation in multi-robot cooperative firefighting. While distributed methods are suitable for large-scale, highly dynamic scenarios with dispersed computing resources, their generated task allocation schemes risk falling into local optima, and frequent inter-agent communication also incurs additional overhead.

Learning-based task allocation methods learn complex scheduling strategies through interaction with the environment, making them particularly suitable for dynamic, uncertain, and difficult-to-model complex environments. Through extensive training, these methods can adapt to changing task requirements and environmental conditions, demonstrating good adaptability and global optimization capabilities. However, they face challenges in terms of data requirements, computational resources, and model generalization ability [2].

The aforementioned methods can, to some extent, solve multi-agent multi-objective task allocation problems. However, when addressing multi-UGV firefighting task allocation in forest fire scenarios, existing research less frequently focuses on the impact of the time-varying nature of fire point states on task allocation effectiveness. Therefore, this paper constructs a multi-objective optimization model with time-varying task states. To cope with challenges in the problem model such as large discrete decision space, complex constraints, and dynamic fire changes, an improved NSGA-II algorithm is proposed. This aims to provide an intelligent multi-UGV firefighting task allocation solution for forest fire rescue, quickly suppressing forest fires while reducing resource consumption.

2 Problem Modeling

2.1 Problem Description

In a forest firefighting scenario, UGVs are responsible for performing water-spraying missions on discrete fire points. The scenario modeling consists of a task set $M = \{m_j, j = 1, 2, ..., m\}$ composed of m fire points to be suppressed and an execution agent set $U = \{u_i, i = 1, 2, ..., n\}$ composed of n UGVs capable of performing firefighting missions ($m > n$).

Each fire point m_j is treated as a task point to be processed, quantitatively quantified by two key parameters: spatial coordinates $l_j = (x_j, y_j)$ and a state quantity W_{mj} representing the intensity of the fire. The water demand (in liters) required to extinguish a task point can be expressed as an exponential function shown in the equation:

$$W_{mj}(t) = W_{mj0} \cdot e^{\alpha_j t} \tag{1}$$

where W_{mj0} is the initial water demand to extinguish task point m_j at time $t = 0$, and α_j is the fire growth rate. When the water needed for a fire point falls below a set minimal threshold, the task is considered complete. In this study, this threshold is set to 10^{-5}(L).

The execution agents are a collection of n homogeneous UGVs. They depart from a dispatch center, complete their assigned tasks, and then return. The model assumes an environment without complex obstacles, with UGVs moving at a constant speed along straight lines, and the distance between two points is represented by Euclidean distance. The capability of each UGV is quantified by parameters such as spatial coordinates $l_i = (x_i, y_i)$, maximum navigable distance spatial coordinates D_i, maximum water storage capacity W_{ui} and a task execution list a_i.

2.2 Decision Variables

The decision variable is a binary matrix $x^i_{j',j}$. If UGV u_i proceeds from task point m'_j to task point m_j to perform a task, then $x^i_{j',j} = 1$; otherwise, $x^i_{j',j} = 0$.

2.3 Objective Function

Minimize Task Completion Time. The latest completion time among all tasks is used as a key indicator to measure the overall task completion efficiency:

$$\min f_1 = \max_{j \in 1,2,...,m} ct_j \tag{2}$$

where ct_j is the actual completion time of task point m_j.

Minimize Total Path of UGVs. To optimize resource utilization and save operational costs, the objective function for minimizing the total UGV travel path is constructed as:

$$\min f_2 = \sum_{i=1}^{n} \sum_{j'=1}^{m} x_{j',j}^{i} \cdot d_{j',j} \tag{3}$$

where $d_{j',j})$ is the Euclidean distance from task point $m_{j'}$ to task point m_j.

2.4 Restrictive Condition

Task Deadline Constraint. Since the fire intensity of task points increases exponentially, a strict task deadline must be set to ensure that UGVs arrive at the task point before this time, thereby effectively avoiding task failure due to excessive fire growth:

$$at_j \leq et_j, \forall j = 1, 2, ..., m \tag{4}$$

where et_j is the deadline for task point m_j, and at_j is the arrival time of UGV at task point m_j, which equals the sum of the previous task completion time and the estimated travel time:

$$at_j = ct_{j'} + t_{j',j} \tag{5}$$

Task Single Matching Constraint. Each task must be executed by only one UGV to avoid redundant operations and scheduling conflicts:

$$\sum_{i=1}^{n} \sum_{j'=1}^{m} x_{j',j}^{i} = 1, \forall j = 1, 2, ..., m \tag{6}$$

Task Execution Sequence Constraint. A UGV must complete its current task before proceeding to the next task point:

$$a_i = \{m_{i,1}, m_{i,2}, ..., m_{i,|a_i|}\}, \forall i = 1, 2, ..., n \tag{7}$$

UGV Firefighting Resource Constraint. The maximum water storage capacity of a UGV must be greater than or equal to the sum of the water demands of all task points in its task sequence:

$$\sum_{j=1}^{|a_i|} W_{mj0} \cdot e^{\alpha_j \cdot at_j} \leq W_{ui}, \forall i = 1, 2, ..., n \tag{8}$$

UGV Travel Distance Constraint. The total travel path of a UGV (including the return trip from the last task point to the dispatch center) cannot exceed its maximum navigable distance:

$$\sum_{j=1}^{|a_i|} d_{j',j} + d_{|a_i|,0} \leq D_i, \forall i = 1, 2, ..., n \tag{9}$$

3 Improved NSGA-II Algorithm

The basic process of NSGA-II includes population initialization, generating offspring through crossover and mutation operations, calculating fitness, performing fast non-dominated sorting and crowding distance sorting, and updating the population via an elitist retention strategy. In multi-objective optimization problems, NSGA-II maintains the superiority structure of solutions through non-dominated sorting while preserving the diversity of population distribution using crowding distance measures.

However, when facing problem scenarios involving task sequence, path and resource coupling constraints in this study, traditional NSGA-II algorithms suffer from issues such as incompatible genetic operations, rapid loss of population diversity, and convergence dependence on parameters. To improve the performance of the NSGA-II algorithm in multi-UGV task allocation problems, this paper proposes an improved NSGA-II algorithm, which mainly focuses on three aspects: problem-adaptive integer encoding, improved genetic operators and adaptive mutation strategies, and dynamic crowding distance sorting.

3.1 Integer Encoding Rules

This study adopts an integer encoding scheme designed specifically for the characteristics of the multi-UGV task allocation problem. A chromosome consists of an integer sequence with $m+n-1$ positions. Among these, encoding values less than or equal to m represent the tasks to be extinguished, while values greater than m represent the UGVs; if a UGV is encoded as A, it corresponds to UGV $u_{i=A-m+1}$.

Starting from the left side of the sequence, the task encoding sequence before the first UGV encoding constitutes the extinguishing task sequence for UGV u_1. Subsequently, the task encoding sequence between every two adjacent UGV encodings forms the extinguishing task sequence for the UGV represented by the UGV encoding on the left, and so on.

For example, when the number of tasks to be extinguished is $m = 10$ and the number of UGVs is $n = 3$, as shown in Fig. 1. For the encoding sequence [2, 5, 1, 11, 4, 6, 3, 12, 9, 7, 10, 8], where "11" corresponds to u_2 (i.e., $11-10+1=2$) and "12" corresponds to u_3 (i.e., $12-10+1=3$). Then, the extinguishing task sequence for u_1 is [2, 5, 1]; the extinguishing task sequence for u_2 is [4, 6, 3]; and the extinguishing task sequence for u_3 is [9, 7, 10, 8].

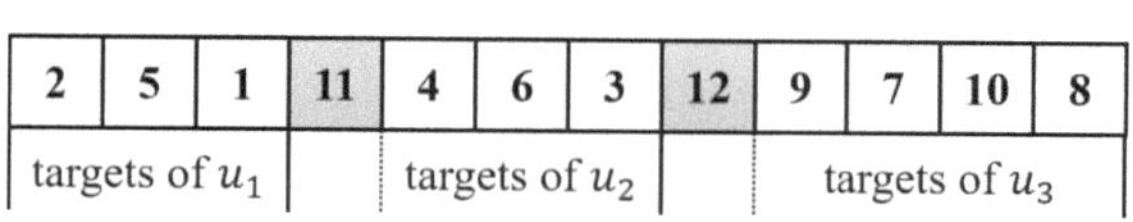

Fig. 1. Encoding Rule Diagram.

This encoding method is scientifically rigorous, satisfies the single matching constraint between tasks and UGVs (avoiding duplicate encodings), and considers the task execution sequence constraint, which helps reduce redundant operations and computational load in subsequent evolutionary computations.

3.2 Partially Mapped Crossover and Adaptive Swap Mutation

To overcome issues like task duplication or omission that simple crossover and mutation operations might cause in task allocation problems, this paper modifies the following genetic operators.

Partially Mapped Crossover (PMX). A segment is randomly selected on the chromosome for exchange, and a mapping relationship is established based on the exchanged region. For duplicate genes appearing in the unexchanged parts of the chromosome, they are replaced according to the mapping relationship [14], thereby ensuring that the chromosome remains a valid feasible solution, avoiding task duplication or omission. An example is shown in Fig. 2.

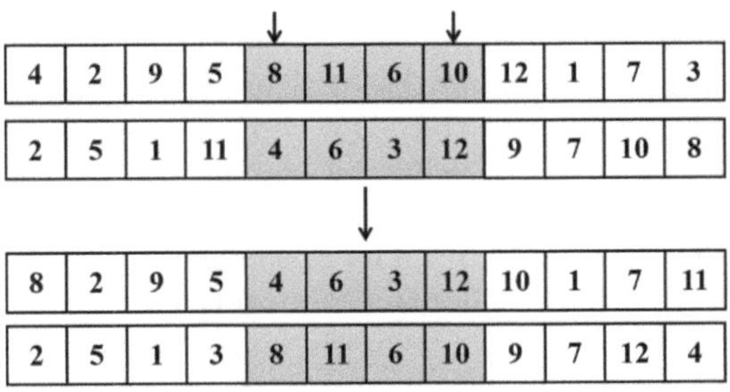

Fig. 2. Partially Mapped Crossover Diagram.

Adaptive Exchange Mutation This method uses exchange mutation by randomly selecting two positions on the chromosome and swapping their genes [15], as shown in Fig. 3.

However, since the demand for mutation differs at various evolutionary stages of an algorithm [14], a fixed mutation rate has obvious limitations. To improve the search efficiency and the ability of the algorithm to escape local optima, the mutation rate is set to adjust adaptively. The mutation rate linearly increases with the number of iterations, allowing the algorithm to perform more robust global exploration with a lower mutation rate in the early stages of evolution, and targeted local search and escape from stagnation with a higher mutation rate in later stages.

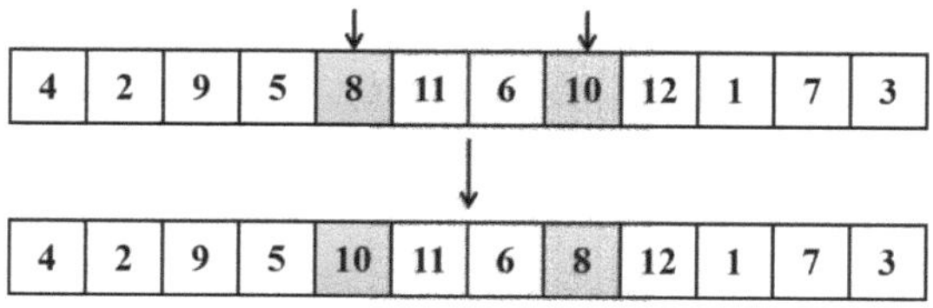

Fig. 3. Exchange Mutation Diagram

The formula for adaptive mutation rate is expressed as:

$$P_m = P_{m\,\min} + 0.5 \times \frac{n_{gen}}{n_{gen\,\max}} \times (1 - P_{m\,\min}) \tag{10}$$

where $P_{m\,\min}$ is the initial mutation rate, $n_{gen\,\max}$ is the set maximum number of iterations, and n_{gen} is the current iteration count of the algorithm.

3.3 Dynamic Crowding Distance Sorting.

In the traditional NSGA-II algorithm, as the population gradually approaches the Pareto front, more and more individuals cluster in the first non-dominated layer, and the criterion for judging the quality of individuals mainly relies on crowding distance sorting. Since the algorithm sorts all individuals based on pre-calculated crowding distance values in one go, directly eliminating several individuals with the smallest crowding distances within a layer, this leads to large gaps in that area, resulting in a loss of solution set diversity.

To address this, this paper adopts an improved crowding distance sorting method based on a dynamic update mechanism. In each iteration, the algorithm no longer eliminates all individuals with small crowding distances at once. Instead, it gradually removes the individual with the smallest current crowding distance and dynamically updates the crowding distance values of its immediate neighbors. This process is repeated until the number of retained individuals meets the requirements. This dynamic update mechanism can more accurately reflect the relative distribution of individuals in the local space, effectively avoiding the problem of discarding entire dense regions that static methods might cause, thereby better maintaining the diversity of the solution set.

3.4 Improved NSGA-II Algorithm Process

The specific process of the improved NSGA-II algorithm is as follows:

Step 1: Randomly generate N_{pop} individuals conforming to the constraints according to the encoding rules to form the initial population.

Step 2: Perform partially mapped crossover and adaptive exchange mutation on the parent population, conduct feasibility checks, and generate the offspring population.

Step 3: Merge the parent population and the offspring population, and perform fast non-dominated sorting on the merged population.

Step 4: According to the elitist retention strategy and the dynamic crowding distance sorting mechanism, select the optimal N_{pop} individuals from the merged $2N_{pop}$ individuals for retention, forming a new population.

Step 5: Determine if the set maximum number of iterations has been reached. If it has, proceed to Step 6; otherwise, set the new population as the parent population and return to Step 2.

Step 6: Output the task allocation results.

4 Simulation Experiment Validation and Analysis

The simulation experiments were conducted on a Windows 11 operating system with a computer configuration of 16.0 GB RAM and an AMD Ryzen 7 8845H wCPU @ 3.80 GHz processor. The simulation software used was Matlab R2019a.

4.1 Parameterization

To comprehensively verify the algorithm's performance, three representative scenario examples, namely small-scale, medium-scale, and large-scale, were set up based on different numbers of UGVs and fire points to be extinguished. The specific numbers of UGVs and task points in these three scenarios are shown in Table 1.

Table 1. Statistics on our resources.

Scenario #	number of UGVs	number of Targets
1	3	10
2	5	15
3	10	30

4.2 Experimental Results and Visualization Analysis

Comparative experiments were conducted under small, medium, and large-scale scenario examples, using the improved NSGA-II algorithm proposed in this paper, Multi-Objective Particle Swarm Optimization (MOPSO), and Strength Pareto Evolutionary Algorithm (SPEA-II). The experimental parameters were uniformly set as follows: the number of iterations for all three algorithms was 100, and the population size was 200.

Pareto Front Distribution. The Pareto front simulation results obtained by the three algorithms are shown in Fig. 4. From the figure, it can be seen that the Pareto front obtained by the improved NSGA-II algorithm is closer to the origin of the objective space and has a wider distribution range, preliminarily indicating its advantage in convergence and distribution diversity in multi-objective optimization.

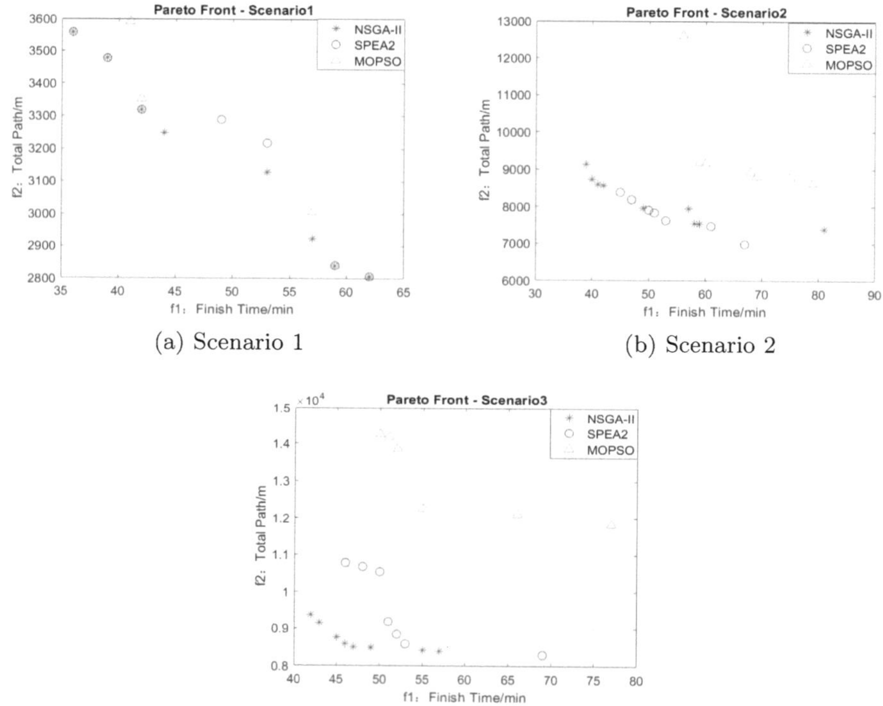

(a) Scenario 1

(b) Scenario 2

(c) Scenario 3

Fig. 4. Pareto fronts of the three algorithms.

Multi-dimensional Performance Radar Chart Comparison. To further evaluate the quality of the solution set, a system containing 5 indicators was constructed from the perspectives of convergence, coverage, and distribution:

Hypervolume (HV): Measures the volume covered by the solution set, reflecting overall superiority and distribution breadth [16].

1/Spacing: Spacing evaluates the uniformity of solutions on the Pareto front by computing the standard deviation of the Euclidean distances between each solution and its nearest neighbor [17]. A smaller value means more uniform distribution. To align with other indicators where larger is better, we use its reciprocal (1/Spacing), so higher values indicate more even spacing of solutions.

Cardinality: The number of solutions on the non-dominated front, reflecting the density of the solution set and the richness of available options.

1/AvgTime: Reflects scheduling efficiency (smaller value is better).

1/AvgPath: Measures resource consumption (smaller value is better).

Figure 5 shows the normalized radar charts for the three algorithms under three scenarios (indicators mapped to the [0,1] range, with higher values indicating better performance). In small-scale (Scenario 1) and medium-scale (Scenario 2) scenarios, the improved NSGA-II leads in most indicators, with only 1/Spacing and average path cost being close to SPEA-II; in the large-scale (Scenario 3) scenario, the improved NSGA-II is significantly superior to the comparison algorithms in all dimensions, validating its comprehensive advantage in complex scenarios.

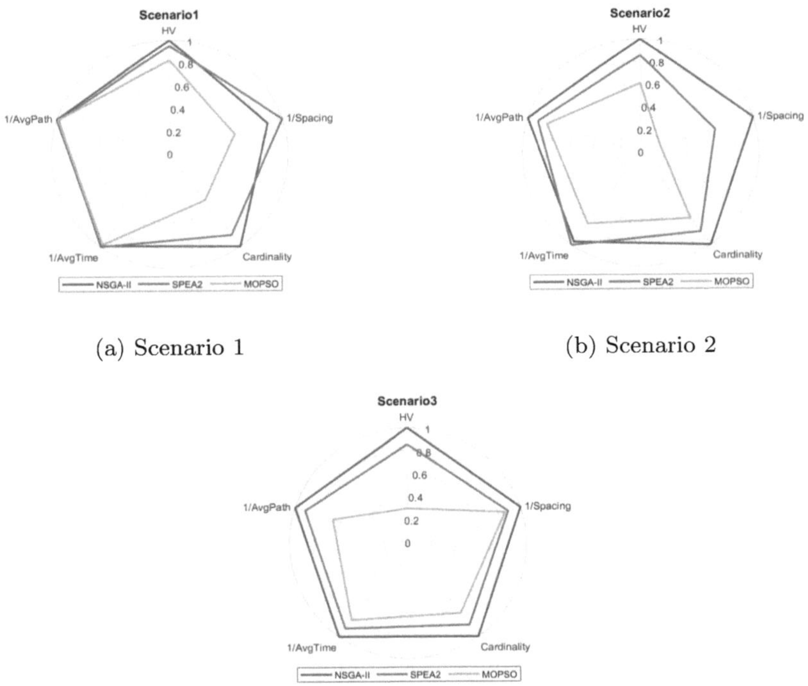

(a) Scenario 1 (b) Scenario 2

(c) Scenario 3

Fig. 5. Algorithm performance comparison radar charts.

Above results indicate that the improved NSGA-II algorithm performs most stably in the current task allocation scenario, demonstrating advantages in obtaining high-quality solution sets, and its relative superiority becomes more pronounced as the problem scale increases. When faced with the complex and dynamic multi-UGV task allocation scenarios serving forest firefighting missions, the NSGA-II algorithm can more effectively plan UGV task allocations and execution sequences, shortening the overall operation time while reducing the total UGV travel path, thereby lowering costs. This verifies its excellent performance in complex scheduling tasks.

4.3 Repeated Experiment Statistical Analysis

Considering that in forest fires, it is desirable to extinguish all fire points as quickly as possible to reduce casualties and economic losses, the task allocation scheme with the minimum task completion time is selected from the Pareto front for output.

To verify algorithm stability, 25 independent repeated experiments were conducted for each of the different scale scenarios, and the average values and standard deviations of the objective functions were statistically analyzed (Table 2, bolded values are the optimal values).

Table 2. Comparison of average objective function values of the three algorithms

Algorithm	Scenario#1		Scenario#2		Scenario#3	
	CT/min(± *std*)	TP/m(± *std*)	CT/min(± *std*)	TP/m(± *std*)	CT/min(± *std*)	TP/m(± *std*)
NSGA-II	**38.80**(±2.38)	**3487.03**(**±100.36**)	**41.72**(**±0.72**)	**5763.77**(**±85.63**)	**40.16**(**±2.03**)	**9382.17**(**±614.94**)
SPEA-II	40.80(**±1.88**)	3558.25(±121.37)	44.08(±2.86)	5919.67(±379.18)	44.40(±2.88)	10021.48(±627.13)
MOPSO	43.04(±3.14)	3843.99(±357.52)	46.88(±3.99)	6264.21(±214.80)	62.92(±6.88)	13953.13(±1345.58)

In terms of task completion time, the improved NSGA-II is optimal under all three scales, with a maximum reduction of 36.17%. In terms of path cost, the improved NSGA-II achieves the shortest paths in all scenarios, reducing path cost by 32.76% compared to MOPSO and 6.38% compared to SPEA-II in the large-scale scenario.

The standard deviation of the improved NSGA-II is generally smaller than that of the comparison algorithms in most cases; only the standard deviation of task completion time in Scenario 1 is slightly higher than SPEA-II, indicating its smaller fluctuation and better robustness across different scales.

In summary, the improved NSGA-II is superior to the comparison algorithms in terms of solution convergence, diversity, and stability, with its advantages becoming more significant as the scale increases, effectively shortening firefighting time and reducing resource consumption.

5 Conclusion

This paper addresses the task allocation problem of multi-UGVs collaborative firefighting in forest fires, constructing a multi-objective optimization model that considers dynamic fire behavior and resource constraints, and proposes an improved NSGA-II algorithm. The algorithm combines integer encoding, partially mapped crossover, adaptive swap mutation, and dynamic crowding distance update strategies to enhance the diversity and convergence of the solution set. Simulation experiments in scenarios of different scales show that this method can not only generate solution sets with higher quality, wider coverage, and more uniform distribution for decision-makers to choose from, but also that the generated task allocation schemes outperform MOPSO and SPEA-II in terms of task completion time and path cost. Additionally, it demonstrates better stability, verifying its practical value in multi-UGV firefighting scenarios for forest fires.

Acknowledgments. This study was funded by the National Natural Science Foundation of China (62088101).

References

1. Liu, Y., Wu, Z., Zhou, Y.: A preliminary study of integrated forest fire prevention and control management in the new situation. J. Wildland Fire Sci. **42**(03), 38–41 (2024)
2. Li, X., Wang, X.: A survey on multi-agent task allocation methods. In: China Conference on Command and Control (2025)
3. Zhao, H.: Research on multi-robot task assignment and pathplanning algorithms in indoor environments. Master's thesis (2023)
4. Moon, S., Oh, E., Shim, D.H.: An integral framework of task assignment and path planning for multiple unmanned aerial vehicles in dynamic environments. J. Intell. Rob. Syst. **70**(1), 303–313 (2013)
5. Rasmussen, S.J., Shima, T.: Tree search algorithm for assigning cooperating UAVs to multiple tasks. Int. J. Robust Nonlinear Control: IFAC-Affili. J. **18**(2), 135–153 (2008)
6. Shima, T., Rasmussen, S.J., Sparks, A.G., Passino, K.M.: Multiple task assignments for cooperating uninhabited aerial vehicles using genetic algorithms. Comput. Oper. Res. **33**(11), 3252–3269 (2006)
7. Li, M., Liu, C., Li, K., Liao, X., Li, K.: Multi-task allocation with an optimized quantum particle swarm method. Appl. Soft Comput. **96**, 106603 (2020)
8. Deb, K., Pratap, A., Agarwal, S., Meyarivan, T.: A fast and elitist multiobjective genetic algorithm: NSGA-II. IEEE Trans. Evol. Comput. **6**(2), 182–197 (2002)
9. Ma, H., Zhang, Y., Sun, S., Liu, T., Shan, Y.: A comprehensive survey on NSGA-II for multi-objective optimization and applications. Artif. Intell. Rev. **56**(12), 15217–15270 (2023)
10. Ma, Y., Li, B., Huang, W., Fan, Q.: An improved NSGA-II based on multi-task optimization for multi-UAV maritime search and rescue under severe weather. J. Marine Sci. Eng. **11**(4), 781 (2023)
11. Liu, C.H., Ma, X., Gao, X., Tang, J.: Distributed energy-efficient multi-uav navigation for long-term communication coverage by deep reinforcement learning. IEEE Trans. Mob. Comput. **19**(6), 1274–1285 (2019)
12. Duan, X., Liu, H., Tang, H., Cai, Q., Zhang, F., Han, X.: A novel hybrid auction algorithm for multi-UAVs dynamic task assignment. IEEE Access **8**, 86207–86222 (2019)

13. Qi, X., Tian, Y., Yang, M., Yang, Y.: Market-based multi-robot task allocation for fire-disaster response. J. Jilin Univ. (Inf. Sci. Ed.) (5), 8 (2009)
14. Xu H: Research on multi-AGV task assignment and path planning in intelligent laboratory. Ph.D. thesis, (2024)
15. Zhang, X., Hu, Y., Li, W., Pang, Q., Yuan, G.: Multi-UAV fire fighting mission planning based on improved artificial bee colony algorithm. J. Chin. Inertial Technol. (004), 028 (2020)
16. Van Veldhuizen, D.A., Lamont, G.B.: Multiobjective evolutionary algorithm test suites. In: Proceedings of the 1999 ACM Symposium on Applied Computing, pp. 351–357 (1999)
17. Knowles, J., Corne, D.: On metrics for comparing nondominated sets. In: Proceedings of the 2002 Congress on Evolutionary Computation, CEC'02 (Cat. No. 02TH8600), vol. 1, pp. 711–716. IEEE (2002)

Research on Cooperative Application of UAV Swarm Based on Cyber Network

Mingqiu Ren[1(✉)], Bingqie Wang[1], and Xin Wang[2]

[1] Air Force Early Warning Academy, Wuhan HB 430019, China
ren19821006@126.com
[2] Unit 94981, Jiangxi JX 330200, China

Abstract. UAV cluster operation is an important operation form in the future battlefield, and network coordination will become the main mode of UAV cluster operation. UAV cluster network situation awareness adopts mature technologies such as node information detection and collection, network routing information detection and acquisition, website application service information search and detection, etc., extracts key information data from massive original data information, and identifies key elements in battlefield electromagnetic situation. Through cyber network cooperative perception and diversified attack technologies, multi domain capabilities are integrated to realize OODA closed loop, providing key technical support for destroying, weakening, destroying or disintegrating the enemy's key information network system.

Keywords: UAV Swarm · Cyber network · information extraction · network situation awareness

1 Introduction

UAV cluster operation is a brand-new operation mode proposed in recent years. It is a typical intelligent cluster operation behavior. It can suppress enemy targets with numerical advantages and complete combat tasks at a lower cost. UAV cluster is a distributed system that integrates a large number of UAVs under the open architecture, based on the collaborative network control between platforms, and aims to improve the ability of collaborative tasks. The US Department of defense released the comprehensive roadmap of unmanned systems for fiscal year 2017–2042, which listed unmanned aerial vehicle cluster as one of the 15 key technologies, and pointed out that it will have the capability of reconnaissance and jamming integration, which will have a significant impact on the composition of future aviation equipment system and combat style. In March 2021, the US Department of the Navy published the unmanned combat framework, which requires the unmanned cooperative system of the Navy and the Marine Corps to become a reliable and sustainable force in the maritime force structure. Through rapid integration with the manned system, the lethality, force scale, survivability and operation speed of the two services are improved, and the vision of transforming from the traditional force structure to the manned unmanned hybrid force is clearly put forward. On February 9, 2024, the US

H. Ma et al. (Eds.): IWACIII 2025, CCIS 2780, pp. 61–72, 2026.
https://doi.org/10.1007/978-981-95-6730-0_5

Special Competitive Research Program Defense Panel (SCSP) released a report titled "Unmanned Systems and the Transformation of US Combat Capability," stating that only when drones are combined with digital command and control systems and military civilian sensor networks can a transformative trinity be formed. Unmanned systems can be used for tasks such as battlefield reconnaissance and artillery firing, and have also evolved into forms such as patrol ammunition, first person unmanned aerial vehicles (FPVs), and one-way attack drones to enhance long-range strike capabilities. Digital command and control will directly transmit battlefield information that is traditionally only accessible to a few high-level decision-makers to grassroots combat personnel, allowing frontline combat personnel and command center personnel to access the same battlefield information and have the same situational awareness capabilities. The military civilian sensor network is spread throughout the battlefield, which can expand the battlefield space to any location where commercial equipment is deployed. By unprecedentedly integrating open source and confidential data, it greatly shortens the time from sensor detection to target strike. When the above three elements are combined, it can provide a rich selection space for combat troops.

At present, domestic universities and scientific research institutes have more research on the relevant technologies of UAV cluster equipment, and less on the networks and typical application scenarios necessary for the efficient operation of UAV cluster. Therefore, it is of great significance to track the development trend of overseas electronic warfare UAV clusters, study the application scenarios of Cyber Network Collaboration Technology UAV clusters, promote the systematization and intellectualization of unmanned systems, and improve the autonomous cooperation performance of sea, land and air unmanned systems, so as to improve the combat capability of UAVs in intelligent wars.

2 Characteristics of UAV Cyber Network Collaboration

2.1 The Number of Large-Scale Cyber Network

In the future intelligent warfare, both the innovative design of intelligent warfare theory and the control of intelligent warfare cannot be separated from the massive scale of cyber network support. Based on the support of the cyber network, operations are carried out under the system architecture of the "combat cloud" to achieve rapid integration and resource allocation of unmanned aerial vehicles, realize networked dispersion and aggregation, and achieve the application effect of precise strikes in all time and all domain. Compared with the unmanned aerial vehicle, the electronic warfare unmanned aerial vehicle can play its unique advantages of good coordination, strong concealment and high cost-effectiveness in the operation against land and sea targets. It will become an important means to suppress and destroy the air defense early warning system outside the defense zone. The individual load of a single unmanned platform is small, the function is simple, and the destructive force is bound to be limited. However, once the scale is formed, the complexity of its function and the destructive force strength can be greatly improved. Therefore, it is important for the cyber collaborative network.

2.2 The Complex Structure Compositon of Large-Scale Cyber Network

The operational functional requirements such as early warning detection, wide area surveillance, forward reconnaissance, electronic countermeasure, saturation attack, tactical decoy, active defense, and network warfare determine the need for an unmanned combat platform to carry different mission systems. In actual combat, according to the different nature and scale of combat tasks, the types and numbers of UAVs that form a "hive" must be determined to build a systematic "hive" that meets the needs of the mission. Unmanned aerial vehicle networked cluster collaborative combat typically consists of heavy unmanned tanks, light unmanned vehicles, robots, and other components. After obtaining air superiority in urban warfare, the army usually scientifically organizes different tasks in different directions, sections, and depths based on the image information transmitted from the front-line positions. Through ground launch, air transport aircraft launch, artillery launch and other forms, dozens, hundreds or even more small and medium-sized unmanned aerial vehicles are deployed to carry out multi wave and cluster precision attacks on the target area. Its formation style and attack method can be "many to many", "many to few", "many to one", or it can launch multi wave and multi-directional attacks. It can also distribute and collaborate to strike multiple targets at similar times.

2.3 The Intelligent Control of Large-Scale Cyber Network

The digital upgrade of future weapons and equipment means that weapons and equipment can directly rely on their sensors to connect to decision-making networks. With the update of algorithms, well-trained artificial intelligence assisted decision-making has reached a level that is no different from or even higher than human level. This application of big data analysis and generative artificial intelligence technology in combat will have a direct impact on the effectiveness of multiple unmanned aerial vehicles. To realize the real-time control of the UAV cluster combat platform, the cyber network platform needs to develop and apply algorithms, and hand a large amount of information processing to the algorithms and the built-in microprocessors of the unmanned platform. The algorithm expresses and specifies the procedures and rules for the command and control of the unmanned aerial vehicle cluster combat platform, so that the large-scale unmanned combat platform can be used as a team for detailed division of labor and effective cooperation.

2.4 Collaborative Resource Scheduling of Unmanned Aerial Vehicles Based on Large-Scale Networks

The problem of interfering with resource allocation can be seen as optimizing a certain objective function under resource or performance constraints. And to measure the interference of the jamming machine against lightning

The interference effect of the drone cluster can guide the allocation of interference resources, and the construction of the objective function for the interference efficiency of the drone cluster is very important. ZHANG, SHIGQ, GENGXT proposed a target selection and power allocation method based on adaptive genetic algorithm with the objective function of minimizing the overall interference loss power. Simulation results showed that the proposed method can reduce the detection range of enemy radar. In practical applications, it is necessary to construct a typical drone cluster collaborative combat scenario, using the speed, angle, altitude, and target threat level of drones to construct different evaluation indicators, and based on this, a multi-objective optimization model is established. Then, an improved non support sorting genetic algorithm and simulated annealing algorithm are attempted to solve the attack task allocation scheme. In summary, the current research findings provide a reference for the allocation of interference resources in the context of formation penetration. However, the interference resource allocation models constructed in relevant literature are mostly non convex, which to some extent face the problems of high computational complexity and easy getting stuck in local optimal solutions. It can be seen that in order to measure the interference impact of jamming aircraft on enemy radar, there is an urgent need for a reasonable and objective evaluation index, which lays the foundation for further achieving rapid allocation of jamming resources for unmanned aerial vehicle formations under breakthrough missions. If the interference resource allocation model based on large-scale cyber networks is a convex optimization problem or easily transformed into a convex optimization problem, the real-time solving ability of unmanned aerial vehicle interference resource allocation will be more easily satisfied, and there will be more flexible choices in the solving algorithm.

2.5 Intelligent Air Combat Decision-Making for Drone Clusters Based on Large-Scale Cyber Networks

Aircraft with autonomous decision-making capabilities outperform humans in terms of reaction speed, without considering the physiological limits of humans during flight. Their supercomputing ability can more accurately predict the development of combat to gain the initiative in confrontation. However, the problem of aerial combat by aircraft is very complex, with high dynamics, strong real-time characteristics, and a larger solution space, which poses a huge challenge to the implementation of intelligent decision-making in aerial combat. Against the backdrop of increasingly complex drone combat modes in contemporary times, the collaboration and adversarial strategies of multi-agent systems have become an important research topic in the field of electromagnetic warfare.. Drones often face challenges from multiple targets and dynamic environments in practical missions, requiring comprehensive consideration of multidimensional functions such as reconnaissance, interference, and strike to gain advantages in hostile environments. However, traditional centralized decision-making methods often lack decision-making efficiency due to environmental uncertainty and information delay, making it difficult to adapt to rapidly changing situations. For this reason, the long cyber network provides the possibility for solving the intelligent air combat decision-making problem of drone swarms. In August 2020, the Defense Advanced Research Projects Agency (DARPA) of the United States

In the "AlphaDog - fight" human-machine aerial combat competition organized by the Projects Agency (DARPA), the design by Egret based on

The deep reinforcement learning based air combat intelligent decision machine defeated human pilots 5–0. By leveraging multi-agent collaboration under a distributed architecture, the flexibility and adaptability of drone swarm air combat decisions can be greatly improved in complex scenarios. Tang Hongsong proposed PSRO: using distributed systems to highly couple reinforcement learning and self game frameworks. Firstly, DPSRO introduces TOP-K truncation, prioritizing the selection of high-quality strategies while limiting the size of the strategy pool to avoid increasing computational costs due to diversity optimization, effectively controlling the consumption cost of intelligent adversarial decision-making for drones.

3 Progress of UAV Cluster Network Collaboration Project

3.1 Locust

Low-Cost UAV Swarming Technology (LOCUST) project was proposed by the US Naval Research Office in April 2015, and its core is autonomous cluster flight technology. Through the compact launch system, multiple folded pocket UAVs can be launched into the air in a short time, and the UAV formation can perform preset tasks or perform tasks independently. The project uses the coyote unmanned aerial vehicle of seonsitel company, with a maximum speed of about 145 km / h, an endurance of one hour and 30 launches per minute. The number of low-cost UAV clusters ranges from tens to hundreds of thousands. The promotion of large-scale UAV Systems will cause the difficulty of command, operation, control and communication of the system to increase exponentially. Therefore, the cooperative control of UAV platforms must be carried out based on the cyberspace network. In July 2016, Racine conducted a cluster flight test of ship-based "Coyote" UAVs in the Gulf of Mexico. The company said that it has basically completed the UAV formation verification phase, and preliminarily tested the UAV launch capability of the C-130H transport aircraft. Later, it will carry out more complex high autonomous flight tests, and continue to carry out launch capability tests on the platforms such as the P-8 "Poseidon" maritime patrol aircraft and the V-22 "Osprey" tilt rotor aircraft. The Naval Research Office is also testing a small low-cost rotor UAV called nomad, which is a very low-cost and consumable design, and can be deployed in a cluster without ensuring the return to the mother ship. The test results show that multiple "Ranger" UAVs can operate safely in the same airspace and fly in a coordinated manner (Fig. 1).

3.2 Offset

Urban combat is an important typical scenario of heterogeneous unmanned system cooperative combat. In 2016, DARPA funded the "offensive swarm enabled tactics (OFFSET) project, which seeks to use up to 250 small unmanned aerial vehicle systems and small unmanned ground systems in the future small-scale infantry force operations, and rapidly develop and deploy unmanned cluster systems and form combat effectiveness by using unmanned swarm technology and autonomous, autonomous and human-machine coordination technologies. In September 2020, the project adopted a multi-platform unmanned cluster system composed of rotary wing UAV, fixed wing UAV and ground unmanned vehicle, focusing on the intelligent interaction technology combining self-organized communication link and natural interaction means of no link. The unmanned

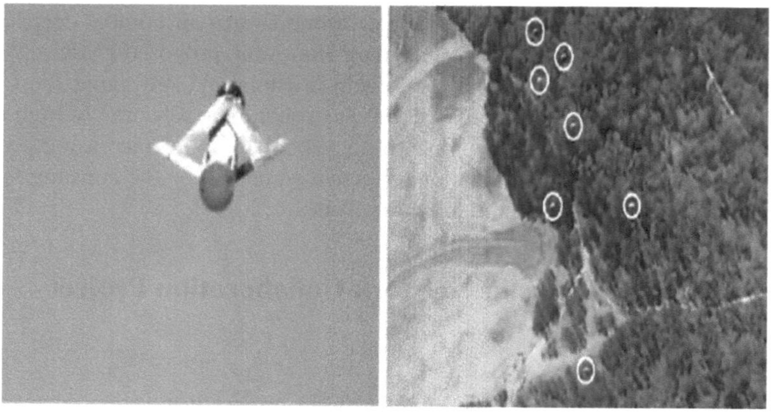

Fig. 1. Schematic diagram of Coyote UAV and formation flight.

cyberspace collaborative network system is affected by tall buildings, closed indoor environment and electromagnetic interference, and the stability of communication links is easily affected. While improving the bandwidth, distance, stability and self-organization of communication links, we should consider indirect natural interaction modes such as special actions and behaviors to achieve efficient communication interaction between unmanned systems and between people and unmanned systems. The unmanned combat equipment of the air ground system has become an indispensable new quality combat force on the modern battlefield. At present, the ground unmanned equipment of armies around the world mainly includes unmanned combat systems such as unmanned combat vehicles, unmanned transport vehicles, unmanned ground reconnaissance vehicles, quadruped robots, etc. There are many types of ground unmanned equipment being developed and tested. The Russian "Uranus 9″ unmanned combat system was unveiled at the 2023 Moscow military parade and can be used for reconnaissance, counter-terrorism, and street combat fire support tasks in urban environments. It can also provide fire support for infantry in the field when necessary (Fig. 2).

3.3 The US "Clone" Program

In August 2023, Deputy Secretary of Defense Katherine Hicks first proposed the "Clone" project at the Washington Defense Emerging Technologies Conference, aiming to make significant adjustments to the future development direction of unmanned combat systems. The goal is to quickly produce thousands of autonomous unmanned combat systems for deployment in sea, land, air and other combat domains in the short term, and to rapidly consume enemy wartime resources through economies of scale, creating conditions for deep strikes on subsequent high combat platforms. With the increasingly obvious characteristics of intelligence, unmanned, autonomous, and distributed combat, emerging concepts such as "swarm" formation combat, manned unmanned collaborative combat, and distributed combat are gradually constructing a blueprint for future combat modes. The battlefield environment in the future information warfare presents

Fig. 2. Cyber reconnoitre of OFFensive Swarm-Enabled Tactics.

the characteristics of multidimensional interconnection, that is, multiple combat dimensions of land, sea, air, space, electricity, and network will form a more compact organic whole, interact and cascade with each other, and form a more complex battlefield environment ecology. At the same time, the research and development of aerial combat equipment and autonomous unmanned technology are accelerating and will be deeply involved in future warfare, promoting the coupling of artificial intelligence, autonomous learning algorithms, and other unmanned systems for "formation application" and "one aircraft for multiple purposes", affecting the complex development of tactical tactics and troop mobilization decisions. The autonomous unmanned system equipment in the "Clone" project does not have a specific type or model, including but not limited to high-altitude long endurance unmanned aerial vehicles, unmanned vehicles, "Swarm" unmanned aerial vehicles, unmanned ships, and satellites. From the selection criteria of unmanned aerial platforms, since August 2023, the US military has gradually determined operational configuration requirements based on the "Clone" project in various combat domains. Among them, the US Air Force plans to focus on developing various types of small autonomous unmanned aerial vehicles, and use the currently in service large unmanned aerial vehicles and manned aerial vehicles as the "mother planes" of small autonomous unmanned aerial vehicles to achieve collaborative formation combat. The US Navy plans to develop a carrier based autonomous unmanned aerial vehicle system to improve the maritime continuous combat system. The US Army, Marine Corps and other land-based forces are advancing the development of new micro and small autonomous reconnaissance drones, which will serve as an important means of tactical reconnaissance.

4 Progress of UAV Cluster Network Collaboration Project

4.1 Design of Multi Domain Joint Cooperative Control Architecture Based on Cyber Network

Cooperative control architecture is the most basic problem in designing unmanned systems. In the future, unmanned systems will conduct military operations in sea, land, air, space, network, electricity and other operational domains. The development goal of the multi domain joint cooperative control architecture is to improve the range of joint operations, connect all sensors and shooters in near real time, fuse a large number of multi-source data at a speed and accuracy far beyond the current level, realize the interconnection and interoperability of sensors, weapons and communication systems, shorten the OODA loop, achieve accurate and efficient operations, and reduce the war loss ratio of our own side. The multi domain joint cooperative control architecture based on cyberspace network is an important guarantee to realize the utilization efficiency of heterogeneous UAV clusters. On the one hand, heterogeneous UAV groups can make use of their complementary advantages, break through the limitations of single UAV and single type UAV, and perform complex combat tasks through self-organization, adaptation and reconfiguration mechanisms. However, in the three typical control architectures of centralized, distributed and limited centralized, many factors such as UAV type, formation number and combat scenario need to be considered. There is no reasonable cooperative control architecture that can comprehensively solve the problems such as communication link stability, computing load balancing and distribution and platform behavior control in heterogeneous unmanned systems. On the other hand, the speed of decision-making and action is improved by incorporating machine control into the command and control process. Commanders need to undertake the task of supervision and control in the interaction with heterogeneous UAVs, and realize distributed command and control with the help of intelligent decision support system. The commander formulates an operational plan reflecting his strategy and superior intention, commands the heterogeneous UAV control system through the computer network interface, assigns operational tasks, and estimates the size and effectiveness of the opposing forces. The cyber collaborative network accurately manages each unmanned combat unit and switches between a large number of high-risk tasks and a small number of low-risk tasks according to the operational objectives For example, the capabilities of drones in traditional mission areas such as intelligence monitoring, surveillance, strike, and transportation continue to expand and explore their potential. Unmanned aerial vehicles such as the MQ-4C and MQ-9B have shown outstanding performance in the maritime field. With their long endurance and advanced sensor systems, they can continuously monitor vast sea areas and collect real-time intelligence information on maritime targets, such as the movements of enemy ships and marine environment information, providing important intelligence support for maritime operations.

4.2 Emitters Signal Acquisition Based on Cyber Network Situation Awareness

Network situation awareness mainly uses mature technologies such as node information detection and collection, network route information detection and acquisition, website application service information search and detection, etc. it extracts all information data related to battlefield confrontation situation from the massive original data information, identifies key elements in the situation information, and conducts correlation analysis and vulnerability assessment, so as to generate electromagnetic space target status The overall distribution of attack capabilities and attack resources. For example, Airbus has collaborated with multiple European companies to carry out helicopter and drone collaboration projects, successfully demonstrating manned helicopter and drone collaboration missions in multiple locations in France and Italy; The US Marine Corps has repeatedly tested the coordination between the XQ-58A drone and the F-35B fighter jet, achieving data transmission between aircraft in the "Cuiqi-2024" exercise, demonstrating its ability as a sensing platform to provide target data for fighter jets, and promoting the development of unmanned collaborative operations; Lockheed Martin Corporation invests in research and development to enable the F-35 fighter jet to control 8 CCA aircraft, develop related pods and control stations, and use tablet computers to configure and control multiple CCA aircraft, promoting innovation in manned/unmanned collaborative air combat tactics. Through the Cyber Network Cooperative perception and diversified attack technologies, the capabilities in multiple fields are integrated to realize the combination of reconnaissance discovery strike or discovery strike, and destroy, weaken, destroy or disintegrate the enemy system. With the promotion and application of active phased array antenna array and integrated radio frequency power control technology, the integrity of electronic reconnaissance signals is getting lower and lower, the difficulty of locating and identifying the enemy's radiation source and judging the key nodes of the enemy's information network system is further increased, and the precision electronic attack efficiency is significantly reduced. In the highly confrontational, heterogeneous UAV clusters must be based on the cyber information network, gather intelligence such as electronic reconnaissance, active detection, remote sensing and telemetry, acquire accurate battlefield information in near real time.

4.3 Cross Domain Integration of Unmanned Formation Based on Cyber Network

The cross domain maritime surveillance and targeting (CDMaST) project initiated by DARPA, which is based on the Cyber Network to realize the integration of the ocean oriented cross domain collaborative system, aims to transform the current maritime force of the US Navy into a system, decompose a variety of maritime combat functions into a large number of low-cost manned unmanned systems that can be upgraded through cyberspace, and disperse them to the wide area of intense confrontation, Build a monitoring and targeting task system that can be implemented across domains. In June 2021, the US Navy held the "unmanned system integrated combat problem 21" (uxsibp 21) exercise to simulate the coordination of unmanned systems in the air and water under real mission scenarios as shown in Fig. 3. The sea and air platforms participating in the exercise include: MQ-8B shipborne unmanned reconnaissance helicopter, MQ-9A long endurance unmanned aerial vehicle (UAV), "vanilla" ultra-long endurance unmanned aerial vehicle, jumwalt class missile destroyer, Arley Burke class missile destroyer, San Antonio class dock transport ship, p-8a maritime patrol aircraft, E-2C shipborne early warning aircraft, ea-18 g shipborne electronic warfare aircraft, etc. Through the compact launch system commanded by the cyber cooperative network, the US naval and air combat platform can launch multiple light UAVs in the air in a short time, and quickly form the maritime distributed combat capability. It can be predicted that the UAV group with small target reflection, relatively low speed, good concealment, and strong sudden attack can actively cruise and ambush in the main activity area of the opponent according to the task instructions by adding electronic attack task load and network control terminal platform, independently perform or preset combat tasks, and conduct latent, detection, positioning and jamming on key targets through network tactical coordination and tactical configuration complex and rejection environment.

Fig. 3. Cyber network coordination of MQ-9A drone coordinated with a moribund combat carrier.

4.4 Adaptive Cross Domain Kill of UAV Based on Cyber Network

The "Adaptive Cross Domain Kill Network" is an important project of DARPA's "Mosaic Warfare". The core of the "mosaic warfare" combat concept is the "system of systems" architecture that enables distributed deployment capabilities between manned and unmanned platforms throughout the entire battlefield space. Combat personnel no longer need to rely on carefully designed fixed killing networks, but can build new killing networks based on these distributed capabilities, creating multiple challenges for the enemy and also having a certain degree of concealment and confusion. The "Adaptive Cross Domain Kill Network" is an important project to achieve the core values mentioned above. It aims to develop auxiliary decision-making software to allocate effective systems to the kill network in real time, realizing the integrated application of reconnaissance, interference, and detection of unmanned aerial vehicle systems. It can also reallocate system resources in real time to adapt to changes in combat. Based on the cyber network system, it assists unmanned units in global combat to select suitable unmanned aerial vehicle combat elements from the shared resource pool to form the kill network, and thus carry out multiple combat tasks concurrently. Currently, the single kill chain combat method based on traditional information links still faces problems such as concentrated risks, difficulty in networking, and insufficient cross domain capabilities. The research on the "Adaptive Cross Domain Kill Network" project can solve the above problems. Based on large-scale cyber networks, it can improve the anti cancellation ability, increase the flexible networking and killing ability of drone clusters, and greatly improve the adaptability of drones to battlefield environments and the adaptive matching ability of combat tasks. The project adopts a "supplier user" architecture, in which each node can be regarded as a potential network "service provider", and each command and control (C2) node can be regarded as a network node "user". The "service provider" provides battlefield resource integration, allocation, and decision-making recommendation services for each "user". Through large-scale network connections, the drone killing network flexibly organizes and utilizes various network services according to specific mission requirements to achieve established combat tasks. The UAV clusters have the ability of multi-source information collection, fusion and processing, unstructured environment perception, complex target learning and reasoning, and achieve accurate efficient electromagnetic situational awareness.

5 Conclusion

Cyber network collaboration is the advanced stage of the development of unmanned systems, and is an important technical approach and application means to cope with increasingly complex environments and missions. With the rapid development of related technologies such as cyber network and artificial intelligence, network collaboration will surely bring more opportunities and challenges to the development of unmanned systems. From the perspective of the development history and technical characteristics of unmanned system network collaboration, the following key issues should be focused on in relevant planning and promotion of relevant research: first, adhere to demand traction. There are often different evaluation methods and characteristics for application effectiveness in different fields. Therefore, when conducting relevant planning and technology research and development of UAV collaborative network, attention should be paid to the close integration with application scenarios. The second is to learn from the development ideas of system theory and system theory, give full play to the characteristics of interdisciplinary, organically combine Cybernet technology with various UAV loads, and give full play to the system integration advantages of mission load function nodes and network application elements.

References

1. Li, T., Lin, W., Ma, R., et al.: CoDetect: cooperative anomaly detection with privacy protection towards UAV swarm. Sci. China Inf. Sci. **67**, 159103 (2024)
2. Xiong, R., Chen, C., Xu, J., Dong, X., Pu, J.: Multi-UAV assisted on-demand charging in dense wireless rechargeable sensor networks. IEEE Internet Things J. **12**, 1–12 (2024)
3. Zheng, Z., Wei, C., Duan, H.: UAV swarm air combat maneuver decision-making method based on multi-agent reinforcement learning and transferring. Sci. China Inf. Sci. **67**, 180204 (2024)

4. Wang Jianfeng, Jia Gaowei, Xin Hongbo: Joint optimization for heterogeneous multi-UAV configuration and mission planning within SEAD scenarios. Guofang Keji Daxue Xuebao/J. Natl. Univ. Def. Technol. **46**(1), 32–41 (2024)
5. Wang, X., Wang, W., Song, K., Wang, M.: Uav air combat decision based on evolutionary expert system tree. Ordnance Industry Automat. **38**(1), 42–47 (2019)
6. Tomar, A., Muduli, L., Jana, P.K.: A fuzzy logic-based on-demand charging algorithm for wireless rechargeable sensor networks with multiple chargers. IEEE Trans. Mob. Comput. **20**(9), 2715–2727 (2020)
7. Zhang, W., Zhao, T., Zhao, Z.: Performance analysis of deep reinforcement learning-based intelligent cooperative jamming method confronting multi-functional networked radar. Signal Process. **7**, 108965 (2023)

A Situation Awareness Method for UAV Swarm Based on Modified Endsley Model

Bi Wu[1], Wei Lu[2], Liang Yu[2], and Zhiqian Zhou[1](✉)

[1] Uint 95899 of PLA, Beijing, China
Zhiqian.zhou13@hotmail.com

[2] AVIC Xi'an Flight Automatic Control Research Institute, Xi'an, China

Abstract. As the basis of decision-making, distributed radar array-based Situation Awareness (SA) plays a critical role in UAV swarm combats. To improve the accuracy of SA, a modified Endsley three-level SA model is proposed, which focus on distributed radar signal processing and fusion. Firstly, a consistency maintenance algorithm is proposed to deal with the inconsistency of the situation information within swarm. Secondly, an identification model of the topological structure and nodes is constructed in SA cognition layer. Finally, the interactive multiple model algorithm and analytic hierarchy process are introduced to estimate the states of units and the threat level of the whole swarm. Besides, the effectiveness of the method proposed in this paper is verified by numerical simulation.

Keywords: UAV swarm · situation awareness · kill web · threat assessment

1 Introduction

With the rapid development of modern warfare towards information-based and intelligent operations, UAV swarms, as a representative unmanned combat system, has become a research hotspot in the military field. Since UAVs within the swarm can carry various sensors and weapons, such as camera, radar systems, and missiles, they can undertake different combat tasks and form a dynamic distributed detection and engagement network through cooperative action. Though recent conflicts have proven that such a swarm is superior to traditional single-node and high-performance armaments, the performance of such swarms is highly dependent on cognitively perceive and understand the battlefield environment and enemies. Therefore, a robust distributed situation awareness (SA) plays a fundamental and critical role in UAV swarm combat systems.

The distributed SA model aims to achieve consistency of situation information. Typical SA models can be classified into three classes: the Endsley SA model, the group SA model, and the distributed SA model [1]. The situation information sharing and prediction method in the group SA model has been studied in [2]. The consistency evaluation algorithm of SA model has been studied in [3, 4] often without explicitly addressing the signal-level challenges of distributed radar arrays. Different nodes within the UAV swarm execute different tasks, such as reconnaissance, communication, jamming, strike, and cooperate with each other to form a kill web. Identifying the nodes functions and the

H. Ma et al. (Eds.): IWACIII 2025, CCIS 2780, pp. 73–86, 2026.
https://doi.org/10.1007/978-981-95-6730-0_6

topological structure of enemy swarm based on the SA model can achieve cognitive superiority, which is conducive to predicting and counter the enemy's tactics, and optimizing the allocation of our resources. The identifying of key nodes in complex network has been researched in [5, 6], leveraging distributed radar signal features (e.g., RCS patterns, Doppler signatures, spatial angles-of-arrival) for functional classification and topology mapping presents distinct challenges. Based on the existing research, it can be seen that the identification task of swarm depends on the situation information model to support decision-making. Due to the limited capacity of individuals, the tactical intentions of UAV swarm are usually manifested in the form of coordinated patterns at the group level. Therefore, the dynamic characteristics of swarm have obvious pattern recognisability. Cui et al. and Xu et al. utilized fuzzy theory to achieve the mapping of intentions for different behaviours, enabling the prediction of combat intentions for aerial targets [7, 8]. Wu et al. and Ou et al. have developed intention prediction algorithm frameworks based on expert knowledge systems, enabling the reasoning from target states rather than raw or fused distributed radar sensor data to tactical intentions [9, 10].

Compared with individual system, distributed UAV swarm SA faces the following problems: it's hard to construct a unified situation information model, difficult to identify the functions and topological structure of the enemy swarm, and hard to predict the enemy's tactical intentions [11].

In response to the above problems, this paper proposes a SA method for UAV swarm based on the Endsley model, specifically tailored for distributed radar array applications. The method explicitly incorporates three layers: situation information acquisition layer based on distributed radar systems, situation information cognition layer and situation information prediction layer. Based on the three layers, the proposed method can maintain the consistency of situation information, identify the function of nodes, and assess the threat level of swarms. Finally, a numerical simulation is used to verify the effectiveness of the method.

2 UAV Swarm SA Model

Following the classic Endsley three-level SA model, the UAV swarm SA model framework can be divided into three layers: the situation information acquisition layer, which acquires information through observation; the situation information cognition layer, which recognizes the current situation through information integration; and the situation information prediction layer, which predicts the tactics of enemy swarm. The SA model framework of units within swarm is shown in Fig. 1.

The situation information acquisition layer achieves the low-level SA consensus and is responsible for the information integration function of various sensors. The information at this layer flows from the physical domain to the information domain. The situation information cognition layer and the situation information prediction layer achieve the high-level SA consensus, and the information in these layers flow from the information domain to the cognitive domain, achieving the perception of the current situation and the future situation, respectively. The three-level model is not a strictly linear continuous relationship. Distributed radar data/information sharing and consistency ensures information consistency within the swarm by situation information sharing mehod, runs through the entire process from investigation to decision-making, and directly affects the coordination accuracy, anti-interference ability, and task success rate of the swarm.

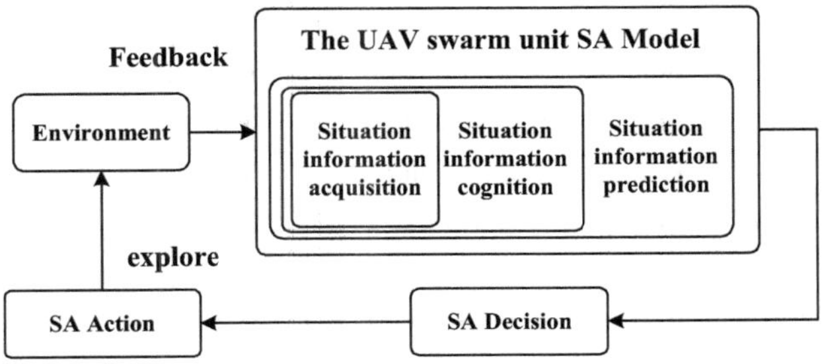

Fig. 1. SA model of unit within swarm

3 Situation Information Acquisition Layer: Distributed Radar Data Fusion

3.1 Distributed Radar Situation Information Model

The situation information acquisition layer is responsible for modelling information and fusing information from the multi-source, distributed radar sensors to form a consistent, swarm-wide situation picture. The information model must accommodate the spatial distribution and potential heterogeneity of the radar nodes. The information model can be represented as a triple in a hyper-network structure, classifying the levels based on the system, equipment, and node as $G = (G_S, G_E, G_N)$, including system layer G_S, equipment layer G_E, and node layer G_N. Crucially, node classification (E) is inferred based on fused radar sensor observations (e.g., RCS magnitude, Doppler spectrum characteristics, emission detection, spatial location). The functions of each layer in the information model are shown in Table 1.

Without considering the communication within the swarm, the functional nodes can be divided into sensing nodes S, target nodes T, decision nodes D and execution nodes A, corresponding to the equipment type information T_E, and there is a link relationship O_E between the nodes. This division method is no longer limited to the information within one's own group. It takes into account the connection of heterogeneous functional nodes and the relationship between mission objectives.

From the C2 organization in UAV swarm, it is known that the decision nodes D is the core of the swarm mission organization responsible for regulating other nodes within the mission subgroup to complete the closure of the kill chain. It should be noted that the enemy information in the information model is unknown at the initial moment and is updated as the swarm confrontation progresses.

Table 1. Functions of each layer in the information model

Layer name	Information Type	Specific functions
System Layer G_S	Organization information O_S	Topological structure of swarm
	Environment information E_S	Battlefield scenario
	Mission Information T_S	Task target of UAV swarm
	Historical information H_S	Trusted perceived information
Equipment Layer G_E	Type Information T_E	Functional type of equipment
	Capability Information C_E	Inherent attributes of equipment
	Relationship Information O_E	Relationships between nodes
Node Layer G_N	Dynamic Information D_N	Position, velocity, etc.
	Task Information T_N	Specific task content

3.2 Distributed Aituation Information Update

Considering the scenario where several sensing nodes jointly investigate the same target, there are overlapping parts of perception information among swarm, and the perception results of the same object may different. Assume the number of sensing nodes is m, and then the set of sensing nodes can be denoted as $S = \{S_1, S_2, \cdots, S_m\}$, $m \geq 2$. When multiple sensing nodes investigate the same target, the resulting information are $X = \{x_1, x_2 \cdots x_n\}$. Due to the heterogeneity of sensing nodes and the differences in perception methods, each node has different preferences for perception results. A fuzzy preference matrix can be established according to [11], and satisfies:

$$P^k = \left\{p_{ij}^k\right\}, i, j \in \{1, 2, \cdot, n\}, k \in \{1, 2, \cdot, m\} \tag{1}$$

For any two sensing nodes $\{S_p, S_q\}, p, q \in \{1, 2, \cdots, m\}, p \neq q$, the similarity matrix can be established as:

$$L = \left(l_{ij}^{pq}\right), l_{ij}^{pq} = 1 - \left|p_{ij}^p - p_{ij}^q\right| \tag{2}$$

The consistency matrix for all sensing nodes of a certain target can be expressed as:

$$Y = (y_{ij}), y_{ij} = \phi\left(l_{ij}^{pq}\right) \tag{3}$$

where $\phi(\cdot)$ is the computed value to measure the degree of deviation of all sensing node preference relationships, and the arithmetic mean or median of each node preference matrix can be taken. The degree of consistency of perception result x_i for all sensing nodes can be solved:

$$\kappa = \frac{\sum_{j=1, j \neq i}^{n} (y_{ij} + y_{ji})}{2(n-1)} \tag{4}$$

The degree of consistency in the preference relationship of all nodes for the result is:

$$\sigma = \sum_{i=1}^{n} \kappa_i / n \tag{5}$$

Defined γ as the preference decision threshold, if the degree of consistency satisfies $\sigma \geq \gamma$, it is determined that all sensing nodes are consistent with the perception results, else the sensing node preference matrix with bias must be repaired.

4 Situation Information Cognition Layer

4.1 Identification Node Type in Enemy Swarm

In the process of confrontation between swarms, although the SA of both swarm combat and anti-swarm combat can be classified as topological identification of the enemy's kill web, the specific requirements are different as follows: (1) In offensive swarm combat scenarios, the core mission is to strike key defence sites. Thus, the principle of swarm is saving the strength of itself as much as possible. Therefore, the combat principle of the offensive swarm SA is to identify the weak nodes in the enemy swarm's kill web and break through by destroying them to create local blind spots. (2) In defensive anti-swarm combat scenarios, the mission objective is to prevent and intercept enemy swarm targets as much as possible. Therefore, the objective of anti-swarm operations is to identify the key nodes within the enemy's kill web whose destruction would maximize the deduction of combat effectiveness.

The function type of enemy unit can be identified by analysing information such as the function mode, dynamic characteristics, physical features, etc. Sensing nodes may hover or patrol over the target area, with a wide range of trajectory coverage, or fly low near critical terrain, usually far from enemy anti-aircraft fire, and fly at a high altitude in high-threat areas to expand reconnaissance range. These nodes normally carry electro-optical/infrared cameras, radars, or electronic reconnaissance equipment to communicate with other nodes. Decision nodes periodically send task instructions to other nodes. It has the highest priority during swarm tactical actions and is usually located at the central node of the swarm to ensure that decision instructions reach all nodes within the swarm. These nodes may carry a high-performance computer for real-time data processing and decision generation. Execution nodes' trajectories are aggressive breakouts, dives or cooperative encirclement routes, and may break away from the swarm to carry out attacks in the later stages of the mission. These nodes apparently carry weapons or jamming equipment.

4.2 Identification of Topological Structure in Enemy Swarm

Because the functional nodes behaviours in the swarm have strong causal correlation with the accusation of the decision node, the Hawkes process can be used to effectively model the dynamic dependencies between events, thereby enabling precise identification of the enemy swarm kill web topology. Define $N(t)$ represents the total number of events that occurred up to time t, where the time point of each event is denoted as t_i, which can represent a sequence of random events as $\Gamma = \{t_1, t_2, \cdots, t_n\}$. The intensity equation used to describe the general pattern of events is defined as:

$$\lambda(t) = \lim_{\Delta t \to 0} \frac{E[N(t + \Delta t) - N(t)]}{\Delta t} \tag{6}$$

Given the possibility that the behaviour of functional nodes may be influenced by multiple decision nodes, we can model it with multivariable Hawkes process, assuming the variable dimension is M, the conditional intensity function of the event m can be defined as

$$\lambda_m(t) = \mu_m(t) + \sum_{m'=1}^{M} \omega_{mm'} \sum_{t_{m'} < t} \gamma(t - t_{m'}) \tag{7}$$

where $\mu_m(t)$ represents the basic intensity or rate of the generation of event m, which can be regarded as a constant μ_m if time changes are not taken into account. M is the total number of event types that $m \in [1, M]$, $m' \in [1, M]$. $\omega_{mm'}$ represents the trigger tendency of event m' to event m, and represents the accusation relationship between decision nodes m' and the functional nodes m. $\gamma(t)$ represents the temporal relationship between current possible events and historical events, usually set as $\gamma(t) = \beta e^{-\beta t}, t \geq 0$.

Define the base event as (t_k, u_k), where t_k is the event trigger time, u_k indicating that the enemy decision node triggered the communication signal. The log-likelihood function can be solved when there is a sequence of events $\Gamma = \{(t_1, u_1), (t_2, u_2), \cdots (t_n, u_n)\}$ within the time interval$(0, T]$:

$$\begin{aligned} L(\boldsymbol{\mu}, \mathbf{W}) = & \sum_{i=1}^{n} \lg\left[u_{u_i} + \sum_{t_k < t_i} \omega_{u_i u_k} \gamma(t_i - t_k)\right] \\ & - \left\{\sum_{j=1}^{M}\left[u_j T + \sum_{i=1}^{n} \omega_{j u_i} \int_0^{T - t_i} \gamma(s) ds\right]\right\} \end{aligned} \tag{8}$$

where $\boldsymbol{\mu} \in \mathbf{R}^{M \times 1}$, $\mathbf{W} \in \mathbf{R}^{M \times M}$, n represents the number of events. Then, the log-likelihood function is decomposed to the multi-dimensional event type m:

$$L_m(\mu_m, \omega_m) = \sum_{k=1}^{n} \lg \lambda_m(t_k) - \int_0^T \lambda_m(t) dt \tag{9}$$

Based on the log-likelihood function, the model parameters are estimated with the maximum likelihood function. The connection strength $\mathbf{W}_{ij}$ between the enemy node u_i and u_j can be obtained through the event sequence Γ. If the strength does not meet the threshold requirement, it is determined that there is no connection between the node u_i and u_j, otherwise, the topological relationship can be set.

4.3 Identification of Key Nodes in Enemy Swarm

A complete kill chain in swarm kill web should be a closed loop starting from the target node T_i, passing through several nodes $\{D_i, S_i, A_i\}$, and returning to the target node T_i. If the reusability count of each node in the closed-loop kill chain during the execution of the strike mission is used as the basis for evaluation, then the importance degree of each node in kill web can be determined.

Assuming the length of the longest kill chain in the enemy's swarm kill web against our node T_i is l, define $val_{T_i}^k$ as the sum of the combat effectiveness of all kill chains with the starting and ending points T_i and the length k. The combat effectiveness matrix of the enemy swarm kill web is $\mathrm{H} = \{H^k| \ k = 1, 2, \cdots, l\}$, where H^k is the combat effectiveness matrix of all kill chains with length k. Define the element h_{ij}^k represents the sum of combat effectiveness of all kill chains of length from the target node i to the target node j, then $val_{T_i}^k$ locates at position i on the main diagonal of H^k. Since the reliability of the kill chain decreases with the increase in the number of nodes, assuming the penalty weight ω_k satisfies $\omega_1 > \omega_2 > \cdots > \omega_k$, the combat effectiveness of the enemy swarm kill web against our node T_i can be defined as:

$$val_{T_i} = \sum_{k=1}^{l} \omega_k val_{T_i}^k = \sum_{k=1}^{l} \omega_k h_{ii}^k \tag{10}$$

Let $q(\mathrm{H}) = \sum_{k=1}^{l} \omega_k \mathrm{H}^k$, $\omega_k h_{ii}^k$ is the element at the position i on the main diagonal of the matrix $q(\mathrm{H})$. Define $\omega_k = 1/k!$, $val_{T_i}^k$ can be rewritten as:

$$val_{T_i} = \sum_{k=1}^{l} \frac{val_{T_i}^k}{k!} = \sum_{k=1}^{l} \frac{h_{ii}^k}{k!} = \sum_{k=1}^{l} \frac{[q(\mathrm{H})]_i}{k!} = \left(e^H\right)_i - 1 \tag{11}$$

The recognition of the importance of nodes in offensive swarm combat and defensive anti-swarm combat is different. The former targets nodes with a high number of reuse times (set performance of the same type nodes to same value), while the latter targets nodes with high capability in kill web.

5 Situation Information Prediction Layer

5.1 Estimation of Tactical Maneuver Mode

The tactical behavior of swarm is often indirectly reflected through swarm maneuver mode, and the maneuver modes of the swarm can be estimated. The Interacting Multiple Model (IMM) algorithm uses several base models to mix and estimate the motion states of complex trajectories. Suppose the IMM algorithm applied contains n basic maneuver models, and the discretized state equation and observation equation of the model i satisfy:

$$\begin{cases} x_i(k) = F_i(k-1)x_i(k-1) + u_i(k-1) \\ z_i(k) = H_i(k-1)x_i(k) + v_i(k) \end{cases}, 1 \le i \le n \tag{12}$$

where $F_i(k-1)$ is the state transition matrix of the model i at the time $k-1$, $H_i(k-1)$ is the state observation matrix of the model i at the time $k-1$, $u_i(k-1)$ and $v_i(k)$ are the random noise of the state transition equation and the observation equation, respectively. The basic maneuver models are set as CV model, CA model, and CSCT model, satisfying in any dimension of XYZ:

$$
\begin{aligned}
&F_{CV} = \begin{bmatrix} 1 & T & 0 \\ 0 & 1 & 0 \\ 0 & 0 & 0 \end{bmatrix}, F_{CA} = \begin{bmatrix} 1 & T & T^2/2 \\ 0 & 1 & T \\ 0 & 0 & 1 \end{bmatrix} \\
&F_{CSCT} = \begin{bmatrix} 1 & \sin(\omega T)/\omega & (1-\cos(\omega T))/\omega^2 \\ 0 & \cos(\omega T) & \sin(\omega T)/\omega \\ 0 & -\omega\sin(\omega T) & \cos(\omega T) \end{bmatrix}
\end{aligned} \tag{13}
$$

where T is the sampling time, and ω is the turning rate. Based on the basic maneuver model, define the confidence probability of the model j at time k is $m_j(k)$, the final state estimation result $\hat{X}(k|k)$ satisfies:

$$
\hat{X}(k|k) = E\Big[X(k)|Z^k\Big] = \sum_i \hat{X}_i(k|k)u_i(k) \tag{14}
$$

And the probability $u_j(k)$ of model j satisfies:

$$
\begin{aligned}
\Lambda_j(k) &= p\big(v_j(k)|m_j(k), Z^{k-1}\big) \\
&= \frac{1}{\sqrt{|2\pi S_j(k)|}} e^{-\frac{v_j^T(k)S_j(k)v_j(k)}{2}} \\
u_j(k) &= \frac{u_j(k|k-1)\Lambda_j(k)}{\sum_i u_i(k|k-1)\Lambda_i(k)}
\end{aligned} \tag{15}
$$

The outputs of each model are combined with the confidence probability to achieve the estimation and prediction of maneuver modes within a certain period of time.

5.2 Threat Assessment of Enemy Swarm Behaviour

In the swarm confrontation scenario, the enemy swarm is composed of multiple functional nodes and can perform several different types of combat actions. Therefore, the threat assessment of the enemy swarm behavior needs to take into account several factors comprehensively. Fuzzy analytic hierarchy process (FAHP) can achieve a combination of multiple evaluation metrics, and is often used in multi-criteria decision-making problems.

Firstly, establish the triangular fuzzy number complementary judgment matrix $\tilde{P} = (\tilde{p}_{ij})_{k\times k}$ of the criterion layer, where $\tilde{p}_{ij}$ represents the importance of the evaluation metric i compared to the evaluation metric j, and k is the number of all metrics. Design $\tilde{p}_{ij}$ as the triangular fuzzy number satisfying $\tilde{p}_{ij} = \left(p_{ij}^L, p_{ij}^M, p_{ij}^H\right)$, where p_{ij}^L is the lower bound of $\tilde{p}_{ij}$, p_{ij}^M is the mean of $\tilde{p}_{ij}$, p_{ij}^H is the upper bound of $\tilde{p}_{ij}$, with $0 \le p_{ij}^L \le p_{ij}^M \le p_{ij}^H \le 1$. Next, get the expectation matrix $E(\tilde{P}) = (e_{ij})_{k\times k}$:

$$
E(\tilde{p}_{ij}) = \left(p_{ij}^L + 2p_{ij}^M + p_{ij}^H\right)/4 \tag{16}
$$

Solve the reciprocal judgment matrix $H(E) = (e_{ij}/e_{ji})_{k \times k}$ and conduct consistency tests on it. Fuzzy evaluation values $\tilde{u}_i$ and expected value $E(\tilde{u}_i)$ can be calculated as:

$$\begin{aligned} \tilde{u}_i &= \left(\frac{\sum_{j=1}^{n} h_{ij}^L}{\sum_{i=1}^{n}\sum_{j=1}^{n} h_{ij}^H}, \frac{\sum_{j=1}^{n} h_{ij}^M}{\sum_{i=1}^{n}\sum_{j=1}^{n} h_{ij}^M}, \frac{\sum_{j=1}^{n} h_{ij}^H}{\sum_{i=1}^{n}\sum_{j=1}^{n} h_{ij}^L} \right) \\ E(\tilde{u}_i) &= \left(\frac{\sum_{j=1}^{n} h_{ij}^L}{\sum_{i=1}^{n}\sum_{j=1}^{n} h_{ij}^H} + 2\frac{\sum_{j=1}^{n} h_{ij}^M}{\sum_{i=1}^{n}\sum_{j=1}^{n} h_{ij}^M} + \frac{\sum_{j=1}^{n} h_{ij}^H}{\sum_{i=1}^{n}\sum_{j=1}^{n} h_{ij}^L} \right) / 4 \end{aligned} \tag{17}$$

Normalize the expected value to obtain the criterion layer weight vector $\alpha^k = (\alpha_1, \alpha_2, \cdots, \alpha_k)$. Similarly, the weights of the metric layer can be calculated as $\beta^k = (\beta_1, \beta_2, \cdots, \beta_k)$, and the final weights of each metric can be obtained by linearly weighting the weights of the two layers.

6 Simulated Experiment

Assuming the topological structure identification result of the enemy swarm at a certain moment is as shown in Fig. 2. To distinguish between swarm and anti-swarm combat scenarios, define the performance values of each node relative to the mission target as capability, and the degree of performance decline after node failure as importance. The results are shown in Table 2 based on the key node identification algorithm.

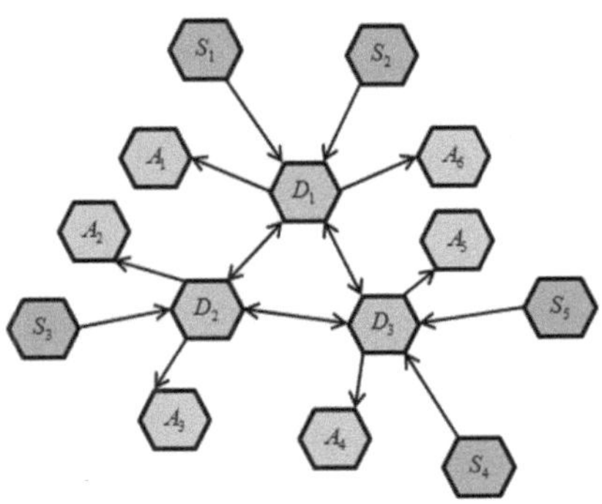

Fig. 2. The topological structure of enemy swarm

As can be seen from Table 2, the evaluation of node importance in the same topology varies in different combat scenarios. In the anti-swarm combat scenario, the final impact of nodes on the performance of the kill web is entirely dependent on the nodes that have a higher degree of reuse, rather than nodes having a greater impact on the performance of the kill web in the swarm combat scenario.

The IMM algorithm is used to achieve the estimation of the target's maneuver mode. The simulation parameters are set as follows: State transition noise variance $q = 10$, observation noise variance $r = 30$, and three types of turning rates in the CSCT model are set as $\omega_1 = 80^*\pi/180$, $\omega_2 = 50^*\pi/180$, and $\omega_3 = 25^*\pi/180$.The three-dimensional trajectory of the target during the simulation is shown in Fig. 3, the state estimation errors during the process are shown in Fig. 4, the confidence probability estimation of models are shown in Fig. 5. From the results we can find that the IMM algorithm achieves state estimation and obtains the confidence probability of the maneuver models accurately.

Table 2. Node capability and importance configuration

Node	Swarm combat	Anti-swarm combat		
	Capability	Importance	Capability	Importance
D_1	0.92	52.52%	1	52.23%
D_2	0.96	51.53%	1	52.23%
D_3	0.94	36.42%	1	36.31%
S_1	0.83	20.34%	0.8	20.00%
S_2	0.87	20.16%	0.8	20.00%
S_3	0.82	20.07%	0.8	20.00%
S_4	0.8	20.07%	0.8	20.00%
S_5	0.78	19.79%	0.8	20.00%
A_1	0.92	18.30%	0.85	17.96%
A_2	0.91	17.99%	0.85	17.96%
A_3	0.84	17.76%	0.85	17.96%
A_4	0.83	17.64%	0.85	17.96%
A_5	0.85	14.31%	0.85	14.08%
A_6	0.87	13.99%	0.85	14.08%

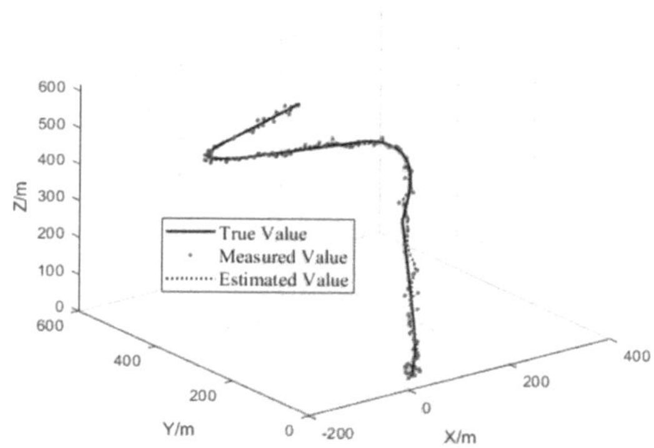

Fig. 3. The three-dimensional trajectory of enemy swarm

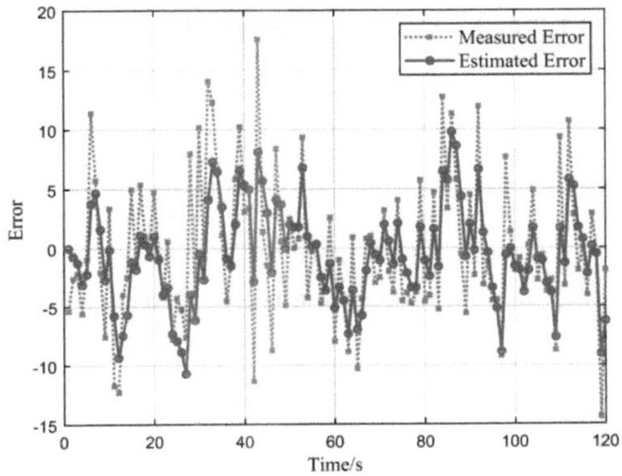

Fig. 4. The state errors in IMM algorithm

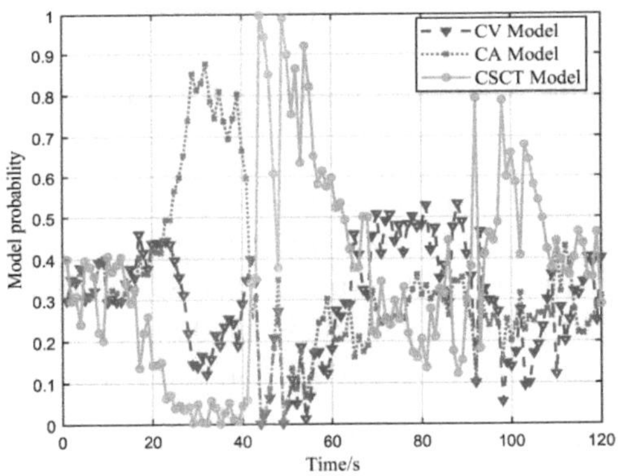

Fig. 5. The estimated model probability of enemy swarm

Suppose there are 5 enemy swarms attacking our key position, the normalized threat characteristics are shown in Table 3.

Table 3. Normalized threat characteristics of enemy

Characteristic	1	2	3	4	5
Kill chain number	0.1702	0.2128	0.2979	0.1914	0.1276
Node scale	0.1812	0.2282	0.2752	0.1879	0.1275
Sense node number	0.2414	0.2069	0.2414	0.1724	0.1379
Decision node number	0.1905	0.2380	0.2380	0.1905	0.1429
Execution node number	0.1967	0.2295	0.2459	0.1639	0.1639
Maneuver mode	0.1	0.3	0.2	0.1	0.3

(continued)

Table 3. (*continued*)

Characteristic	1	2	3	4	5
Swarm speed	0.1938	0.2171	0.2351	0.1550	0.1989
Swarm distance	0.1507	0.2563	0.2110	0.1960	0.1859

Establish a complementary judgment matrix $\tilde{P}$ of triangular fuzzy numbers at the criterion level as:

$$\tilde{P} = \begin{bmatrix} (0.50, 0.50, 0.50) & (0.25, 0.30, 0.35) & (0.35, 0.40, 0.45) \\ (0.65, 0.70, 0.75) & (0.50, 0.50, 0.50) & (0.55, 0.60, 0.65) \\ (0.55, 0.60, 0.65) & (0.35, 0.40, 0.45) & (0.50, 0.50, 0.50) \end{bmatrix}$$

With the threat assessment algorithm in this paper, the criterion layer weight vector α can be calculated as:

$$\alpha = (0.2668, 0.3998, 0.3333)$$

Similarly, establish the index layer triangular fuzzy number complementary judgment matrix $\tilde{P}_i$:

$$\tilde{P}_1 = \begin{bmatrix} (0.5, 0.5, 0.5) & (0.35, 0.4, 0.45) \\ (0.55, 0.6, 0.65) & (0.5, 0.5, 0.5) \end{bmatrix}$$

$$\tilde{P}_2, \tilde{P}_3 = \begin{bmatrix} (0.5, 0.5, 0.5) & (0.75, 0.8, 0.85) & (0.85, 0.9, 0.95) \\ (0.15, 0.2, 0.25) & (0.5, 0.5, 0.5) & (0.55, 0.6, 0.65) \\ (0.05, 0.1, 0.15) & (0.35, 0.4, 0.45) & (0.5, 0.5, 0.5) \end{bmatrix}$$

The index layer weight vector β can be calculated as

$$\beta_1 = (0.4501, 0.5499)$$
$$\beta_2 = (0.4893, 0.2888, 0.2220)$$
$$\beta_3 = (0.4893, 0.2888, 0.2220)$$

Finally, the comprehensive weight γ satisfies:

$$\gamma = \big(0.1201, 0.1467, 0.1956, 0.1155, 0.0888, 0.1631, 0.0963, 0.0740\big)$$

The threat level of each enemy swarm can be predicted based on the weights γ as shown in Table 4.

Table 4. Threat levels of enemy swarms

Swarm index	1	2	3	4	5
Threat level	1.4387	1.8894	1.9485	1.3326	1.3912

According to the threat level values of each swarm, it can be seen that swarm 3 has the highest threat level, followed by swarm 2, which is related to the indicators of each swarm. It is possible to prioritize the attack on the key nodes in the high-threat colony according to the threat level.

7 Conclusion

In this paper, a full-process SA method is constructed for UAV swarm based on the modified Endsley three-level SA model, specifically through the lens of distributed radar array signal processing and its combat applications. An information consistency maintenance algorithm is proposed in the situation information acquisition layer. The identification problem of node type, topological structure and key nodes are addressed in situation information cognition layer. IMM algorithm is introduced to estimate the maneuver mode of swarm and FAHP algorithm is used to assess enemy swarm's threat level in situation information prediction layer. Finally, the experimental results prove the effectiveness of the method proposed in this paper.

Acknowledgments. This work was supported by the National Defense Science and Technology Fund for Outstanding Young Scholars under Grants JCJQ-ZQ-2022-092.

Disclosure of Interests The authors have no competing interests to declare that are relevant to the content of this article.

References

1. Gao, Y., Li, D.S., Cheng, Z.X.: UAV distributed swarm situation awareness model. J. Electron. Inf. Technol. **40**(6), 1271–1278 (2018)
2. Saner, L.D., Bolstad, C.A., Gonzalez, C., et al.: Measuring and predicting shared situation awareness in teams. J. Cognit. Eng. Decis. Making. **3**(3), 280–308 (2009)
3. Zhang, D.G., Meng, H., Zhao, H.Y.: Analytical metric model for situational awareness level. Syst. Eng. Electron. Technol. **38**(8), 1808–1815 (2016)
4. Huang, C., Tao, J.Y., Zhang, D.G., et al.: Research on distributed situation awareness consistency analysis and measurement method. Command Control Simul. **36**(3), 1–8 (2014)
5. Wang, J.C., Wu, X.P., Liao, W., et al.: Improved method of node importance evaluation in weighted complex networks. Comput. Eng. **38**(10), 74–76 (2012)
6. Chen, J., Sun, L.F.: Evaluation of node importance in complex networks. J. Southwest Jiaotong Univ. **44**(3), 426–429 (2009)
7. Cui, Y.P., Wu, Q.X., Chen, M.: Aerial target intention prediction based on adaptive neuro-fuzzy inference system. In: The 15th China Symposium on System Simulation Technology and its Application, pp. 277–281, China, Hefei (2014)

8. Xu, J.P., Zhang, L.F., Han, D.Q.: Air target intention recognition based on fuzzy inference. Command Inf. Syst. Technol. **11**(3), 44–48 (2020)
9. Wu, Z.Q., Li, D.F.: A model for aerial target attacking intention judgment based on reasoning and multi-attribute decision making. Electron. Opt. Control. **17**(5), 10–13 (2010)
10. Ou, W., Liu, S.J., He, S.Y., et al.: Tactical intention recognition algorithm based on encoded temporal features. Command Control Simul. **38**(6), 36–41 (2016)
11. Aniello, G.D., Loia, V., Orciuoli, F.: A multi-agent fuzzy consensus model in a situation awareness framework. Appl. Soft Comput. **30**(5), 430–440 (2015)

Smart Education

Intelligent Detection of Student Classroom Behavior Based on an Improved YOLOv8 Model

Meng Zhou[1], Huifang Zhang[1], Yuchao Peng[1], Jing Wang[1], Zhe Dong[1(✉)], and Jie Fan[2]

[1] The School of Electrical and Control Engineering, North China University of Technology, Beijing 100144, People's Republic of China
dongzhe@ncut.edu.cn

[2] Istituto Italiano di Tecnologia, Genova, Italy

Abstract. This paper proposes an intelligent detection method of student classroom behavior based on an improved YOLOV8 model. First, a dataset comprising five categories of classroom misbehaviors, i.e., "using mobile phones", "sleeping on the desk", "looking down", "whispering" and "looking around" is constructed. Second, an improved YOLOv8 model is designed to detect the student classroom behavior. To improve the model's robustness in complex scenes, a spatial adaptive feature modulation network is introduced, leveraging multi-scale feature fusion. Then, the proposed method is validated on the constructed dataset, and the results demonstrate that the proposed method has improvement in recognition accuracy, i.e., increasing the mean average precision (mAP@0.5) to 81.7%. Furthermore, a visual system interface is developed using PyQt5. Leveraging integrated real-time video analysis and behavioral metrics, this work facilitates optimized teaching management and contributes to the development of a smart classroom ecosystem.

Keywords: Classroom behavior detection · improved YOLOv8 model · educational intelligence · visual system interface · multi-scale feature fusion

1 Introduction

With the rapid advancement of artificial intelligence technology, various industries are exploring the integration of advanced and efficient artificial intelligence solutions. The "Education Informatization 2.0 Action Plan" calls for building a

This paper is funded by the National Key Research and Development Program of China (2023YFC3306401), Yuxiu Innovation Project of NCUT (2024NCUTYXCX107), the Outstanding Talent Development Program in Yanqing District (No. YQRC2024-18).

H. Ma et al. (Eds.): IWACIII 2025, CCIS 2780, pp. 89–101, 2026.
https://doi.org/10.1007/978-981-95-6730-0_7

new "Internet + Education" ecosystem. It also highlights using new technologies like artificial intelligence to transform education models [1]. Furthermore, "China's Education Modernization 2035" lists speeding up educational reform in the information age as a key task. This provides strong policy support for the education sector's digital transformation [2].

The classroom environment is crucial for ensuring teaching quality and student learning efficiency. Well-regulated classroom discipline not only enhances knowledge transfer but also serves as a vital platform for cultivating students' concentration and self-management abilities. However, real-world teaching scenarios frequently encounter classroom misbehaviors such as students using mobile phones, sleeping, and engaging in unauthorized conversations. These actions not only distract individual focus but also create a chain effect, persistently disrupting the learning environment. Traditional classroom supervision, relying primarily on manual observation by teachers, suffers from limitations such as significant blind spots and difficulties in quantification, hindering precise and real-time behavior management. Consequently, classroom behavior detection technology has become a prominent research area in the effort to integrate classrooms with digital intelligence.

The earliest research on automatic face recognition originated in the 1960s [3]. By the late 1980s, the field of automatic face recognition mainly had two research directions: template-based recognition methods and feature-based recognition methods. Among them, template-based recognition methods included pyramid machines and neural network methods. These template-based methods identify faces by matching them with stored templates based on similarity. Feature-based recognition methods included deformable templates. These feature-based methods identify faces by extracting and matching key facial features, which makes the approach more sensitive and flexible to local variations. In the statistical learning recognition framework, the Eigenface method is proposed, in which principal component analysis is employed for face recognition [4]. While the Viola-Jones algorithm significantly enhances face detection speed by integrating AdaBoost, integral images, and cascaded classifiers [5].

In recent years, deep learning technique has developed rapidly. In the 2012 ImageNet image classification competition, the AlexNet deep convolutional neural network was designed by Alex Krizhevsky and Geoffrey Hinton et al., which was achieved a decisive victory [6]. Its key advantages mainly lie in the deep convolutional neural network architecture, GPU parallel computing, ReLU activation function and mini-batch stochastic gradient descent, and it also uses Dropout layers to mitigate overfitting. Human action recognition mainly involves two processes: feature representation, action recognition and understanding. Among them, feature representation extracts features that can represent the key information of the video from the video data, this process plays a crucial role in the entire recognition process, and the quality of the features directly affects the final recognition result, in addition, the action recognition and understanding stage takes the feature vectors obtained in the previous stage as input and uses machine learning algorithms to learn, and inputs the feature vectors obtained in the testing process or application scenarios into the model obtained in the above process for type recognition [7]. Later, a new detection method named YOLO was proposed. From then on, the series of YOLO models have led to significant improvements in both speed and accuracy in object detection [8].

In general, deep learning-based object detection can be divided into two categories: two-stage detection and one-stage detection. Two-stage detection is a "coarse-to-fine" process, while the other is an end-to-end "one-step completion". Usually, one-stage detection attempts to directly classify each region of interest as background or target object. It can directly give the category probability and position coordinate value of the object through only one stage. Typical examples include the YOLO series. In 2018, Redmon and Farhadi used YOLOv3 with optimized feature extraction and multi-scale detection to improve the recognition of small objects, especially rear-row faces in classroom activity detection. Next in 2020, Ultralytics proposed YOLOv5 to improve both detection speed and accuracy, leading to its widespread adoption in student classroom behavior recognition. Then in 2023, Ultralytics designed YOLOv8, which was built on YOLOv5 with further optimizations for speed-accuracy balance, to create an efficient and robust framework for real-time student classroom behavior detection [9]. YOLOv8 is boasting higher detection accuracy and faster inference speed. It supports multi-task learning, such as object detection and pose estimation, and performs excellently in scenarios like real-time video analysis, autonomous driving, medical imaging and agricultural monitoring.

Based on the above analysis of the YOLO series, this paper selects YOLOv8 model to implement intelligent monitoring of student classroom behavior, hoping to provide assistance for the construction of future smart classrooms driven by big data. Currently, big data-driven smart classrooms are an inevitable trend in the future of education [10]. Observing student classroom behavior mainly relies on teachers' manual supervision and video analysis. However, this approach distracts teachers' attention, reduces teaching efficiency, and makes it difficult to comprehensively and accurately record all students' behaviors. Therefore, there is a growing expectation for the application of deep learning and computer vision techniques to analyze student behavior in the classroom. Consequently, the automatic detection and analysis of students' classroom behavior using computer-aided instruction has become a prominent research topic in smart education [11,12].

This paper proposes a computer vision-based method for intelligent student behavior detection, aiming to overcome the inefficiency and subjectivity of manual classroom management. The main contributions of this paper consist of the following aspects:

1. A dataset containing five types of classroom misbehaviors is constructed, which comprises 1089 images from various classroom environments, annotated using the labeling tool, and divided into training and validation sets.
2. An improved YOLOv8 model is constructed, in which a spatial adaptive feature modulation network is introduced to enhance multi-scale feature fusion capabilities. This improves the model's detection performance in complex sce-

narios and enables real-time recognition and statistical analysis of classroom behavior.

The following structure of this paper is organized as follows. In Sect. 2, the dataset is established. In Sect. 3, the main improved model is designed. In Sect. 4, simulation verification is conducted. Finally, some conclusions of this work are given in Sect. 5.

2 Dataset Construction

High-quality data allows models to accurately fit the association between features and targets [13]. Therefore, a large amount of image and video data depicting various student behaviors in the classroom should be collected to train models with high recognition accuracy and strong scene generalization capabilities. In this section, a dataset for student learning behavior detection in natural scenes is constructed to transform the abstract problem of "classroom management" into a quantifiable object detection task, a clear and evaluable optimization direction is provided for subsequent model design.

Considering the diverse and complex nature of student behavior in the classroom, five common behavioral characteristics have been selected for analysis, i.e., using mobile phones, sleeping on the desk, bowing the head, whispering, and looking around. Detailed description is shown in Table 1.

Table 1. Student Misbehavior Recognition

Student Misbehavior	Description
Using a mobile phone	Using mobile phones in various postures
Sleeping on the desk	Lying on their desks
Bowing the head	Bow their heads with a large angle
Whispering	Tilt their heads and are in a speaking state
Looking around	Turn their heads or bodies significantly

The data sources are mainly twofold: obtained from online platforms (such as the Mota community and CSDN) and a self-made dataset, consisting of simulated classroom misbehavior videos filmed by our school's students.

In this paper, we construct a dataset of 1089 images, containing 4374 annotations of "using mobile phone", 1252 of "sleeping on desk", 4071 of "head down", 970 of "whispering", and 1711 of "looking around". Captured from top-down and horizontal perspectives, these images encompass diverse classroom environments and lighting conditions, thereby ensuring the model's generalization ability. In the following, the above annotations will be abbreviated as "phone", "lean", "bow", "talk", "turn head", respectively.

Next, the images are cropped and filtered to highlight feature quality, which enhances the model's robustness during training, enabling it to better adapt to recognition in diverse environments and conditions. Finally, based on the evaluation criteria for students' classroom misbehaviors, the rectangular annotation of the image data using the labelimg toolkit is proposed.

Upon completion of the annotation process, the system generates a JSON-formatted label file, which is then converted into a TXT format file for YOLO training. The annotated dataset is subsequently split into training and validation sets with a 7:3 ratio. These sets are then organized into the following directories: training images, validation images, training labels, and validation labels, facilitating their use during subsequent YOLO model training.

3 Main Results

3.1 YOLOv8 Model Structure

Compared with previous YOLO models, YOLOv8 offers significant improvements in both speed and accuracy, especially excelling in small object detection and complex scenes. Its enhanced network architecture and optimized training strategies make it more efficient for real-time detection tasks, while also increasing robustness. YOLOv8 has strong performance in its efficiency and stability, hence, this paper selects YOLOv8 as the benchmark network [14].

The YOLOv8 structure is mainly composed of two parts: a Backbone for feature extraction and a Head for prediction output. These components synergize through multi-level feature fusion, ensuring real-time performance without compromising detection accuracy. The entire data flow progresses layer by layer from input to output, showcasing a concise and efficient structural design that reflects careful consideration of the balance between computational resources and performance. The Backbone employs a top-down, level-by-level transmission structure, primarily consisting of alternating convolutional (Conv) and C2f layers. The Head begins with an Upsample layer, which receives data from the SPPF layer located at the Backbone's end.

3.2 Model Improvement

Object detection task involves recognizing both the position of objects and categorizing each object within a given image. The YOLO series employs a single-stage approach to address object detection tasks, enabling one-shot detection. This single-stage algorithm also significantly improves computational speed [15–17]. The performance of YOLO models can be enhanced through techniques such as multi-scale training and testing, feature pyramid networks, multi-task learning, and multi-modal information fusion [18]. A direct method to improve YOLO is PDNet, which predicts an object's class and boundary location separately from different locations to improve prediction accuracy [19].

While YOLOv8s exhibits strong performance on certain datasets, its generalization capabilities are limited. In particular, it struggles with recognizing

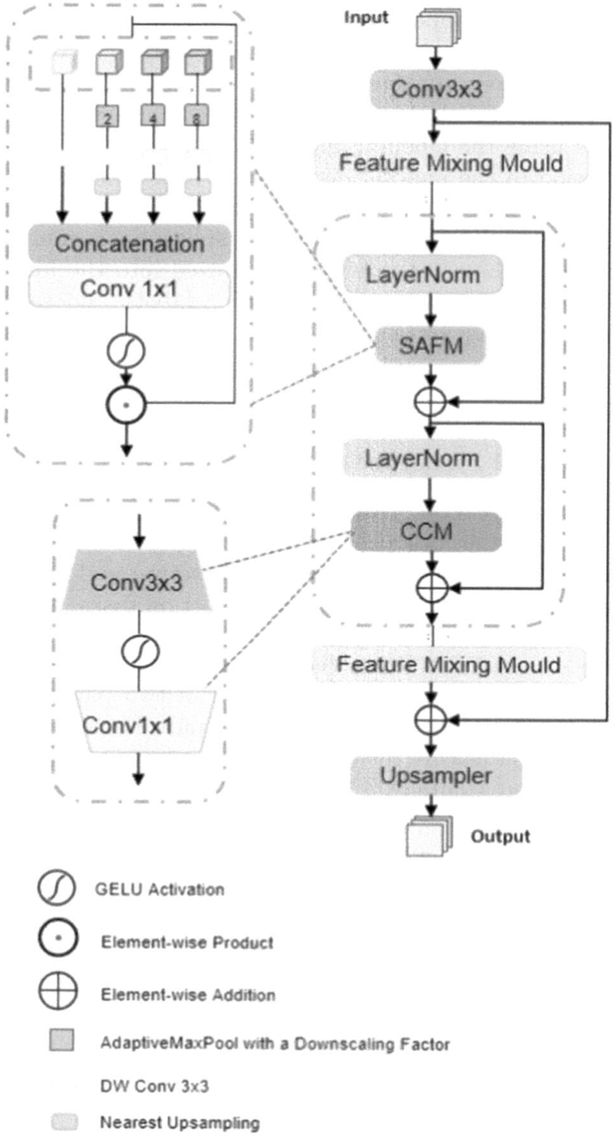

Fig. 1. SAFM Architecture Diagram.

students' behavior under occlusion, and its accuracy can still be improved. To address these issues, in this work a spatial adaptive feature modulation (SAFM) is considered to enhance the model. The SAFM is proposed by Sun Long et al., it is a deep network designed for image super-resolution tasks [20]. Its structure is shown in Fig. 1. By combining local and global features, SAFM processes and fuses data through its components, stabilizes training, and captures details to achieve high-quality image reconstruction.

The SAFM module uses a multi-scale feature interaction mechanism to adaptively integrate visual information, it captures feature information at different levels by employing convolutional kernels or pooling layers of varying sizes. And first multi-angle features are extracted as

$$F_{ms} = \{F_{s1}, F_{s2}, F_{s3}\} \tag{1}$$

where F_{s1}, F_{s2}, and F_{s3} represent feature maps at different scales.

Feature fusion is subsequently performed as

$$F_{fused} = \text{Concat}(F_{s1}, F_{s2}, F_{s3}) \tag{2}$$

By concatenating feature maps of different scales along the channel dimension and integrating multi-scale feature information, the formula allows the network to consider both details and global context, thereby enriching the feature representation.

Finally, the features are modulated as

$$F_{\text{modulated}} = F_{\text{fused}} \odot \sigma(W \cdot F_{\text{fused}}) \tag{3}$$

where σ, W and $\odot$ denote the activation function, weight matrix and element-wise multiplication.

First, the fused feature map F_{fused} undergoes a linear transformation via the weight matrix W. Then, non-linearity is introduced through the activation function σ. Finally, element-wise multiplication ($\odot$) is used to multiply the original fused feature map by the transformed feature map, achieving adaptive modulation of features. This modulation mechanism allows the network to dynamically adjust the importance of different parts of the input features.

The convolutional channel mixer (CCM) module employs a three-stage structure, serially connected by a convolutional layer of size 3, a GELU activation layer, and a convolutional layer of size 1. By sequentially performing spatial feature extraction, non-linear transformation, and channel dimension compression, the module achieves lightweight feature optimization while preserving the input resolution.

The module operates by first applying convolution to the input feature maps, which is formulated as

$$F_{\text{conv}} = \text{Conv}(F_{\text{input}}) \tag{4}$$

where the convolution operation performed on the input feature map F_{input}. Convolutional layers are the fundamental building blocks of Convolutional Neural Networks, which extract local features and generate a new feature map F_{conv}.

Then the GELU activation function is applied to introduce non-linearity as

$$F_{\text{activated}} = \text{GELU}(F_{\text{conv}}) \tag{5}$$

where $F_{\text{activated}}$ represents the application of the GELU activation function to the convolved feature map F_{conv}. GELU is a commonly used activation function with the representation as

$$\text{GELU}(x) = x \cdot \Phi(x) \tag{6}$$

Finally, the channel dimension is reduced through a convolutional layer of size 1, as shown as

$$F_{\text{compressed}} = \text{Conv}(F_{\text{activated}}) \tag{7}$$

where $F_{\text{compressed}}$ describes channel compression of the activated feature map $F_{\text{activated}}$, achieved through a convolutional layer of size 1.

A convolutional layer of size 1 is a specialized convolution operation that preserves the spatial dimensions of the feature map while adjusting the number of channels, enabling dimensionality reduction or expansion.

The SAFM mechanism can be applied to student classroom behavior detection. It effectively fuses multi-scale features and employs dynamic spatial modulation to focus on key regions. Furthermore, the CCM within this mechanism captures local details such as gestures and expressions. And the lightweight design of the SAFM structure, with its reduced parameters and computational cost, makes it well-suited for real-time video stream processing and facilitates easy deployment on classroom monitoring devices. The model's stability across diverse classroom environments is also enhanced by its adaptive multi-scale feature extraction and normalization strategy. This achieves a favorable balance between efficiency and accuracy, offering a practical and reliable improvement for real-time classroom behavior analysis.

4 Simulation Verification

4.1 Experimental Setup

As shown in Fig. 2, a dataset is established in the experiment. For the YOLOv8s model, the depth multiple is set to 0.33, and the channel multiple to 0.50. The pre-labeled materials are then used to train the model via a custom-written train file, where training parameters are configured. Specifically, the epoch is set to 400, although the model typically converges within 300 rounds. With CUDA pre-installed, the device is set to 0 to leverage the GPU for training.

Fig. 2. The Behavior dataset is shown above.

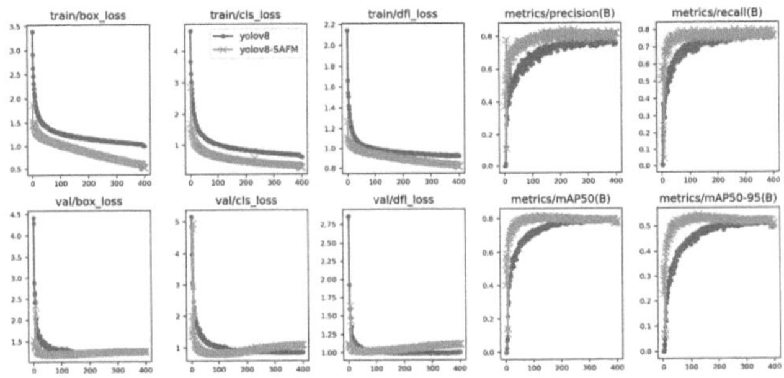

Fig. 3. Performance comparison between YOLOv8 and YOLOv8-SAFM models.

4.2 Model Results and Analysis

To demonstrate the effectiveness of the proposed method, it is compared with Yolov8 model. Through experimental comparison, YOLOv8-SAFM shows a faster decline in loss functions during both training and validation processes. Moreover, it outperforms the original YOLOv8 model in terms of final precision, recall, and mean Average Precision (mAP), demonstrating superior performance, as illustrated in Fig. 3.

By comparing the performance of the baseline model and the improved model on the behavior dataset, the results show that the improved model has shown improvements in multiple key metrics. Specifically, the mAP@0.5 of the improved model increased from 0.809 to 0.817, representing an absolute increase of 0.8% points, the mAP@[0.5:0.95] also increased from 0.526 to 0.544, achieving an absolute improvement of 1.8% points. Detailed comparison results are presented in Table 2. This performance optimization is primarily attributed to the improvement in model design, which significantly enhanced the model's accuracy in recognizing the four behaviors: "phone", "lean", "bow", "talk" and "turn head" thereby meeting the basic standards for high-precision detection. And the results of system identification are shown in Fig. 4.

Table 2. The Performance of the Baseline model and the improved model

Methods	P	R	mAP 0.5	mAP 0.5:0.95
YOLOv8	0.785	0.768	0.809	0.526
YOLOv8s-SAFM	**0.809**	**0.769**	**0.817**	**0.544**

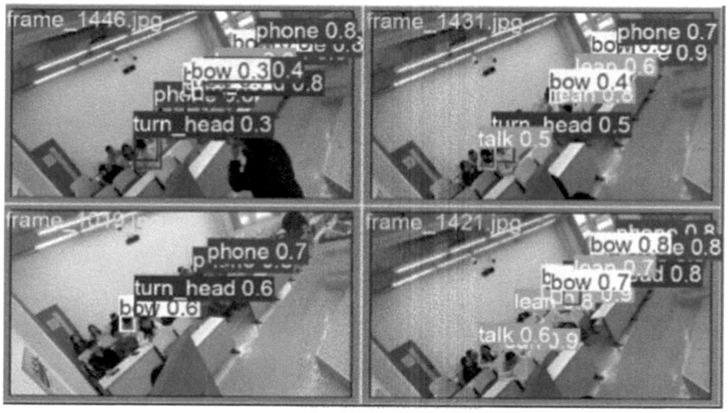

Fig. 4. The recognition results of the overall algorithm for the Behavior dataset.

4.3 Classroom Behavior Detection Application Design

PyQt5 is a GUI development toolkit built on the Python language. By seamlessly integrating the Qt framework with Python, it combines the robustness of Qt with the simplicity of Python, creating an ideal cross-platform development environment.

The system interface is simple and intuitive, adopting the design of left and right split screen. The display area on the left can dynamically present the system analysis picture, and the output window of system text information is below it. Five function keys on the right cover the main functions of the system. The interface of the system running is shown in Fig. 5.

Users can click the "Select File" button to import classroom video and images. The system supports processing videos in MP4, AVI and MOV formats as well as pictures in JPG and PNG formats. After clicking "Start Recognition", the system first processes the input video and images to enhance model recognition accuracy. It then activates the pre-trained model to identify students' head posture, limb movements, and other features in the data. This enables precise detection of classroom misconduct behaviors such as using mobile phones or sleeping at desks, with detection results displayed in the text output window below.

After completing the recognition process, click the "Save Results" function key to save the recognition result image to a designated location for later review and traceability. Clicking the "Classroom Data Extraction" function key generates corresponding statistical data: for images, it produces a category-quantity

Fig. 5. Class behavior analysis system operation interface.

bar chart, for videos, it generates a category-quantity-time line chart. The classroom data images generated from video and image recognition are shown in Fig. 6.

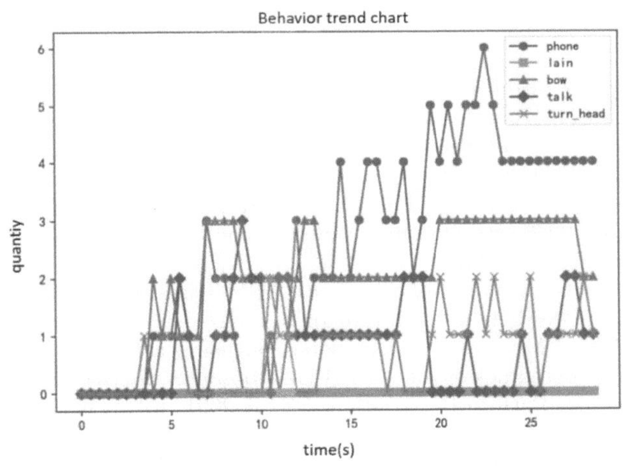

Fig. 6. Category-quantity-time line chart.

The system achieves full-process coverage from data collection to in-depth analysis. The generated quantitative indicators assist teachers in optimizing instructional designs and support educational researchers in studying teaching behavior patterns, providing reliable technical backing for building a smart classroom ecosystem.

5 Conclusion

This paper proposes a student classroom behavior detection system based on computer vision and artificial intelligence technologies, which can effectively detect inattentive behaviors of students in class, such as using mobile phones, sleeping on the desk, looking down, whispering and looking around. The proposed behavior detection algorithm is built based on an improved YOLOv8 model. To enhance the perception ability of the model, a SAFM is introduced, which incorporates adaptive max-pooling to realize flexible regulation of the receptive field range, enabling better perception of the spatial features of target objects. Finally the PyQt5 toolkit is also used to implement a monitoring platform that is simple to operate, easy to use, and has good visual effects.

The currently used dataset has certain limitations in terms of scene coverage, behavior categories, and sample quantity, making it unable to be applied to all classroom environments. In the future, it is necessary to expand the data sources, including collecting classroom behavior data from more classroom environments, lighting conditions, and crowds, to form a more representative classroom behavior dataset and provide possibilities for the generalization of the system.

References

1. Yan, S., Yang, Y.: Education informatization 2.0 in China: motivation, framework, and vision. ECNU Rev. Educ. **4**(2), 410–428 (2021)
2. Zhu, Y.: New national initiatives of modernizing education in China. ECNU Rev. Educ. **2**(3), 353–362 (2019)
3. Gupta, O.P., Agrawal, A.P., Pal, O.: A study on evolution of facial recognition technology. In: 2023 International Conference on Disruptive Technologies (ICDT), pp. 769–775. IEEE (2023)
4. Turk, M.: Over twenty years of eigenfaces. ACM Trans. Multimedia Comput. Commun. Appl. (TOMM) **9**(1s): 1–5 (2013)
5. Viola, P., Jones, M.: Rapid object detection using a boosted cascade of simple features. In: Proceedings of the 2001 IEEE Computer Society Conference on Computer Vision and Pattern Recognition, CVPR 2001, vol. 1, p. I. IEEE (2001)
6. Krizhevsky, A., Sutskever, I., Hinton, G.E.: ImageNet classification with deep convolutional neural networks. Commun. ACM **60**(6), 84–90 (2017)
7. Kong, Y., Fu, Y.: Human action recognition and prediction: a survey. Int. J. Comput. Vision **130**(5), 1366–1401 (2022)
8. Redmon, J., Divvala, S., Girshick, R., et al.: You only look once: unified, real-time object detection. In: Proceedings of the IEEE Conference on Computer Vision and Pattern Recognition, pp. 779–788 (2016)
9. Cao, Y., Cao, Q., Qian, C., et al.: YOLO-AMM: a real-time classroom behavior detection algorithm based on multi-dimensional feature optimization. Sensors **25**(4), 1142 (2025)
10. Singh, H., Miah, S.J.: Smart education literature: a theoretical analysis. Educ. Inf. Technol. **25**(4), 3299–3328 (2020)
11. Zhou, J., Ran, F., Li, G., et al.: Classroom learning status assessment based on deep learning. Math. Prob. Eng. **1**, 7049458 (2022)

12. Hu, M., Wei, Y., Li, M., et al.: Bimodal learning engagement recognition from videos in the classroom. Sensors **22**(16), 5932 (2022)
13. Liu, Q., Jiang, R., Xu, Q., et al.: Yolov8n_bt: research on classroom learning behavior recognition algorithm based on improved YOLOv8n. IEEE Access **12**, 36391–36403 (2024)
14. Han, L., Ma, X., Dai, M., et al.: A WAD-YOLOv8-based method for classroom student behavior detection. Sci. Rep. **15**(1), 9655 (2025)
15. Wang, S., Xu, M., Sun, Y., et al.: Improved single shot detection using DenseNet for tiny target detection. Concurr. Comput. Pract. Exp. **35**(2), e7491 (2023)
16. Diwan, T., Anirudh, G., Tembhurne, J.V.: Object detection using YOLO: challenges, architectural successors, datasets and applications. Multimedia Tools Appl. **82**(6), 9243–9275 (2023)
17. Wang, P., Sun, X., Diao, W., et al.: FMSSD: feature-merged single-shot detection for multiscale objects in large-scale remote sensing imagery. IEEE Trans. Geosci. Remote Sens. **58**(5), 3377–3390 (2019)
18. Zhao, Z.Q., Zheng, P., Xu, S., et al.: Object detection with deep learning: a review. IEEE Trans. Neural Netw. Learn. Syst. **30**(11), 3212–3232 (2019)
19. Yang, L., Xu, Y., Wang, S., et al.: PD Net: toward better one-stage object detection with prediction decoupling. IEEE Trans. Image Process. **31**, 5121–5133 (2022)
20. Sun, L., Dong, J., Tang, J., et al.: Spatially-adaptive feature modulation for efficient image super-resolution. In: Proceedings of the IEEE/ CVF International Conference on Computer Vision, pp. 13190–13199 (2023)

Improving Generative Aspect-Based Sentiment Analysis via Multi-view Data Augmentation and Supervised Contrastive Learning

Yanhong Wang[1,2], Lei Chen[1], Xin Wang[2], and Lin Yao[3(✉)]

[1] Guangdong University of Technology, Guangzhou 510520, China
2112314095@mail2.gdut.edu.cn, chenlei3@gdut.edu.cn
[2] PKU Shenzhen Institute, Shenzhen 518055, China
xin.wang@imsl.org.cn
[3] IMSL Shenzhen Key Lab, PKU-HKUST Shenzhen Hong Kong Institute, Shenzhen 518055, China
lin.yao@imsl.org.cn

Abstract. Aspect-based Sentiment Analysis (ABSA) employing generative pre-trained language models (PLMs) has emerged as a prominent approach in sentiment analysis, particularly in addressing the challenging task of Aspect-Category-Opinion-Sentiment Quadruple Extraction (ACOSQE). The primary challenges of ACOSQE manifest in two key areas: the scarcity and heterogeneity of datasets, and the complexity of the aspect and opinion terminology embedded in texts. To address these limitations, we propose MvGEN-SCL, a novel framework incorporating two techniques for enhanced structured generation in ACOSQE. First, We introduce an efficient multi-view augmentation strategy that fixes the aspect term to reduce computational overhead while maintaining performance, while generating multiple labels from a single input by leveraging human-like problem-solving intuitions from different perspectives. Furthermore, we enhance the model's training objective by incorporating supervised contrastive learning as an auxiliary task, enabling the model to effectively capture both explicit and implicit relationships between target elements. Extensive experimental results demonstrate that the proposed MvGEN-SCL achieves strong performance when implemented on the T5-large model. Additionally, MvGEN-SCL achieved notable improvements on challenging implicit aspect and opinion tasks that have traditionally proven difficult for existing ACOSQE methods.(https://github.com/Rastimus/MvGEN-SCL-for-ACOSQE)

Keywords: Aspect-based sentiment analysis · Pre-trained language models · Data augmentation · Contrastive learning

1 Introduction

ABSA is a crucial fine-grained sentiment analysis task in Natural Language Processing (NLP), which aims to identify and analyze sentiment polarities towards

H. Ma et al. (Eds.): IWACIII 2025, CCIS 2780, pp. 102–116, 2026.
https://doi.org/10.1007/978-981-95-6730-0_8

Table 1. Two examples of generative ABSA. Better viewed in colors. Example-1 is a PARAPHRASE, and Example-2 is GEN-NAT.

Example-1	*The pasta yesterday was delicious!*
Target-1	Food quality is great because pasta is delicious
	⇓ Recovery
Label-1	(c, a, o, p): (food quality, pasta, delicious, POS)
Example-2	*Everything they serve here ... was just very disappointed, I wish they would change next time*
Target-2	Food quality | It is disappointed | negative
	⇓ Recovery
Label-2	(c, a, o, p): (food quality, NULL, disappointed, NEG)

specific aspects in text [1]. One of the critical tasks in ABSA is Aspect-Category-Opinion-Sentiment quadruple extraction, also known as Aspect Sentiment Quad Prediction (ASQP), which involves extracting four elements of information from text, including aspect term (a), aspect category (c), opinion term (o), and sentiment polarity (p). For example, given the sentence *"The pizza is delicious, but the service is terrible."*. Here are two quadruplets, and the target label of our task is (food, pizza, delicious, positive) and (service general, service, terrible, negative).

As in the survey study [2,3], in the field of ACOSQE, numerous recent efforts have made significant breakthroughs by building upon PLMs. Table 1 presents two examples of generative ABSA. The advent of Transformer-based architectures, notably BERT [4], BART [5], and T5 [6], has fundamentally advanced the state-of-the-art performance through their innovative integration of PLMs and advanced fine-tuning methodologies. To be specific, Zhang et al. [7] proposed a novel paradigm that reformulates quadruple extraction as a paraphrasing generation task. They "re-write" the original sentence into target sequences through paraphrasing with a fixed template order. Additionally, others have adopted approaches using natural annotation tree [8], named entity recognition [9], or opinion trees [10] to guide the generative models.

However, several challenges remain [2]. Currently, existing publicly annotated datasets remain limited in scope and are particularly deficient in capturing implicit aspects and opinions, thereby failing to adequately represent the full complexity and diversity inherent in the ACOSQE task. Moreover, ACOSQE is also affected by the ambiguity, implicitness, and complexity of aspects and opinions, as well as the challenges of relationships and consistency between quadruples. For different domains or contexts, the same opinion terms may carry different sentiments. For example, "small" is negative when referring to computer memory, but it could be positive when describing a computer's size.

In response to these challenges, we propose MvGEN-SCL, a novel framework that advances the existing approach through two fundamental enhancements. The multi-view data augmentation mechanism incorporates element ordering prompts, which enhance conventional generation approaches through the aggregation of sentiment elements generated in varying orders. The supervised contrastive loss component employs four fully connected layers to optimize similarity for same-label representations and minimize it for different labels (sentiment, aspect, opinion). It is used to maximize the margin between explicit and implicit terms while minimizing the margin among the same samples.

Our main contributions include: (1) We propose MvGEN-SCL, a unified ACOSQE framework combining multi-view data augmentation and supervised contrastive learning, which demonstrates improvements in implicit aspects and opinions. (2) To further enhance the performance of the ABSA task, we expanded our exploration of T5-large model capabilities in this domain and investigated the impact of multi-dimensional contrastive learning model dimensions. (3) Extensive experiments under various settings (explicit-implicit breakdown) demonstrate that our framework based on T5-large significantly improves the accuracy of implicit aspect and opinion quadruple extraction on benchmark datasets.

2 Related Work

2.1 Generative ABSA

ABSA has drawn significant attention in recent years, particularly driven by the advancement of transformer-based language models, as initially introduced by Vaswani et al. [11]. Recent advances in ABSA demonstrate a clear preference for generative methods over pipeline processing strategies. Numerous studies have extensively investigated generative models and their associated techniques. T5 serves as a foundational framework for ACOSQE research, facilitating various advancements, including prompt engineering, data augmentation, and contrastive learning approaches, and more. Zhang et al. [7] firstly reformulated quadruple extraction as a natural language generation task by converting structured elements into a unified sentence format, which improves semantic representation learning. Hu et al. [12] and Gou et al. [13] proposed a template-based data augmentation approach through element order permutation to enhance model performance. Hu et al. [14] reviewed the template-order data augmentation method, identifying new challenges and proposing Uncertainty-Aware Unlikelihood Learning to enhance ACOSQE performance. Besides, Bao et al. [10] introduced Opinion Tree generation as a novel task that organizes sentiment elements into a hierarchical tree structure for review sentences.

2.2 Supervised Contrastive Learning

Supervised Contrastive Learning (SCL) is a training framework that combines traditional supervised learning with contrastive learning principles. It leverages label information to create more effective representations by pulling together samples from the same class while pushing apart samples from different classes. Numerous supervised contrastive learning-based approaches have been proposed to improve Generative ABSA performance. Several researchers [8,15,16] adopted supervised contrastive learning to capture sentiment-specific features by matching representations with identical sentiment polarities. For multi-aspect sample construction, Wang et al. [17] introduced cross-channel augmentation and in-domain generation techniques. Additionally, Lin et al. [18] incorporated both token and sentence-level augmentation along with sentiment information to facilitate cross-lingual contrastive learning. Further investigation by Xu et al. [19] analyzed and evaluated two prevalent contrastive learning strategies in the context of ABSA.

3 Methodology

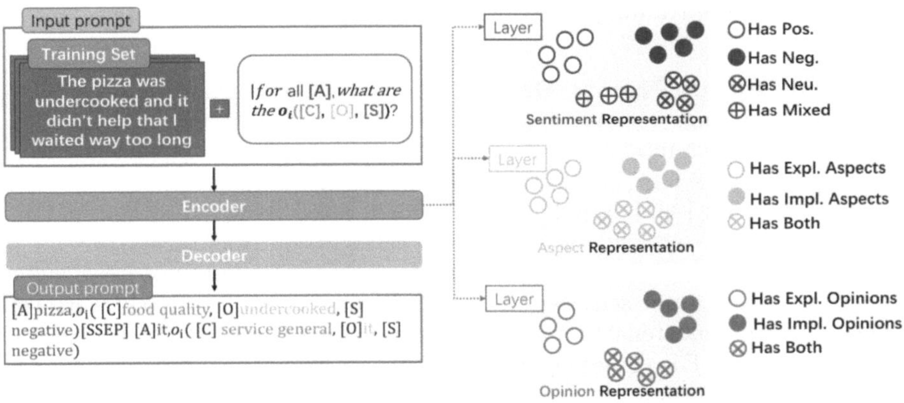

Fig. 1. Framework of the MvGEN-SCL Approach. MvGEN-SCL simultaneously employs multi-view data augmentation and supervised contrastive learning, where the left part illustrates the flow of prompts for input and output after data augmentation through the T5 encoder-decoder model, with O_i denoting the ordering. The right part demonstrates the contrastive learning process, where the layers represent sentiment, aspect, and opinion respectively.

3.1 ACOSQE Task Definition

We follow Zhang et al. [7] and Cai et al. [20] in formulating ABSA as a joint quadruple extraction task, ACOSQE is considered the most challenging task in ABSA as it aims to accurately predict four sentiment elements simultaneously: aspect terms (a), aspect categories (c), opinion terms (o), and sentiment polarity (s). Given an input sentence, the ASQP task aims to extract all sentiment tuples: $T = \{(a,\ c,\ o,\ s)\}$. To facilitate the semantic utilization of labels by generation models, following previous works, we map sentiment labels to their natural language expressions. For example, the sentiment polarity "positive" is mapped to "great", while the opinion term "NULL" is expressed as "it".

3.2 Multi-view Data Augmentation

We find that different element orderings in views significantly affect the performance of sequence-to-sequence learning, consistent with the findings of Hu et al. [12] and Gou et al. [13]. Therefore, we propose to combine multiple template orders through data augmentation to enhance the performance of the ACOSQE task. However, unlike Gou et al., we attempt to fix the elements with minimal impact. As concluded in Hu et al.'s analysis, the performances of each element are connected to its position. Specifically, the aspect term shows the most consistent performance across different positions, demonstrating better stability. Therefore, unlike previous studies, where they conducted selection experiments on all $4! = 24$ views, which was inefficient, we fix the aspect terms at the first position and perform multi-view data augmentation with $3! = 6$ views, significantly reducing the training time, which will be introduced in detail.

Input and Output Prompt Design. To augment data from the perspective of template ordering, we control different prediction sequences by adjusting the positions of element markers in prompts. Specially, given an input sentence x and its quadruplets $\{(x_{ac}, x_{at}, x_{ot}, x_{sp})\}$. We assign different markers to each sentiment element: [A], [C], [O], and [S] are allocated to x_{ac}, x_{at}, x_{ot}, and x_{sp} respectively. Then, by concatenating these markers, we construct a prompt p_i that reflects the desired sequence of sentiment elements. The prompt is appended as a suffix to each input sentence to obtain the final input x_{p_i}. Each sentiment element is then concatenated with its corresponding marker to form the prediction target y_{p_i}. For example, the input and output forms are structured as follows:

Input (x_{p_i}): The wine list yesterday was excellent, but the place is too tiny for me!| For all **[A]**, what are the **[O][C][S]**?

Output (y_{p_i}): **[A]** wine list **[O]** excellent **[C]** drinks style **[S]** great [SSEP] **[A]** place **[O]** too tiny **[C]** ambience general **[S]** bad

Thus, by permuting and combining elements, we can flexibly model different views of the augmented data.

3.3 Constrained Decoding

To enhance the generation quality and maintain consistency in the generative model's output, we implemented constrained decoding (CD) mechanisms during the inference phase. This approach ensures that the generated sequences strictly adhere to predefined vocabulary constraints and structural patterns [21]. The CD mechanism dynamically filters the model's output distribution at each decoding step, only allowing tokens that satisfy the specified constraints to be selected. As shown in Table 2, when generating a '[' token, the decoder restricts the next choice to special terms (A, C, O, S, SSEP) while tracking generation history to guide subsequent token selection.

Table 2. Candidate lists of tokens. "..." denotes arbitrary text.

Generated output	Candidate tokens
	[
...[A/[C/[O/[S/[SSEP	]
...[A]	Input sentence, "it"
...[C]	All categories
...[O]	Input sentence,'it"
...[S]	great,neutral,bad
...[A]...	Input sentence,"it",[
...[C]...	All categories,[
...[O]...	Input sentence,'it",[
...[S]...	great, neutral, bad,[
...[A]...[	C,O,S
...[C]...[	O,S
...[O]...[	C,S
...[S]...[	C,O
...[SSEP]	[
...[SSEP][	A

3.4 Training Loss

After applying our multi-view data augmentation, we construct $3! = 6$ different ordered prompts and targets for each sentence. Since we can fine-tune the pre-trained sequence-to-sequence language model T5 using the training dataset augmented with multi-view data. The loss function is defined as:

$$\mathcal{L}_{\mathrm{CE}} = -\sum_{i=1}^{N}\sum_{j=1}^{k} \log p(y_i^j | x_i^j; \theta) \tag{1}$$

where N denotes the number of sentences in the training set, k denotes the number of views for each sentence, x_i^j and y_i^j represent the input prompt and target output corresponding to the j-th view of the i-th sentence respectively, and θ represents the model parameters.

3.5 Supervised Contrastive Learning

Following previous work, we adopt the method of using a general SCL formula like others [8,22]. This approach encourages the model to distinguish several key features from the input, pulling together sentiment representations with the same emotional polarity, while separating sentiment representations with different emotional polarities. In this study, we propose a model to learn representations of three essential characteristics: **Sentiment**, **Aspect**, and **Opinion** at the example level.

SCL Formulation and Implementation. In Supervised Contrastive Learning (SCL), we strive to maximize the similarity between representations of same-labeled samples, while simultaneously minimizing the similarity between representations of samples with different labels. As shown on the right side of Fig. 1, to generate representations for individual examples x_i, the system employs a three-layer fully connected network where each layer corresponds to a specific characteristic (Sentiment, Aspect, Opinion), taking the sum-pooled encoder representation as input to produce h_i^c. The dimensionality of the fully connected layers varies between T5 variants: T5-large processes 1024-dimensional input and 512-dimensional output, while T5-base transforms 768-dimensional inputs into 128-dimensional representations.

Our SCL objective is to enable the model to distinguish between explicit and implicit terms in complex sentiment expressions. We follow Sedghamiz et al. [23] in their general SCL formulation. It first calculates the similarity between the target sample i and each sample in its corresponding positive sample set $P(i)$, while also calculating the similarity with samples in the negative sample set $B(i)$. These similarities are converted into loss values through the negative logarithm form. Notably, the negative sample set does not include the sample itself, i.e., $b \neq i$. The entire process regulates the similarity distribution through a temperature parameter τ.The formula is as follows:

$$\mathcal{L}_i^c = \sum_{i \in I} \frac{-1}{|P(i)|} \sum_{p \in P(i)} \log \frac{\exp(h_i \cdot h_p / \tau)}{\sum_{b \in B(i)} \exp(h_i \cdot h_b / \tau)} \tag{2}$$

In this experiment, let $I \equiv \{1, 2, \ldots, kN\}$ denote the index of samples after Multi-view augmentation, where k is the views. $P(i)$ is the set of data samples that have the same label as sample i. During training, a dropout probability of 0.1 is used to prevent overfitting. $B \equiv I \backslash \{i\}$ denotes the set of indices excluding index i, and $P(i) \equiv \{p \in B(i) : y_p = y_i\}$ is the set of indices of all positive samples (i.e., samples that have the same label as sample i) in the current batch, excluding index i itself.

Final Training Objective. With our three characteristic-specific losses, the final loss is defined as follows:

$$\mathcal{L}_{total} = \mathcal{L}_{\text{CE}} + \alpha_1 \mathcal{L}_{\text{sentiment}} + \alpha_2 \mathcal{L}_{\text{aspect}} + \alpha_3 \mathcal{L}_{\text{opinion}} \tag{3}$$

4 Experiment Setup

4.1 Datasets

ACOSQE Dataset. We utilize the Restaurant-ACOS and Laptop-ACOS datasets constructed by [20], which originate from the restaurant and e-commerce domains. These datasets contain reviews annotated with ACOSQE, incorporating both explicit and implicit aspects and opinions. The key statistics of these datasets are summarized in Table 3.

Table 3. Dataset statistics. E: explicit, I: implicit, A: aspect, O: opinion. E.g., IAEO refers to "implicit aspect, explicit opinion".

	REST	LAPTOP
#Categories	13	121
#Sentences	2284	4076
#EAEO Quads	2431 (66.40%)	3278 (56.78%)
#IAEO Quads	530 (14.48%)	912 (15.80%)
#EAIO Quads	350 (9.56%)	1241 (21.50%)
#IAIO Quads	350 (9.56%)	342 (5.92%)
#Quads/Sent	1.60	1.42

4.2 Compared Methods

To make an extensive evaluation, we choose the following baseline methods:

BERT-Based Methods. JET-ACOS [20] is an adaptation of JET, designed for ACOSQE by extracting aspect-opinion-sentiment triples and leveraging a BERT-based model for aspect categorization. TAS-BERT-ACOS is an adaptation of TAS-BERT for ACOSQE, utilizing input transformation to enable category-sentiment conditional aspect-opinion co-extraction. It filters out invalid aspect-opinion pairs to form final quadruples. Extract-Classify-ACOS [20]is a two-step approach for ACOSQE, first performing aspect-opinion co-extraction followed by category-sentiment prediction.

T5-Based Methods. GAS [24]presents a unified generative approach that seamlessly integrates various ABSA tasks through the encoding of semantic-rich natural language labels. Paraphrase [7]uses T5 to generate sentiment-containing sentences, then extracts sentiment quadruples. Seq2Path [25] generates tree paths of sentiment tuples to better handle the one-to-many relationships between aspects and opinions. DLO/ILO [12] leverages quadruple order-free property to generate diverse template-based training data for ACOSQE. GEN-SCL-NAT [8] is a T5-Large based generative model that combines supervised contrastive learning and natural annotation tree for ACOSQE task. MvP [13] improves sentiment tuple prediction by aggregating multi-view results. ASQP-ITSCL [22]uses instruction tuning and contrastive learning to enhance its performance in implicit term extraction.

4.3 Experiment Details

We adopt the T5-base and T5-large [6] from the Hugging face Transformers library[1] as the pre-trained generative model, and fine-tune T5 with the constructed training samples. We conducted the training process using different GPU setups: the T5-base model was trained on Nvidia 3090 GPUs for 10 epochs (batch size = 16, learning rate = 9e-5), while the T5-large model was trained on Nvidia A40 GPUs for 20 epochs (batch size = 16). The hyperparameters in Eq. 2 are set as follows: $\alpha_1 = \alpha_2 = \alpha_3 = \alpha$, with $\alpha = 0.05$, and the SCL temperature τ is set to 0.15. These parameters are tuned on the training set.

4.4 Evaluation Metrics

The evaluation metrics for ACOSQE tasks are precision (P), recall (R), F1 score (F1). The experiment uses F1 scores (F1) as the main evaluation metric. A sentiment quad prediction is correct if all predicted elements match the gold labels.

5 Results and Discussions

5.1 Main Performance Results

Table 4. Performance comparison of different methods on the Restaurant-ACOS and Laptop-ACOS datasets.

Method	Model	Restaurant			Laptop		
		P.	R.	F1.	P.	R.	F1.
JET-ACOS	BERT	59.81	28.94	39.01	44.52	16.25	23.81
TAS-BERT-ACOS	BERT	26.29	46.29	33.53	47.15	19.22	27.31
Extract-Classify-ACOS	BERT	38.54	52.96	44.61	45.56	29.48	35.80
PARAPHRASE	T5-base	–	–	60.97	–	–	44.08
GAS	T5-base	57.09	57.51	57.30	43.45	43.29	43.37
Seq2Path	T5-base	–	–	58.41	–	–	42.97
DLO	T5-base	61.03	60.55	60.78	43.78	43.53	43.65
GEN-SCL-NAT	T5-large	–	–	62.62	–	–	45.16
MvP	T5-base	–	–	61.54	–	–	43.92
ASQP-ITSCL	T5-base	61.45	60.92	61.18	44.69	44.19	44.43
MvGEN-SCL	T5-base	61.18	60.32	60.75	44.49	44.59	44.54
MvGEN-SCL	T5-large	**64.38**	**63.57**	**63.97**	**46.44**	**45.82**	**46.13**

[1] https://github.com/huggingface/transformers.

The main results compared with baseline methods are presented in Table 4. Compared with other methods with BERT and T5-base models, although MvGEN-SCL with T5-base showed slight improvement on the Laptop dataset, there was no significant improvement on the Restaurant dataset. However, the T5-large variant achieves the highest F1 scores on these two datasets, 63.97 and 46.13, respectively. Although MvP and DLO (ILO) employ data augmentation based on element order, which mainly focuses on enhancing overall data performance and is effective in low-resource environments, they overlook the implicit elements within the data. Furthermore, our method places emphasis on distinguishing between input aspects, such as implicit and explicit aspects, as well as opinions. We find that our method achieves strong performance on all datasets.

5.2 Explicit and Implicit Sentiment Analysis

Table 5. Performance comparison of explicit and implicit analysis methods on the Restaurant and Laptop datasets. EA, EO, IA, and IO represent explicit aspects, explicit opinions, implicit aspects, and implicit opinions, respectively.

Method	Restaurant (F1.)				Laptop (F1.)			
	EAEO	IAEO	EAIO	IAIO	EAEO	IAEO	EAIO	IAIO
JET-ACOS	52.3	N/A	N/A	N/A	35.7	N/A	N/A	N/A
TAS-BERT-ACOS	33.6	31.8	14.0	39.8	26.1	41.5	10.9	21.2
Extract-Classify-ACOS	45.0	34.7	23.9	33.7	35.4	39.0	16.8	18.6
PARAPHRASE	65.4	53.3	45.6	49.2	45.7	51.0	33.0	39.6
GEN-SCL-NAT	66.5	56.5	**46.2**	50.7	45.8	54.0	34.3	39.6
ASQP-ITSCL(T5-base)	69.8	51.2	31.9	45.6	46.4	59.1	30.3	40.0
ASQP-ITSCL(T5-large)	**71.8**	53.2	44.4	52.2	**47.2**	61.3	**34.4**	39.7
MvGEN-SCL(T5-large)	68.5	**55.6**	43.6	**53.2**	46.2	**62.8**	**34.4**	40.0

To validate the performance of our method on explicit and implicit terms, Table 5 presents the detailed results across four explicit/implicit aspect-opinion combinations (EAEO, IAEO, EAIO, IAIO). Experimental results indicate that our method excels particularly in IA* (Implicit Aspect) subsets. Across both datasets, our approach achieves notable improvements on the IAEO subset. This enhancement can be attributed to our proposed methodology, which first localizes aspect terms and reinforces the model's localization capability through supervised contrastive learning. We observe that improving performance on *IO subsets consistently presents greater challenges than on *EO subsets. However, our method demonstrates robust performance on IAIO subsets for the restaurant dataset and EAIO subsets for the laptop dataset. Although our method shows no improvement on EAEO subtasks, these results still demonstrate the effectiveness of our approach in implicit term extraction.

6 Additional Analyses

6.1 Ablation Study

Ablation of Constrained Decoding. By constraining the decoding strategy to ensure the legitimacy of the generated content, that is, the aspect terms and opinion terms generated by the model must be one or more consecutive words in the input sentence.As shown in Table 6, the experimental results suggest that the constrained decoding module can enhance the performance of the generative prediction framework to a certain degree.

Table 6. Ablation analysis of ITSCL (T5-large) model.

Method	Restaurant	Laptop
PARAPHRASE	60.97	44.08
GEN-SCL-NAT	62.62	45.16
MVP	61.54	43.92
MvGEN-SCL (T5-large)	**63.97**	**46.13**
-w/o Constrained decoding	63.85	46.04
-w/o Sentiment Rep.	62.58	44.70
-w/o Aspect Rep.	62.14	45.90
-w/o Opinion Rep.	63.31	45.25
-w/o All Rep.(MvGEN)	62.47	44.23

We ablate each of the three SCL objectives as shown in the Table 6, and observed that removing individual loss components led to performance degradation across both datasets. Notably, eliminating the aspect term representation had the most significant impact on the restaurant dataset but showed minimal effect on the laptop dataset. Furthermore, when removing all SCL representations and solely utilizing our enhanced multi-view data augmentation method (MvGEN), we observed consistent degradation across all datasets. However, the performance still surpassed that of MvP and PARAPHRASE, which can be attributed to our method's extension on T5-large.

6.2 SNE Representations

To understand the impact of the SCL objective on the model's hidden representations, t-SNE visualizations of the meanpooled final encoder layer were generated [26]. Figure 2 displays results on the Restaurant test set. We have observed that MvGEN-SCL enables the encoder to simultaneously represent sentiment, aspect, and opinion information effectively.

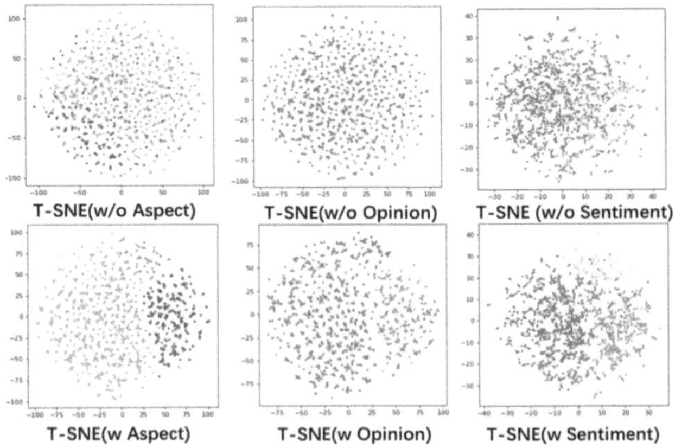

Fig. 2. T-SNE visualization of mean-pooled encoder final layer on the Restaurant dataset. Our MvGEN-SCL objective encourages the encoder to produce distinguishable representations of three key input combinations: aspects, opinions, and sentiment.

6.3 SCL Model Size Effect

The SCL model consists of three fully connected layers to feed forward the sum of encoder representations, where the input dimension is determined by the output dimension of the T5 encoder. To investigate the impact of output dimension variations on model performance, we explored different output dimensions of our proposed method based on the Restaurant. As shown in Fig. 3, the SCL model's performance varies with different output dimensions. As the output dimension

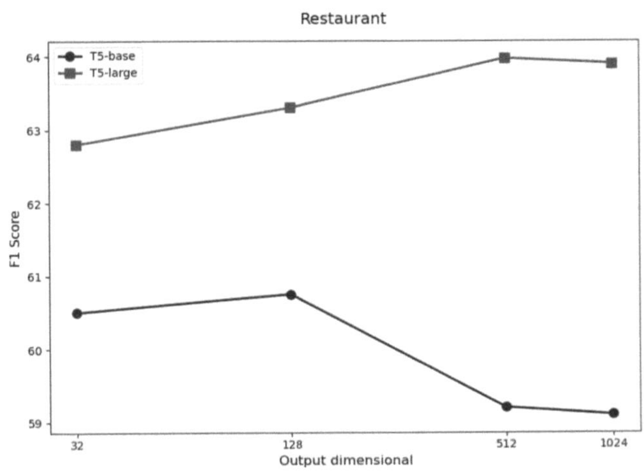

Fig. 3. F1 Score Variation with SCL model output dimensionality in Restaurant Dataset.

increases from 32 to 1024, T5-large shows a general upward trend in F1 score. In contrast, T5-base exhibits performance degradation at higher dimensions due to its limited model capacity, which may result in insufficient representation capabilities when dealing with high-dimensional feature spaces.

7 Conclusions

In this work, we introduce MvGEN-SCL, an ACOSQE approach based on the T5 generative model. By combining multi-view data augmentation and supervised contrastive learning methods, MvGEN-SCL enables the model to acquire knowledge through comprehensive training and identify implicit sentiments. Our multi-view data augmentation not only reduces training overhead but also advances research in generative modeling for tuple structure prediction. The supervised contrastive learning framework further enhances model performance by optimizing the similarity between representations with the same labels while distinguishing between representations with different labels. Our extensive experiments on benchmark datasets demonstrate that MvGEN-SCL achieves significant improvements in extracting implicit aspect terms.

Although our method achieves strong performance, it still has notable limitations. First, the proposed data augmentation approach is highly task-specific, with applicability largely restricted to ACOSQE. Extending it to other ABSA tasks, such as Aspect Sentiment Triplet Extraction (ASTE) and Target Aspect Sentiment Detection (TASD), may require redesigning input prompt templates. Second, our experiments are limited to English sentence-level ABSA with benchmark datasets; future work should examine its effectiveness on document-level and cross-lingual settings.

Acknowledgements. This study was supported by the Shenzhen Science and Technology Program (No:GJHZ20220913144201002) and the Shenzhen Key Research Program (JCYJ20220818101411025)

Disclosure of Interests. Work was done when Yanhong Wang interned at IMSL.

References

1. Wang, J., Xu, B., Zu, Y.: Deep learning for aspect-based sentiment analysis. In: 2021 International Conference on Machine Learning and Intelligent Systems Engineering (MLISE), pp. 267–271 (2021)
2. Zhang, H., Cheah, Y.N., Alyasiri, O.M., An, J.: Exploring aspect-based sentiment quadruple extraction with implicit aspects, opinions, and ChatGPT: a comprehensive survey. Artif. Intell. Rev. **57**(2), 17 (2024)
3. Zhang, W., Li, X., Deng, Y., Bing, L., Lam, W.: A survey on aspect-based sentiment analysis: tasks, methods, and challenges. IEEE Trans. Knowl. Data Eng. **35**(11), 11019–11038 (2023)

4. Devlin, J., Chang, M.W., Lee, K., Toutanova, K.: BERT: pre-training of deep bidirectional transformers for language understanding. In: Burstein, J., Doran, C., Solorio, T. (eds.) Proceedings of the 2019 Conference of the North American Chapter of the Association for Computational Linguistics: Human Language Technologies, Minneapolis, Minnesota, June 2019, vol. 1 (Long and Short Papers), pp. 4171–4186. Association for Computational Linguistics (2019)
5. Lewis, M., et al.: BART: denoising sequence-to-sequence pre-training for natural language generation, translation, and comprehension. In: Jurafsky, D., Chai, J., Schluter, N., Tetreault, J. (eds.) Proceedings of the 58th Annual Meeting of the Association for Computational Linguistics, Online, July 2020, pp. 7871–7880. Association for Computational Linguistics (2020)
6. Roberts, A., et al.: Exploring the limits of transfer learning with a unified text-to-text transformer. Technical report, Google (2019)
7. Zhang, W., Deng, Y., Li, X., Yuan, Y., Bing, L., Lam, W.: Aspect sentiment quad prediction as paraphrase generation. In: Moens, M.F., Huang, X., Specia, L., Yih, S.W., (eds.) Proceedings of the 2021 Conference on Empirical Methods in Natural Language Processing, Online and Punta Cana, Dominican Republic, November 2021, pp. 9209–9219. Association for Computational Linguistics (2021)
8. Peper, J.J., Wang, L.: Generative aspect-based sentiment analysis with contrastive learning and expressive structure. In: Conference on Empirical Methods in Natural Language Processing (2022)
9. Lee, S.K., Kim, J.H.: Sener: sentiment element named entity recognition for aspect-based sentiment analysis. In: ICASSP 2023 - 2023 IEEE International Conference on Acoustics, Speech and Signal Processing (ICASSP), pp. 1–5 (2023)
10. Bao, X., Wang, Z., Jiang, X., Xiao, R., Li, S.: Aspect-based sentiment analysis with opinion tree generation. In: Proceedings of the Thirty-First International Joint Conference on Artificial Intelligence, IJCAI 2022, Vienna, Austria, 2022, pp. 4044–4050. ijcai.org (2022)
11. Vaswani, A., et al.: Attention is all you need. In: Advances in Neural Information Processing Systems, pp. 5998–6008 (2017)
12. Hu, M., Wu, Y., Gao, H., Bai, Y., Zhao, S.: Improving aspect sentiment quad prediction via template-order data augmentation. In: Goldberg, Y., Kozareva, Z., Zhang, Y. (eds.) Proceedings of the 2022 Conference on Empirical Methods in Natural Language Processing, Abu Dhabi, United Arab Emirates, December 2022, pp. 7889–7900. Association for Computational Linguistics (2022)
13. Gou, Z., Guo, Q., Yang, Y.: MvP: multi-view prompting improves aspect sentiment tuple prediction. In: Rogers, A., Boyd-Graber, J., Okazaki, N. (eds.) Proceedings of the 61st Annual Meeting of the Association for Computational Linguistics, vol. 1: Long Papers, Toronto, Canada, July 2023, pp. 4380–4397. Association for Computational Linguistics (2023)
14. Hu, M., et al.: Uncertainty-aware unlikelihood learning improves generative aspect sentiment quad prediction. In: Rogers, A., Boyd-Graber, J., Okazaki, N. (eds.) Findings of the Association for Computational Linguistics: ACL 2023, Toronto, Canada, July 2023, pp. 13481–13494. Association for Computational Linguistics (2023)
15. Li, Z., Zou, Y., Zhang, C., Zhang, Q., Wei, Z.: Learning implicit sentiment in aspect-based sentiment analysis with supervised contrastive pre-training. In: Proceedings of the 2021 Conference on Empirical Methods in Natural Language Processing, EMNLP 2021, pp. 246–256 (2021)

16. Liang, B., et al.: Enhancing aspect-based sentiment analysis with supervised contrastive learning. In: Proceedings of the 30th ACM International Conference on Information & Knowledge Management, CIKM '21, pp. 3242–3247. ACM (2021)
17. Wang, B., Ding, L., Zhong, Q., Li, X., Tao, D.: A contrastive cross-channel data augmentation framework for aspect-based sentiment analysis. In: Proceedings of the 29th International Conference on Computational Linguistics, COLING 2022, pp. 6691–6704 (2022)
18. Lin, N., Yingwen, F., Lin, X., Zhou, D., Yang, A., Jiang, S.: Cl-xabsa: contrastive learning for cross-lingual aspect-based sentiment analysis. IEEE/ACM Trans. Audio Speech Lang. Process. **31**, 2935–2946 (2023)
19. Lingling, X., Wang, W.: Improving aspect-based sentiment analysis with contrastive learning. Nat. Lang. Process. J. **3**, 100009 (2023)
20. Teo, T.W., Choy, B.H.: in. In: Tan, O.S., Low, E.L., Tay, E.G., Yan, Y.K. (eds.) Singapore Math and Science Education Innovation. ETLPPSIP, vol. 1, pp. 43–59. Springer, Singapore (2021). https://doi.org/10.1007/978-981-16-1357-9_3
21. De Cao, N., Izacard, G., Riedel, S., Petroni, F.: Autoregressive entity retrieval. In 9th International Conference on Learning Representations, ICLR 2021, Virtual Event, Austria, 3–7 May 2021. OpenReview.net (2021)
22. Zhang, H., Cheah, Y.N., He, C., Yi, F.: An instruction tuning-based contrastive learning framework for aspect sentiment quad prediction with implicit aspects and opinions. In: Al-Onaizan, Y., Bansal, M., Chen, Y.N. (eds.) Findings of the Association for Computational Linguistics: EMNLP 2024, Miami, Florida, USA, November 2024, pp. 7698–7714. Association for Computational Linguistics (2024)
23. Sedghamiz, H., Raval, S., Santus, E., Alhanai, T., Ghassemi, M.: Supcl-seq: Supervised contrastive learning for downstream optimized sequence representations (2021)
24. Zhang, W., Li, X., Deng, Y., Bing, L., Lam, W.: Towards generative aspect-based sentiment analysis. In: Proceedings of the 59th Annual Meeting of the Association for Computational Linguistics and the 11th International Joint Conference on Natural Language Processing, vol. 2: Short Papers, pp. 504–510 (2021)
25. Mao, Y., Shen, Y., Yang, J., Zhu, X., Cai, L.: Seq2Path: generating sentiment tuples as paths of a tree. In: Muresan, S., Nakov, P., Villavicencio, A. (eds.) Findings of the Association for Computational Linguistics: ACL 2022, Dublin, Ireland, May 2022, pp. 2215–2225. Association for Computational Linguistics (2022)
26. van der Maaten, L., Hinton, G.: Visualizing data using t-sne. J. Mach. Learn. Res. **9**(86), 2579–2605 (2008)

Dual-Path Attention-Enhanced Graph Convolutional Network for Children's Speech Emotion Recognition

Deneng Tang[1(✉)], Liwei Lin[1], Haojie Huang[2], and Qixuan Liu[1]

[1] Beijing institute of technology, Zhuhai, Zhuhai 519000, China
tang1847442546@163.com

[2] Nanyang Technological University, 639798 Singapore, Singapore

Abstract. Emotion is an indispensable part of daily interpersonal communication, and people can communicate through emotional expression. However, the social trend of younger people suffering from psychological problems has attracted widespread attention. Aiming at children's mental health, this paper discusses a proposed dual-path attention-enhanced graph convolutional network for children's speech emotion recognition, and also using the attention mechanism to enhance the model's ability to understand complex speech signals, learn the changes in emotional time information, and classify children's speech emotions. The performance of the proposed Hierarchical Dual-Attention Network classifier is extensively evaluated. The results show that compared with existing models provided by other published studies, it achieves significant recognition improvements, recognizing 86.49% in the integrated dataset with the FAU-Aibo dataset, C-BESD dataset, and MESD dataset. This study provides new ideas and technical support for children's speech emotion recognition and has potential application value in children's mental health assessment, intelligent education, and human-computer interaction.

Keywords: Children's speech emotion recognition · Deep learning · Attention Mechanism · Graph Convolutional Network

1 Introduction

There are various ways to recognize human emotions, including voices, text, and facial expressions. In voice communication, messages contain not only the amount of information but also the way the information is transmitted. Information can be combined through two communication methods, verbal (speech) and non-verbal (facial expressions). To identify emotions in speech, a study [1] confirmed that the speaker's facial expressions accounted for about 55% of the influence, 38% of the influence was conveyed through the tone of voice, and 7% was conveyed through spoken language. Speech is an important channel for studying human emotions, aiming to detect the emotional state of speakers through audio processing analysis, and has a wide range of research applications. However, because clear pronunciation and other acoustic characteristics

H. Ma et al. (Eds.): IWACIII 2025, CCIS 2780, pp. 117–134, 2026.
https://doi.org/10.1007/978-981-95-6730-0_9

of speech can be effectively conveyed in adult speech, most of the current research on emotion recognition is concentrated in the adult field, and there are relatively few studies on children's emotional recognition. Due to the gradual youth of today's social psychological problems, the current social environment, the intensification of academic competition pressure, and the change of family communication patterns lead to the frequent occurrence of children's psychological problems [3]. Existing speech emotion recognition faces challenges in children's emotional recognition, such as the scarcity of public children's emotional speech databases, the difficulty of children's speech collection due to children's more active behavior, and the relative difficulty of capturing children's emotional changes at the same time. Although some progress has been made in the study of children's emotion recognition in recent years, the current situation is still not optimistic. Therefore, in the case of scarcity of children's emotional speech datasets, improving the accuracy of children's emotion recognition algorithms is the mainstream method. In recent years, deep learning has made significant progress in the field of speech processing, providing a new opportunity to solve the problem of children's speech emotion recognition. This study aims to explore an efficient deep learning model to improve the accuracy and robustness of children's speech emotion recognition.

2 Related Work

In order to identify emotions in speech as well as emotions in spoken speech, it also becomes important to study the speech characteristics of the speaker. A wide variety of emotions can be detected from adult speech because clear pronunciation and other acoustic characteristics of speech can be effectively conveyed in adult speech. Therefore, most researchers focus on adult speech emotion recognition (SER) systems. There is very limited research on how to effectively recognize emotions in children's speech, and one reason why there is less research on speech emotion recognition in children is that children's speech characteristics are different from those of adults. Adults have more consistent language expression than children. Children's speech has low clarity and consistency, making it a significant challenge to recognize children's emotions. Most emotion recognition applications target children as end-users, and over the past few decades, many emotion databases based on performing and non-performing corpora have emerged. Some of these databases also include children's emotional speeches, covering English, German, Russian, Mexican, and more.

Children's speech contains numerous lengthy segments, so emotions are not expressed semantically but rather in acoustic aspects. Because of the low clarity and consistency of children's speech, more researchers pay attention to features such as the variance of the audio wave.

In speech emotion recognition applications, selecting appropriate features ensures adequate model training and improves accuracy. Ustubioglu [8] used spectrograms as recognition features to improve emotion recognition accuracy. Yenigalla [9] proposed inputting speech sequences and traditional spectrograms into two convolutional neural networks (CNNs) to extract features. The short-term fluctuations in children's speech require the model to capture both static and dynamic features, which traditional CNN models cannot meet.

The successful application of CNNs in image processing has prompted some researchers to develop end-to-end SER models. For example, Dong Hu proposed a Chinese children's speech emotion recognition model based on DPCNN-SLSTM [2]. Anvarjon [5] developed a CNN-based lightweight speech emotion recognition system that improved speech emotion recognition classification using ordinary rectangular filters. In addition, Li et al. [6] proposed a SER framework that interfaces with CNN and LSTM models to classify speech emotions through end-to-end contextual awareness.

Due to the characteristics of CNN, CNN also lacks the ability to capture temporal dynamic characteristics when simulating the dynamic change of speech dynamic emotions [4].

Graph convolutional networks are a class of deep learning models that are specifically designed to process graph data. GCN can effectively capture the relationship between nodes and the structural information of the graph, and through the layer-by-layer propagation of the graph convolutional layer, GCN can learn the node representation and realize effective feature extraction in the graph data [7]. In emotion recognition tasks, speech audio data can be regard as a timing graph, where each node contains speech features, and each node is connected by edges, representing the dependencies between speech features. GCN is able to consider node correlations in data to understand and analyze complex association patterns more comprehensively, and we can better capture the complex structure and relationships of speech data. For example, Amir and Tanaya [13] proposed a depth graph learning method that uses GCN to learn speech features like the signals, which are extracted as graph data. Although GCN captures the time sequence information lost by CNN in the convolution process, the performance of GCN is not as good as that of CNN in local feature extraction.

Therefore, by examining the advantages and disadvantages of CNN2D and GCN architectures, we can gain a deeper understanding of the advantages of both architectures and explore how they can be combined. CNNs excel in analyzing static features but struggle with dynamic features, particularly in predicting changes in pitch and timbre. Graph convolutional networks (GCNs) are good at predicting dynamic features, and deformable convolutions can adaptively adjust the sampling position according to the input features to better capture irregular, emotion-related temporal changes in speech signals.

The main works of this paper are as follows:

1) Using some dynamic data augmentation strategies, such as adding Gaussian white noise, time stretching, shifting, and pitching to increase data samples, improve model adaptability, and also enhance the model's generalization and robustness.
2) Constructing a dual-path attention-enhanced graph convolutional network and also use an attention mechanism to improve the effective utilization of features.

3 Proposed Solution

3.1 Preprocessing

Samples from the selected dataset are loaded and resampled to 16,000 kHz. All sample audio files are resized to a fixed length of 414 frames, with a fixed feature dimension of two dimensions to maintain a consistent sampling rate across language-independent datasets. Initial and trailing silences are then removed to obtain a signal containing speech information for further processing. Because some audio data in the FAU Aibo dataset is short, a minimum duration of 0.1 s was set to remove these shorter audio clips. All input data was normalized to eliminate dimensionality differences and accelerate model convergence (Fig. 1).

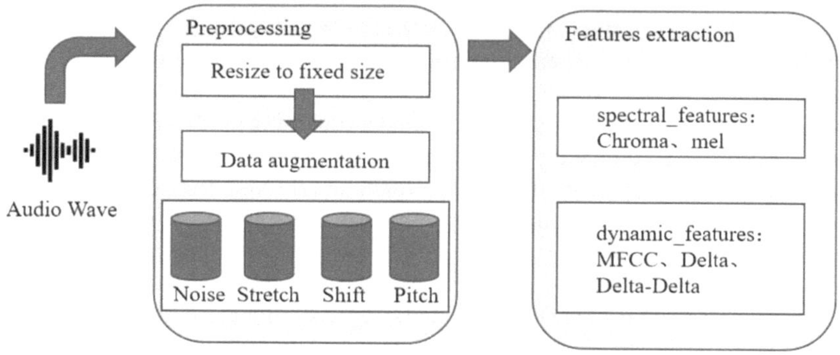

Fig. 1. Preprocessing and feature extraction.

3.2 Data Augmentation

Due to the small number of samples in the children's speech emotion dataset and the imbalance in data across different emotion categories, training results can be biased. To improve the effective utilization of limited audio samples without substantially increasing the data, some dynamic data augmentation strategies are implemented to modify the original audio samples by adding Gaussian white noise, time stretching, shifting, and pitching. This not only ensures that the model extracts more effective information and addresses the data imbalance in the dataset but also improves the model's generalization and robustness while avoiding overfitting.

Gaussian White Noise. For numerical sequence data, a certain amount of noise can be randomly added to each value to generate a new sequence without affecting the overall properties and label information of the sequence. Adding Gaussian white noise to the original audio can change the continuous characteristics of the speech signal and make the sample more resistant to noise, making the model more versatile in different environmental changes. This article sets the signal-to-noise ratio (SNR) to 20 to simulate an interference environment. Standard Gaussian noise is relatively easy to generate. Multiplying the standard Gaussian noise by k can obtain noise with a fixed signal-to-noise ratio, According to the formula (1) and formula (2) to calculate the power of the speech signal P_s and generate standard Gaussian noise P_{n1}.

$$P_s = \frac{\sum_{i=1}^{N} (x_i)^2}{N} \tag{1}$$

$$P_{n1} = \frac{\sum_{i=1}^{N} (n_i)^2}{N} \tag{2}$$

X_i represents the speech signal, n_i represents the noise, and the sound length is N. Assuming that the fixed signal-to-noise ratio is S dB, the signal-to-noise ratio formula is listed as formula (3):

$$S = 10 \log 10 \frac{P_s}{P_n} \tag{3}$$

P_n is the final required Gaussian white noise value. After transforming and sorting Eq. (3), the k value is obtained as formula (4):

$$k = \sqrt{\frac{P_s}{10^{\frac{s}{10}} \times P_{n1}}} \tag{4}$$

The noise is linearly superimposed on the original signal, which conforms to the additive white Gaussian noise (AWGN) model as formula (5):

$$x_{noisy} = x + \varepsilon \tag{5}$$

Stretch. To enhance the model's adaptability to varying speech rhythms, such as children's faster speech than adults and the elderly, a technique was developed to adjust the duration of speech samples without changing their pitch (frequency). This technique involves adjusting the stretching ratio, which adjusts the phase gradient to achieve time-domain stretching, and randomly generates new training data to simulate varying speech rates. This paper sets a random stretching ratio, r, within the range [0.9, 1.1].

Shift. While preserving emotional characteristics, we randomly raise or lower the pitch by two semitones to shift the pitch and increase timbre diversity. This paper sets the random number of semitone steps, n_{step}, within the range $[-2, +2]$. Substituting semitone steps number n_{step} into formula (6), we achieve pitch shifting.

According to the twelve-tone equal temperament, the pitch ratio formula (6) is

$$\text{Scale} = 2^{\frac{n_{step}}{12}} \tag{6}$$

Pitch. While preserving the emotional characteristics, we randomly select an offset N within the range of $[-1000, +1000]$ to perform temporal translation to simulate different environments. Substituting this into formula (7) will achieve pitch shifting.

$$x_{shifted}[n] = x\left[(n - shift) \bmod N\right] \tag{7}$$

3.3 Feature Extraction

After preprocessing, speech samples are augmented using data augmentation strategies to increase the sample size and address data imbalance in emotion classification.

Previous studies have shown that multimodal acoustic feature fusion offers significant advantages in speech emotion recognition. Compared to single-feature models, the integration of multi-dimensional features such as Chroma, Melspectrogram, and MFCC (Multi-Function Combination Criteria) and their differentials improves the robustness of emotion classification through complementary representations.

Chroma. Chroma features represent the energy distribution across the twelve chromatic pitch classes (C, C♯, D, D♯, E, F, F♯, G, G♯, A, A♯, B) within an audio signal, providing octave-invariant harmonic content representation. This feature aggregates spectral information corresponding to each pitch class regardless of octave, resulting in a 12-dimensional feature vector that captures tonal characteristics essential for music information retrieval tasks. Chroma features demonstrate significant utility in chord recognition, key detection, and harmonic similarity analysis, making them fundamental components in computational musicology applications.

Melspectrogram. Melspectrogram constitutes a perceptually motivated time-frequency representation that maps linear frequency components to the Mel scale, which approximates human auditory perception characteristics. The Mel scale exhibits non-linear frequency spacing, providing higher resolution at lower frequencies where human auditory discrimination is more sensitive. The extraction process involves three primary steps:

(1) application of Short-Time Fourier Transform (STFT) to obtain spectral representation
(2) frequency mapping through triangular filter banks distributed according to the Mel scale
(3) logarithmic compression to approximate the dynamic range of human hearing perception.

MFCC. MFCCs represent the spectral envelope of audio signals through cepstral analysis, effectively modeling vocal tract configurations in speech processing applications. The Mel-frequency cepstrum characterizes the short-term power spectrum by applying the Discrete Cosine Transform (DCT) to log-scaled Mel-spectrogram coefficients, resulting in decorrelated features that separate slowly-varying spectral envelope from rapidly-varying harmonic structure. This cepstral representation enhances robustness against noise and channel distortions while maintaining perceptual relevance through Mel-scale frequency distribution that closely approximates human auditory system response characteristics, making MFCCs particularly effective for speech recognition and audio classification tasks.

This study categorizes different features. Chroma describes the tonal content of music by mapping notes to 12 pitch-level energy representations. MFCC, as a core spectral feature, effectively captures acoustic details related to emotion, and its differences can reflect emotional changes during expression.

The feature extraction stage utilizes a multimodal joint modeling approach: First, the short-time Fourier transform (STFT, n_fft = 512) is calculated, followed by the extraction of the following two types of features:

Spectral features: consisting of a 40-dimensional Mel-spectrogram (fmax = 8 kHz) and 12-dimensional chromaticity features;

Dynamic features: 20-dimensional MFCC coefficients and their first and second-order differences are calculated.

After all features are extracted through a sliding window (hop_length = 256), they are uniformly aligned to 414 frames. Features such as Chroma (12), Mel (40), MFCC + Delta + Delta-Delta (60) are horizontally stacked into three vectors through stacking technology, and fused into 52-dimensional spectral features and 60-dimensional dynamic features, forming a 52 × 414 spectral feature matrix and a 60 × 414 dynamic feature matrix.

3.4 Dual-Path Attention-Enhanced Graph Convolutional Network

The children's speech emotion recognition model proposed in this study uses a dual-path deep learning architecture, combining the advantages of spectral and dynamic features. It also fuses and enhances these features through an attention mechanism and graph convolutional networks. The overall model architecture is shown in Fig. 2.

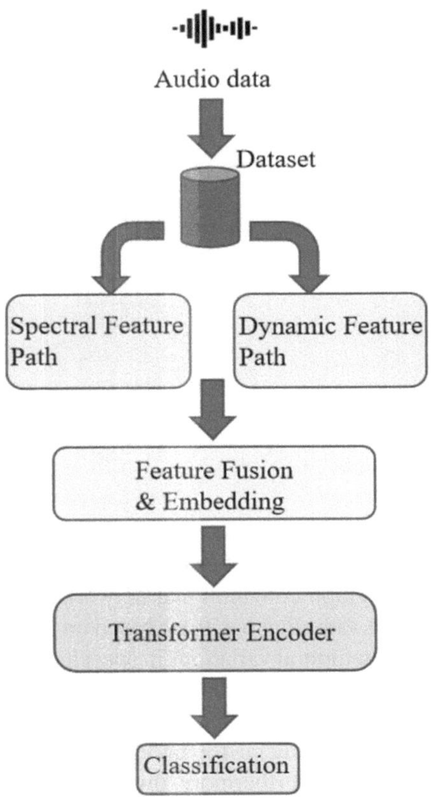

Fig. 2. Dual-Path Attention-Enhanced Graph Convolutional Network

Feature Extraction Module. The feature extraction module is the core component of the model, designed to extract discriminative emotional features from raw speech signals. This module consists of two parallel paths: a spectral feature path and a dynamic feature path.

Spectral Feature Path. The spectral feature path is primarily responsible for capturing the emotional information of speech signals in the frequency domain. This path first performs preliminary processing on the input spectral features through convolutional layers and batch normalization layers. Subsequently, a residual dense layer (DenseLayer) is introduced to enhance feature reuse and information flow, effectively alleviating the vanishing gradient problem in deep networks. Each dense layer concatenates the output of the current layer with the outputs of all previous layers, achieving deep feature fusion. To further enhance the model's focus on important features, this path also integrates a channel attention mechanism (ChannelAttention), enabling it to adaptively learn the importance of different channels. Finally, through residual connections and adaptive average pooling, the spectral features are mapped to a fixed-dimensional feature vector (Fig. 3).

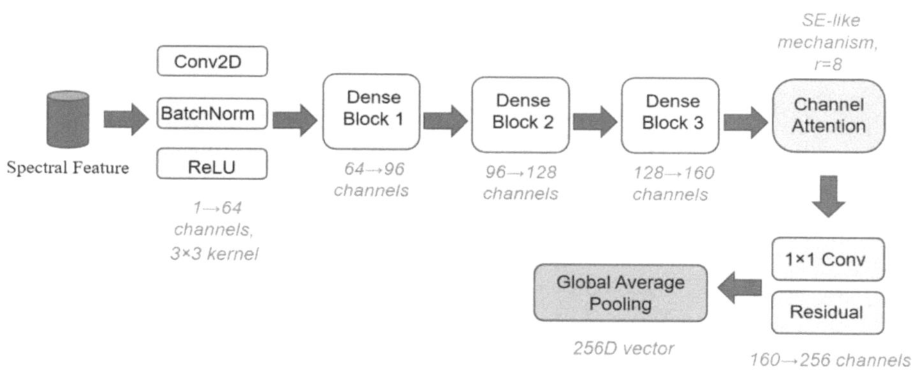

Fig. 3. Spectral Feature Path

Dynamic Feature Path. The dynamic feature path aims to capture emotional variations in speech signals over time. Its innovation lies in the introduction of deformable convolution (DeformConv2D) and graph convolutional networks (GraphConv). Deformable convolution adaptively adjusts sampling positions based on input features, better capturing irregular, emotion-related temporal variations in speech signals. Graph convolutional networks treat speech features as graph-structured data. By propagating and aggregating information across the graph, they effectively model the complex dependencies between speech frames, which is crucial for understanding dynamic emotional cues such as prosody and speaking rate. Furthermore, this path employs depthwise separable convolution to reduce model parameters and improve computational efficiency. Finally, dynamic features are converted into fixed-dimensional feature vectors through adaptive average pooling (Fig. 4).

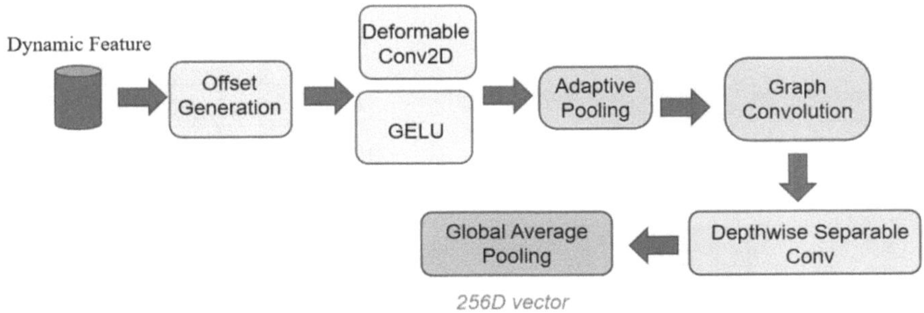

Fig. 4. Dynamic Feature Path

Feature Fusion and Attention Mechanism. The feature vectors extracted from the spectral feature path and the dynamic feature path are concatenated to form a fused feature vector. To further enhance the model's understanding and representation of the fused features, this study introduces the multi-head attention mechanism (MultiHeadAttention) and the masked multi-head attention mechanism (MaskedMultiHeadAttention). The attention mechanism allows the model to assign different weights to features at different positions when processing sequential data, thereby better capturing long-range dependencies in speech emotion. This multi-head mechanism enables the model to learn different attention information from different representation subspaces, further enriching the feature representation. Furthermore, the model employs residual connections and layer normalization (AddNorm) to stabilize training and accelerate convergence. After feature extraction and attention processing, the fused feature vector is ultimately input to a feedforward neural network (FeedForward) and a linear classification layer. The linear classification layer maps the high-dimensional features to a probability distribution of emotion categories and outputs the final emotion recognition result through a softmax function (Fig. 5).

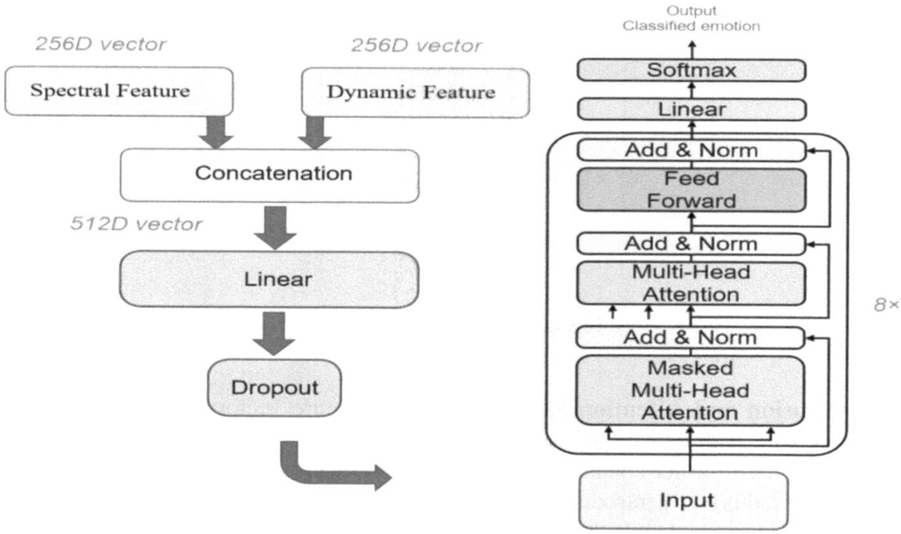

Fig. 5. Feature Fusion and Attention Mechanism

4 Evaluation

4.1 Datasets

This experiment uses publicly available datasets for children's emotional speech recognition: the Children-Bilingual Emotion Speech Dataset (C-BESD) [12], the Mexican Emotional Speech Database (MESD) [14], and the FAU Aibo Emotion Corpus [10, 11].

The FAU Aibo Emotion Corpus is a collection of spontaneous, emotionally charged speech from children interacting with Sony's pet robot, Aibo. Fifty-one children aged 10–13 were recorded using close-range microphones at two schools in Germany. The corpus contains approximately nine hours of German speech and is screened for speech that resembles six emotional categories: anger, happiness, sadness, fear, disgust, and neutral. The C-BESD dataset was released in 2024, providing rich resources for cross-lingual research on emotional development. It includes six different emotions in two languages (Telugu and English), and contains 4,200 emotional voices of six categories (anger, happiness, sadness, fear, disgust, and neutral) of children aged 6–12. It has significant advantages in cross-voice and data balance; The MESD dataset was released in 2021, providing six categories of emotional voices (anger, disgust, fear, happiness, neutrality, and sadness), and selected audio of 6 non-professional child actors with an average age SD $= 9.83 \pm 1.17$ in its dataset through screening.

For the datasets mentioned in this article, all participants and parents of the children signed a written informed consent form.

The recordings in the MESD dataset were conducted in accordance with the Declaration of Helsinki and were approved by the Ethics Committee of the Monterrey Institute of Technology and Higher Education on July 14, 2020 (with national ethics committee registration number CONBIOETICA 19 CEI 011-2016-10-17), with the reference number: P000409-autismoEEG2020-CEICCR002.

4.2 Experimental Settings

Speech emotion recognition is the process of classifying speech signals based on their emotional characteristics. This paper completes feature extraction and model construction by training a designed deep network to learn the emotional characteristics of speech data.

Data Processing Parameter Settings. The number of samples after data augmentation was set to 12,000, with 2,000 for each emotion category. The target sample length for children's speech emotions (anger, happiness, sadness, fear, disgust, and neutrality) was uniformly set to 414 frames. All recordings were filtered for maximum and minimum durations, and the sampling rate for all audio files was 16 kHz. The target samples were randomly divided into a training set (70%) and a test set (30%).

Training Configuration. This study implemented the proposed deep learning model in the PyTorch framework. The experimental configuration parameters are as follows: the experimental hardware configuration is an Intel(R) Xeon(R) Silver 4310 CPU with 74 GB of RAM and an A6000-48G GPU with 48 GB of video memory. The software environment is Python 3.7.0, the deep learning framework is PyTorch 1.13.1, and CUDA 11.7.0. The batch size (batch_size) is 32, the number of emotion classes (num_classes) is 6, the sequence length (seq_len) is 414, the dropout rate (dropout_rate) is 0.1, the number of attention heads (num_heads) is 8, the feedforward network dimension (ff_dim) is 512, and the embedding dimension (embed_dim) is 256. The total number of epochs (epochs) of model training is set to 200, and the initial learning rate (lr) is 1e-4. The Adam optimizer is used, and the loss function is the cross-entropy loss. In order to improve training efficiency and reduce memory consumption, this study also adopted mixed precision training and gradient accumulation strategies.

Evaluation Metrics. Evaluation metrics include Accuracy (Acc), Precision (Pre), Recall (Rec), and F1-Score (F1). Accuracy refers to the model's overall recognition accuracy across all samples. Precision refers to the percentage of actual samples that the model predicts to be positive. Recall refers to the percentage of positive samples correctly identified among all positive samples in the model. The F1-Score represents the model's recognition accuracy and its coverage of positive samples. It is calculated as the harmonic mean of Precision and Recall. In multi-class speech emotion classification, it comprehensively reflects both Accuracy and Recall. A comprehensive analysis of these four metrics forms a multi-class evaluation framework that measures the model's effectiveness in speech emotion recognition tasks.

$$Acc = \frac{TP + TN}{TP + FN + TN + FP} \tag{8}$$

$$\Pr e = \frac{\sum_{i=1}^{k} TP_i}{\sum_{i=1}^{k} TP_i + \sum_{i=1}^{k} FP_i}, \tag{9}$$

$$\operatorname{Re} c = \frac{\sum_{i=1}^{k} TP_i}{\sum_{i=1}^{k} TP_i + \sum_{i=1}^{k} FN_i} \tag{10}$$

$$F1 = \frac{2\sum_{i=1}^{k} pre_i \times 2\sum_{i=1}^{k} \operatorname{Re} c_i}{2\sum_{i=1}^{k} pre_i + 2\sum_{i=1}^{k} \operatorname{Re} c_i} \tag{11}$$

$$WAA = \sum_{i=1}^{k}\left(\frac{TP_i + TN_i}{TP_i + FN_i + TN_i + FP_i}\right)/k \tag{12}$$

Where *TP* is the number of speech samples that are accurately classified into a specific emotion category; *TN* is the number of speech samples that are accurately classified as non-emotional labels but lack specific emotion labels; *FN* is the number of speech samples that have specific emotion labels but are not correctly classified; *FP* is the number of speech samples that are incorrectly identified as specific emotion labels; *WAA* is the weighted average accuracy, which is the average classification accuracy of all categories in the dataset; k is the number of emotion categories. The reason for setting k is that the distribution of different categories is not uniform in some datasets.

4.3 Result Analysis (Tables 1 and 2)

Table 1. Speech emotion recognition results (after performing DA) of the proposed models

classification	Acc/%	Pre/%	Rec/%	F1/%
Angry	84.33	92.00	84.33	88.00
Disgust	90.33	84.82	90.33	87.48
Fear	84.33	92.67	84.33	88.30
Happy	91.33	79.42	91.33	84.96
Neutral	77.83	87.61	77.83	82.43
Sad	90.83	84.89	90.83	87.76

Table 2. Speech emotion recognition results (before performing DA) of the proposed models

classification	Acc/%	Pre/%	Rec/%	F1/%
Angry	60.66	62.50	60.66	61.57
Disgust	57.58	53.52	57.58	55.48
Fear	65.48	58.26	65.48	61.66
Happy	51.11	63.38	51.11	56.79
Neutral	80.24	74.42	80.24	77.22
Sad	62.50	66.03	62.50	64.22

This paper comprehensively investigates various deep learning-based speech emotion recognition (SER) systems for children based on three different datasets covering bilingual, Mexican, and Russian. By fusing two features, the proposed model comprehensively understands emotional information in speech from multiple dimensions. A dual-channel model is used to process the two speech features, and the fused features are introduced into an attention mechanism to extract long-term global contextual dependencies and correlations. Through data augmentation, the proposed model achieves significant performance, achieving accuracy of 84.33%, 90.33%, 84.33%, 91.33%, 77.83%, and 90.83% for the six emotion categories of anger, disgust, fear, happiness, neutrality, and sadness, respectively. The recognition accuracy for "neutral" emotion is relatively low, at approximately 77.83%. This may be due to the less distinct speech characteristics of "neutral" emotion or a degree of confusion with other emotions. Analysis of the accuracy and loss curves before and after data augmentation clearly demonstrates that data augmentation effectively improves the accuracy of emotion recognition categories and also promotes better model convergence (Figs. 6, 7, 8, and 9).

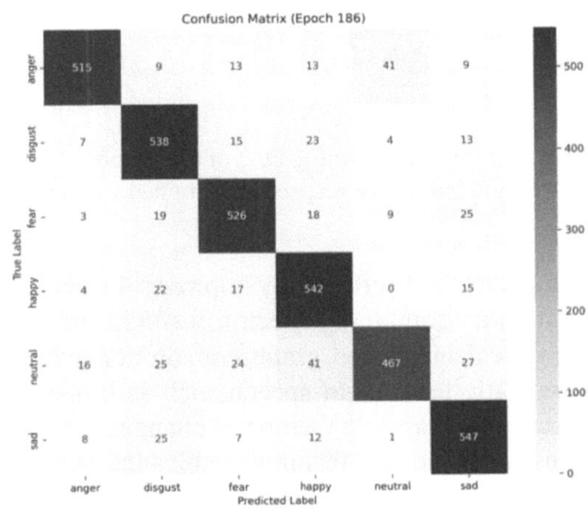

Fig. 6. Confusion matrix (after performing DA) of the proposed models

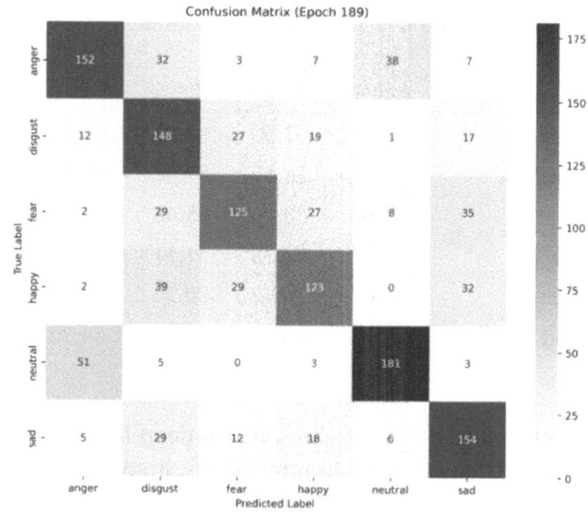

Fig. 7. Confusion matrix (before performing DA) of the proposed models

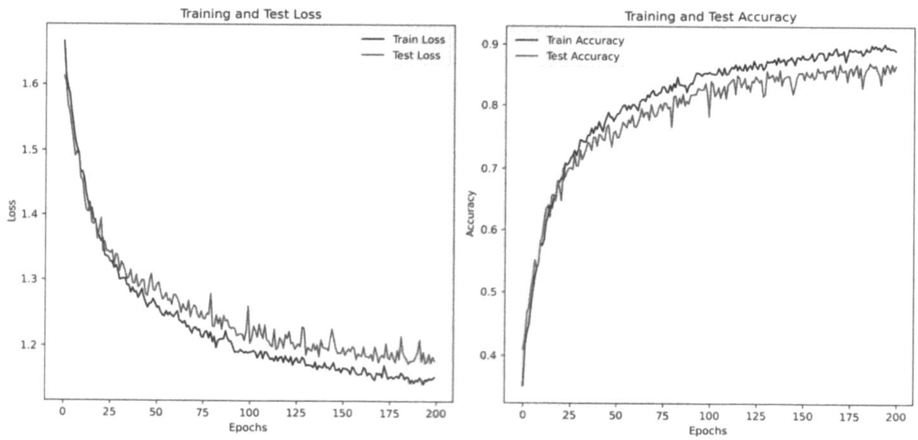

Fig. 8. Performance evaluation (after performing DA) of the proposed models show the training vs. validation accuracy curve and loss curve respectively, trained for 200 epochs.

The model proposed in this study effectively improves the performance of children's speech emotion recognition by combining spectral features and dynamic features, and introducing an attention mechanism and graph convolutional networks. The spectral feature path can capture static features of speech such as timbre and pitch, while the dynamic feature path focuses on capturing temporal changes in speech such as speaking rate and rhythm. The fusion of the two features enables the model to fully understand speech emotion information from different dimensions. The introduction of the attention mechanism enables the model to focus on the most emotion-relevant parts of the speech signal, while the graph convolutional network can better model the complex

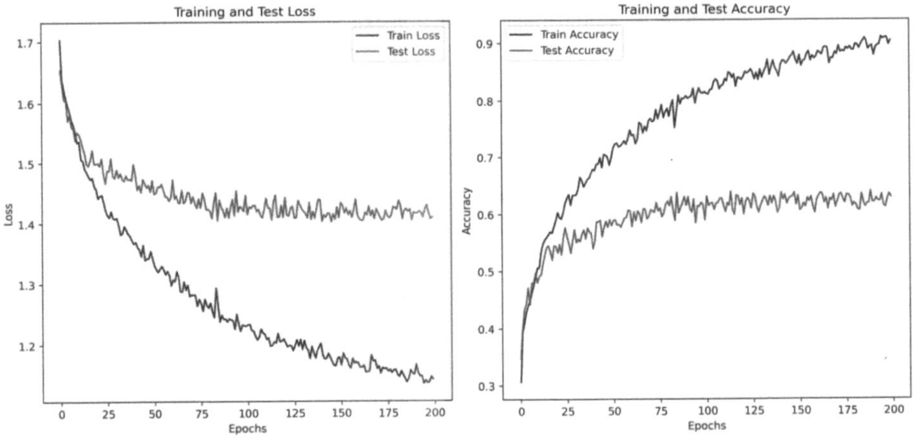

Fig. 9. Performance evaluation (before performing DA) of the proposed models show the training vs. validation accuracy curve and loss curve respectively, trained for 200 epochs

dependencies between speech frames. In addition, the application of mixed precision training and gradient accumulation strategies significantly improves training efficiency while ensuring model performance (Tables 3 and 4).

Table 3. Performance comparison with other algorithms

Datasets	Methodology	Features	Accuracy/%
FAU	CNN + WMRN [18]	MFCC	49.53
FAU	CNN + WMRN [18]	Chroma	42.81
FAU	LSTM [15]	MFCC、RMS、ZCR	76.52
FAU	Attention + LSTM [15]	MFCC、RMS、ZCR	80.55
Belfast、EMO-DB、CASIA、FAU	LSTM + CNN [16]	MFCC	73.68
	SVM [16]	MFCC	56.76
FAU	CNN-BiLSTM [17]	MFCC	61.20
FAU	AttRNN-RNN [19]	MFCC	67.50
FAU	CNN-AttBiLSTM [17]	MFCC	71.60
C-BESD	Conv2D [20]	MFCC、ZCR、Harmonic Ratio, Pitch	75.98
MESD	Conv2D + Conv1D + MLP [21]	ZCR、Chroma、MFCC RMS、Mel	96.00
FAU、C-BESD、MESD	This Work	MFCC、Chroma、Mel	86.49

Table 4. Comparison of ablation experiment results

Methods	Accuracy/%
Only Spectral Feature Path	60.63
Only Dynamic Feature Path	84.98
No Deformable Convolution(Dynamic Feature Path)	82.12
No Graph Convolution(Dynamic Feature Path)	83.78
No Dense block(Spectral Feature Path)	81.26
No channel attention(Spectral Feature Path)	84.88
This Work	86.49

Based on the quantitative ablation study, the proposed model achieves a state-of-the-art accuracy of 86.49%, outperforming all its variants. This gain stems from the effective integration of its core components.

The dynamic feature path (84.98%) contributes more significantly than the spectral path (60.63%), highlighting the primary importance of modeling temporal dynamics. However, the full model's higher performance confirms that spectral features still provide valuable complementary information.

Within the dynamic path, both deformable convolution (82.12% without it) and graph convolution (83.78% without it) are essential, with deformable convolution being more critical for capturing precise motion patterns. In the spectral path, the dense block is extremely important (81.26% without it), enabling complex feature learning, while channel attention also contributes (84.88% without it) by emphasizing informative frequencies (Table 5).

Table 5. Comparison of sample sizes of different datasets in different model applications

Datasets	Methods	Object of the study	Amount
MESD	Begazo, Rolinson, et al. [21]	4 adult males,4 adult females,8 children	1150
MESD	**This work**	8 children (5 girls and 3 boys)	286

In using the MESD dataset, the model by Begazo, Rolinson, et al. [21]. achieved an accuracy of 96.00%, primarily trained and evaluated on adult and child populations. Although its accuracy is higher, its performance may be exaggerated due to learning patterns of adult language, which could limit its specific applicability in children's speech emotion recognition. In contrast, our model, aimed at the child population but trained and tested only on a purified subset of the same data, achieved a slightly lower accuracy of 86.49%. Despite the smaller sample size, this focused approach ensures that the model concentrates on learning the unique acoustic features of children's emotional expressions. Therefore, although the absolute accuracy is lower, our model offers greater effectiveness and reliability for the targeted application of children's speech emotion recognition.

5 Conclusion and Future Works

Insufficient child speech data hinders deep neural network models from reaching their full potential, posing a significant challenge for deep learning-based speech emotion recognition in children. This lack of data samples often leads to overfitting in deep and complex models.

Most children's speech emotion datasets are collected in different environments. Due to the influence of these environments and children's pronunciation, processing real data can be challenging.

To improve the accuracy of emotion recognition in Chinese children's speech, this paper proposes a fusion model based on a Dual-Path Attention-Enhanced Graph Convolutional Network. CNN2D effectively extracts static spectral features of speech signals, while graph convolutional networks capture the temporal relationship between changes in emotional states. By combining the strengths of these two models, the model's ability to recognize the emotional states of children's speech is effectively improved. Experimental results show that the proposed Dual-Path Attention-Enhanced Graph Convolutional Network model achieves excellent performance on the child speech emotion recognition task across three children's speech emotion datasets, with recognition accuracy exceeding that of current typical CNN, LSTM, CNN + LSTM, and Attention+LSTM models. However, its recognition rate is lower than that of the Conv2D + Conv1D + MLP model [21] on the MESD dataset, which can be attributed to the fact that the method targets both adults and children. The MESD dataset used in this paper is entirely derived from children, but the number of samples directed at children in the MESD dataset is relatively small, which results in a higher accuracy compared to other models.

In addition, this paper proposes that, in order to fully train deep learning-based models in the face of insufficient sample size in children's speech emotion datasets, the experiments used to create these datasets are simulated, semi-natural, and far removed from the natural environments found in real-world scenarios. This has raised questions about the ability of the children's speech emotion recognition model to detect correct emotions in noisy real-world scenes. In the future, we will further design relevant hardware structures to collect effective children's speech emotion signals, and build more reasonable and effective children's speech datasets and model fusion strategies to further improve the performance of speech emotion recognition.

References

1. Amsel, T.T.: An urban legend called: "the 7/38/55 Ra o rule". Eur. Polygraph. **13**(2), 95–99 (2019)
2. Hu, D., Gaofeng, P., Wei, C.: Chinese children's speech emotion recognition based on DPCNN-SLSTM [J]. Commun. Technol. **57**(07), 666–671 (2024)
3. Syed, E.U., Hussein, S.A., Mahmud, S.: Screening for emotional and behavioural problems amongst 5–11-year-old school children in Karachi, Pakistan. Soc. Psychiatry Psychiatr. Epidemiol. **42**(5), 421–427 (2007)

4. Han, W., et al.: Towards temporal modelling of categorical speech emotion recognition. In: Interspeech (2018)
5. Anvarjon, T., Mustaqeem, Soonil Kwon: Deep-net: a lightweight CNN-based speech emotion recognition system using deep frequency features. Sensors. **20**(18), 5212 (2020)
6. Li, D., et al.: BLSTM and CNN stacking architecture for speech emotion recognition. Neural. Process. Lett. **53**(6), 4097–4115 (2021)
7. Xia, F., et al.: Graph learning: a survey. IEEE Trans. Artif. Intell. **2**(2), 109–127 (2021)
8. Ustubioglu, A., Ustubioglu, B., Ulutas, G.: Mel spectrogram-based audio forgery detection using CNN. SIViP. **17**(5), 2211–2219 (2023)
9. Yenigalla, P, et al.: Speech emotion recognition using spectrogram & phoneme embedding. Interspeech, vol. 2018, (2018)
10. Steidl, S.: Automatic Classification of Emotion-Related User States in Spontaneous Children's Speech. Logos-Verlag, Berlin, Germany (2009)
11. Batliner, A., et al.: Private emotions versus social interaction: a data-driven approach towards analysing emotion in speech. User Model. User-Adap. Inter. **18**(1), 175–206 (2008)
12. Gudivaka, S.R., Polipogu, P., Penumudi, J.: Child bilingual emotion speech dataset (child-BESD). Kaggle Dataset (2024)
13. Shirian, A., Guha, T.: Compact graph architecture for speech emotion recognition. In: ICASSP 2021-2021 IEEE International Conference on Acoustics, Speech and Signal Processing (ICASSP). IEEE (2021)
14. Duville, M.M., Alonso-Valerdi, L.M., Ibarra-Zarate, D.I.: Mexican emotional speech database based on semantic, frequency, familiarity, concreteness, and cultural shaping of affective prosody. Data. **6**(12), 130 (2021)
15. Chenyang, L., Qiong, Z., Min, H.: Children's emotional speech recognition technology based on attention-LSTM. Acoust. Technol. **44**, 402–409 (2025)
16. Long, L., Liang, T.: Multi-distributed speech emotion recognition based on Mel frequency Cepstogram and parameter transfer. Chin. J. Electron. **31**, 155–167 (2022)
17. Cao, G., Tang, Y., Sheng, J., Cao, W.: Emotion recognition from children speech signals using attention based time series deep learning. In: 2019 IEEE International Conference on Bioinformatics and Biomedicine (BIBM), pp. 1296–1300 (2019)
18. Onwujekwe, Gerald N. "Using Deep Learning-Based Framework for Child Speech Emotion Recognition." (2021).
19. Huang, C.W., Narayanan, S.S.: Deep convolutional recurrent neural network with attention mechanism for robust speech emotion recognition [C]. In: IEEE International Conference on Multimedia and Expo, pp. 583–588 (2017)
20. Rao, D.V., et al.: Children's speech emotion recognition using feature fusion-based deep neural networks. In: IEEE International Conference on Intelligent Signal Processing and Effective Communication Technologies (INSPECT). IEEE (2024)
21. Begazo, R., et al.: A combined CNN architecture for speech emotion recognition. Sensors. **24**(17), 5797 (2024)

Evaluating Learning States in Synchronous Remote Classes via Qwen2.5-Max with RAG and ReAct Agent

Haoyuan He[1], Bemnet Wondimagegnehu Mersha[1(✉)], Yaping Dai[2], Kaoru Hirota[1], Wei Dai[3], and Yumin Lin[3]

[1] School of Automation, Beijing Institute of Technology, Beijing 100081, China
MershaBemnetW@bit.edu.cn

[2] Beijing Institute of Technology Zhuhai, Tangjiawan, Zhuhai 519088, Guangdong, China

[3] River Security Technology Co., LTD., Shanghai 200336, China

Abstract. In a synchronous remote class, where lectures are simultaneously delivered to both local and remote students, evaluating the group learning state of the remote class presents significant challenges. To address this problem, a series of methods for evaluating students' learning states in the synchronous remote class based on Qwen2.5-Max is proposed. First, a behavior recognition model and a facial emotion recognition model were constructed to recognize each student's actions and facial emotions in the class. Subsequently, Qwen2.5-Max with the RAG individual learning state recognition method is proposed. Finally, Qwen2.5-Max with ReAct agent for the synchronous remote class group learning state recognition method is proposed to determine the group learning state and provide instructional suggestions to the teacher. The proposed methods are tested using a custom-made dataset. The results indicate that the methods can help teachers balance local and remote classes by enhancing the quality of synchronous remote teaching.

Keywords: Synchronous Remote Class · Qwen2.5-Max · RAG · ReAct Agent

1 Introduction

With the advancement of educational modernization, the issue of allocating educational resources has garnered widespread attention. Due to the uneven distribution of educational resources, students in remote areas often struggle to access high-quality education. To achieve a balanced distribution of educational resources, the concept of synchronous remote classes has emerged. In a synchronous remote class, the teacher delivers lessons to students in the local class, while students in one or more remote classes can simultaneously view the live broadcast of the teacher's instruction on large screens and interact online

H. Ma et al. (Eds.): IWACIII 2025, CCIS 2780, pp. 135–150, 2026.
https://doi.org/10.1007/978-981-95-6730-0_10

with the teacher. This approach enables students from various locations to participate simultaneously in the same class and share high-quality educational resources. However, synchronous remote teaching has certain limitations. For remote classes, teachers typically rely on internet-based tools to deliver lessons and cannot interact with students face-to-face. Consequently, it becomes challenging to accurately and promptly assess the learning states of remote students, which hinders the balance of instructional progress between local and remote classes [8]. Therefore, there is an urgent need for an intelligent evaluation system that can provide timely feedback on students' learning state in synchronous remote classes.

To ensure the quality of instruction and enhance the effectiveness of synchronous remote teaching, this study proposes a series of novel methods for evaluating students' learning states in a synchronous remote class based on image recognition and Qwen2.5-Max Large Language Models (LLMs) (see Fig. 1). First, a custom-made dataset containing students' behavioral images is developed, and a ResNet34 model based on this dataset is built to recognize six types of students' behaviors (actions). Second, the facial emotion recognition model is built based on MobileNetV2 network to recognize five types of students' facial emotions and is tested on the Emotion-Domestic (Asia) Expression Recognition dataset. Metrics, including accuracy, precision, recall, and F1 score, are applied to assess the performance of the models. Then, the Qwen2.5-Max with RAG individual learning states recognition method is constructed to evaluate the learning state of each student. The individual learning state works using an improved four-dimensional space (4D state space). Its outputs were compared with the results given by the Certainty Factor (CF) model to assess its accuracy. Finally, the Qwen2.5-Max with ReAct-agent group learning state recognition method for the synchronous remote class is constructed to evaluate the learning states of both local and remote classes and provide instructional suggestions to the teacher according to the learning states of the class based on the Qwen and ReAct agent. The outputs were compared with the responses of a human volunteer to assess whether the outputs are precise, specific, and practical.

The remainder of this paper is organized as follows: Sect. 2 introduces the datasets used in the paper. Section 3 introduces the proposed methods, including the behavior (action) and facial emotion recognition models, the 4D state space, the Certainty Factor (CF) model, and the two recognition methods, one for the learning state evaluation of individuals and the other for group learning. Section 4 introduces the experiments and results analysis, including the experiments for behavior recognition, facial emotion recognition, and the two types of learning state evaluation methods. Finally, Sect. 5 concludes the paper.

2 Dataset

2.1 Dataset for Behavior Recognition

A custom-made behavior dataset was created to train the behavior recognition model. First, 20 middle school students were invited to participate in photo

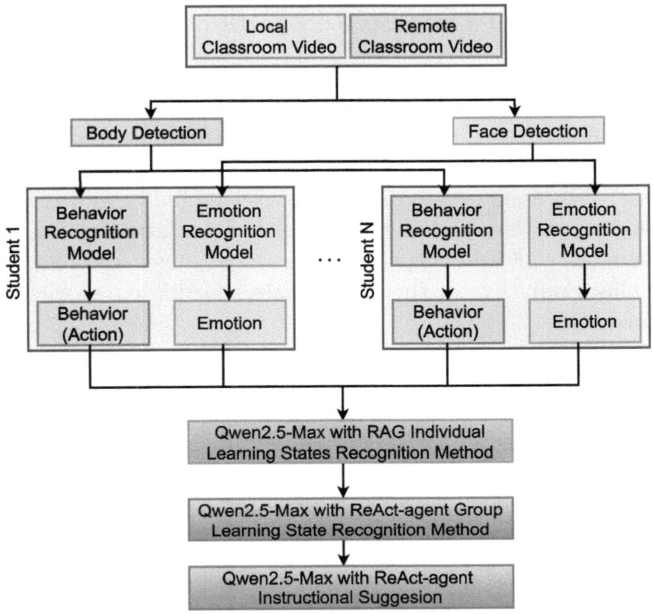

Fig. 1. Proposed group learning states evaluations in a synchronous remote class.

shoots for six specific behavior types: playing phone, raising hand, sitting, sleeping, standing, and writing. The camera was positioned in front of the left side of the students. Second, the Human Body Analysis API developed by Baidu [1], along with functions from OpenCV are used to crop the single-person behavior images. Subsequently, the images were categorized into six behavior types and divided into training sets (80%) and testing sets (20%). Moreover, a related dataset that contains the same six types of students' behaviors was also used in the study [15]. This resulted in a combined student behavior dataset that contains a total of 7,941 images (see Fig. 2).

(a) Custom-made dataset

(b) Related dataset [15]

Fig. 2. Behavior dataset.

2.2 Dataset for Facial Emotion Recognition

Although there are various facial emotion recognition datasets, the Emotion-Domestic (Asia) Expression Recognition Dataset, which primarily consists of images of East Asian faces collected from the Internet, was chosen because the research was conducted for application in Chinese schools. The Emotion-Domestic (Asia) Expression Recognition Dataset includes approximately 54,000 images. These images were classified into seven basic emotions: anger, disgust, fear, happy, neutral, sad, and surprise. However, as anger and fear are relatively infrequent in class, only five of the seven basic emotion categories are included in the final dataset.

2.3 Dataset for Simulated Classroom Scenarios

Five middle school students were invited as volunteers to simulate real classroom scenarios in both local and remote classes. The experimental setup is as follows:

(1) A tripod was placed in front-left of the students, with a camera mounted on the top.
(2) Volunteers were asked to perform specific behaviors and facial emotions, and several video clips of approximately 10 s were recorded. There were variations in the students' behaviors and emotions among each video clips.

To effectively evaluate the proposed method, a distinction was made between the simulated local classroom and the remote classroom. In the simulated local classroom scenarios, the behaviors of "playing phone" and "sleeping" were excluded, whereas these behaviors were incorporated into the simulated remote classroom scenarios (see Fig. 3). This differentiation was necessary due to the challenges teachers face in monitoring remote classes, which often result in poor learning conditions.

(a) Image of local class (b) Image of remote class

Fig. 3. Local and remote custom-made dataset.

3 Proposed Method

3.1 Behavior (Action) and Facial Emotion Recognition Models

The ResNet34 network [3] is used for the recognition of student behavior (action) due to its high performance in feature extraction and classification with residual learning. The training utilized a custom behavioral dataset (see Sect. 2.1). The main training parameters included a batch size of 10, a learning rate of 0.001, and a maximum iteration number of 50. To prevent overfitting, early stopping is implemented in the training process, and patience is set to 5 epochs. The MobileNetV2 [7] network is used for the recognition of student facial emotion because it is lightweight and accurate. The training utilized the Emotion-Domestic (Asia) Expression Recognition Dataset. The specific training parameters included a batch size of 64, a learning rate of 0.001, and a maximum of 100 iterations. Early stopping is also introduced, with an early stopping time of five epochs. Furthermore, due to the significant imbalance in the number of images across different emotion categories, the loss function is modified by employing a weighted loss function. After each emotion and behavior recognition, the number of students in the classroom and every student's recognized behavior and emotion are stored in a Python dictionary for evaluations of students' individual learning states (see Sect. 3.4).

3.2 Four-Dimensional Space of Individual Learning States (4D State Space)

To develop a scientific model to evaluate students' individual learning states, the study proposes a four-dimensional state space model of individual learning states (4D state space). It is constructed based on insights from the literature [12,13]. The definitions of the four dimensions are as follows:

(1) **Engagement (engaged-detached)**: Evaluating whether the student is engaging in the class and focusing on the course content.
(2) **Interactivity (active-silent)**: Evaluating whether the student is active in making statements, and participating in the interaction with the teacher.
(3) **Discipline (disciplined-disruptive)**: Evaluating whether the student complies with rules and regulations.
(4) **Comprehension (clear-confused)**: Evaluating whether the student has understood and grasped the course content.

Specifically, if a student is in an intermediate state without an evident tendency toward either end, the state in this dimension is evaluated as "unknown".

3.3 Certainty Factor (CF) Model

The study introduces the Certainty Factor (CF) model as a benchmark for the proposed Qwen2.5-Max with the RAG individual learning state recognition method. The CF Model is a data fusion model combining probability theory and fuzzy theory [9], offering a reliable method for reasoning. The CF model describes the reliability of inferring a hypothesis given one or more pieces of evidence. Its general form is

$$\text{IF } E \text{ THEN } H \quad CF(H, E), \tag{1}$$

where E is the evidence, H is the hypothesis, and $CF(H, E)$ is the certainty factor for H being true given E ($-1 \leq CF(H, E) \leq 1$). When $0 < CF(H, E) \leq 1$, the evidence E supports the hypothesis H being true. The larger the $CF(H, E)$ value, the greater the certainty that H is true. When $-1 \leq CF(H, E) < 0$, the evidence E supports the hypothesis H being false. The smaller the value of $CF(H, E)$, the greater the certainty that H is false. When $CF(H, E) = 0$, the truth of H is independent of the evidence E. When multiple pieces of evidence are present, combined CF must be calculated based on the CF of the two evidence [4]. Based on the CF model, the study defines the student as displaying a certain kind of behavior or facial emotion in class as evidence E_1 and E_2, respectively. The student's current learning state (the four positive learning states in the 4D state space—Engaged, Active, Disciplined, and Clear) is defined as hypotheses H_1 to H_4. By calculating the combined CF values of behaviors and emotions for each of the four conclusions, and formulating rules to evaluate students' learning states based on the CF values, the learning states corresponding to various behavior-emotion combinations can be determined. To convert the CF values into the evaluation results of the learning state, a judgment rule is developed. Using "Engaged" as an example, the judgment rule based on CF values is defined as

$$\text{Engagement} = \begin{cases} \text{Engaged}, & \text{if } 0.5 \leq CF \leq 1, \\ \text{Unknown}, & \text{if } -0.5 \leq CF < 0.5, \\ \text{Detached}, & \text{if } -1 \leq CF \leq -0.5. \end{cases} \tag{2}$$

The same rule applies to "Active," "Disciplined," and "Clear."

3.4 Qwen2.5-Max with RAG Individual Learning States Recognition Method

Based on Qwen2.5-Max and Retrieval-Augmented Generation (RAG), the study proposes a method called Qwen2.5-Max with RAG individual learning states recognition method (individual learning state recognition) to evaluate the individual learning state of each student in the synchronous remote class. We also proposses a method called Qwen2.5-Max with ReAct-agent group learning state recognition method for the synchronous remote class (group learning state recognition) to evaluate the learning state of the whole class. Qwen2.5-Max is a type of advanced MoE (Mixture-of-Experts), and has manifested obvious superiority in many benchmarks [10]. Qwen2.5-Max is selected because the Qwen series are superior in Mandarin tasks compared to many other LLMs. As the study is carried out in Mandarin, it is suitable to utilize Qwen2.5-Max.

RAG is applied to improve Qwen2.5-Max for individual learning state recognition. It has been widely used for the enhancement of LLMs [6]. The basic working process of RAG is as follows: when a query is given, the RAG system retrieves the most relevant information from the knowledge vector database and integrates it into a prompt for the LLM, which then generates an answer. Based on the working principle of RAG, the workflow of the individual learning state recognition consists of the following three parts:

(1) Data Preprocessing
First, the knowledge base must be preprocessed to convert textual content into a vector form before retrieval. This process mainly includes document chunking and text embedding. The former divides long texts into smaller blocks (chunks) to reduce retrieval load and comply with the context length limitations of large models, whereas the latter transforms text into semantic vectors using pre-trained deep language models. After data preprocessing, the semantic vectors are stored in a vector database for retrieval. In the proposed individual learning state recognition, BGE-m3 [2] is used for embedding, whereas Faiss [5] and LangChain are used to construct the vector database. We also developed two types of individual learning state recognitions. Their working principles and workflow are approximately the same. The main difference is in their knowledge bases. These two knowledge bases share definitions and descriptions of the 4D state space, but differ in the following ways:
1) Knowledge Base 1 contains the evaluation results of students' learning states for all behavior-emotion combinations. Therefore, there is no need to provide the CF values of students to the system; only behaviors and facial emotions are needed.
2) Knowledge Base 2 contains the evaluation rules (as shown in Equation (3)). Therefore, the CF values of students should also be provided.
In our research, the system based on Knowledge Base 1 is referred to as "individual learning state recognition 1" ("recognition 1"), and the one based on Knowledge Base 2 as "individual learning state recognition 2" ("recognition 2").
(2) Data Retrieval
In this step, the RAG system converts the query into a query vector using the same embedding method for data preprocessing and then retrieves the vector database so that the LLM can provide answers based on the retrieved context. The main method used for retrieving data is semantic retrieval, an advanced similarity retrieval method. Unlike keyword-based or word-frequency-based methods like TF-IDF, semantic retrieval can capture and compare the semantic meanings of the semantic vectors and then retrieve relevant documents even without keyword or word frequency overlap. Semantic retrieval compares the similarities of two vectors using metrics such as cosine similarity, Euclidean distance, and Manhattan distance, which is the same as traditional similarity retrieval. In the individual learning state recognition, the knowledge base is a custom-written document named "Correspondence Between Individual Learning States and Their Behaviors and Facial Emotions."

(3) Prompt Construction
The prompts consist of the query, the retrieved context, and other requirements. The LLM replies to the query based on the context and the requirements. The individual learning state recognition iterates through the behavior and emotion of each student based on the total number of students in the class, which is stored in a dictionary (see Sect. 3.1). A query is given about the learning state of the student. For recognition 1, the query includes the student's behavior and emotion, which are also stored in the same dictionary. For recognition 2, the query includes the contents of recognition 1's query. However, the CF values corresponding to the four dimensions in the 4D state space are calculated first based on behavior and emotion, and then the values are included in the query as well. The individual learning state recognition utilizes a kind of prompt engineering called Chain-of-Thought (CoT) [11], which allows LLMs to break down complex tasks into sub-tasks and perform step-by-step reasoning to complete the task. The process of CoT involves the following four steps:
1) Identify question: Each query aims to recognize the learning state of the student by identifying the behavior and emotion, or the four CF values of the student;
2) Find related clues: Find the related clues for individual learning state recognition based on the given information;
3) Reason step by step: For recognition 1, Qwen2.5-Max only needs to find the related behavior-emotion combination. For recognition 2, Qwen2.5-Max needs to evaluate the dimension-level learning states of the four dimensions one by one based on the CF values and the judgment rule given in Sect. 3.3;
4) Give conclusion: For recognition 1, Qwen2.5-Max gives the learning state corresponding to the behavior-emotion combination. For recognition 2, Qwen2.5-Max needs to integrate the dimension-level learning states and give the final evaluation.

3.5 Qwen2.5-Max with ReAct-Agent Group Learning State Recognition Method for the Synchronous Remote Class

After the individual learning state recognition, the Qwen2.5-Max with ReAct-agent group learning state recognition method for the synchronous remote class (group learning state recognition) is proposed. This method is designed to provide the learning states of the whole local and remote classes while also providing instructional suggestions to the teachers based on these evaluation results. Considering the limitations in reasoning and acting capabilities of LLMs, as well as their tendency to generate hallucinations and fail to deliver accurate and detailed analysis, the proposed group learning state recognition method adopts ReAct agent with retrieval capabilities and Qwen2.5-Max to achieve the research objectives. The ReAct agent (reasoning and acting agents) enables LLMs to alternate between reasoning and acting until the final conclusion is drawn [14].

Based on the working principle of ReAct agent, the core workflow of the group learning state recognition consists of the following steps:

(1) Input In this step, the agent can access the query and the chat history. As mentioned in Sect. 3.4, chat history includes all students' behaviors, facial emotions, and individual student learning states.
(2) Thought This step allows the agent to perform reasoning based on current information, mainly including the retrieved context as well as the group learning state and instructional suggestions it has already reasoned, and then plan the next action.
(3) Action Based on its reasoning, the agent selects tools from the toolkit to execute the desired task. In the group learning state recognition, the toolkit mainly consists of a retrieval tool, the principle of which is almost the same as that of the retrieval process in a RAG system. The ReAct agent can retrieve related information from the knowledge base so that it can enable Qwen2.5-Max to better understand the characteristics of the concepts in the study, such as the synchronous remote class and the 4D state space.
(4) Action Input The agent extracts relevant keywords from the query and previous observations as inputs for the next step. For instance, for the query shown in step (1), the action input may be "local/remote," "learning state," "instructional suggestions," etc.
(5) Observation The agent observes the outcome of the performed action, which serves as feedback for further reasoning.

Steps (2)-(5) can repeat multiple times until a final answer is obtained. This mechanism enables ReAct agent to alternate between reasoning and acting effectively. The knowledge base used here is the custom-written document titled "Assistance Document for Group Learning States Recognition and Instructional Suggestions Feedback in synchronous remote classes," which mainly consists of the following four components:

(1) Definitions and Characteristics of Local and Remote Classes: This section first clarifies the definitions of "local class" and "remote class." Second, it elaborates on the characteristics of remote classes, highlighting that students in such settings are more prone to poor learning state because of physical separation from the teacher.
(2) Introduction to the 4D States Space: This section provides definitions for the four dimensions in the 4D state space.
(3) Analysis of Students' Behavior-Emotion Combinations: This section analyzes the likely intent of students based on various behavior-emotion combinations. For instance, if a student raises their hand with a surprised facial expression, it may suggest that the student has a question and is eager to express it.
(4) Instructional Suggestions for the Teacher: This section provides examples of instructional suggestions that can be provided to the teacher under each of the four dimensions in the 4D states space.

4 Experiments and Results Analysis

4.1 Experiments for Behavior (Action) and Facial Emotion Recognition

(1) Experiments for Behavior Recognition

The experimental results of the behavior recognition model indicate that early stopping was applied during the ninth epoch. The training and testing accuracies reached 99.95% and 99.67%, respectively, which were fairly high. The training and testing loss values were 0.0025 and 0.0416, respectively. Also, a report containing the precision, recall, F1-score, and total number (the total number of images in the testing set) for each classification is shown in Table 1.

Table 1. The machine learning-based behavior recognition model result

Label	Class	Precision	Recall	F1 Score	Total Number
0	Playing Phone	100%	99.11%	99.55%	224
1	Raising Hand	100%	99.60%	99.80%	247
2	Sitting	99.33%	100%	99.66%	296
3	Sleeping	100%	99.60%	99.80%	249
4	Standing	99.62%	99.62%	99.62%	262
5	Writing	99.20%	100%	99.60%	249

(2) Experiments for Facial Emotion Recognition

The experimental results indicate that early stopping was implemented during the 21st epoch. The training and testing accuracies were high at 88.10% and 94.11%, respectively. The training and testing loss values were 0.3715 and 0.2054, respectively. A report containing the precision, recall, F1-score, and total image number for each classification is shown in Table 2.

Table 2. The facial emotion recognition model result

Label	Name	Precision	Recall	F1 Score	Total Number
0	Disgust	87.96%	91.05%	89.48%	313
1	Happy	98.58%	97.35%	97.96%	2419
2	Neutral	92.16%	93.03%	92.62%	1112
3	Sad	83.45%	84.03%	83.74%	288
4	Surprise	90.24%	91.74%	90.98%	363

Furthermore, confusion matrices of the recall percentages for the testing sets were also generated (see Fig. 4). The results indicate that the models successfully accomplish the task of recognizing six student behaviors and five student facial emotions, respectively.

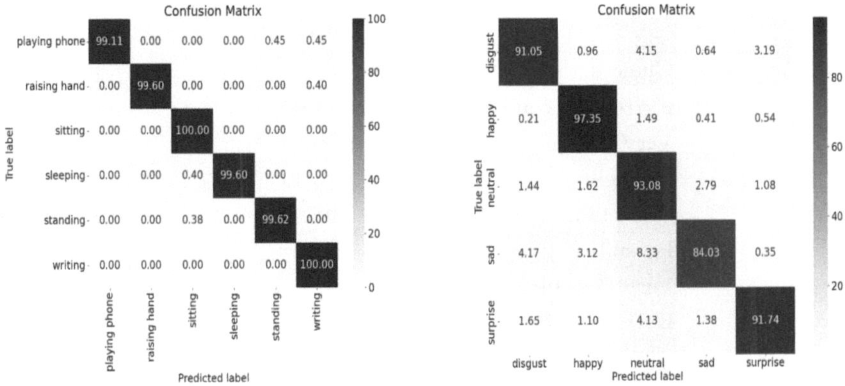

(a) Confusion matrix for behavior recognition

(b) Confusion matrix for emotion recognition

Fig. 4. Confusion matrices.

4.2 Experiments for Qwen2.5-Max with RAG Individual Learning States Recognition Method

(1) Experiments for the Certainty Factor (CF) Model

To determine the CF value of all behaviors and facial emotions, 30 online questionnaires via the 'Wenjuanxing' platform were distributed. The result of the survey for the behaviors part of the "Engaged" learning state is shown in Table 3. The survey questions of the remaining three learning states follow a similar format. The survey results were then converted into CF values. Each option is assigned a certainty factor (CF) value. Two assigning standards were developed. Standard 1 equally divides the $[-1, 1]$ interval into six levels. Standard 2 emphasizes the distinction between clear and uncertain states using ± 1.00 and ± 0.80 for strong tendencies. The exact CF values are presented in Table 4.

Table 3. When students exhibit the following behaviors in class, what are the possibilities that they are "focusing on the course content"?

Behaviors\Choices	Extremely Unlikely	Very Unlikely	Unlikely	Uncertain	Possible	Very Possible	Extremely Possible
Playing Phone	18	3	1	1	5	0	2
Raising Hand	0	2	0	0	2	18	8
Sitting	1	2	0	21	3	2	1
⋮	⋮	⋮	⋮	⋮	⋮	⋮	⋮

Table 4. Standards for mapping questionnaire options to certainty factor values

Choices	Standard 1	Standard 2
Extremely Unlikely	–0.99	–1.00
Very Unlikely	–0.66	–0.80
Unlikely	–0.33	–0.40
Uncertain	0	0
Possible	0.33	0.40
Very Possible	0.66	0.80
Extremely Possible	0.99	1.00

Once a scoring standard is selected, the certainty factor for each behavior/emotion under each of the four dimensions is calculated using

$$CF = \frac{\sum_{i=1}^{N} s_i \cdot n_i}{N}, \quad (3)$$

here, CF is the Certainty Factor, N is the total number of valid survey responses, s_i is the CF value assigned to option i (from Table 4) and n_i is the number of participants who chose that option. Using this approach, the CF values for all behaviors and emotions were computed, and part of the results of behaviors are shown in Table 5.

Table 5. 4D state space CF values of behaviors under different standards

Behaviors	Engaged		Active		Disciplined		Clear	
	Std. 1	Std. 2	Std. 1	Std. 2	Std. 1	Std. 2	Std. 1	Std. 2
Playing Phone	–0.70	–0.71	–0.68	-0.71	–0.67	–0.68	–0.43	–0.41
Raising Hand	0.76	0.86	0.69	0.79	0.68	0.73	0.37	0.44
Sitting	0.04	0.05	–0.08	–0.09	0.27	0.33	0.11	0.13
⋮	⋮	⋮	⋮	⋮	⋮	⋮	⋮	⋮

Under Standard 2, the CF values for"raising hand," "standing," and "writing" exhibited an increase, while other values remained relatively constant. Consequently, Standard 2 is more closely aligned with practical expectations and is thus adopted for behaviors in the CF model. Regarding emotions, CF values under Standard 2 are generally higher than those under Standard 1. To determine an appropriate standard for CF modeling of emotions, the combined CF values of all behavior-emotion combinations were calculated using Standard 2 for behaviors and both standards for emotions. To address scenarios where the model fails to detect behaviors or emotions (due to occlusion, etc.), an "unknown" category was introduced for both behaviors and emotions, with CF values assigned as 0. Furthermore, when students are asleep, emotions are undetectable; hence, only behavior CF values are applicable. In total, 37 behavior-emotion combinations are considered. Subsequently, based on the judgment rule provided in 3.3, the learning state corresponding to each behavior-emotion combination can be determined. The results when the student is playing on the phone are shown below:

Individual Student Learning States for 3 of the 37 students

```
Student 1: The student is playing phone with a disgust expression. The learning
state is: detached, silent, disruptive, and confused.
Student 2: The student is playing phone with a happy expression. The learning
state is: unknown engagement, unknown interactivity, disruptive, and unknown
comprehension.
Student 3: The student is playing phone with a neutral expression. The learning
state is: detached, silent, disruptive, and unknown comprehension.
```

A comparison of the learning states determined using Standards 1 and 2 reveals only minor differences. These differences occur primarily when emotions are "happy" or "surprised." Given that Standard 2 results better match common sense, it is used for emotions in the CF Model. The evaluation results of individual learning states under the 37 behavior-emotion combinations are used as the standard and benchmark of the individual learning state recognition.

(2) Experiments for Two Types of Individual learning state recognition

In our research, to assess the accuracy of recognition 1 and recognition 2, both systems were used to evaluate the individual learning states for all 37 behavior-emotion combinations, and the evaluation results were compared with the benchmark results given by the CF model. The comparison reveals that the results of recognition 1 are exactly the same as the benchmark results, indicating that it is capable of evaluating the individual learning states based on the standards of this study. However, the results of recognition 2 show poor performance. The calculation of its accuracy is shown below and involves 37 behavior-emotion combinations. Each combination is evaluated along four dimensions of the 4D state space, resulting in a total of $37 \times 4 = 148$ dimension-level evaluation outcomes. Upon comparing the results produced by recognition 2 with those derived from the CF Model, it was found that 52 evaluation results were not the same. This corresponds to an accuracy of $(148 - 52)/148 \times 100\% = 64.9\%$. This can be mainly attributed to the nature of LLM training: most models are trained with probabilistic objectives that prioritize fluency and coherence over formal logical accuracy. Consequently, LLMs often understand symbols and equations in terms of linguistic context, rather than formal semantics. This can lead the model to violate strict evaluation criteria, relying instead on semantic intuition. Based on the above experiment, recognition 1 was adopted as the individual learning state recognition.

4.3 Experiments for Qwen2.5-Max with ReAct-Agent Group Learning State Recognition Method

In this study, given that the students' behaviors and emotions remain relatively consistent across frames, only one single frame was selected from each simulated classroom scenario video clip to represent the classroom environment. Both the system and the human volunteer evaluated the group learning states and offered feedback to the teacher. To maintain objectivity, the volunteer was not informed of the specific evaluation criteria or principles employed in the study.As the main focus of this study is on remote classes, the following part only shows the results of the simulated remote class:

Proposed Method Output:

Group Learning States for the Remote Classroom

```
Some students showed signs of distraction, low interactivity, or poor discipline
(e.g., phone use and sleeping), suggesting a dull classroom atmosphere and low
participation. Others had neutral or surprised emotions, hinting at confusion or a
lack of interest. This is common in remote classes where distance reduces engagement
and interactivity.
```

Recommendations for the Teacher

```
(1) Enhance interactivity: Use questions, discussions, or mini-games to engage
    students. Ask those with surprised or neutral emotions if they are confused, and
    offer explanations.
(2) Improve focus: For distracted students, use name calling or assign tasks such as
    note-taking to regain their attention.
(3) Adjust pacing: Allow time for students to digest the content. Encourage
    questions and offer rewards for well-behaved students to set positive examples.

These methods can boost classroom engagement and help students better absorb
materials.
```

Volunteer Evaluations:

Group Learning States

```
Most students seemed to be disengaged. Few participants showed clear participation,
and many looked tired or unmotivated. Although the room remained orderly, overall
energy and attention were low.
```

Recommendations for the Teacher

```
(1) Use live Q&A or short quizzes: These can help refocus students and keep them on
    track during lessons.
(2) Utilize on-site teaching assistants: They can assist in managing classroom
    behavior and offering support where needed.
(3) Collect regular student feedback: Use feedback to adapt teaching strategies and
    improve engagement in low-energy classrooms.
```

By comparing the system results with those provided by the volunteer, it can be observed that the length and keynotes of both the group learning states and instructional suggestions are largely consistent, with only minor differences in detail. This demonstrates that the group learning state recognition is capable of providing precise and specific evaluations and feedback to assist teachers in better coordinating the learning progress of local and remote classes. However, compared with the suggestions provided by the volunteer, the system's teaching suggestions for remote classes might be less practical than the suggestions given by the volunteer, as it did not effectively incorporate the specific characteristics of the remote class. In future work, more suggestions tailored to the characteristics of the remote class should be added to the knowledge base, such as assigning on-site teaching assistants or introducing live Q&A proposed by the volunteer.

5 Conclusion

In this study, a series of methods for evaluating the learning states of students in the synchronous remote class are proposed. First, a student behavior (action) dataset combining custom-made images is used to train the ResNet34 model for behavior (action) recognition, achieving an accuracy of 99.7% across six behavior categories. Also, the Emotion-Domestic (Asia) Expression Recognition dataset is used to train the MobileNetV2 model for emotion recognition, achieving an accuracy of 95% across five facial emotion categories. Second, a 4D state space for evaluating students' individual learning states is proposed, and Qwen2.5-Max with RAG individual learning states recognition using the metrics of the 4D state space is established, resulting in an accuracy of 100%. Finally, Qwen2.5-Max with ReAct-agent group learning state recognition method for the synchronous remote class is established, which can accurately evaluate the learning state of both local and remote classes while providing detailed instructional suggestions to the teacher accordingly. In summary, the methods proposed in this paper help teachers in synchronous remote teaching, especially the teaching of remote classes, and effectively enhance the education quality of synchronous remote classes.

References

1. Baidu AI Cloud: Human body analysis (2022). https://intl.cloud.baidu.com/product/body.html
2. Chen, J., Xiao, S., Zhang, P., Luo, K., Lian, D., Liu, Z.: Bge m3-embedding: multi-lingual, multi-functionality, multi-granularity text embeddings through self-knowledge distillation (2024). https://arxiv.org/abs/2402.03216
3. He, K., Zhang, X., Ren, S., Sun, J.: Deep residual learning for image recognition (2015). https://arxiv.org/abs/1512.03385
4. Heckerman, D.: The certainty-factor model. Technical report, University of Southern California. https://www.microsoft.com/en-us/research/wp-content/uploads/2016/11/The-Certainty-Factor-Model.pdf
5. Johnson, J., Douze, M., Jégou, H.: Billion-scale similarity search with gpus (2017). https://arxiv.org/abs/1702.08734
6. Lewis, P., et al.: Retrieval-augmented generation for knowledge-intensive nlp tasks (2021). https://arxiv.org/abs/2005.11401

7. Sandler, M., Howard, A., Zhu, M., Zhmoginov, A., Chen, L.C.: Mobilenetv2: inverted residuals and linear bottlenecks (2019). https://arxiv.org/abs/1801.04381
8. Shao, S., Li, S., Hirota, K., Dai, Y.: An automatic evaluation method for students' attention in remote classrooms. Trans. Beijing Inst. Technol. **44**(5) (2024)
9. Shortliffe, E.H., Buchanan, B.G.: A model of inexact reasoning in medicine. Math. Biosci. **23**, 351–379 (1975)
10. Team, Q.: Qwen2.5 technical report. arXiv preprint arXiv:2412.15115 (2024)
11. Wei, J., et al.: Chain-of-thought prompting elicits reasoning in large language models (2023). https://arxiv.org/abs/2201.11903
12. Yan, F., Wu, N., Iliyasu, A.M., Kawamoto, K., Hirota, K.: Framework for identifying and visualising emotional atmosphere in online learning environments in the covid-19 era. Appl. Intell. **52**(8), 9406–9422 (2022). https://doi.org/10.1007/s10489-021-02916-z
13. Yang, L., Mersha, B.W., Dai, Y., Hirota, K., Dai, W., Lin, Y.: Integrated fuzzy atmosfield and machine learning to evaluate group learning status in smart classroom. In: ISCIIA and ITCA 2024, Beijing, China (2024)
14. Yao, S., et al.: React: synergizing reasoning and acting in language models (2023). https://arxiv.org/abs/2210.03629
15. Zhao, T., Mersha, B.W., Dai, Y., Hirota, K., Dai, W., Lin, Y.: Classroom engagement evaluation using decision-level fusion of student behavior and emotion recognition. In: ISCIIA and ITCA 2024, Beijing, China (2024)

Smart Healthcare and Rehabilitation

Deep Learning-Based Quantification of Lumbar Disc Herniation on High-Resolution Magnetic Resonance Imaging

Lizhong Ding[1], Wenxin Chen[2], Qina Wu[1], Hongdian Zhu[1], Changsheng Li[1,3,4], Jing Zhao[1,3,4], and Xingguang Duan[1,3,4](✉)

[1] School of Medical Engineering, Beijing Institute of Technology, Zhuhai, Zhuhai 519088, China

[2] School of Medical Technology, Beijing Institute of Technology, Beijing 100081, China

[3] School of Mechatronical Engineering, Beijing Institute of Technology, Beijing 100081, China

[4] Key Laboratory of Biomimetic Robots and Systems, Beijing Institute of Technology, Ministry of Education, Beijing 100081, China

duanstar@bit.edu.cn

Abstract. This paper focuses on the segmentation of lumbar spine high-resolution magnetic resonance imaging. We designed an end-to-end deep learning-based model for automatic segmentation. Additionally, our method includes the automatic quantification of the herniated disc by incorporating acquisition parameters. We collected high-resolution magnetic resonance imaging from 17 patients with lumbar disc herniation. Based on deep learning techniques, we designed a convolutional neural network that accepts 3D inputs. The segmentation results of our model show high similarity to those obtained through manual segmentation, with the mean dice coefficient exceeding 0.8. By incorporating the acquisition parameters of magnetic resonance imaging, the automatic quantification method has an accuracy comparable to that of manual segmentation. Our research demonstrates that the designed deep learning model can reliably extract key features and reconstruct critical structures. Our method has the potential to be allowed for potential routine reporting in the clinical setting.

Keywords: deep learning · lumbar disc herniation · high-resolution magnetic resonance imaging · medical image segmentation

1 Introduction

Spine is one of the most important skeletal structures in the human body, consisting of multiple vertebrae and intervertebral discs. The intervertebral discs, composed of the annulus fibrosus and nucleus pulposus, absorb and distribute

H. Ma et al. (Eds.): IWACIII 2025, CCIS 2780, pp. 153–164, 2026.
https://doi.org/10.1007/978-981-95-6730-0_11

the impacts experienced during various activities. Lumbar disc herniation (LDH) is a common disease with a high prevalence worldwide [1]. LDH occurs when the annulus fibrosus ruptures, allowing the nucleus pulposus to protrude through the fissure and compress nearby nerves [2]. Clinically, the symptoms of LDH are diverse, in addition to the typical low back pain and lower extremity radiating pain (sciatica), some patients may also present with numbness, muscle weakness, and even bowel and bladder dysfunction in severe cases, severely affecting the quality of patient's life.

Medical imaging technology plays an important role in diagnosing lumbar disc diseases [3,4]. X-rays can preliminarily evaluate the structure and stability of the lumbar vertebrae but are less effective in diagnosing LDH. Additionally, although computed tomography (CT) scans offer expedited imaging, magnetic resonance imaging (MRI) provides clearer imaging of the intervertebral discs, nerves, and surrounding soft tissues [5], and can relatively accurately display the location, size, and degree of compression of the herniated disc. Therefore, MRI is generally considered the preferred imaging modality for assessing lumbar disc diseases. However, the main issue with MRI is its lengthy procedure, necessitating an optimal balance between time and effectiveness. Standard MRI typically includes several axial images of the lumbar discs, combined with localizer images to assess severity. This limited number of slices may miss critical parts of the lumbar discs, especially when herniation occurs at different levels, posing a risk of misdiagnosis.

To overcome this limitation, high-resolution MRI (HR-MRI) has been introduced in clinical practice. HR-MRI involves more slices with thinner slice thickness, aiding in more accurate diagnosis [6]. However, HR-MRI takes longer to complete and is not widely adopted, being reserved for specific situations [7]. Despite its detailed imaging, HR-MRI presents challenges. Analysing HR-MRI images requires significant expertise, as surgeons must identify abnormalities from multiple angles and layers, constructing a 3D understanding of the surgical area. This can be particularly challenging for novice surgeons who lack the extensive experience needed to handle and analyze HR-MRI images effectively.

To address the aforementioned challenges, deep learning has shown powerful capabilities in medical image processing. Deep learning solves complex problems by mimicking the cognitive processes of the human brain [8]. Xu et al. designed a deep learning-based model to achieve the grading and location of lumbar disc herniation on magnetic resonance imaging. However, the model only achieved the location of the herniated disc in the image and could not finely divide and quantify its edges [9].

Medical image segmentation is a crucial aspect, aiming to isolate and identify voxel-level positional information of target tissues from images [10]. With adequate training, deep learning models can learn expert-level medical knowledge from a large amount of medical images, enabling the automatic extraction of image features [11]. Compared to natural images, medical images are expensive to obtain and require experienced surgeons to spend considerable time on annotation [12–15]. Despite these challenges, deep learning technology holds great potential for medical image segmentation [16]. In the context of lumbar disc disease, some researchers have also explored deep learning applications. For example, VerSe is a dataset specifically designed for spinal segmentation [17–19]. Similarly, CTSpine1K is a large-scale spinal CT dataset, larger than VerSe, with manually annotated vertebrae [20]. However, these CT datasets lack soft tissue annotations due to the limited capability of CT imaging to capture soft tissues.

SPIDER is a large-scale MRI-based spinal dataset that includes annotations for the spine, intervertebral discs, and spinal canal. However, SPIDER's MRI voxel spacing is relatively large (median 3.3, 0.63, 0.66), which results in insufficient imaging details for specific intervertebral discs. Additionally, SPIDER does not provide annotations for herniation, limiting the ability to quantitatively assess disease severity. Another spinal MRI dataset, MRSpineSeg, contains 172 samples, each of which also includes annotations for the spine, intervertebral discs, and spinal canal. Similarly, MRSpineSeg suffers from relatively large MRI voxel spacing (median 4.4, 0.34, 0.34), making it difficult to quantitatively evaluate the severity of the disease.

Regarding spinal segmentation algorithms, Pang et al. collected 215 T2-weighted MR volumetric images from a local hospital and manually annotated them [21]. They developed a dual-stage spinal segmentation framework based on a 3D graph convolutional segmentation network and a 2D ResUNet. This approach achieved a mean dice similarity coefficient of $87.49 \pm 3.81\%$ across all 19 spinal structures, indicating excellent segmentation performance. Using this MRI dataset, Huang et al. developed a semi-supervised learning method to utilize unlabeled data, achieving higher segmentation accuracy [22]. Additionally, a CTAM module was designed to efficiently fuse intra- and inter-slice features. Chang et al. leveraged global information and trained a spatial graph convolutional network in an end-to-end manner, achieving multi-vertebrae segmentation from MRI in a single step [23]. However, this method only segmented the vertebrae and did not include soft tissue segmentation. In addition, He et al. proposed a lightweight convolutional neural network (CNN) for spinal segmentation, named SALW-Net, which is based on a self-adjusting loss function. This model has application potential in scenarios with low computing power [24]. Furthermore, H. M"oller et al. proposed a two-stage automatic segmentation model for spinal MRI named SPINEPS. They conducted experiments using public datasets and a subset collected from the German National Cohort (NAKO). Benefiting from the sliding window mechanism and the approach of using a second model to generate instance masks from the semantic model, this model demonstrated relatively good segmentation performance in terms of the Dice metric. [25]. However, neither SALW-Net nor SPINEPS has conducted experiments specifically targeting high-resolution MRI (HR-MRI).

Due to the limited adoption of HR-MRI, there are currently no publicly available datasets of this type. Similar MRI datasets lack annotations for herniated discs. Therefore, the goal of this paper is to develop a deep learning-based method to automatically evaluate the volume of herniated discs.

The major contributions of this paper include the following:

- We demonstrated the accuracy of our deep learning-based algorithm for segmenting the tissues surrounding the lumbar intervertebral discs on high-resolution MRI.
- We designed a method to automatically calculate the volume of herniated discs by incorporating image acquisition parameters.
- Experiments on the instrument segmentation dataset demonstrate that our designed automatic segmentation algorithm produces results highly similar to those obtained from manual annotation.

2 Material and Methods

Creating an automatic method for calculating the volume of herniated lumbar intervertebral discs involves several key steps: data collection, data preprocessing, model design, model training, and volume calculation.

2.1 Dataset

This work was approved by the Spine Surgery Department of Peking University People's Hospital. The authors maintained complete control over all patient information and imaging data throughout the study. From July 27, 2023, to October 9, 2023, the Spine Surgery Department of Peking University People's Hospital conducted HR-MRI scans on 17 eligible patients. The MRI sequence used was T2-mx3d-spair, with a magnetic field strength of 1.5 T. The lumbar MRI were stored as Digital Imaging and Communications in Medicine (DICOM) files. Due to the small size of the dataset, the model training process employed cross-validation. The dataset was divided into a training set and a test set, with 13 patients in the training set and 4 patients in the test set. For the MRI of the 13 patients in the training set, experienced clinicians conducted manual segmentation using the ITK-SNAP software [26]. The anatomical tissue segmentation labels of the training sample data used in this paper are manually annotated by the same senior medical expert. Therefore, the sample labels are consistent and valid. This medical expert does not participate in the training and prediction of our model. Four objects were segmented: vertebra, normal Disc, nerve, and Herniated disc. The window width and window level were adjusted according to different image features to improve the accuracy of manual segmentation. The results of manual segmentation, along with the HR-MRI, constituted the dataset used for training the deep learning model. Figure 1 shows sagittal plane images of the surgical area of some patients in the dataset along with the corresponding manual segmentation results.

To automatically extract multiple tissues from HR-MRI, preprocessing of the data is necessary to meet the requirements for model training. Data preprocessing involves three main steps: resampling, image cropping, and normalization.

Different imaging devices and scanning parameters can result in differences in image spacing, potentially impacting the effectiveness of segmentation algorithms. Convolutional neural networks do not consider the spacing between data

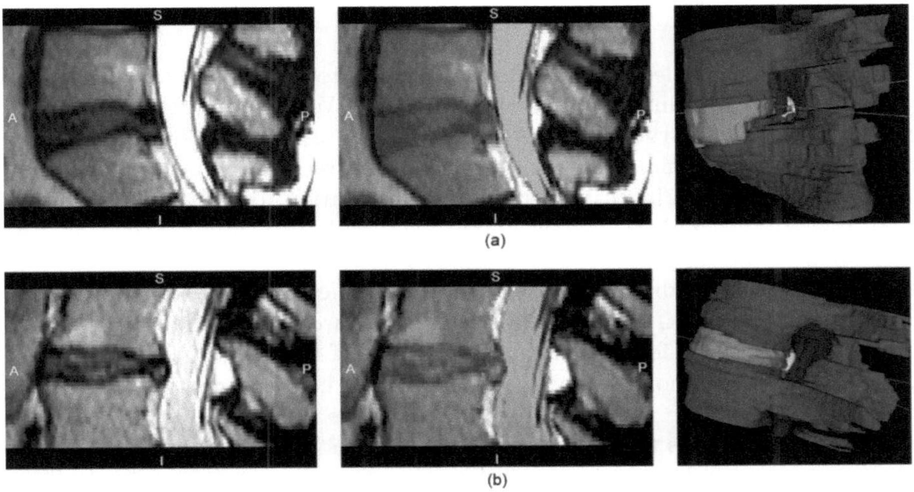

Fig. 1. Two sagittal plane images of the lumbar spine for two patients. The first column depicts the original images. The second column illustrates the results of manual segmentation. The third column displays a 3D representation of the manually segmentation.

when processing data; instead, they directly apply convolution and other operations to the input. Resampling ensures consistent spatial resolution of image data.

In this study, we resampled the image data in the dataset to a median spacing of [1.04, 0.5, 0.5]. To make the processed data more realistic, we utilized trilinear interpolation for resampling. Trilinear interpolation involves constructing piecewise cubic polynomials to fit the resampling target points. This method ensures the continuity of the first and second derivatives of the curve, thereby resulting in smoother fitting results.

The HR-MRI data from scans consist of dozens of DICOM files, and the file sizes are large. For deep learning models, it is not feasible to input the entire high-resolution scan directly. In the data preprocessing process, we divide the image data into multiple small patches. In this work, the shape of our patches is [28, 256, 256]. Although this strategy sacrifices the ability to obtain global information, dividing the patches can accelerate the model training process.

In medical image processing, data standardization is an essential step. Standardization enhances data quality by improving contrast and helps unify the distribution of data across different devices or scanning conditions. Z-score normalization transforms data into a standard normal distribution with a mean of 0 and a standard deviation of 1. The z-score value of each data point directly reflects its position relative to the mean. Z-score normalization ensures the consistency of image data during both model training and inference processes.

2.2 Method

This study utilizes a deep learning approach to automate the calculation of the volume for herniated lumbar intervertebral discs. We achieve automatic feature extraction and reconstruction of key tissues in high-resolution MRI scans by constructing a convolutional neural network. The accuracy of the proposed method is validated by comparing it with manual segmentation results.

In this study, we utilized UNet [27] to train and infer various tissues in HR-MRI of the lumbar spine, as illustrated in Fig. 2. UNet is a widely used deep learning model for medical image segmentation. UNet comprises an encoder and a decoder. In this work, we designed 7 levels of feature extraction in the encoding process. Each level of feature extraction consists of two convolutional modules. Each convolutional module includes convolution, instance normalization, and Leaky ReLU. Instance normalization calculates the mean and variance of each sample and normalizes the sample using these statistics. Acting as an activation function, Leaky ReLU introduces non-linearity into the model. By introducing a non-zero slope, Leaky ReLU ensures that even if the input is negative, the neuron can still receive gradient signals, thus preventing neuron death. In the encoding process, the shape of the feature maps is halved after each level of feature extraction. Due to the specific nature of image segmentation tasks, the decoding stage necessitates the continuous use of transposed convolution to restore the shape of the feature maps. Subsequently, two convolutional modules are employed to refine the upsampled feature maps.

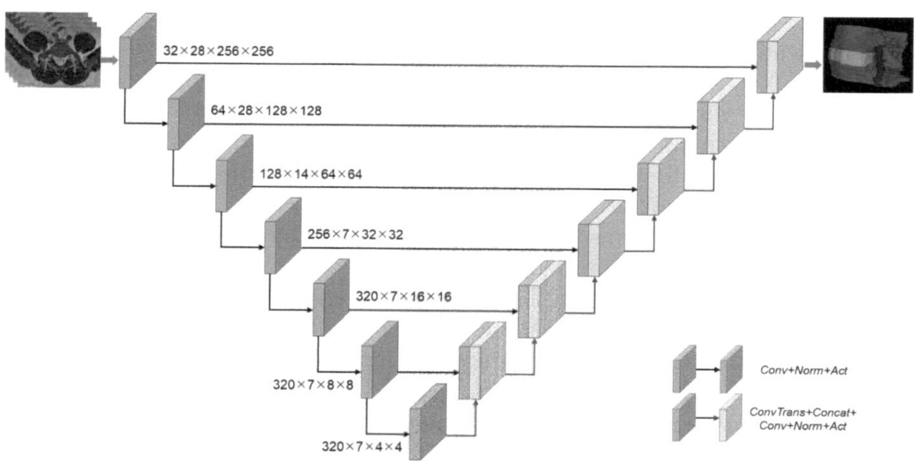

Fig. 2. Deep learning model designed in this study takes lumbar MRI images as input and generates automatic segmentation results as output.

Next, let's delve into the details of model training. The accuracy of the deep learning model developed in this study is validated by comparing it with manual segmentation. Our loss function combines the dice loss and cross-entropy loss. The cross-entropy loss calculates the difference between the probability distributions of the model output and manual segmentation. The dice loss helps alleviate the training difficulties caused by sample imbalance in the dataset. After computing the loss function, an efficient optimizer is selected to minimize the loss value and guide the parameter updates of the model. In this study, we opt for stochastic gradient descent (SGD) as the optimizer. To mitigate overfitting, we incorporate weight decay regularization. By adding a penalty term proportional to the magnitude of the weights, the model is encouraged to learn smaller weight values, thereby enhancing its generalization ability.

Due to the relatively small size of the dataset used for lumbar spine MRI segmentation, the separate validation set was not divided during model training. Instead, we utilize 5-fold cross-validation to evaluate the model's generalization ability. Initially, the 13 sets of data in the training set are randomly divided into subsets with proportions of 3:3:3:2:2. Subsequently, the model undergoes 5 rounds of training and validation. In each round, one subset is selected as the validation set, while the remaining 4 subsets are combined to form the training set for model training. As a result, we obtained five sets of different model parameters and their corresponding performance metrics.

During the inference process, we employ an ensemble approach to consolidate the test results. The test image is inputted into each of the 5-fold cross-validation models separately, and the intersection of the predictions from the five models is taken as the final output. Typically, there is only one main subject for each tissue in medical images. Base on this prior knowledge, we perform post-processing operations on the output results. We filter out all small regions except for the largest connected region to reduce the occurrence of false positives.

After obtaining the segmentation results, we quantify the volume of the herniated lumbar intervertebral disc portion by integrating the acquisition parameters from the HR-MRI. Next, let's introduce the quantification method. Firstly, we extract the acquisition parameters from the HR-MRI and record the spacing in the three key directions most relevant to this study. In the segmentation results, the volume of each annotated region is calculated by multiplying the number of voxels by the voxel spacing (1):

$$V_{voxel} = d_x \times d_y \times d_z \tag{1}$$

where d_x, d_y, d_z are the spacings in the three directions of the data. Next, we iterate through all the lumbar intervertebral disc portions identified as herniated in the segmentation results and calculate the volume contribution of each voxel (2):

$$V(i,j,k) = T(i,j,k) \times V_{voxel} \tag{2}$$

where $T(i,j,k)$ is a boolean function indicating whether the voxel at position (i, j, k) is the part of the herniated disc. Let i, j, k denote the index ranges of the voxels along the x, y, and z axes, respectively. Finally, we calculate the total volume of the herniated lumbar intervertebral disc (3):

$$V = \sum_{i=1}^{I} \sum_{j=1}^{J} \sum_{k=1}^{K} V(i, j, k) \tag{3}$$

By manual segmentation, we have constructed a dataset for automatic volume calculation. Using a deep learning approach, we trained a convolutional neural network to automatically extract and reconstruct the features of lumbar intervertebral disc. Subsequently, we combined the acquisition parameters of the images to calculate the volume of the herniated disc.

3 Results

During the training of the deep learning model, we utilized an RTX 4090 GPU with 24 GB of memory, running the PyTorch framework. The Python version used was 3.10. Throughout the training process, we recorded the dice coefficients and IoU (Intersection over Union) of various tissues in the validation set and calculated the average metrics for each tissue. The model's performance is presented in Table 1, with the top performers highlighted in bold.

Table 1. Validation set performance for 5-fold cross-validation

Fold	Vertebra		Normal Disc		Nerve		Herniation		Mean	
	Dice	IoU	Dice	IoU	Dice	IoU	Dice	IoU	Dice	IoU
0	0.8957	0.8112	0.8215	0.7002	0.8535	0.7460	0.6682	0.5074	**0.8097**	**0.6912**
1	0.8792	0.7844	0.8360	0.7239	0.7833	0.6491	0.3766	0.2782	0.7188	0.6089
2	0.8992	0.8177	0.8863	0.7959	0.8166	0.6908	0.5960	0.4314	0.7995	0.6840
3	0.8553	0.7479	0.7008	0.5408	0.8129	0.6851	**0.6902**	**0.5281**	0.7648	0.6255
4	0.9075	0.8307	0.8468	0.7351	0.8324	0.7131	0.5818	0.4137	0.7921	0.6731

From the data in the table, it can be observed that the top-performing model achieved average dice coefficients of over 0.8 for various tissues, with the mIoU also approaching 0.7. Overall, the model training yielded satisfactory results. Considering the individual tissue, the segmentation performance for vertebrae was the best, with a dice coefficient close to 0.9. Relatively, segmenting the herniated disc proved to be more challenging for the model, with the highest IoU reaching 0.53. In terms of segmentation metrics, the results produced by the deep learning algorithm exhibited good consistency with manual segmentation, establishing a foundation for quantifying the volume of the herniated disc.

To more intuitively evaluate the effectiveness of the automatic segmentation method, we randomly selected image data from the training set for visualization, as shown in Fig. 3. From the images, it can be observed that the segmentation

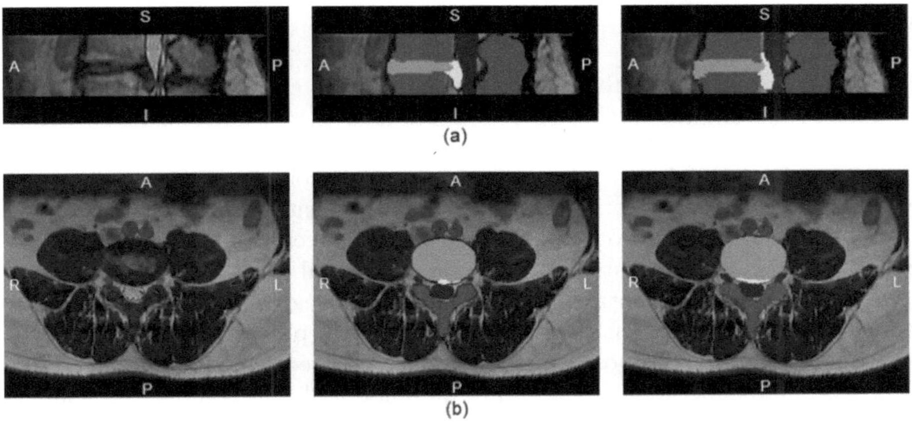

Fig. 3. Visualization of the two segmentation methods in different planes. The first column represents the original image. The second column represents the output from the deep learning model. The third column represents manual segmentation. (a) Sagittal plane. (b) Transverse plane.

method proposed in this study shows a high degree of similarity to manual segmentation, successfully achieving the goal of automation.

As we employed cross-validation for training in this study, we randomly selected 3 pairs of data from the training set. Utilizing the quantification method proposed in this paper, we verified the accuracy of segmentation from the imaging perspective. We recorded the volumes calculated by our method and those from manual segmentation. The absolute error and relative error of our method were computed, and the results are presented in Table 2.

Table 2. Volume comparison between deep learning model and manual segmentation

#	Deep learning model (mm^3)	Manual segmentation (mm^3)	$\Delta(mm^3)$	$E_r(\%)$
1	789.91	852.26	-62.35	-7
2	1312.75	1217.50	95.25	8
3	1324.70	1096.18	228.52	21

Analyzing the data in the table reveals that the automatic quantification method proposed in this paper aligns well with the results of manual segmentation. Among the randomly selected samples, the maximum absolute error is 228 cubic millimeters, and the maximum relative error is 21%. From an imaging perspective, the automatic quantification method proposed in this study shows high reliability.

4 Discussion

High-resolution magnetic resonance imaging enables clearer observation of the tissues surrounding the lumbar spine, but surgeons still need to mentally reconstruct the three-dimensional structure of the surgical area. To address this limitation, we have developed an automatic segmentation method based on deep learning and proposed a method for automatically calculating the volume of lumbar intervertebral discs. We applied deep learning techniques to automatically extract features and reconstruct lumbar intervertebral discs with accuracy comparable to manual segmentation methods. Manual segmentation often requires several hours. However, with the introduction of deep learning technology, the time required for our proposed method can be reduced to minutes or even seconds. Automated methods save a significant amount of time compared to manual methods.

The method proposed in this study has certain limitations. During the manual segmentation process, there is still room for improvement in annotating certain images. In future work, we aim to enhance the accuracy of manual segmentation by incorporating more patient image data. This will provide more precise references for automated segmentation. From a modeling perspective, there is significant room for improvement in the automatic segmentation of protruding lumbar intervertebral discs. Our next steps involve refining the deep learning model to enhance its capability to extract features of lumbar intervertebral discs.

During the experimental phase, this study analyzed imaging data from multiple patients, yielding both intuitive and objective experimental results. However, a potential limitation of this experimental approach is the absence of clinical validation. In future studies, we plan to record the volumes of protrusions removed during surgery and compare them with the results obtained using the automatic quantification method proposed in this paper. This comparison will provide more robust data support.

Despite the aforementioned limitations and the limited scale of HR-MRI, the experimental results demonstrate the effectiveness of the proposed method in this paper. We believe that our research holds significant potential for clinical application in the future.

5 Conclusion

In lumbar discectomy procedures, determining the volume of protrusions to be removed is challenging from a clinical perspective and heavily reliant on the surgeon's judgment. The high recurrence rate of disc herniation is often attributed to incomplete removal of herniated disc. To address this issue, we have developed a method for the automatic quantification of lumbar disc protrusion volume. This involved constructing a dataset for training through manual segmentation, building a convolutional neural network, designing training strategies, and training the model. Finally, we quantified the herniated disc based on imaging acquisition parameters. Through quantitative and qualitative experiments, we validated the accuracy of the deep learning method in quantifying protruding discs. The automated quantification method proposed in this paper provides quantitative references for clinicians, thereby reducing their workload.

Acknowledgments. The data that support the findings of this study are available on request from the corresponding author. The data are not publicly available due to privacy or ethical restrictions. This work was supported in part by the National Natural Science Foundation of China under Grant 62473053, and in part by the Beijing Natural Science Foundation under Grant L252173.

Disclosure of Interests. The authors declare no conflicts of interest.

References

1. Zhang, A.S., et al.: Lumbar disc herniation: diagnosis and management. Am. J. Med. **136**(7), 645–651 (2023)
2. Xia, W., et al.: The compensatory mechanisms for global sagittal balance in degenerative spinal kyphosis patients: a radiological analysis of muscle-skeletal associations. BMC Musculoskelet. Disord. **22**, 1–9 (2021)
3. Valarmathi, G., Devi, S.N.: Automatic localization and classification of intervertebral disc herniation using hybrid classifier. Biomed. Signal Process. Control **86**, 105291 (2023)
4. Cai, R., et al.: Deep-learning based segmentation of ultrasound adipose image for liposuction. Int. J. Med. Robot. Comput. Assis. Sur. **19**(6), e2548 (2023)
5. Gao, X., Zheng, G.: SMILE: Siamese multi-scale interactive-representation LEarning for hierarchical diffeomorphic deformable image registration. Comput. Med. Imaging Graph. **111**, 102322 (2024)
6. Sun, S., et al.: Evaluation of deep learning reconstructed high-resolution 3D lumbar spine MRI. Europ. Radiol. **32**(9), 6167–6177 (2022)
7. Hokamura, M., et al.: Super-resolution deep learning reconstruction approach for enhanced visualization in lumbar spine MR bone imaging. Eur. J. Radiol. **178**, 111587 (2024)
8. Han, Z., et al.: A respiratory motion prediction method based on LSTM-AE with attention mechanism for spine surgery. Cyborg and Bionic Sys. **5**, 0063 (2024)
9. Xu, Y., et al.: Deep learning model for grading and localization of lumbar disc herniation on magnetic resonance imaging. J. Magn. Reson. Imaging. **61**(1), 364-375 (2024)
10. Li, C., et al.: Advances of surgical robotics: image-guided classification and application. Natl. Sci. Rev., **11**(9,) nwae186 (2024)
11. Chen, W., et al.: Dual-stage semantic segmentation of endoscopic surgical instruments. Med. Phys. **51**(12), 9125–9137 (2024)
12. Li, C., Li, X., Wang, K., Chen, W., Liu, Q., Duan, X.: Self-supervised monocular depth estimation for endoscopic imaging. IEEE J. Biomed. Health Inf. (2024)
13. Qin, N., Wu, J., Liu, X., Lin, Z., Wang, Z.: MCRNet: Underwater image enhancement using multi-color space residual network. Biomimetic Intell. Robot. **4**(3), 100169 (2024)
14. Wang R., Zheng, G.: PFMNet: Prototype-based feature mapping network for few-shot domain adaptation in medical image segmentation. Comput. Med. Imaging Graph. 102406 (2024)

15. Nema, S., Vachhani, L.: Unpaired deep adversarial learning for multi-class segmentation of instruments in robot-assisted surgical videos. Int. J. Med. Robot. Comput. Assis. Surg. **19**(4), e2514 (2023)
16. Uesugi, K., Mayama, H., Morishima, K.: Analysis of rowing force of the water strider middle leg by direct measurement using a bio-appropriating probe and by indirect measurement using image analysis. Cyborg and Bionic Sys. **4**, 0061 (2023)
17. Sekuboyina, A., et al.: VerSe: A Vertebrae labelling and segmentation benchmark for multi-detector CT images. Med. Image Anal. **73**, 102166 (2021)
18. Löffler, M.T., et al.: A vertebral segmentation dataset with fracture grading. Radiol. Artif. Intell., **2**(4), e190138 (2020)
19. Liebl, H., et al.: A computed tomography vertebral segmentation dataset with anatomical variations and multi-vendor scanner data. Sci. Data **8**(1), 284 (2021)
20. Deng, Y., et al.: Ctspine1k: A large-scale dataset for spinal vertebrae segmentation in computed tomography. arXiv preprint arXiv:2105.14711 (2021)
21. Pang, S., et al.: SpineParseNet: spine parsing for volumetric MR image by a two-stage segmentation framework with semantic image representation. IEEE Trans. Med. Imaging **40**(1), 262–273 (2020)
22. Huang, M., Zhou, S., Chen, X., Lai, H., Feng, Q.: Semi-supervised hybrid spine network for segmentation of spine MR images. Comput. Med. Imaging Graph. **107**, 102245 (2023)
23. Chang, H., Zhao, S., Zheng, H., Chen, Y., Li, S.: Multi-vertebrae segmentation from arbitrary spine MR images under global view. In: Martel, A.L., et al. (eds.) MICCAI 2020. LNCS, vol. 12266, pp. 702–711. Springer, Cham (2020). https://doi.org/10.1007/978-3-030-59725-2_68
24. He, S., Li, Q., Li, X., Zhang, M.: SALW-Net: a lightweight convolutional neural network based on self-adjusting loss function for spine MR image segmentation. Med. Biol. Eng. Comput. **62**(4), 1247–1264 (2024)
25. Möller, H., et al.: SPINEPS–automatic whole spine segmentation of T2-weighted MR images using a two-phase approach to multi-class semantic and instance segmentation. Eur. Radiol. **35**(3), 1178–1189 (2025)
26. Yushkevich, P.A., et al.: User-guided 3D active contour segmentation of anatomical structures: significantly improved efficiency and reliability. Neuroimage **31**(3), 1116–1128 (2006)
27. Ronneberger, O., Fischer, P., Brox, T.: U-net: Convolutional networks for biomedical image segmentation. In: Medical Image Computing and Computer-Assisted Intervention – MICCAI 2015, Munich, Germany, vol. 18, pp. 234–241 (2015)

A Survey of Deep Learning Models from DCNNs to Transformers for Correcting MRI Artifacts and Assisting Smart Healthcare

Dingxi Liu[1,2], Lichuan Ning[1,2(✉)], and Yuanmin Xie[2,3]

[1] Center for Biomechanics and Intelligent Measurement and Control, Wuhan University of Science and Technology, Wuhan, Hubei 430081, China
ninglichuan@wust.edu.cn

[2] Key Laboratory of Metallurgical Equipment and Control, Ministry of Education, Wuhan University of Science and Technology, Wuhan 430081, Hubei, China
xieyuanmin@wust.edu.cn

[3] Hubei Key Laboratory of Mechanical Transmission and Manufacturing Engineering, Wuhan University of Science and Technology, Wuhan 430081, Hubei, China

Abstract. Artifacts in MRI images are major factors that cause image blurring and interfere with diagnosis. Advanced deep learning models are rapidly evolving and play a crucial role in smart healthcare. In particular, they improve the quality of MRI images corrupted by artifacts. However, only a few studies have systematically analyzed model architectures and usage methods for artifact correction. This study first proposes evaluation methods for correction performance, including both subjective and objective assessments. It then reviews recent progress in six types of state-of-the-art deep learning correction algorithms. The review highlights design strategies, structural innovations, and key aspects for future model development. Finally, the study discusses current technical challenges and outlines research directions for future advancements.

Keywords: Smart Healthcare · Deep Learning · Model Design · Structure Evolution · MRI Artifacts

1 Introduction

Artifacts appear in MRI images are not exist in the original clinical object, and caused by mechanical faults of imaging equipment or imaging-related human factors [1]. These artifacts are manifested in various forms and accompanied with different types of MRI scans. The geometric characteristics were exhibited such as pixel imbalance, rings [2], straight lines, blurriness, abnormal brightness, internal structural deformations. These features make artifacts similar to pathological forms that could lead to misdiagnosis and conceal vital physiological information. The severity of artifacts is proportional to the decrease in image quality.

Motion artifacts primarily appear in MRI due to patient or equipment movement, and present as blurring or distortion [3] of image details and regions. The causes include

H. Ma et al. (Eds.): IWACIII 2025, CCIS 2780, pp. 165–176, 2026.
https://doi.org/10.1007/978-981-95-6730-0_12

movement of organs such as the eyes [4], heart [5], liver [6], the motion of tissues such as brain blood flow [4], blood [7], the movements of the aorta, human respiration [8, 9], head movement [10, 11], arterial pulsation and swallowing [1], and patient body movements [12–14]. Metal artifacts caused by metal prostheses [15, 16] and metal implants [17] rarely appear in MRI as black or white streaks or spots. Ring artifacts appear in MRI because of insufficient high-frequency data sampling in the k-space domain [2], manifesting as concentric circles or arcs centered around the rotation axis.

The complex generation of artifacts and numerous factors pose difficulties for the correction methods. With the development of deep learning (DL) models, it is more operable and practical than modifying equipment, and easy to obtain higher performance. The technological advantages are manifold. First, they enable robust recognition of regions of interest (ROI). Moreover, they facilitate quantitative analysis of image features, which enhances the reliability of assessments. In addition, these technologies can provide partial assistance to experienced professionals, thereby improving efficiency in clinical practice. Finally, they demonstrate low dependence on specific imaging modalities, which increases their generalizability across different applications. Widely, these advantages have allowed deep learning models to assume a growing significance in the field of intelligent healthcare.

The main innovation of this paper are as follows:

1. This survey provides a comprehensive summary of the mechanisms, causes, and effects of MRI artifacts.
2. This survey provides a detailed summary of recent advancements and architectural evolution strategies of six of the most frequently used DL models for MRI artifact correction. The models include deep convolutional neural networks (DCNNs), convolutional autoencoders (CAEs), generative adversarial networks (GANs), attention mechanisms, transformers, and integrated models (IMs).
3. This survey proposes both subjective and objective evaluation methods for artifact correction, and outlines performance assessment metrics.

2 Related Overview of MRI Artifact Correction

2.1 Artifact MRI Correction Principle

Artifact correction is closely related to image reconstruction, and serves as a crucial technical approach to enhancing image quality. The simple linear imaging process can be represented as $y = Ax$, where the target image x passes through a domain transformation operation A to obtain the sensor-sampled data y. The reconstruction problem involves finding the inverse solution to this equation, as shown in Eq. (1), where F represents the mapping of data statistical characteristics. The clean image I_C is obtained by subtracting the reconstructed artifact image I_R by correction model in the image domain from the artifact-corrupted image I_P, as shown in Eq. (2).

$$x^* = \underset{x \in K^{N_x}}{\arg\min} F\left[A(x), y\right] \tag{1}$$

$$I_C = I_P - I_R = X - \underset{x \in K^{N_x}}{\arg\min} F\left[A(x), y\right] \tag{2}$$

2.2 DL-Based Correction Models and Their Structural Evolution

DL models are vital to extract and characterize artifact features by adopting correction principles. This approach enables the decomposition of complex relationships between artifacts and clean images, ultimately achieving automated artifact correction. The goal is to establish a feature mapping function F between an artifact-corrupted image dataset $X = \{x_1, x_2, \ldots, x_N\}$ and a clean image dataset $Y = \{y_1, y_2, \ldots, y_N\}$, such that the cost function F_C is minimized, as shown in Eq. (3).

$$F_C = \arg\min_F \frac{1}{N} \sum_{i=1}^{N} \|F(X) - Y\|_F^2 \tag{3}$$

Where, ||·|| is Frobenius norm, N is the number of data samples.

DCNN models lie in optimizing the combination of convolutional layers, pooling layers, fully connected layers, and activation functions. The dimensions of the model components are set according to the data. The basic convolution blocks (BCBs) adopt structures such as "Convolution + ReLU + Batch Normalization (BN)". The convolutional layers establish local connections with maps through convolution kernels, weight sets $W = \{W_1, W_2, \ldots, W_k\}$ and bias sets $B = \{b_1, b_2, \ldots, b_k\}$, extracting feature maps $X^{l\text{-}1}$, and transporting to next layer $X^l{}_k = f(W_k * X + b_k)$, as shown in Fig. 1 (a). The enhancement of DCNN correction performance can be summarized into three main strategies. ① Optimizing the architecture. ② Appropriate deepening of convolutional layers. ③ Rational design of functional modules.

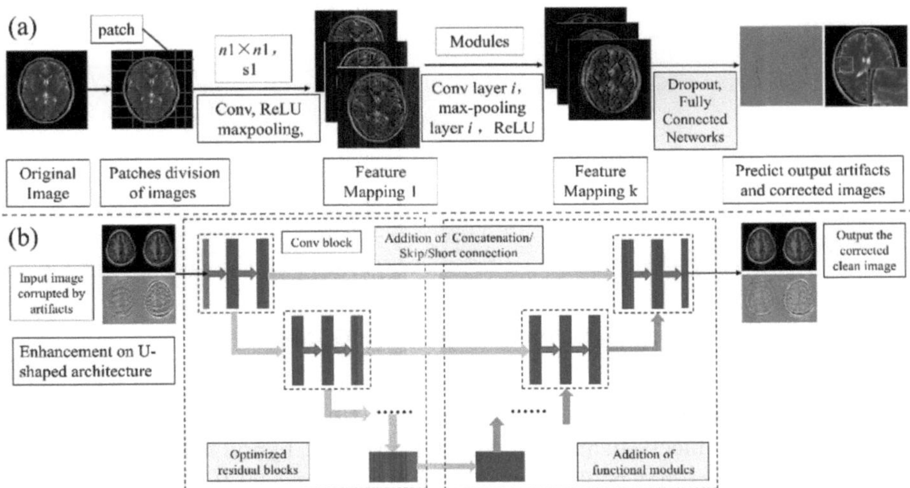

Fig. 1. Process for correcting MRI artifacts using DCNN.

A convolutional autoencoders (CAE) replaces the fully connected networks of a standard autoencoder with convolutional and pooling layers. The encoder and decoder are typically built using BCBs or residual structures. CAE-based correction strategies can be categorized into three types. ①Addition skip connections or concatenation layers (CAE-S1) similar to U-Net [18]. ② Enhanced U-Net-like CAE (CAE-S2). CAE-S2 optimizes the structure and adjusts the number of encoders and decoders. ③ Feature-enhanced CAE (CAE-S3). CAE-S3 incorporates additional functional modules to enhance the scale feature extraction. The correction process is illustrated in Fig. 1 (b).

The GAN-based correction model optimizes the structure of the generator (G) and discriminator (D) by following the optimization strategies of DCNN and CAE. GANs can be categorized into Convolutional GANs (ConvGANs), Conditional GANs (CGANs), and Autoencoder GANs (AGANs) [19]. ConGANs mean both G and D adopt a DCNN-based structure. A CGAN extends a ConvGAN by incorporating specified conditional information to guide image generation. In an AGAN, either G or D is upgraded from a convolutional block structure to a CAE-based structure. G learns the distribution of artifacts and generates reconstructed images and passes them to D. D selects the corrected images with the smallest loss, as illustrated in Fig. 2. GANs demonstrate strong capabilities in transforming medical images from one modality to another [20]. Consequently, they serve as a relatively universal solution for converting artifact-corrupted images into clean ones, with high effectiveness across multiple complex patterns.

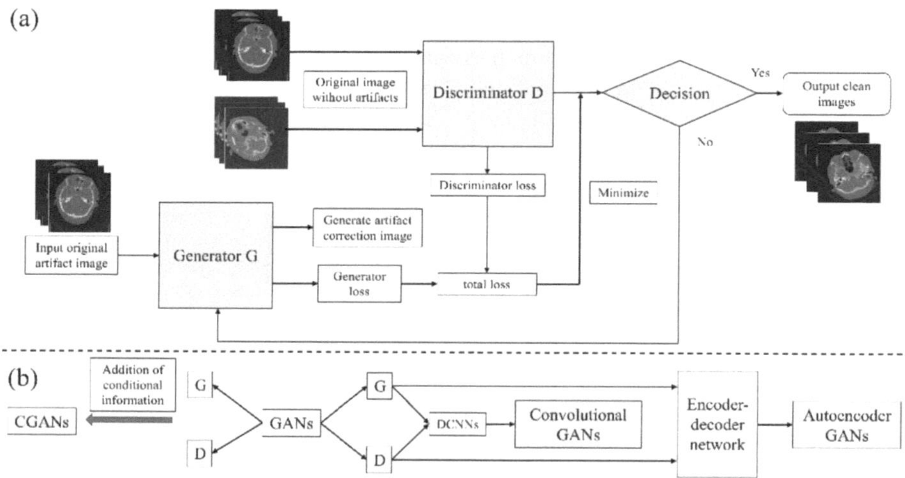

Fig. 2. Artifact correction by GANs.

A basic attention module typically includes spatial attention (SA), channel attention (CA), convolutional block attention module (CBAM) [21], hybrid attention, and so on. The CBAM integrates max pooling, average pooling, shared MLP, convolution layers, and element-wise multiplication to refine feature maps. It is formulated as $F_{CABM} = F_{SA}(F_{CA}(F)F) \otimes (F_{CA}(F) \otimes F)$. Attention mechanism architectures are closely related to residual structures and are often embedded as auxiliary modules in DCNNs, CAEs, and GANs. The model focuses on ROIs at spatial or channel levels. Attention mechanisms provide feature maps for feature fusion, learn the weight distribution of artifact features, and reinforce the model's supervision and focus on artifacts.

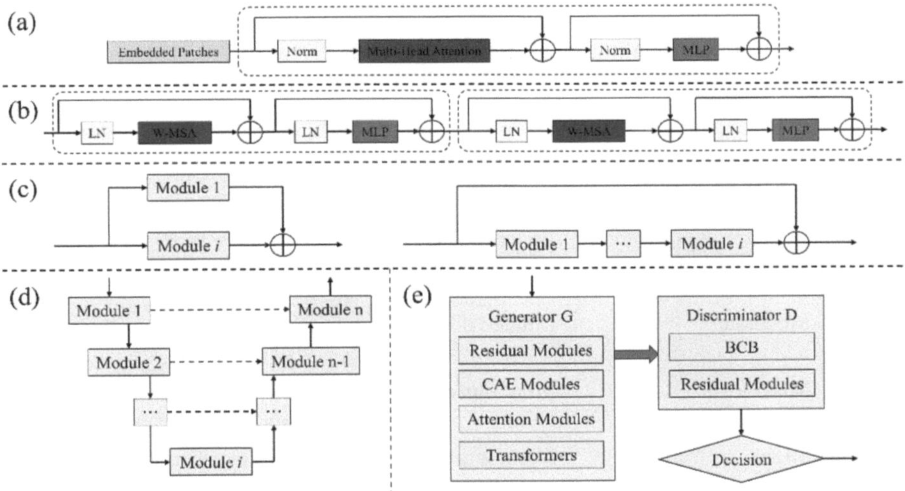

Fig. 3. Architecture of transformers and integrated models.

The Transformer [22], Vision Transformer (VIT) [23], and Swin Transformer Block (STB) [24] have been progressively applied to artifact correction, as shown in Fig. 3 (a) and (b). The core components of encoders and decoders include a Multi-Head Self-Attention (MHSA) and a Feed Forward Network (FFN) built on fully connected layers. By integrating Global Attention (GA) and Local Attention (LA), the Transformer computes the dot product between query vectors (Q) and key vectors (K) using weighted vectors (V). This architecture makes it a crucial module for extracting both global contextual dependencies and specific local features in MRI images, and significantly enhances computational efficiency and optimizes model architecture. For input data X, the self-attention derived feature scores are given in Eq. (4).

$$Attention(Q, K, V) = soft\max\left(\frac{QK^T}{\sqrt{d_k}}\right)V \tag{4}$$

Integrated Models (IM) are built upon the structures of DCNNs, CAEs, and GANs by integrating one or more modules such as BCBs, residual blocks, CAEs, attentions, or transformers within a model. Both the overall architecture of IMs and individual module designs can make full use of residual structures, CAEs, GANs, or their combination to form a comprehensive model. The design strategies of IMs can be categorized as Residual Convolution-IM (RC-IM), Convolutional Autoencoder-IM (CAE-IM), and Generative Adversarial Network-IM (GAN-IM), as shown in Fig. 3 (c), (d) and (e).

2.3 Performance and Evaluation of MRI Artifact Correction Algorithm

The quality of medical MRI images and the performance of artifact correction are assessed by measuring the difference between the artifact-corrupted images and the clean images.

The subjective evaluation was performed by the observer using the Likert scale for visual comparison and qualitative assessment of image quality [10, 13], as shown in Table. 1.

Table. 1. Correspondence between Likert scale score and artifact severity.

Score	Severity of Artifacts	Diagnostic Results	Structural Visibility
1	Prominent/Severe	Underdiagnosis	Little or no visibility of critical structures
2	Moderate	Limited diagnosis	Significantly reduce the visibility of key structures
3	Mild	Mild impact	Moderately reduction of the visibility of some structures
4	Slight	Slight impact	Slightly reduction of the visibility of some structures
5	Non-existent/Almost non-existent	Adequate diagnosis	Visibility of all expected structures

The objective evaluation of correction effectiveness involves acquiring the image ROI, calculating and recording the peak signal noise ratio (PSNR) [25] and structural similarity index measure (SSIM) [7], as shown in Eqs. (5 and 6) .

$$PSNR(p) = \log_{10}\left[\frac{N_x \bullet N_y \bullet \left(\mathrm{Max}\left(I_f(i,j)\right)\right)^2}{\sum_0^{N_x}\sum_0^{N_y}\left(I_f(i,j) - I_c(i,j)\right)^2}\right] \tag{5}$$

$$SSIM\,(p) = \frac{\left(2\mu_x\mu_y + c_1\right)\left(2\sigma_{xy} + c_2\right)}{\left(\mu_x^2 + \mu_y^2 + c_1\right)\left(\sigma_x^2 + \sigma_y^2 + c_2\right)} \tag{6}$$

Where N represents the intensity of regions x and y, μ denotes the HU of regions x and y, σ signifies the SD of regions x and y, and c_1 and c_2 are constants that are set according to the requirements.

3 Applications of DL Models in Artifacts Correction

3.1 Motion Artifacts Correction

The highlights and innovations of DL models for correcting MRI image artifacts are showed in Table. 2. The growth percentages of SSIM and PSNR were calculated and listed to quantify the correction effect of the model.

The dimension of the convolution kernel in a DCNN depends on the input MRI image dimensions [10]. As described in [6, 13], 3D MRI images were divided into patches prior to training. Grad-CAM attention was incorporated to enhance the feature extraction capability of the 3D MS-CNN [8].

Table. 2. Comparison of deep learning models for correcting MRI image artifacts.

Model	Type	Artifact	SSIM (↑)	PSNR (↑)
High-Res 3D Net [10]	DCNN	Motion	0.30	5.07
MSCNN [8]	DCNN	Motion	–	–
UDDN [26]	CAE	Motion	4.21	19.66
U-Net-style CNN [10]	CAE	Motion	0.30	5.07
Residual U-net [27]	CAE	Motion	24.09	61.60
ORBA-DL [14]	ConvGAN	Motion	46.07	13.66
3D cGAN [11]	CGAN	Motion	0.10	2.65
De-artifact network [29]	RC-IM	Motion	29.57	22.58
DCGAN-MS [25]	GAN-IM	Motion	14.67; 13.89; 6.33	19.16; 26.57; 4.99
MACS-Net [30]	CAE-IM	Motion	15.47; 166.77	45.67; 59.78
U-Net structural CNN [15]	CAE	Metal	4.38, 2.54; 4.78, 2.48; 6.55, 3.42	31.28, 28.39; 31.16, 25.82; 18.57, 9.40
GRA-CNN [2]	DCNN	Ring	0.61; 0.62; 0.72; 0.40; 0.20; 0.51	5.11; 3.64; 4.14; 3.67; 3.27; 6.57
FD-Net [31]	CAE	Susceptibility	18.50	6.72

The CAE model [4, 6, 10] uses CAE-S1 and adds skip connections to correct artifacts. An improved U-Net with nine convolutional layers [6] apply CAE-S2 and corrects liver artifacts. As for CAE-S3, UDDN [26] integrates attention-based CDAF module to capture and fuse spatial and frequency domain features. A U-shaped residual network [27] optimizes residual blocks to significantly correct 2D brain artifacts.

GANs are proved particularly effective for processing 3D complex artifacts, ORBA-DL [14] features a U-shaped G with four concatenation layers, and a D based on patch-GAN. 3D MoCo-cGAN [11] improves upon 2D pix2pix by employing a 3D U-Net as the G and using 3D convolutions for both G and D. A two-stage GAN [32] corrects dynamic contrast-enhanced liver artifacts. G employs the DRN-DCMB network, and D uses LPer to update G's weights. DRN-DCMB consists of three U-shaped super-resolution blocks for combining global information. The total loss function is given by Eq. (7).

$$L_{total} = L_{MSE} + \alpha L_G + \beta L_{Per} \quad (7)$$

The integrated model has been gradually applied by focusing on deep edge features and intensity artifact features. MC2-Net [28] employs the Transformer to fuse the features of artifacts. An edge-enhanced network [29] combines EFFE-Net to extract and aggregate edge features, and EFT-Net to focuses on edge information using Transformer.

MACS-Net [30] integrates STBs into every U-Net block, with each STB composed of eight connected Swin Transformer layers.

The technical advantages of the models are shown in Table. 3.

Table. 3. Comparison of the highlights and innovations of DL models.

Model	The Highlights and Innovations
High-res 3D Net [10]	The model excels at correcting mild to moderate Fourier-domain simulated motion artifacts.
MSCNN [8]	Six gradient-weighted CAM attention guide the DCNN to focus on the ROIs.
UDDN [26]	The model undergoes unsupervised training on unpaired data, with the CDAF enhancing the fusion of temporal and frequency-domain features.
U-Net-style CNN [10]	The model has relatively higher memory efficiency, improving correction quality.
Residual U-net [27]	The first proposed using a residual U-Net for correcting brain MRI motion artifacts.
ORBA-DL [14]	The model does not require paired data, preserving organ texture.
3D cGAN [11]	The model's training data consists of clean and artifact-corrupted image pairs, resulting in improved image quality after correction.
De-artifact network [29]	EFFE-Net enhances edge feature extraction, while EFT-Net focuses more on edge features.
DCGAN-MS [25]	Image domain conversion is achieved through two domain generators to reduce artifacts and overcome the insufficiency of real data.
MACS-Net [30]	Effectively extracts both local and global image features, achieving strong correction performance.
U-Net structural CNN [15]	The first DL model to incorporate spatial occlusion for correcting metal artifacts, with excellent performance.
GRA-CNN [2]	The model has a lightweight architecture, subtracting artifact features extracted by CNN.
FD-Net [31]	High computational efficiency and strong generalization capability.

3.2 Other Artifacts Correction

The DL models are also applied to correct metal artifacts, ring artifacts, and aliasing artifacts. With CAE-S1 strategy, MLP [15] and U-shaped convolutional network [15] correct metal artifacts. A four-layer U-shaped convolutional network [33] introduces skip connections and concatenation at the first and last convolutional layers of each encoding-stage convolutional block, preserving multi-scale features. GRA-CNN [2] employs two convolutional layers to extract patches and corrects ring artifacts. With CAE-IM strategy, McSTRA [34] integrates dual-channel Swin-Unet (DS) blocks, and each block consists of STBs embedded in U-Net, to correct aliasing artifacts.

In addition, several deep learning models have also been developed to correct other types of MRI artifacts. With CAE-S2 strategy, U-shaped DAS-Net [35] corrects T1E artifacts in k-t domain by modifying the convolutional layers. U-shaped FD-Net [30] includes four skip connections, and an encoder composed of nine convolutional layers. As for ConvGAN, REGAIN [36] enhances cardiac MRI resolution through 23 residual dense blocks in G, facilitating feature channel fusion and local dense feature extraction.

4 Discussion and Future Research Direction

In the context of MRI images [37], this study focuses on discussing the architectural characteristics and applicable scenarios of the correction model. Regarding artifact types, motion artifact is the most frequently encountered challenges in MRI processing and the most concerned in existing research. From a model perspective, DCNNs have been mainly used for MRI artifact correction. The CAE-IM model achieved the largest growth percentage of SSIM (+166.77%) and PSNR (+59.78%) in the currently released research, and received the best correction and improvement effect. The ConGAN and CAE models also have outstanding performances.

DL-based artifact correction algorithms have achieved significant progress. However, several challenges remain unresolved. ① The interpretability of models, and their deep association with the operational expertise and domain knowledge have yet to be fully explored. ② DL models need improved adaptability and transferability when dealing with modality variations in MRI imaging. ③ There is still room for improvement in the correction effect of artifacts. ④ The reliability of the paired datasets of artifact images and clean images of the same patient is stronger than that of the synthetic datasets, but the collection methods of the paired datasets are limited.

To deal with those challenges, several plans can be explored. ① Imaging techniques. Future research should focus on expanding training datasets, optimizing scanner intensity settings. ② Model innovations. It is necessary to integrate artifact generation mechanisms and characteristics for continuous development of DL models, including innovation of module architectures, optimization of loss functions, and a focus on harmonizing the strengths of different models. ③ Leveraging Large Language Models (LLMs). The emergence of LLMs such as ChatGPT and DeepSeek presents new opportunities for model development, and will serve to enrich the content of artifact correction. Therefore, their potential should be carried on exploring. ④ Field promotion. Advanced DL models can be further extended to more medical application fields, making greater contributions to improving the performance of image segmentation and detection tasks.

5 Conclusion

This study provides a systematic review of DL model architectures and their evolution in the field of MRI artifact correction. After processing with DL models, the artifact-corrupted regions in images are significantly reduced, blurriness is reduced, and evaluation metrics achieve optimal improvements. For the first time, this study proposes a comprehensive evaluation approach that integrates subjective visual assessment with quantitative indicators to assess correction performance. Our study focuses on the architecture and its evolution of DCNNs, CAEs, GANs, attention mechanisms, Transformers, and integrated models in artifact correction. DCNN models, residual structures, CAEs, and GANs serve as the foundational frameworks for model design. Integrated models will become an increasingly powerful choice. Finally, we discuss the remaining challenges in DL-based MRI artifact correction and provide insights and recommendations for future research directions. Meanwhile, these measures will also provide useful experiences for the model development of tasks related to smart healthcare.

References

1. Lavdas, E., Papaioannou, M., Boci, N., et al.: Common and uncommon artifacts in T1 FLAIR SAG sequences of MRI brain. Curr. Probl. Diagn. Radiol. **50**(1), 59–65 (2021). https://doi.org/10.1067/j.cpradiol.2019.08.001
2. Zhang, Q., Ruan, G., Yang, W., et al.: MRI Gibbs-ringing artifact reduction by means of machine learning using convolutional neural networks. Magn. Reson. Med. **82**(6), 2133–2145 (2019). https://doi.org/10.1002/mrm.27894
3. Chen, Z., Pawar, K., Ekanayake, M., et al.: Deep learning for image enhancement and correction in magnetic resonance imaging-state-of-the-art and challenges. J. Digit. Imaging. **36**(1), 204–230 (2023). https://doi.org/10.1007/s10278-022-00721-9
4. Hales, P.W., Pfeuffer, J., Clark, C.A.: Combined denoising and suppression of transient artifacts in arterial spin labeling MRI using deep learning. J. Magn. Reson. Imaging. **52**(5), 1413–1426 (2020). https://doi.org/10.1002/jmri.27255
5. Oksuz, I., Clough, J.R., Ruijsink, B., et al.: Deep learning-based detection and correction of cardiac MR motion artefacts during reconstruction for high-quality segmentation. IEEE Trans. Med. Imaging. **39**(12), 4001–4010 (2020). https://doi.org/10.1109/TMI.2020.3008930
6. Tamada, D., Kromrey, M.L., Ichikawa, S., et al.: Motion artifact reduction using a convolutional neural network for dynamic contrast enhanced MR imaging of the liver. Magn. Reson. Med. Sci. **19**(1), 64–76 (2020). https://doi.org/10.2463/mrms.mp.2018-0156
7. Napoli, A.D., Cheng, S.F., Gregson, J., et al.: Arterial spin labeling MRI in carotid stenosis: arterial transit artifacts may predict symptoms. Radiology. **297**(3), 652–660 (2020). https://doi.org/10.1148/radiol.2020200225
8. Zhang, Q., Hann, E., Werys, K., et al.: Deep learning with attention supervision for automated motion artefact detection in quality control of cardiac T1-mapping. Artif. Intell. Med. **110**, 101955 (2020). https://doi.org/10.1016/j.artmed.2020.101955
9. An, H., Shin, H.G., Ji, S., et al.: DeepResp: deep learning solution for respiration-induced B0 fluctuation artifacts in multi-slice GRE. Neuroimage. **224**, 117432 (2021). https://doi.org/10.1016/j.neuroimage.2020.117432
10. Duffy, B.A., Zhao, L., Sepehrband, F., et al.: Retrospective motion artifact correction of structural MRI images using deep learning improves the quality of cortical surface reconstructions. NeuroImage. **230**, 117756 (2021). https://doi.org/10.1016/j.neuroimage.2021.117756

11. Johnson, P.M., Drangova, M.: Conditional generative adversarial network for 3D rigid-body motion correction in MRI. Magn. Reson. Med. **82**(3), 901–910 (2019). https://doi.org/10.1002/mrm.27772
12. Shaw, R., Sudre, C.H., Varsavsky, T., et al.: A k-space model of movement artefacts: application to segmentation augmentation and artefact removal. IEEE Trans. Med. Imaging. **39**(9), 2881–2892 (2020). https://doi.org/10.1109/TMI.2020.2972547
13. Sommer, K., Saalbach, A., Brosch, T., et al.: Correction of motion artifacts using a multiscale fully convolutional neural network. AJNR Am. J. Neuroradiol. **41**(3), 416–423 (2020). https://doi.org/10.3174/ajnr.A6436
14. Oh, G., Lee, J.E., Ye, J.C.: Unpaired MR motion artifact deep learning using outlier-rejecting bootstrap aggregation. IEEE Trans. Med. Imaging. **40**(11), 3125–3139 (2021). https://doi.org/10.1109/TMI.2021.3089708
15. Seo, S., Do, W., Luu, H.M., et al.: Artificial neural network for slice encoding for metal artifact correction (SEMAC) MRI. Magn. Reson. Med. **84**(1), 263–276 (2020). https://doi.org/10.1002/mrm.28126
16. Galley, J., Sutter, R., Stern, C., et al.: Diagnosis of Periprosthetic hip joint infection using MRI with metal artifact reduction at 1.5 T. Radiology. **296**(1), 98–108 (2020). https://doi.org/10.1148/radiol.2020191901
17. Schwartz, F.R., Tailor, T., Gaca, J.G., et al.: Impact of dual energy cardiac CT for metal artefact reduction post aortic valve replacement. Eur. J. Radiol. **129**, 109135 (2020). https://doi.org/10.1016/j.ejrad.2020.109135
18. Ronneberger, O., Fischer, P., Brox, T.: U-net: convolutional networks for biomedical image segmentation. Medical image computing and computer-assisted intervention-MICCAI 2015. **9351**, 234–241 (2015). https://doi.org/10.1007/978-3-319-24574-4_28
19. Creswell, A., White, T., Dumoulin, V., et al.: Generative adversarial networks: An overview. IEEE Signal Process. Mag. **35**(1), 53–65 (2018). https://doi.org/10.1109/MSP.2017.2765202
20. Blv, A., Jwg, B., Ap, A., et al.: The role of generative adversarial networks in radiation reduction and artifact correction in medical imaging. J. Am. Coll. Radiol. **16**(9), 1273–1278 (2019). https://doi.org/10.1016/j.jacr.2019.05.040
21. Woo, S., Park, J., Lee, J.Y., et al.: CBAM: Convolutional Block Attention Module. In: Proceedings of the European Conference on Computer Vision (ECCV), pp. 3–19 (2018). https://doi.org/10.48550/arXiv.1807.06521
22. Vaswani, A., Shazeer, N., Parmar, N., et al.: Attention is All You Need. 31st Conference on Neural Information Processing Systems (NIPS 2017). arXiv:1706.03762, 6000–6010 (2017). https://doi.org/10.48550/arXiv.1706.03762
23. Dosovitskiy, A., Beyer, L., Kolesnikov, A., et al.: An image is worth 16x16 words: transformers for image recognition at scale. Int Conf Learning Representations. (2021). https://doi.org/10.48550/arXiv.2010.11929
24. Liu, Z., Lin, Y., Cao, Y., et al.: Swin Transformer: hierarchical vision transformer using shifted windows. arXiv:2103.14030, 10012–10022 (2021). https://doi.org/10.48550/arXiv.2103.14030
25. Chen, G., Xie, H., Rao, X., et al.: MRI motion correction through disentangled CycleGAN based on multi-mask K-space subsampling. IEEE Trans. Med. Imaging. **44**(4), 1907–1921 (2025). https://doi.org/10.1109/TMI.2024.3523949
26. Wu, B., Li, C., Zhang, J., et al.: Unsupervised dual-domain disentangled network for removal of rigid motion artifacts in MRI. Comput. Biol. Med. **165**, 107373 (2023). https://doi.org/10.1016/j.compbiomed.2023.107373
27. Oksuz, I.: Brain MRI artefact detection and correction using convolutional neural networks. Comput. Methods Prog. Biomed. **199**, 105909 (2021). https://doi.org/10.1016/j.cmpb.2020.105909

28. Lee, J., Kim, B., Park, H.: MC^2-net: motion correction network for multi-contrast brain MRI. Magn. Reson. Med. **86**(2), 1077–1092 (2021). https://doi.org/10.1002/mrm.28719
29. Jiang, N., Zhang, Y., Li, Q., et al.: A cardiac MRI motion artifact reduction method based on edge enhancement network. Phys. Med. Biol. **69**(9), 095004 (2024). https://doi.org/10.1088/1361-6560/ad3884
30. Hossain, M.-B., Shinde, R.K., Imtiaz, S.M., et al.: Swin transformer and the Unet architecture to correct motion artifacts in magnetic resonance image reconstruction. Int. J. Biomed. Imaging. **2024**(1), 8972980 (2024). https://doi.org/10.1155/2024/8972980
31. Zaid, A.-A., Çukur, T., Saritas, E.U.: FD-net: An unsupervised deep forward-distortion model for susceptibility artifact correction in EPI. Magn. Reson. Med. **91**(1), 280–296 (2024). https://doi.org/10.1002/mrm.29851
32. Wu, Y., Liu, J., White, G.M., et al.: Image-based motion artifact reduction on liver dynamic contrast enhanced MRI. Phys. Med. **105**, 102509 (2023). https://doi.org/10.1016/j.ejmp.2022.12.001
33. Kwon, K., Kim, D., Kim, B., et al.: Unsupervised learning of a deep neural network for metal artifact correction using dual-polarity readout gradients. Magn. Reson. Med. **83**(1), 124–138 (2020). https://doi.org/10.1002/mrm.27917
34. Ekanayake, M., Pawar, K., Harandi, M., et al.: McSTRA: a multi-branch cascaded swin transformer for point spread function-guided robust MRI reconstruction. Comput. Biol. Med. **168**, 107775 (2024). https://doi.org/10.1016/j.compbiomed.2023.107775
35. Sheen, H., Han, Y., Shin, E., et al.: SU-E-J-01: analysis of acquisition parameters that caused artifacts in four-dimensional (4D) CT images of targets undergoing regular motion. Phys. Med. **39**(6), 3652–3652 (2012). https://doi.org/10.1118/1.4734833
36. Yoon, S., Nakamori, S., Amyar, A., et al.: Accelerated cardiac MRI cine with use of resolution enhancement generative adversarial inline neural network. Radiology. **307**(5), e222878 (2023). https://doi.org/10.1148/radiol.222878
37. Spieker, V., Eichhorn, H., Hammernik, K., et al.: Deep learning for retrospective motion correction in MRI: a comprehensive review. IEEE Trans. Med. Imaging. **43**(2), 846–859 (2024). https://doi.org/10.1109/TMI.2023.3323215

LungAide: An Effective 3D Transformer V-Net for Lung Nodule Detection

Xinyuan Gao[1], Lei Dong[1], Xingwang Liu[1], Sijie Yin[1], and Hao Chen[2(✉)]

[1] Beijing Institute of Technology, Beijing 100081, People's Republic of China
[2] Changping No.1 High School, Beijing 102200, People's Republic of China
704243312@qq.com

Abstract. Existing lung nodule detection methods on computed tomography (CT) have the problems of missing small nodules and losing spatial information. This paper proposes LungAide, a novel 3D Transformer V-Net for accurate lung nodule detection in chest CT scans, with a particular focus on small nodules. The proposed architecture integrates the encoder-decoder network and concatenation operation of V-Net with the Transformer blocks featuring a lightweight self-attention mechanism with a Bi-Path downsampling, enabling to preserve spatial information and capture long-range dependencies between lung nodules. To validate the effectiveness of LungAide, experiments are conducted on the LUNA16 dataset, which achieving a Competition Performance Metric of 0.915. Comparative and ablation studies confirm that LungAide effectively reduces false positives while enhancing sensitivity.

Keywords: Lung Nodule Detection · Deep Learning · V-Net · Transformer

1 Introduction

Cancer remains one of the most significant global public health challenges nowadays. In 2022, lung cancer emerged as the most frequently diagnosed cancer, representing 12.5% of all new cancer cases worldwide, maintaining its position as the primary cause of cancer-related deaths at the same time [1]. Lung nodules have the risk of deteriorating into lung cancers. In order to reduce mortality and improve prognosis, CT has been adopted as a standard screening method for lung nodule identification [2]. Compared with X-radiation and magnetic resonance imaging, CT significantly reduces radiation exposure while preserving adequate image resolution for characterizing critical nodule features including calcification patterns, margin definition, sphericity morphology, and spiculation in CT scans, which are key to distinguishing nodules from nodule-like tissues [3]. Figure 1 shows the true lung nodules and false positives in CT scans.

In order to enhance diagnostic precision while minimizing the risk of misdiagnosis and missed nodules due to subjective judgment, Computer-Aided Detection (CADe) systems have been developed for automatically lung nodule identification. Conventional nodule detection systems typically employ intensity-based threshold algorithms, clustering techniques, and morphological operations for preliminary candidate identification

H. Ma et al. (Eds.): IWACIII 2025, CCIS 2780, pp. 177–189, 2026.
https://doi.org/10.1007/978-981-95-6730-0_13

[4]. Lu et al. developed an integrated tree-structured network incorporating multiple well-established algorithms to address the challenge of diverse nodule characteristics [5]. Lo et al. utilized gray-level thresholding techniques to identify three-dimensionally contiguous structures within lung CT scans, followed by a rule-based scheme and a cascaded automated classifier to distinguish true nodules from normal anatomical structures [6]. Conventional nodule detection approaches face two key limitations: (1) their heavy reliance on experience to extract features, leading to poor generalization abilities, and (2) certain techniques such as region growth algorithms demanding manual selection of seed points, lacking full automation.

(a) (b)

Fig. 1. The examples of lung nodules and nodule-like tissues. (a) True lung nodules. (b)False positives.

With the development of data-driven methods, deep learning is also widely used in the field of medical imaging. George et al. introduced the YOLO architecture for lung nodule detection, a widely-used object detection model originally designed for two-dimensional (2D) natural image sets. Their approach employed large input sizes to achieve optimal detection performance [7]. However, using YOLO may result in small nodules' compression or disappearance, making them undetectable. Besides, 2D images fail to preserve complete spatial information. Therefore, three-dimensional (3D) methods can further improve the accuracy of lung nodule detection. Convolutional neural networks (CNNs) have also demonstrated strong performance in lung nodule detection. Xu et al. proposed an improved Faster R-CNN model specifically optimized for CT-based lung nodule identification by integrating advanced techniques aimed at enhancing detection of small, irregular, and difficult-to-classify nodules [8]. However, the two-stage detection results in slow inference speed. Transformers have recently become a powerful architecture in computer vision (CV) tasks, performing well in small target detection due to its capability of long-rang dependencies. Mkindu et al. proposed a scheme based on a 3D multi-scale vision transformer (3D-MSViT) to detect lung nodules from CT scans, relying entirely on attention mechanisms within a local–global transformer architecture to reduce the parameter count and enhance feature extraction [9]. The Swin Transformer preserves spatial continuity by focusing on local regions, thereby addressing the discontinuity caused by non-overlapping window operations —an aspect that can otherwise hinder the accuracy of lung nodule detection. While the Vision Transformer can capture long-range dependencies between surrounding tissues and lung nodules in CT scans through the self-attention mechanisms, which is especially important for detecting lung nodules. Despite the integration of a 3D shifted window-based Swin Transformer, the 3D-MSViT is computationally intensive and data-hungry.

To address the loss of spatial information in 2D images and the high computational cost of Transformer, we propose LungAide, an effective 3D Transformer V-Net architecture for lung nodule detection. Our model leverages the encoder-decoder structure and the concatenation operation of V-Net to avoid the loss of position information. To optimize small object detection, Transformer blocks are introduced within the V-Net, enabling the modeling of global relationships. To reduce computational complexity simultaneously, we introduce a lightweight attention within our Transformer block. Finally, we conduct the experiments on the public lung dataset LUNA16 to evaluate the effectiveness of the proposed architecture, and compare our network with other existing methods.

Our key contributions include:

1) A 3D V-Net architecture integrated with Transformer blocks that capture global contextual dependencies within volumetric CT scans while preserving high-resolution spatial details, thereby enhancing feature representation for small nodules and reducing false negatives.
2) A novel Transformer block that employs a lightweight attention mechanism to reduce the computational cost and parameter count, retaining the crucial global contextual modeling capability while making the network efficient and well-suited for the limited scale of data.
3) A Bi-Path downsampling introduced within the attention block, where one path performs 2× downsampling to preserve local structures, while the other path applies 4× downsampling to obtain more contextual information with a larger receptive field. To mitigate information loss, the 4× branch is followed by a 2× upsampling operation, restoring its resolution before fusion with the 2× branch.

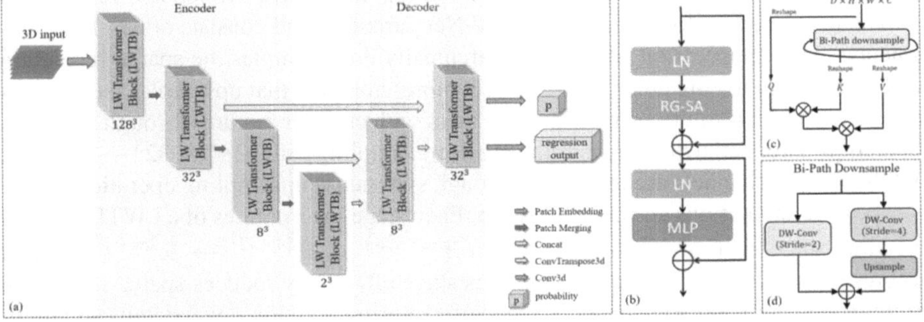

Fig. 2. The architecture of LungAide. (a) A network with LW Transformer Block. Number on the bottom of each block indicates spatial dimensions. (b) LW Transformer Block with RG-SA that is adopted in (a). (c) The RG-SA. (d) The Bi-Path Downsample applied in RG-SA.

2 Method

Our proposed LungAide is a 3D Transformer V-Net for lung nodule detection in CT scans. The model which combines the encoder-decoder structure and the concatenation of V-Net with the self-attention mechanism of Transformer is illustrated in Fig. 2. Specifically, the Transformer block is integrated with a Recursive Generalized Self-Attention (RG-SA) [10] to maintain low computational cost. Meanwhile, a Bi-Path downsampling mechanism applied in RG-SA is designed to reserve more contextual information. Patch merging and transposed convolution operations are used for downsampling and upsampling respectively. The network contains three different sizes of anchors. Finally, the network outputs a probability vector of the likelihood of lung nodule presence at each voxel, and a regression map encoding bounding box.

2.1 LungAide

LungAide features a hybrid 3D Transformer V-Net architecture. This approach integrates two key components: the hierarchical encoder-decoder framework and feature concatenation from V-Net, and attention-based global context modeling implemented using Lightweight Transformer Blocks (LWTB). The network is illustrated as Fig. 2(a).

LungAide: To operate within the LWTB, the input is first passed through patch embedding. Given an input volume $x \in \mathbb{R}^{B \times C \times D \times H \times W}$, we first divide it into several volumetric patches. Each 3D input volume is unfolded into non-overlapping P^3 volumetric patches (where P denotes patch size). Each patch is flattened and linearly projected into d-dimensional feature space, generating a sequence of $x \in \mathbb{R}^{B \times N \times d}$, where $N = \frac{DHW}{P^3}$ denotes the resulting number of patches, which corresponds to the input sequence length for the LWTB. After the embedding step, we obtain the required token $Z \in \mathbb{R}^{B \times N \times d}$.

The network then follows the typical V-Net paradigm and consists of two key components: (1) a compression path that hierarchically downsamples the spatial resolution while increasing feature channels, (2) a decompression path that upsamples the features to recover spatial resolution and combines them with encoder features through skip concatenation. The spatial dimension follows a downsampling path $128^3 \rightarrow 32^3 \rightarrow 8^3 \rightarrow 2^3$, and the feature channels are recovered through symmetric upsampling operations using transposed convolutions and concatenation. Each stage incorporates of a LWTB. During encoding stage, features are transited from $\mathbb{R}^{32\times 128^3} \rightarrow \mathbb{R}^{64\times 32^3} \rightarrow \mathbb{R}^{256\times 2^3}$ through progressive downsampling operations. This stage effectively reduces spatial resolution while increasing feature channels, capable of capturing both low-level details and high-level semantics. The decoder restores spatial resolution by upsampling and concatenating skip-connected encoder features at each level. In the decoder stage, features are upsampled from $\mathbb{R}^{256\times 2^3} \rightarrow \mathbb{R}^{64\times 32^3}$. The network employs three different predefined anchor sizes corresponding to bounding boxes of length 5, 10, and 30. These anchors serve as reference templates for regressing nodule proposals. Finally, the network outputs a probability map p representing the likelihood of nodule presence at each voxel, and a regression map encoding bounding box coordinates (x, y, z, r), where (x, y, z) denotes the center coordinates and r indicates the radius.

2.2 Lightweight Transformer Block (LWTB)

LWTB constitutes the core computational unit within the LungAide network. The input Z undergoes layernorm (LN) before being processed by the RG-SA module. A residual connection adds the original input to the output of the RG-SA module. This result then passes through a second LN, followed by a multi-layer perceptron (MLP) block. Finally, another residual connection adds the output of the MLP output to the previous result, enabling each processing block to establish global dependencies from volumetric input data. The LWTB is depicted in Fig. 2(b).

RG-SA: To efficiently model long-range 3D dependencies while avoiding the quadratic complexity of the standard self-attention, the RG-SA module is introduced. By leveraging local feature interactions and recursive spatial propagation, the RG-SA eliminates explicit attention matrix computation. A Bi-Path downsampling module is introduced in the RG-SA. The RG-SA and the Bi-Path downsampling module is depicted in Fig. 2(c) and Fig. 2(d) respectively.

For the input token $Z \in \mathbb{R}^{B \times N \times d}$, the RG-SA initially reduces the token length to a fixed number of global representative maps $Z_r \in \mathbb{R}^{B \times M \times d_r}$, where $M \ll N$ and $d_r = d \bullet c_r$ is the reduced dimension with scaling factor $c_r \in (0, 1]$. This reduction process is achieved through the depth-wise convolution and the 1×1 point-wise convolution.

Then the query (Q) are computed through the original input features. The key (K) and value (V) are computed through the compressed maps:

$$Q = Z \bullet W_Q,\ \ K = Z_r \bullet W_K,\ \ V = Z_r \bullet W_V, \tag{1}$$

$$Z_M = W_O\left[Softmax\left(\frac{QK^T}{\sqrt{d_r}}\right) \bullet V\right], \tag{2}$$

where $W_Q \in \mathbb{R}^{d \times d_r}$, $W_K \in \mathbb{R}^{d_r \times d_r}$, and $W_V \in \mathbb{R}^{d_r \times d_r}$ are trainable parameters, and $W_O \in \mathbb{R}^{d \times d}$ denotes the projection matrix. Instead of calculating the full attention matrix, RG-SA efficiently approximates global context aggregation through localized attention patterns, which could reduce memory complexity and allow recursive attention across multiple spatial scales, thus maintaining the modeling of global relationships with low computational cost. The total complexity of RG-SA is $\mathcal{O}(BNd)$, which is linear in sequence length N. In the contrast, the complexity of the local self-attention of Swin Transformer can be calculated as $\mathcal{O}(BNs^2d)$, where s is the window size. Thus, the total complexity of Swin Transformer is with a significantly larger constant factor, which is more computationally complex than RG-SA.

Bi-Path Downsample: To further enhance the RG-SA module, we replace the original depth-wise downsampling operation with a Bi-Path Downsampling strategy. This module consists of two parallel paths: the main path performs $2\times$ downsampling, preserving high-resolution local details, while the other path applies $4\times$ downsampling to aggregate low-frequency global context. To reduce information loss, the $4\times$ branch is followed by a $2\times$ upsampling operation, restoring its resolution to match the main path. The outputs of the two paths are then summed, producing a feature map that integrates both fine-grained local information and broader contextual cues. By incorporating the Bi-Path downsampling, RG-SA benefits from an enlarged effective receptive field without increasing the final downsampled feature resolution, allowing the attention mechanism to capture richer multi-scale 3D dependencies efficiently.

MLP: The RG-SA outputs undergo processing through a position-wise MLP containing two fully connected layers with Gaussian Error Linear Unit (GELU) activation. Given the input Z_M after RG-SA and LN, the MLP module strengthens channel-wise representation learning:

$$MLP(Z_M) = [W_2 \bullet \phi(W_1 \bullet Z_M + b_1) + b_2], \tag{3}$$

where $\phi(\cdot)$ denotes the GELU non-linearity function, and $W_1 \in \mathbb{R}^{d \times d'}$ projects each token into a higher-dimensional space with $d' = 4d$. $W_2 \in \mathbb{R}^{d' \times d}$ projects back to the original feature dimension d.

2.3 Loss Function

Class Imbalance: Class imbalance is a prevalent and challenging problem in detection tasks, primarily caused by the disproportionate distribution of samples among various categories. This imbalance can lead to biased model training, where the network becomes more inclined to predict the majority class, resulting in high false negative rates and poor sensitivity to rare but critical targets [11]. The loss function plays a pivotal role in addressing the imbalance problem by quantifying the discrepancy between the model's predictions and the ground truth, and guiding the optimization process accordingly.

The tasks of lung nodule detection often face the problem of the obvious imbalance between nodules and non-nodule parts, since nodules only occupy a small part of a lung CT scan, usually less than 1% of the whole image voxel. This extreme sparsity can severely hinder the model's ability to learn meaningful representations of the minority class. To address this issue and reduce training bias toward the majority class, we adopt the Focal Loss [12] in our model.

Focal Loss: Focal Loss is a loss function specifically designed to focus learning on sparse examples while reducing the impact of well-classified ones, which is defined as:

$$L_{focal}(p_t) = -\alpha_t (1 - p_t)^{\gamma} \log(p_t), \tag{4}$$

where $\alpha_t \in [0, 1]$ is a weighting factor to address class imbalance, and $\gamma \geq 0$ is the focusing parameter that down-weights easy examples. $(1 - p_t)$ and p_t act as ground truth class label $y = 1$ and the otherwise. In our model, The α_t and γ of the Focal Loss function are set as 0.5 and 2 respectively.

Regression Loss: Specifically, the bounding box in our model is parameterized as (x, y, z, r), where (x, y, z) denotes the center coordinates and r represents the radius of the nodule. To supervise the regression of these parameters, a SmoothL1 loss is employed, which combines the robustness of L1 loss with the stability of L2 loss, thereby mitigating the influence of outliers while maintaining sensitivity to small errors. The regression loss is defined as:

$$L_{reg}(n, \hat{n}) = \sum\nolimits_{i \in \{x,y,z,r\}} SmoothL1(n_i - \hat{n}_i), \tag{5}$$

where $\hat{n_i}$ denotes the predicted center coordinates and radius, and n_i is the corresponding ground truth. The SmoothL1 in Eq. (5) is defined as:

$$SmoothL1(x) = \begin{cases} 0.5x^2 & if\ |x| < 1 \\ |x| - 0.5 & otherwise \end{cases}. \tag{6}$$

Total Loss: The final detection loss is formulated as a weighted sum of the classification and regression losses:

$$L_{total} = \lambda_{cls} L_{focal} + \lambda_{reg} L_{reg}, \tag{7}$$

where λ_{cls} and λ_{reg} are balancing coefficients. In our implementation, both coefficients are empirically set to 1.0 to ensure equal contribution from classification and localization.

3 Experiments and Analysis

3.1 Datasets

The Lung Image Database Consortium and Image Database Resource Initiative (LIDC-IDRI) is a public dataset initiated by the National Cancer Institute (NCI) [13]. It comprises 1,018 cases of diagnostic and lung cancer screening thoracic CT scans with annotated lesions. For each subject of the dataset, clinical thoracic CT scans and an associated XML file are included. The XML file contains the results of a two-phase image annotation process performed by a maximum of four and a minimum of two experienced thoracic radiologists. The detected lesions are classified into three categories: 'nodule > or =3 mm', 'nodule <3 mm', and 'non-nodule > or =3 mm'.

LUNA16 dataset represents a subset of LIDC-IDRI, which excludes scans with a slice thickness greater than 2.5 mm from LIDC-IDRI. At last, 888 scans are included in LUNA16 dataset. The dataset contains the annotations for luna detection tasks, including nodules' position, diameter and class labels. Additionally, the dataset provides the automatically generated lung segmentation for all CT images.

3.2 Evaluation Indicators

To quantitatively evaluate the performance of our proposed nodule detection network, we employ several widely-used metrics specifically designed for medical object detection tasks. Sensitivity, which evaluates the model's capability to accurately recognize nodule regions in the input, is defined as:

$$Sensitivity = \frac{TP}{TP + FN}, \tag{8}$$

where *TP* and *FN* represent the number of true positives and false negatives respectively. High sensitivity indicates superior performance in minimizing missed true cases. We also employ the FROC curve, which extends the traditional receiver operating characteristic curve analysis by plotting sensitivity (true positive rate) against the average number of false positives per scan (FPs/scan) instead of false positive rate.

The Competition Performance Metric (CPM) serving as a scalar summary of the FROC curve and representing the official criterion for the LUNA16 nodule detection challenge is also adopted in the experiments. It is mathematically defined as the average sensitivity computed at a predefined set of false positive rates:

$$CPM = \frac{1}{7}\sum\nolimits_{i=FPs} Sensitivity(i), FPs = \left\{\frac{1}{8}, \frac{1}{4}, \frac{1}{2}, 1, 2, 4, 8\right\}. \quad (9)$$

3.3 Experimental Details

For data standardization, all CT scans from LUNA16 were initially converted to Hounsfield Units (HU). A threshold of -600 was then applied to binarize the pixel into a [0,1] range. Subsequently, the LUNA16's provided segmentation ground truth were utilized to remove the background from each CT scan, thereby eliminating redundant information that could interfere with the detection results. To accommodate computational constraints, the CT scans were partitioned into $128 \times 128 \times 128$ patches for network input. The training set was augmented through random flipping, affine transformations, elastic deformation, and Gaussian blurring to enhance data diversity. Anchor sizes were set to [5,10,30] as multiple sizes, with bounding boxes exceeding an Intersection over Union (IoU) threshold of 0.5 being considered as positive samples.

Each model was trained for 150 epochs with a weight decay of $1e^{-4}$. The initial learning rate was set to $1e^{-2}$, then reduced to $1e^{-3}$ after epoch 75, and further decreased to $1e^{-4}$ at epoch 120. All experiments were implemented using PyTorch and conducted on NVIDIA GeForce GTX 1080 Ti GPUs.

3.4 Comparison Experiment

To validate the effectiveness of our LungAide network, we conducted comparative experiments with other competitive 3D detection models on the LUNA16 dataset. The quantitative experiment results are summarized in Table. 1.

Comparative results on the LUNA16 dataset demonstrate that the LungAide network outperforms five state-of-the–art models, achieving the highest average sensitivity score of 0.915, highlighting the model's effectiveness in reducing false positives. The results validate that our proposed model offers an effective and robust solution for handling the lung nodule detection tasks.

Figure 3 represents the visualization results of the LungAide detection method. The middle slice of the CT scan is selected, and the true nodule is automatically detected and highlighted with a red rectangle. 'N1' outside the red rectangle indicates the nodule number for the case.

To further demonstrate the lightweight of LungAide, we conducted comparative experiments against the 3D Swin Transformer. The experiments were conducted on the same GPU. Table. 2 shows the results of the computational cost comparison, where 'FLOPs' indicates the floating-point operations per second, 'Params' denotes the total learnable weights, 'Times' measures the inference latency per sample, and 'GPU memory' represents the amount of GPU memory used. The floating-point operations and the

Table. 1. Comparison results of different models on LUNA16.

Approach	False Positive per Scan							CPM
	0.125	0.25	0.5	1	2	4	8	
Yang et al. [14]	0.698	0.800	0.823	0.844	0.847	0.852	0.861	0.818
Dou et al. [15]	0.659	0.745	0.819	0.865	0.906	0.933	0.946	0.839
Zhu et al. [16]	0.692	0.769	0.824	0.865	0.893	0.917	0.933	0.842
Sont et al. [17]	0.723	0.838	0.887	0.911	0.928	0.934	0.948	0.881
Mkindu et al. [18]	–	–	–	0.941	–	**0.971**	–	0.909
LungAide	**0.781**	**0.865**	**0.922**	**0.943**	**0.957**	0.965	**0.971**	**0.915**

total learnable weights of LungAide are both less than that of 3D Swin Transformer, also leading to the faster running speed and less GPU memory consumption. These results indicate that LungAide reduces the computational burden compared to the 3D Swin Transformer, which is already considered a relatively lightweight network. The main reason for the lower computational cost of LungAide lies in the the LWTB with the RG-SA of Bi-Path downsampling, where the use of depthwise convolutions effectively decreases the number of parameters relative to the hierarchical window-based attention mechanism employed in Swin Transformer.

Table. 2. Comparison results of the computational cost of 3D Swin Transformer and LungAide.

Approach	FLOPs	Params	Times	GPU Memory
3D Swin Transformer	24.64G	10.04 M	23.572 ms	518 M
LungAide (LWTB)	16.23G	9.17 M	18.685 ms	407 M

3.5 Ablation Experiment

To evaluate the superiority of the V-Net with LWTB over the traditional V-Net and to assess the effectiveness of the LWTB with RG-SA compared to other Transformer-based variants, we performed comprehensive ablation experiments. Three models were implemented respectively to systematically validate the performance improvements of our LungAide network: (1) a traditional 3D V-Net network, (2) a 3D V-Net utilizing Swin Transformer (3D Swin T V-Net), and (3) the LungAide network which uses LWTB, denoted as LungAide(LWTB).

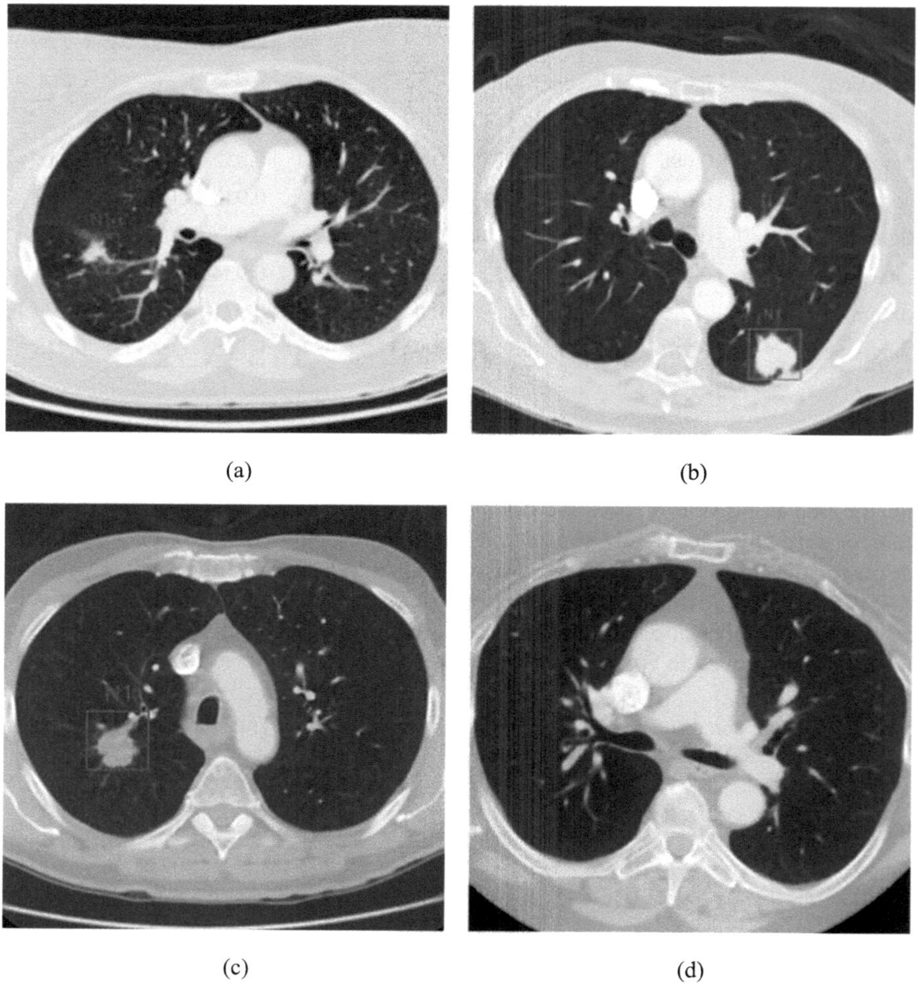

Fig. 3. The detection result of the proposed LungAide. (a), (b), (c) and (d) indicate different nodules of different cases respectively.

Table. 3. Ablation results of various parts of LungAide.

Approach	False Positive per Scan							CPM
	0.125	0.25	0.5	1	2	4	8	
3D V-Net	0.696	0.761	0.792	0.818	0.842	0.870	0.884	0.809
3D Swin T V-Net	0.720	0.792	0.845	0.878	0.904	0.919	0.922	0.854
LungAide (LWTB)	**0.781**	**0.865**	**0.922**	**0.943**	**0.957**	0.965	**0.971**	**0.915**

The FROC curve of the ablation models is presented in Fig. 4. Our LungAide network demonstrates superior performance compared to both the 3D V-Net and 3D Swin T V-Net, across all false positive rates. Quantitatively, the models achieved CPM score of 0.809 (3D V-Net), 0.854 (3D Swin T V-Net), and 0.915 (our proposed LungAide), respectively. These results demonstrate the effectiveness of integrating LWTB with V-Net structure, leading to improved sensitivity in lung nodule detection. The detailed CPM results for the three models are presented in Table. 3.

Table. 4. Ablation results of the computational cost with different attention in LungAide.

Approach	FLOPs	Params	Times	GPU Memory
LungAide (with self-attention)	27.90G	9.62 M	24.116 ms	554 M
LungAide (with RG-SA)	16.23G	9.17 M	18.685 ms	407 M

To further validate the superiority of the RG-SA with the Bi-Path downsampling module in reducing computational complexity, we conducted related ablation experiments, and the results are showed in Table. 4. Specifically, we compared two configurations: (1) the LungAide with RG-SA augmented by Bi-Path downsampling module, and (2) the LungAide with self-attention. The experimental results clearly demonstrate that replacing self-attention with the RG-SA combined with the Bi-Path downsampling, which incorporates depthwise separable convolutions, leads to a significant reduction in both floating-point operations and the total number of learnable parameters. Meanwhile the inference latency per sample and the GPU consumption are also reduced. This efficiency gain is achieved not by adding extra operators, but by substituting part of the standard computation with downsamling achieved by depthwise convolutions and lightweight feature fusion. As a result, LungAide of the RG-SA with the Bi-Path downsampling mechanism achieves lower computational cost while maintaining comparable representational capacity.

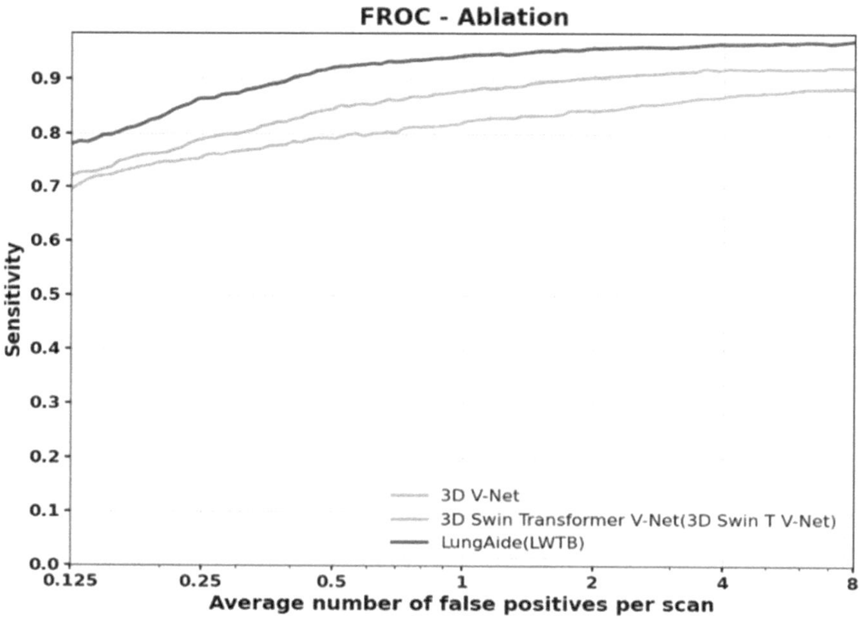

Fig. 4. The FROC curve of the ablation experiment result. The cyan line represents the FROC performance of the 3D V-Net network, the orange line corresponds to 3D Swin T V-Net, and the red line represents our proposed LungAide(LWTB).

4 Conclusion

This paper proposes LungAide, a 3D Transformer V-Net for lung nodule detection on CT images. The network combines V-Net and a self-attention mechanism to reduce information loss in lung nodule detection. By leveraging the encoder-decoder structure and the concatenation operation, the spatial information is preserved to the greatest extent possible. To capture long-range dependencies between lung nodules while reducing the computational cost, a lightweight Transformer block incorporating RG-SA is introduced in the V-Net. To further preserve the global information, the RG-SA introduces a Bi-Path downsampling module which is designed to receive larger receptive field while avoiding the information loss. The experimental results on the LUNA16 dataset demonstrate that LungAide surpasses multiple state-of-the-art approaches, confirming its accuracy in small lung nodule detection. Ablation studies further verify the effectiveness of the combination of V-Net and LWTB, which leverages the information reuse capability of the former and the long-range dependency modeling between lung nodules established by the latter. In the future, the LungAide network may be extended the detection range to include the classification of the lung nodule malignancy, providing more comprehensive computer-aided diagnostic support for early lung cancer screening.

References

1. Chhikara, B.S., Parang, K.: Global cancer statistics 2022: the trends projection analysis. Chem. Biol. Lett. **10**(1), 451–451 (2023)
2. Cao, M., et al.: cancer screening in China: the current status, challenges, and suggestions. Cancer Lett. **506**, 120–127 (2021)
3. Diederich, S., et al.: Screening for early lung cancer with low-dose spiral CT: prevalence in 817 asymptomatic smokers. Radiology. **222**(3), 773–781 (2002)
4. Murphy, K., et al.: A large-scale evaluation of automatic pulmonary nodule detection in chest CT using local image features and k-nearest-neighbour classification. Med. Image Anal. **13**(5), 757–770 (2009)
5. Lu, L., Tan, Y., Schwartz, L.H., Zhao, B.: Hybrid detection of lung nodules on CT scan images. Med. Phys. **42**(9), 5042–5054 (2015)
6. Lo, S.C., et al.: Artificial convolution neural network techniques and applications for lung nodule detection. IEEE Trans. Med. Imaging. **14**(4), 711–718 (1995)
7. George, J., Skaria, S.: Using YOLO based deep learning network for real time detection and localization of lung nodules from low dose CT scans. In: Medical Imaging 2018: Computer-Aided Diagnosis, vol. 10575, pp. 347–355. SPIE (2018)
8. Xu, J., Ren, H., Cai, S., Zhang, X.: An improved faster R-CNN algorithm for assisted detection of lung nodules. Comput. Biol. Med. **153**, 106470 (2023)
9. Mkindu, H., Wu, L., Zhao, Y.: 3D multi-scale vision transformer for lung nodule detection in chest CT images. Signal Image Video Process. SIViP. **17**(5), 2473–2480 (2023)
10. Chen, Z., Zhang, Y., Gu, J., Kong, L., Yang, X.: Recursive generalization transformer for image super-resolution. arXiv preprint arXiv 230., 06373 (2023)
11. Hodson, T.O.: Root mean square error (RMSE) or mean absolute error (MAE): when to use them or not. Geosci. Model Dev. Discussionsv. **15**(14), 5481–5487 (2022)
12. Lin, T.Y., Goyal, P., Girshick, R., He, K., Dollár, P.: Focal loss for dense object detection. In: Proceedings of the IEEE international conference on computer vision, pp. 2980–2988 (2017)
13. Armato III, S.G., et al.: The lung image database consortium (LIDC) and image database resource initiative (IDRI): a completed reference database of lung nodules on CT scans. Med. Phys. **38**(2), 915–931 (2011)
14. Yang, L., et al.: A lightweight neural network for lung nodule detection based on improved ghost module. Quant. Imaging Med. Surg. **13**(7), 4205–4221 (2023)
15. Dou, Q., et al.: automated pulmonary nodule detection via 3d convnets with online sample filtering and hybrid-loss residual learning. In: International Conference on Medical Image Computing and Computer-Assisted Intervention, pp. 630–638. Springer International Publishing, Cham (2017)
16. Zhu, W., Liu, C., Fan, W., Xie, X.: Deeplung: deep 3d dual path nets for automated pulmonary nodule detection and classification. In: In 2018 IEEE Winter Conference on Applications of Computer Vision (WACV), pp. 673–681. IEEE (2018)
17. Song, T., et al.: CPM-net: a 3D center-points matching network for pulmonary nodule detection in CT scans. In: International Conference on Medical Image Computing and Computer-Assisted Intervention, pp. 550–559. Springer International Publishing, Cham (2020)
18. Mkindu, H., Wu, L., Zhao, Y.: Lung nodule detection in chest CT images based on vision transformer network with Bayesian optimization. Biomed Signal Proc Contr. **85**, 104866 (2023)

Concordance-Assisted Learning for Estimating Optimal Individualized Treatment Regimes with Censored Data

Jianlin Zhang(✉), Yinglin Jiao, Lin Luo, Hui Dong, and Jianuo Yang

School of Mathematics and Information Science, Zhongyuan University of Technology, Zhengzhou, China
mathzhangjianlin@hotmail.com

Abstract. Individualized treatment regimes aim to determine the optimal treatment decision based on patient-specific covariates to maximize the clinical benefit for the entire patient population. When the response variable is survival time, the presence of censoring in survival data complicates the direct estimation of the optimal ITR. We propose a novel consistent estimator for the optimal ITR in the context of censored data. Assuming the propensity score model is known, our method adjusts the censored data via a K-S-V transformation, which is essentially a weighting adjustment of the response variable. Subsequently, the optimal decision regimen is estimated by maximizing the concordance function and the value function, respectively. The performance of our proposed method is demonstrated through simulation studies and its application to the ACTG175 clinical trial dataset.

Keywords: Concordance · Individualized treatment regimen · Censored data · Value function

1 Introduction

In recent years, precision medicine has attracted increasing attention due to the heterogeneity in treatment response among patients. A single treatment may benefit most individuals but be suboptimal for specific subgroups [1,4]. For instance, Lu et al. (2013) [11] found that in AIDS patients, ZDV+ddl was more effective for older individuals, while ZDV+zalcitabine benefited younger patients. To address such differences, individualized treatment regimens (ITRs) have emerged, aiming to tailor treatment decisions to patient-specific covariates (e.g., age, genotype), thus optimizing clinical outcomes. Mathematically, an ITR maps patient characteristics to treatment decisions and seeks to maximize the expected treatment benefit while maintaining interpretability [14]. Estimating

This work is supported by National Natural Science Foundation of Henan Youth Program under Grant 242300420670; 252300420939.

H. Ma et al. (Eds.): IWACIII 2025, CCIS 2780, pp. 190–203, 2026.
https://doi.org/10.1007/978-981-95-6730-0_14

optimal ITRs has become a key research focus in both randomized trials and observational studies [3,5].

Estimation approaches are generally categorized into indirect and direct methods. Indirect methods (regression-based) first model the conditional outcome and then select the optimal treatment. Representative approaches include the l_1-penalized regression method [14], as well as Q-learning and A-learning for dynamic regimens [12]. However, indirect methods may suffer from model misspecification and challenges in high-dimensional covariate spaces.Direct methods, such as Zhao et al. proposed outcome weighted learning (OWL) [20], bypass outcome modeling and formulate the estimation as a weighted classification problem. Later developments, including residual weighted learning (RWL) [21] and adaptive outcome weighted learning (AOWL) [10], improved performance and extended applicability to dynamic treatment settings. Additionally, inverse probability weighted estimators (IPWE) and augmented inverse probability weighted estimators (AIPWE) provide double robustness by incorporating both outcome and treatment models [18]. Recent advances, such as Entropy Learning and Proximal Learning, further enrich the field [15].

Current research mainly focuses on extending methods originally developed for binary treatment regimens to other settings, such as multi-category treatment regimens or dynamic treatment regimens [8,16]. For survival outcomes, censoring complicates estimation. Treating censored data as exact outcomes leads to bias [17]. Zhao et al. [19] extended OWL to censored data through inverse censoring outcome weighted learning (ICO) and doubly robust outcome weighted learning (DRO), incorporating survival and censoring models, with DRO achieving double robustness. Other methods include survival tree-based estimations combined with OWL [6].

Considering binary treatment regimens, Fan et al. [7] constructed a concordance function using maximum rank estimation and derived the optimal decision rule based on the concordance function and the value function.In this paper, we extend concordance-assisted learning to solve for the optimal individualized treatment regimens when the response variable is right-censored. We adjust the censored data using a K-S-V transformation, which primarily involves reducing the observed values of censored data to 0 and increasing the observed values of uncensored data. Under the assumption that the propensity score model is known, we derive the concordance function for censored data and estimate the optimal treatment regime by estimating both the concordance function and the value function under right-censored data.

The subsequent sections of this study are arranged as follows: Sect. 2 introduces the basic framework of individualized treatment regimens, reviews concordance-assisted learning, and further applies it to right-censored data. Sections 3 and 4 validate the performance of the proposed method through simulations and the ATCG175 dataset, respectively.

2 Methodological Approach

2.1 Notation and Fundamental Framework

Consider a single-stage treatment regimen with two possible treatment options. For a specific study population of interest, n subjects are randomly selected to participate in a randomized clinical trial or an observational study. The data structure for the i-th subject can be represented as $\{(X, A, T), i = 1, 2, \ldots, n\}$, where $X = (x_1, x_2, \ldots, x_p) \in \mathbb{R}^p$ denotes a p dimensional vector of covariates for the subject. let $A \in \{0, 1\}$ denote the two possible treatment options. Typically, $A = 0$ represents the control group treatment, while $A = 1$ represents the experimental group treatment. The variable T denotes the survival time of the patient, where a larger value of T indicates a better treatment outcome. However, in clinical trials, due to patient withdrawal or deaths unrelated to the disease of interest, the survival time T may not always be directly observed.

Let τ denote the censoring time for the study, and $\widetilde{T}$ represent the time when the event of interest occurs. The observed survival time is thus defined as $T = \min\{\widetilde{T}, \tau\}$. Let C denote the censored time, which may exceed the study cut-off time τ. Define the censoring indicator $\Delta = I(T \leq C)$, where $\Delta = 1$ indicates that the actual survival time is observed, and $\Delta = 0$ indicates that only the censoring time is observed, preventing observation of the true survival time.

Let $Y = \min\{T, C\}$ denote the actually observed survival time (also called the observed outcome). Under this framework with censored data, the observed data structure for estimating individualized treatment strategies can be represented as $\{(X_i, A_i, Y_i, \Delta_i), i = 1, 2, \ldots, n\}$. It is assumed that the observations from different individuals are independent and identically distributed. These data are typically collected from randomized clinical trials or observational studies. The goal of individualized treatment strategy estimation is to utilize such quadruple data structures to estimate the optimal individualized treatment regimen under censored data.

An individualized treatment rule (ITR), denoted by $d(\mathbf{x})$, is a mapping from the covariate space to the treatment space. Here, $d(\mathbf{x}) \in \mathcal{D}$ represents a class of ITRs of interest. In the context of a randomized clinical trial, we assume that the treatment A is assigned probabilistically. To ensure the estimability of the treatment effect, two key assumptions in causal inference are typically stated [2]:

1. **Stable Unit Treatment Value Assumption** :

$$Y_i(d(x)) = Y_i^*(1)d(x) + Y_i^*(0)(1 - d(x))$$

   where $Y_i^*(1)$ and $Y_i^*(0)$ represent the potential survival times of subject i under treatment 1 and treatment 0, respectively. Each patient can only receive one treatment assignment. Thus, we can observe either $Y_i^*(1)$ or $Y_i^*(0)$, but not both for any individual.
2. **Strong Ignorability Assumption**:

$$\{Y^*(0), Y^*(1)\} \perp\!\!\!\perp A \mid X$$

which implies that treatment assignment is independent of potential outcomes given the covariates X. This assumption holds if there is no unobserved confounding, and it is crucial in randomized trials where it is assumed that any treatment assignment is random and independent of the potential outcomes.

These assumptions are fundamental for estimating the optimal individualized treatment regimen in the presence of observational data or randomized trial data.To proceed further, let us denote the propensity score as

$$\pi(A, X) = \Pr(A = a \mid X = x) \tag{1}$$

which represents the conditional probability of an individual with covariates $X = x$ receiving treatment $A = a$. Suppose the propensity score is not equal to zero. In randomized clinical trials, $\pi(A, X)$ is known and is often set to $\pi(A, X) = 0.5$ to achieve balance, or designed to follow a certain dependency structure with the covariates. However, in observational studies, the propensity score $\pi(A, X)$ is typically unknown. It is common practice to assume a parametric model for $\pi(A, X; \gamma)$, and it is often assumed to follow a logistic distribution, that is

$$\pi(A, X; \gamma) = \left[1 + \exp\left(-(\gamma_1 x_1 + \ldots + \gamma_p x_p)\right)\right]^{-1} \tag{2}$$

Using methods such as least squares estimation or maximum likelihood estimation, one can obtain an estimate $\hat{\gamma}$ of the unknown parameter γ, leading to an estimated propensity score model $\pi(A, X; \hat{\gamma})$. Let $S_C^*(Y_i \mid A_i, X_i)$ denote the conditional survival function of the censoring time, given covariates $X = x$ and treatment $A = a$. In the presence of censoring, this function is crucial for appropriately adjusting the estimation.In the presence of censoring, let $S_C^*(Y_i|A_i, X_i)$ denote the conditional survival function of the censoring time, given covariates $X = x$ and treatment $A = a$. The value function of a treatment rule $d(x)$ can then be expressed as

$$V(d(x)) = \mathbb{E}\left[\frac{\Delta Y}{S_C^*(Y_i|A_i, X_i)\pi(A, X; \hat{\gamma})} I(A = d(x))\right] \tag{3}$$

By plugging in a consistent estimator S_C^* of the conditional censoring survival function S^c into the expression above, we obtain a consistent estimator $\hat{V}(d(x))$ of the value function $V(d(x))$, given by

$$\hat{V}(d(x)) = \frac{1}{n}\sum_{i=1}^{n}\left[\frac{\Delta_i Y_i}{\hat{S}_C^*(Y_i \mid A_i, X_i)\pi(A_i, X_i; \hat{\gamma})} I(A_i = d(x_i))\right] \tag{4}$$

Among a class of treatment rules $\mathcal{D}$, the optimal individualized treatment regimen d^{opt} is defined as

$$d^{\text{opt}}(x) = \arg\max_{d \in \mathcal{D}} V_n(d(x)) \tag{5}$$

In other words, for any treatment regime $d \in \mathcal{D}$, the optimal treatment regime d^{opt} satisfies $V(d^{\text{opt}}) \geq V(d)$. Considering a class of linear decision rules of the form $d(x) = I(\beta' X > c)$ (regardless of whether the form of the decision rule is correctly specified), the IPWE estimator $\hat{V}(d(x)) = \hat{V}(\beta, c)$ can be maximized over (β, c) to obtain the optimal treatment regime.

2.2 Concordance-Assisted Learning with Censored Data

Concordance-assisted learning aims to estimate the optimal individualized treatment regime by studying the benefits of different treatment options for each individual. While the true optimal decision rule may be nonlinear, we often assume it to be linear for better interpretability and computational efficiency, as linear rules can still achieve a high degree of accuracy.

Let the optimal linear decision regime be denoted by $I(\beta^{*'}X \geq c^*)$. In the absence of censored data, Fan et al. (2017) [6] proposed a method based on maximum rank correlation estimation for a concordance-assisted learning approach to estimate this optimal regimen. The fundamental concept is that for any two patients, i and j, if the incremental benefit of receiving treatment 1 versus treatment 0 is greater for patient i than for patient j, i.e., $(T_i^*(1) - T_i^*(0)) > (T_j^*(1) - T_j^*(0))$, then patient i is a more suitable candidate for treatment 1.

The Concordance-assisted learning method determines the optimal decision rule in two steps. First, it finds an optimal β^* by maximizing a Concordance function $C(\beta)$, which is defined as

$$C(\beta) = E((G(X_i) - G(X_j))I(\beta^T X_i > \beta^T X_j)) \tag{6}$$

where $G(X_i) = T_i^*(1) - T_i^*(0)$,which represents the benefit of treatment strategy 1 over treatment strategy 0.In the second step, β^* is substituted into the value function $V(d(x))$ (i.e., Eq. 4), and c^* is subsequently obtained by maximizing this function. The value function $V(d(x))$ is essentially a form of inverse propensity score weighting. Therefore, the optimal decision rule can be expressed as $I(\beta^* X > c^*)$.

However, when the outcome of interest is survival time, the presence of right-censoring complicates the estimation, making it impossible to find the optimal decision rule by directly maximizing the consistency function. To address this issue, assuming the propensity score model $\pi(A;X)$ is known, we introduce a K-S-V transformation to adjust the observed event times. The main idea is to up-weight the data from non-censored subjects and set the transformed value for censored subjects to zero. The K-S-V transformation [9] is then defined as

$$\tilde{T}_i^* = \frac{\Delta Y_i}{S(Y_i|A_i, X_i)}, \quad i = 1, 2, ..., n \tag{7}$$

where Δ_i is the event indicator for subject i. If the propensity score model $\pi(A;X)$ is known, based on the principles of A-learning [13], we can obtain the following relationship

$$\begin{aligned} G(X) &= E(\tilde{T}|X, A=1) - E(\tilde{T}|X, A=0) \\ &= E\left\{ \frac{\{\tilde{T} - m(X,\theta)\}\{A - \pi(A,X)\}}{\{1-\pi(A,X)\}\pi(A,X)} \Big| X \right\} \end{aligned} \tag{8}$$

Consequently, the modified consistency function can be expressed as

$$C(\beta) = E\Bigg\{ \bigg(\frac{\{T_i - m(X_i, \theta)\}\{A_i - \pi(A_i, X_i)\}}{\{1 - \pi(A_i, X_i)\}\pi(A_i, X_i)} - \frac{\{T_j - m(X_j, \theta)\}\{A_j - \pi(A_j, X_j)\}}{\{1 - \pi(A_j, X_j)\}\pi(A_j, X_j)} \bigg) \times I(\beta^T X_i > \beta^T X_j) \Bigg\} \tag{9}$$

Here, $m(X, \theta)$ is the parametric model for the conditional mean response under the baseline treatment, i.e., $E(Y|A = 0, X)$. Typically, $m(X, \theta)$ is assumed to be a constant or to follow a linear model. In the case of a linear model,$\hat{\theta}$ can be estimated using maximum likelihood or ordinary least squares with data from the $A = 0$ group.

Due to the presence of censored for some subjects, the true survival time $\tilde{T}$ cannot be fully observed. Therefore, we adapt the K-S-V transformation to deduce the contrast function from Eq. (7) as $G(X) = E[\Gamma(\theta, X)]$,where the transformed outcome function, $\Gamma(\theta, X_i)$, can be expressed as

$$\Gamma(\theta, X_i) = \Bigg\{ \frac{I(A=1)}{S_c^*(Y_i|A=1, X_i)\pi(A=1; X_i)} - \frac{I(A=0)}{S_c^*(Y_i|A=0, X_i)\pi(A=0; X_i)} \Bigg\} \Delta_i\{Y_i - m(X_i, \theta)\} \tag{10}$$

In this expression, $S_c^*(\cdot)$ denotes the estimated survival function of the censored time.

Finally, β^* can be estimated using the censored data by maximizing the following consistency function. The function is defined as:

$$\hat{C}(\beta) = E[\{\Gamma(\hat{\theta}, X_i) - \Gamma(\hat{\theta}, X_j)\} \times I(\beta^T X_i > \beta^T X_j)] \tag{11}$$

If we let $\hat{Z}_{ij}(\hat{\theta}) = \Gamma(\hat{\theta}, X_i) - \Gamma(\hat{\theta}, X_j)$ and plug in $\hat{\theta}$, then Eq. (11) can be written as

$$\hat{C}(\beta, \hat{\theta}) = \frac{1}{n(n-1)} \sum_{i \neq j} \hat{Z}_{ij}(\hat{\theta}) \times I(\beta^T X_i > \beta^T X_j) \tag{12}$$

where the estimation of β is subject to the constraint $||\beta|| = (\beta' \beta)^{\frac{1}{2}} = 1$. Thus, the estimator for β is $\hat{\beta} = \arg\max_{||\beta||=1} \hat{C}(\beta, \hat{\theta})$.

2.3 Estimation of Parameters θ, β, and c

To estimate the parameters β and c for the optimal decision rule, we must first estimate the unknown parameter θ in the baseline regression model, $m(X, \theta)$. In practical applications, $m(X, \theta)$ is often assumed to be linear, i.e., $m(X, \theta) = Z' \theta$, where Z is a set of basis functions derived from the covariates X, potentially including main effects and interaction terms. In some cases, $m(X, \theta)$ might be assumed to be a constant, or even zero. Given the data, $\hat{\theta}$ can be estimated using ordinary least squares (OLS) or maximum likelihood estimation (MLE).

Once $\hat{\theta}$ and the propensity score model have been estimated, these estimates are substituted into the objective function defined in Eq. (12). Maximizing this function yields an initial estimate of the parameter vector, which we denote as $\hat{\beta}^*$. For the purpose of standardization and comparison with other methods, this vector is subsequently normalized. The maximization of the target function to find $\hat{\beta}^*$ can be accomplished through various numerical optimization methods. For instance, a grid search is feasible when the covariate dimension is low. For high-dimensional data, a genetic algorithm is a more suitable approach, which can be implemented using packages such as `optim` package in R.

Next, with the normalized estimate $\hat{\beta}^* = \hat{\beta}^*/||\hat{\beta}^*||^2$, we proceed to estimate the cutoff value $\hat{c}^*$. This is done by maximizing the value function in Eq. (4) with respect to c. Since c is a scalar, this optimization can be easily performed using a simple grid search. The resulting optimal individualized treatment rule is then given by

$$\hat{d}^{\text{opt}}(x) = I(\hat{\beta}^{*\prime} X > \hat{c}^*) \tag{13}$$

It is important to note a key feature of this methodology. The estimation procedure first finds $\hat{\beta}^*$ by maximizing the consistency function (11), and subsequently finds $\hat{c}^*$ by maximizing the value function. Consequently, the optimal decision rule obtained through this two-step approach is not guaranteed to maximize the overall value function. Instead, this procedure is designed to find the decision rule that maximizes the empirical consistency function.

3 Simulations

To validate the performance of our proposed method, we conducted extensive simulation studies. We considered the following two methods for comparison:

- The concordance-assisted learning method, denoted as CAL.
- The inverse probability weighted estimator method, denoted as IPWE.

In our simulation, the observed event time is given by $T = \min(Y, C)$, with a censoring indicator $\Delta = I(Y \leq C)$. The censoring time C was generated from a uniform distribution, $C \sim U(0, M)$, where the parameter M is chosen to control the proportion of censoring.

We defined two data-generating scenarios for the survival time Y. The general model is:

$$Y = u(X) + AD(X) + \varepsilon \tag{14}$$

Here, the covariate vector is $X = (X_1, X_2, X_3, X_4)^T$, where each component is independently generated from a standard normal distribution. The treatment assignment A is drawn from a Bernoulli distribution, $A \sim \text{Bernoulli}(\pi(X))$. The error term is $\varepsilon \sim N(0, 0.1)$. We considered two distinct scenarios for the baseline and treatment effect functions:

Case I $u(X) = 1 + \gamma_1^T X$ and $D(X) = 2\beta_0^T X$.
Case II $u(X) = 1 + \gamma_1^T X$ and $D(X) = \exp(\beta_0^T X) - 1$.

In these scenarios, we set the parameter vectors as $\gamma_1 = (1, -1, 1, 1)^T$. To satisfy the unit norm constraint, we define $\beta_0 = (0.5, 0.5, -0.5, -0.5)^T$. To ensure the non-negativity of the survival time Y, if a generated Y value was negative, both Y and the corresponding covariate vector X were replaced by their negatives.

To evaluate the performance under different levels of censoring, we set the censoring proportions to approximately 10% and 20% for both scenarios. At a 10% censoring rate, the values of M are 24.01, 23.21, 19.2, and 15.36, respectively. For a censoring rate of 20%, the corresponding values are 11.55, 11.55, 9.61, and 7.71, respectively. For both scenarios, the true optimal decision rule is $d^{\text{opt}}(X) = I(\beta_0^T X \geq c_0)$, with $\beta_0 = (0.5, 0.5, -0.5, -0.5)^T$ and $c_0 = 0$. Based on a Monte Carlo simulation with 10^3 replicates under the no-censoring condition, the true values of the value function corresponding to this optimal rule are 2.91 for Case I and 2.83 for Case II.

Our objective is to estimate the optimal individualized treatment regime from the simulated tuple data (T, Δ, A, X) by maximizing the concordance and value functions. To validate the performance of our proposed method, we compare it against an inverse probability weighted estimator method.

Specifically, for the IPWE approach, the first step is to specify a class of ITRs. We assume a linear form for the decision rule, $d(X; \beta, c)$, while acknowledging that this form could be misspecified. The unknown parameters β of the ITR are then found by maximizing the estimated value function (the objective function). Due to the presence of an indicator function within the decision rule, this objective function is discontinuous and non-smooth. Therefore, we employ numerical optimization methods such as grid search or a genetic algorithm to find the optimal parameters.

In our implementation, we use the genetic algorithm provided by the `genoud` package in R. For the optimization procedure across all four simulation scenarios, we consistently used the following settings: `pop.size=3000`, `wait.generations=10`, and `optim.method="Nelder-Mead"`. The search space for the parameters was defined by a 5×2 matrix for the 'Domains' argument, with lower bounds of $(-1, -1, -1, -1, -1)$ and upper bounds of $(1, 1, 1, 1, 1)$. The initial search values were set to $(0, 0, 0, 0, 0)$. All other settings were left to their default values.

The IPWE estimator for the value of a given rule, $\hat{V}(\beta, c)$, is formulated as follows

$$\hat{V}(\beta, c) = \frac{1}{n} \sum_{i=1}^{n} \left[\frac{A_i d(X_i; \beta) \Delta_i Y_i}{\pi(A_i, X; \hat{\gamma}) S_c(Y_i | A, X)} + \frac{(1 - A_i)\{1 - d(X_i; \beta)\} \Delta_i Y_i}{\{1 - \pi(A_i, X; \hat{\gamma})\} S_c(Y_i | A, X)} \right] \tag{15}$$

We conducted 500 simulation replications, with a sample size of $n = 200$ for each run. To evaluate the performance of the estimators, we calculated several metrics. For our proposed CAL estimator, we report the mean of the parameter estimates (Est) for $\hat{\beta}$ and $\hat{c}$, the empirical standard deviation of these estimates across replications (SD), and the average of the estimated standard errors (SE).

To facilitate the calculation of the SE, as described in Sect. 3, we introduced a random perturbation to the consistency function. Specifically, we multiplied the individual contributions within the function by random weights drawn from an Exponential distribution with a mean and variance of one, and then maximized the perturbed objective function (12) to obtain β^*.

For the competing IPWE , we report the mean (Est) and standard deviation (SD) of the estimates for $\hat{\beta}$ and $\hat{c}$.

To assess the overall accuracy of the optimal decision rules derived from each method, we also examined the mean and standard deviation of the estimated value function ($\hat{V}$) and the Percentage of Correct Decisions (PCD) under the different scenarios and censoring rates. The value of an estimated optimal rule, $\hat{d}^{\text{opt}}(X_i)$, was calculated on a large independent test set of size N using the true data-generating functions:

$$\hat{V} = \frac{1}{N}\sum_{i=1}^{N}\{u(X_i) + \hat{d}^{\text{opt}}(X_i)D(X_i)\} \tag{16}$$

where $\hat{d}^{\text{opt}}(X_i) = I(\hat{\beta}^T X_i \geq \hat{c})$. The Percentage of Correct Decisions (PCD) was calculated as the proportion of subjects for whom the estimated rule matches the true optimal regime:

$$PCD = \frac{1}{n}\sum_{i=1}^{n} I\big(I(\hat{\beta}^T X_i \geq \hat{c}) = I(\beta_0^T X_i \geq c_0)\big) \tag{17}$$

The simulation results are presented in Tables 1 and 2. Table 1 displays the results for the IPWE and CAL methods for each scenario under a 10% censoring rate. Table 2 presents the corresponding results under a 20% censoring rate. The results from these simulations are summarized as follows.

When comparing the two methods under identical censoring rates and scenarios, the parameter estimates from the CAL method are consistently closer to the true values, exhibiting smaller bias than those from IPWE. Furthermore, the standard deviation of the CAL estimates is also smaller than that of the IPWE estimates. This demonstrates the stability and accuracy of the CAL method in estimating the optimal decision rule from censored data. While the standard deviation of the parameter estimates increases with the censoring rate, the bias remains small, indicating that the estimates fluctuate closely around the true values. This suggests that the CAL method remains relatively stable even as the amount of censoring increases.

Across both censoring levels, the empirical standard deviation (SD) and the average estimated standard error (SE) for the CAL method are in close agreement. This indicates that our resampling-based method for estimating the standard error of the CAL estimator performs well. Additionally, the Percentage of Correct Decisions (PCD) for CAL remains above 85% in all scenarios. As the censoring rate increases, the PCD shows a slight, acceptable decrease, with a correspondingly slight increase in its standard deviation. In contrast, the PCD

Table 1. Simulation results for four scenarios with a 10% censoring rate.

Method	Statistic	$\hat{\beta}_1$	$\hat{\beta}_2$	$\hat{\beta}_3$	$\hat{\beta}_4$	$\hat{c}$	$\hat{V}(d^{opt})$	PCD
IPWE	Mean	0.4141	0.4609	0.4609	−0.5871	0.4162	2.7414	89.22%
	SD	0.1620	0.1686	0.1294	0.1642	0.1365	0.1292	0.1378
CAL	Mean	0.5057	0.5019	−0.4996	0.4927	−0.0537	2.5825	95.35%
	SD	0.0837	0.0858	0.0831	0.0847	0.1244	0.0995	0.0468
	SE	0.0833	0.0861	0.0825	0.0840			
IPWE	Mean	0.2648	0.3679	−0.4437	0.6497	−0.0336	2.6964	78.82%
	SD	0.2562	0.2745	0.1008	0.1573	0.0781	0.1154	0.0141
CAL	Mean	0.5134	0.5155	−0.4699	0.4999	−0.0853	2.7204	89.94%
	SD	0.1138	0.1166	0.1300	0.1132	0.2502	0.0992	0.1367
	SE	0.1117	0.1164	0.1301	0.1116			

Table 2. Simulation results for four scenarios with a 20% censoring rate.

Method	Statistic	$\hat{\beta}_1$	$\hat{\beta}_2$	$\hat{\beta}_3$	$\hat{\beta}_4$	$\hat{c}$	$\hat{V}(d^{opt})$	PCD
IPWE	Mean	0.4080	0.4434	−0.6037	0.4084	−0.0725	2.7613	88.70%
	SD	0.1752	0.1717	0.1240	0.1743	0.1140	0.1326	0.1292
CAL	Mean	0.5027	0.4967	−0.5164	0.4836	−0.0453	2.5943	94.73%
	SD	0.1122	0.1093	0.0896	0.1103	0.1245	0.1074	0.0611
	SE	0.1119	0.1091	0.0889	0.1091			
IPWE	Mean	0.2548	0.3967	−0.4589	0.5395	0.0763	2.6811	83.85%
	SD	0.3243	0.3380	0.1397	0.1940	0.1959	0.1273	0.2338
CAL	Mean	0.5144	0.5065	−0.5014	0.4768	−0.5469	2.7335	85.93%
	SD	0.1923	0.1914	0.1552	0.1870	0.2527	0.1105	0.2323
	SE	0.1907	0.1897	0.1526	0.1847			

for the IPWE method exhibited greater variability across the different simulation scenarios.

In the specific instance of *case I* with a 10% censoring rate, the value function estimated via the IPWE method was higher than that from the CAL method. This result is consistent with our methodology; as stated in Sect. 2.3, the CAL procedure is designed to maximize the concordance function, which does not guarantee the maximization of the value function. In other scenario, however, the value function produced by the CAL method was higher than that produced by both the IPWE.

Under identical censoring rates and scenarios, the standard deviation of the estimated value function for the CAL method was substantially smaller than the standard deviation for the IPWE method, highlighting the superior stability of our proposed approach.

To account for the complexity that the optimal decision rule may be nonlinear, we consider a setting where the true optimal treatment regime is not linear. Specifically, the models are specified as follows:

Case III $u(X) = 1 + \sin(\gamma_1^\top \mathbf{X}) + 0.5(\gamma_2^\top \mathbf{X})^2$ and
$D(X) = |X_1^2 + X_1X_2 + X_3X_4| \operatorname{sign}(X_2^2 - X_1^2 + X_3 - X_4^2)$.

where γ_1 and γ_2 take the same values as in the previous setting, and the propensity score is set to $\pi(\mathbf{X}) = 0.5$. Under this setup, the corresponding optimal decision rule is $d^{\rm opt}(\mathbf{X}) = I\left(X_2^2 - X_1^2 + X_3 - X_4^2 > 0\right)$

Table 3. Simulation results for four scenarios with a 10% censoring rate.

Method	Statistic	$\hat{\beta}_1$	$\hat{\beta}_2$	$\hat{\beta}_3$	$\hat{\beta}_4$	$\hat{c}$	$\hat{V}(d^{\rm opt})$
IPWE	Mean	0.8312	0.5531	0.3019	0.1021	−0.0325	2.529
	SD	0.1764	0.1731	0.1241	0.1789	0.1126	0.1331
CAL	Mean	0.7027	0.6607	0.2558	0.0653	−0.0710	2.6345
	SD	0.1025	0.1370	0.1641	0.1782	0.0659	0.1599

We compare the performance of CAL and IPWE in this nonlinear setting by evaluating the mean and standard deviation of the estimated value functions. Since a linear model is misspecified under this scenario, it provides a rigorous test of robustness. Table 3 reports the estimated value functions for both methods. The results indicate that while both methods yield decision rules that approximate the true optimal rule, the CAL method achieves superior performance, with larger estimated value functions.

4 Real Data Analysis

To illustrate the application of our proposed method for handling censored data, we analyze data from the AIDS Clinical Trials Group Protocol 175 (ACTG175) study. This randomized clinical trial was designed to investigate the therapeutic effects of different treatment regimens on patients infected with HIV-1. A total of 2139 HIV-1-infected patients were enrolled and randomly assigned to one of four treatment arms:

- 532 patients received zidovudine (ZDV) monotherapy.
- 522 patients received didanosine (ddI) monotherapy.
- 561 patients received combination therapy with ZDV + ddI.
- 524 patients received combination therapy with ZDV + zalcitabine (ZAL).

The outcome of interest was the observed survival time for each patient. A censoring indicator, Δ, was used to denote whether the event of interest was observed, where $\Delta = 1$ indicates an observed event (i.e., no censoring) and $\Delta = 0$ indicates the survival time was right-censored.

For the purpose of estimating the optimal individualized treatment rule, our analysis focuses on the comparison between the two combination therapies. We designate the ZDV + zalcitabine (ZAL) arm as the control group ($A = 0$) and the ZDV + ddI arm as the experimental group ($A = 1$).

Table 4. Comparison of the mean and standard deviation of the estimated coefficient for the covariate *age* under different methods, using the ACTG175 dataset.

	IPWE	CAL
Mean	0.9063	0.9301
SD	0.0278	0.0215

To compare the performance of the different methods in our analysis, we conducted 500 bootstrap replications. For each replication, we applied both the CAL and IPWE methods to estimate the optimal rule and its value. Following the covariate selection approach of Lu (2011) [11], we focused our analysis on the influence of the single covariate, 'age'. The results, summarized in Table 4, show that while the mean estimated value is similar between the two methods, the standard error of the value estimate produced by the CAL procedure is substantially smaller than that from the IPWE method. This highlights the superior stability and efficiency of the CAL approach.

Using the CAL method, the optimal individualized treatment regimen was estimated to be $\hat{d}^{\text{opt}} = I(\text{age} \geq 35.33)$. This finding is in close agreement with the work of Lu et al., who, using CD4 cell count as the outcome variable, derived an optimal decision rule of $\hat{d}^{\text{opt}} = I(\text{age} \geq 35)$. The consistency of these results, despite the different outcome variables, provides validation for our approach.The cutoff value of 35.33 is derived directly from the data properties and the model structure. For the ACTG175 dataset, the mean value of *age* is 35.33 with a standard deviation of 8.75. When considering only the standardized *age* variable, the estimated rule can be expressed as:

$$\hat{d}^{\text{opt}} = I\left(8.91 \times \frac{\text{age} - 35.33}{8.75} > 0\right)$$

Solving the inequality simplifies the rule to its final, interpretable form:

$$\hat{d}^{\text{opt}} = I(\text{age} \geq 35.33)$$

The reason we obtained a nearly identical decision rule despite using a different outcome variable is that with a threshold $c = 0$, the optimal rule depends only on the sign of the standardized covariate, not the specific magnitude of its coefficient. This rule has a clear clinical interpretation: for older patients (age ≥ 35.33), the ZDV + ddI combination therapy ($A = 1$) appears to be more effective, whereas for younger patients, the ZDV + zalcitabine therapy ($A = 0$) is estimated to yield greater benefit.

5 Discussion

In this paper, we extend the framework of CAL to accommodate censored data, proposing a novel concordance-assisted learning method specifically for right-censored outcomes. Under the assumption of a correctly specified propensity score model, our method sequentially utilizes concordance function and value function to estimate the optimal treatment regimen. When compared with existing inverse probability weighting estimation methods, our approach yields more accurate value function estimates and achieves a higher percentage of correct decisions.

Survival data analysis often involves various types of censoring, including left, right, and interval censored, with right-censored being the most common. To date, the majority of research on estimating optimal individualized treatment regimens from censored data has concentrated on right-censored scenarios. However, our proposed CAL methodology is not inherently limited to this data type. Through the development of appropriate re-weighting techniques for left-censored and interval-censored data, the CAL framework can be readily adapted to these settings as well.

Furthermore, identifying the optimal regimens from multiple treatment options is a popular topic in contemporary ITR research. While a simple approach involves performing pairwise comparisons, the CAL framework can be extended to address the multi-category treatment estimation problem directly, building upon many well-established techniques in this area. This allows for the estimation of decision rules for multiple treatments where survival time is the outcome of interest.

References

1. Ashley, E.A.: Towards precision medicine. Nat. Rev. Genet. **17**(9), 507–522 (2016)
2. Basu, D.: Randomization analysis of experimental data: the fisher randomization test. In: Selected Works of Debabrata Basu. SWPS, pp. 305–325. Springer, New York (2011). https://doi.org/10.1007/978-1-4419-5825-9_28
3. Chakraborty, B., Murphy, S.A.: Dynamic treatment regimes. Annu. R. Stat. Appl. **1**(1), 447–464 (2014)
4. Collins, F.S., Varmus, H.: A new initiative on precision medicine. N. Engl. J. Med. **372**(9), 793–795 (2015)
5. Crutcher, M., Waldman, S.: Biomarkers in the development of individualized treatment regimens for colorectal cancer. Front. Med. **9**, 1062423 (2022)
6. Cui, Y., Zhu, R., Kosorok, M.: Tree based weighted learning for estimating individualized treatment rules with censored data. Electr. J. Stat. **11**(2), 3927 (2017)
7. Fan, C., Lu, W., Song, R., Zhou, Y.: Concordance-assisted learning for estimating optimal individualized treatment regimes. J. R. Stat. Soc. Ser. B Stat Methodol. **79**(5), 1565–1582 (2017)
8. Huang, X., Goldberg, Y., Xu, J.: Multicategory individualized treatment regime using outcome weighted learning. Biometrics **75**(4), 1216–1227 (2019)
9. Koul, H., Susarla, V., Van Ryzin, J.: Regression analysis with randomly right-censored data. Annals Stat. pp. 1276–1288 (1981)

10. Liu, Y., Wang, Y., Kosorok, M.R., Zhao, Y., Zeng, D.: Augmented outcome-weighted learning for estimating optimal dynamic treatment regimens. Stat. Med. **37**(26), 3776–3788 (2018)
11. Lu, W., Zhang, H.H., Zeng, D.: Variable selection for optimal treatment decision. Stat. Methods Med. Res. **22**(5), 493–504 (2013)
12. Moodie, E.E., Chakraborty, B., Kramer, M.S.: Q-learning for estimating optimal dynamic treatment rules from observational data. Can. J. Stat. **40**(4), 629–645 (2012)
13. Murphy, S.A.: Optimal dynamic treatment regimes. J. R. Stat. Soc. Ser. B Stat Methodol. **65**(2), 331–355 (2003)
14. Qian, M., Murphy, S.A.: Performance guarantees for individualized treatment rules. Ann. Stat. **39**(2), 1180 (2011)
15. Shen, T., Cui, Y.: Optimal treatment regimes for proximal causal learning. Adv. Neural. Inf. Process. Syst. **36**, 47735–47748 (2023)
16. Wang, C., Tom, B.D.: A tutorial on optimal dynamic treatment regimes. arXiv preprint arXiv:2502.16988 (2025)
17. Wang, P., Li, Y., Reddy, C.K.: Machine learning for survival analysis: a survey. ACM Comput. Sur. (CSUR) **51**(6), 1–36 (2019)
18. Zhang, B., Tsiatis, A.A., Laber, E.B., Davidian, M.: A robust method for estimating optimal treatment regimes. Biometrics **68**(4), 1010–1018 (2012)
19. Zhao, Y.Q., Zeng, D., Laber, E.B., Song, R., Yuan, M., Kosorok, M.R.: Doubly robust learning for estimating individualized treatment with censored data. Biometrika **102**(1), 151–168 (2015)
20. Zhao, Y., Zeng, D., Rush, A.J., Kosorok, M.R.: Estimating individualized treatment rules using outcome weighted learning. J. Am. Stat. Assoc. **107**(499), 1106–1118 (2012)
21. Zhou, X., Mayer-Hamblett, N., Khan, U., Kosorok, M.R.: Residual weighted learning for estimating individualized treatment rules. J. Am. Stat. Assoc. **112**(517), 169–187 (2017)

PID Control with Pedal Torque Estimation for Safety Monitoring in Lower-Limb Rehabilitation Robots

Yue Jing[1,2,3], Jinhua She[4](✉), Seiichi Kawata[1,2,3], and Zewen Wang[5]

[1] School of Automation, China University of Geosciences, Wuhan 430074, China
[2] Hubei Key Laboratory of Advanced Control and Intelligent Automation for Complex Systems, Wuhan 430074, China
[3] Engineering Research Center of Intelligent Technology for Geo-Exploration, Ministry of Education, Wuhan 430074, China
[4] Graduate School of Engineering, Tokyo University of Technology, Hachioji, Tokyo 192-0982, Japan
she@stf.teu.ac.jp
[5] School of Mechanical Engineering and Electronic Information, China University of Geosciences, Wuhan 430074, China

Abstract. This article presents a PID control system for a lower-limb rehabilitation robot that not only features satisfactory control performance for pedal angle but also provides us with the function of pedal torque estimation. A PID controller guarantees that the pedal angle tracks a reference trajectory. The integration of the equivalent-input-disturbance (EID) approach in the control system precisely estimates pedal torque in a real-time fashion. And the method achieves safety monitoring by detecting abnormal torque variations such as those caused by spasms or excessive effort. Simulation results and comparisons with active disturbance estimation methods validate the effectiveness of the method.

Keywords: Equivalent input disturbance (EID) · Human–robot interaction · Lower-limb rehabilitation robot (LLRR) · Torque estimation

1 Introduction

The global population suffering from neurological injuries and physical disabilities continues to grow annually, with stroke being the main cause of lower-limb motor dysfunction [1]. Lower-limb rehabilitation places a heavy burden on physiatrists. Moreover, currently there are not enough physiatrists [2].This situation accelerated the development of lower-limb rehabilitation robots (LLRRs) [3,4].

Effective human-robot interaction (HRI) requires accurate estimation of pedal torque for rehabilitation assessment and safety assurance [5,6], particularly across different rehabilitation stages. Current approaches using torque sensors or electromyography are limited by cost, susceptibility to noise, and an inability to address specific requirements during the all-stage rehabilitation [7,8].

H. Ma et al. (Eds.): IWACIII 2025, CCIS 2780, pp. 204–215, 2026.
https://doi.org/10.1007/978-981-95-6730-0_15

Active disturbance-estimation methods, such as the disturbance observer (DOB) [9] and the extended state observer (ESO) [10], are applied to torque estimation. Sebastian et al. [11] designed a linear ESO to estimate time-varying interaction force. Although active disturbance-estimation methods are useful for estimating torque, the DOB requires an inverse model of a plant [12], and the ESO relies on assumptions about a disturbance. The challenge of accurate pedal torque estimation is further compounded by the complex neuromuscular state of stroke patients throughout the rehabilitation process [13]. Since the frequency ranges of pedal torque and tremor torque are close, how to accurately estimate pedal torque is a difficult task [14].

As one of the active disturbance-estimation methods, the equivalent-input-disturbance (EID) approach [15,16] overcomes the shortcomings of the DOB and ESO methods. We present a strategy of estimating pedal torque in this paper. The strategy combines the PID controller and the EID estimator in a pedal rehabilitation robot system that regards the overall torque as a total disturbance and uses the EID approach to estimate the pedal torque of patients. Compared to the aforementioned methods, this strategy has two advantages:

1. The EID approach with a high-order low-pass filter ensures high-precision extraction of pedal torque.
2. The EID approach is effective in achieving the safety monitoring of torque.

2 Modeling of Lower-Limb Rehabilitation Robot

Since an ankle joint is tied to the pedal using a bandage, the ankle joint and pedal can be considered as a whole. This provides us with a four-link model of the human-robot system for one lower limb (Fig. 1(a)).

Set the hip joint be the origin of a coordinate system. Let A be the knee joint, B be the pedal pivot and C be the crank pivot (Fig. 1(b)). l_i $(i = 1, 2, 3)$ is the length of the thigh rod, calf rod, and crank rod, respectively. l_4 is the fixed length from the hip joint to the crank pivot after adjusting the sitting position, and h is the corresponding vertical distance from the hip joint to the crank pivot. θ_i $(i = 1, 2, 3)$ are the rotation angles of the hip joint, knee joint, and crank, respectively; and θ_4 is the angle between the thigh of and the horizontal line at $\theta_3 = 0$. The geometric relationship between a lower limb and the LLRR is

$$\overrightarrow{OA} + \overrightarrow{AB} = \overrightarrow{OC} + \overrightarrow{CB}, \tag{1}$$

or in the exponential expression

$$l_1 \mathrm{e}^{i\theta_1} + l_2 \mathrm{e}^{i\theta_2} = l_3 \mathrm{e}^{i\theta_3} + l_4 \mathrm{e}^{i\theta_4}. \tag{2}$$

Deriving θ_1 in (2) yields

$$\theta_1 = 2 \arctan \frac{\varphi_2 + \sqrt{\varphi_1^2 + \varphi_2^2 - \varphi_3^2}}{\varphi_1 + \varphi_3}, \tag{3}$$

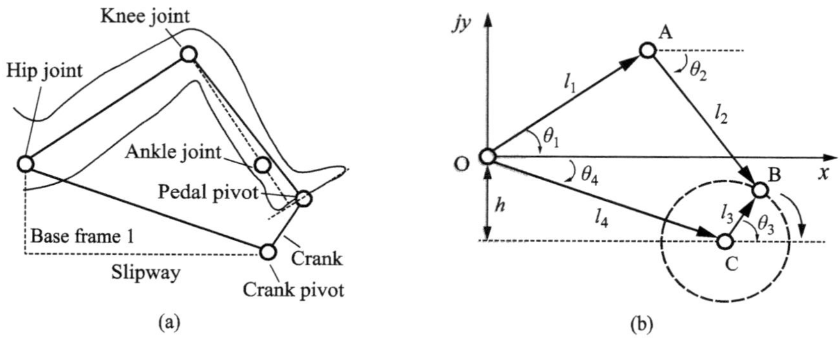

Fig. 1. Four-link model of human-robot system: (a) Schema and (b) four-link model.

where

$$\begin{cases} \varphi_1 = l_3 \sin\theta_3 + l_4 \sin\theta_4, \\ \varphi_2 = l_3 \cos\theta_3 + l_4 \cos\theta_4, \\ \varphi_3 = \dfrac{l_1^2 - l_2^2 + \varphi_1^2 + \varphi_2^2}{2l_1}. \end{cases} \tag{4}$$

The Lagrangian equation of the system is

$$\frac{\mathrm{d}}{\mathrm{d}t}\left(\frac{\partial L'}{\partial \dot{\theta}}\right) - \frac{\partial L'}{\partial \theta} = \tau, \tag{5}$$

where $L' = T - V$ is the Lagrangian function, θ is the generalized coordinate, $\dot{\theta}$ is the velocity of the system, and τ is the generalized torque. For $i = 1$ (OA), 2 (AB), and 3 (BC), m_i is the mass of Link i, I_i is the moment of inertia of Link i, and $\dot{\theta}_i$ is the angular velocity of the center of mass of Link i.

Choosing θ_3 as the generalized coordinate and solving (5) yields

$$\left(\frac{1}{4}m_3 l_3^2 + I_3\right)\ddot{\theta}_3 + \frac{1}{2}m_3 g l_3 \cos\theta_3 = \tau_{\theta_3}, \tag{6}$$

where

$$\begin{cases} \tau_{\theta_3} = \tau_f + \tau_h, \\ \tau_h = \tau_p + \tau_t + \tau_g, \end{cases} \tag{7}$$

τ_f is the driving torque of the pedal, τ_h is the total torque, τ_p is the pedal torque, τ_g is the gravity torque, and τ_t is the tremor torque of a user. Rewriting (6) gives

$$\ddot{\theta}_3 = \frac{1}{\frac{1}{4}m_3 l_3^2 + I_3}(\tau_f + \tau_h) - \frac{\frac{1}{2}m_3 g l_3}{\frac{1}{4}m_3 l_3^2 + I_3}\cos\theta_3. \tag{8}$$

The state-space representation of (8) is

$$\begin{cases} \dot{x} = Ax + B_f f(x) + B(\tau_f + \tau_h), \\ y = Cx, \end{cases} \tag{9}$$

where

$$x = \left[\theta_3 \; \dot{\theta}_3\right]^{\mathrm{T}}, \; f(x) = \cos\theta_3, \tag{10}$$

$$\begin{cases} A = \begin{bmatrix} 0 & 1 \\ 0 & 0 \end{bmatrix}, \; B = \left[0 \; \dfrac{4}{m_3 l_3^2 + 4I_3}\right]^{\mathrm{T}}, \\ B_f = \left[0 \; \dfrac{2m_3 g l_3}{m_3 l_3^2 + 4I_3}\right]^{\mathrm{T}}, \; C = \left[1 \; 0\right]. \end{cases} \tag{11}$$

3 Configuration of PID-EID Control System

A PID-EID control system (Fig. 2) consists of seven parts: a plant, a feedforward, a nonlinear state feedback, a reference trajectory planner, a PID controller, a state observer, and an EID estimator.

The EID contains a mearsurable part, that is the gravity of the user. A feedforward produces a τ_g that compensates for the gravity of the user.

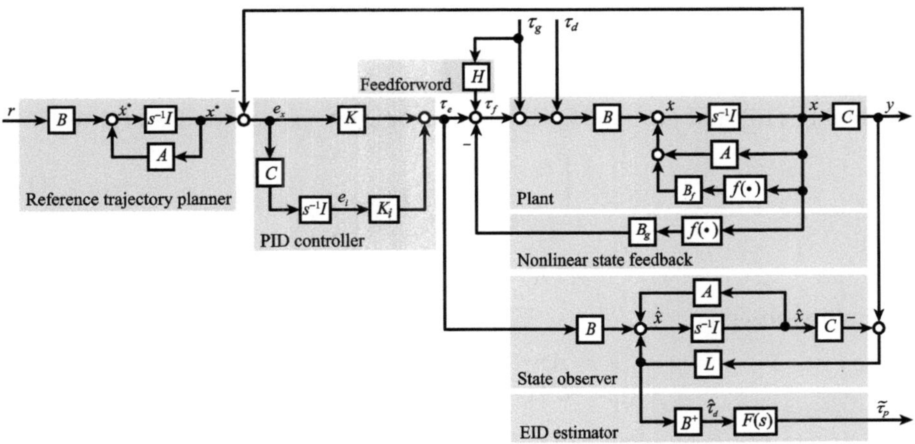

Fig. 2. Configuration of PID-EID control system.

The nonlinear behavior in (9) is related to the gravity of the crank. A nonlinear state feedback is $\tau_n = B_g f(x)$ that compensates for the gravity of the crank, where

$$\begin{cases} B_g = B^+ B_f, \\ B^+ = \left(B^{\mathrm{T}} B\right)^{-1} B^{\mathrm{T}}. \end{cases} \tag{12}$$

Nonlinear state feedback and feedforward form a control law

$$\tau_f = \tau_e + H\tau_g - \tau_n, \tag{13}$$

where $H = -I$ represents the measurement matrix, and τ_e is the output of the PID controller. Let $\tau_d = \tau_p + \tau_t$, where τ_d is the total disturbance including the pedal torque and tremor torque. The plant (9) becomes a linear one

$$\begin{cases} \dot{x} = Ax + B(\tau_e + \tau_d), \\ y = Cx. \end{cases} \tag{14}$$

by using the control law (13).

The reference trajectory is planned by $\dot{x}^* = Ax^* + Br$, where r is a torque signal that is used to generate the desired trajectory, $x^* = \left[\theta_3^* \ \dot{\theta}_3^*\right]^{\mathrm{T}}$, and θ_3^* and $\dot{\theta}_3^*$ devote the desired trajectory for angle and velocity respectively.

The PID controller is

$$\tau_e = K_p e + K_i \int_0^t e \ \mathrm{d}t + K_d \frac{\mathrm{d}e}{\mathrm{d}t}, \tag{15}$$

where $e = \theta_3^* - y$ is the error of trajectory tracking. Define

$$e_x = x^* - x = \left[e \ \ \dot{e}\right]^{\mathrm{T}}, \quad e_i = \int_0^t e \ \mathrm{d}t. \tag{16}$$

Combining (15) and (16) yields the implementation expression for (15) as

$$\begin{cases} \dot{e}_i = Ce_x, \\ \tau_e = Ke_x + K_i e_i, \end{cases} \tag{17}$$

where

$$K = \left[K_p \ K_d\right]. \tag{18}$$

A state observer

$$\begin{cases} \dot{\hat{x}} = A\hat{x} + B\tau_e + L(y - C\hat{x}), \\ \hat{y} = C\hat{x}, \end{cases} \tag{19}$$

reconstructs the liear plant (14), L is an observer gain, and $\hat{x}$ and $\hat{y}$ are the estimate state and output respectively. Let $\Delta x = x - \hat{x}$. An EID estimate of the torque including the pedal torque and the tremor torque is

$$\hat{\tau} = B^+ L(y - C\hat{x}) = B^+ LC\Delta x. \tag{20}$$

Since pedal torque and tremor torque have different frequencies, a low-pass filter $F(s)$ in the EID estimator selects the frequency band for the pedal torque estimation and filters out the tremor torque. The filter satisfies

$$|F(j\omega)| \approx 1, \ \forall \omega \in [0, \omega_c], \tag{21}$$

where ω_c is the cutoff frequency of $F(s)$. The filtered disturbance estimate, $\tilde{\tau}_p$, is

$$\tilde{T}_p(s) = F(s)\hat{T}(s), \tag{22}$$

where $\tilde{T}_p(s)$ and $\hat{T}(s)$ are the Laplace transforms of $\tilde{\tau}_p$ and $\hat{\tau}$, respectively.

4 Simulation Verification

This section verifies the effectiveness of the PID-EID method by simulation. The simulations are designed to evaluate the performance of the proposed control and estimation method under various rehabilitation scenarios, including early-stage and mid-to-late-stage rehabilitation conditions with different patient groups.

4.1 Design of Parameters

Experiments were conducted using Simulink Desktop Real-Time (R2022b) with a sampling frequency of 1000 Hz. Table 1 lists the parameters of the human–robot system model for simulating the process of pedal training for patients.

To meet the rehabilitation training requirements of patients, the reference angular velocity was $\dot{\theta}_3^* = \pi\,(1 - \cos\pi t)$. This simulated the process of acceleration and deceleration during normal pedaling. Thus, we get the reference input of the crank angle $\theta_3^* = \int \pi\,(1 - \cos\pi t)\;\mathrm{d}t$. And the external input r is obtained to drive the reference trajectory generator for producing the reference trajectory:

$$r = \frac{\pi^2\left(m_3 l_3^2 + 4I_3\right)\sin\pi t}{4}. \tag{23}$$

The parameters, K_p, K_i, and K_d, in the PID controller are designed by the optimal control method. So we get $K_p = 1.31,\ K_i = 31.62,\ K_d = 9.15$.

For different stages of rehabilitation, the patient's pedal torque and tremor also vary. In the early stage of rehabilitation, the patient's muscle strength is severely weakened. Since τ_g was compensated, the actual disturbance τ_{d1} added to the control system was generated by τ_{p1} and τ_{t1}

$$\begin{cases} \tau_{p1} = 0.6\sin\pi t, \\ \tau_{t1} = 0.2\sin 8\pi t, \end{cases} \tag{24}$$

where τ_{p1} and τ_{t1} are sinusoidal signals with 0.5 Hz and 4 Hz, respectively (Fig. 3(a)).

Table 1. Parameters of the human–robot system model.

Parameter	Thigh	Calf	Crank
l_i (m)	0.434	0.447	0.200
m_i (kg)	9.843	4.808	2.500
I_i $(\mathrm{kg}\cdot\mathrm{m}^2)$	0.464	0.241	0.025

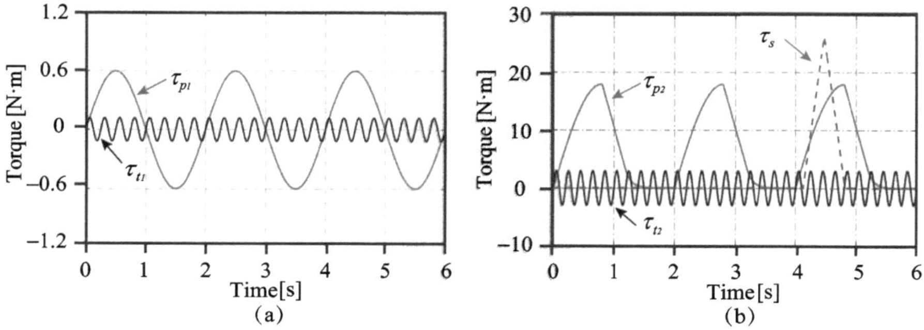

Fig. 3. Input of the disturbance: (a) τ_{p1} and τ_{t1}, (b) τ_{p2} , τ_{t2}, and τ_s.

The patient's pedal torque in the mid-to-late stage of rehabilitation was designed as follows:

$$\tau_{base} = \begin{cases} 30\sin\left(\dfrac{\pi}{2}\cdot\dfrac{\phi}{0.4T}\right), & 0 \le \phi < 0.4T \\ 30 - 25\left(\dfrac{\phi - 0.4T}{0.2T}\right), & 0.4T \le \phi < 0.6T \\ 5\exp\left(-8\dfrac{\phi - 0.6T}{0.4T}\right), & 0.6T \le \phi < T, \end{cases} \tag{25}$$

with $\phi = t \bmod T$ ($T = 2$ s cycle), while tremor torque increases to $\tau_{t2} = 3\sin 10\pi t$. Let $\tau_{p2} = \beta \cdot \tau_{base}$, where $\beta \in \{1.0, 0.6, 0.5\}$ for healthy, stroke, and Parkinson's patients respectively (Fig. 3(b)).

To achieve safety monitoring of patient's voluntary torque, the spasm torque τ_s with triangular wave characteristics was introduced. Therefore, the total disturbance τ_{d2} was

$$\tau_{d2} = \tau_{p2} + \tau_{t2} + \tau_s. \tag{26}$$

A third-order Butterworth low-pass filter was constructed as

$$F(s) = \frac{\omega_c^3}{s^3 + 2\omega_c s^2 + 2\omega_c^2\, s + \omega_c^3}, \tag{27}$$

where $\omega_c = 2\pi$ rad/s is the cutoff angular frequency range for torque estimation. The design of $F(s)$ ensures that the magnitude of $F(s)$ is approximately 0 dB in Ω_d (Fig. 4)

To achieve safety monitoring of abnormal torque variations such as those caused by spasms in stroke patients, a patient-specific safety threshold for the torque change rate ($\frac{\mathrm{d}\tau}{\mathrm{d}t}$) was established. This threshold is designed to distinguish between high voluntary effort and involuntary spasms, thereby preventing potential injuries caused by sudden abnormal torque variations.

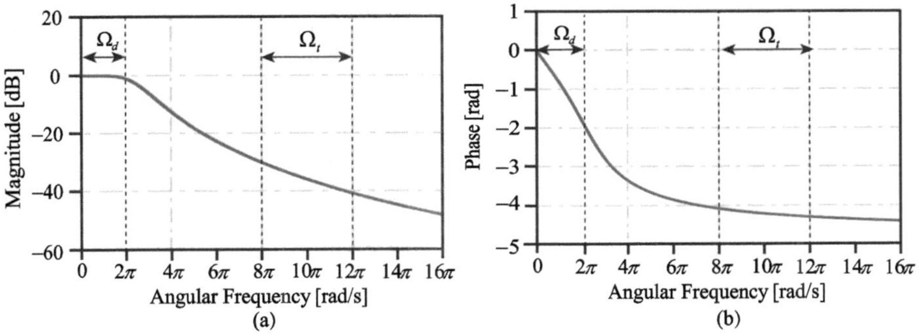

Fig. 4. Bode plots of $F(s)$: (a) magnitude and (b) phase.

The threshold is derived from the patient's own maximum voluntary torque capability. For the mid-to-late-stage stroke patients defined in the simulation (τ_{p2} with $\beta = 0.6$), the peak pedal torque is 18 $\mathrm{N \cdot m}$, which is achieved in approximately 0.8 s. The corresponding peak torque change rate during normal pedaling is estimated as the foundation for threshold setting:

$$\left(\frac{\mathrm{d}\tau}{\mathrm{d}t}\right)_{\mathrm{normal\_peak}} \approx 35\ \mathrm{N \cdot m/s}. \tag{28}$$

Since spasms are characterized by a rapid, involuntary muscle contraction with a torque change rate significantly exceeding that of voluntary movement, the safety threshold must be set above this normal peak. The system response time $\Delta t_{\mathrm{response}} = 0.05$ s is considered to ensure timely intervention. A safety margin factor α is introduced to account for the intensified effort patients may exhibit during recovery. The threshold is thus calculated as:

$$\left(\frac{\mathrm{d}\tau}{\mathrm{d}t}\right)_{\mathrm{safe}} = \alpha \cdot \left(\frac{\mathrm{d}\tau}{\mathrm{d}t}\right)_{\mathrm{normal\_peak}} + \frac{\Delta\tau_{\mathrm{base}}}{\Delta t_{\mathrm{response}}}. \tag{29}$$

Here, $\Delta\tau_{\mathrm{base}} = 2.5\ \mathrm{N \cdot m}$ represents a baseline torque variation considered safe within the response window, and $\alpha = 1.5$ is the safety margin factor. This yields the specific safety threshold for stroke patients:

$$\left(\frac{\mathrm{d}\tau}{\mathrm{d}t}\right)_{\mathrm{safe}} = 1.5 \times 35 + \frac{2.5}{0.05} = 102.5\ \mathrm{N \cdot m/s} \approx 100\ \mathrm{N \cdot m/s}. \tag{30}$$

This patient-specific threshold, rounded to 100 $\mathrm{N \cdot m/s}$ for implementation, ensures sensitive detection of abnormal spasms while maintaining robustness against false alarms during effective rehabilitation training.

4.2 Simulation Results

The experimental work was composed of two parts. In the first part, the trajectory tracking performance of the PID controller was evaluated for the pedal

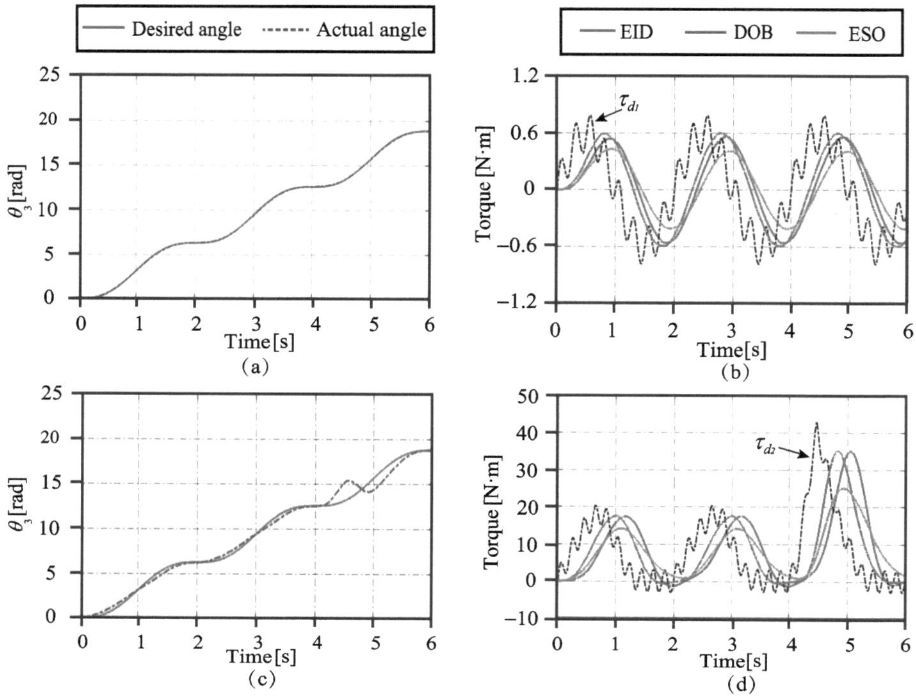

Fig. 5. Simulation results: (a) Trajectory tracking of θ_3 (early-stage), (b) Estimation of τ_{p1} (early-stage), (c) Trajectory tracking of θ_3 (mid-to-late-stage) and (d) Estimation of $\tau_{p2} + \tau_s$ (mid-to-late-stage).

rehabilitation robot system. Since the pedal and tremor torques of the patient were minimal in the early rehabilitation stage, the PID-controlled crank angle trajectory (blue dashed line) can well track the desired trajectory (red solid line), showing high tracking accuracy and stability (Fig. 5(a)). However, the pedal and tremor torques of patient significantly increased in the mid-to-late rehabilitation stage, and there may be abnormal torque caused by spasms. Such high-amplitude, dynamically severe abnormal torque disturbances lead to a significant decline in system tracking performance, with a noticeable increase in the deviation of the crank angle trajectory (Fig. 5(c)).

In the second part, the performance evaluation of the EID method for pedal torque estimation and its application to safety monitoring were presented. To evaluate the effectiveness and superiority of the EID approach, comparative experiments were conducted against ESO and DOB approaches. Table 2 summarizes the performance comparison of the three estimation methods in different rehabilitation stages. In the early rehabilitation stage (Fig. 5(b)), the EID method achieved an RMSE of 0.37 $\text{N} \cdot \text{m}$, representing a 47.6 % improvement over DOB (0.70 $\text{N} \cdot \text{m}$) and a 51.8 % improvement over ESO (0.77 $\text{N} \cdot \text{m}$). In the mid-to-late rehabilitation stage (Fig. 5(d)), the EID method maintained sig-

Table 2. Performance comparison of torque estimation methods.

Method	RMSE (N · m)		Phase Lag (rad)	
	Early Stage	Mid-to-Late Stage	Early Stage	Mid-to-Late Stage
EID	0.37	7.61	0.52	1.58
DOB	0.70	15.26	1.04	2.64
ESO	0.77	12.62	1.33	2.42

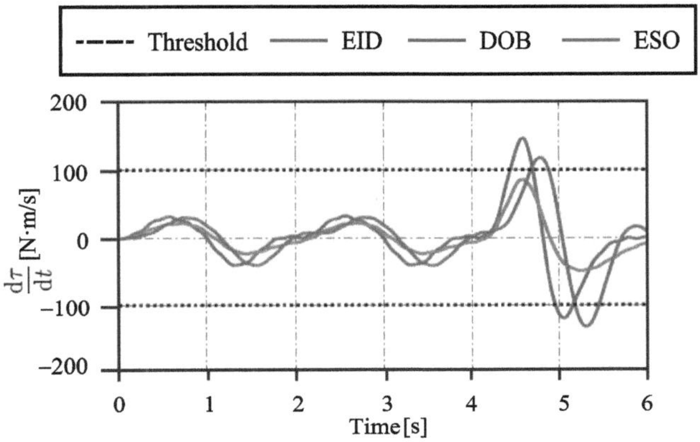

Fig. 6. Simulation result of safety monitoring.

nificantly lower estimation error (7.61 N · m) compared to DOB (15.26 N · m) and ESO (12.62 N · m), corresponding to improvements of 50.1 % and 39.7 %, respectively. The EID method also exhibited the smallest phase lag in both rehabilitation stages, measuring 0.52 rad compared to 1.04 rad for DOB and 1.33 rad for ESO in the early stage, and 1.58 rad versus 2.64 rad for DOB and 2.42 rad for ESO in the mid-to-late stage. The EID method successfully extracted the pedal torque while effectively rejecting high-frequency tremor torque, demonstrating its frequency-selective estimation capability. This is particularly important for rehabilitation applications because accurately estimating the patient's voluntary effort is crucial for treatment assessment and adjustment.

The analysis of $\frac{d\tau}{dt}$ played a critical role in achieving real-time safety monitoring of pedal torque. With a safety threshold of 100 N · m/s based on clinical safety standards, the EID and DOB methods successfully detected abnormal torque exceeding the safety threshold (Fig. 6). The ESO method failed to detect abnormal torque in time due to its slow response characteristics and low estimation accuracy. In addition, the EID method detected approximately 2.53 s earlier than the DOB method, significantly improving response time in safety monitoring. Rapid safety monitoring provides valuable time for safety interventions, preventing patients from feeling uncomfortable or getting injured during rehabilitation training.

5 Conclusion

This paper presents a new pedal torque estimation method based on PID-EID for pedal rehabilitation robots. The key contributions of this work include the development of a human-robot system model of the pedal rehabilitation robot and the analysis in terms of kinematics and dynamics, the design of a nonlinear state feedback controller using measurable nonlinear terms, and the experimental validation of the method for the pedal rehabilitation robot. Simulation results show that the method can effectively extract the pedal torque while monitoring the abnormal torque.

Future work will focus on experimental validation with human subjects and further optimizing safety thresholds based on clinical data collected during actual rehabilitation training. Additional considerations for implementation include the integration of patient-specific adaptive algorithms and the development of user-friendly interfaces for clinical therapists.

Acknowledgments. This work was supported in part by the Natural Science Foundation of Hubei Province, China, under Grant 2020CFA031; Fundamental Research Funds for the Central Universities, China University of Geosciences (Wuhan), under Grants CUG250652 and CUG240635; the "CUG Scholar" Scientific Research Funds at China University of Geosciences (Wuhan) under Grants 2025030 and 2022029; JSPS (Japan Society for the Promotion of Science) KAKENHI under Grants 24K03325 and 25K07807; and Foundation of Shanghai Academy of Spaceflight Technology (SAST) under Grant SAST2023-034.

References

1. Chen, R., Zhang, Y., Wang, X.: Treatment effects of low-frequency repetitive transcranial magnetic stimulation combined with motor relearning procedure on spasticity and limb motor function in stroke patients. Front. Neurol. **14**, 1213624 (2023)
2. Wu, Q., Chen, Y.: Variable admittance time-delay control of an upper limb rehabilitation robot based on human stiffness estimation. Mechatronics **90**, 102935 (2023)
3. Kuroda, S., She, J., Nakamuro, S., Wang, R.: Selecting pedal load for lower-limb rehabilitation based on the combination of muscle synergy and Fourier series. J. Adv. Comput. Intell. Intell. Inform. **28**(3), 595–605 (2024)
4. Zhu, Z., Liu, L., Zhang, W., Jiang, C., Li, J.: Design and motion control of exoskeleton robot for paralyzed lower limb rehabilitation. Front. Neurosci. **18**, 1355052 (2024)
5. Jiang, L., Zhao, J., Wang, F., Zhou, Y., Ge, W., She, J.: Rehabilitation evaluation system for lower-limb rehabilitation robot. Front. Neurosci. **27**(4), 691–699 (2023)
6. Wang, Y., Tian, Y., Guo, Y., Wang, H.: Active torque-based gait adjustment multilevel control strategy for lower limb patient-exoskeleton coupling system in rehabilitation training. Math. Comput. Simul. **215**, 357–381 (2023)
7. Jiang, J., Gao, Z., Li, G.: A miniature elastic torque sensor for index finger exoskeletons. IEEE Trans. Instrum. Meas. **72**, 1–10 (2023)

8. Li, S., Zhang, L., Meng, Q., Yu, H.: A real-time control method for upper limb exoskeleton based on active torque prediction model. Bioengineering **10**(12), 1441 (2023)
9. Mohammadi, A.: and Dallali, H: Chapter 5 - Disturbance observer applications in rehabilitation robotics: an overview. In: Powered Prostheses, pp. 113–133. Academic Press (2020)
10. She, J., Miyamoto, K., Han, Q., Wu, M., Hashimoto, H., Wang, Q.: Generalized-extended-state-observer and equivalent-input-disturbance methods for active disturbance rejection: deep observation and comparison. IEEE/CAA J. Autom. Sinica **10**(4), 957–968 (2023)
11. Sebastian, G., Li, Z., Crocher, V., Kremers, D., Tan, Y., Oetomo, D.: Interaction force estimation using extended state observers: an application to impedance-based assistive and rehabilitation robotics. IEEE Robot. Autom. Lett. **4**(2), 1156–1161 (2019)
12. Aoun, C. and Moncayo, H: Disturbance observer-based reinforcement learning control and the application to a nonlinear dynamic system. In: AIAA SCITECH (2022)
13. He, R., Yang, J., Tang, B., Sun, Q.: Research advances in the treatment of essential tremor. Chin. J. Neurol. **54**(4), 404–408 (2021)
14. Cortez, J., Lee, J., Kim, J., Chang, S.: Thalamo-cortical network is associated with harmaline-induced tremor in rodent model. Exp. Neurol. **358**, 114210 (2022)
15. Wang, H., She, J., Wang, H., Kawata, S., Iwasaki, M.: Disturbance rejection and performance improvement for control systems using a finite-time equivalent-input-disturbance approach. IEEE Trans. Industr. Electron. **72**(5), 5365–5375 (2025)
16. Wang, H., She, J., Wang, F.: Disturbance rejection for nonlinear systems using nonlinear-state-observer-based equivalent-input-disturbance approach. IEEE Trans. Industr. Inform. **20**(1), 953–962 (2024)

Machine Learning

Optimization of Traction Transformer Status Prediction Based on Missing Data Imputation

Guanlin Qu[1], Wei Wei[1(✉)], and Weifan Wang[2]

[1] School of Intelligent Engineering and Automation, Beijing University of Posts and Telecommunications, Beijing 102206, China
weiwell@bupt.edu.cn

[2] Infrastructure Inspection Research Institute, China Academy of Railway Sciences Co., Ltd., Beijing 100081, China

Abstract. Traction transformers face challenges, such as complex operating environments, communication interference, and sensor malfunctions, which result in missing or inconsistent data, thereby affecting the equipment condition prediction models. To address it, a Wasserstein generative adversarial imputation network with gradient penalty (WGAIN-GP) is employed to impute missing data, which is then utilized in the model prediction. Imputed data generated by various imputation methods were added as augmentation data to a Sequence-to-sequence (Seq2seq) model for predicting traction transformer status. Results indicate that the WGAIN-GP demonstrates superior imputation performance, and the imputed data significantly improve the prediction accuracy of traction transformer status.

Keywords: Traction transformers · Data imputation · Time series forecasting · Generative adversarial networks

1 Introduction

Traction transformers, as critical components of railway power supply systems, require effective monitoring and prediction of their operational status to ensure the safe and stable operation of the entire system [1, 2]. With the development of the industrial internet of things (IIoT), increasing amounts of operational data from equipment are recorded and stored in real time via sensors connected to computer systems, enabling the analysis and identification of operational patterns [3, 4]. However, due to the complex operating environment of the equipment and issues such as communication interference and sensor malfunctions, the collected operational data contains a significant number of missing values and outliers, resulting in an insufficient number of valid samples for the model to fully learn useful information [5]. Therefore, it is necessary to impute the missing data and use it as augmented data for traction transformer condition prediction.

Traditional imputation methods rely on statistical principles, such as mean imputation [6] and multiple imputation of chained equations (MICE) [7], which are simple, reliable, and yield good results when dealing with simple linear time series. However,

H. Ma et al. (Eds.): IWACIII 2025, CCIS 2780, pp. 219–228, 2026.
https://doi.org/10.1007/978-981-95-6730-0_16

these methods have significant limitations when applied to complex nonlinear multi-dimensional time series. With the development of artificial intelligence, an increasing number of algorithms have been adapted for use in missing value imputation. These include classical machine learning algorithms [8, 9] and subsequent advancements in deep learning algorithms [10, 11]. Generative adversarial networks (GANs) are generative models that employ adversarial learning mechanisms. In [12], an algorithm combining deep and multi-channel convolutional networks for data imputation was proposed, and experimental results demonstrated its strong performance. In [13], generative adversarial imputation nets (GAIN) were proposed for imputation, where the input to the generator is not simple random noise but missing data and corresponding mask vectors. This lays an important foundation for GAN-based imputation algorithms. However, like other GAN-based approaches, the GAIN still faces challenges such as training instability and mode collapse. In [14], a cue Wasserstein generative adversarial net with gradient penalty (CWGAIN-GP) was proposed, which extracts hidden latent data evolution features using a context information cue matrix. It also addresses training instability and mode collapse through a gradient penalty mechanism and the Wasserstein distance.

To address the issue of missing data in traction transformer datasets, this paper adopts the Wasserstein generative adversarial net with gradient penalty (WGAIN-GP), which mitigates training difficulties and mode collapse through its adversarial learning mechanism, thus enabling the generation of high-quality imputed data. This paper selects the Sequence-to-sequence (Seq2seq) [15] as the traction transformer state prediction algorithm and demonstrates through a comparison of prediction performance before and after adding imputed data that imputed data can provide more useful information for the prediction model.

2 Dataset Construction

The dataset used in this study was collected from traction transformers at a substation in southern China. Variations in sensor configurations led to inconsistent sampling timestamps and intervals across variables, producing misaligned time series requiring preprocessing before assembly. Visualization analysis further indicated that changes in most variables occurred far less frequently than the sampling rate. This not only increased storage demands but also diminishing the proportion of informative data.

To address these issues comprehensively, we implemented a downsampling approach to resolve the data misalignment problem. Specifically, the calculation formula is showed as

$$\mathrm{var}_{T'}(i) = median\left(\left[\mathrm{var}_T\left(j > (i-1) * T'\right), \ldots, \mathrm{var}_T\left(j < i^* T'\right)\right]\right) \quad (1)$$

where $var_{T'}(i)$ is the i-th sample in the downsampled dataset with a sampling period of T', $var_T(j)$ is the j-th sample in the dataset with a sampling period of T, $median(\cdot)$ is the function computes the median value of the input dataset.

We established a unified reference timestamp for all variables and implemented a fixed 1-h sampling interval. Following downsampling with (1), the variables were merged to construct the final dataset.

Since no faults occurred in the traction transformers during data collection, we identified normal operating ranges for each variable based on expert knowledge and industry standards.

$$\begin{cases} \text{var}_i^{new}(j) = \text{var}_i^{old}(j) & \text{var}_i^{old}(j) \text{ within normal range} \\ \text{var}_i^{new}(j) = NA & \text{otherwise} \end{cases} \quad (2)$$

Variables with severe missing values or constant values distort the true data distribution and are therefore selected for deletion. The final retained variables also have missing values, but these are mostly concentrated in a few intervals. After removing these intervals, the remaining intervals form eight time-continuous datasets without missing values. Table 1 shows the time intervals and uses of each dataset.

Table 1. Dataset specifications.

Data number	Time Intervals	Applications
Data1	2023.10.19–2023.11.15	Test imputation model/ Augmented data for prediction
Data2	2023.11.22–2024.01.28	Train imputation model/ Train prediction model
Data3	2024.11.02–2024.11.13	Test imputation model/ Augmented data for prediction
Data4	2024.11.14–2024.11.21	Train imputation model/ Train prediction model
Data5	2024.11.22–2025.1.09	Train imputation model/ Train prediction model
Data6	2025.01.10–2025.02.09	Test imputation model/ Augmented data for prediction
Data7	2025.05.08–2025.05.31	Train imputation model/ Train prediction model
Data8	2025.06.01–2025.06.30	test prediction model

To enhance the generalization of the experiment, this paper will also validate the algorithm on the ETTh1 dataset, containing variables such as transformer load and oil temperature. The ETTh1 dataset will be partitioned in a manner similar to the dataset used in this study. To distinguish the data used in this study from ETTh1, the dataset presented here is named Transformers Data (TD).

3 Imputation

The imputation experiment consists of two stages. In the first stage, missing data is artificially generated to train the model and test its imputation performance. In the second stage, the trained model is used to generate imputation data, which provides imputation data for subsequent prediction experiments.

3.1 Data Partitioning Strategy

This paper uses the sliding window sampling method to create samples. In order to evaluate the imputation performance, it is necessary to artificially create missing samples. We constructed the mask matrix sets $M = \{M_1, M_2, \ldots, M_n\}$ of the same dimensions as the samples to indicate the missing status at each position, The elements of the mask matrix are defined as follows:

$$m_{ij} = \begin{cases} 0, & x_{ij} = NA \\ 1, & \text{otherwise} \end{cases} \tag{3}$$

where m_{ij} denotes the element at the i-th row and j-th column of the mask matrix, while x_{ij} represents the corresponding element in the sample matrix. We first generate a set of mask matrices, then artificially create missing samples according to

$$\tilde{X} = X \odot M \tag{4}$$

where X denotes non-missing samples, $\tilde{X}$ denotes the missing samples, $\odot$ is the Hadamard product.

3.2 WGAIN-GP

We employ the WGAIN-GP algorithm for an imputation experiment, and the algorithm retains the same structural framework as GAIN.

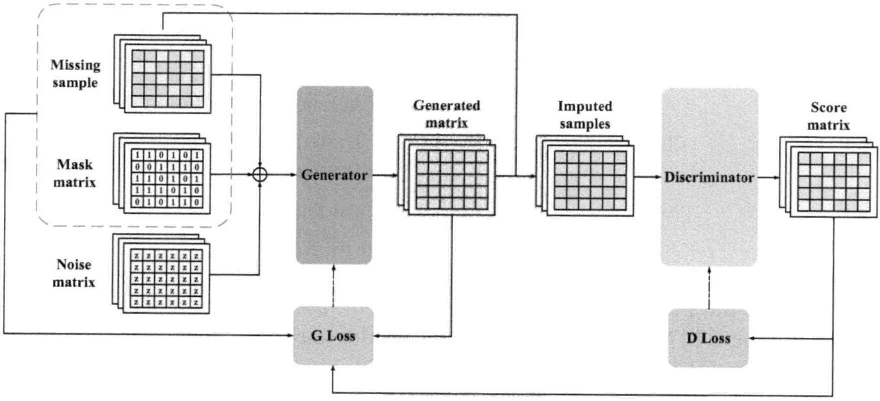

Fig. 1. Schematic diagram of the WGAIN-GP model.

As shown in Fig. 1, the generator is designed to produce imputed data samples. The function of the generator is

$$\begin{aligned} \overline{X} &= G(\tilde{X}, M, (1 - M) \odot Z) \\ \hat{X} &= (1 - M) \odot \overline{X} + M \odot \tilde{X} \end{aligned} \tag{5}$$

where $\overline{X}$ denotes the generated data matrix, Z is a random noise matrix of identical dimensions, $\hat{X}$ denotes the final imputed data matrix.

Similar to GAIN, the generator loss function of WGAIN-GP also consists of reconstruction loss and generation loss. The reconstruction loss is the mean squared error (MSE) between the non-missing parts of the missing samples and the corresponding positions of the generated samples. The generation loss differs from GAIN because this paper adopts the Wasserstein distance as the discriminator's loss criterion. The generator loss function L_G is defined as follows:

$$L_G = -(1 - M) \odot D_{score}\left(\overline{X}\right) + \lambda \left\| \tilde{X} \odot M - \overline{X} \odot M \right\|_2 \tag{6}$$

where D_{score} represents the discriminator's output, specifically the score matrix shown in Fig. 1, and λ is the scaling coefficient serving as a hyperparameter to regulate the proportion of reconstruction loss.

Unlike the GAIN discriminator that outputs a probability matrix, WGAIN-GP produces a score matrix D_{score} that simultaneously participates in the loss computations of both generator and discriminator. The generator aims to maximize the scores for generated samples in (6), while the discriminator seeks to minimize the scores for generated samples and maximize the scores for non-generated samples.

When using the Wasserstein distance metrics, traditional weight clipping, which is used to ensure Lipschitz continuity, inevitably sacrifices model performance. To address this, we introduce a gradient penalty mechanism that maintains model performance while ensuring Lipschitz constraints. Accordingly, the discriminator's loss function is formulated as:

$$L_D = D_{score}\left(\overline{X}\right) - D_{score}(X) + \eta GP \tag{7}$$

$$GP = \left(\left\| \nabla_{\breve{X}} D\left(\breve{X}\right) \right\|_2 - 1 \right)^2 \tag{8}$$

$$\breve{X} = \varepsilon X + (1 - \varepsilon)\overline{X} \tag{9}$$

where GP denotes the gradient penalty term, with its specific mathematical formulation provided in (8), η serves as the scaling coefficient to regulate the magnitude of the gradient penalty term, ∇ is the differentiation factor, ε is a random number from 0 to 1.

3.3 Experiments and Results

The generator and discriminator of the WGAIN-GP algorithm are both constructed using fully connected layers, and the Adam optimizer is selected as the optimizer. The batch = 64, hyperparameters λ =120 and η =20. All data underwent normalization to eliminate the influence of varying scales and units, with the root mean square error (RMSE) being adopted as the evaluation metric for assessing imputation performance. To demonstrate the imputation capabilities of WGAIN-GP, we conducted comparative experiments with the Mean method and the denoising autoencoder (DAE) method. Since data with excessively high missing rates lose a significant amount of information, we

Table 2. Comparison of imputation results on different missing rates.

Data	Model	Error	Missing rate				
			20%	30%	40%	50%	60%
TD	Mean	RMSE	**0.1128**	**0.1187**	**0.1193**	**0.1204**	**0.1220**
	DAE		0.2203	0.2232	0.2235	0.2287	0.2123
	WGAIN-GP		0.2010	0.2087	0.2104	0.2185	0.2137
ETTh1	Mean		**0.1071**	**0.1061**	**0.1062**	**0.1073**	**0.1085**
	DAE		0.2049	0.2068	0.2072	0.2082	0.2097
	WGAIN-GP		0.1880	0.1894	0.1902	0.1912	0.1926

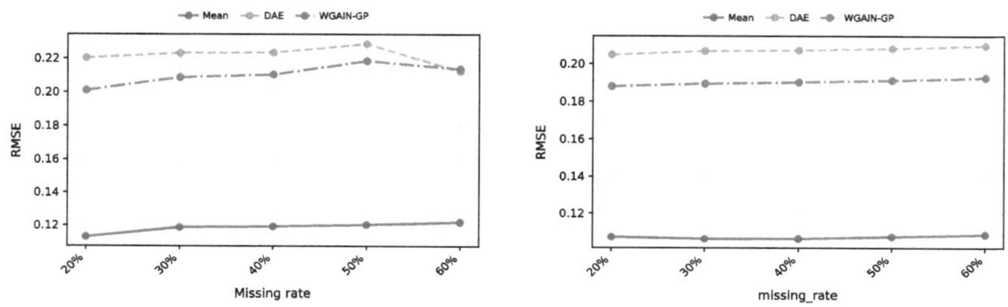

Fig. 2. Imputation results of the dataset (left: TD dataset; right: ETTh1 dataset).

used data with missing rates between 20% and 60% to evaluate model performance (Fig. 2).

Table 2 presents the performance evaluation results of the Mean method, DAE model, and WGAN-GP model based on the RMSE evaluation metric, with the optimal results highlighted in bold. As shown in the table, the imputation performance of the models generally decreases with increasing missing rate, which is consistent with empirical knowledge. Compared to artificial intelligence algorithms, the Mean method achieves the best performance under different missing rate conditions. When comparing the DAE and WGAN-GP models, the WGAN-GP algorithm shows slightly lower performance compared to the DAE at a 60% missing rate on the TD dataset, but outperforms the DAE in all other cases. shows a visualization of the results presented in Table 2, allowing for a more intuitive observation of the model's relative performance. To investigate the superior performance of the mean method, we visualize the imputation resuts, as shown in Fig. 3.

Figure 3 visualizes the imputation results of a specific variable from the Data 3 subset of the TD dataset. It demonstrates that the mean method achieves the smallest deviation from ground truth among the three approaches, but incurs the most severe disruption to temporal patterns. Further impacts will be evaluated in forecasting experiments.

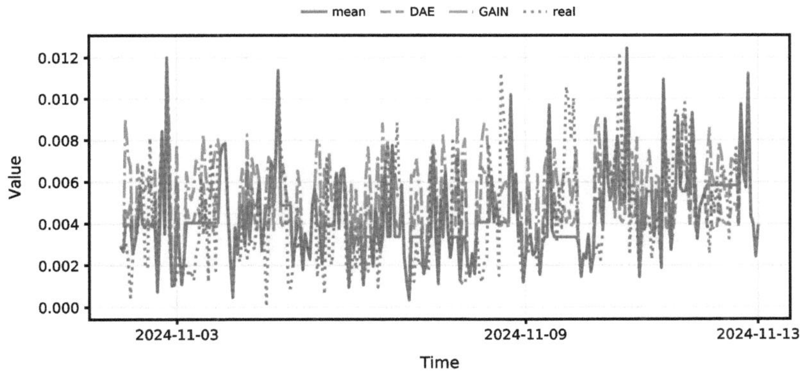

Fig. 3. Visualization of imputation results for the TD dataset.

4 Prediction

In addition to verifying the impact of imputed data on the prediction model, prediction experiments are also conducted to achieve traction transformer status prediction and provide guidance for traction transformer status maintenance.

4.1 Seq2Seq

As shown in Fig. 4, Seq2Seq consists of an encoder and a decoder. The encoder takes historical time series data as input and propagates information through its cell and hidden states. The final cell and hidden states of the encoder are then provided as the initial states for the decoder. Subsequently, the last data point of the historical sequence is used as input for the decoder, and the output serves as the predicted value for the current time step. This predicted value is then used as the input for the next time step, with the process iterating until the predefined sequence length is reached.

Additionally, we incorporate an attention mechanism into the architecture, enabling the decoder to focus on the most informative parts of the historical window at each prediction step, thereby improving forecasting performance.

4.2 Experiments and Results

We use MSE as the loss function and the Adam optimizer for training. The batch size is 64, with a historical window size of 64 and a prediction length of 12. All data undergo normalization preprocessing, and RMSE is selected as the evaluation metric.

The experiment first trains the model using Data 2, Data 4, Data 5, and Data 7, with Data 8 serving to evaluate the model's performance. Subsequently, the incrementally imputed data from the WGAIN-GP, DAE, and Mean method in the imputation experiment are separately incorporated into model training, and Data 8 is reused to assess the training outcomes. Detailed results are presented in Table 3.

Table 3 shows the impact of imputed data on the performance of different models in predicting the timing of traction transformers under different missing rates. First,

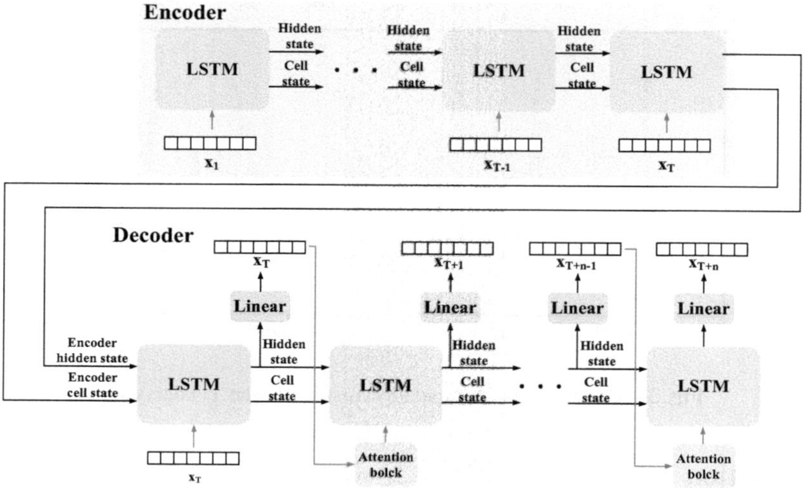

Fig. 4. Seq2Seq architecture diagram.

Table 3. Comparison of prediction results on different missing rates.

Data	Model	Missing rate					
		0%	20%	30%	40%	50%	60%
TD	No incremental data	0.1391	/	/	/	/	/
	Complete data	**0.1192**	/	/	/	/	/
	Mean	/	0.1233	0.1272	0.1211	0.1398	0.1333
	DAE	/	0.1246	0.1393	0.1251	0.1328	**0.1287**
	WGAIN-GP	/	**0.1166**	**0.1228**	**0.1188**	**0.1318**	0.1323
ETTh1	No incremental data	0.0903	/	/	/	/	/
	Complete data	**0.0858**	/	/	/	/	/
	Mean	/	0.0899	0.0884	0.0887	0.0895	0.0908
	DAE	/	**0.0878**	0.0874	0.0879	0.0899	0.0903
	WGAIN-GP	/	0.0879	**0.0866**	**0.0877**	**0.0886**	**0.0895**

the Seq2Seq model generally performs better when imputed data is added than without imputed data. This may be because the initial training dataset is insufficient for the model to fully learn the data information. When comparing the effects of imputed data generated by different imputation models on the prediction model, we found that the mean method performed well in the imputation experiments. However, it did not yield the optimal imputation results. Instead, the WGAIN-GP algorithm performed best in most cases.

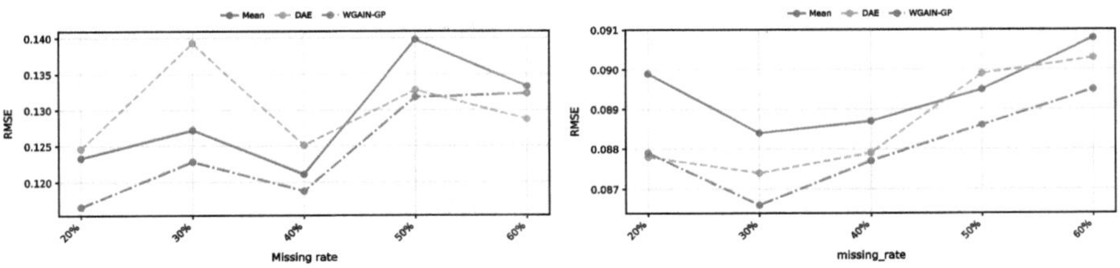

Fig. 5. Prediction results of the traction transformer dataset (left: TD dataset; right: ETTh1 dataset).

This is because, although the Mean method has smaller deviations between imputation values and actual values, its variable imputation values are identical, which disrupts the data structure and obscures some useful information. Compared to the WGAIN-GP model, the Mean method's imputation effect is inferior, highlighting the superiority of WGAIN-GP over traditional imputation algorithms. As the missing rate increases, the improvement in prediction model performance due to imputed data gradually diminishes, consistent with empirical knowledge. Figure 5 visualizes part of the results in Table 3, allowing an intuitive comparison of the effects of imputed data from different algorithms on prediction performance.

In addition, we directly input complete data without artificially introduced missing values as augmentation data into the prediction model to evaluate its performance. As shown in Table 3, the inclusion of complete data enables the model to achieve state-of-the-art performance, which also demonstrates the significant role of augmentation data in enhancing the predictive model's capabilities.

5 Conclusion

This paper uses expert consultation and a series of data preprocessing algorithms to filter out invalid variables and outliers from real-world traction transformer data, thereby creating a time-series dataset. The Seq2Seq algorithm is then employed to predict the time-series behavior of traction transformer variables. The WGAIN-GP imputation algorithm is employed to impute missing values, and the resulting data are incorporated into the training set as augmentation data. The results indicate that WGAIN-GP generates imputed data that better aligns with the distribution of real data. When the number of valid samples is limited, imputing missing values and incorporating them as augmentation data can enhance the accuracy and reliability of the prediction model.

Acknowledgments. This work is supported by the Project of China Railway Research Institute Group Co., Ltd. Infrastructure Testing Institute, grant number 2023YJ331.

Disclosure of Interests The authors have no competing interests to declare that are relevant to the content of this article.

References

1. May, M.C., Glatter, D., Arnold, D., et al.: IIoT system canvas—from architecture patterns towards an IIoT development framework. J. Manuf. Syst. **72**(1), 437–459 (2024)
2. Salim, M.M., Azzaoui, A.E., Deng, X., et al.: FL-CTIF: a federated learning based CTI framework based on information fusion for secure IIoT. Inf. Fusion. **102**(1), 1–12 (2024)
3. Aslan, B., Balci, S., Kayabasi, A.: Fault diagnosis in thermal images of transformer and asynchronous motor through semantic segmentation and different CNN models. Appl. Therm. Eng. **265**(1), 1–13 (2025)
4. Lyu, Z., Wan, Z., Bian, Z., et al.: Integrated digital twins system for oil temperature prediction of power transformer based on Internet of Things. IEEE Internet Things J. **12**(10), 13746–13756 (2025)
5. Xu, L., Xu, L., Yu, J.: Time series imputation with GAN inversion and decay connection. Inf. Sci. **643**(1), 1–13 (2023)
6. Little, R.J.A., Rubin, D.B.: Statistical Analysis with Missing Data, 3rd edn. John Wiley & Sons, Hoboken (2019)
7. Samad, M.D., Abrar, S., Diawara, N.: Missing value estimation using clustering and deep learning within multiple imputation framework. Knowl.-Based Syst. **249**(1), 1–12 (2022)
8. Zhang, S.: Nearest neighbor selection for iteratively kNN imputation. J. Syst. Softw. **85**(11), 2541–2552 (2012)
9. Stekhoven, D.J., Bühlmann, P.: MissForest—non-parametric missing value imputation for mixed-type data. Bioinformatics. **28**(1), 112–118 (2012)
10. Du, W., Côté, D., Yan, L.: Saits: self-attention-based imputation for time series. Expert Syst. Appl. **219**(1), 1–15 (2023)
11. Xu, D., Peng, H., Tang, Y., et al.: Hierarchical spatio-temporal graph convolutional neural networks for traffic data imputation. Inf. Fusion. **106**(1), 1–11 (2024)
12. Guo, Z., Wan, Y., Ye, H.: A data imputation method for multivariate time series based on generative adversarial network. Neurocomputing. **360**(1), 185–197 (2019)
13. Yoon, J., Jordon, J., Schaar, M.: Gain: missing data imputation using generative adversarial nets. In: Dy, J., Krause, A. (eds.) The 35th International Conference on Machine Learning 2018, PMLR, vol. 80, pp. 5689–5698. PMLR, Stockholm (2018)
14. Wang, Y., Xu, X., Hu, L., et al.: A time series continuous missing values imputation method based on generative adversarial networks. Knowl. Based Syst. **283**(1), 1–13 (2024)
15. Zhong, J., Liu, J.: Seq2seq modelling for cross-site temporal forecasting of urban air pollutant concentrations leveraging sensor data. Build. Environ. **269**(1), 1–13 (2025)

Online Fine-Tuning for Robustness in Offline Reinforcement Learning: Actor-Critic vs Decision Transformer

Shingo Ayabe, Hiroshi Kera, and Kazuhiko Kawamoto(✉)

Chiba University, 1-33, Yayoi-cho, Inage-ku, Chiba 263-8522, Japan
{ayabe.shingo,kera}@chiba-u.jp, kawa@faculty.chiba-u.jp

Abstract. Offline reinforcement learning enables agents to learn from fixed datasets without requiring online interaction, offering clear advantages in scenarios where collecting real-world data is costly or risky. However, policies trained solely offline often lack robustness to action-space perturbations due to the absence of exploration during training. This study explores whether online fine-tuning with randomly perturbed actions can enhance robustness. We consider two types of models: an actor-critic method and a Transformer-based model. We evaluate the models in simulated the legged robot environment, where perturbations are applied to joint torque outputs to emulate control faults. Fine-tuning significantly improves robustness in the actor-critic method, whereas only marginal gains are observed in the Transformer-based model, which primarily relies on trajectory imitation. These results reveal the difficulty of adapting Transformer-based models to perturbed environments. Furthermore, they suggest that integrating actor-critic mechanisms into such models may provide a promising direction for improving robustness within the offline-to-online RL framework.

Keywords: Offline reinforcement learning · Offline-to-online reinforcement learning · Action-space perturbation · Robustness evaluation

1 Introduction

Offline reinforcement learning (offline RL) enables policy learning from fixed datasets without online interaction, making it attractive for safety-critical domains such as robotics, healthcare, and energy systems [5,13,21,36]. In contrast to online RL, which relies on real-time interaction with the environment, offline RL learns policies from previously collected datasets. This paradigm reduces both the cost and safety risk of frequent online interactions [6,21].

However, applying standard RL algorithms in the offline setting often overestimates Q-values due to distributional shift. To address this issue, conservative approaches have been developed to constrain the learned policy to remain close to the dataset behavior [8,10,17,19]. Decision Transformer (DT) [4] offers a different formulation by casting offline RL as a sequence modeling problem using

H. Ma et al. (Eds.): IWACIII 2025, CCIS 2780, pp. 229–243, 2026.
https://doi.org/10.1007/978-981-95-6730-0_17

supervised learning. While this approach differs from conventional offline RL methods, DT remains conservative, as it imitates actions from offline trajectories.

The conservatism of these offline RL methods stems from their lack of exploration, which limits adaptability to environmental changes and uncertainties. This is particularly critical in real-world applications, where robustness to unexpected dynamics—such as external disturbances or actuator faults—is essential. While robustness to state-space perturbations has been studied in offline RL [23,35,42], vulnerabilities to action-space perturbations remain relatively underexplored despite being observed [1]. Improving robustness in this setting requires policies to actively adapt their actions to such perturbations, which is inherently challenging in offline RL due to the absence of exploration. This limitation also applies to Transformer-based models, which are trained through supervised action prediction from offline trajectories. Since their action selection is constrained to reproducing observed sequences, they lack the flexibility to adapt to perturbed environments.

One promising approach to address the lack of adaptability is offline-to-online RL, which combines offline pretraining with online fine-tuning. This framework enables policies to adapt to deployment-time environmental changes and uncertainties not covered in the offline dataset, thereby improving performance [20,22,37,38]. In this study, we investigate whether fine-tuning an offline-trained policy under action-space perturbations improves robustness. We evaluate this approach on both a conventional actor-critic (AC) approach and a Transformer-based sequence modeling approach using legged robot tasks from OpenAI Gym [2].

Our contributions are summarized as follows:

- We propose an online fine-tuning method for offline-trained policies in environments with action-space perturbations.
- We show that fine-tuning significantly enhances robustness in a conventional actor-critic algorithm, even under severe perturbations.
- We apply the fine-tuning method to a Transformer-based model and find that robustness gains are limited, highlighting the difficulty of adapting such sequence models to perturbed environments.

2 Related Work

2.1 Offline RL

Offline RL is a framework that trains policies using datasets without environment interaction. However, due to the absence of exploration, conventional online RL algorithms tend to overestimate Q-values for out-of-distribution state-action pairs, which can lead to learning failure [6]. To address this issue, conservative approaches have been proposed to keep the learned policy close to the data distribution [8,10,17–19]. For example, TD3+BC [8] achieves efficient offline learning by adding a simple behavior cloning loss to the actor update in the

Twin Delayed Deterministic Policy Gradient (TD3) [9]. Nevertheless, offline RL algorithms with such conservative designs are often vulnerable to perturbations in the action space [1]. Training solely in offline settings makes it difficult to acquire sufficient robustness.

This study employs an offline-to-online RL framework to improve policy robustness against action-space perturbations. The offline phase uses TD3+BC to initialize the policy. This algorithm is selected not only for its learning efficiency but also for its architectural consistency with TD3 used in the online phase. Such a unified design enables a smooth transition from offline to online learning and facilitates consistent comparisons across different training stages.

2.2 Robust RL

Robustness in RL for robotic control primarily tackles discrepancies arising from mismatches in physical parameters such as friction coefficients or terrain geometry, as well as mechanical failures. One common approach to handling these discrepancies involves introducing perturbations during the training process. Domain randomization [3,16] generates diverse environments by randomizing parameters such as friction, enabling the policy to adapt to a wide range of variations. Adversarial training [27,41] enhances robustness by incorporating adversarial perturbations designed to degrade rewards. For instance, differential evolution has been used to generate adversarial perturbations that significantly reduce performance in locomotion scenarios [24].

In offline RL, robustness against unexpected perturbations encountered during testing-time has also been studied. RORL [35] improves robustness against state-space perturbations by smoothing the value distribution and applying conservative value estimation. Nguyen [23] proposed a method that generates adversarial perturbations based on KL divergence and introduces a defense mechanism against them.

In contrast, robustness against action-space perturbations remains relatively unexplored. Recent studies have pointed out that offline-trained policies are vulnerable against action-space perturbations [1]. No effective countermeasures have yet been established. Offline RL alone is insufficient for acquiring robustness to such disturbances due to the lack of active exploration. This study addresses this challenge through the offline-to-online RL framework.

2.3 Offline-to-Online RL

Offline-to-online RL has gained attention as a method for overcoming the limitations of offline RL, which depends on the quality of the dataset [12,20,22, 30,37,38]. A major challenge in this setting is performance degradation during the early stages of fine-tuning. This issue stems from distributional shifts between offline data and online experiences [20,22]. Various approaches have been proposed to address this problem. These include relaxed or adaptive policy constraints [22,38,39], sample selection methods that prioritize near-on-policy data [20], dynamic switching between multiple policies depending on the state [30,37], and exploration strategies combined with meta-adaptation [12]. This study proposes a method for improving policy robustness against action-space perturbations through online fine-tuning.

2.4 Sequence Modeling for RL

Sequence modeling methods for offline RL have attracted significant attention, particularly those based on Decision Transformer (DT) and its variants [4,15,31,33]. These methods do not explicitly learn Q-functions or policy gradients. Instead, they use Transformer architectures [25,29] to predict sequences of actions conditioned on task specifications such as target states or return-to-go. DT relies on trajectory imitation and lacks the ability to stitch novel trajectories [11,31,33]. Performance degradation under stochastic environments has also been reported. Since action-space perturbations introduced in this study can be considered a form of stochasticity, DT may be vulnerable under such conditions.

To address this issue, the offline-to-online RL framework is applied in this study. Several prior studies have extended DT for online fine-tuning. For example, Online Decision Transformer (ODT) [40] introduces stochastic policies and entropy regularization—similar to Soft Actor-Critic (SAC) [14]—to enhance exploration. Other approaches replace return-to-go with latent representations to improve downstream task adaptability [32]. This study adopts ODT as the simplest variant among these extensions. Robustness of the learned policy against action-space perturbations is evaluated and compared across both pretraining and online fine-tuning phases.

3 Preliminaries

This section provides background on the methods used in this study. We first introduce the offline reinforcement learning setting and describe its limitations under action-space perturbations. We then present the offline-to-online RL framework as a strategy to address these limitations. Finally, we explain sequence modeling approaches, particularly the Decision Transformer and its extension, Online Decision Transformer, which are evaluated in our experiments.

3.1 Offline RL

RL is formulated as a Markov decision process (MDP), defined by the tuple $(\mathcal{S}, \mathcal{A}, P, P_0, r, \gamma)$. Here, $\mathcal{S}$ denotes the state space, $\mathcal{A}$ the action space, $P(\boldsymbol{s}'|\boldsymbol{s}, \boldsymbol{a})$ the state transition probability, P_0 the initial state distribution, $r(\boldsymbol{s}, \boldsymbol{a})$ the reward function, and γ the discount factor. The goal is to find an optimal policy π^* that maximizes expected return.

Offline RL learns policies using a fixed dataset $\mathcal{D} = \{(\boldsymbol{s}_t, \boldsymbol{a}_t, \boldsymbol{s}_{t+1}, r_t)_i\}$ without further interaction with the environment. This dataset is typically collected from expert demonstrations, past policy rollouts, or random actions, and is assumed to follow some behavior policy π_β. Offline settings often suffer from overestimated Q-values for out-of-distribution state-action pairs. Without exploration, these overestimated values cannot be corrected, leading to poor generalization. To mitigate this issue, conservative policy learning constrains the learned policy to stay close to the behavior policy. The learning objective is formulated as:

$$\pi = \arg\max_{\pi} \mathbb{E}_{\boldsymbol{s}_t \sim \mathcal{D}} \left[Q^{\pi}(\boldsymbol{s}_t, \pi(\boldsymbol{s}_t)) \right] \quad \text{s.t. } D(\pi(\cdot|\boldsymbol{s}_t), \pi_\beta(\cdot|\boldsymbol{s}_t)) < \epsilon, \tag{1}$$

where D represents a divergence metric such as KL divergence [6]. This strategy introduces potential vulnerabilities when action-space perturbations are present in the environment. In real-world systems, actions are often altered by disturbances or noise before execution. The actual action becomes $\boldsymbol{a} + \boldsymbol{\delta}$, where $\boldsymbol{\delta}$ is a perturbation applied to the policy output $\boldsymbol{a}$. As a result, the reward r and next state $\boldsymbol{s}_{t+1}$ depend on the perturbed action.

However, datasets typically contain only unperturbed actions $\boldsymbol{a}$. This prevents the learner from evaluating whether the chosen actions are effective under the perturbation condition. Additionally, acquiring new data that reflects successful outcomes under such conditions is not feasible in the offline setting. This mismatch highlights a fundamental limitation of offline RL in learning robust behaviors. Conservative constraints may further reinforce vulnerable actions present in the dataset. These issues reveal a structural vulnerability of offline RL to action-space perturbations, arising from both the lack of robust data and the restrictive nature of the learning paradigm.

3.2 Offline-to-Online RL

Offline-to-online RL is effective when offline training alone is insufficient, or when a mismatch exists between the dataset distribution and the actual deployment environment [20,22]. This approach first performs offline pretraining and then applies online fine-tuning to improve adaptability and robustness. Online fine-tuning allows the agent to accumulate new experiences under perturbed conditions. As a result, the policy can adapt to the perturbations and is expected to acquire robust behaviors.

In this study, offline pretraining is conducted using TD3+BC [8], a conservative actor-critic RL algorithm. Online fine-tuning is then performed using TD3 [9]. The policy update rule for TD3+BC is defined as:

$$\pi = \arg\max_{\pi} \mathbb{E}_{(\boldsymbol{s}_t, \boldsymbol{a}_t) \sim \mathcal{D}} \left[Q^{\pi}(\boldsymbol{s}_t, \pi(\boldsymbol{s}_t)) - \|\pi(\boldsymbol{s}_t) - \boldsymbol{a}_t\|^2 \right], \tag{2}$$

where the second term penalizes deviation from the behavior policy. In the fine-tuning phase, the TD3 policy update is defined as:

$$\pi = \arg\max_{\pi} \mathbb{E}_{\boldsymbol{s}_t \sim \mathcal{R}} \left[Q^{\pi}(\boldsymbol{s}_t, \pi(\boldsymbol{s}_t)) \right], \tag{3}$$

where $\mathcal{R}$ represents the replay buffer containing online experience. Some prior work continues to apply the TD3+BC constraint during fine-tuning [38,39]. In this study setting, perturbations are introduced during online interaction. It is unclear whether the actions in the replay buffer remain valid under perturbed dynamics. Applying constraints in such cases may hinder exploration. Therefore, the regularization term is excluded to enable more flexible policy adaptation.

3.3 Sequence Modeling for RL

DT formulates RL as a sequence modeling problem using Transformer architecture [4,15]. The model treats an agent's experience as a trajectory $\tau = \{(\boldsymbol{s}_0, \boldsymbol{a}_0, \hat{R}_0), (\boldsymbol{s}_1, \boldsymbol{a}_1, \hat{R}_1), \ldots, (\boldsymbol{s}_T, \boldsymbol{a}_T, \hat{R}_T)\}$, where $\boldsymbol{s}_t$ is the state, $\boldsymbol{a}_t$ is the action, and $\hat{R}_t$ is the target return-to-go at time t. Instead of learning a value function or policy directly, DT autoregressively predicts actions conditioned on the trajectory:

$$a_t = \text{Transformer}(\boldsymbol{s}_0, \hat{R}_0, \boldsymbol{a}_0, \boldsymbol{s}_1, \hat{R}_1, \boldsymbol{a}_1, \ldots, \boldsymbol{s}_t, \hat{R}_t). \quad (4)$$

This structure enables supervised learning on offline datasets. However, DT lacks online interaction capabilities, which limits its ability to explore and adapt to new environments. The objective of DT is to minimize the prediction error on actions observed in offline trajectories [4]:

$$\mathcal{L}_{\text{DT}} = \mathbb{E}_{\tau \sim \mathcal{D}} \left[\sum_t \| a_t - \pi_{\text{DT}}(s_{\leq t}, a_{<t}, \hat{R}_{\leq t}) \|^2 \right]. \quad (5)$$

ODT extends DT by enabling online interaction while keeping the same trajectory-based structure. To encourage exploration and policy diversity, ODT introduces trajectory-level entropy regularization. The objective of ODT [40] is defined as:

$$\mathcal{L}_{\text{ODT}} = \mathbb{E}_{\tau \sim \mathcal{D}} \left[-\log \pi_\theta(\tau) + \beta \cdot \mathcal{H}(\pi_\theta(\cdot|\tau)) \right], \quad (6)$$

where $\mathcal{H}(\pi_\theta(\cdot|\tau))$ is the entropy of the policy conditioned on the trajectory. This term promotes action diversity during online adaptation. During fine-tuning, ODT also applies hindsight relabeling to update the return targets $\hat{R}_t$ using actual returns observed from the environment. This mechanism allows the model to adapt to environment dynamics and task variations.

In this study, ODT is employed as a fine-tuning framework for DT. Random perturbations are introduced to the policy output during the fine-tuning phase. The effectiveness of this approach is evaluated in terms of robustness under perturbed conditions.

4 Proposed Method

This section presents our offline-to-online RL method for enhancing robustness to action-space perturbations. We fine-tune offline-trained policies under perturbed conditions using TD3+BC/TD3 and ODT, and evaluate robustness based on episodic rewards.

4.1 Problem Setting

This study addresses the problem of improving the robustness of policies against action-space perturbations in offline reinforcement learning settings. Specifically, the goal is to fine-tune a policy pretrained with offline data in an environment where random perturbations are applied to the executed actions. The central objective is to evaluate whether such fine-tuning enables the policy to adapt to the perturbations and achieve higher performance in perturbed environments.

4.2 Proposed Fine-Tuning Method

This study performs online fine-tuning of RL models pretrained using offline RL in the legged robot environment from OpenAI Gym [2]. Two types of models are considered: TD3+BC/TD3 (TD3+BC: pretrain, TD3: fine-tuning), a conventional actor-critic approach, and ODT, a sequence modeling approach.

The fine-tuning algorithm differs depending on the type of pretrained model. For TD3+BC, the procedure is described in Algorithm 1. The pretrained policy and critic are first initialized.

To stabilize fine-tuning, a subset of offline data is added to the replay buffer $\mathcal{R}$ at the beginning of training, following prior studies [20,38,39]. The number of offline samples M_{off} is computed as:

$$M_{\text{off}} = r_{\text{off}} \cdot S_{\mathcal{R}}, \tag{7}$$

where r_{off} is the sampling ratio, and $S_{\mathcal{R}}$ is the buffer capacity.

At the start of each episode, a perturbation vector $\boldsymbol{\delta}$ is generated according to Eq. (9). At each interaction step, an action is sampled from the policy and perturbed with probability q as follows [1]:

$$\boldsymbol{a}_t' = \boldsymbol{a}_t + \boldsymbol{\delta} \odot \boldsymbol{a}_t, \tag{8}$$

where t is the timestep within an episode and $\odot$ denotes the element-wise (Hadamard) product. The perturbed action is executed in the environment, and the resulting transition is stored in the buffer. The policy and critic are then updated using TD3.

Algorithm 2 follows the standard training framework of ODT. The only modification introduced in this study is the injection of perturbations into the predicted actions during interaction, applied in the same manner as in the TD3+BC case.

4.3 Perturbation Conditions

Perturbation $\boldsymbol{\delta}$ follows the definition in [1]. Fine-tuning is conducted under two perturbation conditions: normal and random.

$$\boldsymbol{\delta} = \begin{cases} \mathbf{0} & \text{normal}, \\ \boldsymbol{\delta}_r \sim U([-\epsilon, \epsilon]^{N_a}) & \text{random perturbations}. \end{cases} \tag{9}$$

Algorithm 1. Online Fine-tuning with TD3

Require: $\pi_{\text{off}}, Q_{\text{off}}$: policy and critic pretrained with an offline RL method
Require: $\mathcal{D}_{\text{off}}$: offline dataset
Require: $q \in \mathbb{R}$: perturbation probability
Require: $\mathcal{R}$: replay buffer
Require: T_{ft}: total fine-tuning steps

Initialize $\pi_{\text{ft}}, Q \leftarrow \pi_{\text{off}}, Q_{\text{off}}$
Determine the number of offline samples using Eq. (7)
Initialize $\mathcal{R} \leftarrow$ sampled offline data
$T_{\text{total}} \leftarrow 0$
while $T_{\text{total}} < T_{\text{ft}}$ **do**
  Generate perturbation $\boldsymbol{\delta}$ using Eq. (9)
  Sample $u \sim U([0, 1])$
  $t \leftarrow 0$
  Sample initial state $\boldsymbol{s}_0 \sim P_0(\cdot)$
  while episode not terminated **do**
    Sample action $\boldsymbol{a}_t \sim \pi_{\text{ft}}(\cdot|\boldsymbol{s}_t)$
    if $u < q$ **then**
      $\boldsymbol{a}'_t \leftarrow \boldsymbol{a}_t + \boldsymbol{\delta} \odot \boldsymbol{a}_t$
    else
      $\boldsymbol{a}'_t \leftarrow \boldsymbol{a}_t$
    end if
    $(r_t, \boldsymbol{s}_{t+1}) \sim \text{Env}(\boldsymbol{a}'_t)$
    Append $(\boldsymbol{s}_t, \boldsymbol{a}_t, r_t, \boldsymbol{s}_{t+1})$ to $\mathcal{R}$
    Optimize π_{ft} and Q using TD3
    $T_{\text{total}} \leftarrow T_{\text{total}} + 1$
    $t \leftarrow t + 1$
  end while
end while

In the normal condition, no perturbation is applied ($\boldsymbol{\delta} = \mathbf{0}$). Random perturbation $\boldsymbol{\delta}_r$ is sampled from a uniform distribution over $[-\epsilon, \epsilon]^{N_a}$, where ϵ controls the perturbation strength and N_a is the action dimension.

4.4 Testing-Time Robustness Evaluation

The robustness of fine-tuned policies is evaluated through forward walking tasks using legged robots. Each policy is tested under two perturbation conditions defined in Eq. (9). The primary evaluation metric is the average episodic reward, computed as:

$$\bar{R}_{\text{episode}} = \frac{1}{M} \sum_{m=1}^{M} \sum_{t=0}^{T-1} r_t^{(m)}, \tag{10}$$

where $r_t^{(m)}$ is the reward at timestep t in episode m, T is the episode length, and M is the number of evaluation episodes. In practice, the average episodic reward is normalized based on the D4RL normalized score.

Algorithm 2. Online Fine-tuning with ODT

Require: π_{ODT}: Transformer-based policy pretrained with offline trajectories

Initialize π_{ODT} and trajectory buffer $\mathcal{R}$ following the ODT procedure
$T_{\text{total}} \leftarrow 0$
while $T_{\text{total}} < T_{\text{ft}}$ **do**
  Generate perturbation $\boldsymbol{\delta}$ using Eq. (9)
  Sample $u \sim U([0, 1])$
  $t \leftarrow 0$
  while episode not terminated **do**
    Sample trajectory prefix $\tau_{0:t}$ from $\mathcal{R}$
    Sample action $\boldsymbol{a}_t \sim \pi_{\text{ODT}}(\tau_{0:t})$
    if $u < q$ **then**
      $\boldsymbol{a}'_t \leftarrow \boldsymbol{a}_t + \boldsymbol{\delta} \odot \boldsymbol{a}_t$
    else
      $\boldsymbol{a}'_t \leftarrow \boldsymbol{a}_t$
    end if
    $(r_t, \boldsymbol{s}_{t+1}) \sim \text{Env}(\boldsymbol{a}'_t)$
    Append $(\boldsymbol{s}_t, \boldsymbol{a}_t, r_t, \boldsymbol{s}_{t+1})$ to $\mathcal{R}$
    Optimize π_{ODT} following the standard ODT procedure
    $T_{\text{total}} \leftarrow T_{\text{total}} + 1$
    $t \leftarrow t + 1$
  end while
end while

5 Experiments

5.1 Experimental Setup

Pretrained Model This study uses two types of pretrained models for fine-tuning. The first is a model trained with the offline reinforcement learning algorithm TD3+BC for 5 million steps. We use the implementation provided in the Data-Driven Deep Reinforcement Learning library for Python (d3rlpy) [26]. The training dataset is the expert dataset from Datasets for Deep Data-Driven Reinforcement Learning (D4RL) [7].

The second model is pretrained using the original implementation of the ODT. The model is pretrained for 100,000 steps. All other hyperparameters follow the original ODT settings [40].

Online Fine-tuning Setup. Online fine-tuning for both TD3 and ODT is conducted over 1 million steps. For TD3, fine-tuning is performed in five independent runs under two different conditions: a normal condition with no perturbation and a random perturbation condition. The perturbation probability is set to $q = 1$. The perturbation strength is set to $\epsilon = 0.5$ following [1]. The replay buffer size is set to one million, and offline data is inserted into the buffer with a ratio of $r_{\text{off}} = 1$.

For ODT, fine-tuning is also performed for 1 million steps under the same basic training conditions as in TD3. However, the offline data insertion followed the default implementation of ODT [40], in which 1 million samples from the offline dataset are inserted into the replay buffer at the beginning. This setting results in the same buffer size and number of inserted samples as in TD3.

Legged Robot Environment. This study uses the Ant-v2 environment from OpenAI Gym, running in the MuJoCo physics simulator [28]. The robot is trained to perform a forward walking task. The reward function is defined as:

$$r(\boldsymbol{s}, \boldsymbol{a}) = v_{\mathrm{fwd}} - 0.5\|\boldsymbol{a}\|^2 - 0.5 \cdot 10^{-3} \left\|\boldsymbol{f}\right\|^2 + 1, \tag{11}$$

where v_{fwd} denotes the forward walking velocity, and $\boldsymbol{f}$ represents the contact force exerted on the robot's body.

Testing-time Robustness Evaluation. For robustness evaluation, each fine-tuned policy is tested on 100 episodes under both normal and random perturbation conditions. The final results are reported as the average over five independent fine-tuning runs.

5.2 Robustness Evaluation Results

Table 1 summarizes the evaluation results of both pretrained and fine-tuned models over one million steps under two perturbation conditions: normal (no perturbation) and random perturbation. TD3+BC/TD3 refers to the setting where the model is pretrained using TD3+BC and fine-tuned with TD3. ODT represents the setting where both pretraining and fine-tuning are conducted using the ODT algorithm. The *Pretrained* column shows the performance before fine-tuning. *Fine-tuned (Normal)* indicates the results after fine-tuning under the normal condition. *Fine-tuned (Random)* corresponds to the results after fine-tuning under the random perturbation condition. Each value represents the D4RL normalized average episodic reward with standard deviation.

Both TD3+BC/TD3 and ODT exhibit significant performance degradation under the random perturbation condition before fine-tuning. Fine-tuning under the normal condition improves performance accordingly; however, little or no improvement occurs under the random perturbation. Fine-tuning under the random perturbation condition leads to marked improvements for TD3+BC/TD3 in both normal and perturbed conditions.

These results confirm the effectiveness of the proposed fine-tuning approach. In comparison, ODT exhibits only minor improvements, and the gains are substantially limited compared to TD3+BC/TD3.

Figure 1 presents the learning curves. The x-axis indicates the number of fine-tuning steps, and the y-axis shows the D4RL normalized average episodic reward. The left plot corresponds to evaluations in the normal condition, while the right plot illustrates the performance under random perturbations.

Under the normal condition, TD3+BC/TD3 exhibits an initial performance drop, followed by rapid recovery and eventual stabilization. In contrast, ODT maintains stable performance from the beginning without such degradation. Under the random perturbation condition, TD3+BC/TD3 shows consistent improvement throughout training. While ODT achieves relatively high returns

Table 1. Robustness evaluation results of fine-tuned models under different perturbation conditions. The table reports D4RL normalized average episodic rewards (± standard deviation) for TD3+BC/TD3 and ODT, evaluated under both normal and random perturbations.

RL Algorithms	Training	Evaluation Condition	
		Normal	Random
TD3+BC/TD3	Pretrained	119.9 ± 39.9	96.0 ± 43.3
	Fine-tuned (Normal)	133.1 ± 23.6	103.5 ± 25.7
	Fine-tuned (Random)	130.8 ± 29.3	120.2 ± 32.0
ODT	Pretrained	104.5 ± 41.5	80.5 ± 45.7
	Fine-tuned (Normal)	114.5 ± 36.3	79.6 ± 41.8
	Fine-tuned (Random)	112.9 ± 38.7	91.5 ± 45.0

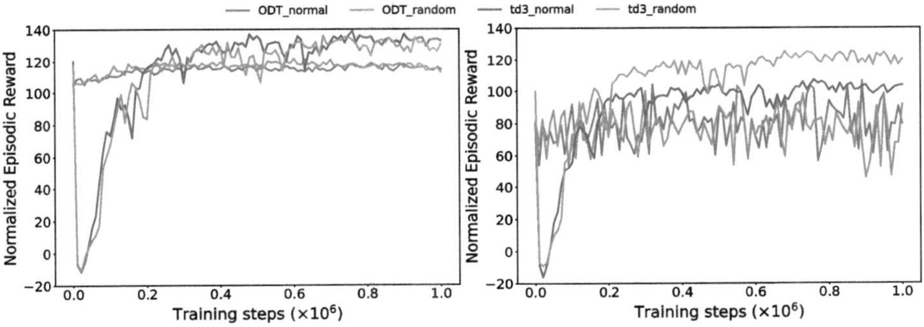

Fig. 1. Learning curves of fine-tuned models under different perturbation conditions. The left panel shows evaluation results under the normal condition, and the right panel shows results under the random perturbation condition. Each line represents the normalized episodic reward obtained by evaluating the model during training.

at times, it exhibits high variance and an overall flat trend. Although Table 1 reports slight numerical improvement for ODT in the *Fine-tuned (Random)*, Fig. 1 suggests the actual improvement is limited. These findings indicate that the proposed method has limited impact on the robustness of ODT.

This discrepancy can be attributed to differences in policy optimization mechanisms. TD3+BC/TD3 leverages Q-value estimation to assess the value of each action, enabling learning of perturbation-resilient policies. On the other hand, ODT is based on trajectory imitation through sequence modeling. Despite incorporating entropy regularization during fine-tuning, the training objective remains trajectory prediction rather than explicit value estimation. This limitation likely impedes adaptation to perturbed environments. Prior studies have identified similar weaknesses in imitation-based models, including limited stitching capability and vulnerability to stochastic environments. These limitations stem from the absence of value-based decision mechanisms [11,33,34]. Improving robustness in such models requires design strategies that go beyond pure imitation frameworks.

6 Conclusion

This study addressed the robustness of offline-pretrained reinforcement learning policies against action-space perturbations. A simple yet effective fine-tuning strategy was introduced, where action-space perturbations are applied during online interaction. The proposed method was evaluated using both a conventional actor-critic RL model (TD3+BC/TD3) and a Transformer-based sequence model (Online Decision Transformer, ODT) in the simulated legged robot environment. Experimental results demonstrated that TD3+BC/TD3 benefits significantly from fine-tuning under perturbation conditions, achieving robust performance across both perturbed and unperturbed settings. In contrast, ODT exhibited only limited improvements. This discrepancy highlights fundamental differences between the actor-critic and sequence-modeling approaches. Actor-critic methods leverage Q-value estimation to adjust actions in response to perturbations, while ODT relies on trajectory prediction and lacks an explicit mechanism to evaluate action utility. As a result, ODT struggles to modify its behavior under unexpected disturbances.

Improving robustness in Transformer-based policies remains an open challenge. Future work includes incorporating Q-value estimation into sequence models during training or inference, allowing the model to evaluate the utility of candidate actions. Such integration may enable sequence-based policies to adapt their behavior based on action quality, thereby improving robustness under perturbed environments. Addressing this limitation is critical for deploying such models in real-world robot applications.

Acknowledgements. This work was supported by JSPS KAKENHI Grant Numbers JP23K24914 and the Telecommunications Advancement Foundation.

References

1. Ayabe, S., Otomo, T., Kera, H., Kawamoto, K.: Robustness evaluation of offline reinforcement learning for robot control against action perturbations. arXiv:2412.18781 (2024)
2. Brockman, G., et al.: OpenAI gym. arXiv:1606.01540 (2016)
3. Chebotar, Y., et al.: Closing the sim-to-real loop: adapting simulation randomization with real world experience. In: 2019 IEEE International Conference on Robotics and Automation (ICRA 2019), pp. 8973–8979 (2018)
4. Chen, L., et al.: Decision transformer: reinforcement learning via sequence modeling. In: Advances in Neural Information Processing Systems, vol. 34, pp. 15084–15097 (2021)
5. Emerson, H., Guy, M., McConville, R.: Offline reinforcement learning for safer blood glucose control in people with type 1 diabetes. J. Biomed. Inform. **142**, 104376 (2023)

6. Figueiredo Prudencio, R., Maximo, Mardos, R.O.A., Colombini, E.L.: A survey on offline reinforcement learning: taxonomy, review, and open problems. IEEE Trans. Neural Netw. Learn. Syst. **35**(8), 10237–10257 (2024)
7. Fu, J., Kumar, A., Nachum, O., Tucker, G., Levine, S.: D4RL: datasets for deep data-driven reinforcement learning. arXiv:2004.07219 (2020)
8. Fujimoto, S., Gu, S.S.: A minimalist approach to offline reinforcement learning. In: the 35th Conference on Neural Information Processing Systems (NeurIPS 2021), vol. 34, pp. 20132–20145 (2021)
9. Fujimoto, S., Hoof, H., Meger, D.: Addressing function approximation error in actor-critic methods. In: the 35th International Conference on Machine Learning (ICML 2018), pp. 1582–1591 (2018)
10. Fujimoto, S., Meger, D., Precup, D.: Off-policy deep reinforcement learning without exploration. In: the 36th International Conference on Machine Learning (ICML 2019), vol. 97, pp. 2052–2062 (2019)
11. Gao, C.X., Wu, C., Cao, Mingjun ant Kong, R., Zhang, Z., Yu, Y.: Act: empowering decision transformer with dynamic programming via advantage conditioning. In: the 38th AAAI Conference on Artificial Intelligence (AAAI 2024) (2024)
12. Guo, S., Zou, L., Chen, H., Qu, B., Chi, H., Yu, P.S., Chang, Y.: Sample efficient offline-to-online reinforcement learning. IEEE Trans. Knowl. Data Eng. **36**(3), 1299–1310 (2024)
13. Gürtler, N., et al.: Benchmarking offline reinforcement learning on real-robot hardware. In: the 11th International Conference on Learning Representations (ICLR 2023) (2023)
14. Haarnoja, T., Zhou, A., Abbeel, P., Levine, S.: Soft actor-critic: off-policy maximum entropy deep reinforcement learning with a stochastic actor. In: the 35th International Conference on Machine Learning (ICML 2018), vol. 80, pp. 1861–1870 (2018)
15. Janner, M., Li, Q., Levine, S.: Offline reinforcement learning as one big sequence modeling problem. In: the 35th Conference on Neural Information Processing Systems (NeurIPS 2021), vol. 34, pp. 1273–1286 (2021)
16. Kadokawa, Y., Zhu, L., Tsurumine, Y., Matsubara, T.: Cyclic policy distillation: sample-efficient sim-to-real reinforcement learning with domain randomization. Robot. Auton. Syst. **165**, 104425 (2023)
17. Kostrikov, I., Nair, A.N., Levine, S.: Offline reinforcement learning with implicit Q-learning. In: the 10th International Conference on Learning Representations (ICLR 2022) (2022)
18. Kostrikov, I., Tompson, J., Fergus, R., Nachum, O.: Offline reinforcement learning with fisher divergence critic regularization. In: the 38th International Conference on Machine Learning (ICML 2021), vol. 139, pp. 5774–5783 (2021)
19. Kumar, A., Zhou, A., Tucker, G., Levine, S.: Conservative Q-learning for offline reinforcement learning. In: the 34th Conference on Neural Information Processing Systems (NeurIPS 2020), vol. 33, pp. 1179–1191 (2020)
20. Lee, S., Seo, Y., Lee, K., Abbeel, P., Shin, J.: Offline-to-online reinforcement learning via balanced replay and pessimistic Q-ensemble. In: 5th Annual Conference on Robot Learning (CoRL 2021) (2021)
21. Levine, S., Kumar, A., Tucker, G., Fu, J.: Offline reinforcement learning: tutorial, review, and perspectives on open problems. arXiv:2005.01643 (2020)
22. Nair, A., Gupta, A., Dalal, M., Levine, S.: AWAC: accelerating online reinforcement learning with offline datasets (2021)

23. Nguyen, T., Luu, M.T., Ton, T., Yoo, D.C.: Towards robust policy: enhancing offline reinforcement learning with adversarial attacks and defenses. arXiv:2405.11206 (2024)
24. Otomo, T., Kera, H., Kawamoto, K.: Adversarial joint attacks on legged robots. In: 2022 IEEE International Conference on Systems, Man, and Cybernetics, pp. 676–681 (2022)
25. Radford, A., Narasimhan, K., Salimans, T., Sutskever, I.: Improving language understanding by generative pre-training (2018)
26. Seno, T., Imai, M.: d3rlpy: an offline deep reinforcement learning library. J. Mach. Learn. Res. **23**, 1–20 (2022)
27. Sheng, J., Zhai, P., Dong, Z., Kang, X., Chen, C., Zhang, L.: Curriculum adversarial training for robust reinforcement learning. In: 2022 International Joint Conference on Neural Networks (IJCNN 2022), pp. 1–8 (2022)
28. Todorov, E., Erez, T., Tassa, Y.: Mujoco: a physics engine for model-based control. In: 2012 IEEE/RSJ International Conference on Intelligent Robots and Systems (IROS 2012), pp. 5026–5033 (2012)
29. Vaswani, A., et al.: Attention is all you need. In: the 31st Conference on Neural Information Processing Systems (NIPS 2017), vol. 30. Curran Associates, Inc. (2017)
30. Wang, S., et al.: Train once, get a family: State-adaptive balances for offline-to-online reinforcement learning. In: the 37th Conference on Neural Information Processing Systems (NeurIPS 2023) (2023)
31. Wu, Y.H., Wang, X., Hamaya, M.: Elastic decision transformer. In: the 37th Conference on Neural Information Processing Systems (NeurIPS 2023) (2023)
32. Xie, Z., Lin, Z., Ye, D., Fu, Q., Yang, W., Li, S.: Future-conditioned unsupervised pretraining for decision transformer. In: the 40th International Conference on Machine Learning (ICML 2023), pp. 38187–38203 (2023)
33. Yamagata, T., Khalil, A., Santos-Rodriguez, R.: Q-learning decision transformer: leveraging dynamic programming for conditional sequence modelling in offline RL. In: the 40th International Conference on Machine Learning (ICML 2023) (2023)
34. Yan, K., Schwing, A., Wang, Y.X.: Reinforcement learning gradients as vitamin for online finetuning decision transformers. In: the 38th Annual Conference on Neural Information Processing Systems (NeurIPS 2024) (2024)
35. Yang, R., Bai, C., Ma, X., Wang, Z., Zhang, C., Han, L.: RORL: robust offline reinforcement learning via conservative smoothing. In: the 36th Conference on Neural Information Processing System (NeurIPS 2022), vol. 35, pp. 23851–23866 (2022)
36. Zhan, X., Xu, H., Zhang, Y., Zhu, X., Yin, H., Zheng, Y.: DeepThermal: combustion optimization for thermal power generating units using offline reinforcement learning. In: the 36th AAAI Conference on Artificial Intelligence (AAAI 2022), vol. 36, pp. 4680–4688 (2022)
37. Zhang, H., Wei, X., Yu, H.: Policy expansion for bridging offline-to-online reinforcement learning. In: the 11th International Conference on Learning Representations (ICLR 2023) (2023)
38. Zhao, Y., Boney, R., Ilin, A., Kannala, J., Pajarinen, J.: Adaptive behavior cloning regularization for stable offline-to-online reinforcement learning. arXiv:2210.13846 (2022)
39. Zheng, H., Luo, X., Wei, P., Song, X., Li, D., Jiang, J.: Adaptive policy learning for offline-to-online reinforcement learning. In: the 37th AAAI Conference on Artificial Intelligence (AAAI 2023), vol. 37, pp. 11372–11380 (2023)

40. Zheng, Q., Zhang, A., Grover, A.: Online decision transformer. In: the 39th International Conference on Machine Learning (ICML 2022), pp. 27042–27059 (2022)
41. Zheng, X., Ma, X., Wang, S., Wang, X., Shen, C., Wang, C.: Toward evaluating robustness of reinforcement learning with adversarial policy. In: the 54th Annual IEEE/IFIP International Conference on Dependable Systems and Networks (DSN 2024), pp. 288–301 (2024)
42. Zhou, Z., Zhou, Z., Bai, Q., Qiu, L., Blanchet, J., Glynn, P.: Finite-sample regret bound for distributionally robust offline tabular reinforcement learning. In: the 24th International Conference on Artificial Intelligence and Statistics (AISTATS 2021), vol. 130, pp. 3331–3339 (2021)

Feature Expansion Based Graph Transformer for Node Classification

Yun Chen[1], Canxin Chen[1], Jiaqing Tong[1], and Xiangyuan Ma[2(✉)]

[1] School of Mathematics and Computational Science, Xiangtan University, Xiangtan 411105, People's Republic of China
[2] Department of Biomedical Engineering, Shantou University, Shantou 515063, People's Republic of China
maxiangyuan@stu.edu.cn

Abstract. Graph Transformer (GT) models have demonstrated superior performance in various graph-related tasks due to their ability to model long-range node dependencies. However, current GT-based methods are constrained by low-dimensional node attributes and insufficient exploitation of global graph structure, leaving their feature embedding space underexplored. In this study, we propose a Feature Expansion based Graph Transformer (FEGT) for effective graph data processing, aiming to improve node classification accuracy. FEGT integrates node, edge, and structural embeddings to extend the original feature space of Graph Transformers, employing a fusion architecture that combines message-passing mechanisms and Transformer modules for node and edge feature extraction. Importantly, we capture structural information by decomposing the graph adjacency matrix and extracting principal component features, which are incorporated into the feature space as structural embeddings. We present numerical experiments on benchmark graph datasets (Cora, Citeseer, Pubmed, PATTERN, CLUSTER) to verify the effectiveness of FEGT. The results demonstrate that FEGT achieves superior performance in node classification compared to seven competitive models (GCN, GIN, GAT, GatedGCN, SAN, GT, GPS). Ablation studies further validate the effectiveness of the feature expansion with structural embeddings.

Keywords: Graph Transformer · Feature Expansion · Feature Embedding · MPNN · Matrix Decomposition

1 Introduction

As a complex non-Euclidean structure, graph data is prevalent in real-world applications such as social networks and transportation systems [1,2]. Its rapid growth in scale and expanding application have made graph data research a key focus in artificial intelligence. A fundamental task in this domain is node classification, which aims to predict class labels for unlabeled nodes within a graph [3]. Graph Neural Networks (GNNs) offer an effective approach to processing this

H. Ma et al. (Eds.): IWACIII 2025, CCIS 2780, pp. 244–258, 2026.
https://doi.org/10.1007/978-981-95-6730-0_18

task, enabling better capture of structural dependencies, mitigation of noise, and improved scalability [4].

GNNs are commonly divided into spectral, spatial, and message-passing methods. Valued for their simplicity and effectiveness, they have been successfully applied to tasks such as prediction of molecular properties and social recommendation [4]. However, current GNNs still suffer from over-smoothing and over-squeezing, which impede long-range dependency modeling and limit their scalability to large graphs [5]. Moreover, being constrained by the first-order Weisfeiler–Lehman Isomorphism Test (WL), they fail to capture global graph information [6].

To address the limitations of traditional GNNs in capturing long-range dependencies and scaling to large graphs, the Graph Transformer framework was introduced [7]. Its core idea is to model node relationships through positional encodings combined with self-attention. To more effectively capture complex global node interactions, researchers have proposed various extensions. For instance, the Spectral Attention Network (SAN) [8] leverages the graph Laplacian spectrum to encode positional information, while Graphormer [9] incorporates centrality, spatial, and edge encodings to enhance long-range reasoning. Other models introduce edge channels for structural parsing [10] and embed structural context directly into self-attention mechanisms [11]. More recent advances include subgraph-aware embeddings learned via self-supervised methods [12] and local structure encoders for enhanced neighborhood aggregation [13]. Collectively, these methods demonstrate the strong performance and adaptability of the Graph Transformer framework across a variety of graph learning tasks and benchmarks.

Although existing Graph Transformer models have demonstrated superior capability for modeling long-range node dependencies, their feature embedding space remains underexplored. Specifically, current approaches are constrained by low-dimensional node attributes and insufficient exploitation of global graph structure [14]. In this study, recognizing that performance hinges on feature-space design, we draw inspiration from Sun et al.'s feature expansion approach [15] and propose the Feature Expansion based Graph Transformer (FEGT). The proposed FEGT integrates node, edge, and structural embeddings to extend the original feature space of Graph Transformers. In particular, through a systematic analysis of the self-attention mechanism, we develop a message-passing mechanism and a Transformer-based fusion architecture. Importantly, structural principal component features are extracted through decomposition of the graph adjacency matrix and incorporated into the aforementioned fusion architecture, further enhancing feature representations.

We organize this paper in four sections. In Sect. 2, we describe the developed FEGT model, including the update of local-global features and the expansion of feature space. We compare in Sect. 3 the FEGT with several existing baseline models in terms of node classification tasks. Section 4 is devoted to the conclusion of the paper.

2 Feature Expansion Based Graph Transformer

We propose in this section a feature expansion based graph transformer (FEGT) model, aiming at improving the accuracy of node classification. The developed FEGT integrates node, edge, and structural embeddings to extend the original feature space of Graph Transformers. In particular, through a systematic analysis of the self-attention mechanism, we develop a fusion architecture that combines message-passing mechanisms and Transformer modules for node and edge feature extraction. Importantly, we capture structural information by decomposing the graph adjacency matrix and extracting principal component features, which are incorporated into the feature space as structural embeddings. The resulting architecture of FEGT is given in the Fig. 1.

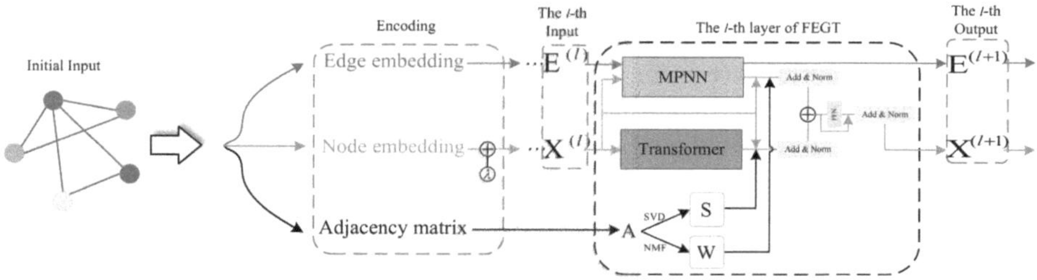

Fig. 1. Architecture of the proposed FEGT

In Fig. 1, we decouple the node, edge and structure embedding pathways. Firstly, the edge embedding is computed exclusively through the Message Passing Neural Network (MPNN) to generate the output of the corresponding layer. The node embedding is augmented with Laplacian positional (λ) encoding, followed by parallel processing through MPNN and Transformer modules. Subsequently, the outputs of the MPNN is combined with the non-negative matrix factorization (NMF) derived principal component feature W, and is normalized via a normalization (Norm) layer. This normalized output is then summed with the Transformer output, which has been augmented with the singular value decomposition (SVD) based principal component features S. The resulting representation is further refined through a Feed-Forward Network (FFN) module equipped with skip connections, and finally normalized through another norm layer to yield the ultimate output.

2.1 Update Strategy of Local-Global Features

This section describes the update strategy of local-global features for FEGT. Here, the update of local-global features is based on the designed fusion framework of MPNN and Transformer with respect to the node embedding.

During the data preprocessing phase, the model initially incorporates position embeddings into the nodes. These position embeddings are designed to capture the spatial positioning of individual nodes within the graph structure, effectively encoding their global structural information. We employs Laplacian positional encoding by selecting the k smallest non-trivial eigenvectors $\lambda_i = [U_{i1}, U_{i2}, \ldots, U_{ik}]^T \in \mathbb{R}^k$ (for $i = 1, 2, \ldots, n$) from Λ in the following formula

$$\Delta = I - D^{-1/2} A D^{-1/2} = U^T \Lambda U. \tag{1}$$

These eigenvectors collectively form the matrix $\Lambda \in \mathbb{R}^{n \times k}$, where k corresponds to the embedding dimension of the nodes. Subsequently, this positional encoding matrix is added to the initial node feature matrix X

$$X^{(0)} = X + \lambda. \tag{2}$$

FEGT employs the Graph Isomorphism Network (GIN) as the MPNN in this paper. GIN [6] is a message-passing-based graph neural network specifically designed to maximize the representational capacity of GNN, thereby enabling the model to effectively distinguish between graph structures. The detailed update procedure of GIN is outlined as follows. Let $X^{(l)} = (h_1^{(l)}, h_2^{(l)}, \ldots, h_k^{(l)})^\top$. The update mechanism of node representation in GIN can be formulated as follows

$$h_v^{(l)} = \mathrm{MLP}^{(l)} \left((1 + \epsilon^{(l)}) \cdot h_v^{(l-1)} + \sum_{u \in \mathcal{N}(v)} h_u^{(l-1)} \right), \tag{3}$$

where $h_v^{(l)}$ denotes the representation of node v at the l-th layer, $h_v^{(l-1)}$ represents the representation of node v at the $(l-1)$-th layer, $\mathcal{N}(v)$ indicates the set of neighboring nodes of v, $\epsilon^{(l)}$ is a learnable parameter that adjusts the relative weights between the node's own representation and its neighbors' representations, and $\mathrm{MLP}^{(l)}$ refers to a multi-layer perceptron employed to apply nonlinear transformations to the node representations.

Hence, the updated node features by MPNN and Transformer can be denoted by $X_M^{(l)}$ and $X_T^{(l)}$, respectively. The resulting representations are defined as

$$X_M^{(l)} = \mathrm{MPNN}(X^{(l-1)}) = \mathrm{GIN}(X^{(l-1)}), \tag{4}$$

$$X_T^{(l)} = \mathrm{Transformer}(X^{(l-1)}). \tag{5}$$

2.2 Feature Expansion Based on Matrix Decomposition

To address the limitation of insufficient dimensionality in node attribute features, this section extend the original feature space through the incorporation of additional feature subspace. Here, the feature subspace is based on the decomposition of the adjacency matrix with structure information.

In this paper, FEGT incorporates the principal component features by the NMF and SVD of the adjacency matrix to the outputs of MPNN and Transformer, respectively. By the message-passing mechanism of MPNN, the non-negative constraint enhances the node representations by only considering a limited set of related communities, which may mitigate the over-smoothing issue of

GNNs. With Transformer model, its global attention mechanism inherently captures long-range dependencies among nodes. Moreover, the dominant low-rank principal components of the graph-structure data are extracted by the SVD of the adjacency matrix. This may lead to the suppression of high-frequency noise and the preservation of the essential low-frequency structural features. With Eqs. (4) and (5), the output by feature expansion is described as the following formulations

$$X_M^{(l)} = W + \text{MPNN}(X_M^{(l-1)}), \tag{6}$$

$$X_T^{(l)} = S + \text{Transformer}(X_T^{(l-1)}), \tag{7}$$

where the non-negative basis matrix W and the basis matrix S are obtained by the NMF and SVD, respectively. NMF and SVD are described as follows.

(1) NMF

For the input adjacency matrix A, the following row normalization procedure is applied

$$A_{ij} \leftarrow \frac{A_{ij}}{\sum_j A_{ij}}. \tag{8}$$

Subsequently, the non-negative matrices W and H are initialized and iteratively updated through alternating optimization rules to converge to a solution.

- With H fixed, the matrix W is subsequently updated

$$W_{ij} \leftarrow W_{ij} \cdot \frac{(AH^\top)_{ij}}{(WHH^\top)_{ij} + \epsilon}, \tag{9}$$

- With W fixed, the matrix H is subsequently updated

$$H_{ij} \leftarrow H_{ij} \cdot \frac{(W^\top A)_{ij}}{(W^\top WH)_{ij} + \epsilon}. \tag{10}$$

During the update process, it is essential to maintain the non-negativity of both matrices W and H

$$W \leftarrow \max(W, \epsilon), H \leftarrow \max(H, \epsilon). \tag{11}$$

Finally, the column-normalized form of the obtained non-negative basis matrix is output

$$W_{:,i} \leftarrow \frac{W_{:,i}}{\|W_{:,i}\|_2}, \quad \forall i = 1, \ldots, k. \tag{12}$$

(2) SVD

By Eq. (8), the SVD of A is given by

$$A = U\Sigma V^\top. \tag{13}$$

We choose the first z singular values and their corresponding singular vectors, and define the structural principal component feature S as

$$S = U_z \cdot \Sigma_z. \tag{14}$$

To ensure the stability of the basis vectors, S is normalized as

$$S_{:,i} \leftarrow \frac{S_{:,i}}{\|S_{:,i}\|_2}, \quad \forall i = 1, \ldots, k. \tag{15}$$

2.3 Skip Connections and Feed-Forward Network Layers

Skip connections are designed to further simplify the model architecture and improve the training efficiency across multiple network layers, thereby mitigating the over-smoothing phenomenon. As illustrated in Fig. 2, the original feature X^{l-1} is incorporated into both MPNN(X^l) and Transformer(X^l) as skip connections. By the independently weighted nature of both MPNN and Transformer, the independence of the original feature is maximally preserved. Furthermore, skip connections are also implemented following the final feed-forward network (FFN) layer of the model. From the perspective of model architecture, skip connections enable the direct propagation of untransformed features to subsequent modules, thereby preserving information integrity via multiple transformation layers. This mechanism facilitates direct gradient backpropagation to shallow layers, effectively mitigating the vanishing or exploding gradient problem and enhancing the overall training performance of deep network structures. Particularly in graph neural networks, this architectural design effectively prevents critical information from being excessively smoothed during multi-layer propagation. The above analysis can be formulated as

$$X_M^{(l)'} = \text{BatchNorm}(X_M^{(l)} + X^{(l-1)}), \tag{16}$$

$$X_T^{(l)'} = \text{BatchNorm}(X_T^{(l)} + X^{(l-1)}), \tag{17}$$

$$X^{(l)'} = X_M^{(l)'} + X_T^{(l)'}. \tag{18}$$

Furthermore, the FFN layer is specifically designed to transform node features and enhance node diversity, thereby mitigating the over-smoothing phenomenon. It is a basic MLP consisting of two fully connected layers

$$X^{(l)} = \sigma(X^{(l)'} W_1) W_2 + X^{(l)'}, \tag{19}$$

where W_1 and W_2 denote the weight matrices of the fully connected layers without bias terms. The hidden dimension of FFN is typically set to be larger than that of $X(l)'$. Normalization is applied subsequent to the fully connected layers, and the normalization operation is omitted in Eq. (19) for notational simplicity. The FFN layer serves as a critical component within the model architecture, as evidenced by the experimental results presented in the subsequent section.

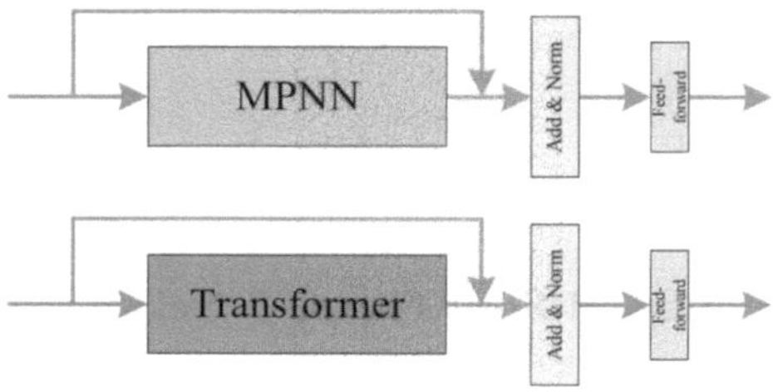

Fig. 2. Skip connection architecture.

2.4 Loss Function

At the output of the network's final layer, the model feeds the node representation h^L into an MLP to produce a prediction vector for each class of sample v. This vector is represented in one-hot encoding, where the position corresponding to the true class is set to 1 and all other positions are 0. We denote the prediction vector by $p_v \in \mathbb{R}^C$, where C represents the total number of classes. To address class imbalance by reducing the influence of classes with a larger number of samples, this section defines a weight for each class c

$$w_c = \frac{V - |c|}{V} \cdot \mathbb{I}(|c| > 0). \tag{20}$$

where $|c|$ denotes the number of samples in class c, V represents the total number of samples in the current batch, and the indicator function $\mathbb{I}(|c| > 0)$ evaluates to 1 if and only if at least one sample from class c is present in the current batch (i.e., $|c| > 0$); otherwise, it is 0. This function ensures that no weight is assigned to empty classes, thereby avoiding unnecessary or invalid computations. As observed from Eq. (20), the weight w_c is inversely proportional to the number of samples $|c|$ in class c. Specifically, when $|c|$ is large, w_c becomes small, thereby suppressing the loss contribution of that class; conversely, when $|c|$ is small, w_c increases, amplifying the loss contribution. This mechanism effectively mitigates the issue of severe class imbalance in model predictions. Based on the aforementioned definition of w_c, the multi-class weighted cross-entropy is defined as

$$Loss = -\frac{1}{V} \sum_{v=1}^{V} \sum_{c=1}^{C} w_c y_v[c] \log p_v[c], \tag{21}$$

where $y_v[c]$ denotes the c-th component in the true label of sample v.

2.5 Workflow of FEGT

Drawing upon the preceding overview of FEGT's overall architecture and detailed description of its individual modules, this section provides a comprehensive summary of the computational workflow of FEGT, as elaborated in the following algorithm.

Algorithm 1. Computational Workflow of the FEGT Model

Require: Graph $G(V, E)$, adjacency matrix A of graph G, input node feature matrix X of graph G, number of graph layers $l-1$, and node feature matrix $X^{(l-1)}$ at layer l.

Ensure: The output of the node at the l-th layer is denoted as $X^{(l)}$.

1: **if** $l - 1 = 0$ **then**
2: By incorporating positional encoding λ into the input node features X through Eq. (2), the resulting node representation at the initial layer is denoted as $X^{(0)}$;
3: **else**
4: The updated node feature matrices $X_M^{(l)}$ and $X_T^{(l)}$ are computed using Eqs. (4) and (5), respectively;
5: Based on Eqs. (6) and(7), the node feature matrices $X_M^{(l)}$ and $X_T^{(l)}$ are obtained by incorporating the principal component features W and the principal component features S, respectively;
6: Specifically, W is computed using Eqs. (8)–(12), while S is derived from equations (13)–(15);
7: According to Eqs. (16) and (17), the normalized node feature matrices $X_M^{(l)'}$ and $X_T^{(l)'}$ are obtained after applying Batch Normalization to the outputs of the skip connection;
8: Based on Eq. (18), the combined feature representation $X^{(l)'}$ is derived by integrating $X_M^{(l)'}$ and $X_T^{(l)'}$;
9: The output of the l-th layer, denoted as $X^{(l)}$, is obtained after forward propagation through the network using Eq. (19).
10: **end if**

3 Experimental Validation

In this section, we conduct numerical experiments to evaluate the proposed FEGT, covering node classification performance, ablation analyses, and hyperparameter tuning. We first compare FEGT with seven competitive models through five benchmark datasets. Ablation studies are carried out to assess the contribution of the structural principal component features. Furthermore, we conduct hyperparameter tuning experiments for efficiency check. The environment parameters of experiments in this study are summarized as Table 1. Four quantitative metrics, including Accuracy (Acc.), Precision (Prec.), Recall (Rec.) and F1, are used to assess the experiment results.

Table 1. Experimental Environment Configuration

Parameter	Experimental Environment Configuration
OS	Ubuntu 20.04.6 LTS
RAM	128 GB DDR5
CPU	Intel(R) Core(TM) i9-14900K
GPU	GeForce RTX 4090
Tool	Python 3.9.19 + Pytorch 1.13.0

3.1 Datasets and Baseline Models

Five benchmark datasets are described as follows. They include Cora, Citeseer, Pubmed, PATTERN and CLUSTER. In particular, as mentioned in [16], Cora is applied for the validation of learning models, Citeseer with sparsity and noise is employed to assess the robustness of models, and Pubmed serves to evaluate the scalability of the models on large-scale graphs. Furthermore, PATTERN and CLUSTER are specifically designed to assess the ability of node classification and the structural interpretability for models [17]. The detailed parameters of these datasets in the experiments are listed in Table 2.

Table 2. Setup of datasets, where AvNo and AvEd are the average number of nodes and edges in graphs, respectively, DimNo is the dimension of nodes.

Dataset	Graph	AvNo	AvEd	Classes	DimNo
Cora [16]	1	2708	5429	7	1433
Citeseer [16]	1	3327	4732	6	3703
Pubmed [16]	1	19717	44338	3	500
PATTERN [17]	14000	117	4301	2	7
CLUSTER [17]	12000	117	4749	6	3

Seven baseline models contain four GNNs and three graph transformer models. GCN [18] is a type of graph neural network that is grounded in the architecture of message-passing neural networks. GIN [6] is a spatial-domain graph convolutional network that operates on the basis of the message-passing mechanism. GAT [19] is an anisotropic graph neural network that dynamically computes the edge weights between nodes through an attention mechanism. GatedGCN [20] is an anisotropic graph neural network that incorporates a gating mechanism to regulate information propagation. SAN [8] is able to accurately determine the positional information of each node within a given graph, which is subsequently fed into a fully connected transformer architecture. GT [7] is a transformer-based architecture designed for graph-structured data. GPS [21] is a general-purpose, robust, and scalable graph transformer architecture. Here, the first four models are graph networks, and the last three are graph transformer models. In this experiment, the dropout rate is set to 0.01. The optimal number of network layers for achieving the best performance on Cora, Citeseer, Pubmed, PATTERN, and CLUSTER are 3, 4, 4, 4, and 12, respectively. The number of attention heads per layer is varied between 1 and 10. The initial learning rate for the AdamW [22] optimizer employed in FEGT is fixed at 10^{-3} across all datasets, with a maximum of 100 training epochs allowed.

3.2 Results and Analysis

A. Node Classification

We now compare the developed FEGT with the seven competitive models on five datasets for the task of node classification. The results are summarized in Table 3, reporting the high accuracy. With the results, it shows that FEGT has the best performance on four datasets: Cora (83.1%), Pubmed (81.9%), PATTERN (89.7%), and CLUSTER (78.9%). Compared to the second-best models, FEGT achieves 0.1%, 1.7%, 3%, and 0.9% accuracy improvements, respectively. Although FEGT did not achieve the best performance on Citeseer (70.9%), it still outperforms three baseline graph transformer models.

Table 3. Evaluation of results for models on five datasets.

Model	Cora	Citeseer	Pubmed	PATTERN	CLUSTER
	Accuracy(%)				
GCN [18]	81.5	70.3	79.0	71.9	68.5
GIN [6]	80.2	69.8	78.1	85.4	66.4
GAT [19]	83.0	**72.5**	79.0	78.3	70.6
GatedGCN [20]	82.7	72.1	80.2	85.6	75.0
SAN [8]	79.0	68.1	79.8	86.6	76.7
GT [7]	80.7	68.5	78.1	84.8	73.2
GPS [21]	80.0	64.2	78.6	86.7	78.0
FEGT(ours)	**83.1**	70.9	**81.9**	**89.7**	**78.9**

We also observe that FEGT demonstrates superior performance compared to traditional graph networks on Pubmed, PATTERN, and CLUSTER. These results imply that FEGT can preserve the local aggregation properties of graph networks through MPNN, while effectively capturing long-range dependencies by the attention mechanism of transformer. Furthermore, from Table 3, FEGT achieves the best performance on all datasets, compared with three graph transformer models. By expanding the feature space, FEGT uses both the SVD and the NMF to collaboratively enhance the discriminative capability of node embeddings. This enhancement facilitates more accurate community boundary partitioning, thereby improving overall model performance.

We further evaluate these models through four metrics in terms of CLUSTER, reporting the results as Table 4. By the results and model characteristics, it is observed that FEGT outperforms all models with an accuracy of 78.9%, demonstrating the strong overall correctness of the MPNN-Transformer fusion framework and feature expansion. In particular, the principal component features derived from matrix decomposition maybe suppress noise nodes, thereby enabling more accurate classification.

Table 4. Four metrics based evaluation for results of models on CLUSTER.

Model	CLUSTER (%)			
	Acc.	Prec.	Rec.	F1
GCN	68.5	67.7	68.6	68.6
GIN	66.4	67.0	66.1	66.4
GAT	70.6	70.1	70.8	70.8
GatedGCN	75.0	74.8	74.7	74.8
SAN	76.7	76.9	76.7	76.8
GT	73.2	73.4	73.3	73.3
GPS	78.0	76.5	**79.6**	**78.0**
FEGT (ours)	**78.9**	**78.7**	76.5	77.2

B. Ablation Studies

This section investigates the impact of two feature extension components in FEGT as well as the final FNN layer on the overall model performance through systematic ablation experiments. In this experiment, five distinct models were designed in Table 5, labeled A, B, C, D, and FEGT. As shown in Fig. 3, five models on all datasets exhibits a consistent trend in terms of relative effectiveness. The superior performance of Model B over Model C on three datasets provides indirect evidence that truncated SVD is more effective in feature extension within the computationally complex Transformer architecture. The comparison between Models A and D further confirms the efficacy of feature expansion. Moreover, the comparison between Models D and FEGT clearly demonstrates that the FFN layer plays a critical role in model training, which may be attributed to its capacity to enhance node diversity and substantially mitigate the over-smoothing issue.

C. Hyperparameter Tuning

This section presents hyperparameter experiments conducted from two perspectives, followed by the analysis of the experimental results.

Impact of Attention Heads for FEGT. From Table 6, the number of attention heads has a direct impact on the performance of FEGT. To investigate this effect, this study establishes multiple experimental groups with the number of attention heads set to 1, 2, 4, 8, and 10, respectively. The results reveal that the

Table 5. Model setup in ablation studies.

Model	SVD	NMF	FFN
A	×	×	×
B	✓	×	✓
C	×	✓	✓
D	✓	✓	×
FEGT	✓	✓	✓

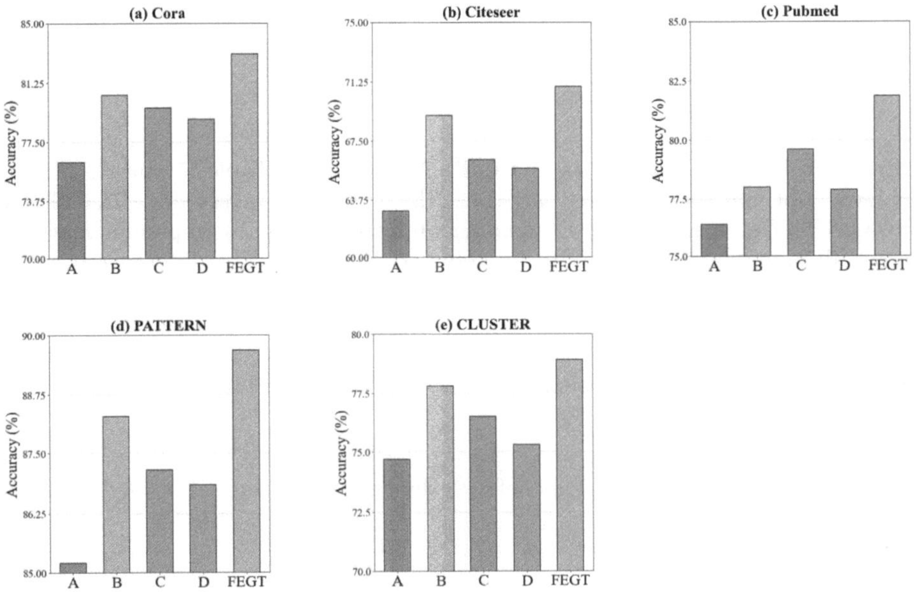

Fig. 3. Results of ablation studies on five datasets.

FEGT performance does not exhibit a positive correlation with the number of attention heads. Specifically, on the PATTERN dataset, the optimal performance is achieved when H = 4, whereas on the CLUSTER dataset, the best results are obtained when H = 8. From the perspective of model mechanism, multi-head attention facilitates the establishment of associations within distinct feature subspaces for graph-structure data. However, when the number of attention heads becomes excessively large, redundant information may be introduced, potentially leading to conflicts among the information captured by different heads. When the number of attention heads is small, the ability of capturing local-to-local relationships diminishes, thereby limiting its capacity to explore subgraph structures effectively. From the perspective of data characteristics, the task of PATTERN requires determining the existence of a specific subgraph, which heavily depends on precise matching of local structural patterns. In such cases, an excessively large head maybe result in excessive decomposition of local features and the introduction of noise. In contrast, the task of CLUSTER demands both strong intra-subgraph cohesion and inter-subgraph separation, requiring model to capture multi-scale features. Consequently, it benefits from a larger number of head.

Table 6. Evaluation of attention heads for FEGT through two datasets.

heads	1	2	4	8	10
Acc.(PATTERN)	87.1	87.4	**89.7**	86.5	86.0
Acc.(CLUSTER)	77.2	77.3	78.4	**79.0**	78.7

Effect of Model Layers for FEGT. As shown in Tables 7 and 8, FEGT achieves optimal performance at the 4th and 12th layers for PATTERN and CLUSTER, respectively. Expanding uniform principal component features across all model layers is likely to introduce redundancy. The difference in optimal layer depth between the two datasets stems from inherent characteristics of the datasets themselves. PATTERN emphasizes local matching and benefits from shallow layers to achieve rapid convergence, whereas CLUSTER depends on multi-level feature representation, necessitating deeper architectures for effective feature extraction.

Table 7. Evaluation of FEGT model with different layersin terms of PATTERN.

Model Layers	1	2	3	4	5	6
Acc.	88.5	89.1	89.3	**89.7**	89.4	88.9

Table 8. Evaluation of FEGT model with different layers in terms of CLUSTER.

Model Layers	6	8	10	12	14	16
Acc.	76.9	78.1	78.4	**78.9**	77.7	76.5

4 Conclusion

This paper proposed the FEGT model and conducts experiments to evaluate its performance. First, considering the lack of systematic research on the feature embedding space of Transformers in existing graph models, this work analyzes the attention mechanism and improves the graph convolutional architecture by incorporating Transformer components. The proposed framework, which integrates a message passing mechanism with a Transformer-based fusion structure, effectively combines local and global information to enable long-range dependency modeling of graph-structured data. Furthermore, to mitigate the limitations imposed by the dimensionality of node features, this work introduces the SVD and NMF applied to the graph adjacency matrix, thereby expanding the principal component feature space and enhancing the diversity of node representations.

The experimental results presented in this chapter demonstrate that FEGT achieves competitive performance on several key evaluation metrics when compared with other competitive models across the Cora, Citeseer, Pubmed, PATTERN, and CLUSTER datasets. Additionally, ablation studies and hyperparameter tuning experiments were conducted to validate the effectiveness of the model design and further enhance its performance.

Acknowledgment. This study is supported in part by the National Natural Science Foundation of China under Grant 62388101, in part by the Natural Science Foundation of Hunan Province under Grant 2022JJ40419, in part by the Youth Project of Hunan Provincial Department of Education under Grant 24B0176, and in part by the National Natural Science Foundation of China under Grant 62371476.

Disclosure of Interests. The authors have no competing interests to declare that are relevant to the content of this article.

References

1. Fan, W., Ma, Y., Li, Q., et al.: Graph neural networks for social recommendation. In: The World Wide Web Conference, pp. 417–426 (2019)
2. Wu, Z., Pan, S., Chen, F., et al.: A comprehensive survey on graph neural networks. IEEE Trans. Neural Netw. Learn. Syst. **32**(1), 4–24 (2020)
3. Xiao, S., Wang, S., Dai, Y., Guo, W.: Graph neural networks in node classification: survey and evaluation. Mach. Vis. Appl. **33**, 4 (2022)
4. Gori, M., Monfardini, G., Scarselli, F.: A new model for learning in graph domains. In: 2005 Proceedings of the IEEE International Joint Conference on Neural Networks, pp. 729–734 (2005)
5. Min, E., Chen, R., Bian, Y., et al.: Transformer for graphs: An overview from architecture perspective. arXiv preprint arXiv:2202.08455 (2022)
6. Xu, K., Hu, W., Leskovec, J., et al.: How powerful are graph neural networks? arXiv preprint arXiv:1810.00826 (2018)
7. Dwivedi, V.P., Bresson, X.: A generalization of transformer networks to graphs. arXiv preprint arXiv:2012.09699 (2020)
8. Kreuzer, D., Beaini, D., Hamilton, W., et al.: Rethinking graph transformers with spectral attention. Adv. Neural. Inf. Process. Syst. **34**, 21618–21629 (2021)
9. Ying, C., Cai, T., Luo, S., et al.: Do transformers really perform badly for graph representation? Adv. Neural. Inf. Process. Syst. **34**, 28877–28888 (2021)

10. Hussain, M.S., Zaki, M.J., Subramanian, D.: Global self-attention as a replacement for graph convolution. In: Proceedings of the 28th ACM SIGKDD Conference on Knowledge Discovery and Data Mining, pp. 655–665 (2022)
11. Chen, D., O'Bray, L., Borgwardt, K.: Structure-aware transformer for graph representation learning. In: International Conference on Machine Learning, pp. 3469–3489. PMLR (2022)
12. Ko, Y., Yang, H., Kim, T., et al.: Subgraph-aware training of language models for knowledge graph completion using structure-aware contrastive learning. In: Proceedings of the ACM on Web Conference, pp. 72–85 (2025)
13. Mao, Q., Liu, Z., Liu, C., et al.: HINormer: representation learning on heterogeneous information networks with graph transformer. In: Proceedings of the ACM Web Conference 2023, pp. 599–610 (2023)
14. Chen, F., Wang, Y., Wang, B., et al.: Graph representation learning: a survey. APSIPA Trans. Signal Inf. Process. **9**, e15 (2020)
15. Sun, J., Zhang, L., Chen, G., et al.: Feature expansion for graph neural networks. In: International Conference on Machine Learning, pp. 33156–33176 (2023)
16. Sen, P., Namata, G., Bilgic, M., et al.: Collective classification in network data. AI Mag. **29**(3), 93 (2008)
17. Dwivedi, V., Joshi, C., Luu, A., et al.: Benchmarking graph neural networks. J. Mach. Learn. Res. **24**(43), 1–48 (2023)
18. Kipf, T., Welling, M.: Semi-supervised classification with graph convolutional networks. arXiv preprint arXiv:1609.02907 (2016)
19. Veličković, P., Cucurull, G., Casanova, A., et al.: Graph attention networks. arXiv preprint arXiv:1710.10903 (2017)
20. Bresson, X., Laurent, T.: Residual gated graph convnets. arXiv preprint arXiv:1711.07553 (2017)
21. Rampášek, L., Galkin, M., Dwivedi, V., et al.: Recipe for a general, powerful, scalable graph transformer. Adv. Neural. Inf. Process. Syst. **35**, 14501–14515 (2022)
22. Loshchilov, I., Hutter, F.: Decoupled weight decay regularization. arXiv preprint arXiv:1711.05101 (2017)

API Sequence Representation via Stochastic Distribution Prediction

Qiaohua Deng(✉), Wei Dai, Yumin Lin, Weiyan Ma, Yan Yu, and Luguang Huang

RiverSecurity Co., Ltd., Shanghai, China
qiaohua.deng@riversecurity.com

Abstract. With the increasing adoption of APIs across diverse applications, API security has become one of the most critical concerns within the broader field of cybersecurity. This paper focuses on modeling API call sequences to address security challenges associated with API usage. Specifically, we model the sequential behavior of the API by predicting the probabilistic distribution of subsequent API calls, enabling the extraction of rich semantic representations of execution patterns. These pretrained representations are then utilized in downstream tasks such as bot detection and anomaly detection to identify bot-like behavior without relying on labeled data. Our method captures subtle differences in API usage between benign and malicious users, making it effective against stealthy or previously unseen bots. Experimental results demonstrate that our pretraining method consistently outperforms traditional sequence modeling paradigms in security detection tasks.

Keywords: representation learning · sequential data mining · API security

1 Introduction

Application Programming Interfaces (APIs) have become a fundamental component in modern software systems, enabling seamless interaction between diverse services and applications across distributed architectures. The ubiquity of APIs in contemporary digital ecosystems, from web services and mobile applications to cloud computing platforms and Internet of Things (IoT) devices, has fundamentally transformed how software components communicate and share data. As APIs continue to proliferate across various domains and become increasingly sophisticated in their functionality, their security has emerged as a critical concern within the broader landscape of cybersecurity.

The growing reliance on API-driven architectures has unfortunately been accompanied by a corresponding increase in security threats specifically targeting these interfaces. API-related attacks, including automated scraping, credential stuffing, distributed denial-of-service (DDoS) attacks, injection vulnerabilities, and other sophisticated malicious behaviors, are increasingly common and pose

H. Ma et al. (Eds.): IWACIII 2025, CCIS 2780, pp. 259–271, 2026.
https://doi.org/10.1007/978-981-95-6730-0_19

significant risks to the integrity, availability, and privacy of user data. These attacks often exploit the programmatic nature of APIs, utilizing automated tools and bots to conduct large-scale operations that can overwhelm systems, extract sensitive information, or compromise user accounts. The challenge is further compounded by the difficulty in distinguishing between legitimate automated API usage and malicious bot activities, as both may exhibit similar technical characteristics in their interaction patterns.

Traditional sequence modeling techniques, widely explored in natural language processing (NLP) and user behavior analysis, have shown promise in capturing temporal dependencies and patterns in sequential data [1,2]. However, directly applying these methods to API access sequences presents unique challenges due to the complexity and heterogeneity of API calls. To address this gap, this work draws inspiration from NLP and behavioral modeling to propose a novel pretraining framework based on random distribution prediction of API sequences. By dividing the API access timeline into uniform time windows and modeling the distribution of API calls within these windows, our approach captures temporal dynamics and usage patterns more effectively than conventional behavior-level prediction methods.

The key innovation of our approach lies in its ability to learn robust representations that are invariant to minor variations in API call sequences while remaining sensitive to meaningful behavioral differences that distinguish legitimate users from malicious actors. This is achieved through an objective function that combines distribution prediction with consistency regularization, ensuring that the learned representations are accurate in predictions and consistent internally.

We validate the effectiveness of the proposed method on multiple downstream tasks, including bot detection and anomaly detection, using both real-world CRM system data and synthetically constructed API sequences. Experimental results demonstrate that our approach consistently outperforms state-of-the-art baselines, such as masked behavior prediction and next-behavior prediction models, in terms of detection accuracy and robustness.

Additionally, an auxiliary consistency loss further enhances the quality of learned representations by aligning embeddings of original and masked sequences. This regularization technique prevents the model from overfitting to specific sequence patterns and promotes the learning of more generalizable representations that capture the essential characteristics of user behavior while remaining robust to minor variations and noise in the input data. The consistency loss also contributes to improved model stability during training and better convergence properties, making the overall approach more practical for real-world deployment scenarios.

2 Related Work

API Sequence Modeling. API sequence modeling has been extensively explored in software engineering and cybersecurity domains. Traditional approaches often rely on static analysis or keyword-based matching, which may not capture the dynamic and contextual nature of API usage. Deep learning techniques, particularly Recurrent Neural Networks (RNNs) and Transformer models, have shown promise in learning sequential patterns in API calls. DeepAPI, introduced by Gu et al. [2] addresses this challenge by employing a deep learning-based approach to generate API usage sequences from natural language queries.

The application of RNNs in malware detection through API sequences has demonstrated significant potential. Research on extracting representative API call patterns of malware families using recurrent neural networks has shown that RNNs can effectively capture the sequential dependencies inherent in malicious API call patterns, enabling the identification of family-specific behavioral signatures. This approach leverages the temporal nature of API calls to distinguish between different malware variants and their evolutionary patterns.

More recent advancements include the application of Transformer-based models [3], which have been widely adopted in natural language processing tasks due to their ability to capture long-range dependencies in sequences. These models have been adapted for API sequence modeling, offering improved performance over traditional RNN-based approaches.

User Behavior Modeling. Understanding user behavior is crucial for various applications, including personalized services and security detection. Traditional user modeling methods often require task-specific labeled data, which may be insufficient for training accurate models. To address this, pre-training user models on unlabeled user behavior data has been proposed as a potential solution. For example, UserBERT [4] introduces a contrastive user model pre-training method that incorporates self-supervision tasks to empower user modeling. Similarly, the Self-supervised User Modeling Network (SUMN) [5] and Multi-scale Stochastic Distribution Prediction(MSDP) [6] exploits behavioral consistency to develop universal user representations applicable to various downstream tasks.

Sequence Modeling in API Security. Integrating sequence modeling techniques into API security detection has gained significant traction in recent research. By modeling the temporal and contextual relationships between API calls, these approaches can identify patterns indicative of malicious activities. The development of deep frameworks for dynamic malware detection based on API sequence intrinsic features represents a major advancement in this field. These frameworks focus on extracting inherent characteristics from API call sequences that are invariant to superficial modifications, thereby improving the robustness of detection systems against evasion techniques.

The extraction of intrinsic features from API sequences has emerged as a critical research direction. Unlike traditional approaches that rely on hand-crafted features, Iltaek et al. [7] use Recurrent Neural Network (RNN) automatically learn representative features that capture the essential characteristics of malware behavior. This capability is particularly valuable when dealing with polymorphic malware that employs code obfuscation and other evasion techniques.

With the exceptional performance demonstrated by Transformer architectures on sequential data, the application of Transformer-based models for sequence modeling has become increasingly prevalent across various domains. In the context of malware detection, the adoption of corresponding model architectures has enabled more sophisticated analysis and modeling of API call sequences. Li et al. [8] proposed a Transformer-based model framework that integrates the intrinsic structural characteristics of multiple API features, demonstrating enhanced capability in capturing complex dependencies within malicious behavior patterns. Similarly, Maniriho et al. [9] introduced a comprehensive deep learning framework specifically designed for API call sequence analysis in Windows environments, leveraging advanced neural architectures to improve detection accuracy and robustness. These Transformer-based approaches represent a significant advancement over previous RNN-based methodologies, exhibiting substantial improvements in both computational performance and detection efficacy.

However, it is important to note that these existing studies primarily focus on malware API sequence construction, where malicious software exhibits relatively predictable and deterministic sequential patterns. Malware typically follows specific execution flows and API calling conventions that are characteristic of their underlying attack strategies and implementation approaches. In contrast, this work addresses API call sequence modeling in web applications, where user behavior is inherently more random and unpredictable. Web application users exhibit diverse interaction patterns, spontaneous decision-making, and highly variable session characteristics that require fundamentally different modeling approaches for effective behavior mining and analysis.

3 Method

This study aims to capture the long-term behavioral patterns embedded within API call sequences, which are crucial for tasks such as bot detection and anomaly analysis. Instead of building specialized models for each specific application, we propose learning a general-purpose encoder that transforms API sequences into universal representations. These representations effectively reflect consistent user or bot activity over extended periods. For instance, frequent and characteristic API calls made by a bot or user reveal intrinsic usage habits that can be distilled into these embeddings. Once trained, the universal encoder enables direct application of these representations to various downstream tasks without additional fine-tuning, greatly simplifying the deployment process and improving scalability.

As shown in Fig. 1, The pre-training tasks include distribution prediction for learning a robust sequence representation, and consistency loss task for model regularization that enforces representation stability under input perturbations. The intermediate layer vectors obtained from the pre-trained sequence model are subsequently utilized to support a variety of downstream tasks.

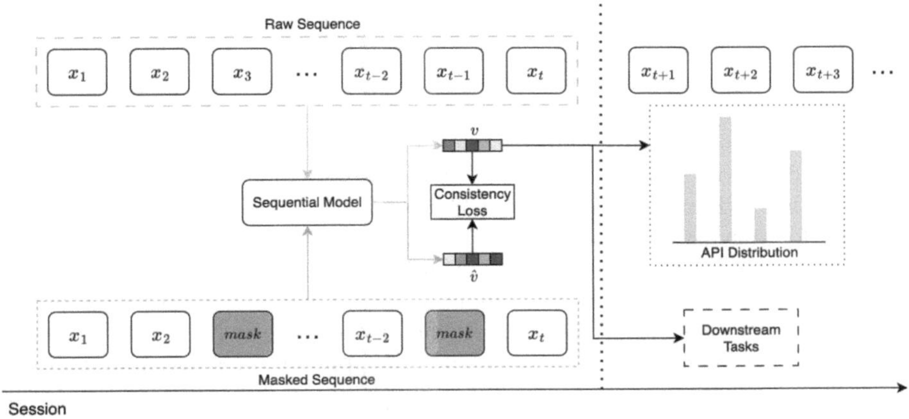

Fig. 1. Illustration of our approach for learning API sequence representation. Given a session of API call sequence, we pre-train the Sequential Model to get a sequence representation v. The inferred representations can be applied to various downstream tasks.

3.1 Preliminaries

The initial stage in distinguishing between a human visitor and a web bot involves extracting user sessions from log data. Session identification is a fundamental preprocessing step that significantly impacts the quality of subsequent analysis. A common approach in existing studies is to consider a session terminated if there is a period of inactivity exceeding 30 min. This threshold is based on empirical observations of typical user behavior patterns, where extended periods of inactivity often indicate the end of a coherent browsing session. In other words, if no requests are made within a 30-minute window, any subsequent request is treated as the beginning of a new session.

Let $S = x_1, ..., x_t, x_{t+1}, ...$ be a set of API call sequences in a session, where x_t denotes the specific API of the t-th call. Each API call is characterized by multiple attributes including the request timestamp, endpoint, parameters, and response characteristics. The sequential nature of these calls carries important behavioral information that distinguishes legitimate user interactions from automated bot activities.

The choice of session segmentation directly influences the model's ability to capture meaningful behavioral patterns. Too short sessions may not contain sufficient context for accurate prediction, while overly long sessions may introduce noise and dilute the coherence of behavioral patterns. The 30-minute threshold represents a balance between capturing complete user interactions and maintaining session coherence.

Window Setting. To ensure the effectiveness of the training sequences and that the predicted API distribution within the target window has statistical significance, two segmentation strategies can be considered: dividing sessions based on uniform time intervals or based on the number of API calls. Each approach has distinct advantages and limitations that must be carefully evaluated based on the specific characteristics of the dataset and the intended application.

Time-based segmentation ensures that each window represents a consistent temporal duration, which is particularly valuable when analyzing behavioral patterns that exhibit time-dependent characteristics. This approach naturally accounts for varying activity levels across different time periods and provides a uniform temporal context for prediction tasks.

Count-based segmentation ensures that each window contains a consistent number of API calls, which can be beneficial when the focus is on sequential patterns rather than temporal dynamics. This approach may be more suitable for scenarios where the density of API calls varies significantly across different sessions or user types.

In this study, based on comprehensive analysis of data characteristics and extensive experimental evaluation, we adopt the approach of splitting each session evenly into n time windows according to its total time span (n is set to 10 in this work). The choice of n=10 represents an optimal balance between providing sufficient historical context and maintaining computational efficiency.

Specifically, the entire duration of a session is divided into n equal-length segments, each representing a distinct time window. This uniform temporal segmentation ensures that the model receives consistent temporal context across different sessions, regardless of their total duration. The API calls occurring within the final (n-th) time window are designated as the prediction target window, serving as the ground truth for evaluating the model's predictive capabilities.

The preceding n-1 windows serve as historical context, forming the input sequence used to predict the API distribution in the final window. This temporal hierarchy allows the model to learn both short-term and long-term dependencies in user behavior patterns. The segmentation strategy enables the model to capture the temporal dynamics of API usage throughout the session and supports short-term forecasting of future behavior patterns based on prior activity.

The effectiveness of this windowing approach is further enhanced by the inclusion of temporal features that capture the relative timing of API calls within each window. These features help the model understand not only what APIs are called but also when they are invoked relative to the session timeline.

3.2 Stochastic Distribution Prediction

The core of our approach lies in formulating the prediction task as a stochastic distribution learning problem. This formulation acknowledges the inherent uncertainty in user behavior and the stochastic nature of API call patterns. Rather than attempting to predict exact sequences, which would be overly restrictive and prone to overfitting, we focus on learning the underlying probability distributions that govern API usage patterns.

Given the API sequence s_t over a period of time window (0, n-1], we use the API over the last window as the self-supervision signal. This self-supervised approach eliminates the need for manually labeled data while providing a natural learning objective that aligns with the predictive task. The ground-truth distribution is computed from the actual API calls observed in the target window, providing a robust training signal that captures the true behavioral patterns.

The ground-truth distribution is defined as the normalized occurrences of APIs:

$$p(x|v) = \frac{log(1 + count(x))}{\sum_{x \in \nu} log(1 + count(x))} \quad (1)$$

where ν denotes the set of all unique APIs in the target window, and $count(x)$ represents the number of occurrences of API x in the target window. The logarithmic transformation serves multiple important purposes:

i) Smoothing effect: The log(1+count) transformation prevents zero probabilities for unseen APIs and provides a smoother probability distribution.

ii) Reducing dominance: High-frequency APIs do not completely dominate the distribution, allowing the model to learn patterns from both common and rare API calls.

iii) Numerical stability: The logarithmic transformation improves numerical stability during training and prevents gradient explosion issues.

The pre-training task is to predict the API distribution in the target window based on the historical context provided by the preceding windows. The Kullback-Leibler (KL) divergence is employed to measure the differences between the predicted and ground-truth distributions, which defines the primary loss function:

$$L_{sim} = D_{KL}(p||\hat{p}) \propto -\sum_{x \in \nu} p(x|v) log \hat{p}(x|v) \quad (2)$$

where $\hat{p}$ is the predicted distribution produced by the sequence representation learning model. The KL divergence provides a principled way to measure the dissimilarity between probability distributions and has desirable properties for optimization, including convexity and differentiability.

The feature representation vector v is derived from the embedding of an intermediate hidden layer within the model, capturing contextualized sequential information. This representation encodes both local and global patterns in the API sequence, enabling the model to make informed predictions about future behavior based on historical context.

The choice of intermediate layer embeddings, rather than final layer outputs, is motivated by the need to capture rich, multi-level representations that can be effectively utilized for both the primary prediction task and auxiliary objectives. These embeddings typically contain more nuanced information about the input sequences compared to final layer outputs, which are often optimized specifically for the primary task.

3.3 Consistency Loss

Besides leveraging self-supervision signals derived from the behavioral distribution within the future window to facilitate learning of sequence representations over the past, we additionally propose a behavior reconstruction objective. This auxiliary task serves as a critical regularization mechanism that prevents the model from developing overly specialized representations that only work well for the primary prediction task.

The behavior reconstruction objective operates by randomly masking portions of the input behavioral sequence and requiring the model to maintain consistent representations between the original and masked sequences. This approach is inspired by successful masked language modeling techniques in natural language processing, adapted specifically for the sequential nature of API call data.

Masking Strategy: We employ a dynamic masking strategy where approximately 15% of the API calls in the input sequence are randomly selected for masking. The masked positions are replaced with special tokens, and the model must learn to generate representations that are robust to these missing elements. This masking ratio is chosen based on empirical evaluation to balance between providing sufficient challenge for the model while maintaining enough context for meaningful representation learning.

The auxiliary task reconstructs the masked portion of the behavioral sequence, acting as a regularizer to prevent the model from excessively fitting future behavior patterns at the expense of adequately modeling the observed sequence itself. This dual objective ensures that the learned representations capture both forward-looking predictive patterns and backward-looking reconstructive patterns.

We employ cosine similarity to enforce a consistency loss between the representation vectors of the original sequence and the randomly masked sequence, thereby enhancing the overall robustness of the model. Cosine similarity is particularly well-suited for this task as it measures the angle between vectors, focusing on directional similarity rather than magnitude differences.

The similarity is defined by:

$$\Omega(v, \hat{v}) = \frac{v}{\|v\|_2} \cdot \frac{\hat{v}}{\|\hat{v}\|_2} \tag{3}$$

where $\|\cdot\|_2$ is the $\ell_2 - norm$, $\hat{v}$ is the API sequence representation of the masked sequence.

The consistency loss encourages the model to learn representations that are invariant to small perturbations in the input sequence, which is crucial for generalization to unseen data. This invariance property is particularly important in cybersecurity applications where attackers may attempt to evade detection by introducing minor modifications to their behavior patterns.

3.4 Objective Function

In order to balance between accurate prediction of the target distribution and maintaining robust sequence representation, the overall objective function is formulated as a combination of the KL divergence loss in Eq. 2 and the consistency loss in Eq. 3 as

$$L = L_{sim} - \lambda \Omega(v, \hat{v}) \tag{4}$$

where λ is a non-negative coefficient to control the strength of the consistency loss and its influence on the final learning objective.

The parameter λ plays a crucial role in balancing the two objectives and requires careful tuning based on the specific characteristics of the dataset and the intended application. A small value of λ emphasizes predictive accuracy, while a larger value prioritizes representation robustness.

4 Experiments

To verify the effectiveness of the proposed method in API security scenarios, we report extensive experimental results on both real-world and synthetic datasets, comparing its performance against baseline approaches.

4.1 Datasets

Real-world Dataset. The real-world dataset is primarily collected from traffic logs of a CRM system server operating in a production environment. From this traffic data, API calls related to functionalities are extracted, such as */login*, */query*, and */download*, etc. The dataset includes a total of 544 APIs, which are also the basis for distribution prediction during pretraining. The bot traffic in the dataset is generated by automated tools, characterized mainly by complex patterns. The downstream task focuses on leveraging the learned sequence representations to accurately identify these bots.

Synthetic Dataset. Based on the 544 APIs present in the training dataset, we constructed synthetic API call sequences that do not appear in the original training data. Using the pretrained representation model, these sequences are vectorized and then subjected to anomaly detection methods to identify sequences that are out-of-distribution. The synthetic sequences are generated by composing n-grams that are entirely absent from the training sequences. For example, a constructed sequence such as "ABCD" is considered novel if none of its 2-grams (e.g., "AB", "BC", "CD") or 3-grams (e.g., "ABC", "BCD") appear in the original dataset.

4.2 Baselines

We conduct comparative experiments against representative pretraining strategies that focus on different behavioral prediction paradigms. The selected baselines include: a representative model based on the Masked Behavior Prediction (MBP) task [11], which learns to predict randomly masked behaviors within a sequence, and a method that incorporates both MBP and Next Behavior Prediction (NBP) [10], combining the tasks of predicting masked behaviors and the subsequent k behaviors to enhance sequence representation learning.

Ablation on Consistency Loss. To assess the impact of the consistency loss on model performance, we conduct ablation experiments by comparing the full model with a variant that excludes the consistency constraint. The version incorporating the consistency loss serves as a baseline for evaluating the contribution of representation alignment between original and masked sequences. This comparison allows us to isolate and quantify the effect of the consistency objective on the quality of learned representations and the final task performance.

Parameter Setting. For all experiments, we use a Transformer encoder as the sequence model, maintaining consistent hyperparameter settings across different runs. The embedding dimension is set to 256, and the factor λ in the objective function is fixed at 0.5, except in the ablation study where λ is set to 0 to disable the consistency loss. Optimization is performed using the Adam optimizer with a learning rate of 0.001 and a batch size of 256.

4.3 Evaluation and Result

Bot Detection. Bot detection is one of the most common tasks in the field of API security. Compared to normal users, bots typically exhibit significantly different access patterns. While simple bots can often be accurately identified using superficial traffic features such as the User-Agent field, advanced bots employ various evasion techniques that make them much harder to detect through rule-based or shallow inspection methods. Nonetheless, bots are generally designed with specific objectives, and their behavioral patterns—though often subtle—tend to reflect these intents.

To capture such nuanced characteristics, we adopt a deep learning-based sequence modeling approach to learn latent representations of API access behaviors. This enables the model to identify hidden patterns that distinguish bots from legitimate users, even when the bots are designed to mimic normal traffic.

For the bot detection task, we use traffic data collected from a real-world production system. The dataset includes normal samples representing typical business operations initiated by legitimate users, and bot samples generated by automated tools such as scanners, scrapers, and other malicious crawlers. Given the highly imbalanced nature of the original dataset—where bot traffic constitutes a small fraction—we apply the following strategies to mitigate class imbalance:

i) Data Augmentation: We simulate a portion of bot activities using automated tools to enrich the minority class.

ii) Oversampling: Attack samples are oversampled during training to ensure more balanced exposure.

iii) Focal Loss: We apply the Focal Loss function to adjust the training objective, reducing the impact of dominant normal samples and improving sensitivity to minority-class instances.

These combined strategies enable the model to better generalize and maintain robust performance in detecting various forms of bot activities in real-world API environments.

For bot detection, we use accuracy and AUC as evaluation metrics. After pretraining, the behavioral sequence representations are fed into an identical MLP classifier for downstream classification. The downstream task training settings and hyperparameters are kept consistent across all pretrained models to ensure a fair comparison. As shown in Table 1, our proposed method achieves noticeable improvements over previous approaches across multiple metrics. Furthermore, the ablation study demonstrates that incorporating the consistency loss yields a modest but consistent enhancement in performance.

Table 1. Performance comparison on real-world dataset

Method	Accuracy	AUC
MBP	67.23	0.7453
NBP&MBP	67.43	0.7465
Our Method(w/o CL)	68.05	0.7489
Our Method	**68.23**	**0.7506**

Anomaly Detection. Anomalies in API call sequences represent another important scenario in the domain of API security. In well-designed systems, API workflows typically follow predefined and logical patterns—for instance, a user is expected to log in before accessing any data-related services. However, due to flaws or oversights in system design, it is sometimes possible to bypass certain APIs and directly access sensitive or unauthorized endpoints. Such deviations from the expected sequence indicate potential vulnerabilities or malicious behavior, making API sequence anomaly detection a critical security task.

In our anomaly detection setting, we assume that the API sequences present in the existing dataset represent normal behavior. To simulate abnormal access patterns, we synthetically generate anomalous API sequences that violate expected logical or business process flows. These include skipping essential steps (e.g., querying without authentication) or introducing rare and unrealistic combinations of API calls.

We then apply a range of unsupervised anomaly detection algorithms—such as Isolation Forest, DBSCAN, and Autoencoder-based detectors—on the representations learned by the pre-trained sequence model. The goal is to evaluate the effectiveness of the model's behavioral representations in distinguishing between normal and anomalous sequences. By comparing performance across different detection algorithms, we assess the quality and expressiveness of the learned embeddings in capturing meaningful behavioral structures.

In the downstream anomaly detection task, we randomly sample 10% of the real-world dataset as a validation set and combine it with the synthetic anomaly dataset. The remaining 90% of the real data is used to train the sequence representation model. Once trained, we apply various unsupervised anomaly detection algorithms to the learned representations. To evaluate detection performance, we use accuracy as the primary metric. As shown in Table 2, our proposed method achieves consistent improvements over other approaches across all detection settings. Moreover, the variant with consistency loss shows only marginal performance differences compared to the one without it, suggesting that while the consistency objective contributes to representation stability, its influence on anomaly detection performance is relatively limited.

Table 2. Performance comparison in terms of accuracy on synthetic dataset

Method	IForest	DBSCAN	AutoEncoder
MBP	0.5223	0.5053	0.6123
NBP&MBP	0.5298	0.5215	0.6243
Our Method(w/o CL)	0.5332	**0.5273**	0.6590
Our Method	**0.5334**	0.5206	**0.6621**

5 Conclusion

This work draws inspiration from NLP and user behavior representation modeling to propose a novel approach for API sequence representation based on random distribution prediction. The method leverages temporal windowing and probabilistic modeling to capture the inherent variability in user behavior patterns, addressing the unique challenges posed by API sequence analysis in web application environments. The proposed framework is systematically applied to various downstream tasks, including API security detection, bot identification, and anomaly detection, with experimental results demonstrating its substantial advantage over traditional sequence modeling paradigms commonly used in NLP and cybersecurity domains.

This study represents an initial exploration into probabilistic sequence modeling for API security applications. Future research directions will focus on incorporating a broader spectrum of characteristics inherent in API data, including parameter semantics, temporal dynamics, and contextual dependencies, to design more comprehensive and theoretically grounded modeling frameworks. Additionally, we plan to investigate the integration of multi-modal features and develop adaptive mechanisms that can dynamically adjust to evolving attack patterns, thereby enhancing the long-term effectiveness and practical deployment viability of API security solutions in production environments.

References

1. Wang, X., Yiu, S.M.: A multi-task learning model for malware classification with useful file access pattern from API call sequence. arXiv preprint arXiv:1610.05945 (2016)

2. Gu, X., Zhang, H., Zhang, D., et al.: Deep API learning. In: Proceedings of the 2016 24th ACM SIGSOFT International Symposium on Foundations of Software Engineering, pp. 631–642 (2016)
3. Jiang, H., Li, Q.: Approximation rate of the transformer architecture for sequence modeling. Adv. Neural. Inf. Process. Syst. **37**, 68926–68955 (2024)
4. Wu, C., Wu, F., Yu, Y., et al.: Userbert: contrastive user model pre-training. arXiv preprint arXiv:2109.01274 (2021)
5. Gu, J., Wang, F., Sun, Q., et al.: Exploiting behavioral consistence for universal user representation. In: Proceedings of the AAAI Conference on Artificial Intelligence, vol. 35, no. 5, pp. 4063–4071 (2021)
6. Fu, C., Wu, W., Zhang, X., Hu, J., Wang, J., Zhou, J.: Robust user behavioral sequence representation via multi-scale stochastic distribution prediction. In: Proceedings of the 32nd ACM International Conference on Information and Knowledge Management (CIKM '23), pp. 4567–4573. Association for Computing Machinery, New York (2023). https://doi.org/10.1145/3583780.3614714
7. Kwon, I., Im, E.G.: Extracting the representative API call patterns of malware families using recurrent neural network. In: Proceedings of the International Conference on Research in Adaptive and Convergent Systems (RACS '17), pp. 202–207. Association for Computing Machinery, New York (2017). https://doi.org/10.1145/3129676.3129712
8. Li, C., Lv, Q., Li, N., et al.: A novel deep framework for dynamic malware detection based on API sequence intrinsic features. Comput. Secur. **116**, 102686 (2022)
9. Maniriho, P., Mahmood, A.N., Chowdhury, M.J.M.: API-MalDetect: automated malware detection framework for windows based on API calls and deep learning techniques. J. Netw. Comput. Appl. **218**, 103704 (2023)
10. Chuhan, W., Fangzhao, W., Qi, T., Lian, J., Huang, Y., Xie, X.: PTUM: pre-training user model from unlabeled user behaviors via self-supervision. In: Findings of the Association for Computational Linguistics: EMNLP 2020, pp. 1939–1944 (2020)
11. Zhao, Q.: RESETBERT4Rec: a pre-training model integrating time and user historical behavior for sequential recommendation. In: Proceedings of the 45th International ACM SIGIR Conference on Research and Development in Information Retrieval, pp. 1812–1816 (2022)

Cascade Adaptive Phase Optimized Active Disturbance Rejection Control for Ultra-supercritical Units

Jiangyu Jia and Wei Wei(✉)

School of Intelligent Engineering and Automation, Beijing University of Posts and Telecommunications, Beijing 102206, China
weiwell@bupt.edu.cn

Abstract. The coordinated control system of ultra-supercritical (USC) units often faces multi-source uncertain disturbances, such as coal quality variations and load fluctuations. Those factors weaken the system's response and stability. In addition, the boiler's thermal inertia and the turbine's mechanical inertia add complexities to the control task. To address it, a cascade adaptive phase optimized active disturbance rejection control (CAPOADRC) is proposed. An adaptive phase optimized extended state observer (APOESO) is constructed to enhance the disturbance estimation. To overcome the bandwidth limitations of ESO, a cascade structure is introduced, the first layer estimates partial disturbances and filters the system output, while the second layer utilizes the estimation information from the first layer to get the remaining disturbances. The CAPOADRC can estimate the total disturbance rapidly and accurately. Numerical results validate the proposed method in suppressing system disturbances and improving response speed. It may provide a feasible solution for the coordinated control of USC units.

Keywords: Ultra-Supercritical Units · Coordinated Control · Active Disturbance Rejection Control · Adaptive Phase Optimized Extended State Observer

1 Introduction

With increasing demands for stability, economic efficiency, and environmental protection in power systems, ultra-supercritical (USC) units have emerged as the dominant technology for large-scale power plants due to their high thermal efficiency and low emissions. However, under complex and variable operating conditions, the coordinated control of USC units faces increasingly stringent challenges. Those systems are frequently subjected to multiple sources of uncertainties, such as variations in coal quality, load disturbances, and environmental fluctuations, which significantly degrade their dynamic performance or break the stability of the system [1].

In recent years, worldwide researchers have studied the coordinated control of USC units. Traditional methods primarily rely on PID control or multivariable decoupling control [2, 3]. Those approaches are simple in structure and easy to be implemented.

H. Ma et al. (Eds.): IWACIII 2025, CCIS 2780, pp. 272–280, 2026.
https://doi.org/10.1007/978-981-95-6730-0_20

However, their performance deteriorates significantly under strong disturbances and couplings conditions. To enhance performance, control strategies such as predictive control [4–7] and neural network control [8, 9] have been introduced. They can effectively address nonlinearities and time-varying disturbances. However, they are largely dependent on model accuracy and training data. It poses challenges in terms of computational burden and implementation.

Active disturbance rejection control (ADRC) attracts much attention due to its strong robustness to external disturbances and system uncertainties [10]. The coordinated control system (CCS) of USC units is subject to multiple unknown disturbances. However, an extended state observer (ESO), a key part of the ADRC, still has limitations on estimation in terms of accuracy and speed. Although increasing the bandwidth can accelerate estimation, it cannot be increased arbitrarily due to physical constraints such as sampling frequency, actuator bandwidth, and sensor noises. To address those undesired issues, a hybrid control incorporating error-based ADRC, Kalman filtering, and reinforcement learning was proposed [11]. However, filters inevitably reduce the estimation speed. To balance estimation and noise suppression, a cascaded ESO structure was introduced [12]. To further enhance the robustness of the system, an improved cascaded ESO design was proposed [13]. Nevertheless, those improvements still come at the cost of reduced estimation speed.

Rapid disturbance estimation ensures that a system can promptly adjust control actions. To overcome the limitations imposed by a high-bandwidth ESO, it has been shown that optimizing estimation speed from a phase perspective is effective [14]. In this paper, a cascaded adaptive phase optimized active disturbance rejection control (CAPOADRC) is proposed to improve the coordinated control of USC units.

2 Nonlinear Model of an Ultra-supercritical OTB Unit

A third-order nonlinear model is utilized to describe the dynamics of a 1000 MW USC unit [15].

$$\begin{cases} \dot{x}_1 = -0.0056x_1 + 0.0056u_1(t-17) \\ \dot{x}_2 = \frac{f_{x_2}\times(500-1.31x_3)}{1060000(1.31x_3-1205)}u_3 + 0.0157x_1^{1.031} + 0.000665u_2 \\ \dot{x}_3 = \frac{f_{x_2}\times(3000-1.31x_3)}{59830(1.31x_3-1205)}u_3 + 0.278x_1^{1.031} - 0.03u_2 \\ y_1 = x_2 - 0.13x_2^{0.882} \\ y_2 = x_3 \\ y_3 = 0.00055u_3f_{x_2} \\ f_{x_2} = \left(43.22x_2 - 5.62x_2^{0.882} - 31.84\right)\times\left(-8.96x_2 + 1.165x_2^{0.882} + 2512.4\right) \end{cases} \tag{1}$$

where x_1, x_2, and x_3 are the coal feed rate into the boiler, the steam pressure at the steam-water separator, and the outlet enthalpy, respectively. The input variables u_1, u_2, and u_3 are coal feed command u_B, feedwater flow rate D_e, and valve opening u_T, respectively. Output variables y_1, y_2, and y_3 are the main steam pressure P_{st}, the separator outlet enthalpy H_m, and the unit load N_e, respectively.

3 The Cascade Adaptive Phase Optimized Active Disturbance Rejection Control

3.1 Active Disturbance Rejection Control

To illustrate the ADRC, a first-order nonlinear system is taken.

$$\dot{y} = f + bu + d \tag{2}$$

where y is the system output, u is the control signal, b is the unknown control gain, $f(\cdot)$ is the nonlinear part of the system, and d is the external disturbance.

$$\dot{y} = f + bu + d = f + (b - b_0)u + d + b_0u = F + b_0u \tag{3}$$

here, b_0 is an adjustable value. F is the total disturbance, and$F = f + (b - b_0)u + d$.

A second-order linear ESO can be described as

$$\begin{cases} \dot{z}_1 = z_2 + b_0u + \beta_1(y - z_1) \\ \dot{z}_2 = \beta_2(y - z_1) \end{cases} \tag{4}$$

where z_1, z_2 estimate y and F, respectively. β_1, β_2 are observer gains.

Based on bandwidth-parameterized approach in [10], $\beta_1 = 2\omega_o, \beta_2 = {\omega_o}^2$, where ω_o is an observer bandwidth. The designed control law is

$$u = \frac{K_p(r - z_1) - z_2}{b_0} \tag{5}$$

where K_p is the controller gain.

3.2 Cascade Adaptive Phase Optimized Extended State Observer

Under varying operating conditions, USC units are often subjected to various disturbances. To achieve fast and accurate disturbance estimation, a cascaded adaptive phase optimized extended state observer (CAPOESO) is constructed.

The first layer of CAPOESO is designed as

$$\begin{cases} e = y - z_1, S(e) = (\tanh(e))^c \\ \dot{z}_1 = z_2 + b_0u + \beta_1 e \\ \dot{z}_2 = \beta_2 e \\ \dot{z}_{2u} = \dot{z}_2 + \frac{dS(e)}{dt} = \beta_2 e + \lambda(e)\dot{e} \end{cases} \tag{6}$$

where e is the observation error, and $S(e)$ is a constructed regulatory term. c is an adjustable parameter, and it is set to be 5/7 in this study. z_1 and z_{2u} are the outputs of the first layer. $\lambda(e)$ is an adaptive coefficient, $\lambda(e) = c(\tanh(e))^{c-1} \cdot (1 - \tanh^2(e))$, taking Laplace transformation on z_{2u}, one has $z_{2u} = \frac{\lambda s + \beta_2}{s^2 + \beta_1 s + \beta_2}F$. β_1 and β_2 are gain coefficients, $\beta_1 = 2\omega_{o1}, \beta_2 = {\omega_{o1}}^2$.

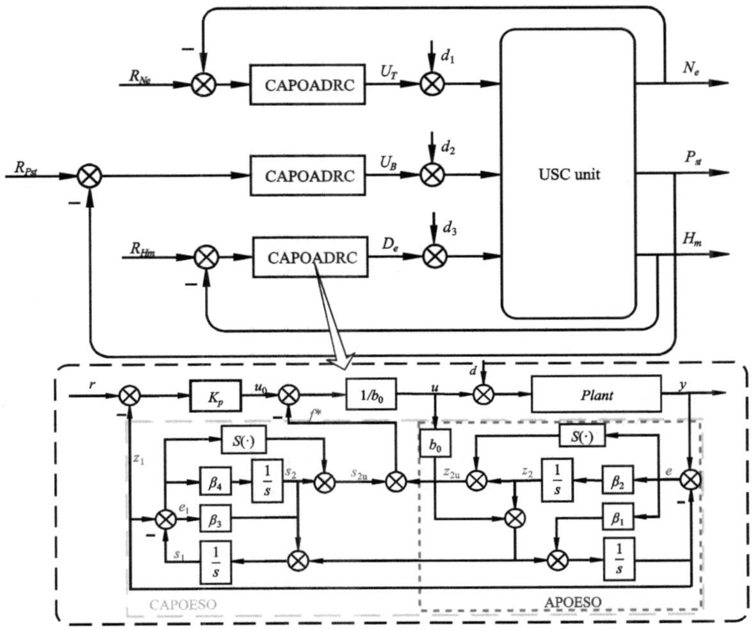

Fig. 1. Scheme of the CCS and the CAPOADRC

The first layer of CAPOESO primarily estimates partial disturbances and filters the system output. The second layer utilizes the estimation information from the first layer to obtain the remaining disturbances. Similarly, the second layer of CAPOESO is

$$\begin{cases} e_1 = z_1 - s_1, S(e_1) = (\tanh(e_1))^c \\ \dot{s}_1 = s_2 + b_0 u + \beta_3 e_1 + z_2 \\ \dot{s}_2 = \beta_4 e_1 \\ \dot{s}_{2u} = \beta_4 e_1 + \lambda(e_1)\dot{e}_1 \end{cases} \tag{7}$$

where s_{2u} is the output of the second layer. β_3 and β_4 are gain coefficients, $\beta_3 = 2\omega_{o2}$, $\beta_4 = {\omega_{o2}}^2$. The total disturbance estimated by CAPOESO is

$$\mathrm{f}^* = \mathrm{z}_{2\mathrm{u}} + \mathrm{s}_{2\mathrm{u}} \tag{8}$$

Based on the above, the control law for CAPOADRC is

$$\mathrm{u} = \frac{\mathrm{K_p}(\mathrm{r} - \mathrm{z}_1) - \mathrm{f}^*}{\mathrm{b}_0} \tag{9}$$

where $K_p = \omega_c$, ω_c is the controller bandwidth. Scheme of the coordinated control system and CAPOADRC is shown in Fig. 1.

4 Simulation Results

As illustrated in Fig. 1, the coordinated control system consists of three control loops: S_{Ne}, S_{Pst} and S_{Hm}. A reference signal is set to simulate the unit's sliding pressure operation. Initially, the load setpoint increases from 450 MW to 1000 MW within 660 s. Subsequently, the load decreases from 1000 MW to 700 MW at the same rate, and the load increases again from 700 MW to 900 MW. The simulation step size is set to be 1×10^{-2} s. At $t = 6000$ s, a constant disturbance, whose amplitude is 9, is introduced.

During the experiments, ADRC and CPOADRC are designed to make a comparison. Parameters are shown in Table 1. Tracking responses and tracking errors of N_e, P_{st} and H_m are shown in Fig. 2. and Fig. 3. As shown in Fig. 2, with lower observer bandwidth and identical controller parameters, CPOADRC outperforms ADRC in tracking performance. Furthermore, with the same observer bandwidth and controller parameters, CAPOADRC achieves a faster response and significantly lower maximum overshoot than CPOADRC. Under disturbances, CAPOADRC exhibits the smallest deviation amplitude and the shortest convergence time. This demonstrate CAPOADRC is robust enough to disturbances, uncertainties, and strong nonlinear couplings.

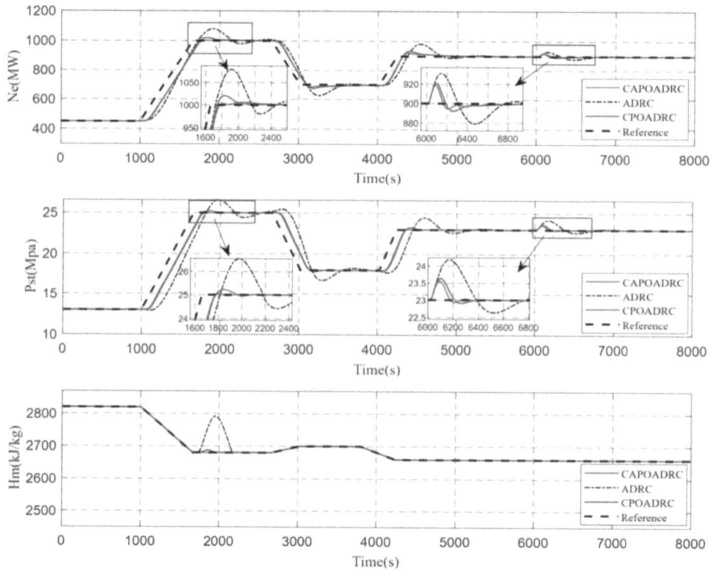

Fig. 2. Outputs of the CAPOADRC, ADRC and CPOADRC

In Table 2, the integrated time absolute error ($ITAE = \int t|e_1|dt$) and control energy ($E = \int u^2 dt$) are calculated. It shows that CAPOADRC achieves the best tracking accuracy with least energy.

Figure 4 illustrates total disturbance estimations. Generally, due to the time required for each layer to process and transmit information, the cascaded ESO tends to estimate total disturbances more slowly than a single ESO. However, by dynamically optimizing

Table 1. Simulation parameters of the ADRC, CPOADRC and CAPOADRC

Controllers	Parameter	S_{Ne}	S_{Pst}	S_{Hm}
ADRC	b_0	3000	9	-1.5
	ω_o	0.09	99	5
	ω_c	10	0.013	80
POADRC	b_0	3000	9	-1.5
	ω_{o1}	0.08	60	2.5
	ω_{o2}	0.06	40	2.5
	ω_c	10	0.013	80
CAPOADRC	b_0	3000	9	-1.5
	ω_{o1}	0.08	60	2.5
	ω_{o2}	0.06	40	2.5
	ω_c	10	0.013	80

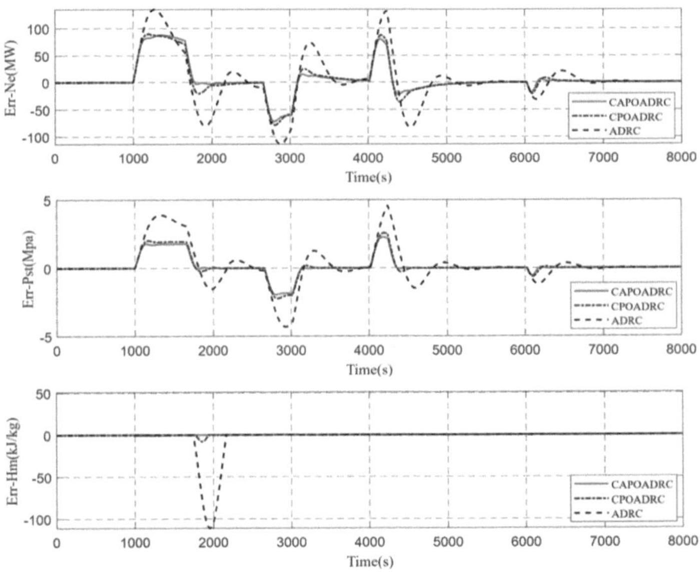

Fig. 3. Tracking errors

the phase of the observer, the CAPOESO achieved faster estimation speeds. Thus, the CAPOADRC responds more promptly than the other two methods. It also can be found in Fig. 5.

Table 2. Performance indexes of the ADRC, CPOADRC and CAPOADRC

Controllers	E			ITAE		
	S_{Ne}	S_{Pst}	S_{Hm}	S_{Ne}	S_{Pst}	S_{Hm}
ADRC	4.84×10^7	3.41×10^9	4.24×10^3	6.23×10^8	1.85×10^7	5.60×10^7
CPOADRC	4.81×10^7	3.41×10^9	4.20×10^3	3.31×10^8	7.03×10^6	2.49×10^6
CAPOADRC	4.80×10^7	3.40×10^9	4.19×10^3	2.96×10^8	6.11×10^6	9.31×10^5
Improvement (CPOADRC)	0.6%	0	0.9%	46.9%	62.0%	95.6%
Improvement (CAPOADRC)	0.8%	0.3%	1.2%	52.5%	67.0%	98.3%

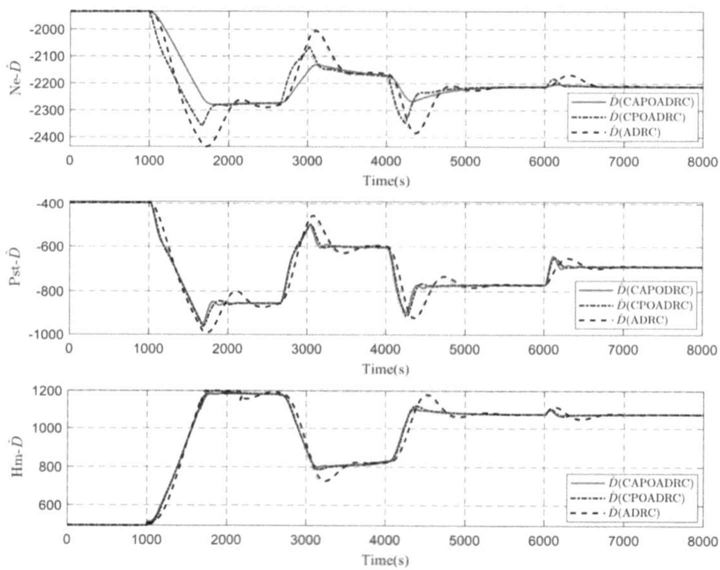

Fig. 4. The total disturbance estimations

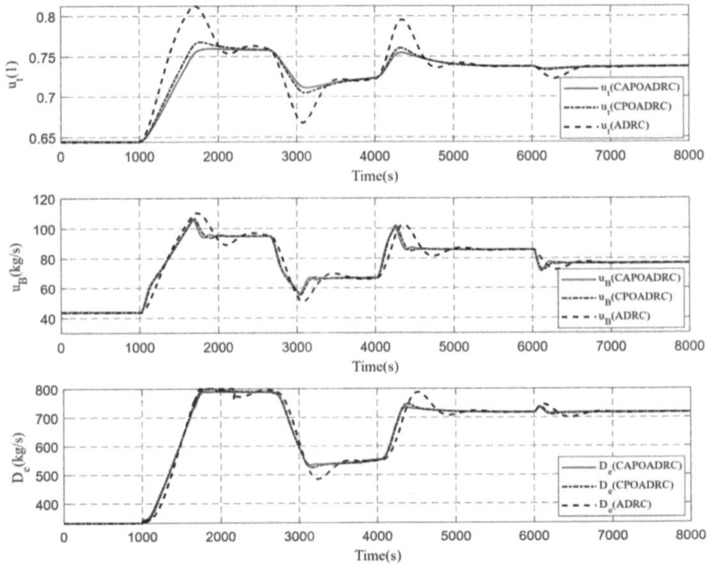

Fig. 5. Control signals under the ADRC, CPOADRC and CAPOADRC

5 Conclusions

In this article, the coordinated control for USC units is considered. To enhance the system's disturbance rejection capability, a cascaded adaptive phase optimized active disturbance rejection control is proposed. An APOESO is constructed to adaptively optimize the phase of disturbance estimation. Concurrently, a cascaded structure is introduced to resolve the fundamental bandwidth constraints. Simulation results validate the superiority of the proposed method in both tracking accuracy and disturbance rejection capability. It may be a promising way to improve the USC units' coordinated control. However, experimental results are necessary, and it is the work in the near future.

Acknowledgments. This work is supported by the Open Project Program of State Key Laboratory of Low-carbon Smart Coal-fired Power Generation and Ultra-clean Emission, grant number D2023FK112.

Disclosure of Interests The authors have no competing interests to declare that are relevant to the content of this article.

References

1. Hou, G., Zhou, Y., Zhang, F.: Data-driven modeling of ultra-supercritical unit based on improved PatchTST network with multi-task balancing. Energy. **330**(1), 136958–136971 (2025)

2. Magzoub, M.A., Alquthami, T.: Optimal design of automatic generation control based on simulated annealing in interconnected two-area power system using hybrid PID—fuzzy control. Energies. **15**(1), 1540–1555 (2022)
3. Yang, Q., Chen, G., Guo, M., Chen, T., Luo, L., Sun, L.: Model predictive hybrid PID control and energy-saving performance analysis of supercritical unit. Energies. **17**(24), 6356–6569 (2024)
4. Yang, C., Zhang, T., Sun, L., et al.: Piecewise affine model identification and predictive control for ultra-supercritical circulating fluidized bed boiler unit. Comput. Chem. Eng. **174**(1), 108257–108273 (2023)
5. Zhu, Y., Zhang, K., Zhu, Y.: Development of a three-term MPC and its application to an ultra-supercritical coal fired power plant. Control. Eng. Pract. **143**(1), 105787–105801 (2024)
6. Cheng, C., Peng, C., Zhang, T.: An optimized nonlinear generalized predictive control for steam temperature in an ultra-supercritical unit. ISA Trans. **139**(1), 13–23 (2023)
7. Hou, G., Huang, T., Zheng, F., Huang, C.: A hierarchical reinforcement learning GPC for flexible operation of ultra-supercritical unit considering economy. Energy. **289**(1), 129936–129956 (2024)
8. Ping, B., Zeng, D., Hu, Y., Xie, Y.: Neural network predictive controller based on the improved TPA-LSTM model for ultra-supercritical units. Heliyon. **10**(12), 31997–32016 (2024)
9. Zhou, Z., Zhang, J., Wang, Y., Yang, D., Liu, Z.: Adaptive neural control of superheated steam system in ultra-supercritical units with output constraints based on disturbance observer. IEEE Trans. Circ. Syst. I Regul. Pap. **1**(1), 1–11 (2025)
10. Gao, Z.Q.: Scaling and bandwidth-parameterization based controller tuning. Proceedings of the 2003 American Control Conference. **6**(1), 4989–4996 (2003)
11. Hou, G., Huang, T., Zheng, F.: Application of multi-agent EADRC in flexible operation of combined heat and power plant considering carbon emission and economy. Energy. **263**(1), 125711–125726 (2023)
12. Łakomy, K., Madonski, R.: Cascade extended state observer for active disturbance rejection control applications under measurement noise. ISA Trans. **109**(1), 1–10 (2021)
13. Rojas, H.D., Cortés-Romero, J., Rojas, H.E.: Active disturbance rejection control based on a cascade estimator composed of reduced-order and full-order extended state observers. ISA Trans. **151**(1), 296–311 (2024)
14. Wei, W., Xia, P., Zuo, M.: Active disturbance rejection control using a phase optimized extended state observer for a nanopositioner. Int. J. Automot. Manuf. Mater. **3**(2), 6–25 (2024)
15. Zhou, H., et al.: Affine nonlinear control for an ultra-supercritical coal fired once-through boiler-turbine unit. Energy. **153**(1), 638–649 (2018)

AI for Science And Culture

Research on Decision Support for Cultural Heritage Conservation Using Agent-based AI Systems with Prompt Engineering—A Case Study of the "Heritage Echo Planet" World Heritage Large Language Model

Ximeng Wang[1], Yuzhe Liu[1], Shuai Shao[2], and Lijie Wang[1](✉)

[1] Beijing Information Science and Technology University, 55 Taihang Road Changping District, 10096 Beijing, China
Wanglijie_0203@126.com

[2] Beijing Zhuangguancun Academy, No. 5 Building No. 17 Daniaofang 2nd Ring Road Haidian District, 10094 Beijing, China

Abstract. Amid the global digital transformation of cultural heritage, this study addresses critical challenges such as data fragmentation, the lack of intelligent response mechanisms, and the limited capacity of general Large Language Models (LLMs) in parsing domain-specific semantics. We propose a prompt engineering-based agent AI system that enhances professional task performance through deep knowledge embedding, structured output specification, and dynamic decision reinforcement. This approach effectively mitigates issues including terminology misunderstanding and the absence of specialized agents in general LLMs. Comparative experiments were conducted between the proposed agent (World Heritage LLM) and baseline models on tasks such as professional Q&A and strategic report generation. The results demonstrate that the World Heritage LLM significantly outperforms baseline models in case coverage, systematic strategy formulation, temporal planning, risk responsiveness, innovation capability, and support for international collaboration. These findings underscore the efficacy of prompt engineering in advancing intelligent decision-making for cultural heritage conservation. The study offers a reusable and scalable paradigm of intelligent decision support for the field.

Keywords: Prompt Engineering · Agent-Based Artificial Intelligence · Cultural Heritage Conservation · World Heritage Large Language Model

1 Research Background and Problems

In the wave of global digital transformation, prompt engineering has rapidly risen since 2023 and has become one of the key means for large language models(LLMs) to be implemented in vertical fields. As general LLMs have demonstrated powerful language understanding and generation capabilities, researchers have found that carefully designed

H. Ma et al. (Eds.): IWACIII 2025, CCIS 2780, pp. 283–293, 2026.
https://doi.org/10.1007/978-981-95-6730-0_21

prompt can significantly enhance the LLMs' performance in professional scenarios without changing the LLMs' parameters. Moreover, prompt can improve the LLMs' efficiency in extracting and utilizing domain knowledge. By constructing prompt that include typical domain scenarios, professional terminology, and task requirements, the LLMs can be guided to focus on relevant information when processing data, allowing them to accumulate domain knowledge more targetedly during pre-training. Prompt can also enhance the LLMs' generalization capabilities in specific fields. Through diverse prompt designs covering different tasks and scenarios within a domain, the LLMs can form a more comprehensive understanding of the domain during pre-training. When faced with new tasks or data within the domain, they can adapt more quickly and perform better, while reducing the dependence on large amounts of domain-labeled data, lowering the threshold and cost for domain applications. This has led prompt engineering to evolve from simple instruction templates in its early stages to a complex technical system that includes zero-shot prompting, few-shot prompting, chain-of-thought prompting, tree-of-thought prompting, retrieval-augmented generation, and multi-agent collaboration, etc.

Cultural heritage conservation is also undergoing a profound transformation from traditional manual modes to intelligent paradigms. Data from United Nations Educational, Scientific and Cultural Organization (UNESCO) indicates that the field faces issues such as data fragmentation and lagging policy responses. The 2015–2025 "Digital Heritage" global strategy has planned a digital transformation path for member states. The European Union, Japan, Africa, and other regions have been promoting smart heritage projects, and China has also included digital transformation in the "The 14th Five-Year Plan for Cultural Heritage Conservation and Science and Technology Innovation". The newly revised Cultural Relics Protection Law has even incorporated digital protection into the legal framework. Despite these efforts, the fragmentation of cross-source data and insufficient platform interoperability have led to prominent "information silos," increasing the cost of data integration. Existing policies also respond slowly to emerging risks. The heritage conservation industry currently faces multiple challenges in technology application and data management: Universal LLMs exhibit significant deviations in parsing specialized terminology, historical contexts, and interdisciplinary knowledge associations, such as the "cave temple shape system". They lack domain-specific logical reasoning capabilities, which makes it difficult for them to meet the demands for innovative policy-making and high-precision decision-making. There is a lack of specialized intelligent agent systems that are compatible with world heritage conservation and capable of functions such as data fusion analysis. This deficiency hinders the proposal of innovative conservation policies. Moreover, world heritage data is widely distributed and comes from diverse sources. The insufficient sharing mechanisms between platforms and the high costs of data acquisition severely affect the collection and organization of data. When constructing heritage conservation scenarios based on DeepSeek, the defects in professional semantic processing are prominent. Misunderstandings of domain terminology and the absence of historical context lead to logical gaps, making it difficult to parse the implicit semantics and deep logic of the text. Knowledge graphs have sparse semantic associations and chaotic hierarchical structures, and dynamic semantic updates lag behind the iteration of domain knowledge and policy interpretation needs. The lack of structured responses results in fragmented outputs, which are not conducive to professional institutions formulating implementable decisions. The decision-making process is frequently obstructed by data distortion and frequent fabrication, affecting the rigor of research. Moreover, the system lacks key data, and updates on policy dynamics and academic research progress are not timely, resulting in decision-making risks due to the exclusion of the latest provisions and low efficiency in the transformation of academic achievements.

2 Related Research

As a highly complex professional field, world heritage protection has long relied on decision-making models that integrate expert experience and multi-source information. Moreover, in the face of a vast amount of mixed real and false information on the Internet, it is essential to ensure the accuracy and reliability of the information obtained, and to widely refer to and compare the practices and achievements of other heritage sites around the world in terms of protection, management, and monitoring. For example, CyArk has achieved remarkable results in the 3D modeling and structural preservation of cultural heritage, but is weak in semantic construction and intelligent reasoning. Culturalink relies on LLMs to build knowledge graphs, enhancing cultural understanding and generation capabilities. However, its functions are more focused on entertainment and popular science for the general public, and it cannot support professional protection decision - making scenarios. In China, large - model services are also gradually being introduced into the field of world cultural heritage. The "Chuantou Zhisheng AI system" and the "Nanjing Agricultural University Non - material Cultural Heritage LLMs" have made preliminary progress in semantic aggregation and automatic question - and - answer, but the former mainly focuses on classification retrieval and display functions, and the latter focuses on intangible cultural heritage and does not target world heritage. Overall, existing systems are mostly limited to static retrieval, single - task processing, or mass - service levels, and it is difficult to achieve intelligent protection management and in - depth decision - making support under cross - domain and multi - constraint conditions.

In recent years, with the rapid development of prompt engineering technology, intelligent systems based on LLMs have shown significant potential in many professional tasks. This type of technology uses natural - language interaction to call up expert - level knowledge and conduct complex reasoning, providing a feasible path for building intelligent agents with high domain adaptability. Specifically, in vertical fields such as telemedicine, customer service, and industrial maintenance, researchers have used prompt frameworks such as "Conversation Routines" to achieve precise collection of structured data, conversion of natural language to structured instructions, and interactive fault diagnosis and report generation based on knowledge bases, demonstrating the extensive value of prompt engineering in improving task accuracy, interaction naturalness, and system scalability. Although the above - mentioned achievements highlight the application results of prompt engineering in many fields, its systematic integration and in - depth application in specific complex scenarios such as cultural heritage protection are still in their infancy. The field of world heritage involves complex needs such as accurate content reporting, long - text understanding, cultural context restoration, and cross - language interaction. Current research has not yet fully explored how to deeply integrate cutting - edge prompt design methods with domain knowledge to build professional agents with cultural perception and heritage interpretation capabilities. Therefore, effectively applying prompt engineering to the protection, interpretation, and dissemination of world heritage is still a research direction that is urgent and of great value.

3 Proposed Solution in This Paper

To address these challenges, this paper proposes a agent-based artificial intelligence system architecture for the world heritage LLM based on prompt engineering called "Heritage Echo Planet" . This LLM is based on the locally deployed DeepSeek. By constructing a knowledge base system that crawls authoritative platforms and employing pre-training optimization (leveraging prompt engineering to focus the LLM on specific domains), the LLM addresses a series of issues in general LLMs and ultimately provides decision-making support for professionals.

This paper integrates the ideas of Structured Prompting, Iterative Prompt Engineering, Contextualization and Parameter Setting, Task-Oriented Prompt Design, and Chain-of-Thought Prompting to design prompts for world heritage. These prompts effectively enhance the performance of AI in complex tasks such as heritage conservation, management, and research.

The overall optimization approach for prompts in this paper adopts the Iterative Prompt Engineering method. Based on the output of the LLM the prompts designed are continuously optimized. Subsequently, Structured prompting is employed. Structured descriptions are more accurate than natural language and can more clearly and fully provide context, clarify requirements, anchor expectations, and standardize output formats. They also enhance repeatability and ensure the stability of the output. When designing prompts, several sections are structured, including role, background, skill, goal, constraints, output format, workflow, and example, to guide the LLM in understanding task requirements and avoid understanding deviations caused by the ambiguity of natural language prompts. To ensure consistency and authority when the AI simulates a professional identity, the Contextualization and Parameter Setting method is used. The role of the AI is explicitly set as "Senior World Heritage Research Scholar and Consultant." The background is set to require that users need structured weekly or monthly reports that meet professional standards in the field of world heritage, and they may also pose professional questions that require in-depth analysis and guiding opinions based on the knowledge base. By inputting strong contextual constraints to the LLM, it can activate token sequences that match the professional attributes, language style, and task background of the role, enabling the LLM to support the complete construction of logical chains and avoid historical context loss and reasoning logical gaps. In the Skills section, by clearly stating "rigorous academic analytical skills, policy interpretation abilities, data processing capabilities, and cross-cultural communication skills," the LLM is essentially prompted to invoke the pre-trained knowledge clusters and task processing paradigms corresponding to these skills. During training, the LLM has already learned the knowledge associations and logical processing patterns corresponding to these skills. This setting guides the LLM to prioritize and integrate these pre-stored professional capability modules, thereby naturally presenting analysis depth and recommendation quality that align with the skill descriptions. In the prompt design, the goal module is set, in which three core tasks are explicitly defined. First, generate accurate, objective, and timely weekly and monthly reports in the prescribed format. Second, provide in-depth analysis and decision-referential guidance for users' professional questions based on the knowledge base. Third, ensure that all output content is strictly based on the knowledge base, complies with the law, and avoids subjective assumptions and unverified opinions. By guiding the AI to advance tasks step by step and module by module, strong goal signals are input to the LLM. The LLMs attention mechanism is guided to focus on key task dimensions, and the goal distribution is anchored during conditional generation to reduce output deviation. This ensures that the LLM maintains reasoning coherence and logical clarity in complex scenarios. Through Constraints, the sources of the knowledge base are restricted, illegal tokens are suppressed, the distinction between facts and inferences is guided and labeled, reasonable innovative token combinations based on the knowledge base are retained, and token features matching the input language are aligned to intervene in the probability of token generation. The Constraints explicitly limit the sources of information to the knowledge base, prohibiting fabrication and fiction. It requires distinguishing facts from inferences and labeling the basis to enhance content credibility. It prohibits suggestions that exceed legal provisions to ensure the compliance of the output. However, this design does not restrict the LLM from proposing creative opinions. The prompt includes the statement: "If creative opinions are needed, ensure that they are appropriately extended from the knowledge base and are realistically feasible." This allows the LLM to propose creative opinions while ensuring their feasibility and legality. These constraint rules construct a rigorous quality control system, ensuring that the LLMs output withstands professional scrutiny. To ensure the consistency of input and output languages, the setting "Answer in the same language as the question posed by the user" is used to ensure that the LLM responds in the same language as the questioner. To enhance the readability of the output format, the Task-Oriented Prompt Design method is employed. In the output format, it is clearly stated: "If the user requests a weekly or monthly report, output the report content in rigorous academic language according to the provided report format." This achieves a structured upgrade of the weekly and monthly report content. Legal citation rules and logical chain construction requirements are continuously emphasized to ensure the theoretical basis and practical feasibility of the plans and reports. Based on the two core needs of professionals—first, collecting and obtaining information on events and new regulations that occurred during the week/month, and second, answering professional questions and providing decision-referential opinions—the workflow module is designed. The AI tasks are explicitly divided into two categories: "regular report writing" and "professional question answering," with clear output structures and workflows set for each category. This design helps the LLM quickly switch modes when handling different tasks, ensuring the systematic and professional nature of the output content. The weekly and monthly report formats mentioned in the output format are set through templates in the example section, allowing the LLM to output specified content by calling the templates during output.

4 Experimental Results and Comparison

4.1 Experimental Design and Results

To verify the effectiveness of the aforementioned prompt architecture, this study designed comparative experiments based on DeepSeek-R1 and Qwen2.5-7B-Instruct, and established a control group without prompt instructions (ordinary DeepSeek-R1 and Qwen2.5-7B-Instruct). Utilizing the prompt design scheme mentioned in proposed solution in this paper, experiments were conducted in the same scenario for tasks such as asking professional questions in the field of world heritage and generating weekly and monthly reports. The stability of the answers was tested through repeated questioning. The prompt was continuously optimized based on the comparison of output results, and the optimal prompt that could achieve the best interactive effect was finally determined.

We designed the question: "How to balance tourism development and ecological protection in the Three Parallel Rivers World Natural Heritage Site, based on existing cases and data, to formulate a sustainable management plan?" as an example, combined with the task of generating weekly and monthly reports, the performance of the models was compared from the dimensions of the systematicness of heritage protection policies, interdisciplinary integration capabilities, and data - driven decision - making. The evaluation was based on the actual needs of professors from the Planning Bureau of Peking University, focusing on the authority and accuracy of cases, the precision and preliminary analysis ability of answers, and the integration ability of recent materials.

Taking Chinese as the final generated language as an example (Fig. 1):

Fig. 1. Left: Output of the World Heritage LLM; Right: Output of the Ordinary DeepSeek-R1

4.2 Experimental Analysis

In terms of the authority and accuracy of cases in answering questions, the agent based on DeepSeek-R1 cited various types of world heritage cases, such as the Great Barrier Reef in Australia, Venice in Italy, and the Rock Islands in Palau, Pacific Ocean. It provided experience references from multiple dimensions, including ecological monitoring and community participation, and attached specific literature links for reference. The agent based on Qwen also cited similar cases, showing that the prompt has a certain degree of precision and case consistency across different models, but its analysis was less sufficient. The ordinary DeepSeek mainly cited domestic cases (such as Jiuzhaigou and Lijiang Ancient Town) and surrounding cases (such as the Fjords of New Zealand), with insufficient coverage of international cases and no provision of original sources, which is not conducive to scholars verifying authority. Qwen did not explicitly list cases. In terms of the systematicness and comprehensiveness of strategies, the agent based on DeepSeek constructed a complete system from five dimensions: visitor management, climate - adaptive conservation, resource allocation, community participation rights protection, and educational experience. It proposed specific measures such as disaster early warning and mandatory low - carbon standards in the aspect of climate adaptability, with a clear structure and in - depth analysis. The agent based on Qwen covered the above five dimensions in some answers, but its analysis was more concise, and it once divided into seven aspects, with weaker answer structure. The ordinary DeepSeek only mentioned traditional strategies such as low - carbon transportation, lacking a systematic framework. Qwen's answers were more inclined to macro - level aspects such as law, management, infrastructure, and community coordination, with insufficient systematicness. In terms of innovation and sustainability, the intelligent agent based on DeepSeek proposed innovative mechanisms such as "tourism enterprise ecological performance evaluation" and "extracting 20% of ticket revenue to set up a heritage protection fund", and suggested using intelligent monitoring to dynamically adjust the number of visitors. The agent based on Qwen also introduced big data and artificial intelligence for visitor behavior analysis, but lacked specific implementation content. The ordinary DeepSeek mainly relied on traditional methods such as ecological compensation and concession operation, with insufficient application of innovative tools. Qwen's answers were still more inclined to legal provisions, without innovative and sustainable responses (Table 1).

Table 1. Comparative Analysis of AI Agents in Generating Heritage Conservation Strategies

Dimension	DeepSeek based Agent	Standard DeepSeek	Qwen based Agent	Standard Qwen
Case Quality	International cases, sources linked	Domestic focus, no sources	International cases, brief analysis	No cases cited
Strategy & Framework	Some pillars, weak structure	Traditional, no system	Some pillars, weak structure	Macro, lacks system
Innovation	Vague innovative concepts (e.g., AI)	Traditional methods	Vague innovative concepts (e.g., AI)	Legal focus, no innovation

For the generation of weekly and monthly reports, the agent based on DeepSeek-R1 can output content in the required format, with integrated data sourced exclusively from designated websites and specified domains, effectively mitigating the issue of large models being insensitive to temporal information. This agent not only consolidates the collected materials but also provides preliminary analysis and offers recommendations for key focus areas in the upcoming week or month based on available information. However, since its knowledge base is limited to authoritative websites, the diversity of content is somewhat insufficient. In comparison, the agent based on Qwen is also capable of generating corresponding outputs according to requirements and provided materials, with content largely consistent with that produced by the DeepSeek-R1 agent, differing only slightly in the depth of analysis and predictive details. On the other hand, the standard DeepSeek model exhibits significant shortcomings in generating weekly and monthly reports: it fails to accurately reflect temporal information based on user needs, produces inconsistent content across multiple generations, relies on disorganized and often unverifiable sources of information, and lacks a uniform report format from month to month, making it difficult for researchers to conduct systematic comparisons between weeks or months. The Qwen model shares similar issues with the standard DeepSeek in this regard, as its outputs are generally limited to factual statements without in-depth analysis of the content.

The experimental results of this study demonstrate that the proposed prompt architecture effectively mitigates several limitations of large models in the domain of heritage conservation. In addressing issues such as professional semantic processing deficiencies—for instance, confusing "ecological buffer zones" with "tourism development zones," or logical gaps caused by missing historical context—the model exhibits improved accuracy in terminology understanding and contextual reasoning. The previously sparse semantic associations (e.g., inability to link "climate change" with "heritage conservation policies") and disorganized hierarchies in knowledge graph construction have been enhanced, with dynamic updating capabilities better keeping pace with new domain knowledge and policy revisions. The structural quality of responses has also significantly improved, overcoming previous issues of fragmented and incoherent output (e.g., disorganized content lacking a clear focus), which hindered direct decision-making support. Furthermore, instances of data fabrication and distortion (such as inventing non-existent cases or statistics) have been reduced. The system now incorporates the latest convention developments, academic findings, and policy provisions (e.g., timely reflecting updated UNESCO guidelines), thereby mitigating decision-making risks and improving the efficiency of academic translation.

The agent based on DeepSeek-R1, when applied with the prompt designed in this study, demonstrates strong case coverage breadth and strategic systematicity, effectively guiding the model to produce authoritative and structured content. Its advantages are primarily reflected in its ability to accurately invoke multiple typical World Heritage case studies from around the world based on instructions and provide specific literature sources, thereby enhancing the credibility and reference value of the content. In terms of strategy construction, the agent can systematically develop a comprehensive framework from multiple dimensions and propose quantifiable, actionable recommendations, showcasing strong logical organization and cross-domain integration capabilities. When generating weekly and monthly reports, the agent adheres to specified formats and produces content based on designated data sources, while also exhibiting preliminary analysis and forecasting abilities, to some extent mitigating the large model's insensitivity to temporal information. In multiple Q&A experiments, its responses remained largely consistent, demonstrating a degree of stability. However, there is still room for improvement in the depth of analysis for certain complex issues, such as the integration of interdisciplinary methods and the elaboration of specific implementation pathways, which remain relatively brief. Compared to the Qwen-based agent, this solution exhibits better case consistency and strategy generalization across different models. The prompt design proposed in this study is applicable to various large models, allowing users to make selections based on factors such as budget and model characteristics. Without prompt guidance, the standard DeepSeek and Qwen models produce content lacking structure and authority, indicating that the proposed prompt significantly enhances response quality. Nevertheless, its performance ceiling still depends on the model's inherent capabilities and the coverage of its knowledge base.

5 Conclusion and Outlook

This study addresses core issues in cultural heritage preservation, such as data fragmentation and the lack of intelligent response mechanisms, by proposing and validating a prompt engineering-based agent-type artificial intelligence system architecture. Through the development of the "Heritage Echo Planet" world heritage large language model, this architecture significantly enhances the adaptability and practical performance of the model in complex heritage conservation tasks. Experimental results demonstrate that the introduction of prompt mechanisms effectively improves the large model's professional capability in the field of world heritage, showing optimizations in case coverage, strategy systematicity, time planning rationality, risk response capability, and international collaboration support, highlighting the important potential of prompt engineering in driving intelligent transformation within vertical domains. Future work will involve continuously optimizing prompt design based on output results and introducing multi-model comparisons to facilitate iterative improvements. Further research directions include: firstly, developing a hierarchical prompt template that dynamically adjusts the prompt structure according to task complexity to strengthen the model's ability to process high-level semantic information such as legal provisions and policy logic; secondly, constructing a cross-modal prompt update mechanism that converts non-textual information such as images and geographic coordinates into embeddable semantic descriptors, thereby enhancing the collaborative reasoning capability for multi-source heterogeneous data; thirdly, optimizing a context-aware prompt adaptation mechanism that automatically streamlines redundant prompt units based on dialogue history to improve interactive efficiency and intent recognition accuracy in complex tasks. By continuously refining the prompt engineering system, this research aims to promote the design innovation and application of prompts in cultural heritage preservation scenarios.

References

1. Eason, G., Noble, B., Sneddon, I.N.: On certain integrals of Lipschitz-Hankel type involving products of Bessel functions. Phil. Trans. R. Soc. Lond. **A247**, 529–551 (1955)
2. Schmidt, D.C., Spencer-Smith, J., Fu, Q., White, J.: Cataloging Prompt Patterns to Enhance the Discipline of Prompt Engineering. Vanderbilt University, Nashville, TN (2023)
3. Fei, W., et al.: Extending context window of large language models via semantic compression. arXiv preprint arXiv, 2312.09571 (2023)
4. Fernando, C., Banarse, D.S., Michalewski, H., Osindero, S., Rocktäschel, T.: Promptbreeder: Self-referential self-improvement via prompt evolution. arXiv preprint arXiv, 2309.16797 (2023)
5. Gal, R., et al.: An image is worth one word: Personalizing text-to-image generation using textual inversion. arXiv preprint arXiv, 2208.01618 (2022)
6. Ge, T., et al.: Extensible prompts for language models on zero-shot language style customization. In: Proceedings of the 2022 Conference on Empirical Methods in Natural Language Processing, pp. 1–15 (2022)
7. Guo, Q., et al.: Connecting large language models with evolutionary algorithms yields powerful prompt optimizers. arXiv preprint arXiv, 2309.08532 (2023)
8. Hinton, G.E., Vinyals, O., Dean, J.: Distilling the knowledge in a neural network. arXiv preprint arXiv, 1503.02531 (2015)
9. Hsieh, C.-J., Si, S., Yu, F.X., Dhillon, I.S.: Automatic engineering of long prompts. arXiv preprint arXiv, 2311.10117 (2023)
10. Hsieh, C.-Y., et al.: Distilling step-by-step! Outperforming larger language models with less training data and smaller model sizes. In: Findings of the Association for Computational Linguistics: ACL 2023, pp. 1–15 (2023)
11. Huang, J., et al.: Large language models can self-improve. In: Proceedings of the 2023 Conference on Empirical Methods in Natural Language Processing, pp. 1–12 (2023)
12. Huang, X., Zhang, L.L., Cheng, K.-T., Yang, F., Yang, M.: Fewer is more: boosting LLM reasoning with reinforced context pruning. arXiv preprint arXiv, 2310.12345 (2023)
13. Jiang, H., Wu, Q., Lin, C.-Y., Yang, Y., Qiu, L.: LLM Lingua: compressing prompts for accelerated inference of large language models. In: Proceedings of the 2023 Conference on Empirical Methods in Natural Language Processing, pp. 1–14 (2023)
14. Jiang, H., et al.: Long LLM lingua: accelerating and enhancing LLMs in long context scenarios via prompt compression. arXiv preprint arXiv, 2310.06839 (2023)
15. Jin, C., et al.: APEER: automatic prompt engineering enhances large language model reranking. arXiv preprint arXiv, 2406.14449 (2024)
16. Juneja, G., Natarajan, N., Li, H., Jiao, J., Sharma, A.: Task facet learning: a structured approach to prompt optimization. arXiv preprint arXiv, 2406.10504 (2024)
17. Jung, H., Kim, K.-J.: Discrete prompt compression with reinforcement learning. IEEE Access. **12**, 72578–72587 (2023)
18. Lester, B., Al-Rfou, R., Constant, N.: The power of scale for parameter-efficient prompt tuning. In: Proceedings of the 2021 Conference on Empirical Methods in Natural Language Processing, pp. 1–12 (2021)
19. Li, B., et al.: Deliberate then generate: enhanced prompting framework for text generation. arXiv preprint arXiv, 2305.19835 (2023)
20. Li, L., Zhang, Y., Chen, L.: Prompt distillation for efficient LLM-based recommendation. In: Proceedings of the 32nd ACM International Conference on Information and Knowledge Management (CIKM 2023), pp. 1–10 (2023)
21. Li, M., Wang, W., Feng, F., Cao, Y., Zhang, J., Chua, T.-S.: Robust prompt optimization for large language models against distribution shifts. In: Proceedings of the 2023 Conference on Empirical Methods in Natural Language Processing, pp. 1–15 (2023)

22. Li, X., et al.: Say more with less: understanding prompt learning behaviors through gist compression. arXiv preprint arXiv, 2402.16058 (2024)
23. Li, X.L., Liang, P.: Prefix-tuning: Optimizing continuous prompts for generation. In: Proceedings of the 59th Annual Meeting of the Association for Computational Linguistics and the 11th International Joint Conference on Natural Language Processing (Volume 1: Long Papers), pp. 1–12 (2021)
24. Li, Y., Dong, B., Guerin, F., Lin, C.: Compressing context to enhance inference efficiency of large language models. In: Proceedings of the 2023 Conference on Empirical Methods in Natural Language Processing, pp. 1–14 (2023)
25. Li, Z., Peng, B., He, P., Galley, M., Gao, J., Yan, X.: Guiding large language models via directional stimulus prompting. In: Advances in Neural Information Processing Systems 36: Annual Conference on Neural Information Processing Systems 2023 (NeurIPS 2023), pp. 1–20 (2023)
26. Li, Z., Su, Y., Collier, N.: 500xCompressor: generalized prompt compression for large language models. arXiv preprint arXiv, 2408.03094 (2024)

Building an AI Knowledge Infrastructure for the World Cultural Heritage Vertical: Methodology and Challenges from Local Databases

Yuzhe Liu[1], Ximeng Wang[1], Shuai Shao[2], and Lijie Wang[1](✉)

[1] School of Automation, Beijing Information Science & Technology University, Beijing, China
wanglijie_0203@126.com

[2] Zhongguancun College, Beijing, China
shaoshuai@bit.edu.cn

Abstract. In order to promote the intelligent protection of world cultural heritage, this paper focuses on the construction of local databases in its vertical AI knowledge infrastructure. Professional dispersion characteristics of cultural heritage data method and general adaptation shortage problem, put forward building scheme based on MySQL: the directional crawl and preprocessing multi-source data, design optimization of table structure associated storage, build a three-tier architecture of "database - retrieval module - large model". Ultimately, a large model in the vertical field of world cultural heritage is formed. Compared with 25 professional questions, the experiment verifies that the big model connected to the database is better in accuracy, completeness and pertinacity, and can accurately call authoritative data. This scheme provides data support for the intelligent protection of cultural heritage. In the future, it will continue to improve from the dimensions of expanding storage, optimizing retrieval, and enriching data to help the protection and inheritance of cultural heritage.

Keywords: World cultural heritage · Vertical AI · Local database · Large vertical model

1 Introduction

As an important carrier of human civilization, it is of great significance to protect and inherit the world cultural heritage. With the rapid development of technology, especially the continuous advancement of information technology and digital technology, the methods of protecting and inheriting cultural heritage have gradually shifted from the traditional physical preservation and display models to a digital and intelligent direction. The application and development of cultural heritage digital technology can not only effectively enhance the preservation efficiency and accuracy of cultural heritage, but also provide the public with more abundant interactive experiences and display effects, thereby promoting the development of innovative application technologies for cultural

H. Ma et al. (Eds.): IWACIII 2025, CCIS 2780, pp. 294–308, 2026.
https://doi.org/10.1007/978-981-95-6730-0_22

heritage [1]. From digital documentation to preventive protection, AI will be applied to the field of world cultural heritage, which has become a new research direction and practice hotspot. Build a vertical AI knowledge infrastructure, become a key link in the process of intelligent push cultural heritage protection. As an important part of the knowledge infrastructure, the construction quality of local database directly affects the storage, management and call efficiency of the AI system for cultural heritage related knowledge, and then determines the application efficiency of the AI system in answering professional questions, predicting damage and other functions.

At present, although there have been many studies on database construction and AI knowledge infrastructure construction in other fields, there are still insufficient related studies in the vertical field of world cultural heritage. On the one hand, the data in the field of cultural heritage have the characteristics of strong professionalism, dispersed data sources, and diverse formats, so the existing general database construction methods are difficult to be directly applicable. On the other hand, when using local databases to support AI to realize specific functions such as cultural heritage protection, it faces many challenges such as data quality and data security. Therefore, it is of great theoretical value and practical significance to deeply study the construction methodology and challenges of local databases in AI knowledge infrastructure in the vertical field of world cultural heritage.

2 Related Work

In the field of database construction and AI knowledge infrastructure construction, many scholars have carried out research. In the fields of database construction and the construction of A1 knowledge infrastructure, many scholars have conducted research, resulting in the emergence of several representative platforms and projects.

At the international level, CyArk is one of the most representative non-profit projects in the field of cultural heritage digital protection. The CyArk project was initiated by the Kacyra Family Foundation in 2002. It collects data about cultural sites through 3D laser scanning and related technologies, and then uses computers to process the digital models to preserve and reproduce the conditions of the sites, providing a basis for the management and related research of the sites as well as future restoration work [2]. The platform has long cooperated with World Heritage organizations to use laser scanning, photogrammetry and 3D modeling technologies to digitally archive landmark cultural sites such as Angkor Wat and Pompeii in high definition. The core advantages of the platform are its high accuracy of data acquisition, strong 3D reconstruction ability, and good retention of architectural microstructure, which has high academic research value. However, the technical application of the system is relatively focused on preservation modeling, and its data is mainly for display and archive purposes, lacking semantic structural processing and artificial intelligence reasoning mechanism, and insufficient support for the knowledge construction, value analysis and protection strategy formulation of heritage.

In China, the "Zhisheng AI" platform, developed by Sichuan Investment Group, focuses on semantic aggregation and intelligent recommendation of intangible cultural heritage resources, and constructs a digital knowledge system covering thousands of intangible cultural heritage projects in Sichuan Province. Take the All-County Tourism Big Data Center of Dujiangyan City as an example. This data center was jointly established by the Qiongcheng Mountain-Dujiangyan Scenic Area Administration Bureau and Zhisheng Integration. Relying on the "Digital Qiongcheng Mountain, Cloudy Dujiangyan" smart scenic area management service system, it realizes "one machine for management and services", "one machine for tourists' experience", and "one brain for the command center", providing operational support for the scenic area and cultural tourism enterprises, digital service support for local tourists' experiences, and decision-making support for government management [3]. Platform for the integration of speech recognition, text semantic analysis and automatic label generation capacity, which can realize intangible process, inheritance characters and geographical distribution of dimensions of interactive query. However, the capabilities of the platform mainly focus on the classification, labeling and retrieval of data, and lack the semantic structuring ability based on knowledge graph, which cannot realize the automated reasoning of cultural content. At the same time, its platform logic is mainly based on "display service", and it fails to build an intelligent response mechanism, which is difficult to support the collaborative processing of complex tasks such as research, decision making and public education.

As an official knowledge base of UNESCO, the UNESCO Digital Library covers academic literature and reports in the field of global cultural heritage. It has a high degree of data openness, but its search method is still mainly based on traditional keyword matching, lacking semantic understanding and intelligent reasoning capabilities. The CyArk project stands out in high-precision 3D modeling and digital archiving, but its data organization is oriented towards display and archives, making it difficult to support in-depth knowledge mining and cross-source association analysis. The domestic "ZhiSheng AI" platform has made progress in semantic search and recommendation of intangible cultural heritage resources, but it mostly remains at the level of content classification and tagging, lacking deep collaboration with large models.

In contrast, the vertical domain AI knowledge infrastructure based on MySQL proposed in this paper has the following unique advantages:

1. Structured integration of multi-source heterogeneous data: Text, tables, and semi-structured information are uniformly stored in a relational database, and efficient retrieval is achieved through optimized table structures and foreign key associations.
2. Deep collaboration between retrieval and large models: Combining RAG and SQL queries, natural language questions are transformed into precise structured retrieval, ensuring the accuracy and traceability of answers.
3. Domain-adapted vertical large models: Customized models are developed based on the characteristics of the world cultural heritage field, significantly improving the accuracy and specificity in answering professional questions.

In terms of data processing technology, RAG (Retrieval-Augmented Generation) method is widely used in knowledge base construction and Retrieval tasks, which improves the timeliness and accuracy of information acquisition through text segmentation, vectorization and similarity calculation, and becomes one of the mainstream technologies of knowledge retrieval in vertical fields. Especially suitable for structured data scenarios.

The basic process of RAG can be divided into three stages: knowledge base construction stage, retrieval stage and generation stage. In the knowledge base construction phase, all kinds of data in different formats need to be converted into a unified plain text format. Due to the limited ability of language model to process text, it is necessary to perform text segmentation operation to divide the text into blocks. Then, in order to transform the text into a language that the computer can process, the text needs to be vectorized using the embedding model, and the text is stored in the vector database, where the preliminary work of the data is completed. In retrieval phase, the first input by the user questions, RAG using embedding in the form of a question can be converted to vector, and vector data in the database for similarity calculation, return and question vector is the most similar K before a text block, finally, returns the text block added to the prompt with questions. In the generation stage, the large language model generates answers according to the prompt and the internal knowledge of the model [4].

In the field of cultural heritage, existing research focuses on the integration method of multi-source heterogeneous data, but there is still a lack of research on the construction of AI knowledge infrastructure in vertical fields based on a single relational database (such as MySQL) and the collaborative application of databases and big models. Traditional big data integration platforms mostly use relational databases for system architecture. The system architecture of traditional big data integration platform is slightly insufficient in terms of storage space, computing power, scalability, flexibility, and operating efficiency [5]. It is urgent to propose a targeted scheme combining the characteristics of relational database.

3 Method of This Paper

3.1 Data Acquisition and Preprocessing

There are a wide range of sources of world cultural heritage related data, including the official website of the UNESCO World Heritage Center, the websites of National Cultural Heritage Administration departments, academic journal databases, and professional forums of cultural heritage, covering basic information (name, geographical location, construction age, damage status, etc.), structured and semi-structured data such as protection status reports, research paper abstracts, etc. The main modalities are text and table.

The data collection module is the core part, responsible for obtaining the required data from designated websites and APIs. The main functions of this module include sending HTTP requests to obtain web page content and parsing out the information that users are interested in, such as article titles and links. To achieve this function, two popular Python libraries, Requests and BeautifulSoup, were used [6].

For text and table data, a targeted crawling strategy was used, the requests library was used to send requests, and BeautifulSoup was combined to parse the structure of the web page and extract structured information. For long texts such as academic papers, the text content is extracted by PDF parsing tools and converted into structured fields. To ensure the stability and compliance of the data collection process, this paper adopts multi-level anti-crawling strategies in the targeted crawling stage. Firstly, it simulates a browser by setting the User-Agent. Secondly, it uses a proxy pool to rotate IP addresses, builds and maintains a dynamic IP proxy pool, and eliminates invalid IP addresses through regular health checks to ensure that each request is initiated by an available proxy, effectively reducing the risk of a single IP being banned. Meanwhile, it adds delays using time.sleep() for requests and dynamically adjusts the access strategy based on the website's response status. When receiving error codes such as 403/503, it automatically extends the waiting time and switches to a different proxy IP. Additionally, it strictly adheres to the target website's robots.txt protocol to ensure the legality and compliance of the data collection process.

In the process of collecting and organizing actual data, various types of noise data, data loss, sample duplication and outliers are inevitably generated. However, noise data will affect the quality of data and the results of predictive analysis to varying degrees in different predictive scenarios [7]. Need to cleaning and pretreatment, data collected using Python Pandas library to complete data to heavy (number) based on a unique field such as heritage, missing value filling (through the legacy data mean or same type expert rules forecast), outlier detection and processing (based on the field threshold selection and correct unreasonable data). For the detected outliers, if they can be inferred and repaired through the context, the median of the same type of data is used for filling; if they cannot be repaired, they are marked as "awaiting manual review" or directly removed. At the same time, the text data is structurally labeled according to the attribute (title, date, heritage type, etc.), and stored in a mySQL-compatible field format, which lays a foundation for database storage and retrieval.

3.2 Database Architecture Design

Considering the structural characteristics of world cultural heritage data and the maturity of relational database, this paper uses MySQL as the only local database, and realizes the efficient storage and management of cultural heritage data by optimizing the design of table structure. SQL is an open-source relational database, which is a widely used database system in Web development. It is characterized by reliability, efficiency and ease of use [8].

In the relational database, multiple associated tables are designed to store cultural heritage data. For example, create a "Cultural Heritage Basic Information table" to store core information such as the name, type, and geographical location of cultural heritage; "Protection and Restoration Record table", which records protection measures, maintenance time, maintenance personnel and other protection-related data of cultural heritage; And "research paper table", which stores information such as titles, authors, published journals and abstracts of academic papers related to cultural heritage. By foreign key link each table, set up the logical relationship between data, easy to improve the efficiency of subsequent large retrieves information model and improve the degree of correlation between the data.

The core table structure of the database includes:

- The basic information table of cultural heritage: it stores the core attributes of heritage such as name, number, type, construction date, and time of being listed as World Heritage;
- to protect and repair record: record heritage protection, maintenance time and effect of implementing agencies, maintenance information;

- research literature tables: the title of the store related academic papers, published time, abstract, key words, text, etc. Table 1;

Table 1. Research literature tables

Title	Shijian	Jianjie	Link	All text
UNESCO Strengthens Capacity of the Government of Sindh in Preserving the Historical Monuments at Makli, Thatta in Pakistan	Thursday, 22 May 2025	To enhance the capacity of professionals from the…	https://whc.unesco.org/en/news/2771	["To enhance the capacity of professionals from the Directorate
Strengthening protection of World Heritage Archives with JFIT support	Wednesday, 21 May 2025	World Heritage site managers and archive specialists…	https://whc.unesco.org/en/news/2768	["World Heritage site managers and archive specialists…
Call for Applications: World Heritage Young Professionals Forum 2025	Wednesday, 2 April 2025	UNESCO Headquarters in Paris, 1–8 July 2025 – "World Heritage	https://whc.unesco.org/en/news/2765	["UNESCO Headquarters in Paris, 1–8 July 2025 - "World Heritage in

3.3 Design of Pendant Large Model System

A single universal model is difficult to directly meet the needs of multiple scenarios. Industry-specific large models, however, can integrate industry data, computing power, and human resources to empower specific vertical fields and promote the digital and intelligent industrial upgrading. Compared with universal large models, vertical large models have lower development costs [9]. Universal large models have strong language understanding capabilities and can expand the breadth of application scope, while vertical large models target specific industries or needs and can better meet users' actual requirements in depth [10]. The vertical large model system uses MySQL database as the core data support, and constructs a three-layer architecture of "databases-retrieval module-large model" to achieve efficient collaboration between structured data and large model. MySQL is a relational database server that supports the SQL database language, is simple to use and open source, and is reliable and compatible with all major hosting providers. MySQL relational database is different and superior to other databases because it saves data in different tables, improves the speed of reading and writing data in the database, and improves the flexibility of the database system, so it is widely used at home and abroad. The MySQL database support for multiple operating systems and multiple languages, has a strong universality [11].

The bottom for the MySQL database layer, through structured data table structure design store all cultural heritage, to provide data to add and delete based interface; The middle layer is the retrieval module, implemented based on the RAG method, which converts user questions into structured query conditions (such as matching the literature table by keywords, filtering the basic information table by heritage types), and extracts relevant data for text integration. The top layer is the large model layer, which receives the structured data text returned by the retrieval module and generates professional answers combined with user questions.

4 Experiment

4.1 Experimental Objectives

Through the large model, it was verified that after building based on the local MySQL database, the professional capabilities of the vertical large model in the field of world cultural heritage have been significantly enhanced. Specifically, it is manifested in the accuracy of answering professional questions, the completeness of information, and the domain specificity being superior to the vertical domain large model that is not connected to the database.

4.2 Experimental Setup

The local MySQL database and the large model were docked with the help of "Cherry Studio framework + API interface", and the specific process was as follows: After users input professional questions in the field of world cultural heritage, the retrieval module converts natural language questions into SQL query statements through the SQL Agent of Cherry Studio. The statement invokes MySQL database to perform retrieval and returns structured results. The retrieval module further integrated the results into natural language text, encapsulated as Prompt according to the preset format, and passed it to the large model. The big model generates the final answer based on Prompt.

In the experiment, 25 professional questions (covering basic information query of heritage, protection technology analysis, damage cause traceability and other types) were selected, and the large model without database connection was used as the control group to verify the improvement effect of database access on the quality of answers. Here are 25 specific experimental questions:

1. As the oldest independent stone-built structure in the world, what are the unsolved mysteries of the transportation and processing technology of the building stones at the Ggantija Temples in Malta?
2. What is the height of the Christ the Redeemer statue in Rio de Janeiro, Brazil? What is the main building material used for the statue? How many years did the construction of the project last?

3. What is the weight of the largest sandstone boulder (referred to as the "heel stone") in the British Stonehenge site? How far is the original origin of these stones from the site of the ruins?
4. What was the original height of the famous "Ishtar Gate" in the ancient Babylonian city ruins? What were the three main colors of the glazed tiles on the door?
5. How many buildings in the core protection area of the historical center of St. Petersburg, Russia, are included in the category of World Cultural Heritage? What proportion of them were built in the 18th century?
6. What architectural style does the Independence Hall (in Philadelphia) of the United States, where the Declaration of Independence was signed, belong to? What is the material of the original wooden tables and chairs used for signing documents in the hall?
7. What is the remaining height of the "Pantheon Gate" (also known as the Xerxes Gate) in the Persian Palace ruins? What are the prototypes of the giant stone sculptures on both sides of the gate?
8. In which year was the Robben Island (where Mandela was imprisoned) listed as a World Cultural Heritage site? How many cells are currently remaining on the island?
9. What is the altitude of Uluru (Ayers Rock) in the Uluru-Kata Tjuta National Park of Australia? What is the ratio of the exposed part to the underground part of the rock?
10. In which year was the most intact "Mahatma Temple" in the ruins of the Ayutthaya Kingdom built? What is the height of the Buddha head of the most famous "Tree-encircling Buddha Head" statue in the temple?
11. When did the construction of the Cologne Cathedral in Germany begin? How many years did it take to complete the final project? Was the height of the cathedral's spire the highest building in the world at the time of its completion?
12. What is the circumference of the imperial city wall of the ancient Hue City in Vietnam? How many city gates are there? What is the name of the largest city gate?
13. As a natural and cultural dual heritage site, what traditional activity does the cultural heritage part of Argentina's Valdes Peninsula mainly represent? How many years has this activity been around?
14. What is the area in square meters of the "Great Mosque" in Sana'a, Yemen, built in the 11th century? What is the most ancient artifact preserved in the temple?
15. In the Gyeongju Historical Area of South Korea, what dynasty's tomb is the "Tianma Mausoleum" (also known as the "Eastern Pyramid")? What is the most representative artifact unearthed from the tomb? Where is it currently housed?
16. The wooden structure of the Todaiji Temple in Nara, Japan has developed decay. What are the two types of fungi (specific species names) that caused this? What is the reproduction speed of these fungi in the wood (how many fungi per square meter of wood increase per month)?
17. The stone columns at the Olympia site in Greece have developed inclination. What is the main reason for this? What is the difference in the settlement amount of the foundation in different areas of the site (such as the east and west)? What is the average annual settlement speed in the past five years?
18. The adobe walls of the ancient city of Chan Chan in Peru have collapsed. What are the main reasons for this? What are the contributions of rain erosion and seismic activities in the three recent wall collapses?

19. During the restoration of the mosaic mosaics at the Alhambra Palace, what is the shear strength required for the bonding agent used in the "fragment assembly" technique? What is the material used for the "protection treatment" after the assembly is completed?
20. During the cleaning of the Byzantine-style mosaics in the Hagia Sophia in Istanbul, what temperature range is the melting point of the "microcrystalline wax protection" technology controlled at? What is the thickness of the wax layer?
21. What is the total length of the relatively well-preserved sections of the Great Wall (Ming Great Wall) in China? What proportion of it is located in Hebei Province?
22. During the anti-corrosion treatment of the sandstone walls of the Luxor Temple in Egypt, what is the concentration ratio of the "silane immersion" technology used? What percentage must the water absorption rate of the wall after the immersion treatment be reduced by?
23. During the restoration of the red sandstone building at the Agra Fort in India, what three natural mineral pigments are mainly used in the "mineral pigment mixing" technique to match the original color of the stone? What is the mixing ratio of the pigment and the binder?
24. During the restoration of the brick and stone walls of the Castle of Carcassonne in France, what is the ratio of lime to sand in the "traditional lime mortar masonry" technique? What is the relative humidity required during the mortar curing process?
25. In the stone weathering damage of the Roman Colosseum in Italy, which factor, acid rain erosion or industrial dust adhesion, has a greater impact on the porosity of the stone surface? How many percentage points will the porosity of the stone increase?

4.3 Selection Criteria for Experimental Questions

To ensure the validity and comparability of the experimental results, this study selected 25 professional questions derived from actual work tasks in the field of heritage protection as the test set. The selection process adhered to the following principles: completeness of domain coverage, representativeness of task types, gradient design of difficulty, objectivity of evaluation, and variability in data dependency. The test questions encompass core task types such as basic information queries (e.g., age and geographical location), protection technology analysis (e.g., material composition detection and reinforcement scheme evaluation), and damage cause tracing (e.g., environmental factors, biological erosion, and human destruction), thereby comprehensively reflecting the model's performance across diverse task scenarios. All questions were drawn from real work contexts to ensure they represent common and practically significant information needs in the field, enhancing the external validity of the results. The difficulty gradient ranges from simple factual queries that can be directly retrieved to complex problems requiring multi-step reasoning and cross-domain knowledge integration, allowing for an examination of the role of database access under varying cognitive complexities. Each question was assigned clear standard answers or evaluation criteria to facilitate quantitative comparisons of the accuracy, completeness, and timeliness of responses and minimize subjective bias. Furthermore, the question set includes cases with differing degrees of reliance on external databases—some answerable through general knowledge alone, others requiring access to specialized databases for up-to-date or specific information—enabling a quantified assessment of the performance improvements attributable to database access.

4.4 Experimental Results

For example, The Megalithic Temples of Malta (Ġgantija Temples) are the oldest independent stone-built structures in the world. What are the unsolved mysteries regarding the transportation and processing techniques of the building stones? (Figs. 1 and 2).

The Ġgantija Temples on Gozo, Malta (c. 3600-3200 BCE) truly push the boundaries of Neolithic engineering. Being among the world's oldest free-standing stone structures, their massive scale and precise construction with limited technology pose fascinating unsolved mysteries, particularly concerning stone transport and shaping:

1. **Transportation of Megaliths:**
   - **The Scale:** Moving limestone blocks weighing up to **50 tonnes** (and many in the 10-20 tonne range) over distances of up to 1 km from local quarries, across Neolithic Gozo's landscape.
   - **The Mechanisms (Mystery):**
     - **Rollers vs. Sledges:** The standard explanation involves wooden rollers and ropes. However, Gozo's terrain isn't flat, and stone rollers capable of handling such weight haven't been found. Sledges (possibly on wooden tracks lubricated with water, clay, or animal fat) are another hypothesis, but again, hard evidence (like wear patterns or timber fragments) is scarce.
     - **Human Power vs. Ingenuity:** Calculations suggest moving a single megalith could require hundreds of people pulling simultaneously with ropes – an immense logistical feat requiring sophisticated coordination, robust ropes (possibly made from local plant fibres whose durability is questioned), and cleared paths. Were techniques used that we haven't considered?
     - **Absence of Evidence:** No clear haul paths, specialized vehicles, or significant quantities of surviving wooden artifacts relating to transport have been definitively linked to the construction phase. How were obstacles overcome? How were slopes managed?
2. **Lifting and Precise Placement:**
   - **The Challenge:** Getting massive slabs upright and placing them precisely on top of each other, creating the impressive high walls and complex internal chambers, often with inward sloping walls for stability (corbelling).
   - **The Methods (Mystery):**

Fig. 1. The response of the large model when not accessing the database

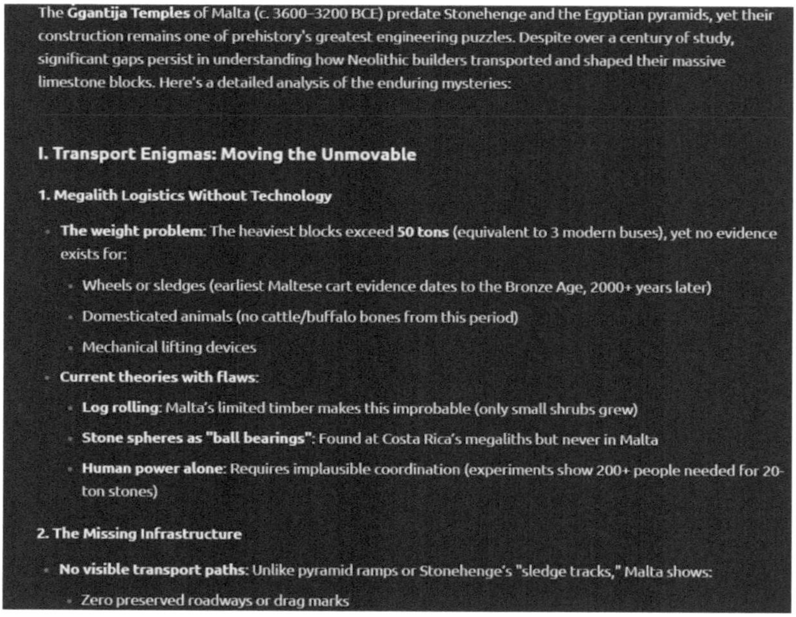

The **Ggantija Temples** of Malta (c. 3600–3200 BCE) predate Stonehenge and the Egyptian pyramids, yet their construction remains one of prehistory's greatest engineering puzzles. Despite over a century of study, significant gaps persist in understanding how Neolithic builders transported and shaped their massive limestone blocks. Here's a detailed analysis of the enduring mysteries:

I. Transport Enigmas: Moving the Unmovable

1. Megalith Logistics Without Technology

- **The weight problem**: The heaviest blocks exceed **50 tons** (equivalent to 3 modern buses), yet no evidence exists for:
  - Wheels or sledges (earliest Maltese cart evidence dates to the Bronze Age, 2000+ years later)
  - Domesticated animals (no cattle/buffalo bones from this period)
  - Mechanical lifting devices
- **Current theories with flaws**:
  - **Log rolling**: Malta's limited timber makes this improbable (only small shrubs grew)
  - **Stone spheres as "ball bearings"**: Found at Costa Rica's megaliths but never in Malta
  - **Human power alone**: Requires implausible coordination (experiments show 200+ people needed for 20-ton stones)

2. The Missing Infrastructure

- **No visible transport paths**: Unlike pyramid ramps or Stonehenge's "sledge tracks," Malta shows:
  - Zero preserved roadways or drag marks

Fig. 2. The response of the large model when accessing the database

From the comparison of the discussions related to the Ġgantija temples in Malta, the improvement in the quality of answers brought about by database access is reflected in five core dimensions: In terms of accuracy, the content without database access contains ambiguous expressions and excessive speculation, such as mentioning that "ropes might have been made from local plant fibers" without specifying the type or basis. In contrast, the answers with database access anchor to specific archaeological evidence, clearly stating verifiable information such as the weight of the stones "exceeding 50 tons" and the transportation distance "about 500 meters to 1 kilometer", avoiding subjective judgments. In terms of completeness, the content without database access focuses on the "transportation, hoisting, and processing" partial links, neglecting key background information such as the construction period and the connection with other sites. The answers with database access, however, construct a systematic framework of "transportation, processing, labor organization, timeline, and emerging research", supplementing information such as "centuries-long construction breaks", achieving a connection between technical issues and social history, and covering a more comprehensive range. In terms of authenticity, the content without database access contains "reasonable imagination" without evidence support, such as speculating that "animal fat lubricated sledges" without explaining the basis. The answers with database access strictly adhere to archaeological evidence as the boundary, clearly distinguishing facts from hypotheses, such as noting that "stone ball bearings have been found in other sites but not confirmed in Ġgantija", avoiding fictional information. In terms of verifiability, the content without database access lacks source indications, making it difficult for readers to verify conclusions such as "hundreds of people were needed to move the stones". The answers with database access provide clear paths for information verification by marking sources such as "archaeological journals, UNESCO reports", and research methods such as "geochemical tracing". In terms of logic, the content without database access piles up questions with weak connections between each link. The answers with database access form a closed loop of "classification → sub-problems → hypotheses → evidence gaps", such as the transportation link through a progressive analysis of "doubts about log rollers → unconfirmed stone ball bearings → water transport dependent on ancient coastlines", strengthening the rationality of the argument. In summary, database access upgrades answers from "common sense speculation" to "structured research constrained by evidence", and its core value lies in establishing a rigorous framework of "problems - evidence - hypotheses", which not only conforms to known facts but also clearly marks the space for exploration, highlighting the rigor of academic answers.

4.5 Experimental Analysis

The experimental results show that the local database based on MySQL significantly improves the professional ability of vertical large model. When the database is not connected, the large model relies on general training data, which is difficult to cover the segmentation data in the field of cultural heritage, resulting in the lack of pertinacity of the answers. After the database is connected, the structured data stored in MySQL is accurately retrieved through SQL query, which provides authoritative and specific domain knowledge support for the big model, so that its answers are more in line with professional needs.

4.6 Discussion

In the research of large model applications, connecting external databases is often regarded as an important way to enhance the accuracy and coverage of responses. However, the experimental results show that when dealing with basic questions in the field of heritage, there is no significant difference in the performance of the model between the condition of connecting the database and the condition of not connecting the database. This phenomenon mainly stems from the characteristics of the task - the knowledge involved in basic questions is usually fully covered during the model training stage, and the external information provided by the database has limited impact on the quality of the answers.

Taking the questions about the height (including the base) of the Christ the Redeemer statue in Rio de Janeiro, Brazil, the primary construction material used for the statue, and the number of years its construction took to complete as an example, Table 2 presents the responses of the large model to these questions, both with and without database connectivity.

Table 2. Large Models' Responses to Basic Questions

连接数据库	未连接数据库
Here are the precise details regarding Rio de Janeiro's iconic Christ the Redeemer statue (Cristo Redentor): 1. Height: -Statue itself: 30 m (98 ft) -Statue on top of the pedestal/building base: The total height including its massive Art Deco-style pedestal/base is 38 m (125 ft). 2. Main Building Material: -Primary structure: Reinforced concrete. This was a relatively new and innovative material for large sculptures at the time (early 20th century). -Outer layer: The surface of the statue is clad with soapstone (steatite). Tens of thousands of small, triangular tiles of this pale, durable stone were cut and affixed to the concrete framework like a mosaic skin. This gives it the smooth appearance and weather resistance. 3. Construction Duration: -Construction lasted 9 years.-It officially began in 1922 (with the cornerstone laid) and was opened to the public on October 12, 1931 -Key Additional Fact:-The statue incorporates extensive lightning rods within its reinforced concrete frame and outer soapstone "skin," a crucial feature given its exposed mountaintop location atop Mount Corcovado. It is struck by lightning multiple times each year, sometimes requiring repairs to the soapstone tiles.	1. Height of the Statue -Total height (with pedestal): 38 m (125 ft) - Statue alone: 30 m (98 ft) - Pedestal (base): 8 m (26 ft) - Arms pan: 28 m (92 ft) 2. Main Building Material -Reinforced concrete (core structure)- Exterior cladding: Soapstone (6,000+ triangular tiles, originally quarried from Minas Gerais)- Modern restorations have replaced some tiles with composite materials for durability. 3. Construction Duration -Total project timeline: 9 years (1922–1931) -Design phase: 1922–1924 (initial sketches by Carlos Oswald, finalized by Heitor da Silva Costa) -Sculpting: 1926–1931 (created by French sculptor Paul Landowski in Paris) -Assembly on-site: 1926–1931 (engineer Albert Caquot oversaw structural work) Additional Notable Facts -Weight: 635 metric tons -Location: Summit of Corcovado Mountain (710 m / 2,330 ft. above sea level) -UNESCO Status: Declared a World Heritage Site in 2012 (as part of Rio's cultural landscape) Sources -Official records from the Archdiocese of Rio de Janeiro -UNESCO World Heritage documents -Instituto Brasileiro de Geografia e Estatística (IBGE) This iconic statue remains the largest Art Deco-style sculpture in the world and a symbol of Brazilian cultural heritage.

5 Conclusion and Prospect

5.1 Conclusion

Building a local database in the AI knowledge infrastructure for the vertical field of world cultural heritage is a key link to promote the intelligent development of cultural heritage protection and utilization. This paper proposes a mySQL-based methodology for the construction of AI knowledge infrastructure in the vertical field of world cultural heritage, which provides a complete and feasible technical solution for the construction of large model of world heritage and agent-based artificial intelligence system through data collection and preprocessing, optimized table structure design and "database retrieval module-large model" system architecture. The experimental results show that the local database constructed by this method can significantly improve the accuracy, completeness and pertinacity of the vertical large model in answering professional questions, and effectively solve the problems of storage, association and efficient invocation of structured data in the field of cultural heritage. The combination of RAG and SQL query realizes the accurate retrieval of domain knowledge, and provides reliable data support for the intelligent protection of cultural heritage.

5.2 Outlook

In the future, we will optimize and improve the scheme from three aspects. First, we will expand the storage capacity of MySQL, and improve the scalability of the database by processing larger scale of cultural heritage data through split-table and split-database technology. The second is to optimize the conversion accuracy of natural language to SQL in the retrieval module, and reduce the query error by combining the domain dictionary and rule base. The third is to enrich the content dimension of the database, incorporate more structured conversion results of unstructured data, including free text, images, audio, video, etc., and explore the docking with other data sources to further improve the knowledge coverage of the large vertical model, and provide more comprehensive intelligent support for the protection and inheritance of world cultural heritage. In the future, with the continuous progress of technology and the deepening of research, the local database construction scheme in the AI knowledge infrastructure in the vertical field of world cultural heritage will be continuously optimized to provide more powerful support for the development of cultural heritage.

References

1. Li, Q.S., Ni, T., Luo, X., et al.: Research progress on digitalization of cultural heritage and visualization service technology for digital cultural tourism. J. Image Graph. **30**(06), 2304–2324 (2025)
2. Computer Graphics Processing technology for cultural heritage sites - CyArk. Computer & Telecommunication (01), 2 (2008)
3. Zuo, Q.: Seize the Outbreak Period of Artificial Intelligence and Develop New Productivity: "Intelligent Transformation and Digital Transformation" Enters the Fast Lane. Sichuan Release (2024-04-22)
4. Wu, X., Fu, T.: Review of research on retrieval-augmented generation technology. Comput. Eng. Appl., 1–20 (2025)
5. Jiao, Y., Yang, X.: Big data integration platform based on hybrid architecture of Hadoop and MPP database. J. Inf. Eng. Univ. **25**(04), 435–440 (2024)
6. Wang, G.Y., Liu, X.N.: Development of web data crawling and analysis tool based on Python. Comput. Programming Skills Maint. (06), 56–58 (2025)
7. Chen, S.: Data cleaning system based on expert knowledge and optimized interpolation. Inf. Technol. (06), 125–130 (2025)
8. Xiong, Q.Y.: Application analysis of MySQL database in the era of big data. Inf. Comput. (Theory Edition). **35**(14), 209–212 (2023)

9. Lin, B.J.: AI large model competition in the ascendant: accurate application in vertical fields becomes key competitiveness. Communications Information News (2025-06-11)
10. Ji, X.M., Sheng, H.Y., Cui, T.T., Zhang, K.L.: Current status and challenges of large model applications in vertical fields. Comput. Eng. Appl. **61**(12), 1–11 (2025)
11. Wang, H.: Design and Implementation of Multi-Source Heterogeneous Data Integration System. Master's thesis, University of Electronic Science and Technology of China (2020)

Sustainable Tourism Water Management Research in Bali Based on Grey Correlation Analysis and Intelligent Optimization Algorithms

Yanran Wei(✉)

Shanghai University, Shanghai 200444, China
wyr-7_@shu.edu.cn

Abstract. Bali is confronting severe water stress exacerbated by tourism expansion, climate change, and population growth, with around 60% of the island facing seasonal water scarcity. Traditional management methods fail to reconcile tourism revenue with freshwater conservation. This study proposes an integrated computational framework combining Grey Relational Analysis (GRA) and Multi-Objective Decision-Making (MODM) to optimize water governance. The GRA module quantifies critical correlations between water use and tourism indicators under data-scarce conditions, identifying tourist volume as the primary driver. The MODM model then pursues dual objectives—maximizing tourism income while minimizing freshwater consumption—through optimized visitor numbers, progressive water taxes, and strategic promotion of low-water-intensity activities. The solution achieves a 23.7% reduction in water use while maintaining robust economic returns. Supported by IoT and remote sensing, this approach provides a dynamic, scalable strategy for sustainable water management in tourism-intensive regions.

Keywords: Tourism Water Sustainability · Grey Relational Analysis (GRA) · Multi-Objective Optimization · Computational Intelligence

1 Introduction

1.1 Background and Crisis

In recent decades, global tourism development has created significant pressure on natural resources in popular destinations, with water scarcity emerging as one of the most critical challenges. Bali, as a globally renowned tourist destination, faces a triple challenge to its water resource system due to climate change, population growth, and tourism expansion. Statistics show approximately 60% of Bali experiences seasonal water scarcity, with increasingly longer dry seasons. More critically, excessive groundwater extraction for tourism has caused coastal saltwater intrusion, degrading water quality and threatening public health and agriculture.

H. Ma et al. (Eds.): IWACIII 2025, CCIS 2780, pp. 309–322, 2026.
https://doi.org/10.1007/978-981-95-6730-0_23

The traditional water resource management approach in Bali usually employs a single indicator for assessment, which is insufficient to cope with the complex systems presented by popular tourist destinations worldwide. The current methods often neglect the dynamic interactions among the environment, tourism growth, and community needs. This fragmented approach hinders the formation of sustainable solutions.

1.2 Characteristics and Systemic Interdependencies in Bali's Tourism Water Management

The water resource management issues in Bali not only involve physical shortages of water resources, but are also closely related to the island's economic structure, seasonal tourism patterns, and geographical limitations. The water crisis caused by tourism in Bali has several unique characteristics, which together exacerbate the water resource pressure and make traditional management methods more complex.

Firstly, there is a significant imbalance in water usage among different industries. Luxury tourism facilities such as golf courses and high-end resorts consume an excessive amount of water, accounting for an estimated 65%–75% of water usage related to tourism, and also contribute significantly to economic income. This creates a fundamental contradiction between economic gains and resource sustainability. Secondly, the mismatch in time and space exacerbates this problem: tourist numbers often peak during the dry season, while water supply is at its lowest, and tourists mainly concentrate in the already over-extracted groundwater-dry coastal areas. Thirdly, reliance on groundwater leads to serious environmental consequences, including depletion of aquifers and seawater intrusion, which deteriorates water quality and threatens the safety of agricultural and household water use. Finally, the lack of data and decentralized governance hinder the implementation of effective policies, so analytical methods that can robustly deal with uncertainties are needed.

These characteristics are supported by the close and often contradictory systematic interdependence relationships among key variables (tourist numbers, economic income, and water consumption). The relationships among these factors form a vicious cycle: an increase in tourist numbers promotes economic growth, but also leads to higher water demand, especially from water-intensive industries. This demand in turn exacerbates the decline of groundwater and environmental degradation, which could eventually undermine the natural and cultural charm that initially attracted tourists to Bali. Therefore, this system presents a feedback loop, where economic success unintentionally intensifies resource consumption, thereby creating a difficult trade-off between short-term income and long-term sustainability.

Understanding these interrelationships is crucial for formulating effective management strategies. Simply reducing the number of tourists is not sufficient, as it would entail significant economic costs. Similarly, if only water-saving tourism is emphasized without considering economic incentives, the effect may not be satisfactory. Therefore, a comprehensive approach is needed - by strategically shifting tourism activities to options with lower water consumption and introducing economic measures that reflect the true environmental cost of water usage, to achieve decoupling of economic growth from water usage.

1.3 Research Significance and Objectives

The water crisis in Bali highlights the critical need for sustainable management, as competing demands from tourism, agriculture, and residential sectors create inequitable distribution and ecological strain. Addressing these challenges requires innovative solutions, particularly through intelligent technologies like IoT sensing, satellite remote sensing, and deep learning for real-time monitoring, combined with computational methods such as Grey Relational Analysis (GRA) and Multi-Objective Decision-Making (MODM) to balance conflicting priorities—ecological conservation, tourism revenue, and agricultural needs—despite limited data availability.

The primary objective of this study is to develop an integrated GRA-MODM framework that optimizes water governance by quantifying linkages between tourism indicators (e.g., visitor numbers, resort water use) and freshwater consumption, while pursuing dual goals: maximizing tourism income and minimizing water usage. This approach incorporates AI-driven policies, including progressive water taxes and visitor flow redirection to low-impact activities, supported by IoT and satellite-based monitoring. By leveraging computational intelligence in data-scarce environments, the framework not only addresses Bali's immediate water stress but also provides a scalable model for other tropical destinations facing similar tourism-induced resource pressures.

1.4 Article Frame

This paper is structured as follows. Section 2 comprehensively reviews modern water resource management strategies and methodologies for balancing tourism development with environmental protection, while Sect. 3 details the application of multi-objective optimization models in tourism. The preliminary model assumptions and mathematical notations are systematically presented in Sect. 3. Section 4 develops the integrated Grey Relational Analysis and Multi-Objective Decision-Making framework, including model construction, solution algorithms, and sensitivity analysis. Finally, Sect. 5 concludes with policy recommendations and discusses the framework's transfer-ability to other island destinations.

2 Literature Review

2.1 Evolution of Water Resource Management Strategies

The field of water resource management has undergone significant changes, driven by increasingly complex water issues and advancements in computing technology. Early strategies often relied on single-objective optimization and departmental management, but it has been proven that these methods are insufficient to address the interrelated systemic challenges.

The first major shift was towards an integrated management framework. These approaches, such as the multi-factor coupling method [1], aim to balance ecological, social and environmental goals through weighted optimization models in electronic decision support tools. This lays a crucial foundation for considering numerous priorities simultaneously.

Meanwhile, this field has made a significant leap in computational capabilities for solving these complex problems. The development of optimization techniques based on artificial intelligence, especially evolutionary algorithms like genetic algorithms, provides the necessary tools for effectively coordinating water resource allocation under industrial, agricultural, and ecological demands [2]. This trend towards algorithmic complexity continues, and so does the recent trend in hybrid intelligent models. For instance, a hybrid simulation-optimization model for joint surface water and groundwater management has been proposed, innovatively combining moth-flame algorithm, symbiotic biological search algorithm, and artificial neural network [3]. This approach represents the cutting-edge integration of bio-inspired optimization and machine learning to achieve more outstanding decision-making.

Based on these algorithm improvements, the current advanced level is defined by a comprehensive data-driven artificial intelligence framework. These systems combine hydrological modeling, remote sensing, and machine learning to create adaptive operation systems that can simultaneously reduce water shortages and minimize environmental impacts. This method has been successfully verified in practical applications in the Segura River Basin in Spain, which has successfully optimized the allocation of 642 million cubic meters of water resources and controlled the water supply gap to below 10% [4].

However, there remains a crucial gap: These advanced frameworks are usually designed to address a wide range of regional-scale issues (such as agriculture, general urban uses), rather than being specifically tailored to the intense and localized pressures imposed by a particular industry (such as tourism). The unique characteristics of water resource pressure driven by tourism, such as extreme seasonal fluctuations, high per capita consumption, and the direct conflict between economic gains and resource conservation, are not adequately covered by these general models.

2.2 Multi-objective Optimization in Sustainable Tourism Management

Meanwhile, the multi-objective optimization (MOO) model has made significant progress in addressing the trade-off issues within the tourism industry. Most of these studies aim to enhance the tourist experience and operational efficiency. For instance, models based on fuzzy theory have been developed to optimize tourism packages, aiming to reduce costs while improving customer satisfaction [5]. Further research has also created frameworks such as MOEA/D-HRR to balance the accuracy and diversity of recommendations [6], as well as the WAEA algorithm, which generates personalized travel routes by optimizing preferences, time, and budget [7].

An emerging branch of this research field applies multi-objective optimization methods to sustainable development. The SA-NSGAII model was proposed for environmentally friendly itinerary planning, aiming to coordinate environmental impact, travel cost, and preferences [8]. It is notable that the enhanced NSGA-II algorithm has been used to optimize the urban tourism carrying capacity in different seasons, balancing economic development and environmental protection [9]. Supplementary non-technical research emphasizes the importance of community-led initiatives [10], ecosystem-based models [11], and strong policy frameworks [12] for achieving sustainable tourism.

Despite these advancements, a discernible research gap persists. Existing MOO applications in tourism primarily focus on micro-level planning (e.g., itineraries, recommendations) or single-sector carrying capacity. However, there is a lack of research that combines multi-objective optimization with tourism water resource management, especially in addressing the conflict between economic growth and resource sustainability. This study aims to fill this gap and proposes a comprehensive analytical framework that combines grey correlation analysis (GRA) with multi-objective decision-making (MODM). The GRA module is used to identify the key drivers of water consumption in situations where data is scarce, while the MODM model is used to simultaneously optimize the number of tourists, economic income, and water allocation strategies for various industries. This approach provides a macro-level decision-making tool to balance the tourism development of island destinations like Bali with water resource sustainability, being both methodologically innovative and having practical policy significance.

3 Preliminary Model Analysis and Mathematical Notations

3.1 Assumptions and Justifications

Considering those practical problems always contain many complex factors, first of all, we need to make reasonable assumptions to simplify the model, and each hypothesis is closely followed by its corresponding explanation.

Assumption 1: Annual freshwater consumption of golf courses remains constant.
Justification: The turf area and irrigation systems of golf courses remain relatively stable within the 1–3 year planning period.

Assumption 2: Estimate the total area of golf courses on Bali by summing up the areas of four large and representative golf courses.
Justification: Their combined area basically reflects the island's total golf course coverage.

Assumption 3: Per tourist daily water consumption of tourists is set at 0.175 m^3.
Justification: Research indicates tourist water consumption in Bali ranges between 150-200 L per day. The median value of 0.175 m^3not only represents the average but also mitigates outlier effects. This range has been verified through water usage records from five major hotels in Bali.

Assumption 4: Daily freshwater consumption per occupied hotel room is 0.8 m^3.
Justification: According to WAHLI (Bali Branch of the Indonesian Environmental Association), standard hotel rooms consume approximately 0.8 m^3per room per day.

Assumption 5: Annual freshwater consumption of high-water-intensity tourism projects equals the sum of all hotel room consumption and total golf course irrigation.
Justification: Annual freshwater consumption of low-water-intensity tourism activities (e.g., hiking, eco-farms) is substituted by total tourist water consumption.

Assumption 6: Annual freshwater consumption of low-water-intensity tourism activities (e.g., hiking, eco-farms) is substituted by total tourist water consumption.
The infrastructure water consumption of ecotourism projects is significantly lower than that of tourists' personal water usage.

Assumption 7: Bali's annual tourism revenue is estimated as half of Indonesia's total tourism revenue.

Justification: Data from Statistics Indonesia (BPS) shows Bali's tourism revenue consistently accounted for 48%–52% of national tourism income. As Bali remains Indonesia's largest tourist destination with minimal percentage fluctuations, the 50% estimation demonstrates statistical significance.

Assumption 8: In terms of water consumption, high-water-intensity tourism assigns a higher weight of 70% while the weight of low-water-intensity tourism is set at 30%.

Justification: The Bali Water Resources Report indicates that luxury hotels and golf courses account for 65–75% of the total water consumption.

3.2 Notations

The key mathematical notations used in this paper are listed in Table 1.

Table 1. Notations used in this paper

Symbol	Description	Unit
W	Annual freshwater consumption	million m^3/year
T	Annual tourist numbers	person
R	Annual tourism revenue	million dollars
W_H	Annual water consumption of high-water-intensity tourism	million m^3/year
W_L	Annual water consumption of low-water-intensity tourism	million m^3/year
W_g	Annual water consumption for golf course irrigation	million m^3/year
W_h	Annual water consumption of hotel rooms	million m^3/year
r_H	Annual hotel room occupancy rate	%
H	Annual total number of hotel rooms	room
L	Average length of stay for tourists	day
r_t	Tourist tax rate	%
α	The proportion of tourists participating in low-water-intensity tourism	/
C_H	High-water-intensity tourism per capita consumption	dollars
C_L	Low-water-intensity tourism per capita consumption	dollars
w_H	High-water-intensity tourism per capita water consumption	million m^3/year
w_L	Low-water-intensity tourism per capita water consumption	million m^3/year
C_E	Environmental governance costs	million dollars
k	The coefficient of treatment cost per unit volume of water	/

4 Modelling

4.1 Grey Relational Analysis Model

4.1.1 The Establishment of GRA

The grey relational analysis model refers to the measurement of the degree of correlation between the factors that change over time or among different objects between two systems, which is called the correlation degree. During the development of the system, if the changing trends of two factors are consistent, then it is considered that the correlation degree between the two factors is higher; conversely, it is lower. The advantage of this method lies in handling small sample sizes and high uncertainty inherent to environmental data, without demanding large datasets [8]. Therefore, based on the grey correlation theory, this paper analyzes and calculates the grey relational coefficients between annual freshwater consumption and annual tourist numbers, annual tourism revenue, annual freshwater consumption of high-water-intensity tourism, and annual freshwater consumption of low-water-intensity tourism (total tourist freshwater consumption).

The first step is to calculate the grey relational degree [13]. The formula is as follows:

$$\begin{aligned} a &= \min_{s}\min_{t}|X_0(t) - X_s(t)| \\ b &= \max_{s}\max_{t}|X_0(t) - X_s(t)| \\ r &= \frac{a+\rho b}{|X_0(k)-X_i(k)|+\rho b} \end{aligned} \tag{1}$$

Here, a represents the minimum absolute difference, b represents the maximum absolute difference, ρ is the resolution coefficient, which takes a value between 0 and 1. Here, it is set to 0.5.

The second step is to calculate the degree of correlation, and the formula is as follows:

$$\gamma(X_0, X_i) = \frac{1}{n}\sum_{k=1}^{n} r(X_0(k), X_i(k)) \tag{2}$$

The value of γ ranges from 0 to 1. The larger the value, the higher the data correlation and the stronger the coupling.

4.1.2 Data Processing

Annual water consumption of high-water-intensity tourism W_H is the sum of Annual water consumption for golf course irrigation Wg and Annual water consumption of hotel rooms W_h. Related formulas are as follows:

$$W_g = 415 \div 120 \times 108758 \div 10000 = 37.61 \cdot \left(million \cdot m^3/year\right) \tag{3}$$

$$W_h = H \times r_H \times 0.8 \times 365 \div 10000 \tag{4}$$

$$W_H = W_g + W_h \tag{5}$$

Here, 415 is the sum area of four large and representative golf courses (unit is hectare), 120 is the area of Handara Golf & Resort Bali (unit is hectare) and 108758 is the annual irrigation water consumption of Handara Golf & Resort Bali (unit is m^3/year).

Due to the assumptions mentioned above, annual water consumption of low-water-intensity tourism W_L is substituted by total tourist water consumption and the formula is below:

$$W_L = T \times L \times 0.175 \div 10000 \tag{6}$$

4.1.3 The Solution of GRA

This article uses data from 2016 to 2024 for calculation (CEIC Database, 2025). The data from 2020 to 2022 were excluded because of the impact of the COVID-19 pandemic, which made the figures significantly different from the normal ones. The results are shown in Table 2 as follows:

Table 2. The solution of GRA

	Reference	Comparison sequences			
	W	T	R	W_H	W_L
γ	/	0.7278	0.6922	0.5427	0.6955

The highest grey relational degree between tourist numbers (T) and freshwater consumption (W) ($\gamma = 0.7278$) demonstrates that visitor volume is the primary driver of water consumption in Bali. This strong correlation indicates that reducing tourist inflows would be the most effective direct strategy for decreasing water stress. The relevant departments should therefore prioritize visitor flow control and carrying capacity policies as essential components of sustainable water resource management in tourism development.

4.2 Multi-objective Optimization Model

4.2.1 Model Framework

Decision Variables

Annual tourist numbers T, tourist tax rate rt—a policy lever proven effective in modulating water-intensive tourism demand [14] and the proportion of tourists participating in low-water-intensity tourism α.

Objective Functions

Maximizing tourism revenue:

$$f_1 = R \cdot (1 - r_t) - C_E \tag{7}$$

Minimizing freshwater consumption:

$$f_2 = W = 0.7 \cdot W_H + 0.3 \cdot W_L \tag{8}$$

Constraint Conditions

Tourist capacity constraint: $T_{\min} \leq T \leq T_{\max}$
Tax rate constraint: $0 \leq r_t \leq r_t^{\max}$
Water consumption ratio constraint: $\alpha \geq \alpha_{\min}$
Water resource constraints: $W \leq W_{\max}$
Economic constraints: $R \geq R_{\min}$

4.2.2 Specific Expression Derivation

Tourism revenue model

$$R = T \cdot [(1 - \alpha) \cdot C_H + \alpha \cdot C_L] \tag{9}$$

Freshwater consumption model

$$w_H = 0.8 \div 10^4 \tag{10}$$

$$w_L = 0.175 \div 10^4 \tag{11}$$

$$W_H = T \cdot (1 - \alpha) \cdot w_H + W_g \tag{12}$$

$$W_L = T \cdot \alpha \cdot w_L \tag{13}$$

Environmental governance costs

$$C_E = k \cdot W \tag{14}$$

4.2.3 Algorithm Implementation

This multi-objective optimization model is solved using the Non-dominated Sorting Genetic Algorithm II (NSGA-II), which is implemented through MATLAB's Global Optimization Toolbox. Based on the preliminary sensitivity test, the algorithm parameters were selected to balance the computational efficiency and the quality of the solutions.

This optimization problem involves 3 decision variables (the number of tourists T, the tax rate r_t, and the proportion of low water intensity α) as well as 5 nonlinear inequality constraints representing the availability of water resources, hotel capacity, revenue requirements, and the proportion of water allocation among departments. The NSGA-II algorithm successfully found a set of Pareto-optimal solutions, which effectively balanced the three competing objectives. The optimization process underwent 102 iterations and 10,202 function evaluations, and finally a Pareto front solution was obtained.

4.2.4 Optimization Results

The Pareto frontier solution was obtained through the genetic algorithm. The recommended solution with the lowest environmental cost is as follows:

Tourist Numbers: 5.00 million,
Tourism Tax Rate: 4.77%,
Low-Water-Intensity Proportion: 65.00%,
Tourism Revenue: 547.07 Million$,
Total Water Consumption: 1058.39 (10 k m^3),
Environmental Cost: 0.53 Million$.

Compared to the conventional situation, this solution reduces water usage by 19%. At the same time, it can maintain stable tourism revenue. The strategic shift towards low water-intensity activities is quite different from the current tourism model, but it demonstrates the great potential for water resource protection without sacrificing economic gains. This allocation effectively achieves a balance between environmental sustainability and economic goals, providing a feasible approach for sustainable tourism development in water-stressed regions.

4.3 Experimental Validation

4.3.1 Comparison with Baseline Models

In order to rigorously evaluate the performance of the proposed GRA-MODM framework, we compared its optimal solution with two benchmark scenarios:

1. **Business-as-Usual (BAU) Scenario:** This scenario assumes no policy interventions, with tourist numbers fixed at the 2023 level (5.27 million), a tourism tax rate of 0%, and the current proportion of low-water-intensity activities (estimated at 35% based on sectoral water usage reports).
2. **Single-Objective Revenue Maximization Scenario:** This model employs the same constraint set as the proposed framework but optimizes for a single objective: maximizing tourism revenue, disregarding water consumption and environmental costs.

The comparative results are summarized in Table 3.

Table 3. Comparative Results

Metric	BAU	Revenue Maximization	GRA-MODM Framework	Improvement over BAU
Tourist Numbers (million)	5.27	6.88	5.03	−4.6%
Low-Water-Intensity Proportion	35%	22%	65%	+85.7%
Tourism Revenue ($ million)	700.0	855.2	686.0	−2.0%

(continued)

Table 3. (*continued*)

Metric	BAU	Revenue Maximization	GRA-MODM Framework	Improvement over BAU
Water Consumption (10 k m^3)	1387.4	1720.1	1058.7	−23.7%
Environmental Cost ($ million)	0.69	0.86	0.53	−23.2%

Note: Revenue for the BAU scenario is projected based on 2023 data.

Compared with the BAU scenario, the proposed model achieved a significant 23.7% reduction in water consumption and a 23.2% decrease in environmental costs, while still retaining 98% of economic benefits (only a 2% reduction in income). This highlights the ability of this framework to separate economic growth from resource consumption.

Compared with the single-objective profit maximization model, our solution sacrificed only a small amount of economic output (approximately a 19.8% reduction in profits), but achieved significant resource savings (a 38.5% reduction in water consumption). This comparison highlights the key limitations of traditional economic-driven methods and confirms the necessity of multi-objective optimization for sustainable governance.

Therefore, the GRA-MODM framework provides a Pareto-optimal solution, successfully achieving an effective balance between economic development and environmental sustainability.

4.3.2 Sensitivity Analysis

Based on the above model establishment, this paper has conducted corresponding sensitivity analysis. The results are as follows:

Water sensitivity to tourist numbers: 10.50 (10 k m^3)/million tourists,
Water sensitivity to low-water proportion: 5250.00 (10 k m^3)/%,
Revenue sensitivity to tax rate: 65000.00 $M/%.

The specific results are shown in the following figures (Figs. 1, 2 and 3):

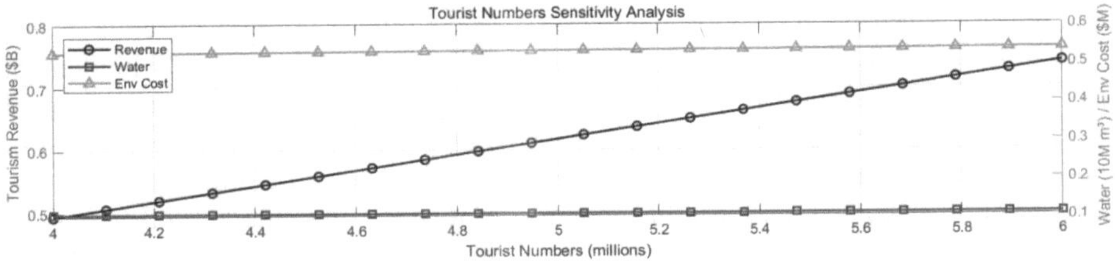

Fig. 1. Tourist Numbers Sensitivity Analysis

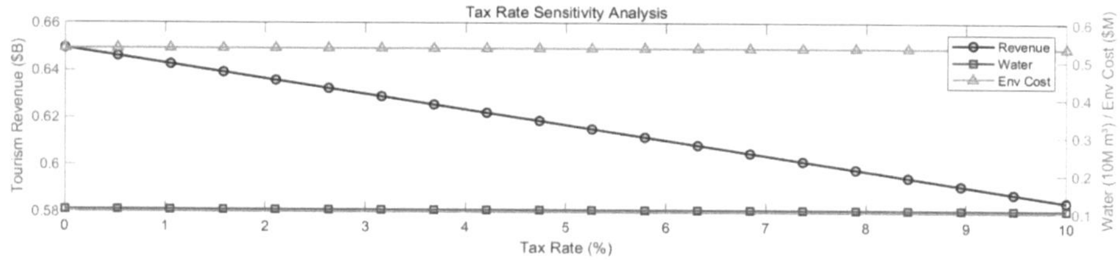

Fig. 2. Tax Rate Sensitivity Analysis

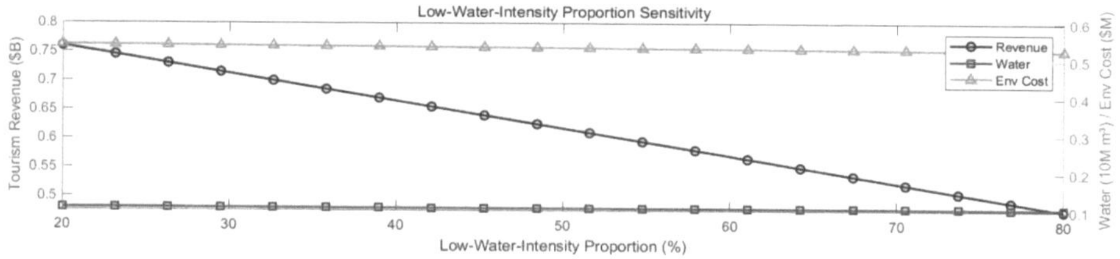

Fig. 3. Low-Water-Intensity Proportion

According to the sensitivity analysis, the number of tourists has been identified as the main driver of water demand. For every increase of 1 million tourists, water consumption will increase by 10.5 (100,000 m^3). Most notably, reallocating tourists to activities with lower water consumption proves to be an excellent policy. For every 1% increase in the proportion of tourists, the total water consumption will decrease by 52.5 (100,000 m^3). Therefore, although controlling the number of tourists is crucial, strategically redistributing tourists provides the most effective way to achieve significant water resource savings without significantly affecting economic benefits.

5 Conclusion

This study establishes an integrated computational intelligence framework to address Bali's tourism-driven water crisis, yielding actionable insights for governance. The analysis confirms tourism's disproportionate hydrological impact through systemic water-revenue linkages, necessitating targeted interventions. Governments should implement dual policies: a progressive water tax calibrated to consumption patterns, coupled with strategic visitor redistribution toward low-water-intensity activities (ecotourism). These measures demonstrably reduce water stress while maintaining economic viability. Supported by IoT and satellite monitoring technologies, the approach enables dynamic resource governance that balances tourism revenue, freshwater conservation, and ecological resilience.

Although this study has made contributions, it also has some limitations, which are worthy of recognition and provide opportunities for future research. Firstly, due to the lack of data, the model relies on several simplified assumptions (such as the annual water consumption of the golf course remaining constant and the water consumption per room being fixed), which may not fully reflect the operational dynamics in real-life situations. Secondly, the genetic algorithm encountered a formal feasibility issue, indicating that although this solution is feasible in practical operations, the mathematical convergence under strict constraints still needs further improvement. Thirdly, this framework currently prioritizes economic and environmental goals, but does not incorporate explicit fairness considerations regarding the impact of policies on the distribution of local communities.

These limitations have directly led to the future research directions. Firstly, integrating artificial intelligence-driven predictions with state-space models can break away from static assumptions and take into account the variability of the climate and market shocks. Secondly, exploring hybrid optimization algorithms or robust optimization techniques will enhance the framework's ability to handle complex real-world constraints. Finally, and most importantly, subsequent research must incorporate social and economic equity indicators to ensure that water resource governance strategies are not only efficient but also fair and inclusive, and to assess their impact on local industries, employment, and community access to water resources.

References

1. Zhang, J., Meng, W., Cui, S., et al.: Method for evaluating water resource management optimization scheme based on multi-factor coupling by using electronic device, involves solving water resource management evaluation optimization model using scene corresponding to maximum value of total target function. Chinese Patent CN116415410-A, 11 July 2023
2. Lu, Y.Q., Gan, L., Chen, Y.H., et al.: Research on the allocation of water resources engineering projects based on multi-objective optimization. Water Supply. **23**(8), 3297–3308 (2023)
3. Akbarifard, S., Madadi, M.R., Zounemat-Kermani, M.: An artificial intelligence-based model for optimal conjunctive operation of surface and groundwater resources. Nat. Commun. **15**, 553 (2024)
4. Otamendi, U., Maiza, M., Olaizola, I.G., et al.: Integrated water resource management in the Segura Hydrographic Basin: an artificial intelligence approach. J. Environ. Manag. **370**, 122526 (2024)
5. Shojatalab, G., Nasseri, S.H., Mahdavi, I.: New multi-objective optimization model for tourism systems with fuzzy data and new algorithm for solving this model. Opsearch. **59**(3), 1018–1037 (2022)
6. Cao, R., Li, Z., Wei, P., Tian, Y., Zheng, C.: A Hybrid tourism recommendation system based on multi-objective evolutionary algorithm and re-ranking. In: Huang, D.S., Premaratne, P., Jin, B., et al. (eds.) Advanced Intelligent Computing Technology and Applications, ICIC 2023, LNCS, vol. 14090, pp. 363–372. Springer, Heidelberg (2023)
7. Du, J.: Application of improved multi-objective evolutionary algorithm in intelligent tourism interest point recommendation and itinerary planning. Mechatron. Syst. Control. **53**(2), 126–133 (2025)
8. Allahmadi, S.Z., Jabbarzadeh, A., Hof, L.A.: A multi-objective optimization approach for sustainable and personalized trip planning: a self-adaptive evolutionary algorithm with case study. Expert Syst. Appl. **261**, 125412 (2025)
9. Tan, Q.H.: Research on sustainable carrying capacity of urban tourism environment based on multi-objective optimization algorithm. Results Eng. **19**, 101344 (2023)
10. Anindhita, T.A., Zielinski, S., Milanes, C.B., et al.: The protection of natural and cultural landscapes through community-based tourism: the case of the indigenous Kamoro tribe in West Papua, Indonesia. Land. **13**(8), 1237 (2024)

11. Tian, J., Li, J.: Analysis and treatment of the conflict between sustainable development and environmental protection based on the ecotourism concept. Front. Environ. Sci. **10**, 1056643 (2022)
12. Guo, Z., Li, Y.: Analysis of the decisive factors of government attracting tourists in public management from the of environmental protection. Probl. Ekorozw. **19**(1), 285–295 (2024)
13. Liu, S., Forrest, J., Yang, Y.: A brief introduction to grey systems theory. Grey Syst. Theory Appl. **2**(2), 89–104 (2012)
14. Goktas, L.S., Çetin, G.: Tourist tax for sustainability: determining willingness to pay. Eur. J. Tourism Res. **35**, 3503 (2023)

Weak N-Best POAFD for Second Order Initial Value Problems of Ordinary Differential Equation

Hongfang Bai and Jianwei Dong(✉)

Department of Mathematics, College of Medical Information and Engineering, Guangdong Pharmaceutical University, Guangzhou 510006, China
177762438@qq.com

Abstract. In this paper, we propose the weak N-best pre-orthogonal adaptive Fourier decomposition (weak N-best POAFD) method for solving second-order initial value problems (IVPs). By introducing a novel weak N-best maximal selection principle, the method constructs orthogonal bases that maximize energy capture at each iterative step. Theoretical analysis demonstrates that the proposed approach achieves a higher convergence rate compared to standard POAFD. Numerical experiments confirm an order-of-magnitude improvement in accuracy.

Keywords: Weak N-best POAFD · Weak N-best maximal selection principle · Second order initial value problem · Reproducing kernel Hilbert space

1 Introduction

We consider the following initial value problem

$$(Lu)(x) = a(x)u'' + b(x)u' + c(x)u = f(x), \quad 0 < x < 1, \tag{1}$$

$$u(0) = 0, u'(0) = 0, \tag{2}$$

where the coefficients $a, b, c \in W^1[0,1]$, $f \in W^1[0,1]$ and $u \in W_0^3[0,1]$. $W^1[0,1]$, $W^3[0,1]$ are reproducing kernel Hilbert spaces defined in Sect. 2.

Initial value problems (IVPs) of ordinary differential equation serve as mathematical models in many scientific and engineering fields to describe the evolution of physical quantities. These problems are characterized by conditions specified at a single point, which determine the unknown function. Obtaining analytical and numerical solutions to IVPs is a fundamental task [3,4].

Within the framework of reproducing kernel Hilbert spaces, Cui et al. proposed the Reproducing Kernel Method (RKM) to express and approximate the solution of differential equations, integral equations and so on [1]. On the other hand, Qian et al. proposed adaptive Fourier decomposition (AFD) type methods, including AFD, pre-orthogonal adaptive Fourier decomposition (POAFD)

H. Ma et al. (Eds.): IWACIII 2025, CCIS 2780, pp. 323–346, 2026.
https://doi.org/10.1007/978-981-95-6730-0_24

and several other variations [5–7] to give a sparse representation of the given function with fast convergence rate. This class of methods has a wide range of applications, such as signal and image processing [8,9], numerical solutions of the operator equations [10,11].

Building on the work of Cui et al. and Qian et al., this paper aims to derive the analytical solution and approximate solutions of IVPs by weak N-best POAFD. Different from weak pre-orthogonal adaptive Fourier decomposition method, our method employs a weak N-best maximal selection principle at each step to choose a sequence of distinct parameters. This leads to an explicit analytical representation of the solution and improved convergence behavior.

The organization of this paper is as follows. In Sect. 2, we introduce some important properties of the reproducing kernel Hilbert spaces $W^1[0,1]$ and $W_0^3[0,1]$, and discuss the boundedness of the inverse of the operator L. It is noteworthy that while the proposed "weak N-best" framework is general, the detailed theoretical derivations and algorithmic implementations in Sects. 3.1 to 3.3 primarily focus on the case of N=2 for clarity. The extension to arbitrary N will be briefly discussed at the end of Sect. 3.3. Thus, in Sect. 3, we construct the sequence $\{a_n, b_n\}_{n\geq 1}$ in $[0,1]$ of distinct points by weak 2-best maximal selection principle (42) in Lemma 5. We then prove the series (46) obtained by weak N-best POAFD is the solution of IVP (1)–(2) in Theorem 4. Also, the convergence properties are analyzed. In Sect. 4, numerical experiments are conducted to verify the feasibility of weak N-best POAFD in solving second-order IVPs. Finally, Sect. 5 ends this work with a brief conclusion.

2 Preliminaries

In this paper, we work on two reproducing kernel Hilbert spaces $W^1[0,1]$ and $W_0^3[0,1]$ and introduce the inner products and the norms of these two spaces as follows.

Definition 1 [1] *The reproducing kernel Hilbert space $W_0^3[0,1]$ is defined as*

$$W_0^3[0,1] = \left\{u \mid u^{(2)} \text{ exists and is absolutely continuous on } [0,1], u^{(3)} \in L^2[0,1], u(0)=0,\ u'(0)=0.\right\}$$

with the inner product and the norm given by

$$\begin{aligned}\langle u, v\rangle_{W_0^3[0,1]} &= u''(0)v''(0) + \int_0^1 u^{(3)}(x)v^{(3)}(x)\mathrm{d}x, \quad u,v \in W_0^3[0,1],\\ \|u\|_{W_0^3[0,1]} &= \sqrt{\langle u,u\rangle_{W_0^3}}, \quad u \in W_0^3[0,1].\end{aligned} \tag{3}$$

The reproducing kernel R of $W_0^3[0,1]$ is

$$R(x,y) = R_y(x) = \begin{cases} \frac{1}{120}y^2(-5xy^2+y^3+10x^2(3+y)), & y \leq x,\\ \frac{1}{120}x^2(-5x^2y+x^3+10(3+x)y^2), & y > x.\end{cases} \tag{4}$$

For any $x, y \in [0,1]$ and any $u \in W_0^3[0,1]$, it is easy to check that R in $W_0^3[0,1]$ satisfies the following:
(i) reproducing property: $\langle u, R_x\rangle_{W_0^3[0,1]} = u(x)$; (ii) symmetric property: $R(x,y) = R(y,x)$.

Definition 2 [1]. *The Sobolev space* $W^1[0,1]$ *is defined as*

$$W^1[0,1] = \left\{ u \;\middle|\; u : [0,1] \to \text{ is absolutely continuous, and } u' \in L^2[0,1]. \right\}$$

with the inner product and the norm:

$$\begin{aligned} \langle u, v\rangle_{W^1[0,1]} &= u(0)v(0) + \int_0^1 u'(\xi)v'(\xi)\, d\xi, \quad u, v \in W^1[0,1], \\ \|u\|_{W^1[0,1]} &= \sqrt{\langle u, u\rangle_{W^1[0,1]}}, \quad u \in W^1[0,1]. \end{aligned} \tag{5}$$

The space $W^1[0,1]$ is a reproducing kernel Hilbert space with the reproducing kernel

$$K(x,y) = K_y(x) = \begin{cases} 1+y, & y \le x, \\ 1+x, & y > x. \end{cases} \tag{6}$$

The reproducing kernel K of $W^1[0,1]$ has the same reproducing property and symmetric property as $R \in W_0^3[0,1]$ discussed above.

Throughout this paper, $\|a\|_\infty = \sup\{\, |a(x)| \mid x \in [0,1]\,\}$. For brevity, we use the notations $\|\cdot\|_{W^1[0,1]} = \|\cdot\|_{W^1}$, and $\|\cdot\|_{W_0^3[0,1]} = \|\cdot\|_{W_0^3}$ along with their related inner products.

Theorem 1 [1]. *Let* R *be the reproducing kernel of RKHS* $W_0^3[0,1]$ *in* (4). *For any* $x, y \in [0,1]$, *and non-negative integers* i, j, *one has*

$$\frac{\partial^{i+j}}{\partial x^i \partial y^j} R(x,y) \in L^2[0,1], \quad i+j = 5; \tag{7}$$

$$\frac{\partial^{i+j}}{\partial x^i \partial y^j} R(x,y), \quad 0 \le i+j \le 4, \tag{8}$$

is absolutely continuous on $[0,1]$ *with respect to* x *or with respect to* y.

Remark 1. (i) By (8), there exists $M_0 > 0$ such that for any x, $y \in [0,1]$, we have $|R(x,y)| \le M_0$.

(ii) Define $\partial^i R_a(x) := \frac{\partial^i}{\partial x^i} R(x,a)$ for $x, a \in [0,1]$. It follows from the reproducing property of R in $W_0^3[0,1]$, (7) and (8) that for any $a \in [0,1]$,

$$u^{(i)}(a) = \langle u, \partial^i R_a\rangle_{W_0^3}, \quad 0 \le i \le 2. \tag{9}$$

(iii) It follows from (7), (8) and (9) that for any $a \in [0,1]$

$$|u^{(i)}(a)| = |\langle u, \partial^i R_a\rangle_{W_0^3}| \le \|\partial^i R_a\|_{W_0^3} \|u\|_{W_0^3} \le M_i \|u\|_{W_0^3}, \qquad 0 \le i \le 2,$$

where $M_i = \max\{\|\partial^i R_a\|_{W_0^3} \mid a \in [0,1]\}$.

Proposition 1. *The linear operator* $L : W_0^3[0,1] \to W^1[0,1]$ *defined in* (1)–(2) *is bounded.*

Proof. It follows from (6), (3) and Remark 1 (ii)–(iii) that

$$\begin{aligned}
\|Lu\|_{W^1}^2 &= [a(0)u''(0)+b(0)u'(0)+c(0)u(0)]^2 + \int_0^1 [(a(x)u''+b(x)u'(x)+c(x)u)']^2 \mathrm{d}x \\
&\le 3[a(0)]^2[u''(0)]^2) + 3[b(0)]^2[u'(0)]^2 + 3[c(0)]^2[u(0)]^2 \\
&+ \int_0^1 [a(x)u'''(x) + [a'(x)+b(x)+c(x)]u''(x) + [b'(x)+c'(x)]u'(x)]^2 \mathrm{d}x \\
&\le 3[a(0)]^2 M_2^2\|u\|_{W_0^3}^2 + 3[b(0)]^2 M_1^2\|u\|_{W_0^3}^2 + 3[c(0)]^2 M_0^2\|u\|_{W_0^3}^2 \\
&+ 3\int_0^1 [a(x)u'''(x)]^2 \mathrm{d}x + 3\int_0^1 [a'(x)+b(x)+c(x)]^2[u''(x)]^2 \mathrm{d}x + 3\int_0^1 [b'(x)+c'(x)]^2[u'(x)]^2 \mathrm{d}x \\
&\le 3[a(0)]^2 M_2^2\|u\|_{W_0^3}^2 + 3[b(0)]^2 M_1^2\|u\|_{W_0^3}^2 + 3[c(0)]^2 M_0^2\|u\|_{W_0^3}^2 \\
&+ 3\int_0^1 [u'''(x)]^2 \mathrm{d}x \|a\|_\infty^2 + 3\int_0^1 [a'(x)+b(x)+c(x)]^2 \mathrm{d}x M_2^2\|u\|_{W_0^3}^2 + 3\int_0^1 [b'(x)+c'(x)]^2 \mathrm{d}x M_1^2\|u\|_{W_0^3}^2 \\
&\le \Big[3[a(0)]^2M_2^2 + 3[b(0)]^2M_1^2 + 3[c(0)]^2M_0^2 + 3\|a\|_\infty^2 + 9M_2^2\|b\|_\infty^2 + 9M_2^2\|c\|_\infty^2 \\
&+ 9M_2^2\int_0^1 [a'(x)]^2 \mathrm{d}x + 6M_1^2\int_0^1 [b'(x)]^2 \mathrm{d}x + 6M_1^2\int_0^1 [c'(x)]^2 \mathrm{d}x\Big] \|u\|_{W_0^3}^2 \le M^2\|u\|_{W_0^3}^2
\end{aligned}$$

and hence $\|Lu\|_{W^1} \le M\|u\|_{W_0^3}$.

Lemma 1. *For any $f \in W^1[0,1]$ and $u \in W_0^3[0,1]$, we have the following inequalities:*

$$|f(x)|^2 \le 2\|f\|_{W^1}^2; \tag{10}$$

$$\|f\|_{L^2[(0,1)]} < \infty; \tag{11}$$

$$\|u^{(i)}\|_\infty < \infty, \quad 0 \le i \le 2; \tag{12}$$

$$\|u^{(i)}\|_{L^2[(0,1)]} < \infty, \quad 0 \le i \le 3. \tag{13}$$

Proof. Since $f \in W^1[0,1]$ is absolutely continuous on $[0,1]$, then $f(x) = f(0) + \int_0^x f'(t)\mathrm{d}t$ for any $x \in [0,1]$. By the Cauchy-Schwarz inequality, we have the following estimates:

$$\begin{aligned}
|f(x)|^2 = \left|f(0) + \int_0^x f'(t)\,\mathrm{d}t\right|^2 &\le 2|f(0)|^2 + 2\left(\int_0^x f'(t)\,\mathrm{d}t\right)^2 \\
&\le 2|f(0)|^2 + 2\left(\int_0^1 1^2\,\mathrm{d}t\right)\left(\int_0^1 |f'(t)|^2\,\mathrm{d}t\right) = 2\|f\|_{W^1}^2.
\end{aligned}$$

Moreover, as $f \in W^1[0,1]$, $u^{(2)}$ are absolutely continuous on $[0,1]$, thus $\|f\|_{L^2[(0,1)]} < \infty$, $\|u^{(i)}\|_\infty < \infty$ $(0 \le i \le 2)$. Noticing that $u''' \in L^2[(0,1)]$, we have $\|u^{(i)}\|_{L^2[(0,1)]} < \infty$ $(0 \le i \le 3)$.

Theorem 2. *Denote by $L : W_0^3[0,1] \to W^1[0,1]$ the linear bounded operator defined in (1)–(2). Let a_0 be a positive constant. If $|a| > a_0 > 0$, then L has a bounded inverse L^{-1}.*

Proof. First, it follows that $|a| > a_0 > 0$ on $[0,1]$ and $f \in W^1[0,1]$ that the initial value problem (IVP) (1)–(2) has one and only one classical solution (see from [3]). Thus $Ker(L) = \{0\}$.

Suppose that u_1, u_2 in $W_0^3[0,1]$ are two particular solutions of the homogeneous differential equation of (1)

$$a(x)u'' + b(x)u' + c(x)u = 0, \quad 0 < x < 1$$

such that the Wronskian

$$W(x) = \begin{vmatrix} u_1(x) & u_2(x) \\ u_1'(x) & u_2'(x) \end{vmatrix} = u_1(x)u_2'(x) - u_1'(x)u_2(x) \neq 0.$$

It follows from Abel's formula for the Wronskian that there exists a constant C such that

$$W(x) = C\mathrm{e}^{-m(x)} \neq 0, \tag{14}$$

where $m(x)$ is a particular solution of $m'(x) = \frac{b(x)}{a(x)}$.

As $|a| > a_0 > 0$ on $[0,1]$ and $b \in W^1[0,1]$, then $\frac{b(x)}{a(x)}$ is continuous on $[0,1]$. Thus the function $m(x) = \int_0^x \frac{b(t)}{a(t)}\mathrm{d}t$ is continuous on $[0,1]$. This implies that there exists a constant m_0 such that $m(x) \leq m_0$ for all $x \in [0,1]$. Consequently, it follows from (14) that there exists a positive constant C_0 such that

$$|W(x)| = \left|C\mathrm{e}^{-m(x)}\right| \geq \left|C\mathrm{e}^{-m_0}\right| = C_0. \tag{15}$$

Let

$$u(x) = c_1(x)u_1(x) + c_2(x)u_2(x) = -u_1(x)\int_0^x \frac{f(t)u_2(t)}{a(t)W(t)}\mathrm{d}t + u_2(x)\int_0^x \frac{f(t)u_1(t)}{a(t)W(t)}\mathrm{d}t \tag{16}$$

be the general solution of the non-homogeneous differential equation (1)–(2) (see from [3]).

Moreover, one can check from (16) that

$$u'(x) = -u_1'(x)\int_0^x \frac{f(t)u_2(t)}{a(t)W(t)}\mathrm{d}t + u_2'(x)\int_0^x \frac{f(t)u_1(t)}{a(t)W(t)}\mathrm{d}t, \tag{17}$$

$$u''(x) = -u_1''(x)\int_0^x \frac{f(t)u_2(t)}{a(t)W(t)}\mathrm{d}t + u_2''(x)\int_0^x \frac{f(t)u_1(t)}{a(t)W(t)}\mathrm{d}t + \frac{f(x)}{a(x)} \tag{18}$$

$$\begin{aligned} u'''(x) = &-u_1'''(x)\int_0^x \frac{f(t)u_2(t)}{a(t)W(t)}\mathrm{d}t + u_2'''(x)\int_0^x \frac{f(t)u_1(t)}{a(t)W(t)}\mathrm{d}t - u_1''(x)\frac{f(x)u_2(x)}{a(x)W(x)} \\ &+ u_2''(x)\frac{f(x)u_1(x)}{a(x)W(x)} + \frac{f'(x)a(x) - f(x)a'(x)}{a^2(x)}. \end{aligned} \tag{19}$$

Clearly, it follows from u_1, u_2 in $W_0^3[0,1]$, (16) and (17)–(19) that $u \in W_0^3[0,1]$. Moreover, it follows from conditions $|a| > a_0 > 0$ on $[0,1]$ and (15) that

$$\left\|\frac{1}{a}\right\|_{W^1} < \frac{1}{a_0}, \quad \left\|\frac{1}{aW}\right\|_\infty < \frac{1}{a_0C_0}, \quad \left\|\frac{1}{aW}\right\|_{L^2([0,1])} < \frac{1}{a_0C_0}. \tag{20}$$

From (16), (3), (17)–(19), (10)–(13), (20), the Cauchy-Schwarz inequality, f, a in $W^1[0,1]$ and u_1, u_2 in $W_0^3[0,1]$, we have

$$\begin{aligned}
\|u\|_{W_0^3}^2 &= |u''(0)|^2 + \int_0^1 [u'''(x)]^2 \mathrm{d}x \\
&= \left[\frac{f(0)}{a(0)}\right]^2 + \int_0^1 [c_1'(x)u_1''(x) + c_2'(x)u_2''(x) + c_1(x)u_1'''(x) + c_2(x)u_2'''(x) + \left[\left(\frac{f(x)}{a(x)}\right)'\right]^2]\mathrm{d}x \\
&\le \left[\frac{f(0)}{a(0)}\right]^2 + 5\int_0^1 [c_1'(x)u_1''(x)]^2 \mathrm{d}x + 5\int_0^1 [c_2'(x)u_2''(x)]^2 \mathrm{d}x + 5\int_0^1 [c_1(x)u_1'''(x)]^2 \mathrm{d}x \\
&+ 5\int_0^1 [c_2(x)u_2'''(x)]^2 \mathrm{d}x + 5\int_0^1 \left[\left(\frac{f(x)}{a(x)}\right)'\right]^2]\mathrm{d}x \\
&\le 5\left\|\frac{f}{a}\right\|_{W^1}^2 + 5\int_0^1 [c_1'(x)]^2 \mathrm{d}x \|u_1''\|_\infty^2 + 5\int_0^1 [c_2'(x)]^2 \mathrm{d}x \|u_2''\|_\infty^2 + 5\int_0^1 [c_1(x)]^2 \mathrm{d}x \|u_1'''\|_{L^2([0,1])}^2 \\
&+ 5\int_0^1 [c_2(x)]^2 \mathrm{d}x \|u_2'''\|_{L^2([0,1])}^2 \\
&\le 5\left\|\frac{f}{a}\right\|_{W^1}^2 + 5\int_0^1 \left[\frac{f(x)u_2(x)}{a(x)W(x)}\right]^2 \mathrm{d}x \|u_1''\|_\infty^2 + 5\int_0^1 \left[\frac{f(x)u_1(x)}{a(x)W(x)}\right]^2 \mathrm{d}x \|u_2''\|_\infty^2 \\
&+ 5\int_0^1 \left[\int_{x_0}^x \frac{f(t)u_2(t)}{a(t)W(t)}\mathrm{d}t\right]^2 \mathrm{d}x \|u_1'''\|_{L^2([0,1])}^2 + 5\int_0^1 \left[\int_{x_0}^x \frac{f(t)u_1(t)}{a(t)W(t)}\mathrm{d}t\right]^2 \mathrm{d}x \|u_2'''\|_{L^2([0,1])}^2 \\
&\le 5\left\|\frac{f}{a}\right\|_{W^1}^2 + 5\|u_1''\|_\infty^2 \left\|\frac{u_2(x)}{a(x)W(x)}\right\|_\infty^2 \int_0^1 [f(x)]^2 \mathrm{d}x + 5\|u_2''\|_\infty^2 \left\|\frac{u_1(x)}{a(x)W(x)}\right\|_\infty^2 \int_0^1 [f(x)]^2 \mathrm{d}x \\
&+ 5\int_0^1 \left[\int_0^1 \frac{f(t)u_2(t)}{a(t)W(t)}\mathrm{d}t\right]^2 \mathrm{d}x \|u_1'''\|_{L^2([0,1])} + 5\int_0^1 \left[\int_0^1 \frac{f(t)u_1(t)}{a(t)W(t)}\mathrm{d}t\right]^2 \mathrm{d}x \|u_2'''\|_{L^2([0,1])} \\
&\le 5\left\|\frac{f}{a}\right\|_{W^1}^2 + 5\|u_1''\|_\infty^2 \left\|\frac{u_2}{aW}\right\|_\infty^2 \|f\|_{L^2([0,1])}^2 + 5\|u_2''\|_\infty^2 \left\|\frac{u_1}{aW}\right\|_\infty^2 \|f\|_{L^2([0,1])}^2 \\
&+ 5\int_0^1 \int_0^1 [f(t)]^2 \mathrm{d}t \int_0^1 \left[\frac{f(t)u_2(t)}{a(t)W(t)}\right]^2 \mathrm{d}t\mathrm{d}x \|u_1'''\|_{L^2([0,1])} \\
&+ 5\int_0^1 \int_0^1 [f(t)]^2 \mathrm{d}t \left[\int_0^1 \frac{f(t)u_1(t)}{a(t)W(t)}\mathrm{d}t\right]^2 \mathrm{d}x \|u_2'''\|_{L^2([0,1])} \\
&\le 5\left\|\frac{f}{a}\right\|_{W^1}^2 + 5\|u_1''\|_\infty^2 \left\|\frac{u_2}{aW}\right\|_\infty^2 \|f\|_{L^2([0,1])}^2 + 5\|u_2''\|_\infty^2 \left\|\frac{u_1}{aW}\right\|_\infty^2 \|f\|_{L^2([0,1])}^2 \\
&+ 5\|f\|_{L^2([0,1])}^2 \left\|\frac{fu_2}{aW}\right\|_{L^2([0,1])}^2 \|u_1'''\|_{L^2([0,1])} + 5\|f\|_{L^2([0,1])}^2 \left\|\frac{fu_1}{aW}\right\|_{L^2([0,1])}^2 \|u_2'''\|_{L^2([0,1])} \\
&\le 5\Big[\left\|\frac{1}{a}\right\|_{W^1}^2 + 2\|u_1''\|_\infty^2 \left\|\frac{u_2}{aW}\right\|_\infty^2 + 2\|u_2''\|_\infty^2 \left\|\frac{u_1}{aW}\right\|_\infty^2 + 2\left\|\frac{fu_2}{aW}\right\|_{L^2([0,1])}^2 \|u_1'''\|_{L^2([0,1])} \\
&+ 2\left\|\frac{fu_1}{aW}\right\|_{L^2([0,1])}^2 \|u_2'''\|_{L^2([0,1])}\Big] \|f\|_{W^1}^2 \le M^2 \|f\|_{W^1}^2.
\end{aligned} \tag{21}$$

This implies that $\|u\|_{W_0^3} = \|L^{-1}f\|_{W_0^3} \le M\|f\|_{W^1}$. Thus L has a bounded inverse L^{-1}.

3 Method for Solutions and Error Estimate

To apply weak N-best POAFD to approximate the solution u of $Lu = f$ (1)–(2), we need to construct a dictionary of $W_0^3[0,1]$.

Lemma 2. *Denote by* $L^* : W^1[0,1] \to W_0^3[0,1]$ *the adjoint operator of* L. *Let* K *be reproducing kernel of* $W^1[0,1]$. *Define* $\mathbf{h}(x) = \phi_x = L^* K_x$, $\psi(x) = \|\mathbf{h}_x\|_{W_0^3}$. *Then*
(i) for any $x \in [0,1]$, *any* $f \in W^1[0,1]$,

$$f(x) = Lu(x) = \langle u, \phi_x \rangle_{W_0^3}; \tag{22}$$

(ii) the function $\mathbf{h}(x)$, $\psi(x)$ *are continuous on* $[0,1]$;
(iii) the function $\mathbf{h}(x) \neq 0$, $\psi(x) > 0$ *for all* $x \in [0,1]$.

Proof. By reproducing property of $K, Lu = f$ and the adjoint property of L^*, we have

$$f(x) = \langle f, K_x\rangle_{W^1} = \langle Lu, K_x\rangle_{W^1} = \langle u, L^* K_x\rangle_{W_0^3} = \langle u, \phi_x\rangle_{W_0^3}.$$

We see that this equation establish the relation between the f and u by the inner product with L^*K_x. Since $K(x,y)$ is a continuous function on $[0,1]^2$, so for any $x_0 \in [0,1]$, we have

$$\begin{aligned}\|K_x - K_{x_0}\|_{W^1}^2 =& \langle K_x - K_{x_0}, K_x - K_{x_0}\rangle_{W^1} = \langle K_x, K_x\rangle_{W^1} + \langle K_{x_0}, K_{x_0}\rangle_{W^1} - 2\langle K_x - K_{x_0}\rangle_{W^1}\\ =& K(x,x) + K(x_0,x_0) - 2K(x,x_0) \to 0, \quad \text{as} x \to x_0.\end{aligned}$$

It follows from L and L^* are continuous operators on $W^1[0,1]$ and $W_0^3[0,1]$ respectively that $\lim_{x\to x_0} \phi_x = \lim_{x\to x_0} L^*K_x = L^*K_{x_0} = \phi_{x_0}$. The continuity of ψ is followed from that of $\mathbf{h}$ on $[0,1]$.

Moreover, since $K_x \neq 0$ for any $x \in E$. Therefore $\phi_x = L^*K_x \neq 0$, and also $\psi = \|\phi_x\|_{W_0^3} > 0$.

3.1 The Dictionary of Weak N-Best POAFD

Theorem 3. *Let* $X = \{x_i\}_{i\geq 1}$ *be a sequence in* $[0,1]$, *and let* $\phi_i = L^*K_{x_i}$ *for all* $i \geq 1$. *If the sequence* X *is dense in* $[0,1]$, *then the linear span of* $\{\phi_i\}_{i\geq 1}$ *is dense in the Hilbert space* $W_0^3[0,1]$. *If* X *consists of distinct points, then the set* $\{\phi_i\}_{i\geq 1}$ *is linearly independent in* $W_0^3[0,1]$.

Remark 2. Theorem 3 shows that $\{\phi_i\}_{i\geq 1}$ can be considered as a dictionary $\mathcal{D}$ in $W_0^3[0,1]$, which is the basis of applying weak N-best POAFD to approximate the solution of Eq. 16 (1)–(2).

We now present a simplified description of $\phi_x = L^*K_x$. We consider the reproducing kernel $K(x,y)$ as a function of two variables (x,y) on the unit square $[0,1]^2$. Then $L^{(1)}$ and $L^{(2)}$ are operators on the space of functions of 2 variables defined on $[0,1]^2$, in which L acts the function $K(x,y)$ on the first and second variable respectively. It follows from the symmetric property $K(x,y) = K(y,x)$ on $[0,1]^2$ that $L^{(1)}K(x,y) = L^{(2)}K(y,x)$ for all $(x,y) \in [0,1]^2$. With this set-up, we have the following two important relations.

Lemma 3. *For any* $x \in [0,1]$, *define* $\mathbf{h}(x) = \phi_x = L^*K_x \in W_0^3[0,1]$. *Then*

$$\mathbf{h}(a) = \phi_a(x) = (LR_x)(a), \quad \textit{for any} \;\; a, x \in [0,1], \tag{23}$$

$$\langle \phi_a, \phi_b\rangle_{W_0^3} = (L^{(1)}L^{(2)}R)(a,b), \quad \textit{for any} \;\; a, b \in [0,1]. \tag{24}$$

Proof. From reproducing property of K and symmetric property of K on $[0,1]^2$, we have

$$\begin{aligned}\phi_b(x) =& L^*K_b(x) = \langle L^*K_b, R_x\rangle_{W_0^3} = \langle K_b, LR_x\rangle_{W^1} = LR_x(b)\\ =& L^{(1)}R(b,x) = L^{(2)}R(x,b);\\ L\phi_b(x) =& (L^{(1)})(L^{(2)}R)(x,b) = (L^{(1)}L^{(2)})R(x,b)\\ \langle \phi_a, \phi_b\rangle_{W_0^3} =& \langle L^*K_a, \phi_b\rangle_{W_0^3} = \langle K_a, L\phi_b\rangle_{W^1} = L\phi_b(a) = (L^{(1)}L^{(2)}R)(a,b)\end{aligned}$$

for any $a, b, x \in [0,1]$.

Remark 3. The formula (24) is used to express $\langle u, B_n^a\rangle_{W_0^3}$ in (31) and (33) later.

Below, we describe how weak N-best POAFD can be used to solve IVP (1)–(2). To characterize weak N-best POAFD, we first outline the structure of weak 2-best POAFD.

3.2 Algorithm Implementation

The key idea of weak 2-best POAFD is to choose distinct pairs of points $\{a_i, b_i\}_{i\geq 1}$ in $[0,1]^2$ successively by weak 2-best maximal selection principle in Lemma 5 on minimizing the norm of the residue terms step by step. Then the solution of (1)–(2) can be expressed as a specified linear combination of $\{\phi_{a_i}, \phi_{b_i}\}_{i\geq 1}$.

For any 2 different points $a, b \in [0,1]$, we have a pair of elements (ϕ_a, ϕ_b) in dictionary $\mathcal{D}$. From Theorem 3, ϕ_a and ϕ_b are linearly independent. Denote

$$E_1(b) = \phi_b - \langle \phi_b, B_1^a\rangle_{W_0^3} B_1^a \neq 0, \tag{25}$$

by Gram-Smith orthogonalization process (G-S), we obtain a pair of unit vector (B_1^a, B_1^b), where

$$B_1^a = \frac{\phi_a}{\|\phi_a\|_{W_0^3}} = \beta_{11}^a\phi_a,\ B_1^b = \frac{E_1(b)}{\|E_1(b)\|_{W_0^3}} = \frac{\phi_b - \langle \phi_b, B_1^a\rangle_{W_0^3} B_1^a}{\|\phi_b - \langle \phi_b, B_1^a\rangle_{W_0^3} B_1^a\|_{W_0^3}} = \beta_{21}^b\phi_a + \beta_{22}^b\phi_b, \tag{26}$$

and $\langle B_1^a, B_1^b\rangle_{W_0^3} = 0$.

Define $u = u_1$ and the residual $u_2 = u_1 - \langle u_1, B_1^a\rangle_{W_0^3} B_1^a - \langle u_1, B_1^b\rangle_{W_0^3} B_1^b$.

Let $C_1(a,b) = |\langle u_1, B_1^a\rangle_{W_0^3}|^2 + |\langle u_1, B_1^b\rangle_{W_0^3}|^2$. By (22) and (26), we have

$$\begin{aligned}C_1(a,b) =& \left|\left\langle u_1, \frac{\phi_a}{\|\phi_a\|_{W_0^3}}\right\rangle_{W_0^3}\right|^2 + \left|\left\langle u_1, \frac{E_1(b)}{\|E_1(b)\|_{W_0^3}}\right\rangle_{W_0^3}\right|^2 = |\langle u_1, \beta_{11}^a\phi_a\rangle|^2 + |\langle u_1, \beta_{21}^b\phi_a + \beta_{22}^b\phi_b|^2\\ =& [\beta_{11}^a f(a)]^2 + [\beta_{21}^b f(a) + \beta_{22}^b f(b)]^2.\end{aligned} \tag{27}$$

From (27), we see that $C_1(a,b)$ can be express by the known value $f(a)$, $f(b)$, ϕ_a and $\langle \phi_b, B_1^a\rangle_{W_0^3}$. Then we do the first decomposition.

Lemma 4. *(i) For any different points a, $b \in [0,1]$, the vector $\|B_1^a\|_{W_0^3} = \|B_1^b\|_{W_0^3} = 1$ are orthonormal vectors, and $\langle u_2, B_1^a\rangle_{W_0^3} = \langle u_2, B_1^b\rangle_{W_0^3} = 0$.*

(ii) $u_1 = u_2 + \langle u_1, B_1^a\rangle_{W_0^3} B_1^a + \langle u_1, B_1^b\rangle_{W_0^3} B_1^b$, *and*

$$\|u_1\|_{W_0^3}^2 = \|u_2\|_{W_0^3}^2 + |\langle u_1, B_1^a\rangle_{W_0^3}|^2 + |\langle u_1, B_1^b\rangle_{W_0^3}|^2.$$

(iii) Let $\rho \in (0,1)$ *be a constant. Define* $C_1(a,b) = |\langle u, B_1^a\rangle_{W_0^3}|^2 + |\langle u, B_1^b\rangle_{W_0^3}|^2$. *Then there exists distinct points* $a_1, b_1 \in [0,1]^2$ *such that*

$$|\langle u, B_{a_1}\rangle_{W_0^3}|^2 + |\langle u, B_{b_1}\rangle_{W_0^3}|^2 \geq \rho \sup\{\langle u, B_1^a\rangle|_{W_0^3}^2 + |\langle u, B_1^b\rangle|_{W_0^3}^2 \mid a, b \in [0,1]^2\}. \quad (28)$$

Proof. (i) It follows from $\langle B_1^a, B_1^b\rangle_{W_0^3} = 0$ that

$$\langle u_2, B_1^a\rangle_{W_0^3} = \langle u_1 - \langle u_1, B_1^a\rangle_{W_0^3} B_1^a - \langle u_1, B_1^b\rangle_{W_0^3} B_1^b, B_1^a\rangle_{W_0^3} = \langle u_1, B_1^a\rangle_{W_0^3} - \langle u_1, B_1^a\rangle_{W_0^3} = 0,$$
$$\langle u_2, B_1^b\rangle_{W_0^3} = 0.$$

(ii) From (i), it is easy to check that $\|u_1\|_{W_0^3}^2 = \|u_2\|_{W_0^3}^2 + |\langle u_1, B_1^a\rangle_{W_0^3}|^2 + |\langle u_1, B_1^b\rangle_{W_0^3}|^2$.

(iii) It follows from Lemma 2 (ii) and (iii) that ϕ_x is continuous on $[0,1]$ and $\|\phi_x\| \neq 0$. Moreover, from the Cauchy-Schwarz inequality, we have $\lim_{x\to x_0} \langle u, \phi_x\rangle = \langle u, \phi_{x_0}\rangle$.

On the other hand, for distinct points $b \neq a$, we notice that $E_1(b) \neq 0$ for any $b \in [0,1] \setminus \{a\}$ from Theorem 3. By definition of $E_1(b)$ in (25), the triangle inequality and the Cauchy-Schwarz inequality, we have

$$\begin{aligned}
&\|E_1(x) - E_1(x_0)\|_{W_0^3}^2 = \|\phi_x - \langle \phi_x, B_1^a\rangle_{W_0^3} B_1^a - (\phi_{x_0} - \langle \phi_{x_0}, B_1^a\rangle_{W_0^3} B_1^a)\|_{W_0^3}^2 \\
=&\|(\phi_x - \phi_{x_0}) - \langle(\phi_x - \phi_{x_0}), B_1^a\rangle_{W_0^3} B_1^a\|^2 = \|\phi_x - \phi_{x_0}\|_{W_0^3}^2 - |\langle(\phi_x - \phi_{x_0}), B_1^a\rangle_{W_0^3}|^2 \\
\leq&\|\phi_x - \phi_{x_0}\|_{W_0^3}^2.
\end{aligned}$$

By Lemma 2, then $\lim_{x\to x_0} E_1(x) = E_1(x_0)$ and $\lim_{x\to x_0} \|E_1(x)\|_{W_0^3} = \|E_1(x_0)\|_{W_0^3} > 0$. Thus, $E_1(b)$ is continuous.

Moreover, from the Cauchy-Schwarz inequality, we get

$$|\langle u_1, E_1(x) - E_1(x_0)\rangle_{W_0^3}| \leq \|u_1\|_{W_0^3} \|E_1(x) - E_1(x_0)\|_{W_0^3}, \quad (29)$$

which implies $\lim_{x\to x_0} \langle u_1, E_1(x)\rangle_{W_0^3} = \langle u_1, E_1(x_0)\rangle_{W_0^3}$. And consequently, $\lim_{x\to x_0} \langle u_1, B_1^x\rangle = \langle u_1, B_1^{x_0}\rangle$. This means $\langle u_1, B_1^b\rangle_{W_0^3}$ is continuous on $[0,1] \setminus \{a\}$, and $[0,1] \setminus \{a\}$ is a union of two disjoint sub-intervals of $[0,1]$. From intermediate value theorem, we see that the range of $\langle u_1, B_1^b\rangle_{W_0^3}$ on $[0,1] \setminus \{a\}$ is also a union of two disjoint bounded intervals, and that $\sup\{\ |\langle u_1, B_1^b\rangle_{W_0^3}|\ \big| b \in [0,1] \setminus \{a\}\ \}$ is finite.

In addition, since $u_1 \neq 0$ and L is bounded inverse, then $L(u_1) = f \neq 0$, and hence $|L(u_1)(x)| = |f(x)| > 0$ for some $x \in [0,1]$. By (26), (22), we have

$$\begin{aligned} C_1(a,b) &= |\langle u_1, B_1^a\rangle_{W_0^3}|^2 + |\langle u_1, B_1^b\rangle_{W_0^3}|^2 = \frac{|\langle u,\phi_a\rangle_{W_0^3}|^2}{\|\phi_a\|^2_{W_0^3}} + \frac{|\langle u_1, E_1(b)\rangle_{W_0^3}|^2}{\|E_1(b)\|^2_{W_0^3}} \\ &= \frac{|f(a)|^2}{\|\phi_a\|^2_{W_0^3}} + \frac{|\langle u_1, \phi_b - \beta_{11}^a\langle\phi_b, B_1^a\rangle_{W_0^3}\phi_a\rangle_{W_0^3}|^2}{\|E_1(b)\|^2_{W_0^3}} \\ &= \frac{|f(a)|^2}{\|\phi_a\|^2_{W_0^3}} + \frac{|\langle u_1, L^*K_b - \beta_{11}^a\langle\phi_b, B_1^a\rangle_{W_0^3}L^*K_a\rangle_{W_0^3}|^2}{\|E_1(b)\|^2_{W_0^3}} \\ &\geq \frac{|f(a)|^2}{\|\phi_a\|^2_{W_0^3}} + \frac{||f(b)| - |\beta_{11}^a\langle\phi_b, B_1^a\rangle_{W_0^3}f(a)||^2}{\|E_1(b)\|^2_{W_0^3}} > 0. \end{aligned}$$

Recall that $0 < \rho < 1$, so for $a, b \in [0,1]$ $(a < b)$,

$$0 < \rho \sup\{\, |\langle u_1, B_1^a\rangle_{W_0^3}|^2 + |\langle u_1, B_1^b\rangle_{W_0^3}|^2\} < \sup\{\, |\langle u_1, B_1^a\rangle_{W_0^3}|^2 + |\langle u_1, B_1^b\rangle_{W_0^3}|^2 \,\}.$$

It follows from definition of supremum, there exists 2 different points $a_1, b_1 \in [0,1]^2$ such that

$$|\langle u, B_{a_1}\rangle_{W_0^3}|^2 + |\langle u, B_{b_1}\rangle_{W_0^3}|^2 \geq \rho \sup\{\langle|\langle u, B_1^a\rangle_{W_0^3}|^2 + |\langle u, B_1^b\rangle_{W_0^3}|^2 | a, b \in [0,1]^2\}.$$

This completes the proof.

After successive $n-1$ steps with $n \geq 2$, suppose that we have chosen

(i) the sequence $(a_1, b_1, a_2, b_2, \cdots, a_{n-1}, b_{n-1})$ of $2(n-1)$ distinct points.
(ii) $2(n-1)$ orthonormal vectors

$$B_{a_1}, B_{b_1}, \cdots, B_{a_{n-1}}, B_{b_{n-1}} \in W_0^3[0,1]$$

obtained by applying Gram-Schmidt orthonormalization to $\phi_{a_1}, \phi_{b_1}, \cdots, \phi_{a_{n-1}}, \phi_{b_{n-1}}$.
Define

$$O_i(a_i) = \phi_{a_i} - \sum_{k=1}^{i-1}(\langle\phi_{a_i}, B_{a_k}\rangle_{W_0^3}B_{a_k} + \langle\phi_{a_i}, B_{b_k}\rangle_{W_0^3}B_{b_k}), \tag{30}$$

then

$$\begin{aligned} B_{a_i} &= \frac{O_i(a_i)}{\|O_i(a_i)\|_{W_0^3}} = \frac{\phi_{a_i} - \sum\limits_{k=1}^{i-1}(\langle\phi_{a_i}, B_{a_k}\rangle_{W_0^3}B_{a_k} + \langle\phi_{a_i}, B_{b_k}\rangle_{W_0^3}B_{b_k})}{\|\phi_{a_i} - \sum\limits_{k=1}^{i-1}(\langle\phi_{a_i}, B_{a_k}\rangle_{W_0^3}B_{a_k} + \langle\phi_{a_i}, B_{b_k}\rangle_{W_0^3}B_{b_k})\|_{W_0^3}} \\ &= \sum_{k=1}^{i-1}\left(\beta_{ik}^{(a_i,a_k)}\phi_{a_k} + \beta_{ik}^{(a_i,b_k)}\phi_{b_k}\right) + \beta_{ii}^{(a_i,a_i)}\phi_{a_i}, \end{aligned} \tag{31}$$

and

$$E_i(b_i) = \phi_{b_i} - \sum_{k=1}^{i-1}(\langle\phi_{b_i}, B_{a_k}\rangle_{W_0^3} B_{a_k} + \langle\phi_{b_i}, B_{b_k}\rangle_{W_0^3} B_{b_k}) - \langle\phi_{b_i}, B_{a_i}\rangle_{W_0^3} B_{a_i}, \tag{32}$$

$$\begin{aligned} B_{b_i} &= \frac{E_i(b_i)}{\|E_i(b_i)\|_{W_0^3}} = \frac{\phi_{b_i} - \sum\limits_{k=1}^{i-1}(\langle\phi_{b_i}, B_{a_k}\rangle_{W_0^3} B_{a_k} + \langle\phi_{b_i}, B_{b_k}\rangle_{W_0^3} B_{b_k}) - \langle\phi_{b_i}, B_{a_i}\rangle_{W_0^3} B_{a_i}}{\|\phi_{b_i} - \sum\limits_{k=1}^{i-1}(\langle\phi_{b_i}, B_{a_k}\rangle_{W_0^3} B_{a_k} + \langle\phi_{b_i}, B_{b_k}\rangle_{W_0^3} B_{b_k}) - \langle\phi_{b_i}, B_{a_i}\rangle_{W_0^3} B_{a_i}\|_{W_0^3}} \\ &= \sum_{k=1}^{i}(\beta_{ik}^{(b_i,a_k)}\phi_{a_k} + \beta_{ik}^{(b_i,b_k)}\phi_{b_k}). \end{aligned} \tag{33}$$

Thus the sequence $\{B_{a_1}, B_{b_1}, \cdots, B_{a_{n-1}}, B_{b_{n-1}}\}$ is orthonormal set.

(iii) For all $k = 1, \cdots, n-1$, $u_1 = u$, we have

$$\begin{aligned} u &= \sum_{i=1}^{k}(\langle u_i, B_{a_i}\rangle_{W_0^3} B_{a_i} + \langle u_i, B_{b_i}\rangle_{W_0^3} B_{b_i}) + u_{k+1}, \\ u_{k+1} &= u - \sum_{i=1}^{k}(\langle u_i, B_{a_i}\rangle_{W_0^3} B_{a_i} + \langle u_i, B_{b_i}\rangle_{W_0^3} B_{b_i}), \end{aligned} \tag{34}$$

where u_{k+1} is the *k–th residual* of u at the k–th step. □

For all $1 \leq k \leq n-1$, from (34) and (ii), we have

$$\begin{aligned} \langle u_k, B_{a_k}\rangle_{W_0^3} &= \langle u - \sum_{i=1}^{k-1}(\langle u_i, B_{a_i}\rangle_{W_0^3} B_{a_i} + \langle u_i, B_{b_i}\rangle_{W_0^3} B_{a_i}), B_{a_k}\rangle_{W_0^3} = \langle u, B_{a_k}\rangle_{W_0^3}, \\ \langle u_k, B_{b_k}\rangle_{W_0^3} &= \langle u, B_{b_k}\rangle_{W_0^3}. \end{aligned} \tag{35}$$

And for $1 \leq k \leq j-1$,

$$\begin{aligned} \langle u_j, B_{a_k}\rangle_{W_0^3} &= \langle u - \sum_{i=1}^{j-1}(\langle u_i, B_{a_i}\rangle_{W_0^3} B_{a_i} + \langle u_i, B_{b_i}\rangle_{W_0^3} B_{a_i}), B_{a_k}\rangle = \langle u, B_{a_k}\rangle_{W_0^3} - \langle u_k, B_{a_k}\rangle_{W_0^3} = 0, \\ \langle u_j, B_{b_k}\rangle_{W_0^3} &= 0. \end{aligned} \tag{36}$$

Thus

$$\begin{aligned} &\langle u_i, B_{a_i}\rangle_{W_0^3} B_{a_i} + \langle u_i, B_{b_i}\rangle_{W_0^3} B_{b_i} = \langle u, B_{a_i}\rangle_{W_0^3} B_{a_i} + \langle u, B_{b_i}\rangle_{W_0^3} B_{b_i}, \\ &C_i(a,b) = |\langle u_i, B_{a_i}\rangle_{W_0^3}|^2 + |\langle u_i, B_{b_i}\rangle_{W_0^3}|^2 = |\langle u, B_{a_i}\rangle_{W_0^3}|^2 + |\langle u, B_{b_i}\rangle_{W_0^3}|^2. \end{aligned}$$

If $u_n = 0$, stop the decomposition; otherwise, we continue to work out the n–th decomposition of u. That is, we shall choose $P_n \in [0,1]$ by the following strategy.

Denote by $I_n := [0,1] \setminus \{a_i, b_i\}_{i=1}^{n-1}$ the punctured interval. For any fixed points $a, b \in I_n$ $(a \neq b)$, we get a new point sequence $(a_1, b_1, \cdots, a_{n-1}, b_{n-1}, a, b)$ of distinct points. Then from Theorem 3, the new vector

$$(\phi_{a_1}, \phi_{b_1}, \cdots, \phi_{a_{n-1}}, \phi_{b_{n-1}}, \phi_a, \phi_b)$$

is linearly independent. We can obtain the orthonormal set $(B_{a_1}, B_{b_1}, \cdots, B_{a_{n-1}}, B_{b_{n-1}}, B_n^a, B_n^b)$ by apply Gram-Schmidt orthonormalization to $(\phi_{a_1}, \phi_{b_1}, \cdots, \phi_{a_{n-1}}, \phi_{b_{n-1}}, \phi_a, \phi_b)$.

Denote the pre-orthogonalization at n–th step.

$$O_n(a) = \phi_a - \sum_{j=1}^{n-1}(\langle\phi_a, B_{a_j}\rangle_{W_0^3} B_{a_j} + \langle\phi_a, B_{b_j}\rangle_{W_0^3} B_{b_j}) \neq 0, \tag{37}$$

$$B_n^a = \frac{O_n(a)}{\|O_n(a)\|_{W_0^3}}, \tag{38}$$

$$E_n(b) = \phi_b - \sum_{j=1}^{n-1}(\langle\phi_b, B_{a_j}\rangle_{W_0^3} B_{a_j} + \langle\phi_a, B_{b_j}\rangle_{W_0^3} B_{b_j}) - \langle\phi_b, B_n^a\rangle_{W_0^3} B_n^a \neq 0, \tag{39}$$

$$B_n^b = \frac{E_n(b)}{\|E_n(b)\|_{W_0^3}}. \tag{40}$$

Then from (37)– (40) and (36), we have

$$\begin{aligned} C_n(a,b) &= |\langle u_n, B_n^a\rangle_{W_0^3}|^2 + |\langle u_n, B_n^b\rangle_{W_0^3}|^2 = \frac{|\langle u_n, O_n(a)\rangle_{W_0^3}|^2}{\|O_n(a)\|_{W_0^3}^2} + \frac{|\langle u_n, E_n(b)\rangle_{W_0^3}|^2}{\|E_n(b)\|_{W_0^3}^2} \\ &= \frac{|\langle u_n, \phi_a - \sum_{j=1}^{n-1}(\langle\phi_a, B_{a_j}\rangle_{W_0^3} B_{a_j} + \langle\phi_a, B_{b_j}\rangle_{W_0^3} B_{b_j})\rangle_{W_0^3}|^2}{\|O_n(a)\|_{W_0^3}^2} \\ &+ \frac{|\langle u_n, \phi_b - \sum_{j=1}^{n-1}(\langle\phi_b, B_{a_j}\rangle_{W_0^3} B_{a_j} + \langle\phi_a, B_{b_j}\rangle_{W_0^3} B_{b_j}) - \langle\phi_b, B_n^a\rangle_{W_0^3} B_n^a\rangle_{W_0^3}|^2}{\|E_n(b)\|_{W_0^3}^2} \\ &= \frac{|\langle u_n, \phi_a\rangle|^2}{\|O_n(a)\|_{W_0^3}^2} + \frac{|\langle u_n, \phi_b - \langle\phi_b, B_n^a\rangle_{W_0^3} B_n^a\rangle_{W_0^3}|^2}{\|E_n(b)\|_{W_0^3}^2} \\ &= \frac{|Lu_n(a)|^2}{\|O_n(a)\|_{W_0^3}^2} + \frac{|Lu_n(b) - \langle\phi_b, B_n^a\rangle_{W_0^3}\langle u_n, B_n^a\rangle_{W_0^3}|^2}{\|E_n(b)\|_{W_0^3}^2} \geq \frac{|Lu_n(a)|^2}{\|O_n(a)\|_{W_0^3}^2} > 0. \end{aligned} \tag{41}$$

Then, we use the following weak 2-best maximal selection principle to establish the existence of the n–th pair of different parameters $a_n, b_n \in I_n$.

Lemma 5. (Weak 2-best Maximal Selection Principle). *For any $n \geq 2$, let $I_n := [0,1] \setminus \{a_i, b_i\}_{i=1}^{n-1}$ be the punctured interval. Suppose that $u_n \neq 0$, then for any 2 different points $a, b \in I_n$*
(i) $\sup\{C_n(a,b) = |\langle u_n, B_n^b\rangle_{W_0^3}|^2 + |\langle u_n, B_n^a\rangle_{W_0^3}|^2 \mid a, b \in I_n\}$ is finite.
(ii) For any fixed $0 < \rho < 1$, there exists 2 different points $a_n, b_n \in I_n$ such that

$$\begin{aligned} C_n(a_n, b_n) &= |\langle u, B_{a_n}\rangle_{W_0^3}|^2 + |\langle u, B_{b_n}\rangle_{W_0^3}|^2 \\ &\geq \rho \sup\{\, |\langle u, B_n^a\rangle_{W_0^3}|^2 + |\langle u, B_n^b\rangle_{W_0^3}|^2 \mid a, b \in [0,1] \setminus \{a_i, b_i\}_{i=1}^{n-1}\,\} > 0. \end{aligned} \tag{42}$$

Proof. (i) First, from definition of $C_n(a,b)$ in (41) and the Cauchy-Schwarz inequality, we have

$$0 \le C_n(a,b) = |\langle u, B_n^a\rangle_{W_0^3}|^2 + |\langle u, B_n^b\rangle_{W_0^3}|^2 \le \|u\|_{W_0^3}^2 \|B_n^a\|_{W_0^3}^2 + \|u_n\|_{W_0^3}^2 \|B_n^b\|_{W_0^3}^2 \le 2\|u\|_{W_0^3}^2,$$

which shows that C_n is bounded on I_n.

To ensure the finiteness of supremum of $C_n(a,b)$ on I_n, it needs to show that C_n is continuous on I_n. First, we notice that $O_n^a \neq 0$ and $E_n^b \neq 0$ for any $P=(a,b)\in I_n$ from Theorem 3. On one hand, by definition of $O_n(a)$ in (37), the triangle inequality and the Cauchy-Schwarz inequality, we have

$$\begin{aligned}\|O_n(a)-O_n(a_0)\|_{W_0^3}^2 =& \|(\phi_a-\phi_{a_0}) - \sum_{j=1}^{n-1}(\langle(\phi_a-\phi_{a_0}),B_{a_j}\rangle_{W_0^3}B_{a_j} + \langle(\phi_a-\phi_{a_0}),B_{b_j}\rangle_{W_0^3}B_{b_j})\|_{W_0^3}^2\\ =& \|(\phi_a-\phi_{a_0})\|_{W_0^3}^2 - \sum_{j=1}^{n-1}(|\langle(\phi_a-\phi_{a_0}),B_{a_j}\rangle_{W_0^3}|^2 + |\langle(\phi_a-\phi_{a_0}),B_{b_j}\rangle_{W_0^3}|^2)\\ \le& \|\phi_a-\phi_{a_0}\|_{W_0^3}^2.\end{aligned} \tag{43}$$

It follows from Lemma 2 (ii) and (43) $\lim_{a\to a_0} O_n(a) = O_n(a_0)$, and $\lim_{a\to a_0}\|O_n(a)\|^2 = \|O_n(a_0)\|^2 > 0$ as well as $E_n(b)$. Thus $O_n(a)$ and $E_n(b)$ are non-zero continuous on I_n.

In addition, by the Cauchy-Schwarz inequality, we have

$$|\langle u_n, O_n(a)-O_n(a_0)\rangle_{W_0^3}| \le \|u_n\|_{W_0^3}\|O_n(a)-O_n(a_0)\|_{W_0^3},$$

which implies $\lim_{a\to a_0}\langle u_n, O_n(a)\rangle_{W_0^3} = \langle u_n, O_n(a_0)\rangle_{W_0^3}$. Thus $\langle u, O_n(a)\rangle_{W_0^3}$ and $\langle u, E_n(b)\rangle_{W_0^3}$ are continuous on $[0,1]$. Noticing that $C_n(a,b) = \dfrac{|\langle u, O_n(a)\rangle_{W_0^3}|^2}{\|O_n(a)\|_{W_0^3}^2} + \dfrac{|\langle u, E_n(b)\rangle_{W_0^3}|^2}{\|E_n(b)\|_{W_0^3}^2}$, consequently, $\lim_{(a,b)\to(a_0,b_0)} C_n(a,b) = C_n(a_0,b_0)$. As C_n is continuous on I_n, and I_n is a finite union of disjoint sub-intervals of $[0,1]$. From intermediate value theorem, we see that the range of C_n on I_n is also a finite union of disjoint bounded intervals, and that

$$\sup\{\ |\langle u, B_n^a\rangle_{W_0^3}|^2 + |\langle u, B_n^b\rangle_{W_0^3}|^2 \mid a,b\in[0,1]\setminus\{a_i,b_i\}_{i=1}^{n-1}\}$$

is finite.

(ii) Since $u_n \neq 0$ and L is bounded inverse, then $L(u_n)\neq 0$, and hence $|L(u_n)(a)|>0$. From (41), we see that

$$C_n(a,b) \ge \frac{|Lu_n(a)|^2}{\|O_n(a)\|_{W_0^3}^2} > 0.$$

Recall that $0<\rho<1$, so

$$\begin{aligned}0 <& \rho\sup\{\ C_n(a,b) = |\langle u, B_n^a\rangle_{W_0^3}|^2 + |\langle u, B_n^b\rangle_{W_0^3}|^2 \mid a,b\in I_n\ \}\\ <& \sup\{\ C_n(a,b) = |\langle u, B_n^a\rangle_{W_0^3}|^2 + |\langle u, B_n^b\rangle_{W_0^3}|^2 \mid a,b\in I_n\ \}.\end{aligned}$$

It follows from definition of supremum, there exists 2 different points $a_n, b_n \in I_n$ such that

$$\begin{aligned} C_n(a_n, b_n) &= |\langle u, B_{a_n}\rangle_{W_0^3}|^2 + |\langle u, B_{b_n}\rangle_{W_0^3}|^2 \\ &\geq \rho \sup\left\{ \ |\langle u, B_n^a\rangle_{W_0^3}|^2 + |\langle u, B_n^b\rangle_{W_0^3}|^2 \ |a, b \in [0,1] \setminus \{a_i, b_i\}_{i=1}^{n-1} \right\} > 0, \end{aligned}$$

which is called the Weak N-best Maximal Selection Principle (42).

Thus we have the n–th orthonormal decomposition of u in terms of

$$(B_{a_1}, B_{b_1}, \cdots, B_{a_{n-1}}, B_{b_{n-1}}, B_{a_n}, B_{b_n})$$

and the n–th residual $u_{n+1} = u_n - \langle u, B_{a_n}\rangle_{W_0^3} B_{a_n} - \langle u, B_{b_n}\rangle_{W_0^3} B_{b_n}$:

$$u = \sum_{i=1}^{n-1}(\langle u, B_{a_i}\rangle_{W_0^3} B_{a_i} + \langle u, B_{b_i}\rangle_{W_0^3} B_{b_i}) + u_n = \sum_{i=1}^{n}(\langle u, B_{a_i}\rangle_{W_0^3} B_{a_i} + \langle u, B_{b_i}\rangle_{W_0^3} B_{b_i}) + u_{n+1}; \tag{44}$$

$$\|u_{n+1}\|_{W_0^3}^2 = \|u\|_{W_0^3}^2 - \sum_{i=1}^{n}(|\langle u, B_{a_i}\rangle_{W_0^3}|^2 + |\langle u, B_{b_i}\rangle_{W_0^3}|^2). \tag{45}$$

3.3 Convergence Analysis

Theorem 4. *Suppose that* $|a| > a_0 > 0$ *on* $[0,1]$*, then IVP* (1)–(2) *has a unique solution* $u \in W_0^3[0,1]$ *for any* $f \in W^1[0,1]$*. For the given dictionary* $\mathcal{D}$ *constructed in Theorem 3, select pairs of parameters* $\{a_i, b_i\}_{i\geq 1}$ *by weak 2-best maximal selection principle in* (42)*, meanwhile, the normalized dictionary elements* $\{B_{a_i}, B_{b_i}\}_{i\geq 1}$ *are obtained by applying Gram-Schmidt orthonormalization to* $\{\phi_{a_i}, \phi_{b_i}\}_{i\geq 1}$*. Then the solution* u *can be represented in* $W_0^3[0,1]$ *by*

$$u = \lim_{n\to\infty} S_n = \sum_{i=1}^{\infty}(\langle u, B_{a_i}\rangle_{W_0^3} B_{a_i} + \langle u, B_{b_i}\rangle_{W_0^3} B_{b_i}), \tag{46}$$

where n*–th partial sum* S_n *is defined by*

$$\begin{aligned} S_n &= \sum_{i=1}^{n}(\langle u, B_{a_i}\rangle_{W_0^3} B_{a_i} + \langle u, B_{b_i}\rangle_{W_0^3} B_{b_i}) \\ &= \sum_{i=1}^{n}\Big(\sum_{k=1}^{n-1}(\beta_{ik}^{(a_i,a_k)} f(a_k) + \beta_{ik}^{(a_i,b_k)} f(b_k) + \beta_{ii}^{a_i} f(a_i))\, B_{a_i} + \sum_{k=1}^{n}(\beta_{ik}^{(b_i,a_k)} f(a_k) + \beta_{ik}^{(b_i,b_k)} f(b_k))\, B_{b_i}\Big). \end{aligned} \tag{47}$$

Proof. It follows from Theorem 2 that L^{-1} exists, and hence the solution $u = L^{-1}f \in W_0^3[0,1]$. Let $\{a_i, b_i\}_{i\geq 1}$ be a sequence of distinct points in $[0,1]$ for the solution u satisfying the weak N-best maximal selection principle (42). It remains to prove that the solution of (46) can be represented by the series $u = \sum_{i\geq 1}(\langle u, B_{a_i}\rangle_{W_0^3} B_{a_i} + \langle u, B_{b_i}\rangle_{W_0^3} B_{b_i})$. The proof is rather long, we divide it into 3 parts.

(I) We first prove the series $\sum_{i\geq 1}(\langle u, B_{a_i}\rangle_{W_0^3} B_{a_i} + \langle u, B_{b_i}\rangle_{W_0^3} B_{b_i}) \in W_0^3[0,1]$. As the corresponding sequence $(B_{a_1}, B_{b_1}, \cdots, B_{a_i}, B_{b_i}, \cdots)$ is an orthonormal set in $W_0^3[0,1]$, so Bessel's inequality implies that

$$\sum_{i\geq 1} C_i(a_i, b_i) = \sum_{i\geq 1}(|\langle u, B_{a_i}\rangle_{W_0^3}|^2 + |\langle u, B_{b_i}\rangle_{W_0^3}|^2) \leq \|u\|_{W_0^3}^2, \tag{48}$$

and hence the series $\sum_{i\geq 1}(\langle u, B_{a_i}\rangle_{W_0^3} B_{a_i} + \langle u, B_{b_i}\rangle_{W_0^3} B_{b_i}) \in W_0^3$.

(II) We are going to prove that this series converges to the solution u of IVP (1)–(2). Suppose to the contrary that the residual

$$g = u - \sum_{i\geq 1}(\langle u, B_{a_i}\rangle_{W_0^3} B_{a_i} + \langle u, B_{b_i}\rangle_{W_0^3} B_{b_i}) \in W_0^3[0,1] \tag{49}$$

is non-zero. Then the weak 2-best maximal selection principle in Lemma 5 implies that the sequence $\{a_i, b_i\}_{i\geq 1}$ is infinite.

For any different $x \in [0,1]$ and $y \in [0,1] \setminus \{a_i, b_i\}_{i=1}^{N-1}$, we define two unit vectors

$$e_x = \frac{\phi_x}{\|\phi_x\|_{W_0^3}}, \tag{50}$$

$$e_y = \frac{\phi_y - \langle \phi_y, B_N^x\rangle_{W_0^3} B_N^x}{\|\phi_y - \langle \phi_y, B_N^x\rangle_{W_0^3} B_N^x\|_{W_0^3}}, \tag{51}$$

where $B_N^x = \frac{\phi_x - \sum_{j=1}^{n-1}(\langle \phi_x, B_{a_j}\rangle_{W_0^3} B_{a_j} + \langle \phi_x, B_{b_j}\rangle_{W_0^3} B_{b_j})}{\|\phi_x - \sum_{j=1}^{n-1}(\langle \phi_x, B_{a_j}\rangle_{W_0^3} B_{a_j} + \langle \phi_x, B_{b_j}\rangle_{W_0^3} B_{b_j})\|}$ defined in (38).

It follows from Lemma 2 (ii) that ϕ_x, $\|\phi_x\|_{W_0^3}$ are continuous on $[0,1]$. And $\|\phi_x\|_{W_0^3} \neq 0$ follows from Lemma 2 (iii). Thus e_x (50) is continuous on $[0,1]$.

On the other hand, it follows from Theorem 3 that $\|\phi_y - \langle \phi_y, B_N^x\rangle_{W_0^3} B_N^x\|_{W_0^3} \neq 0$. As ϕ_y and $\|\phi_y\|_{W_0^3}$ are continuous on $[0,1]$, then $\phi_y - \langle \phi_y, B_N^x\rangle_{W_0^3} B_N^x$ is continuous in $[0,1] \setminus \{a_i, b_i\}_{i=1}^{N-1}$. Since

$$\begin{aligned}\|\phi_y - \langle \phi_y, B_N^x\rangle_{W_0^3} B_N^x\|_{W_0^3}^2 &= \langle \phi_y, \phi_y\rangle_{W_0^3} + |\langle \phi_y, B_N^x\rangle_{W_0^3}|^2 - 2|\langle \phi_y, B_N^x\rangle_{W_0^3}|^2 \\ &= \|\phi_y\|_{W_0^3}^2 - |\langle \phi_y, B_N^x\rangle_{W_0^3}|^2,\end{aligned}$$

from continuity of ϕ_y and $\|\phi_y\|_{W_0^3}$, we see that $\|\phi_y - \langle \phi_y, B_N^x\rangle_{W_0^3} B_N^x\|_{W_0^3}$ is continuous in $[0,1] \setminus \{a_i, b_i\}_{i=1}^{N-1}$. Thus e_y in (51) is continuous in $[0,1] \setminus \{a_i, b_i\}_{i=1}^{N-1}$. By Cauchy-Schwarz inequality, we have

$$|\langle g, e_x\rangle_{W_0^3}| \leq \|g\|_{W_0^3} \|e_x\|_{W_0^3}, \tag{52}$$

$$|\langle g, e_y\rangle_{W_0^3}| \leq \|g\|_{W_0^3} \|e_y\|_{W_0^3}, \tag{53}$$

which imply that $|\langle g, e_x\rangle_{W_0^3}|$ is continuous on $[0,1]$, surely bounded and $|\langle g, e_y\rangle_{W_0^3}|$ is continuous in $[0,1] \setminus \{a_i, b_i\}_{i=1}^{N-1}$. As $[0,1] \setminus \{a_i, b_i\}_{i=1}^{N-1}$ is a finite union of disjoint sub-intervals of $[0,1]$, from intermediate value theorem, we see that the range of $|\langle g, e_y\rangle_{W_0^3}|$ in $[0,1]\setminus\{a_i,b_i\}_{i=1}^{N-1}$ is also a finite union of disjoint bounded intervals, and $\sup\{|\langle g, e_y\rangle_{W_0^3}| | y\in[0,1]\setminus\{a_i,b_i\}_{i=1}^{N-1}\}$ is finite.

Hence there exists a closed interval $[c,d]\subset[0,1]\setminus\{a_i,b_i\}_{i=1}^{N-1}$ with $c<d$ such that

$$\frac{3}{2}C_0=\min_{x,y\in[c,d]}|\langle g,e_x\rangle_{W_0^3}|^2+|\langle g,e_y\rangle_{W_0^3}|^2>0. \tag{54}$$

Recall that $u_N=u-\sum_{i=1}^{N-1}(\langle u,B_{a_i}\rangle_{W_0^3}B_{a_i}+\langle u,B_{b_i}\rangle_{W_0^3}B_{b_i})$ is the $(N-1)$–th residual of u. We are going to estimate $|\langle u_N,B_{a_N}\rangle_{W_0^3}|^2+|\langle u_N,B_{b_N}\rangle_{W_0^3}|^2$ in two different ways.

Firstly, it follows from (48) and $\rho, C_0>0$ that there exists a positive integer N_0 such that for all $N\geq N_0$, one has

$$|\langle u,B_{a_N}\rangle_{W_0^3}|^2+|\langle u_N,B_{b_N}\rangle_{W_0^3}|^2\leq\sum_{i=N}^{\infty}(|\langle u,B_{a_i}\rangle_{W_0^3}|^2+|\langle u,B_{b_i}\rangle_{W_0^3}|^2)<\frac{\rho C_0}{2} \tag{55}$$

Secondly, for any fixed $N\geq N_0$, we select different points $a,b\in[c,d]$. Considering another sequence $(a_1,b_1,\ldots,a_{N-1},b_{N-1},a,b)\in[0,1]^{2N}$ of distinct point, then $(\phi_{a_1},\phi_{b_1},\cdots,\phi_{a_{N-1}},\phi_{b_{N-1}},\phi_a,\phi_b)$ is linearly independent. Let $(B_{a_1},B_{b_1},\cdots,B_{a_{N-1}},B_{b_{N-1}},B_N^a,B_N^b)$ in (38) be the Gram-Schmidt orthonormalization of $(B_{a_1},B_{b_1},\cdots,B_{a_{N-1}},B_{b_{N-1}},\phi_a,\phi_b)$, where

$$\begin{aligned}B_N^a&=\frac{\phi_a-\sum\limits_{i=1}^{N-1}(\langle\phi_a,B_{a_i}\rangle_{W_0^3}B_{a_i}+\langle\phi_a,B_{b_i}\rangle_{W_0^3}B_{b_i})}{\|\phi_a-\sum\limits_{i=1}^{N-1}(\langle\phi_a,B_{a_i}\rangle_{W_0^3}B_{a_i}+\langle\phi_a,B_{b_i}\rangle_{W_0^3}B_{b_i})\|_{W_0^3}},\\ B_N^b&=\frac{\phi_b-\sum\limits_{i=1}^{N-1}(\langle\phi_b,B_{a_i}\rangle_{W_0^3}B_{a_i}+\langle\phi_b,B_{b_i}\rangle_{W_0^3}B_{b_i})-\langle\phi_b,B_N^a\rangle_{W_0^3}B_N^a}{\|\phi_b-\sum\limits_{i=1}^{N-1}(\langle\phi_b,B_{a_i}\rangle_{W_0^3}B_{a_i}+\langle\phi_b,B_{b_i}\rangle_{W_0^3}B_{b_i})-\langle\phi_b,B_N^a\rangle_{W_0^3}B_N^a\|_{W_0^3}}.\end{aligned} \tag{56}$$

Since a_N, b_N are selected according to the weak 2-best maximal selection principle (42) in Lemma 5, we have

$$\begin{aligned}|\langle u_N,B_{a_N}\rangle_{W_0^3}|^2+|\langle u_N,B_{b_N}\rangle_{W_0^3}|&\geq\rho\sup\left\{|\langle u_N,B_N^a\rangle_{W_0^3}|^2+|\langle u_N,B_N^b\rangle_{W_0^3}|^2\;\middle|\;a,b\in[c,d]\right\}\\&\geq\rho[|\langle u_N,B_N^a\rangle_{W_0^3}|^2+|\langle u_N,B_N^b\rangle_{W_0^3}|^2].\end{aligned} \tag{57}$$

In order to arrive at a contradiction, we can consider 2 unit vectors

$$e_a=\frac{\phi_a}{\|\phi_a\|_{W_0^3}},\quad e_b=\frac{\phi_b-\langle\phi_b,B_N^a\rangle_{W_0^3}B_N^a}{\|\phi_b-\langle\phi_b,B_N^a\rangle_{W_0^3}B_N^a\|_{W_0^3}} \tag{58}$$

in $W_0^3[0,1]$, where B_N^a defined in (56). And note that

$$\|\phi_a - \sum_{i=1}^{N-1}(\langle\phi_a, B_{a_i}\rangle_{W_0^3}B_{a_i} + \langle\phi_a, B_{b_i}\rangle_{W_0^3}B_{b_i})\|_{W_0^3}^2 = \|\phi_a\|_{W_0^3}^2 - \sum_{i=1}^{N-1}(|\langle\phi_a, B_{a_i}\rangle|^2 + |\langle\phi_a, B_{b_i}\rangle|^2)$$
$$\leq \|\phi_a\|_{W_0^3}^2,$$
$$\|\phi_b - \sum_{i=1}^{N-1}(\langle\phi_b, B_{a_i}\rangle_{W_0^3}B_{a_i} + \langle\phi_b, B_{b_i}\rangle_{W_0^3}B_{b_i}) - \langle\phi_b, B_N^a\rangle_{W_0^3}B_N^a\|_{W_0^3}^2 \leq \|\phi_b - \langle\phi_b, B_N^a\rangle_{W_0^3}B_N^a\|_{W_0^3}^2. \tag{59}$$

It follows from (36) that for $1 \leq i \leq N-1$, we have

$$\langle u_N, B_{a_i}\rangle_{W_0^3} = \langle u_N, B_{b_i}\rangle_{W_0^3} = 0. \tag{60}$$

Then, (56) and (58)–(60) imply that

$$|\langle u_N, e_a\rangle_{W_0^3}| = |\langle u_N, \frac{\phi_a}{\|\phi_a\|_{W_0^3}}\rangle_{W_0^3}| = \frac{\left|\left\langle u_N, \phi_a - \sum_{i=1}^{N-1}(\langle\phi_a, B_{a_i}\rangle_{W_0^3}B_{a_i} + \langle\phi_a, B_{b_i}\rangle_{W_0^3}B_{b_i})\right\rangle_{W_0^3}\right|}{\|\phi_a\|}$$
$$\leq \frac{\left|\left\langle u_N, \phi_a - \sum_{i=1}^{N-1}(\langle\phi_a, B_{a_i}\rangle_{W_0^3}B_{a_i} + \langle\phi_a, B_{b_i}\rangle_{W_0^3}B_{b_i})\right\rangle_{W_0^3}\right|}{\left\|\phi_a - \sum_{i=1}^{N-1}(\langle\phi_a, B_{a_i}\rangle_{W_0^3}B_{a_i} + \langle\phi_a, B_{b_i}\rangle_{W_0^3}B_{b_i})\right\|_{W_0^3}} = |\langle u_N, B_N^a\rangle_{W_0^3}|, \tag{61}$$

$$|\langle u_N, e_b\rangle_{W_0^3}| = \frac{|\langle u_N, \phi_b - \langle\phi_b, B_N^a\rangle_{W_0^3}B_N^a\rangle|}{\|\phi_b - \langle\phi_b, B_N^a\rangle_{W_0^3}B_N^a\|_{W_0^3}}$$
$$= \frac{\left|\left\langle u_N, \phi_b - \langle\phi_b, B_N^a\rangle_{W_0^3}B_N^a - \sum_{i=1}^{N-1}(\langle\phi_b, B_{a_i}\rangle_{W_0^3}B_{a_i} + \langle\phi_b, B_{b_i}\rangle_{W_0^3}B_{b_i})\right\rangle_{W_0^3}\right|}{\|\phi_b - \langle\phi_b, B_N^a\rangle_{W_0^3}B_N^a\|_{W_0^3}}$$
$$\leq \frac{\left|\left\langle u_N, \phi_b - \langle\phi_b, B_N^a\rangle_{W_0^3}B_N^a - \sum_{i=1}^{N-1}(\langle\phi_b, B_{a_i}\rangle_{W_0^3}B_{a_i} + \langle\phi_b, B_{b_i}\rangle_{W_0^3}B_{b_i})\right\rangle_{W_0^3}\right|}{\left\|\phi_b - \langle\phi_b, B_N^a\rangle_{W_0^3}B_N^a - \sum_{i=1}^{N-1}(\langle\phi_b, B_{a_i}\rangle_{W_0^3}B_{a_i} + \langle\phi_b, B_{b_i}\rangle_{W_0^3}B_{b_i})\right\|_{W_0^3}}$$
$$= |\langle u_N, B_N^b\rangle_{W_0^3}|. \tag{62}$$

Thus, it follows from (61), (62), (57) and (55) that

$$|\langle u_N, e_a\rangle_{W_0^3}|^2 + |\langle u_N, e_b\rangle_{W_0^3}|^2 \leq |\langle u_N, B_N^a\rangle_{W_0^3}|^2$$
$$+ |\langle u_N, B_N^b\rangle_{W_0^3}|^2 \leq \frac{1}{\rho}[|\langle u_N, B_{a_N}\rangle_{W_0^3}|^2 + \langle u_N, B_{b_N}\rangle_{W_0^3}|^2] \tag{63}$$
$$< \frac{1}{\rho}\frac{\rho C_0}{2} = \frac{C_0}{2}.$$

On the other hand, it follows from (49) that one has

$$u_N = u - \sum_{i=1}^{N-1}(\langle u, B_{a_i}\rangle_{W_0^3}B_{a_i} + \langle u, B_{b_i}\rangle_{W_0^3}B_{b_i}) = g + \sum_{i=N}^{\infty}(\langle u, B_{a_i}\rangle_{W_0^3}B_{a_i} + \langle u, B_{b_i}\rangle_{W_0^3}B_{b_i}). \tag{64}$$

Recall that $0 < \rho < 1$, it follows from the triangle inequality, the Cauchy-Schwarz inequality, (54), (55) and (64) that

$$
\begin{aligned}
|\langle u_N, e_a\rangle_{W_0^3}|^2 + |\langle u_N, e_b\rangle_{W_0^3}|^2 &= \left|\left\langle g + \sum_{i=N}^{\infty}(\langle u, B_{a_i}\rangle_{W_0^3} B_{a_i} + \langle u, B_{b_i}\rangle_{W_0^3} B_{b_i}), e_a \right\rangle_{W_0^3}\right|^2 \\
&+ \left|\left\langle g + \sum_{i=N}^{\infty}(\langle u, B_{a_i}\rangle_{W_0^3} B_{a_i} + \langle u, B_{b_i}\rangle_{W_0^3} B_{b_i}), e_b \right\rangle_{W_0^3}\right|^2 \\
&\geq |\langle g, e_a\rangle_{W_0^3}|^2 - \left|\left\langle \sum_{i=N}^{\infty}(\langle u, B_{a_i}\rangle_{W_0^3} B_{a_i} + \langle u, B_{b_i}\rangle_{W_0^3} B_{b_i}), e_a \right\rangle_{W_0^3}\right|^2 \\
&+ |\langle g, e_b\rangle_{W_0^3}|^2 - \left|\left\langle \sum_{i=N}^{\infty}(\langle u, B_{a_i}\rangle_{W_0^3} B_{a_i} + \langle u, B_{b_i}\rangle_{W_0^3} B_{b_i}), e_b \right\rangle_{W_0^3}\right|^2 \\
&\geq |\langle g, e_a\rangle_{W_0^3}|^2 - \left\|\sum_{i=N}^{\infty}(\langle u, B_{a_i}\rangle_{W_0^3} B_{a_i} + \langle u, B_{b_i}\rangle_{W_0^3} B_{b_i})\right\|_{W_0^3}^2 \|e_a\|_{W_0^3}|^2 \\
&+ |\langle g, e_b\rangle_{W_0^3}|^2 - \left\|\sum_{i=N}^{\infty}(\langle u, B_{a_i}\rangle_{W_0^3} B_{a_i} + \langle u, B_{b_i}\rangle_{W_0^3} B_{b_i})\right\|_{W_0^3}^2 \|e_b\|_{W_0^3}|^2 \\
&= |\langle g, e_a\rangle_{W_0^3}|^2 + |\langle g, e_b\rangle_{W_0^3}|^2 - 2\left\|\sum_{i=N}^{\infty}(\langle u, B_{a_i}\rangle_{W_0^3} B_{a_i} + \langle u, B_{b_i}\rangle_{W_0^3} B_{b_i})\right\|_{W_0^3}^2 \\
&\geq \min_{x,y\in[c,d]} |\langle g, e_x\rangle_{W_0^3}|^2 + |\langle g, e_y\rangle_{W_0^3}|^2 - 2\left\|\sum_{i=N}^{\infty}(\langle u, B_{a_i}\rangle_{W_0^3} B_{a_i} + \langle u, B_{b_i}\rangle_{W_0^3} B_{b_i})\right\|_{W_0^3}^2 \\
&= \frac{3C_0}{2} - 2\frac{C_0}{2} = \frac{C_0}{2},
\end{aligned}
$$

which contradicts to (63). Consequently, the residual $g = 0$. And this completes the proof of (II).

(III) We are going to derive another analytic formula (46) of the solution u of IVP $Lu = f$. We may assume that $u_i \neq 0$ for all i; otherwise the series in (46) is a finite sum. For this, we prove that the sequence $S_n = \sum_{i=1}^{n}(\langle u, B_{a_i}\rangle_{W_0^3} B_{a_i} + \langle u, B_{b_i}\rangle_{W_0^3} B_{b_i})$ satisfies $\lim_{n\to\infty} L(S_n) = f$. The result of part (II) implies that $\lim_{n\to\infty} \|u - S_n\|_{W_0^3} = 0$. As L is bounded and

$$\|f - LS_n\|_{W^1} = \|Lu - LS_n\|_{W^1} = \|L(u - S_n)\|_{W^1} \leq \|L\|\,\|u - S_n\|_{W_0^3},$$

we know that the series $LS_n = \sum_{i=1}^{n}(\langle u, B_{a_i}\rangle_{W_0^3} LB_{a_i} + \langle u, B_{b_i}\rangle_{W_0^3} LB_{b_i})$ converges to f in $W^1[0, 1]$.

Then it follows from (2), (31) and (33) that

$$
\begin{aligned}
u &= \sum_{n\geq 1}(\langle u, B_{a_n}\rangle_{W_0^3} B_{a_n} + \langle u, B_{b_n}\rangle_{W_0^3} B_{b_n}) \\
&= \sum_{n\geq 1}(\langle\, u, \sum_{k=1}^{n-1}(\beta_{ik}^{(a_i,a_k)}\phi_{a_k} + \beta_{ik}^{(a_i,b_k)}\phi_{b_k}) + \beta_{ii}^{a_i}\phi_{a_i}\,\rangle_{W_0^3} B_{a_n} \\
&+ \langle\, u, \sum_{k=1}^{n}(\beta_{ik}^{(b_i,a_k)}\phi_{a_k} + \beta_{ik}^{(b_i,b_k)}\phi_{b_k})\,\rangle_{W_0^3} B_{b_n}) \\
&= \sum_{n\geq 1}(\sum_{k=1}^{n-1}(\beta_{ik}^{(a_i,a_k)} f(a_k) + \beta_{ik}^{(a_i,b_k)} f(b_k) + \beta_{ii}^{a_i} f(a_i))\, B_{a_n} \\
&+ \sum_{k=1}^{n}(\beta_{ik}^{(b_i,a_k)} f(a_k) + \beta_{ik}^{(b_i,b_k)} f(b_k)\, B_{b_n}).
\end{aligned}
$$

Similarly, we have the numerical solution

$$
s_n = \sum_{i=1}^{n}\Big(\sum_{k=1}^{n-1}(\beta_{ik}^{(a_i,a_k)} f(a_k) + \beta_{ik}^{(a_i,b_k)} f(b_k) + \beta_{ii}^{a_i} f(a_i))\, B_{a_i} + \sum_{k=1}^{n}(\beta_{ik}^{(b_i,a_k)} f(a_k) + \beta_{ik}^{(b_i,b_k)} f(b_k))\, B_{b_i}\Big). \tag{65}
$$

which completes the proof of Theorem 4.

Remark 4. As $W_0^3[0,1]$ is a RKHS, norm-convergence implies pointwise convergence. In fact, we can even prove uniform convergence in this case.

Corollary 1. *The sequence $\{S_n\}_{n\geq 1}$ in (47) converges uniformly to solution u of (1)–(2) on $[0,1]$.*

To deal with the convergence rate estimation of the weak 2-best POAFD, for any $M > 0$, we introduce the subclass $W_M^3[0,1]$ of $W_0^3[0,1]$ defined as

$$
W_M^3[0,1] = \left\{ u \;\middle|\; \begin{array}{l} u \in W_0^3[0,1],\ \text{exists } (c_i, d_i)_{i\geq 1} \text{ in } {}^2 \text{ and } (\phi_{a_i}, \phi_{b_i})_{i\geq 1} \text{ in } W_0^3[0,1] \text{ such that} \\ u = \sum_{i\geq 1}(c_i\phi_{a_i} + d_i\phi_{b_i}), \text{ and } \sum_{i\geq 1}(|c_i| + |d_i|) \leq M. \end{array} \right\}
$$

One can follow the same argument of the proof of the paper [2,10] to deduce the following

Theorem 5. (Convergence rate [2,10]. *Suppose that weak 2-best POAFD solution $u = \sum_{i\geq 1} c_i\phi_{a_i}$ of IVP (1)–(2) is in the class $W_M^3[0,1]$, and let $u_n = \sum_{i\geq n} c_i\phi_{a_i}$ be the $(n-1)$–th residual of u. Then we have*

$$
\|u_n\|_{W_0^3} \leq \frac{MR_n}{\rho\sqrt{2n}},
$$

where $R_n = \max_{1\leq i\leq k\leq n}\{\|G_k^{a_i}\|_{W_0^3}, \|G_k^{b_i}\|_{W_0^3}\}$, and $G_k^{a_i}$ and $G_k^{b_i}$ in (30) and (32) depend on the first $2(k-1)$ functions ϕ_{a_j}, ϕ_{b_j} $(1 \leq j \leq 2(n-1))$. □

Building upon the analysis of convergence and convergence rate of weak 2-best POAFD, the generalization from weak 2-best POAFD to weak N-best POAFD is straightforward, as detailed below.

For any sequence $\{P_1, P_2, \cdots, P_N\}$ of distinct points in $[0,1]^N$, there relates a sequence of dictionary elements $\{\mathbf{h}_{P_1}, \mathbf{h}_{P_2}, \cdots, \mathbf{h}_{P_N}\}$. Applying G-S orthonormalization to $\{\mathbf{h}_{P_1}, \mathbf{h}_{P_2}, \cdots, \mathbf{h}_{P_N}\}$, we obtain the orthonormal set $\{B_{P_1}, B_{P_2}, \cdots, B_{P_N}\}$, where

$$B_{P_i} = \frac{\mathbf{h}_{P_i} - \sum_{j=1}^{i-1} \langle \mathbf{h}_{P_i}, B_{P_j} \rangle B_{P_j}}{\|\mathbf{h}_{P_i} - \sum_{j=1}^{i-1} \langle \mathbf{h}_{P_i}, B_{P_j} \rangle B_{P_j} \|_{W_0^3[0,1]}}, \quad 1 \le i \le N.$$

For any fixed $0 < \rho < 1$, we determine a sequence $\{Q_1, Q_2, \cdots, Q_N\}$ of distinct points by the following weak N–best maximal selection principle:

$$\sum_{i=1}^{N} \left|\langle u, B_{Q_i} \rangle\right|^2 \ge \rho \sup\{ \sum_{i=1}^{N} \left|\langle u, B_{P_i} \rangle\right|^2 | \{P_i\}_{i=1}^N \in [0,1]^N \}. \tag{66}$$

Denote $B_i = B_{Q_i}$, $1 \le i \le N$, then we can approximate the solution $u \in W_0^3[0,1]$ of IVP(1)–(2) by weak N–best POAFD as follows:

$$u = \lim_{N \to \infty} S_N = \lim_{N \to \infty} \sum_{i=1}^{N} \langle u, B_i \rangle B_i. \tag{67}$$

The existence of these selected parameters (28) and the convergence properties(Theorem 4–5) can be similarly discussed as shown in 2-best case.

4 Numerical Experiments

In this last section, we have established that weak N-best POAFD can produce convergent numerical solutions of linear IVP (1)–(2). Through two examples, we compare the performance of numerical solutions obtained by our weak N-best POAFD method and the Reproducing Kernel Methods (RKM) discussed in Chap. 4 of Cui [1]. The computations of both methods are implemented by the Matlab codes.

For clarity and reproducibility, the pseudocode for weak 2-best POAFD, which solves Initial Value Problems (IVPs), is provided below to allow for straightforward implementation.

Algorithm 1. Numerical solution representation of IVP (1)–(2) by weak 2-best POAFD

Require: The point set $X = \{x_i\}_{i=1}^n \subset [0,1]$, reproducing kernel $R(x,y)$ of image space $W_0^3[0,1]$, the operator L, the maximum number of iterations m $(2m \leq n)$, iteration tolerance ϵ, maximal selection property parameter $0 < \rho < 1$.
Ensure: Approximate solution S_n(65).
Step 0: Compute the dictionary elements $\{\mathbf{h}_i\}_{i=1}^n$ by (23).
Step 1: Normalize $\{\mathbf{h}_a, \mathbf{h}_b\}$ to obtain $\{B_1^a, B_1^b\}$ by (26) for any $a, b \in [0,1]$.
(1) Choose distinct points $a_1, b_1 \in [0,1]^2$ such that
$|\langle u, B_{a_1}\rangle_{W_0^3}|^2 + |\langle u, B_{b_1}\rangle_{W_0^3}|^2 \geq \rho \sup\{|\langle u, B_1^a\rangle|^2_{W_0^3} + |\langle u, B_1^b\rangle|^2_{W_0^3} \mid a, b \in [0,1]^2\}$.
Then $S_1 = \langle u, B_1^a\rangle_{W_0^3} B_1^a + \langle u, B_1^b\rangle_{W_0^3} B_1^b + g_1$, and the residual $g_1 = u - S_1$.
(2) **while** $(1 < i < m$ **or** $|\langle u_i, B_i\rangle| > \epsilon)$ **do**
Orthonormalize $\{B_{a_1}, B_{b_1}, \cdots, B_{a_{i-1}}, B_{b_{i-1}}, \mathbf{h}_a, \mathbf{h}_b \ : \ a, b \in X \setminus \{a_j, b_j\}_{j=1}^{i-1}\}$ using
the Gram-Schmidt process.
Choose a_i, b_i such that

$$\begin{aligned} C_i(a_i, b_i) &= |\langle u, B_{a_i}\rangle_{W_0^3}|^2 + |\langle u, B_{b_i}\rangle_{W_0^3}|^2 \\ &\geq \rho \sup\{|\langle u, B_i^a\rangle_{W_0^3}|^2 + |\langle u, B_i^b\rangle_{W_0^3}|^2 \mid a, b \in [0,1] \setminus \{a_j, b_j\}_{j=1}^{i-1}\}. \end{aligned}$$

Then we obtain
$u = \sum_{j=1}^{i} (\langle u, B_{a_j}\rangle_{W_0^3} B_{a_j} + \langle u, B_{b_j}\rangle_{W_0^3} B_{b_j}) + g_i$, where
$g_i = u - \sum_{j=1}^{i} (\langle u, B_{a_j}\rangle_{W_0^3} B_{a_j} + \langle u, B_{b_j}\rangle_{W_0^3} B_{b_j})$ and set $i = i + 1$.
end while
Step 2: Output approximate solution: $S_m = \sum_{i=1}^{m} (\langle u, B_{a_i}\rangle_{W_0^3} B_{a_i} + \langle u, B_{b_i}\rangle_{W_0^3} B_{b_i})$.

For a fixed positive integer $m > 0$, any integer $0 < 2n \leq m$, we want to compare numerical solutions S_n obtained by weak N-best POAFD in (47) and numerical solutions S_{2n} obtained by RKM in [1] as follows. Here, we should notice that the RKM solution is constructed using $2n$ discrete points. The weak N-best POAFD solution S_n is constructed after n iteration steps, resulting in a linear combination of $2n$ basis functions.

Example 4.1. Consider the following initial value problem of ordinary differential equation

$$\begin{aligned} &u''(x) + u'(x) + x^3 u(x) = f(x), \quad \text{for} \quad 0 \leq x \leq 1, \\ &u(0) = 1, u'(0) = 1, \end{aligned} \tag{68}$$

where $f(x) = (2+x)\mathrm{e}^x + x^3\mathrm{e}^x$. The exact solution is $u(x) = \mathrm{e}^x$. Let $g(x) = -1-x$ and $v(x) = u(x)+g(x) = \mathrm{e}^x - x - 1$. We can homogenize the boundary conditions in (68).

Taking $m = 100$ and $n = 2, 4, 6, 8, 9, 10, 15, 20, 25$, we obtain the approximate solutions S_n which is a linear combination of $2n$ terms. In Table 1 and Fig. 1, we list the maximum absolute errors, the maximum relative errors and norms of the remainders produced by RKM and weak N-best POAFD for various n. It shows that the accuracy of partial sum S_n obtained by weak N-best POAFD method is higher than that of RKM. Also, one observes that the larger n is, the higher accuracy weak N-best POAFD can achieve than the one RKM does.

Table 1. In **Example** 4.1, we compare the maximum absolute errors, the maximum relative errors and norms of the remainders of u obtained by weak N-best POAFD and the RKM. The RKM solution uses $2n$ points, the weak N-best POAFD solution S_n uses n steps ($2n$ basis functions).

	RKM [1]			weak N-best POAFD		
	$\lvert u - S_{2n}\rvert_{max}$	$\left\lvert\frac{u-S_{2n}}{u}\right\rvert_{max}$	$\lVert u - S_{2n}\rVert$	$\lvert u - S_n\rvert_{max}$	$\left\lvert\frac{u-S_n}{u}\right\rvert_{max}$	$\lVert u - S_n\rVert$
$n=2$	4.0153E-01	5.5901E-01	2.5548×10^0	6.6870E-03	2.2435E-02	2.0143E-01
$n=4$	3.7973E-01	5.4519E-01	2.5129×10^0	1.4417E-03	3.4613E-03	9.1519E-02
$n=6$	3.6729E-01	5.2866E-01	2.4695×10^0	9.7938E-04	3.4703E-03	5.8631E-02
$n=8$	3.6091E-01	5.1134E-01	2.4238×10^0	3.4923E-04	5.0576E-04	3.9173E-02
$n=10$	3.2071E-01	4.9346E-01	2.3756×10^0	2.9093E-04	4.8103E-04	3.2087E-02
$n=15$	2.7420E-01	4.4649E-01	2.2417×10^0	1.0036E-04	2.9195E-04	1.4564E-02

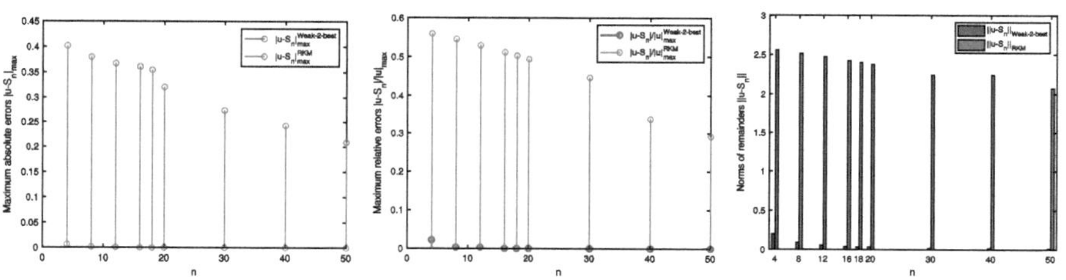

(a) The maximum absolute errors of u (b) The maximum relative errors of u (c) The norms of remainders $\lVert u - S_n\rVert_{W_0^3}$

Fig. 1. Comparison of numerical results for Example 4.1.

Example 4.2. Consider the following initial value problem of ordinary differential equation

$$\begin{aligned} &u''(x) + \cos(x)u'(x) + xu(x) = f(x), 0 < x < 1,\\ &u(0) = 0, u'(0) = 0, \end{aligned} \tag{69}$$

where $f = 2\cos(x) - x\sin(x) + x^2\sin(x) + \cos(x)(x\cos(x) + \sin(x))$. The exact solution is $u = x\sin(x)$.

Taking $m = 100, n = 2, 4, 6, 8, 10, 15$, we obtain the approximate solutions S_n. Table 2 and Fig. 2 list the numerical results obtained by the two methods. These results show the good convergence of the weak N-best POAFD.

Table 2. In Example 4.2, we compare the maximum absolute errors, the maximum relative errors and the norms of the remainders of u obtained by weak N-best POAFD and the RKM. The RKM solution uses $2n$ points, the weak N-best POAFD solution S_n uses n steps ($2n$ basis functions).

	RKM [1]			weak N-best POAFD		
	$\lvert u - S_{2n}\rvert_{max}$	$\left\lvert\frac{u - S_{2n}}{u}\right\rvert_{max}$	$\lVert u - S_{2n}\rVert$	$\lvert u - S_n\rvert_{max}$	$\left\lvert\frac{u - S_n}{u}\right\rvert_{max}$	$\lVert u - S_n\rVert$
n =2	1.5364E-01	3.3081E-01	1.4647×10^0	2.4342E-02	2.8927E-02	5.2164E-01
n =4	1.1086E-01	1.9680E-01	1.2818×10^0	2.9530E-03	6.5904E-03	1.9237E-01
n =6	9.3734E-02	1.3966E-01	1.2365×10^0	1.1990E-03	6.5905E-03	1.0477E-01
n =8	8.1674E-02	1.1182E-01	1.2148×10^0	6.2794E-04	6.5906E-03	7.4780E-02
n =10	7.1548E-02	1.0185E-01	1.1982×10^0	4.2758E-04	6.5906E-03	5.5047E-02
n =15	5.0726E-02	8.2219E-02	1.1475×10^0	2.4330E-04	6.5906E-03	3.9921E-02

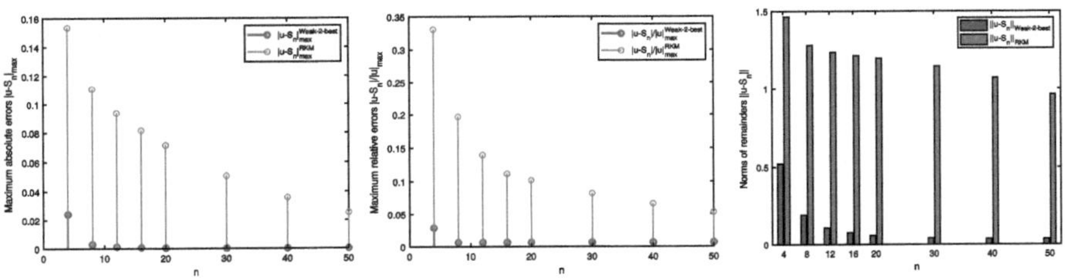

(a) The maximum absolute errors of u (b) The maximum relative errors of u (c) The norms of remainders $\|u - S_n\|_{W_0^3}$

Fig. 2. Comparison of numerical results for Example 4.2.

5 Conclusion

This paper presents the novel application of the weak N-best POAFD algorithm for solving initial value problems (IVPs) in reproducing kernel Hilbert spaces. The primary advantage of our approach lies in its superior accuracy within the space $W_0^3[0,1]$, as demonstrated by our numerical experiments compared to the reproducing kernel method (RKM). This enhancement is attributed to the algorithm's parameter selection mechanism–the weak N-best maximal selection strategy–which systematically chooses n pairs of distinct parameters $a_i, b_{i i\geq 1}$ to improve the approximation.

The numerical results affirm the effectiveness of the method in terms of solution accuracy. However, this comes at the cost of increased computational demands due to the maximization procedure at each step. Thus, the method is presently most suited for problems where accuracy is critical. Future research will focus on developing more efficient search strategies to alleviate the associated computational overhead, thereby extending the applicability to larger-scale problems.

Acknowledgments. This work was supported by Guangdong Provincial Characteristic Innovation Project for Colleges and Universities (Grant No. 2024KTSCX172) and Guangdong Provincial Postgraduate Education Innovation Program Project (Grant No. 2024JGXM_098).

Disclosure of Interests. The authors have no competing interests to declare that are relevant to the content of this article.

References

1. Cui, M.G., Lin, Y.Z.: Nonlinear Numerical Analysis in the Reproducing Kernel Space. Nova Science Publishers, New York (2009)
2. Chen, Q., Qian, T., Tan, L.: A theory on non-constant frequency decompositions and applications. In: Breaz, D., Rassias, M.T. (eds.) Advancements in Complex Analysis, pp. 1–37. Springer, Cham (2020). https://doi.org/10.1007/978-3-030-40120-7_1
3. Stakgold, I.: Green's Functions and Boundary Value Problems, pp. 102–105. Wiley, Hoboken (1979)
4. Willian, G.C.: Numerical Initial Value Problems in Ordinary Differential Equations. Prentice-Hall, Inc., Englewood Cliffs, New Jersey (1971)
5. Qian, T., Ho, I.T., Leong, I.T., Wang, Y.B.: Adaptive decomposition of functions into pieces of non-negative instantaneous frequencies. Int. J. Wavelets Multiresolut. Inf. Process. **34**(8), 813–833 (2010). https://doi.org/10.1142/S0219691310003791
6. Qian, T., Wang, Y.B.: Adaptive Fourier series-a variation of greedy algorithm. Adv. Comput. Math. **34**(3), 279–293 (2011). https://doi.org/10.1007/s10444-010-9153-4
7. Qian, T.: Two-dimensional adaptive Fourier decomposition. Math. Methods Appl. Sci. **39**(10), 2431–2448 (2016). https://doi.org/10.1002/mma.3649
8. Qu, W., Qian, T., Deng, G.T.: A stochastic sparse representation: n-best approximation to random signals and computation. Appl. Comput. Harmon. Anal. **55**(1), 185–198 (2021)
9. Li, Y.T., Zhang, L.M., Qian, T.: 2D partial unwinding-a novel non-linear phase decomposition of images. IEEE Trans. Image Process. **28**(10), 4762–4773 (2019)
10. Qian, T.: Reproducing kernel sparse representations in relation to operator equations. Complex Anal. Oper. Theory **14**(2), 1–15 (2020). https://doi.org/10.1007/s11785-020-00994-4
11. Bai, H.F., Leong, I.T., Dang, P.: Reproducing kernel representation of the solution of second order linear three-point boundary value Problem. Math. Methods Appl. Sci. **45**(17), 11181–11205 (2022)

Backstepping Stabilization of Stochastic Hyperbolic PDE-ODE Systems

Xuejiao Sun(✉), Junmin Wang, and Kaijing Lyu

School of Mathematics and Statistics, Beijing Institute of Technology, Beijing, China
{xjsun,jmwang,kjlv}@bit.edu.cn

Abstract. This paper establishes a constructive boundary control framework for stabilizing coupled hyperbolic PDE-ODE systems subject to stochastic parameter variations. The system under consideration exhibits mixed-directional dynamics, with three positive and one negative characteristic speeds in its PDE subsystem. Building upon nominal backstepping designs for deterministic systems, we develop a parameter-adaptive control law that maintains stabilization capability when system parameters undergo random switching. Through a novel composite Lyapunov analysis, we derive explicit stability conditions relating the controller's robustness to both the magnitude of parameter variations and their transition statistics. The theoretical results demonstrate that mean-square exponential stability can be preserved when the stochastic parameters remain within probability-dependent neighborhoods of their nominal values. A distinctive feature of our approach lies in the unified treatment of PDE-PDE and PDE-ODE coupling terms under stochastic perturbations, which has not been addressed in existing Markov-jumping PDE literature. The control design is validated through numerical experiments, showing consistent stabilization performance across different parameter switching scenarios.

Keywords: Stochastic hyperbolic systems · PDE-ODE cascades · PDE backstepping · Boundary control · Robust stabilization

1 Introduction

The stabilization of linear ordinary differential equations (ODEs) coupled with hyperbolic partial differential equations (PDEs) in the actuation path has been well established for deterministic systems, with backstepping emerging as a systematic solution [6]. For instance, [6] developed a boundary control framework for first-order hyperbolic PDEs with delay components, demonstrating backstepping's effectiveness in handling coupled dynamics.

However, significant gaps remain in stochastic settings where system parameters follow Markov processes or boundary conditions are subject to random disturbances. While Wang et al. [7] addressed robust stabilization of hyperbolic PDEs with Markov jump parameters through linear matrix inequalities (LMIs),

H. Ma et al. (Eds.): IWACIII 2025, CCIS 2780, pp. 347–361, 2026.
https://doi.org/10.1007/978-981-95-6730-0_25

their approach is inherently conservative and cannot be directly extended to PDE-ODE cascades. Recent work by Deutscher [8] on heterodirectional PDE-ODE systems with spatially varying coefficients provides a comprehensive backstepping framework, but remains limited to deterministic parameter configurations. The most relevant advancement comes from Zhang and Auriol [4], who established backstepping control for 4 × 4 stochastic hyperbolic PDEs with Markov jumps. However, their framework excludes the critical case of ODE coupling - a significant limitation for applications requiring actuator dynamics integration or distributed sensing architectures.

This paper develops a backstepping control method for stabilizing coupled hyperbolic PDE-ODE systems with Markov jump parameters. By constructing a novel Volterra transformation that accommodates both the PDE-ODE coupling and stochastic parameter variations, we design a boundary control law that achieves mean-square exponential stability. The proposed controller maintains robustness against Markovian parameter jumps without requiring exact parameter matching, while explicitly accounting for the bidirectional interaction between the PDE and ODE states. Section 2 formulates the stochastic PDE-ODE system. Section 3 presents the backstepping control design. Section 4 provides the stability analysis, and Sect. 5 demonstrates the results through numerical simulations of a flexible structure with random parameter variations.

2 Problem Statement

In this paper, we consider the following stochastic PDE-ODE coupled system

$$\dot{X}(t) = AX(t) + Bv(0,t), \tag{1}$$
$$u_t(x,t) = -\Lambda_1(t)u_x(x,t) + \Sigma_1(x,t)u(x,t) + \Sigma_2(x,t)v(x,t), \tag{2}$$
$$v_t(x,t) = \Lambda_2(t)v_x(x,t) + \Sigma_3(x,t)u(x,t), \tag{3}$$

with the boundary conditions

$$u(0,t) = Q(t)v(0,t) + CX(t), \tag{4}$$
$$v(1,t) = R(t)u(1,t) + U(t) \tag{5}$$

where $t > 0$ and $x \in [0,1]$ are respectively the time and space variable. $X(t) \in \mathbb{R}$ is the ODE state. $u(t,x) \in \mathbb{R}^3$ and $v(t,x) \in \mathbb{R}$ are the PDE states. And $U(t) \in R$ is the control input. The matrices $\Lambda_1(t)$, $\Lambda_2(t)$ are defined as

$$\Lambda_1(t) = \begin{bmatrix} \lambda_1(t) & 0 & 0 \\ 0 & \lambda_2(t) & 0 \\ 0 & 0 & \lambda_3(t) \end{bmatrix}, \quad \Lambda_2(t) = \lambda_4(t). \tag{6}$$

The coefficient matrices including $\Lambda_1(t)$, $\Lambda_2(t)$, $\Sigma_1(x,t) \in \mathbb{R}^{3\times3}$, $\Sigma_2(x,t) \in \mathbb{R}^{3\times1}$, $\Sigma_3(x,t) \in \mathbb{R}^{1\times3}$, $Q(t) \in \mathbb{R}^{3\times1}$ and $R(t) \in \mathbb{R}^{1\times3}$ are time-varying. We denote the set of random variables $\mathscr{X} = \{\Lambda_1, \Lambda_2, \Sigma_1, \Sigma_2, \Sigma_3, Q, R\}$. We consider a homogeneous continuous-time Markov process $\mathcal{X}(t)$ defined for $t \in \mathbb{R}^+$, taking values in a finite state space $\mathcal{S} = \{\mathscr{X}_1, \mathscr{X}_2, \ldots, \mathscr{X}_r\}$. The sample paths of this process are right-continuous functions. Each state $\mathscr{X}_j$ can be represented as: $\mathscr{X}_j = \{\Lambda_{1j}, \Lambda_{2j}, \Sigma_{1j}(x), \Sigma_{2j}(x), \Sigma_{3j}(x), Q_j, R_j\}$. The transition probabilities $P_{ij}(t_1, t_2)$ represent the probability of transitioning from state $\mathscr{X}_i$ at time t_1 to state $\mathscr{X}_j$ at time t_2, where $(i, j) \in \{1, \ldots, r\}^2$ and $0 \leq t_1 \leq t_2$. These probabilities satisfy: $P_{ij} : \mathbb{R}^2 \to [0, 1]$ with $\sum_{j=1}^{r} P_{ij}(t_1, t_2) = 1$. For $\varrho < t$, the transition probabilities obey the Kolmogorov forward equation:

$$\frac{\partial}{\partial t} P_{ij}(\varrho, t) = -c_j(t) P_{ij}(\varrho, t) + \sum_{k=1}^{r} P_{ik}(\varrho, t) \tau_{kj}(t),$$

with initial conditions $P_{ii}(\varrho, \varrho) = 1$ and $P_{ij}(\varrho, \varrho) = 0$ for $i \neq j$. Here, $\tau_{ij}(t)$ and $c_j(t) = \sum_{k=1, k\neq j}^{r} \tau_{jk}(t)$ are nonnegative functions of $\tau_{ii}(t) = 0$. The functions τ_{ij} are upper bounded by a constant $\tau^\star$. and given known bounds $\underline{\mathscr{X}}$ and $\overline{\mathscr{X}}$ satisfying $\underline{\mathscr{X}} \leq \mathscr{X}_1 < \ldots < \mathscr{X}_r \leq \overline{\mathscr{X}}$, and we further assume strictly positive lower bounds on the stochastic velocities. Specifically, for each i we have:

- $\Lambda_{1i} > 0$ with lower bound $\underline{\Lambda}_{1i} > 0$
- $\Lambda_{2i} > 0$ with lower bound $\underline{\Lambda}_{2i} > 0$

3 Backstepping Controller Design

3.1 Nominal System and Backstepping Transformation

We first consider the following PDE system without stochastic uncertanity,called nominal system

$$\begin{aligned}
\dot{X}_{nom}(t) =& A X_{nom}(t) + B v_{nom}(0, t), & (7)\\
\partial_t u_{nom}(x, t) =& -\Lambda_1 \partial_x u_{nom}(x, t) + \Sigma_1 u_{nom}(x, t) + \Sigma_2 v_{nom}(x, t), & (8)\\
\partial_t v_{nom}(x, t) =& \Lambda_2 \partial_x v_{nom}(x, t) + \Sigma_3 u_{nom}(x, t), & (9)
\end{aligned}$$

with boundary conditions

$$\begin{aligned}
u_{nom}(0, t) =& Q_0 v_{nom}(0, t) + C X_{nom}(t), & (10)\\
v_{nom}(1, t) =& R_0 u_{nom}(1, t) + \bar{U}(t) & (11)
\end{aligned}$$

where $\Lambda_1 > 0$, $\Lambda_2 > 0$, Σ_1, Σ_2, Σ_3, Q_0, R_0 are constant matrices, and the function $\bar{U}(t)$ is the boundary control input. We design the target system as follows:

$$\begin{aligned}
\dot{X}_{nom}(t) =& (A + BK) X_{nom}(t) + B \beta_{nom}(0, t), \qquad (12)\\
\partial_t \mathscr{Q}_{nom}(x, t) =& -\Lambda_1 \partial_x \mathscr{Q}_{nom}(x, t) + \Sigma_1(x) \mathscr{Q}_{nom}(x, t) + \Sigma_2(x) \beta_{nom}(x, t)\\
&+ \int_0^x \mathbf{C}_1(x, \xi) \mathscr{Q}_{nom}(\xi, t)\, d\xi + \int_0^x \mathbf{C}_2(x, \xi) \beta_{nom}(\xi, t)\, d\xi
\end{aligned}$$

$$+ D(x)X_{nom}(t), \tag{13}$$

$$\partial_t \beta_{nom}(x,t) = \Lambda_2 \partial_x \beta_{nom}(x,t), \tag{14}$$

$$\mathscr{Q}_{nom}(0,t) = Q_0 \beta_{nom}(0,t) + C_0 X_{nom}(t), \tag{15}$$

$$\beta_{nom}(1,t) = R_0 u_{nom}(1,t) - \int_0^1 (K_0(1,\xi)u_{nom}(\xi,t) + N_0(1,\xi)v_{nom}(\xi,t))\, d\xi - \gamma(1)X(t) + \bar{U}(t). \tag{16}$$

To map the original system (8)–(12)to the target system (13)–(17), we use the following Volterra transformation $\mathcal{K}_0$

$$\begin{cases} \mathscr{Q}_{nom} = u_{nom}, \\ \beta_{nom} = v_{nom} - \int_0^x \mathbf{K}_0(x,\xi)u_{nom}(\xi,t) + N_0(x,\xi)v_{nom}(\xi,t)d\xi - \gamma(x)X_{nom}(t) \end{cases} \tag{17}$$

where the matrices are defined by

$$C_1(x,y) = \Sigma_2 K_0(x,y) + \int_y^x C_2(x,s)K(s,y)\, ds \tag{18}$$

$$C_2(x,y) = \Sigma_2 N_0(x,y) + \int_y^x C_2(x,s)L(s,y)\, ds \tag{19}$$

$$D(x) = \Sigma_2 \gamma(x) + \int_0^x C_2(x,y)\gamma(y)\, dy \tag{20}$$

$$C_0 = C + Q_0 \gamma(0) \tag{21}$$

and the kernels K_0, N_0 and γ are defined as follow

$$\Lambda_2 (\mathbf{K}_0)_x (x,\xi) - (\mathbf{K}_0)_\xi (x,\xi)\Lambda_1 = \mathbf{K}_0(x,\xi)\Sigma_1(\xi) - \Sigma_3(\xi)N_0(x,\xi) \tag{22}$$

$$\Lambda_2 (N_0)_x (x,\xi) + \Lambda_2 (N_0)_\xi (x,\xi) = \mathbf{K}_0(x,\xi)\Sigma_2(\xi), \tag{23}$$

$$(\Lambda_2 \mathbf{I}_3 + \Lambda_1)\, \mathbf{K}_0^\top (x,x) = -\Sigma_3^\top (x) \tag{24}$$

$$\Lambda_2 N_0(x,0) + \mathbf{K}_0(x,0)\Lambda_1 Q_0 = \gamma(x)B \tag{25}$$

$$\Lambda_2 \gamma'(x) - A\gamma(x) = C\Lambda_1 k_0(x,0) \tag{26}$$

The kernel Eqs. (23)–(27) admit the triangular domain a unique solution on $\mathcal{T}$ [1]. We denote $w = [u,v]$ and $\vartheta = [\mathscr{Q},\beta]$. Since the Volterra transformation $\mathcal{K}0$ is boundedly invertible, the L^2 norms of the states w and ϑ are equivalent. Specifically, there exist positive constants m and M satisfying:

$$m\|[u,v]\|_{L^2}^2 \leq \|[\mathscr{Q},\beta]\|_{L^2}^2 \leq M\|[u,v]\|_{L^2}^2. \tag{27}$$

3.2 Nominal Control Law and The L^2-Stability of Nominal System

From the nominal target system (13)–(17), we can easily design a stabilizing control law as [4]

$$\overline{U}(t) = -R_0 u(1,t) + \int_0^1 \mathbf{K}_0(1,\xi)u(\xi,t) + N_0(1,\xi)v(\xi,t)d\xi + \gamma(1)X(t) \tag{28}$$

For convenience of notation, we denote the system variables of the nominal system $\mathscr{Q}_{nom}$, β_{nom} and X_{nom} by $\mathscr{Q}$, β and X respectively. We consider the Lyapunov functional V_0 defined by

$$V_0(t) = \int_0^1 (\mathcal{K}_0 w(x,t))^T D_0(x) \mathcal{K}_0 w(x,t) dx + X(t)^T P X(t) \tag{29}$$

where

$$D_0(x) = \text{Diag}\left\{ \frac{e^{-\frac{\nu}{\lambda_1^0}x}}{\lambda_1^0}, \frac{e^{-\frac{\nu}{\lambda_2^0}x}}{\lambda_2^0}, \frac{e^{-\frac{\nu}{\lambda_3^0}x}}{\lambda_3^0}, a\frac{e^{\frac{\nu}{\lambda_4^0}x}}{\lambda_4^0} \right\}. \tag{30}$$

Besides, $P = P^T > 0$ and $A + BK$ is Huriwitz, so there exists the symmetric positive definite matrix S, s.t. $P(A+BK)+(A+BK)P^T = -S$. This Lyapunov functional is equivalent to the L^2 norm of the system, that is, there exist two constants $k_1 > 0$ and $k_2 > 0$, such that

$$k_1 \|\mathcal{K}_0 w\|_{L^2}^2 \le \int_0^1 (\mathcal{K}_0 w)^T D_0(x) \mathcal{K}_0 w dx \le k_2 \|\mathcal{K}_0 w\|_{L^2}^2 \tag{31}$$

To prove L^2-exponential stability for systems (13)–(17), we differentiate the Lyapunov function

$$\begin{aligned}
\dot{V}_0(t) &\le -\nu \int_0^1 (\mathcal{K}_0 w(x,t))^T D_0(x) \mathcal{K}_0 w(x,t) dx + 2\int_0^1 \mathscr{Q}^T(\Sigma_1 u + \Sigma_2 v) dx \\
&- D_{01}(1)\Lambda_1 \mathscr{Q}^2(1,t) + \mathscr{Q}^2(0,t) - a\beta^2(0,t) + 2X^T[(A+BK)X + B\beta(0,t)] \\
&\le -\eta_1 \int_0^1 (\mathcal{K}_0 w(x,t))^T D_0(x) \mathcal{K}_0 w(x,t) dx + (2q_{10}^2 + 2q_{20}^2 + 2q_{30}^2 + \epsilon - a)\beta^2(0,t) \\
&- SX^T X + \gamma^2(0) X^T X + (C_0 + PB)^T (C_0 + PB) X^T X \\
&\le -\eta V_0 + (2q_{10}^2 + 2q_{20}^2 + 2q_{30}^2 + \epsilon - a)\beta^2(0,t)
\end{aligned} \tag{32}$$

where

$$\begin{aligned}
\eta_1 &= \nu - \frac{1}{\underline{\Lambda}_1 k_1}(2 \max_{x\in[0,1]} \|\Sigma_1(x)\| + \left(1 + \frac{1}{m}\right) \max_{x\in[0,1]} \|\Sigma_2(x)\|), \\
\eta &= min(\eta_1, \frac{S - \gamma^2(0) - (C_0 + PB)^T(C_0 + PB)}{\lambda_{min}(P)}) \\
D_{01}(x) &= \text{Diag}\left\{ \frac{e^{-\frac{\nu}{\lambda_1^0}x}}{\lambda_1^0}, \frac{e^{-\frac{\nu}{\lambda_2^0}x}}{\lambda_2^0}, \frac{e^{-\frac{\nu}{\lambda_3^0}x}}{\lambda_3^0} \right\}
\end{aligned} \tag{33}$$

We select positive constants $a > 0$ and $\nu > 0$ satisfying

$$2q_{10}^2 + 2q_{20}^2 + 2q_{30}^2 + \epsilon - a \le 0, \eta > 0, \tag{34}$$

where q_{10}, q_{20}, and q_{30} are the elements of matrix Q_0. This leads to the inequality

$$\dot{V}_0(t) \le -\eta V_0(t),$$

which establishes the L^2-exponential stability of the system.

3.3 Mean-Square Exponential Stabilization

Lemma 1. *For any initial conditions of the Markov system* $[X(t), w(x,t)]$ *and any initial states* $\mathcal{X}(t) = \mathcal{X}(0)$ *for the stochastic parameters, the system (1)–(5) with the nominal control law (28) has a unique solution such that for any* t,

$$\mathbb{E}\big[\|w(x,t)\| + \|X(t)\|\big] < \infty,$$

where the $\mathbb{E}\{\cdot\}$ *denotes the mathematical expectation.*

Proof. This result extends the lemma established by Zhang and Prieur [3]. The sample paths of our stochastic processes are almost surely right-continuous step functions exhibiting finitely many jumps within any bounded time interval. Consequently, there exists an increasing sequence of stopping times $\{t_k\}_{k=0}^{\infty}$ satisfying $t_0 = 0$, $\lim_{k\to\infty} t_k = \infty$, and $\mathcal{X}(t) \equiv \mathcal{X}(t_k)$ for $t \in [t_k, t_{k+1})$. Using Theorem A.6 from [2], we can successively construct the solution on each interval $[t_k, t_{k+1}]$. The subsequent arguments follow similar reasoning to Proposition 1 in [3].

The primary objective of this work is to demonstrate that the control law (29) maintains stabilization capability for the stochastic system (1)–(5), under the condition that the nominal parameter $\mathcal{X}_0$ remains sufficiently close in expectation to its stochastic counterparts. Formally, we establish the following sufficient condition for robust stabilization:

Theorem 1. *There exists a constant* $\varepsilon^* > 0$*, such that if, for all time* $t > 0$,

$$\sum_{\mathscr{X} \in \mathcal{S}} \mathbb{E}\big(|\mathscr{X}(t) - \mathscr{X}_0|\big) \le \varepsilon^*,$$

then the closed-loop system (1)–(4) with the control law (29) is mean-square exponentially stable, namely, there exist $\zeta > 0$ *such that:*

$$\mathbb{E}_{(0,(\mathscr{X}(0),p(0))}\big(p(t)\big) \le e^{-\zeta t} p(0),$$

where $p(t) = \int_0^1 \|w(x,t)\|_2^2 dx + \|X(t)\|_2^2$*, while* $\mathbb{E}_{\left[0,(\mathscr{X}(0),p(0))\right]}$ *denotes the conditional expectation at time* t *on the initial condition of* $p(t) = p(0)$, $\mathscr{X}(t) = \mathscr{X}(0)$.

This theorem will be proved in the next section.

4 Lyapunov Analysis

4.1 Target System in Stochastic Mode $\mathscr{X}_j$

In this section, we consider that $\mathscr{X}(t) = \mathscr{X}_j$ at time t. The following transformation, similar to Eq. (17), is applied:

$$\begin{cases} \mathscr{Q} = u, \\ \beta = v - \int_0^x \mathbf{K}_0(x,\xi)u(\xi,t) + N_0(x,\xi)v(\xi,t)d\xi - \gamma(x)X(t) \end{cases} \tag{35}$$

we consider the stochastic Lyapunov functional V defined by

$$V(t) = \int_0^1 (K_0 w(x,t))^T D(x) K_0 w(x,t) dx + X(t)^T P X(t) \tag{36}$$

where

$$D(x) = \text{Diag}\left\{ \frac{\mathrm{e}^{-\frac{\nu}{\lambda_1^j}x}}{\lambda_1^j}, \frac{\mathrm{e}^{-\frac{\nu}{\lambda_2^j}x}}{\lambda_2^j}, \frac{\mathrm{e}^{-\frac{\nu}{\lambda_3^j}x}}{\lambda_3^j}, a\frac{\mathrm{e}^{\frac{\nu}{\lambda_4^j}x}}{\lambda_4^j} \right\}. \tag{37}$$

Our objective is first to obtain the equations verified by the state ϑ that appears in the Lyapunov functional. It verifies the following set of equations

$$\dot{X}(t) =(A + BK)X(t) + B\beta(0,t), \tag{38}$$

$$\begin{aligned} \mathscr{Q}_t(x,t) =& - \Lambda_{1j}\mathscr{Q}_x(x,t) + \Sigma_{1j}(x)\mathscr{Q}(x,t) + \Sigma_{2j}(x)\beta(x,t) + \int_0^1 C_1\mathscr{Q}(\xi,t)d\xi \\ &+ \int_0^1 C_2\beta(\xi,t)d\xi + D(x)X(t) \end{aligned} \tag{39}$$

$$\begin{aligned} \beta_t(x,t) =& \Lambda_{2j}\beta_x(x,t) + f_{1j}(x)\mathscr{Q}(x,t) + f_{2j}(x)\beta(0,t) + X(t)f_{3j}(x) \\ &+ \int_0^x f_{4j}(x,\xi)u(\xi,t)d\xi + \int_0^x f_{5j}(x,\xi)v(\xi,t)d\xi, \end{aligned} \tag{40}$$

$$\mathscr{Q}(0,t) =Q_0\beta(0,t) + C_0X(t) \tag{41}$$

$$\beta(1,t) =0 \tag{42}$$

where

$$f_{1j}(x) = \Sigma_{3j}(x) + \Lambda_{2j}K_0(x,x) + K_0(x,x)\Lambda_{1j}, \tag{43}$$

$$f_{2j}(x) = -K_0(x,0)\Lambda_{1j}Q_j + N_0(x,0)\Lambda_{2j} - \gamma(x)B, \tag{44}$$

$$\begin{aligned} f_{3j}(x) =& -\Lambda_{2j}\gamma'(x) - \gamma(x)(A + B\gamma(0)) - \Lambda_{1j}K_0(x,0)C - N_0(x,0)\Lambda_{2j}\gamma(0) \\ &- \Lambda_{1j}k_0(x,0)Q(t)\gamma(0), \end{aligned} \tag{45}$$

$$f_{4j}(x,\xi) = \Lambda_{2j}(K_0)_x(x,\xi) - (K_0)_\xi(x,\xi)\Lambda_{1j} - K_0(x,\xi)\Sigma_{1j}(\xi) - N_0(x,\xi)\Sigma_{3j}(\xi), \tag{46}$$

$$f_{5j}(x,\xi) = \Lambda_{2j}(N_0)_x(x,\xi) + \Lambda_{2j}(N_0)_\xi(x,\xi) - K_0(x,\xi)\Sigma_{2j}(\xi). \tag{47}$$

All w-dependent terms in the target system (38)–(42) can be represented via $[\mathscr{Q},\beta]$ using (35), though this complicates computations unnecessarily for stability analysis. Crucially, when stochastic parameters approach nominal values, all right-hand side terms in the equation become sufficiently small, as formalized in the following lemma.

Lemma 2. *There exists a constant M_0, such that for any realization $\mathscr{X}(t) = \mathscr{X}_j \in \mathcal{S}$ for any $(x,\xi) \in \mathcal{T}$, for all $i \in \{1,2,3,4,5\}$*

$$|f_{ij}| < M_0 \sum_{\mathscr{X} \in \mathcal{S}} |\mathscr{X}_0 - \mathscr{X}_j|.$$

Proof. Let us consider the function $f_{1j}(x)$. For all $x \in [0,1]$, it can be expressed as

$$\begin{aligned} f_{1j}(x) &= \Sigma_{3j}(x) + \Lambda_{2j} K_0(x,x) + K_0(x,x)\Lambda_{1j}, \\ &= (\Sigma_{3j}(x) - \Sigma_{30}(x)) + (\Lambda_{2j} - \Lambda_{20})\, K_0(x,x) \\ &\quad + K_0(x,x)\,(\Lambda_{1j} - \Lambda_{10})\,. \end{aligned} \tag{48}$$

As a result, there exists a constant $K_1 > 0$ such that

$$|f_{1j}| \leq K_1 \sum_{\mathscr{X} \in \mathcal{S}} |\mathscr{X}_0 - \mathscr{X}_j|.$$

Analogous estimates for the functions $f_{2j}(x)$, $f_{3j}(x)$, $f_{4j}(x,\xi)$, and $f_{5j}(x,\xi)$ can be obtained by applying the same reasoning. This concludes the proof.

4.2 Derivation of the Lyapunov Function

Let us denote the state of the target system (38)–(40) as $z(t,\cdot) = (w(t,\cdot), X(t))$. As the solution to (38)–(40) is unique (due to the unicity of the solution of (1)–(3)), $(z, \mathscr{X})$ defines a continuous-time Markov process. Let us consider the Lyapunov functional V defined in Eq. (36). Its infinitesimal generator L is defined as [5]

$$LV(z,\mathscr{X}) = \limsup_{\Delta t \to 0^+} \frac{1}{\Delta t} \times \mathbb{E}(V(z(x+\Delta t), \mathscr{X}(t+\Delta t)) - V(z(x,t), \mathscr{X}(t)))$$

We define L_j, the infinitesimal generator of V obtained by fixing $\mathscr{X}(t) = \mathscr{X}_j \in \mathcal{S}$. We have

$$L_j V(z) = \frac{dV}{dz}(z, \mathscr{X}_j)\, h_j(z) + \sum_{\ell \in \mathscr{S}} (V_\ell(z) - V_j(z))\, \tau_{j\ell}, \tag{49}$$

where $V_\ell(z) = V\left(z, \mathscr{X}^\ell\right)$, and where the operator h_j is defined by

$$h_j = \begin{pmatrix} (A+BK)X(t) + B\beta(0,t) \\ -\Lambda_{1j}\mathscr{Q}_x(x,t) + \Sigma_{1j}(x)\mathscr{Q}(x,t) + \Sigma_{2j}(x)v(x,t) \\ \Lambda_{2j}\beta_x(x,t) + f_{1j}(x)\mathscr{Q}(x,t) + f_{2j}(x)\beta(0,t) \\ +X(t)f_{3j}(x) + \int_0^x f_{4j}(x,\xi)u(\xi,t)d\xi + \int_0^x f_{5j}(x,\xi)v(\xi,t)d\xi \end{pmatrix}$$

To simplify notation, we write:

$$\begin{aligned} V(t) &\triangleq V(z, \mathscr{X}(t)), LV(t) \triangleq LV(z, \mathscr{X}(t)), \\ V_j(t) &\triangleq V(z, \mathscr{X}_j), L_jV(t) \triangleq L_j(V(z)). \end{aligned}$$

Throughout the analysis, we assume the initial condition $\mathscr{X}(0) = \mathscr{X}_{i0} \in \mathcal{S}$. The following lemma holds.

Lemma 3. *There exists $\eta_2 > 0$, $\eta_3 > 0$ and $d_1 > 0$ such that the Lyapunov functional $V(t)$ satisfies*

$$\sum_{j=1}^{r} P_{ij}(0,t)L_j V(t) \leq -V(t)\Big(\eta_3 - d_1 \mathcal{Z}(t) - (\eta_2 M_0 + d_1 r\tau^\star) \sum_{\mathcal{X} \in \mathcal{S}} \mathbb{E}\left(|\mathcal{X}(t) - \mathcal{X}_0|\right)\Big) - \sum_{k=1}^{3} \mathrm{e}^{-\frac{\nu}{\lambda}} \alpha_k^2(1,t). \tag{50}$$

where the function $\mathcal{Z}(t)$ is defined as:

$$\mathcal{Z}(t) = \sum_{l=1}^{r} \sum_{\mathcal{X} \in \mathcal{S}} |\mathcal{X}_l(t) - \mathcal{X}_0| \left(\partial_t P_{ij}(0,t) + c_j P_{ij}(0,t)\right) \tag{51}$$

Proof. Consider that $\mathcal{X} = \mathcal{X}_j$. Let us compute the first term of Eq. (52), namely $\frac{dV}{dz}(z, \mathcal{X}_j)\, h_j(z)$. Applying integration by parts, we obtain

$$\begin{aligned}
&\frac{dV}{dz}(z, \mathcal{X}_j)\, h_j = -\nu \int_0^1 [\mathcal{Q}, \beta]^T D_j(x)[\mathcal{Q}, \beta] dx \\
&+ X^T(t)[P(A + B\gamma(0)) + (A + B\gamma(0))P^T]X(t) + 2X^T(t)PB\beta(0,t) \\
&+ 2\int_0^1 \mathcal{Q}(x,t)D_{1j}(x)\Sigma_{1j}\mathcal{Q}(x,t) + \mathcal{Q}(x,t)D_{1j}(x)\Sigma_{2j}v(x,t)dx + 2\int_0^1 \beta^T(t)D_{2j} \\
&[f1j\mathcal{Q} + f_{2j}\beta(0,t) + X(t)f_{3j}(x) + \int_0^x f_{4j}(x,\xi)u(\xi,t)d\xi + \int_0^x f_{5j}(x,\xi)v(\xi,t)d\xi]dx \\
&+ 2X^T PB\beta(0,t) + \mathcal{Q}^2(0,t) - a\beta^2(0,t) - D_{1j}\Lambda_{1j}\mathcal{Q}^2(1,t)
\end{aligned} \tag{52}$$

where

$D_{1j}(x) = \mathrm{Diag}\left\{ \frac{\mathrm{e}^{-\frac{\nu}{\lambda_j^j}x}}{\lambda_1^j}, \frac{\mathrm{e}^{-\frac{\nu}{\lambda_2^j}x}}{\lambda_2^j}, \frac{\mathrm{e}^{-\frac{\nu}{\lambda_3^j}x}}{\lambda_3^j} \right\}, D_{2j} = a\frac{\mathrm{e}^{\frac{\nu}{\lambda_4^j}x}}{\lambda_4^j}$

Then we consider the term $2\int_0^1 \beta^T(t)D_{2j}f_{2j}\beta(0,t)$.

$$\begin{aligned}
&2\int_0^1 \beta^T(t)D_{2j}f_{2j}\beta(0,t)d\xi \\
&\leq a\frac{e^{\frac{\nu}{\underline{\Lambda_2}}}}{\underline{\Lambda_2}} M_0 \sum_{\mathcal{X} \in \mathcal{S}} |\mathcal{X}_j(t) - \mathcal{X}_0| \int_0^1 \frac{1}{\epsilon_1}\beta^2(\xi,t) + \epsilon_1\beta^2(0,t)d\xi \\
&\leq a\frac{e^{\frac{\nu}{\underline{\Lambda_2}}}}{\epsilon_1\underline{\Lambda_2}} M_0 \sum_{\mathcal{X} \in \mathcal{S}} |\mathcal{X}_j(t) - \mathcal{X}_0| \int_0^1 [\mathcal{Q}, \beta]^T D(x)[\mathcal{Q}, \beta] dx
\end{aligned} \tag{53}$$

$$+ a\frac{e^{\frac{\nu}{\underline{\Lambda_2}}}}{\underline{\Lambda_2}} M_0 \sum_{\mathcal{X} \in \mathcal{S}} |\mathcal{X}_j(t) - \mathcal{X}_0|\epsilon_1\beta^2(0,t) \tag{54}$$

And performing analogous computations on all the terms that appear in the expression of $\frac{dV}{dz}(z)h_j(z)$, we obtain

$$\begin{aligned}
&\frac{dV}{dz}(z)h_j(z) \leq -\eta V_j + (2q_{1j}^2 + 2q_{2j}^2 + 2q_{3j}^2 + \epsilon - a)\beta^2(0,t) \\
&+ 2\int_0^1 \beta^T(t) D_{2j}[f1j\mathscr{Q} + f_{2j}\beta(0,t) + X(t) f_{3j}(x) + \int_0^x f_{4j}(x,\xi)u(\xi,t)d\xi \\
&+ \int_0^x f_{5j}(x,\xi)v(\xi,t)d\xi]dx - a\sum_{k=1}^{3} e^{-\frac{\nu}{\lambda_k}}\alpha_k^2(1,t) \\
&\leq -\eta_1' V_j + (M_0 \sum_{\mathscr{X}\in\mathcal{S}} |\mathscr{X}_j(t) - \mathscr{X}_0| c_1\epsilon_1 + 2q_{1j}^2 + 2q_{2j}^2 + 2q_{3j}^2 \\
&+ \epsilon - a)\beta^2(0,t) + c_2 M_0 \sum_{\mathscr{X}\in\mathcal{S}} |\mathscr{X}_j(t) - \mathscr{X}_0| \int_0^1 [\mathscr{Q},\beta]^T D(x)[\mathscr{Q},\beta]dx \\
&+ \frac{M_0\sum_{\mathscr{X}\in\mathcal{S}} |\mathscr{X}_j(t) - \mathscr{X}_0|}{\lambda_{min}(P)} X^T P X - a\sum_{k=1}^{3} e^{-\frac{\nu}{\lambda_k}}\alpha_k^2(1,t) \\
&\leq -\eta_1' V_j + (M_0 \sum_{\mathscr{X}\in\mathcal{S}} |\mathscr{X}_j(t) - \mathscr{X}_0| c_1\epsilon_1 + 2q_{1j}^2 + 2q_{2j}^2 + 2q_{3j}^2 \\
&+ \epsilon - a)\beta^2(0,t) + \eta_2 M_0 \sum_{\mathscr{X}\in\mathcal{S}} |\mathscr{X}_j(t) - \mathscr{X}_0| V - a\sum_{k=1}^{3} e^{-\frac{\nu}{\lambda_k}}\alpha_k^2(1,t) \qquad (55)
\end{aligned}$$

where η is defined by (33), and

$$\begin{aligned}
\eta_1' &= \nu - \frac{1}{\underline{\Lambda_1} k_1}\left(2\max_{x\in[0,1]} \|\Sigma_1(x)\| + \left(1 + \frac{1}{m}\right)\max_{x\in[0,1]} \|\Sigma_2(x)\|\right) - 2c_2\frac{M_0\overline{\mathscr{X}}}{\epsilon_0 k_1} \\
\eta_2 &= max(c_2, \frac{1}{\lambda_{min}(P)})
\end{aligned}$$

and c_i are positive constants. Notice first that there exists $\eta_3 > 0$, such that for all j

$$-\eta_1' \int_0^1 [\mathscr{Q},\beta]^T D_j(x)[\mathscr{Q},\beta]dx \leq -\eta_3 \int_0^1 [\mathscr{Q},\beta]^T D(x)[\mathscr{Q},\beta]dx. \qquad (56)$$

then

$$\begin{aligned}
&\frac{dV}{dz}(z)h_j(z) \\
&\leq -\eta_3 V(t) + (M_0 \sum_{\mathscr{X}\in\mathcal{S}} |\mathscr{X}_j(t) - \mathscr{X}_0| c_1\epsilon_1 + 2q_{1j}^2 + 2q_{2j}^2 + 2q_{3j}^2 \\
&+ \epsilon - a)\beta^2(0,t) + \eta_2 M_0 \sum_{\mathscr{X}\in\mathcal{S}} |\mathscr{X}_j(t) - \mathscr{X}_0| V - a\sum_{k=1}^{3} e^{-\frac{\nu}{\lambda_k}}\alpha_k^2(1,t) \qquad (57)
\end{aligned}$$

Now, we calculate the second term of L_j, we have:

$$\sum_{\ell\in\mathscr{S}}\left(V_\ell(w)-V_j(w)\right)\tau_{j\ell}=\sum_{\ell=1}^{r}\tau_{j\ell}\left(\int_0^1 (K_0 w(x,t))^\top D_\ell(x)K_0 w(x,t)dx\right.$$
$$\left.-\int_0^1 (K_0 w(x,t))^\top D_j(x)K_0 w(x,t)dx\right)$$
$$\le d_1\sum_{\ell=1}^{r}\tau_{j\ell}\sum_{\mathscr{X}\in\mathcal{S}}|\mathscr{X}_j(t)-\mathscr{X}_0|V(t), \tag{58}$$

Using the property of the expectation and we get

$$\sum_{j=1}^{r}P_{ij}(0,t)L_jV(t)\le -V(t)\Big(\eta_3-d_1Z(t)-(\eta_2M_0+d_1r\tau^*)\sum_{\mathscr{X}\in\mathcal{S}}\mathbb{E}(|\mathscr{X}(t)-\mathscr{X}_0|)\Big)$$
$$-\sum_{k=1}^{3}e^{-\frac{\nu}{\lambda}}\alpha_k^2(1,t) \tag{59}$$

4.3 Proof of Theorem 1

Based on Lemma 3, we introduce the time-varying decay rate function:

$$\phi(t):=\eta_3-d_1\mathcal{Z}(t)-\Gamma\sum_{\mathscr{X}\in\mathcal{S}}\mathbb{E}(|\mathscr{X}(t)-\mathscr{X}_0|)),$$

where $\Gamma=\eta_2M_0+d_1r\tau^\star$ aggregates the system parameters. The exponential Lyapunov functional is constructed as:

$$\Psi(t)=e^{\int_0^t\phi(y)dy}V(t).$$

Applying Dynkin's formula to $\Psi(t)$ yields:

$$\mathbb{E}[\Psi(t)]=\Psi(0)+\mathbb{E}\left[\int_0^t\mathcal{L}\Psi(s)ds\right].$$

From the generator inequality, for some $\tilde{\eta}>0$,

$$\mathcal{L}\Psi(t)=\phi(t)\Psi(t)+e^{\int_0^t\phi(y)dy}\mathcal{L}V(t)\le-\tilde{\eta}\Psi(t),$$

where we used:

$$\mathcal{L}V(t)\le-\phi(t)V(t)-\sum_{k=1}^{3}e^{-\frac{\nu}{\lambda}}\alpha_k^2(1,t)$$
$$\le-\phi(t)V(t)$$

Substituting into Dynkin's formula:

$$\mathbb{E}[\Psi(t)] \leq \Psi(0) - \tilde{\eta} \int_0^t \mathbb{E}[\Psi(s)] ds.$$

By Grönwall's inequality, this implies:

$$\mathbb{E}[\Psi(t)] \leq \Psi(0) e^{-\tilde{\eta} t}.$$

The critical parameter selection:

$$\epsilon^\star = \frac{\eta_3}{2(2d_1 r\tau^\star + \eta_2 M_0)}$$

ensures $\phi(t) \geq \frac{\eta_3}{2} > 0$ for all $t \geq 0$, leading to:

$$\mathbb{E}[V(t)] \leq e^{d_1 \epsilon^\star} e^{-\zeta t} V(0), \quad \zeta = \frac{\eta_3}{2}.$$

The equivalence between $V(t)$ and the system's L^2-norm concludes the mean-square exponential stability proof.

5 Numerical Validation

We simulate the coupled ODE-PDE system with spatial discretization $\Delta x = 0.01$ and time step $\Delta t = 0.005$ over $t \in [0, 300]$. The system parameters are configured as:

$$A = \begin{bmatrix} 0 & 2 \\ -2 & 0 \end{bmatrix}, \quad B = [2, 1]^\top$$

$$\Lambda_{10} = \text{diag}([1, 1, 1]), \quad \Lambda_{20} = 1$$

$$Q_0 = \begin{bmatrix} 1 \\ 1 \\ 1 \end{bmatrix}, \quad R_0 = \begin{bmatrix} 1 & 1 & 1 \end{bmatrix}, \quad C_0 = \begin{bmatrix} 1 & 0 \\ 0 & 0 \\ 0 & 0 \end{bmatrix}$$

for the PDE subsystem, and for the ODE subsystem. The five other possible values are $\in \{0.8, 1.0, 1.1, 1.2, 1.5\}$. And the transition rates τ_{ij} are defined as the same as in [20]. The numerical implementation solves the Kolmogorov forward equations to obtain the time-varying state probabilities, depicted in Fig. 1. These results confirm the Markov process predominantly remains near its nominal operational mode throughout the simulation horizon. Figure 2 demonstrates the closed-loop system response, where all states exhibit asymptotic convergence to zero equilibrium. This numerical behavior rigorously validates the theoretical stability guarantees established in Theorem 1.

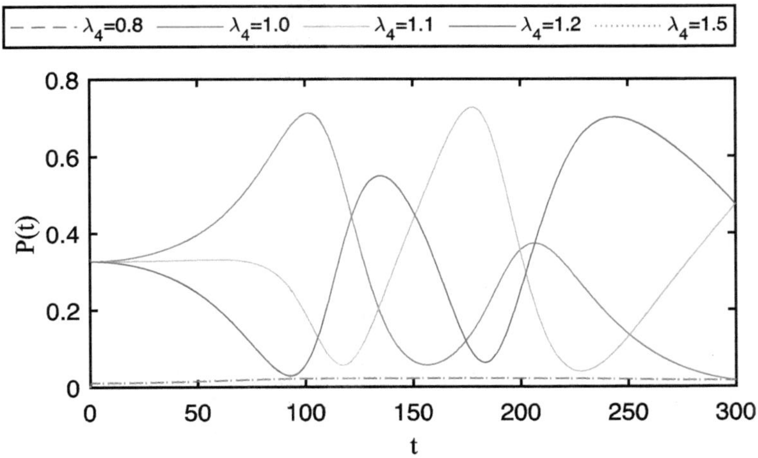

Fig. 1. The probability of Markov states

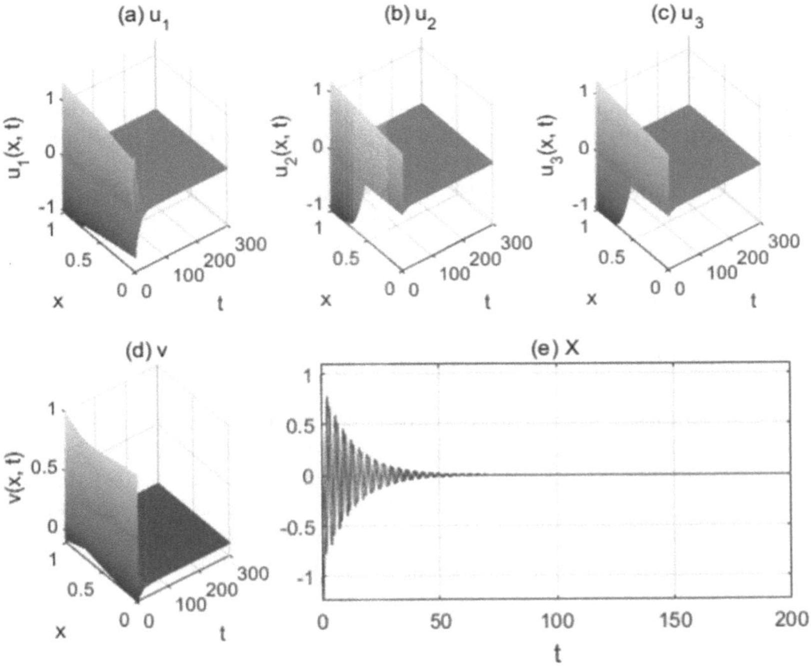

Fig. 2. Closed-loop results of the stochastic system

Acknowledgments. This work is supported by the National Natural Science Foundation of China under Grant 92471108 and 12131008.

Disclosure of Interests. The authors have no competing interests to declare that are relevant to the content of this article.

References

1. Meglio, F., Vazquez, R., Krstic, M.: Stabilization of coupled linear heterodirectional hyperbolic PDE-ODE systems. Automatica **76**, 281–289 (2017). https://doi.org/10.1016/j.automatica.2017.09.027
2. Bastin, G., Coron, J.M.: Stability and Boundary Stabilization of 1-D Hyperbolic Systems. Birkhäuser (2016). https://doi.org/10.1007/978-3-319-32062-5
3. Zhang, L., Prieur, C.: Stochastic stability of Markov jump hyperbolic systems with application to traffic flow control. Automatica **86**, 29–37 (2017). https://doi.org/10.1016/j.automatica.2017.08.007
4. Zhang, Y., Auriol, A., Di Meglio, F.: Robust boundary stabilization of stochastic hyperbolic PDEs, arXiv preprint arXiv:2312.16636 (2023). https://doi.org/10.48550/arXiv.2312.16636
5. Ross, S.M.: Introduction to Probability Models, 10th ed. Academic Press (2010)
6. Krstic, M., Smyshlyaev, A.: Backstepping boundary control for first order hyperbolic PDEs and application to systems with actuator and sensor delays. Syst. Control Lett. **57**(9), 750–758 (2008). https://doi.org/10.1016/j.sysconle.2008.02.005
7. Wang, J.W., Yang, C.Y., Qiu, J.B.: Stochastically exponential stability and stabilization of uncertain linear hyperbolic PDE systems with Markov jumping parameters. Automatica **48**(3), 569–576 (2012). https://doi.org/10.1016/j.automatica.2012.01.006
8. Deutscher, J., Gehring, N.: Output feedback control of general linear heterodirectional hyperbolic PDE-ODE systems with spatially-varying coefficients. Int. J. Control **92**(10), 2274–2290 (2019). https://doi.org/10.1080/00207179.2018.1436770
9. Auriol, J., Bribiesca Argomedo, F.: Output-feedback stabilization of n + m linear hyperbolic ODE-PDE-ODE systems. In: Proceedings of IFAC World Congress, vol. 58, no. 21, pp. 202–207 (2024). https://doi.org/10.1016/j.ifacol.2024.10.213
10. Fischer, F., Deutscher, J., Kurniawan, A.: Fault diagnosis for linear heterodirectional hyperbolic ODE-PDE systems using backstepping-based trajectory planning. Automatica **135**, 109952 (2022). https://doi.org/10.1016/j.automatica.2021.109952
11. Aarabi, E., Ghadiri-Modarres, M., Mojiri, M.: Adaptive backstepping stabilization of thermoacoustic instability in a linearized ODE-PDE Rijke tube model. IEEE Access **11**, 121766–121785 (2023). https://doi.org/10.1109/ACCESS.2023.3329055
12. Xu, X., Liu, Y., Krstic, M.: Backstepping-forwarding designs for doubly-distributed delay-PDE-ODE systems. IEEE Trans. Autom. Control **69**(11), 1–15 (2024). https://doi.org/10.1109/TAC.2024.3383802
13. Zhao, D., Jiang, B., Yang, H.: Backstepping-based decentralized fault-tolerant control of hypersonic vehicles in PDE-ODE form. IEEE Trans. Autom. Control **67**(3), 1210–1225 (2022). https://doi.org/10.1109/TAC.2021.3059689

14. Aarabi, E., Ghadiri-Modarres, M., Mojiri, M.: Adaptive vibration control of the moving cage in the 4×4 hyperbolic PDE-ODE model of the dual-cable mining elevator. IET Control Theory Appl. **19**(1), e70007 (2025). https://doi.org/10.1049/cth2.70007
15. Auriol, J., Pereira, M., Kulcsar, B.: Mean-square exponential stabilization of coupled hyperbolic systems with random parameters. Proc. IFAC World Congress **56**(2), 8153–8158 (2023). https://doi.org/10.1016/j.ifacol.2023.10.991
16. Nikdel, P., Sheikholeslam, F., Krstic, M.: Backstepping boundary control design for a cascaded viscous Hamilton-Jacobi PDE-ODE system. IET Control Theory Appl. **15**(14), 1832–1848 (2021). https://doi.org/10.1049/cth2.12163
17. Xiao, Y., Xu, X., Luo, B., et al.: Adaptive event-triggered tracking control for strict-feedback nonlinear ODE cascaded coupled hyperbolic PDE systems. ISA Trans. (2025, in press). https://doi.org/10.1016/j.isatra.2025.05.019
18. Li, T., Shang, P., Krstic, M.: Control of the transport PDE-ODE cascade with integral-type propagation speed. Math. Control Signals Syst. **37**(2), 451–478 (2025). https://doi.org/10.1007/s00498-025-00407-x
19. Teng, A., Wen, C., Li, Z.: Dynamic event-triggered optimized backstepping containment control of a class of stochastic nonlinear MASs using reinforcement learning. IEEE Internet Things J. (2025, in press). https://doi.org/10.1109/JIOT.2025.3569625
20. De Andrade, G.A.: Backstepping stabilization of a linearized ODE-PDE Rijke tube model. Automatica **96**, 98–109 (2018). https://doi.org/10.1016/j.automatica.2018.06.034
21. Lv, K., Wang, J., Zhang, Y., Yu, H.: Neural operators for adaptive control of freeway traffic. Automatica **182**, 112553 (2025). https://doi.org/10.1016/j.automatica.2025.112553
22. Lv, K., Wang, J., Cao, Y.: Neural operator approximations for boundary stabilization of cascaded parabolic PDEs. Int. J. Adapt. Control Signal Process. **39**(7), 1503–1520 (2025). https://doi.org/10.1002/acs.3902
23. Lyu, K., Biccari, U., Wang, J.: Robust stabilization of hyperbolic PDE-ODE systems via neural operator-approximated gain kernels, arXiv preprint arXiv:2508.03242 (2025). https://doi.org/10.48550/arXiv.2508.03242
24. Lyu, K., Wang, J., Zhang, Y., Yu, H.: Neural operators for adaptive control of traffic flow models, arXiv preprint arXiv:2505.07353 (2025). https://doi.org/10.48550/arXiv.2505.07353

Physically Inspired Intelligent Optimization: from Theory to Real-World Systems

A 3D Gaussian Splatting-Based Framework for Occlusion-Aware Reconstruction of Immovable Cultural Heritage from Crowd-Sourced Data

Jianing Qin, Lan Cheng(✉), Zixi Qi, Zhe Zhang, and Wanchen Liu

Taiyuan University of Technology, Taiyuan 030024, China
18335236230@163.com, taolan_1983@126.com

Abstract. In recent years, the digital preservation of immovable cultural heritage has increasingly relied on 3D reconstruction technologies, with 3D Gaussian Splatting (3DGS) favored for its high-efficiency and high-fidelity rendering quality. However, due to the challenges of performing systematic and controlled image capture at immovable heritage sites, crowd-sourced data offer a low-cost alternative with broad, diverse coverage, albeit often affected by occlusions that degrade 3D reconstruction quality. To address this challenge, we propose an occlusion-aware 3D reconstruction pipeline based on 3DGS and crowd-sourced image collections. Our method integrates an object detection module to identify dynamic occlusions and generate binary masks, followed by a LaMa-based image inpainting model to semantically restore occluded regions. These refined images effectively enhance point cloud quality and are used for high-fidelity reconstruction. Additionally, we incorporate DINOv2 features for pixel-level refinement, improving geometric consistency and the overall accuracy of the reconstructed 3D models. Extensive experiments conducted on multiple public datasets containing immovable cultural heritage demonstrate that our method achieves superior performance in both occlusion removal and reconstruction quality compared to existing approaches, highlighting its strong generalization capability and practical value.

Keywords: Crowd-sourced Data · Occlusion Removal · Immovable Cultural Heritage · 3D Gaussian Splatting (3DGS)

1 Introduction

Digital technologies have opened new possibilities for the accurate documentation and permanent preservation of cultural heritage. These technologies not only enable faithful restoration of the original appearance of artifacts but also provide robust support for subsequent research, restoration, and exhibition. In particular, the development of digital technologies has significantly enhanced the protection and utilization of immovable cultural heritage. Traditional digitization methods, such as 3D scanning [1] and photogrammetry [2], perform well in terms of reconstruction quality. However, these methods still face challenges when applied to immovable heritage. On the one hand,

H. Ma et al. (Eds.): IWACIII 2025, CCIS 2780, pp. 365–376, 2026.
https://doi.org/10.1007/978-981-95-6730-0_26

architectural heritage is often located in open environments, where image capture is inevitably affected by occlusions from crowds, vegetation, and other objects. On the other hand, immovable heritage sites typically involve large volumes of data, making traditional digitization workflows complex, inefficient, and prone to redundancy.

In recent years, Neural Radiance Fields (NeRF) [3] have opened up new directions for cultural heritage digitization, offering high-fidelity rendering and overcoming some limitations of traditional methods in large-scale scenes. However, NeRF relies on implicit neural representations that require hundreds of sampled points per ray, with neural networks predicting color and density for each point—leading to significant computational costs and slow reconstruction. In contrast, 3D Gaussian Splatting (3DGS) [4] explicitly represents scenes with millions of anisotropic Gaussians and employs rasterization-based rendering, avoiding per-point neural inference. As a result, 3DGS achieves comparable rendering quality at rendering speed tens of times faster than NeRF.

While both NeRF and 3DGS have demonstrated excellent performance in static scene reconstruction, they still face significant challenges in real-world applications—particularly in reconstructing immovable heritage that are open to the public. The presence of occlusions such as tourists and vegetation limits camera viewpoints and degrades reconstruction completeness and quality. NeRF-W [5] was the first method to leverage NeRF for scene reconstruction from crowd-sourced images. It introduced a transient head to model occlusion-related density and color, along with an uncertainty field to focus learning on static pixels. However, NeRF-W struggles to recover detailed information under heavy occlusion and often produces artifacts. Although several subsequent NeRF-based methods [6, 7] have improved occlusion handling and reconstruction accuracy, they remain constrained by the inherent inefficiencies of NeRF.

In contrast, 3DGS provides real-time rendering capabilities and holds greater potential for complex scene reconstruction. Recent works [8–10] have attempted to reduce occlusion effects by optimizing the 3DGS training process. However, these methods primarily focus on filtering occlusions during training, overlooking the fact that 3DGS reconstruction quality heavily depends on the distribution of the initial point cloud. As a result, noisy or erroneous points caused by occlusions in the input data still lead to residual artifacts in the final model.

To address this issue, we propose a 3DGS reconstruction pipeline integrated with image inpainting, which removes occlusions and filters noise before training begins, thereby reducing the impact of erroneous points in the initial point cloud. Specifically, we first adopt the Resolution-Robust Large Mask Inpainting (LaMa) [11] to remove prominent occlusions from the images, recovering the basic structure of the heritage scene. Then, we employ the DINOv2 [12] visual representation model to perform pixel-level occlusion refinement, forming a coarse-to-fine occlusion removal pipeline.

Our contributions are summarized as follows:

(1) We propose a novel occlusion removal pipeline based on crowd-sourced data, which progressively eliminates occlusions and restores the occluded details of immovable heritage sites.
(2) We employ the LaMa model to inpaint large occluded regions and use DINOv2 features to refine residual occlusions, achieving fine-grained occlusion removal.

(3) We validate the effectiveness and robustness of our method on public datasets, such as Photo Tourism dataset and NeRF-OSR dataset.

2 Preliminaries

We implement our method based on the 3D Gaussian Splatting pipeline. 3DGS represents a scene using a set of anisotropic 3D Gaussians, each of which is derived from an initial sparse point cloud through a Gaussian function. Each Gaussian is characterized by the following parameters:

$$G_i : \{\mu_i, \Sigma_i, \alpha_i, SH_i\} \tag{1}$$

Here, μ_i denotes the center position of the Gaussian, Σ_i is a semi-definite covariance matrix that controls the shape and orientation, α_i represents the opacity, and SH_i spherical harmonics coefficients encoding the color.

To ensure numerical stability and maintain the positive semi-definite property of the covariance matrix Σ_i, it is decomposed into a rotation matrix R_i and a scaling matrix S_i as follows:

$$\Sigma_i = R_i S_i {S_i}^T {R_i}^T \tag{2}$$

During the rendering stage, each 3D Gaussian is projected onto the 2D image plane using the intrinsic camera matrix K and a viewing transformation matrix W. This yields the 2D mean position $\mu_i^{'}$ and projected 2D covariance matrix $\Sigma_i^{'}$:

$$\mu_i' = KW\mu_i, \quad \Sigma_i' = JW\Sigma_i W^T J^T \tag{3}$$

where J represents the Jacobian of the affine approximation for the projective transformation. Subsequently, rasterization techniques are applied to compute the color of each pixel C_{pixel} in the rendered image:

$$\begin{gathered} C_{pixel} = \sum_{i \in N} T_i \alpha_i' C_i \\ where\ T_i = \prod_{j=1}^{i-1} \left(1 - \alpha_j'\right), \quad \alpha_i' = \alpha_i \times e^{-\frac{1}{2}(x' - \mu_i')^T \Sigma_i'^{-1} (x' - \mu_i')} \end{gathered} \tag{4}$$

Finally, we compute the reconstruction error between the rendered image C_{pixel} and the reference image C_{gt} using a combination of L1 loss and SSIM loss:

$$L_{total} = L_1\left(C_{pixel}, C_{gt}\right) + L_{SSIM}\left(C_{pixel}, C_{gt}\right) \tag{5}$$

This loss serves as the optimization objective, based on which the parameters of Gaussians are iteratively updated, thereby progressively improving the reconstruction quality.

3 Methodology

We propose an occlusion-aware 3D reconstruction framework for immovable cultural heritage sites, aiming to preprocess and optimize reconstruction in occluded scenes using crowd-sourced image collections. The proposed method comprises four main components:

(1) pedestrian detection and the generation of corresponding occlusion masks using an object detection model;
(2) image inpainting to restore occluded regions and improve point cloud quality;
(3) high-fidelity 3D reconstruction based on the 3D Gaussian Splatting (3DGS) algorithm;
(4) pixel-level refinement of the reconstruction results using DINOv2 features to further enhance occlusion removal and reconstruction accuracy.

Figure 1 illustrates the overall pipeline of the proposed method.

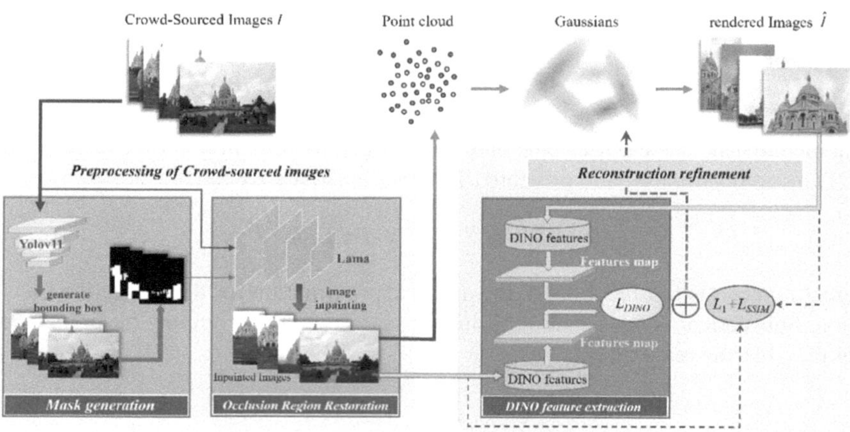

Fig. 1. Overview over the occlusion removal pipeline for immovable cultural heritage. The framework includes: (1) occlusion detection and mask generation via object detection; (2) image inpainting to restore occluded regions; and (3) high-fidelity 3D reconstruction and refinement using 3D Gaussian Splatting (3DGS) enhanced with DINOv2 features.

3.1 Preprocessing of Crowd-Sourced Images

Crowd-sourced images are typically unstructured and heterogeneous due to variations in users, devices, times, and viewpoints, leading to inconsistent geometry, floating artifacts, and incomplete 3D models. To address these issues, we design a preprocessing strategy that employs a robust object detection algorithm, YOLOv11, to automatically identify occlusions in the images, regardless of viewpoint or camera type. The resulting bounding boxes are converted into binary masks that guide the occlusion removal process. We then adopt the LaMa model, which is highly effective in inpainting large missing regions, particularly when provided with regular-shaped masks like rectangles. This model enables semantically consistent and scalable restoration without reliance on camera poses or user prompts. The preprocessing comprises mask generation and occlusion region restoration, which effectively mitigates the impact of occlusions and enhances the quality of crowd-sourced inputs for 3D reconstruction.

Mask Generation. In crowd-sourced image collections of immovable cultural heritage, pedestrians often constitute major and large-scale occlusions. To identify such occlusions, we employ the YOLOv11 object detection model trained on the COCO dataset to detect objects labeled as "person." Each detection is represented by a bounding box, which is then converted into a binary mask: occluded regions are marked as 1 (white), and non-occluded regions are marked as 0 (black).

Let $I \in \mathbb{R}^{H \times W \times 3}$ be an input RGB image. The object detector O produces a set of bounding boxes $B = \{b_i\}_{i=1}^{n}$. Each b_i corresponds to a human in the image. From B, we construct a mask $M\{0, 1\}^{H \times W}$ such that pixels within any b_i are set to 1.

Occlusion Region Restoration. To effectively remove the detected occlusions and recover the underlying heritage structures, we integrate the LaMa image inpainting model. Compared with other inpainting approaches [13, 14], LaMa can generate semantically plausible and structurally consistent results even in the presence of large missing regions, as shown in Fig. 2. Moreover, LaMa exhibits strong adaptability to images of varying resolutions, making it particularly well-suited for crowd-sourced datasets composed of heterogeneous image qualities and resolutions.

Fig. 2. Inpainting results from different models. Given the original occluded image and its corresponding mask, we compare the outputs of LDM [13], DiffIR [14], and LaMa [11]. LaMa produces more semantically coherent and structurally consistent restorations, particularly in regions with large occlusions.

The inpainting model F takes the original image I and the occlusion mask M, and outputs the inpainted image I':

$$I' = F(I, M) \tag{6}$$

3.2 Reconstruction Refinement

Once occluded regions are removed and semantically plausible content is inpainted, the restored image set is fed into the 3D Gaussian Splatting (3DGS) pipeline to enable high-quality 3D reconstruction. To facilitate 3DGS training, we employ COLMAP to estimate camera poses from the inpainted images. These poses, together with the corresponding restored images, serve as inputs to the 3DGS optimization process.

Traditional 3D reconstruction pipelines typically rely on low-level photometric consistency losses (e.g., L1, MSE [15], SSIM [16]). However, under non-ideal conditions such as occlusions, lighting variations, or incomplete image restoration, these loss functions often fail to provide sufficient constraints, leading to degraded reconstruction quality. Given the significant visual variability commonly present in crowdsourced image collections, as well as the potential lack of complete semantic information in the restored images, we introduce a feature-guided optimization module based on DINOv2 features to enhance the model's robustness in handling occlusions.

Specifically, for each image $\hat{I}$ rendered by the 3DGS pipeline and its corresponding original training image I, we extract their corresponding DINOv2 feature maps $\Phi(\hat{I})$, $\Phi(I)$, which are subsequently used to compute a perceptual consistency loss and guide the reconstruction process at a semantic level. The perceptual consistency loss at each pixel block (i, j) be defined as the cosine similarity between the corresponding DINOv2 feature vectors from the rendered image and the original input image:

$$L_{feature}^{(i,j)} = 1 - \frac{\Phi_{i,j}(\hat{I}) \cdot \Phi_{i,j}(I)}{\left\| \Phi_{i,j}(\hat{I}) \right\|_2 \cdot \left\| \Phi_{i,j}(I) \right\|_2} \tag{7}$$

This loss encourages the rendered image to be semantically consistent with the original one. The overall feature loss for an image is computed as the average perceptual loss across all pixel blocks:

$$L_{DINO} = \frac{1}{hw} \sum_{i=1}^{h} \sum_{j=1}^{w} L_{feature}^{(i,j)} \tag{8}$$

Finally, this feature loss is jointly optimized with the photometric rendering loss during the 3DGS training phase:

$$L_{ALL} = L_{total} + L_{DINO} \tag{9}$$

That ensures that the reconstructed results are consistent with the input images both visually and semantically.

4 Experiments

To evaluate the effectiveness of our proposed pipeline in handling occlusions in crowd-sourced data, we conducted experiments on two datasets with strong crowd-sourced characteristics. The Photo Tourism dataset [17] consists of a large number of unstructured Internet images covering various well-known landmarks, featuring realistic and

Fig. 3. Qualitative comparisons on the Photo Tourism dataset across three representative scenes: *Brandenburg Gate*, *Sacre Coeur*, and *Trevi Fountain*. Our method reconstructs finer structural details and better preserves geometry.

complex scenes. The NeRF-OSR dataset [18], specifically designed for outdoor neural radiance field reconstruction, contains high-resolution, real multi-view images captured at different times of day. We compare our method against NeRF [3], 3DGS [4] and two recent state-of-the-art methods for occlusion handling: GS-W [9] and WildGaussians [8].

In our experiments, we selected three scenes from the Photo Tourism dataset (*Sacre Coeur*, *Trevi Fountain* and *Brandenburg Gate*) and three scenes from the NeRF-OSR dataset (*europa*, *lwp*, and *st*). For each scene, we divided the available images into a training and test set according to the standard practice in prior work [8]. Specifically, approximately 80% of the images were used for training and 20% for testing, with exact numbers varying across scenes depending on image availability. The detailed counts for each scene are reported in the supplementary material. All models were trained on a single Nvidia RTX A6000 GPU for 30,000 steps per scene. We quantitatively evaluated the performance of our model using three commonly used standard metrics: PSNR, SSIM [16], and LPIPS [19].

Fig. 4. Qualitative results on the NeRF-OSR dataset. Compared to baseline methods, our pipeline offers sharper and more complete reconstructions, especially in occluded areas.

4.1 Comparison on the Photo Tourism

Table 1 and Fig. 3 present the quantitative and qualitative comparisons on the Photo Tourism dataset. Following the same data split protocol as NeRF-W, our method demonstrates consistently strong performance across all evaluated scenes. Compared to NeRF and 3DGS, the proposed approach yields significant improvements in reconstruction accuracy. Furthermore, it achieves comparable or superior performance to recent state-of-the-art methods designed for in-the-wild reconstruction.

As illustrated in Fig. 3, our method produces more accurate and structurally consistent reconstructions. For instance, in the *Trevi Fountain* scene, fine-grained features such as cracks in stone surfaces are more faithfully restored. Similarly, in the *Brandenburg Gate* scene, structural details like the spokes of the statue's wheels are reconstructed with higher fidelity compared to GS-W and WildGaussians. By employing YOLOv11-based occlusion detection and LaMa-based inpainting, our method ensures that only clean and semantically consistent content contributes to the reconstruction pipeline. This preprocessing reduces artifacts, improves scene completeness, and enhances geometric consistency, particularly in regions heavily affected by occlusions.

Table 1. Comparison on the Photo Tourism Dataset with NeRF [3], 3DGS [4], GS-W [9], WildGaussians [8]. Adopting the same data split as NeRF-W [5], our method achieves superior reconstruction quality across all scenes.

	Sacre Coeur			Trevi Fountain			Brandenburg Gate		
	PSNR↑	SSIM↑	LPIPS↓	PSNR↑	SSIM↑	LPIPS↓	PSNR↑	SSIM↑	LPIPS↓
NeRF	15.60	0.72	0.29	16.14	0.60	0.37	18.90	0.82	0.23
3DGS	16.68	0.82	0.22	15.12	0.56	0.46	19.87	0.88	0.18
GS-W	22.79	**0.86**	0.18	20.87	0.72	0.30	25.40	0.90	**0.17**
WildGaussians	22.47	0.85	0.19	20.43	0.70	0.33	25.48	0.90	0.18
Ours	**22.90**	**0.86**	**0.17**	**21.18**	**0.73**	**0.29**	**25.55**	**0.91**	**0.17**

4.2 Comparison on the NeRF-OSR Dataset

As shown in Table 2 and Fig. 4, our method outperforms the baselines in terms of reconstruction accuracy while effectively mitigating the visual artifacts introduced by dynamic occluders in the NeRF-OSR dataset. Although 3DGS benefits from the use of high-resolution inputs and achieves relatively sharp reconstructions, it exhibits noticeable ghosting artifacts in occluded areas. In contrast, our pipeline restores occluded content with high consistency while preserving fine geometric details, as in the *st* scene where occluded structures such as windows are accurately reconstructed without visual distortions.

Table 2. Comparison on the NeRF-OSR dataset. Our method achieves superior reconstruction quality across all scenes.

	europa			lwp			st		
	PSNR↑	SSIM↑	LPIPS↓	PSNR↑	SSIM↑	LPIPS↓	PSNR↑	SSIM↑	LPIPS↓
3DGS	15.17	0.60	0.47	14.59	0.58	0.41	13.39	**0.62**	0.42
GS-W	14.93	0.66	0.46	15.03	0.52	0.50	12.57	0.54	0.43
WildGaussians	15.76	0.65	0.46	15.16	0.56	0.49	12.64	0.58	0.55
Ours	**16.95**	**0.67**	**0.44**	**15.20**	**0.60**	**0.40**	**13.48**	0.58	**0.41**

However, in the *st* scene, our method slightly underperforms 3DGS. We attribute this phenomenon to the fact that the scene contains minimal occlusions, limiting the benefits of our occlusion-aware pipeline. In addition, the reduced image resolution used in our method slightly affects visual fidelity, whereas 3DGS fully utilizes high-resolution inputs. Nevertheless, our approach remains robust across most scenes. Its current limitation lies in handling severe lighting inconsistencies due to the lack of explicit illumination modeling, occasionally leading to incomplete reconstructions in regions such as the sky. Incorporating lighting-aware priors is a promising direction for future improvement.

4.3 Qualitative Evaluation of Occlusion Removal

Compared to state-of-the-art outdoor occlusion-aware reconstruction methods, our approach demonstrates superior performance in recovering occluded architectural structures while maintaining structural consistency in Fig. 5. In the *europa* scene, although some fine details are missing, our method effectively restores buildings occluded by vegetation, outperforming other baselines. In the *Sacre Coeur* and *lwp* scenes, while other methods approximately reconstruct the target structures, the results often exhibit significant blurring. In contrast, our method yields clearer reconstructions and achieves more accurate geometric details. These results confirm that removing occlusions prior to 3D modeling enables 3D Gaussian Splatting to utilize cleaner inputs, improving robustness on real-world crowd-sourced scenes with complex occlusions.

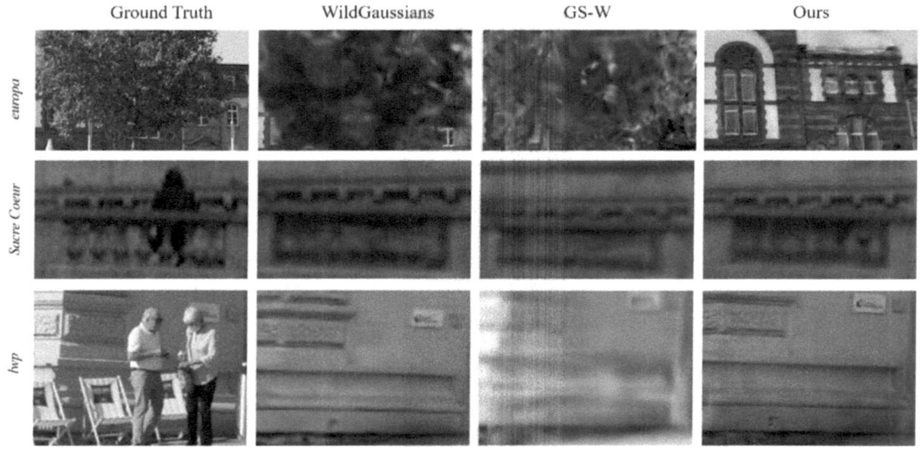

Fig. 5. Compared to the state-of-the-art outdoor occlusion-aware reconstruction methods GS-W [9] and WildGaussians [8].

5 Conclusion

In this paper, we propose a novel occlusion removal pipeline based on 3DGS for the task of immovable cultural heritage reconstruction using crowd-sourced data. Our approach integrates object detection and image inpainting techniques to automatically identify and repair large occluded regions in images, significantly improving the quality of input data and providing richer information for subsequent 3D modeling. Experimental results demonstrate that our method achieves favorable reconstruction performance across multiple challenging scenes, with particularly notable improvements in occlusion region restoration.

While the proposed pipeline effectively removes occlusions and improves reconstruction quality, it introduces additional preprocessing steps that increase computational overhead compared to standard 3DGS. Since our primary objective is to enhance robustness against occlusions, we deliberately prioritize reconstruction accuracy over runtime efficiency. Future work will therefore focus on integrating lightweight detection models and more efficient inpainting strategies to reduce computational cost without compromising reconstruction fidelity. In addition, inherent lighting variations in crowd-sourced images remain a challenge for occlusion restoration, indicating that incorporating lighting-aware priors could further improve the robustness and accuracy of the proposed method.

Acknowledgments. The work was supported by the National Natural Science Foundation of China (62073232) and Natural Science Foundation of Shanxi Province (202403021221054).

Disclosure of Interests The authors have no competing interests to declare that are relevant to the content of this article.

References

1. Kantaros, A., Ganetsos, T., Petrescu, F.I.T.: Three-dimensional printing and 3D scanning: emerging technologies exhibiting high potential in the field of cultural heritage. Appl. Sci. **13**(8), 4777 (2023)
2. Shan, J., Li, Z., Lercel, D., et al.: Democratizing photogrammetry: an accuracy perspective. Geo-spat. Inf. Sci. **26**(2), 175–188 (2023)
3. Mildenhall, B., Srinivasan, P.P., Tancik, M., et al.: Nerf: representing scenes as neural radiance fields for view synthesis. Commun. ACM. **65**(1), 99–106 (2021)
4. Kerbl, B., Kopanas, G., Leimkühler, T., et al.: 3D Gaussian splatting for real-time radiance field rendering. ACM Trans. Graph. **42**(4), 139:1–139:14 (2023)
5. Martin-Brualla, R., Radwan, N., Sajjadi, M.S.M., et al.: NeRF in the wild: Neural radiance fields for unconstrained photo collections. In: Proceedings of the IEEE/CVF Conference on Computer Vision and Pattern Recognition, pp. 7210–7219 (2021)
6. Chen, X., Zhang, Q., Li, X., et al.: Hallucinated neural radiance fields in the wild. In: Proceedings of the IEEE/CVF Conference on Computer Vision and Pattern Recognition, pp. 12943–12952 (2022)
7. Yang, Y., Zhang, S., Huang, Z., et al.: Cross-ray neural radiance fields for novel-view synthesis from unconstrained image collections. In: Proceedings of the IEEE/CVF International Conference on Computer Vision, pp. 15901–15911 (2023)
8. Kulhanek, J., Peng, S., Kukelova, Z., et al: Wildgaussians: 3d gaussian splatting in the wild. arXiv preprint arXiv:2407.08447 (2024)
9. Zhang, D., Wang, C., Wang, W., et al.: Gaussian in the wild: 3D Gaussian splatting for unconstrained image collections. In: European Conference on Computer Vision, pp. 341–359. Springer, Cham (2024)
10. Xu, C., Kerr, J., Kanazawa, A.: Splatfacto-W: a Nerfstudio implementation of Gaussian splatting for unconstrained photo collections. arXiv preprint arXiv:2407.12306 (2024)
11. Suvorov, R., Logacheva, E., Mashikhin, A., et al.: Resolution-robust large mask inpainting with Fourier convolutions. In: Proceedings of the IEEE/CVF Winter Conference on Applications of Computer Vision, pp. 2149–2159 (2022)
12. Oquab, M., Darcet, T., Moutakanni, T., et al.: DINOv2: Learning robust visual features without supervision. arXiv preprint arXiv:2304.07193 (2023)
13. Rombach, R., Blattmann, A., Lorenz, D., et al.: High-resolution image synthesis with latent diffusion models. In: Proceedings of the IEEE/CVF Conference on Computer Vision and Pattern Recognition, pp. 10684–10695 (2022)

14. Xia, B., Zhang, Y., Wang, S., et al.: DiffIR: efficient diffusion model for image restoration. In: Proceedings of the IEEE/CVF International Conference on Computer Vision, pp. 13095–13105 (2023)
15. Wang, Z., Bovik, A.C.: Mean squared error: love it or leave it? A new look at signal fidelity measures. IEEE Signal Process. Mag. **26**(1), 98–117 (2009)
16. Wang, Z., Bovik, A.C., Sheikh, H.R., et al.: Image quality assessment: from error visibility to structural similarity. IEEE Trans. Image Process. **13**(4), 600–612 (2009)
17. Snavely, N., Seitz, S.M., Szeliski, R.: Photo tourism: exploring photo collections in 3D. In: ACM SIGGRAPH 2006 Papers, pp. 835–846 (2006)
18. Rudnev, V., Elgharib, M., Smith, W., et al.: NeRF for outdoor scene relighting. In: European Conference on Computer Vision, pp. 615–631. Springer, Cham (2022)
19. Zhang, R., Isola, P., Efros, A.A., et al.: The unreasonable effectiveness of deep features as a perceptual metric. In: Proceedings of the IEEE Conference on Computer Vision and Pattern Recognition, pp. 586–595 (2009)

G-COA: An Optimization Algorithm Inspired by Celestial Motion Behavior

Kai Wang[1], Kuo Yang[2(✉)], Mifeng Ren[1], Wenjie Zhang[1], and Zhile Yang[3]

[1] College of Electrical Power and Engineering, Taiyuan University of Technology, 79 Yingze West Street, Taiyuan 030024, China

[2] PLA, Beijing 100043, China

yangkuo1985@163.com

[3] Shenzhen Institutes of Advanced Technology, Chinese Academy of Sciences, No. 1068, Xueyuan Avenue, Nanshan District, Shenzhen 518055, China

Abstract. This study proposes an intelligent optimization algorithm inspired by astrophysical kinematics, called gravitational-collision optimization algorithm (G-COA). In a galaxy, the mass of the star accounts for more than 90% of the whole galaxy. According to the size of gravity, it is in the absolute dominant position of the star system. Several planets revolve around them in different orbits. In addition, most of the planets are surrounded by several satellites orbiting the planets. Satellites often encounter the collision of small celestial bodies, resulting in changes in their speed and position, or getting rid of the gravity of the central celestial body, or falling into the central celestial body. The optimization algorithm proposed in this study is based on the motion behavior of these celestial bodies to update the solution, and then find the optimal solution after multiple iterations. Each celestial body represents a feasible solution. G-COA can dynamically adjust exploration and development functions at different stages of iteration. Based on some existing astrophysics-inspired optimization algorithms, this study improves the gravitational search strategy and integrates more celestial motion behaviors into the algorithm. In order to verify the effectiveness of each search strategy, ablation experiments are carried out. Based on the CEC2022 test function set, several latest algorithms and highly cited algorithms are compared. The optimization results of 12 of the 16 test functions are optimal. Finally, two engineering constraint problems are listed, which further proves that G-COA is a promising algorithm.

Keywords: Heuristic algorithm · Astrophysics · Gravity operator · Collision operator

1 Introduction

In the field of engineering optimization, it is often hoped that some of the objectives concerned can achieve the optimal value, so as to achieve the purpose of saving costs and improving the quality of the project [1]. The expression of the objective function of

H. Ma et al. (Eds.): IWACIII 2025, CCIS 2780, pp. 377–394, 2026.
https://doi.org/10.1007/978-981-95-6730-0_27

the traditional optimization method based on mathematical programming is often simple, and the constraints are less. Most of them are continuous derivable functions, and the extreme value of the objective function can be obtained by derivation [2]. However, most engineering optimization problems have complex objective functions and many constraints, and their gradients cannot be obtained explicitly. The heuristic optimization algorithm makes the current solution reach the neighborhood solution by imitating the operating mechanism of the natural law, and then the optimal solution is finally retained by the survival of the fittest [3]. The heuristic algorithm has great advantages in solving multi-objective and variable discontinuous optimization problems. The algorithm is simple and easy to implement. Heuristic optimization algorithms can be divided into several categories: based on biological group behavior, based on biological evolution theory, based on human social behavior and based on the laws of physical chemistry [4].

The inspiration of biological group behavior comes from social activities such as foraging and migration of some animals such as birds. For example, Pavel Trojovský et al. [5] simulated the natural behaviors such as foraging, migration, escape and anti-predator of walrus, and proposed Walrus Optimization Algorithm (WaOA). Fevzi Tugrul Varna et al. [6] proposed two particle swarm optimization algorithms, Biased Eavesdropping PSO (BEPSO) and Altruistic Heterogeneous PSO (AHPSO), based on the modeling of "eavesdropping behavior " and "altruistic behavior "in nature. Dler O. Hasan et al. [7] proposed the Griffon Vultures Optimization Algorithm (GVOA) inspired by the intelligent foraging behavior of vultures, which combines four strategies: following, group foraging, independent search and wreck detection.

The optimization algorithm based on biological behavior has strong population cooperation and adaptive ability, and has a natural diversity maintenance mechanism. The disadvantage is that although there are various variants of biological behavior algorithm, the flexibility of integrating new mechanisms is poor due to the limitation of simulation objects. The algorithm based on biological evolution theory is easy to flexibly integrate with local search, constraint processing, machine learning and other methods, resulting in 'hybridization' or multi-level optimization framework, corresponding to different optimization problems. There are different variants. Xavier Sécheresse et al. [8] combined genetic algorithm with a variety of prompt generation strategies, and automatically optimized prompt through multi-generation evolution, which effectively improved the performance of LLM under multi-task and multi-data sets, a new algorithm SparseEA-AGDS for large-scale sparse multi-objective optimization is proposed. Xia Wang et al. [9] significantly improved the convergence and sparse solution quality of the algorithm on high-dimensional sparse multi-objective problems by combining adaptive genetic operators and dynamic scoring mechanisms.

Although the above two types of algorithms have a certain degree of adaptation and collaboration, most of them are based on simple mechanisms such as pheromones and neighborhood following, mainly rely on global optimization or local neighbors, and the collaboration and division of labor are relatively simple. The human-based social behavior algorithm emphasizes dynamic adaptation and role switching, and can flexibly adjust strategies according to individual or group performance to support multi-role division of labor and collaboration. Ning Fang et al. [10] proposed a new heuristic optimization algorithm for human-like game-Offensive Defensive Optimization (ODO) by transforming a variety of offensive and defensive strategies in chess games into mixed search behaviors. Xiang Li et al. [11] proposed Dragon Boat Optimization (DBO), which is inspired by the team collaboration of dragon boat race. It models the role collaboration and social psychological mechanism of drummer, oar and helmsman. Through a variety of dynamic state update strategies, it effectively improves the diversity, adaptability and global optimization ability of the algorithm.

In the optimization algorithm, there are often some parameters that need to be set manually, and setting the parameter values only by experience may not achieve the best results. The algorithm based on the laws of physical chemistry has a strong theoretical basis in parameter setting, and there is less possibility of subjectivity. In this kind of algorithm, the optimization algorithm inspired by the law of celestial motion is one of the branches, and many algorithms with excellent performance are derived. Mohamed Abdel-Basset et al. [12] proposed Kepler optimization algorithm (KOA) by abstracting the elliptical orbit, gravity, mass and velocity of the planets around the sun as the dynamic update mechanism of the optimization operator. Gang Hu et al. [13] integrated a variety of search strategies on the basis of [12] to further improve the performance of KOA. Amr A. Abd El Mageed et al. [14] proposed an enhanced binary Kepler optimization algorithm (BKOA-MUT) integrated with DE mutation mechanism, which is specially used for feature selection tasks. Aiming at the problem of feature classification in the field of medical imaging, Essam H. Houssein et al. [15] improved KOA by adding integrated opposition learning and local escape operator, which effectively improved the quality of feature selection.

The contributions of this study are as follows :

1. According to the law of kinematics and the idea of infinitesimal, the calculation method of celestial body motion speed is reconstructed, so that the number of self-setting parameters is controlled, and the variable step size strategy is used to adaptively search different stages to balance exploration and development.
2. More celestial bodies with different attributes are added to the population, and celestial bodies with different attributes are generated according to the fitness, and no new individuals are generated. In order to explore the solution space widely, inspired by the collision phenomenon of celestial bodies, the collision mechanism is established as a disturbance.
3. Compared with several newly proposed algorithms or highly cited algorithms, the convergence performance is compared, and the degree of action of each operator is proved by ablation experiments. The practicability of the proposed algorithm is further proved by substituting it into practical engineering problems.

2 C-G Optimization Algorithm

2.1 Define the Solution Space

For any objective function $F(X^N)$, the independent variable X^N is a vector with N degrees of freedom. The definition of X^N is as follows :

$$X^N = (x_1, x_2, \ldots, x_N), x_1 \in U_1, x_2 \in U_2, \ldots, x_N \in U_N \tag{1}$$

where x_i, $i \in \{1, 2, \cdots N\}$ is the variable of X^N on the i-th degree of freedom, and U_i, $i \in \{1, 2, \cdots N\}$ is the space of values of x_i. Then the solution space of variable X^N is:

$$U_{X^N} = (U_1, U_2, \cdots, U_N) \tag{2}$$

It is assumed that the constraints of the optimization problem constitute a set $D(\psi)$. When $D(\psi) = \emptyset$, that is, the optimization problem has no constraints, the value space of the variable x_i is $U_{\mathrm{i}}^{D(\psi)=\varnothing}$. When $D(\psi) \neq \emptyset$, the value space of the variable x_i satisfying the constraint condition is $\arg_{x_i} D(\psi)$. In summary, the value space of x_i is expressed as follows :

$$U_i = U_i^{D(\psi)=\varnothing} \cap \arg_{x_i} D(\psi) \tag{3}$$

2.2 Formation of the Initial Population

By arbitrarily selecting M feasible solutions in the solution space, a set X_M^N is defined such that $X_M^N = \{X_m^N | m = 1, 2, \cdots, M\}$, where X_M^N denotes the m-th feasible solution selected. The objective function value $F(X_m^N)$ is obtained by substituting the $\forall X_m^N \in X_M^N$ into the objective function $F(X^N)$. The M feasible solutions are regarded as M celestial bodies in the solution space, and the objective function value is regarded as the quality of the corresponding celestial bodies. Among them, the objective function value with the best fitness for the optimization problem is recorded as $F_{best}(X_{m'}^N)$, and the objective function value with the worst fitness is recorded as $F_{worst}(X_{m''}^N)$, $m' \neq m''$. In order to avoid the large difference in fitness, some local optimal values occupy a dominant position for a long time in the previous iteration process, and the population fitness value needs to be normalized. The expression of fitness normalization is as follows:

$$\tilde{F}(X_m^N) = \frac{F(X_m^N)}{\sum_{m=1}^{M} F(X_m^N)} \tag{4}$$

$\tilde{F}(X_m^N)$ is the normalized fitness value. The $\tilde{F}(X_m^N)$ is arranged in order of advantages and disadvantages. The solution with the best fitness is regarded as the central celestial body of the galaxy, and the solutions of the remaining fitness are divided into two groups. One of the two sets of solutions with better fitness is used as a planet to orbit around the fixed star, which is convenient for exploring new solutions in the spatial range affected by the gravitational field of the fixed star and developing better solutions around the fixed star. Although the global optimal solution may appear in the vicinity of the current optimal solution, it may also exist in the vicinity of those current non-optimal solutions, especially those feasible solutions with a small gap between the fitness value and the current optimal solution, just because these feasible solutions are far away from the global optimal solution in the current explored range, and there is no potential information that the optimal solution exists in the vicinity. In order to explore the surrounding space of some non-optimal solutions, a set of solutions with poor fitness in the two sets of solutions is regarded as a satellite orbiting the planet. The principle of assigning satellites to planets is as follows :

$$\min \sum \left[\tilde{F}\left(X_{m_p}^N\right) - \tilde{F}\left(X_{m_s}^N\right) \right] \tag{5}$$

The normalized planetary fitness value is denoted by $\tilde{F}\left(X_{m_p}^N\right)$, and the normalized satellite fitness value is denoted by $\tilde{F}\left(X_{m_s}^N\right)$. The above formula shows that the planets with better fitness should also match the satellites with better fitness. This is because the position of individuals with similar fitness in the solution space is also similar. If the distance between the satellite and the planet is too close, the range of the orbit formed by the gravitational operator is very small, and the function of exploring in a larger space range is lost. If the distance is too far, the vast range of solution space between the central celestial body and the satellite orbit will not be explored. Therefore, it is necessary to control the distance between the planet and the satellite through an appropriate fitness matching mechanism. Taking the objective function with two degrees of freedom as an example, the positions and attributes of the generated initial population individuals are shown in Fig. 1:

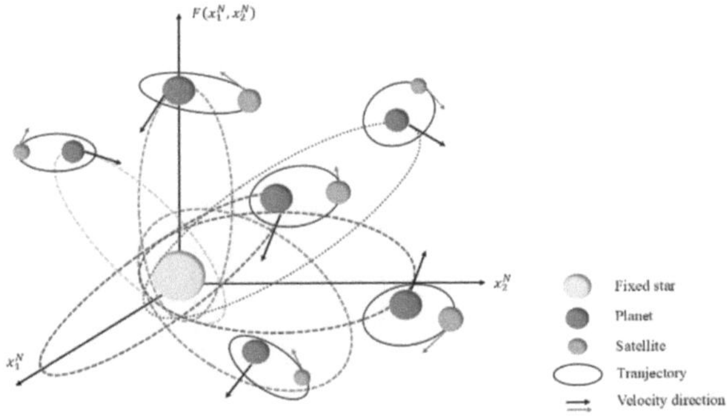

Fig. 1. The schematic diagram of initial population formation

2.3 Gravity Operator

The size and quality of gravity are closely related. This feature can guide all individuals to move towards excellent solutions, and has strong global optimization ability, which is suitable for multi-peak and high-dimensional complex problems [15]. In the intelligent optimization algorithm inspired by Kepler's law of celestial motion, the gravitational constant is a quantity that changes with time, and its value depends on the setting of parameters μ_0 and γ [12]. The values of μ_0 and γ will directly affect the effect of algorithm optimization, and the optimal parameter values need to be determined through multiple experiments. The purpose of setting the dynamic change of the gravitational constant is to adjust the accuracy of the search. In the early stage of exploration, the gravitational constant is large, which increases the search step and the semi-major axis of the orbit. In the later stage, the gravitational constant is reduced, so that the search operator updates the neighborhood solution in a small step in the region closer to the current optimal solution. The algorithm designed in this paper avoids the artificial setting of parameters as much as possible to reduce the influence of subjective factors on the performance of the algorithm. Assume that all the planets or satellites at the initial time start from the near-central celestial point of the orbit, and randomly assign the eccentricity e_m and the revolution period T_m of the orbit for X_m^N, $m \in \{m_p, m_s\}$, where $e_m \sim U(0, 1)$, $T_m \sim U(0, 1)$. The expression of the distance from the planet $X_{m_p}^N$ to the fixed star X_{best}^N or the distance from the satellite $X_{m_s}^N$ to its central planet $X_{m_p}^N$ is as follows :

$$\overrightarrow{R}_{X_{m_p|m_s}^N}^{X_{best|m_p}^N} = X_{best|m_p}^N - X_{m_p|m_s}^N \tag{6}$$

$\overrightarrow{R}_{X_{m_p|m_s}^N}^{X_{best|m_p}^N}$ denotes the Euclidean distance from a planet or satellite to its central object. The normalized distance expression is:

$$\tilde{\overrightarrow{R}}_{X_{m_p|m_s}^N}^{X_{best|m_s}^N} = \frac{\overrightarrow{R}_{X_{m_p|m_s}^N}^{X_{best|m_s}^N} - \min\left(\overrightarrow{R}_{X_{m_p|m_s}^N}^{X_{best|m_s}^N}\right)}{\max\left(\overrightarrow{R}_{X_{m_p|m_s}^N}^{X_{best|m_s}^N}\right) - \min\left(\overrightarrow{R}_{X_{m_p|m_s}^N}^{X_{best|m_s}^N}\right)} \tag{7}$$

According to Kepler's third law, the expression of the initial gravitational constant is obtained as follows:

$$\mu_{m_p|m_s}^0 = 4\pi^2 \frac{\left(\frac{\left\|\tilde{\overrightarrow{R}}_{X_{m_p|m_s}^N}^{X_{best|m_p}^N}\right\|_2}{1-e_{m_p|m_s}}\right)^3}{\tilde{F}\left(X_{best|m_p}^N\right)T_{m_p|m_s}^2} \tag{8}$$

where $\mu_{m_p|m_s}^0$ is the initial value of the gravitational constant of the planet m_p or the satellite m_s. Since the population should explore the solution space as much as possible in the early stage, and focus on developing new optimal solutions near the current optimal solution in the later stage, the gravitational constant is set to a dynamic value that changes with time. In the early stage, the advantages of the population are concentrated on the exploration function of the solution space, so the early value of $\mu_{m_p|m_s}^0$ should be small, so that the traction of the central celestial body to its surrounding celestial body is weak, which is conducive to the exploration of the search operator in a far area. In the later period, the value should be larger, which means that the surrounding celestial body is pulled back to the vicinity of the central celestial body in the later period. At this time, the search operator can develop a better solution near the current optimal solution. The expression of the variation of the universal gravitational constant with the number of iterations is as follows :

$$\mu_{m_p|m_s}^t = \mu_{m_p|m_s}^0 \cdot \ln\left(\frac{t}{T_{\max}}\right) \tag{9}$$

$\mu^{t}_{m_p|m_s}$ represents the gravitational constant at the t-th iteration. The expression of the gravitational force $\overrightarrow{G}^{t}_{m_p|m_s}$ of the central celestial body on the planet or satellite at the t-th iteration is :

$$\overrightarrow{G}^{t}_{m_p|m_s} = \mu^{t}_{m_p|m_s} \frac{\tilde{F}\left(X^{N}_{m_p|m_s}\right)\tilde{F}\left(X^{N}_{m_{best}|m_p}\right)}{\left(\overrightarrow{R}^{X^{N}_{best|}m_p}_{X^{N}_{m_p|m_s}}\right)^2} \cdot \frac{\left(X^{N}_{m_{best}|m_p} - X^{N}_{m_p|m_s}\right)}{\left\| \overrightarrow{R}^{X^{N}_{best|m_p}}_{X^{N}_{m_p|m_s}} \right\|_2} \tag{10}$$

According to the initial value of the gravitational constant and the size of the eccentricity, the expression of the initial linear velocity of the planet or satellite $\left|\overrightarrow{v}^{0}_{m_p|m_s}\right|$ is derived as follows :

$$\left|\overrightarrow{v}^{0}_{m_p|m_s}\right| = \sqrt{\frac{\mu^{0}_{m_p|m_s}\tilde{F}\left(X^{N}_{m_{best}|m_p}\right)\left(1 + e_{m_p|m_s}\right)}{1 - e_{m_p|m_s}}} \tag{11}$$

The direction of the initial velocity is defined as follows :

$$\frac{\overrightarrow{v}^{0}_{m_p|m_s}}{\left|\overrightarrow{v}^{0}_{m_p|m_s}\right|} = \cos\theta \frac{\overrightarrow{\delta}_{m_p|m_s}}{\left|\overrightarrow{\delta}_{m_p|m_s}\right|} + \sin\theta \frac{\overrightarrow{\delta}'_{m_p|m_s}}{\left|\overrightarrow{\delta}'_{m_p|m_s}\right|}, \theta \in (0, 2\pi) \tag{12}$$

Here, $\overrightarrow{\delta}_{m_p|m_s}$, $\overrightarrow{\delta}'_{m_p|m_s}$ satisfies :

$$\overrightarrow{\delta}_{m_p|m_s} \perp \overrightarrow{\delta}'_{m_p|m_s}, \overrightarrow{\delta}_{m_p|m_s} \perp \overrightarrow{R}^{X^{N}_{best|}m_p}_{X^{N}_{m_p|m_s}}, \overrightarrow{\delta}'_{m_p|m_s} \perp \overrightarrow{R}^{X^{N}_{best|}m_p}_{X^{N}_{m_p|m_s}} \tag{13}$$

The above formula indicates that the direction of the initial velocity is in any direction on the plane perpendicular to gravity, $\overrightarrow{\delta}_{m_p|m_s}$, $\overrightarrow{\delta}'_{m_p|m_s}$ represents two vectors that can determine this plane and are not related to each other. In a very short period of time Δt, the surrounding motion of the celestial body can be seen as a uniform variable motion under the action of gravity under the premise of tangential velocity along the orbit of revolution, and the direction and size of gravity are constant. By dividing the time from the t-th iteration to the $t+1$-th iteration into N^t subintervals, the position expression of the solution space reached by the search particle in the $t+1$-th iteration is as follows :

$$X^{N}_{m_p|m_s}(t+1) = X^{N}_{m_p|m_s}(t) + \sum_{i=0}^{N^t-1} \overrightarrow{v}_{m_p|m_s}\left(t + \frac{i}{N^t}\right) \cdot \frac{1}{N^t} + \frac{1}{2}\frac{\overrightarrow{G}^{t}_{m_p|m_s}}{\tilde{F}\left(X^{N}_{m_p|m_s}\right)}\left(\frac{1}{N^t}\right)^2 \tag{14}$$

where $\overrightarrow{v}_{m_p|m_s}\left(t + \frac{i}{N^t}\right)$ denotes the i-th subinterval from the t-th iteration to the $t+1$-th iteration. The recurrence relations of $\overrightarrow{v}_{m_p|m_s}\left(t + \frac{i+1}{N^t}\right)$ and $\overrightarrow{v}_{m_p|m_s}\left(t + \frac{i}{N^t}\right)$ are as follows :

$$\vec{v}_{m_p|m_s}\left(t+\frac{i+1}{N^t}\right)=\vec{v}_{m_p|m_s}\left(t+\frac{i}{N^t}\right)+\frac{\vec{G}^{t}_{m_p|m_s}}{\tilde{F}\left(X^N_{m_p|m_s}\right)}\cdot\frac{1}{N^t} \tag{15}$$

In the iterative process, the search operator gradually changes from the initial rough exploration to the accurate development in the local area. When the value of N^t is small, it means that the step size of the search operator is larger. As N^t increases, the two adjacent iterations are divided into smaller intervals. The step size of the search operator is smaller, and the obtained neighborhood solution is closer to the current optimal solution. Therefore, the value of N^t should be set as a dynamic value that increases with the iteration of the population. The expression of N^t is as follows :

$$N^t = \gamma t \tag{16}$$

γ is an adjustable parameter. The optimal value of γ will be obtained according to the experiment. Taking F5 in CEC2022 test function set as an example, the algorithm convergence values corresponding to different complete γ values are tested, as shown in Fig. 2. According to the optimal value of the graphic γ, take 8.

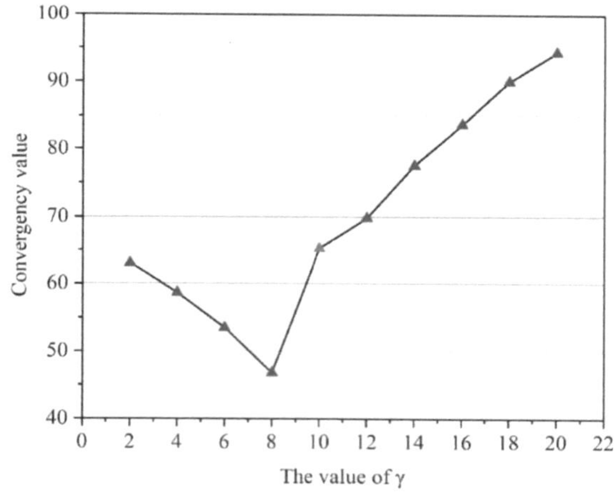

Fig. 2. The curve of convergence value with γ value (CEC2022, F5)

2.4 Collision Operator

Under normal circumstances, both the planet and the satellite run stably on their orbits. If there is a random event, when the random event occurs, any two satellites around the planet's revolution collide, resulting in the change of the speed and direction of the satellite. If the speed decreases after the collision, the semi-major axis of the orbit shrinks, and the search operator develops a better solution near the current optimal solution. If the speed increases after the collision, the semi-major axis of the orbit expands outward, which is conducive to the search operator to explore unknown feasible solutions. The mass of the two satellites that collided is unequal. According to the kinetic energy theorem, it can be seen that the speed of one search operator will decrease and the speed of the other search operator will increase. As shown in Fig. 3.

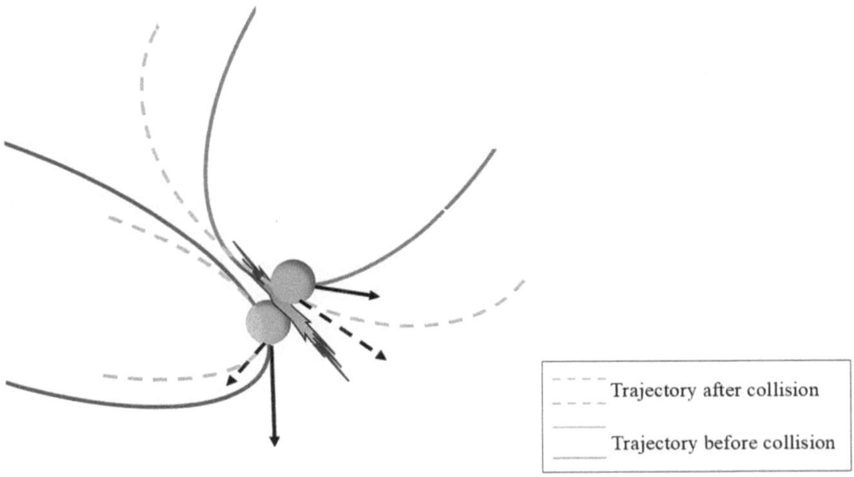

Fig. 3. Schematic diagram of celestial body orbit change after collision

This search mechanism makes it possible that some search operators are still out of the vicinity of the current optimal solution when the algorithm focuses on development in the later stage of the algorithm, which is conducive to jumping out of the local optimal solution. According to the kinematics analysis of the collision process of an object in multidimensional space [16], for any two satellites $X_{m_s}^N$ and $X_{m_s'}^N$ in the solution space, the velocities are $\vec{v}_{m_s}(t)$ and $\vec{v}_{m_s'}(t)$, respectively. The velocity expression after collision is as follows :

$$\vec{v}_{m_s}(t) \leftarrow \vec{v}_{m_s}(t) - \frac{2\tilde{F}\left(X_{m_s'}^N\right)}{\tilde{F}\left(X_{m_s}^N\right) + \tilde{F}\left(X_{m_s'}^N\right)} \cdot \left[\frac{\left(\vec{v}_{m_s}(t) - \vec{v}_{m_s'}(t)\right) \cdot \left(X_{m_s}^N - X_{m_s'}^N\right)}{\left\|X_{m_s}^N - X_{m_s'}^N\right\|^2}\right]\left(X_{m_s}^N - X_{m_s'}^N\right) \tag{17}$$

$$\vec{v}_{m_s'}(t) \leftarrow \vec{v}_{m_s'}(t) - \frac{2\tilde{F}\left(X_{m_s}^N\right)}{\tilde{F}\left(X_{m_s}^N\right) + \tilde{F}\left(X_{m_s'}^N\right)} \cdot \left[\frac{\left(\vec{v}_{m_s}(t) - \vec{v}_{m_s'}(t)\right) \cdot \left(X_{m_s}^N - X_{m_s'}^N\right)}{\left\|X_{m_s}^N - X_{m_s'}^N\right\|^2}\right]\left(X_{m_s}^N - X_{m_s'}^N\right) \tag{18}$$

After the collision, the direction of the velocity will also change, and the whole population will gradually transition to a new search orbit. According to the general design idea of the particle swarm search algorithm, the search operator should have inertia. In the previous iterations, some information of the previous step should be inherited. In the collision operator, the velocity update of the satellite inherits the velocity information when there is no collision. Therefore, the position update of the search operator is a step-by-step change with a basis, rather than a random mutation without a basis. The solution update based on the velocity change is compared with the new solution generated by the original solution of several randomly selected particles through some operation mechanism. Due to the accumulation of time, even if the information of too many particles is not fused, it can change greatly. Reducing random particles to participate in the operation of solution update is conducive to generating stable search results.

The collision operator does not involve collisions between planets or between planets or satellites. The reasons are as follows :

1 Too many search operators participate in the operation of the collision operator, which increases the time complexity of the algorithm, and the number of particles participating in the collision operation reaches a certain number. It will produce saturation, and the participation of more particles is redundant, and the performance of the algorithm will not be greatly improved.
2 The mass gap between the planet and the satellite is large, and the speed of the satellite will change sharply after the collision, while the orbit of the planet does not change significantly compared with the original. If the search step size is too large or too small, the performance of the algorithm will be reduced, so it needs to be avoided.

2.5 Adaptive Strategy

In order to dynamically adjust the position of the search trajectory in the solution space, an adaptive trajectory change mechanism is introduced. In the early stage, searching particles requires a comprehensive exploration of the solution space. As an additional disturbance, the collision operator is an effective means to make the search operator get rid of the current trajectory and enter a new orbit, which means that the search operator goes to a new unknown space for exploration. Therefore, the probability of collision should be increased in the early stage of the algorithm, and the search operator should be guided to basically cover the solution space area. In the later stage of the algorithm, it focuses on finding better solutions near the current optimal solution and several better solutions. The gravitational operator has the function of constraining the search operator near the better solution. Therefore, the probability of collision in this stage is reduced. Each search operator is basically safe and moves normally in its orbit. However, if the probability of collision is 0, it will fall into local optimum in the later stage. In order to avoid such problems, a random variable r_1, $r_1 \sim U(0, 1)$ is defined. The search speed of the satellite is updated according to the following rules :

- If $r_1 \leq \beta(t)$, the value of $\vec{v}_{m_s}(t)$ is calculated according to formulas (11, 12) and (15).
- If $r_1 > \beta(t)$, the value of $\vec{v}_{m_s}(t)$ is calculated according to formulas (17, 18).

Here, $\beta(t)$ is the control factor of the frequency of collision events, and the expression of $\beta(t)$ is as follows :

$$\beta(t) = \beta_{\max} + (\beta_{\max} - \beta_{\min}) \ln\left(\frac{t}{T_{\max}}\right) \tag{19}$$

where $\beta_{max} = 1$, $\beta_{min} = 0$, t is the current iteration, T_{max} is the maximum number of iterations. In the early stage, the value of $\beta(t)$ is closer to 0, and the sampling value of r_1 has a great probability to be greater than $\beta(t)$. At this time, the collision operator is executed; as the number of iterations increases, the value of $\beta(t)$ approaches 1, and the probability that the sampling value of r_1 is less than $\beta(t)$ increases. Most search operators tend to perform gravitational operators.

2.6 Update the Solution

For $\forall X^N(t)$, after updating to $X^N(t+1)$, the solution replacement at the individual level is first performed. Based on the greedy strategy, if the fitness of the newly generated solution is better than the current solution, the current solution is replaced by the new solution, otherwise the current solution is maintained; after completing the solution replacement at the individual level, a new population has been generated, and then the solution replacement at the population level is performed. The solution with the best fitness in the population becomes a new star, and the remaining solution is divided into new planets and satellites after fitness sorting, and the search strategy is started again.

2.7 Algorithm Flow Summary

Firstly, the objective function is defined, and the solution space range is limited according to the optimization problem. A set of initial populations are generated in the solution space, which are defined as stars, planets and satellites according to the size of fitness. The planets optimize around the stars by performing gravitational operators. The satellites not only optimize around the planets by performing gravitational operators, but also collide between any two satellites according to the collision triggering mechanism to optimize in a new orbit. When it is updated to the next generation solution, the replacement of the individual solution and the replacement at the population level are successively performed. After the replacement is completed, the search mechanism of the gravitational operator and the collision operator is repeated until the maximum number of iterations is satisfied. The complete flow chart of the algorithm is shown in Table 1.

Table 1. G-COA algorithm pseudo-code

Algorithm1: The optimal solution generation process of the gravitational-collision algorithm
Input : the number of celestial bodies M, the dimension N of the search space, the maximum number of iterations $T_{\max}$, the step coefficient γ, the maximum value of the probability control factor of the collision event $\beta_{\max}$ and $\beta_{\min}$.   **Output** : The global optimal solution of the objective function.
1.Initialization t = 0 ;   2.Define the solution space according to Equation (3) ;   3.The initial position of each celestial body is randomly generated, and the initial velocity of the celestial body is randomly generated according to equations (11) and (12).   4.Evaluate the initial fitness of each celestial body in the population.   5.The solution with the best fitness in the population is regarded as a fixed star, denoted as $X_{best}^{N}(t)$.   6.The remaining feasible solutions are sorted according to the fitness and divided into two groups. The group with better fitness is regarded as a planet, denoted by $X_{m_p}^{N}(t)$.   7.The group with poor fitness is regarded as a satellite, denoted by $X_{m_s}^{N}(t)$, and assigned to each planet according to formula (5).   8.Normalized solution fitness ; normalize the distance between the planet and the fixed star, and the distance between the satellite and the planet.   **While** $t \geq 0$ and $t < T_{max}$,**do**:   **For** m in 1:M,**do**:   According to Eq. (11), Eq. (12) and Eq. (15), the velocity of the planet $\vec{v}_{m_p}(t)$ is updated.   **IF** $r_1 \leq \beta(t)$,**do:**   According to formula (17), formula (18), the velocity of the satellite $\vec{v}_{m_s}(t)$ is updated.   **ELSE do:**

3 Experiment

3.1 Ablation Experiment

In order to verify the role of the collision operator in the algorithm, the ablation experiment was carried out with the more complex test function F16 in CEC2022 as an example. The experimental results are shown in Fig. 4. Among them, A1 represents the convergence performance of the algorithm only under the action of the gravitational operator, and A2 represents the performance of adding the collision operator on the basis of the gravitational operator. Each case runs 20 times, and the maximum convergence value, minimum convergence value and mean value are counted. When only the gravitational operator is used, the maximum value is 16.4, the minimum value is 11.08, and the mean value is 14.628. After adding the collision operator, the maximum value is 11.67, the minimum value is 7.134, and the average value is 9.788, which is 33.1% lower than the same period last year. From the difference of the convergence value, the variance of the convergence result after adding the collision operator is smaller, which proves that the convergence value is more stable.

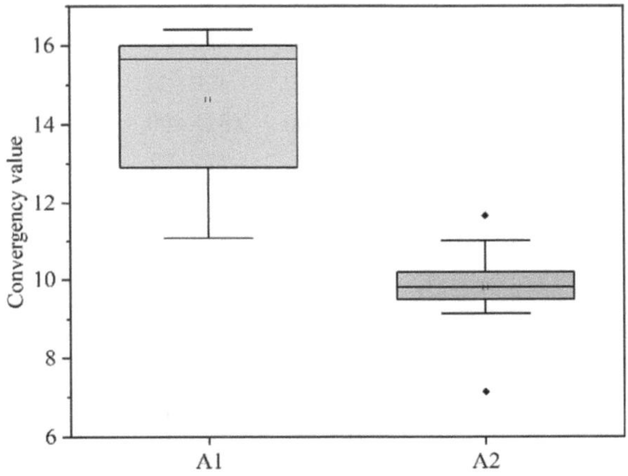

Fig. 4. Operator-convergence value box line graph

3.2 Contrast Experiment

In this paper, CEC2022 test function set is selected to verify the performance of G-COA algorithm. CEC2022 contains 16 test functions, which can be classified into unimodal function, multimodal function, combination of basic elementary functions and combination of several functions in the test set. Five optimization algorithms, Sparrow Search Algorithm (SSA) [18], Spider Wasp Optimizer (SWO) [19], Artificial hummingbird algorithm (AHA) [19], Kepler optimization algorithm (KOA) [12] and CGKOA [13], are selected as comparison objects. Among them, SSA, SWO and AHA are heuristic algorithms based on biological behavior. KOA and CGKOA are algorithms inspired by the laws of astrophysical motion. Each test function runs 20 times, and the maximum number of iterations is unified to 1000 times. The maximum, minimum and average values of each function convergence are counted. The statistical results are shown in Table 2.

Table 2. The convergence value comparison of the algorithm on the CEC2022 function set

		G-COA	SSA	SWO	AHA	KOA	CGKOA
F1	Max	4.01E+04	11.44E+04	7.83E+04	7.61E+04	5.63E+04	4.58E+04
	Min	3.65E+04	8.69E+04	7.31E+04	7.03E+04	5.20E+04	3.89E+04
	Ave	**3.74E+04**	10.06E+04	7.56E+04	7.25E+04	5.33E+04	4.11E+04
F2	Max	4.38E+07	15.12E+07	13.40E+07	12.10E+07	9.16E+07	6.23E+07
	Min	4.10E+07	11.83E+07	10.15E+07	10.35E+07	6.01E+07	5.74E+07
	Ave	**4.29E+07**	13.86E+07	12.74E+07	11.39E+07	8.13E+07	5.97E+07
F3	Max	13.12E+00	14.75E+00	15.16E+00	5.64E+00	13.31E+00	9.53E+00
	Min	13.07E+00	13.67E+00	14.49E+00	1.42E+00	13.06E+00	5.62E+00
	Ave	13.09E+00	14.21E+00	14.83E+00	**3.51E+00**	13.17E+00	7.14E+00
F4	Max	3.02E+02	4.16E+02	3.85E+02	3.71E+02	3.21E+02	2.85E+02
	Min	2.71E+02	4.07E+02	3.56E+02	3.39E+02	3.23E+02	2.33E+02
	Ave	2.85E+02	4.10E+02	3.63E+02	3.52E+02	3.22E+02	**2.42E+02**
F5	Max	75.46E+00	173.56E+00	113.29E+00	152,79E+00	124.23E+00	70.29E+00
	Min	69.92E+00	153.79E+00	99.37E+00	134.59E+00	109.62E+00	62.37E+00
	Ave	72.53E+00	161.87E+00	107.30E+00	146.62E+00	120.93E+00	**64.78E+00**
F6	Max	2.37E+10	5.66E+10	5.58E+10	4.47E+10	3.35E+10	3.07E+10
	Min	1.72E+10	5.23E+10	4.82E+10	3.11E+10	2.87E+10	2.46E+10
	Ave	**2.08E+10**	5.45E+10	5.13E+10	3.78E+10	3.26E+10	2.63E+10
F7	Max	2.26E+04	5.79E+10	6.49E+10	4.57E+04	3.91E+04	2.74E+04
	Min	1.86E+04	5.92E+10	4.47E+10	3.53E+04	3.10E+04	2.27E+04
	Ave	**1.94E+04**	5.83E+10	5.37E+10	4.17E+04	3.73E+04	2.58E+04
F8	Max	12.20E+07	36.5E+07	11.4E+07	47.5E+07	31.3E+07	17.12E+07
	Min	8.33E+07	30.25E+07	7.95E+07	28.8E+07	22.0E+07	11.09E+07
	Ave	**9.72E+07**	32.74E+07	10.87E+07	34.27E+07	24.65E+07	13.45E+07
F9	Max	2.66E+00	3.72E+00	3.06E+00	3.42E+00	3.13E+00	2.93E+00
	Min	1.54E+00	3.19E+00	2.91E+00	3.27E+00	3.08E+00	2.37E+00
	Ave	**2,24E+00**	3.40E+00	2.98E+00	3.35E+00	3.10E+00	2.44E+00
F10	Max	338.92E+00	455.36E+00	732.52E+00	692.41E+00	649.24E+00	372.18E+00
	Min	319.24E+00	424.98E+00	718.37E+00	605.82E+00	545.67E+00	353.76E+00
	Ave	**328.67E+00**	436.57E+00	722.16E+00	673.61E+00	614,46E+00	360.28E+00
F11	Max	8.55E+14	39.62E+14	24.45E+14	47.3E+14	42.7E+14	18.8E+14
	Min	3.57E+14	31.07E+14	20.79E+14	35.2E+14	37.8E+14	12.92E+14
	Ave	**5.87E+14**	34.39E+14	22.53E+14	42.6E+14	40.9E+14	13.36E+14
F12	Max	6.14E+03	9.34E+03	10.02E+03	8.84E+03	7.30E+03	6.65E+03
	Min	5.79E+03	8.03E+03	8.32E+03	8.26E+03	7.19E+03	6.20E+03

(*continued*)

Table 2. (*continued*)

		G-COA	SSA	SWO	AHA	KOA	CGKOA
	Ave	**5.92E+03**	8.76E+03	9.19E+03	8.60E+03	7.23E+03	6.38E+03
F13	Max	19.01E+00	22.80E+00	20.17E+00	18.47E+00	19.55E+00	19.37E+00
	Min	17.86E+00	21.98E+00	19.54E+00	16.25E+00	19.17E+00	17.35E+00
	Ave	**18.02E+00**	22.27E+00	20.04E+00	17.04E+00	19.37E+00	18.13E+00
F14	Max	5.94E+04	8.53E+04	8.87E+04	9.02E+04	7.29E+04	6.32E+04
	Min	5.80E+04	6.88E+04	7.12E+04	7.31E+04	6.89E+04	6.14E+04
	Ave	**5.85E+04**	7.28E+04	8.24E+04	7.82E+04	7.07E+04	6.21E+04
F15	Max	65.06E+00	175.35E+04	232.29E+04	430.71E+04	378.61E+00	93.77E+00
	Min	60.34E+00	194.47E+04	200.76E+04	387.37E+04	278.70E+00	86.29E+00
	Ave	**62.07E+00**	186.33E+04	227.65E+04	414.72E+04	343.18E+00	90.10E+00
F16	Max	10.55E+00	13.34E+00	17.62E+00	19.05E+00	15.03E+00	10.32E+00
	Min	9.52E+00	10.09E+00	14.36E+00	17.62E+00	13.52E+00	3.22E+00
	Ave	9.97E+00	12.03E+00	15.55E+00	18.32E+00	14.46E+00	**5.15E+00**

The statistical results show that among the 16 test functions, the convergence results of 12 test functions are optimal. Compared with the original KOA algorithm, G-COA significantly improves the convergence performance. Although CGKOA does not show obvious advantages, it is close to the optimal solution, and the convergence value is better than G-COA in F3, F4 and F5, indicating that CGKOA has more advantages in solving the optimal value of bounded sine and cosine oscillation function.

3.3 Application Experiment

In order to further prove the practicability of the proposed optimization algorithm in solving engineering problems, taking the design optimization problem of welded beam [20] and the design optimization problem of tension-compression spring [13] as examples, the convergence performance of the six optimization algorithms mentioned in Sect. 3.2 in solving the optimization problem is compared. The experimental results are shown in Fig. 5(a) and (b).

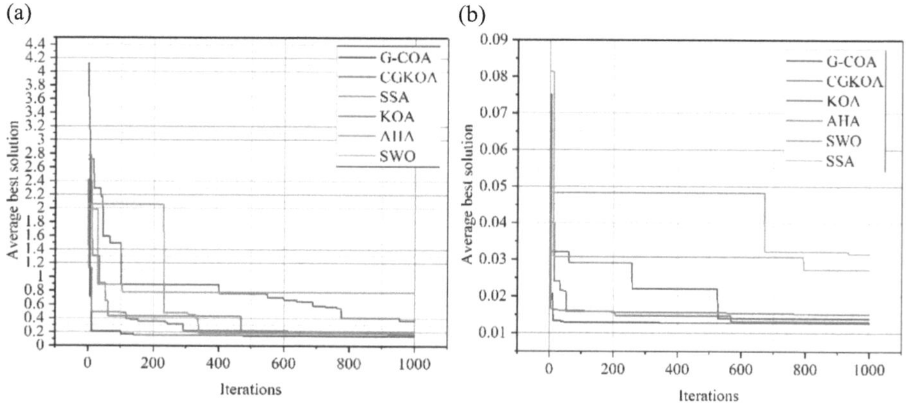

Fig. 5. (a). The welded beam design problem: G-COA's Convergence curve. **(b)** Tension/compression spring design: G-COA's Convergence curve.

The experimental results show that G-COA is superior to other algorithms in terms of convergence speed and convergence value. CGKOA is comparable to G-COA in terms of convergence value performance, but the convergence speed is slightly worse than G-COA. AHA and SWO are relatively poor. AHA falls into local optimum for a long time in the early stage of the optimization process and cannot jump out. SWO converges prematurely in the design optimization of welded beams. However, in the optimal design of tension-compression spring, it stagnates at the local optimum for a long time.

4 Conclusion

This study proposes a search algorithm inspired by the motion behavior of celestial bodies, named G-COA, which is used to solve general single-objective continuous optimization problems. The gravity of the celestial body, the initial velocity of the celestial body and the intervention of the outer celestial body will lead to the change of the trajectory and velocity of the celestial body. The physical model of the motion behavior of these celestial bodies is the basis for establishing the G-COA algorithm. The effectiveness of the G-COA algorithm is verified on the latest test function set CEC2022. Two engineering optimization problems with constraints are listed to prove its ability to solve practical optimization problems. Compared with some latest algorithms and highly cited algorithms, it shows that the G-COA algorithm is closer to the optimal solution in most cases. However, when the iteration enters the later stage, the time for generating a new generation of solutions becomes longer, that is, the time complexity of the operation becomes larger, which may pose a challenge to the automated production of the process industry and is the direction that needs to be improved in the future.

Acknowledgments. This research was supported by the National Natural Science Foundation of China (52307247) and the Fundamental Research Program of Shanxi Province (202203021222124).

Disclosure of Interests The authors have no competing interests to declare that are relevant to the content of this article.

References

1. Kan, S., He, Z., Cen, Y., Li, Y., Mladenovic, V., He, Z.: Contrastive Bayesian analysis for deep metric learning. IEEE Trans. Pattern Anal. Mach. Intell. **45**(6), 7220–7238 (2023). https://doi.org/10.1109/TPAMI.2022.3221486
2. Morquecho, E.G., et al.: Comparison of an improved metaheuristic and mathematical optimization based methods to solve the static AC TNEP problem. IEEE Trans. Power Syst. **39**(2), 3240–3256 (2024). https://doi.org/10.1109/TPWRS.2023.3305431
3. Shehab, M., et al.: Nature-inspired metaheuristic algorithms: a comprehensive review. Int. Arab J. Inf. Technol. **21**, 815–831 (2024)
4. Moshtaghi, H.R., et al.: A Comprehensive Review on Meta-Heuristic Algorithms and their Classification with Novel Approach. J. Appl. Res. Ind. Eng. **8**(1), 63–89 (2021)
5. Trojovský, P., Dehghani, M.: A new bio-inspired metaheuristic algorithm for solving optimization problems based on walruses behavior. Sci. Rep. **13**, 8775 (2023). https://doi.org/10.1038/s41598-023-35863-5
6. Varna, F.T., Husbands, P.: Two new bio-inspired particle swarm optimisation algorithms for single-objective continuous variable problems based on eavesdropping and altruistic animal Behaviours. Biomimetics. **9**, 538 (2024). https://doi.org/10.3390/biomimetics9090538
7. Hasan, D.O., Mohammed, H.M., Abdul, Z.K.: Griffon vultures optimization algorithm for solving optimization problems. Expert Syst. Appl. **276**, 127206.,ISSN 0957–4174 (2025). https://doi.org/10.1016/j.eswa.2025.127206
8. Sécheresse, X., Ly, G., Yves, J., Torcy, A.: GAAPO: genetic algorithmic applied to prompt optimization. Front. Artif. Intell. **8**, 1613007 (2025). https://doi.org/10.48550/arXiv.2504.07157
9. Wang, X., Zhao, W., Tang, J.N., et al.: Evolution algorithm with adaptive genetic operator and dynamic scoring mechanism for large-scale sparse many-objective optimization. Sci. Rep. **15**, 9267 (2025). https://doi.org/10.1038/s41598-025-91245-z
10. Fang, N., Xu, C., Gong, X., et al.: A new human-based offensive defensive optimization algorithm for solving optimization problems. Sci. Rep. **15**, 12119 (2025). https://doi.org/10.1038/s41598-025-96559-6
11. Li, X., et al.: Dragon boat optimization: a meta-heuristic for intelligent systems. Expert. Syst. **42**(2), e13785 (2025). https://doi.org/10.1111/exsy.13785
12. Abdel-Basset, M., Mohamed, R., Abdel Azeem, S.A., Jameel, M., Abouhawwash, M.: Kepler optimization algorithm: a new metaheuristic algorithm inspired by Kepler's laws of planetary motion. Knowl.-Based Syst. **268**, 110454.,ISSN 0950–7051 (2023). https://doi.org/10.1016/j.knosys.2023.110454
13. Hu, G., Gong, C., Li, X., Xu, Z.: CGKOA: an enhanced Kepler optimization algorithm for multi-domain optimization problems. Comput. Methods Appl. Mech. Eng. **425**, 116964.,ISSN 0045-7825 (2024). https://doi.org/10.1016/j.cma.2024.116964
14. El-Mageed, A.A.A., Abohany, A.A., Hosny, K.M.: Enhanced Binary Kepler Optimization Algorithm for effective feature selection of supervised learning classification. J. Big Data. **12**, 93 (2025). https://doi.org/10.1186/s40537-025-01125-6
15. Houssein, E.H., Abdalkarim, N., Abdel Samee, N., Alabdulhafith, M., Mohamed, E.: Improved Kepler optimization algorithm for enhanced feature selection in liver disease classification. Knowl.-Based Syst. **297**, 111960.,ISSN 0950–7051 (2024). https://doi.org/10.1016/j.knosys.2024.111960

16. Fan, C., Yang, L.T., Xiao, L.: A step gravitational search algorithm for function optimization and STTM's synchronous feature selection-parameter optimization. Artif. Intell. Rev. **58**, 179 (2025). https://doi.org/10.1007/s10462-025-11193-y
17. Cai, S.L., Ding, G., Duan, G.H., Jing, X.H., Dai, L.H., Jiang, M.Q.: Scaling laws for two-metallic spheres in a head-on collision. Int. J. Impact Eng. **200**, 105257.,ISSN 0734-743X (2025). https://doi.org/10.1016/j.ijimpeng.2025.105257
18. Cai, Y., Chen, H.: An improved salp swarm algorithm for permutation flow shop vehicle routing problem. Sci. Rep. **15**, 6704 (2025). https://doi.org/10.1038/s41598-025-86054-3
19. Chen, X., Ye, C., Zhang, Y.: Strengthened grey wolf optimization algorithms for numerical optimization tasks and AutoML. Swarm Evol. Comput. **94**, 101891.,ISSN 2210-6502 (2025). https://doi.org/10.1016/j.swevo.2025.101891
20. Houssein, E.H., Hossam Abdel Gafar, M., Fawzy, N., et al.: Recent metaheuristic algorithms for solving some civil engineering optimization problems. Sci. Rep. **15**, 7929 (2025). https://doi.org/10.1038/s41598-025-90000-8
21. Rashid, M.F.F.A.: Tiki-taka algorithm: a novel metaheuristic inspired by football playing style. Eng. Comput. **38**(1), 313–343.,ISSN 0264-4401 (2020). https://doi.org/10.1108/EC-03-2020-0137

A Study on the Force Mechanism of a Many-Objective Artificial Physics Optimization Algorithm Based on Clustering and Information Entropy

Weizhe Li[1] and Liping Xie[1,2](✉)

[1] School of Computer Science and Technology, Taiyuan University of Science and Technology, Taiyuan 030024, China
s202320110942@stu.tyust.edu.cn

[2] Shanxi Key Laboratory of Complex Systems and Computational Intelligence, Taiyuan University of Science and Technology, Taiyuan 030024, Shanxi, China
xieliping@tyust.edu.cn

Abstract. In the many-objective Artificial Physics Optimization (APO) algorithm, the proposed attraction-guided population optimization framework exhibits good convergence properties in the early stages of iteration but suffers from instability in the later stages. This is specifically manifested by the Inverted Generational Distance (IGD) metric oscillating slightly around a suboptimal value, indicating that the particles can neither escape the attraction of suboptimal solutions nor can the algorithm enhance its convergence precision. To overcome these issues, this study proposes a Tri-Cluster Entropy-Guided many-objective Artificial Physics Optimization, abbreviated as TCEG-APO. TCEG-APO partitions the objective space using K-means clustering and, based on the IGD metric, constructs a dynamic hierarchical architecture composed of three cluster layers: elite, intermediate, and exploration. The elite cluster serves as the population's guiding core, shielded from external forces to maintain the search direction towards the global optimum. The intermediate cluster dynamically decides whether to be guided by the force from the nearest elite cluster based on the decreasing trend of the IGD. The exploration cluster, through a distance-weighting mechanism, prevents ineffective guidance from distant particles, thereby enhancing population diversity. Furthermore, the algorithm incorporates an information entropy mechanism to dynamically adjust the interaction forces within each cluster, preventing particle aggregation and precisely balancing convergence and diversity in the high-dimensional space. To comprehensively evaluate the performance of TCEG-APO, this paper conducts a comparative analysis against several many-objective evolutionary algorithms (MaOEAs) on the DTLZ benchmark test suite with 3, 5, 8, and 15 objectives. The experimental results demonstrate that the TCEG-APO algorithm exhibits superior performance.

Keywords: Many-objective Optimization · Artificial Physics Optimization · Clustering · Information Entropy · Hierarchical Learning

H. Ma et al. (Eds.): IWACIII 2025, CCIS 2780, pp. 395–408, 2026.
https://doi.org/10.1007/978-981-95-6730-0_28

1 Introduction

Artificial Physics Optimization (APO) models Newtonian forces between better and worse individuals to steer populations toward the global optimum [1]. For multi-objective optimization, Wang et al. [2] introduced MOAPO via an aggregation-function framework. Sun et al. proposed R2-ICRMOAPO, using the R2 indicator to manage an external archive and maintain diversity [3]; PEMOAPO to mitigate local trapping and poor spread [4]; and IFNS-MOAPO, which incorporates improved fast non-dominated sorting and multiple strategies to balance global–ocal search and speed elite selection [5].

Early in MOAPO, aggregated attraction from multiple non-dominated solutions speeds movement toward the Pareto front, but later the same overlapping forces cause oscillation and hinder focused local search. Clustering splits the population into subpopulations, converting conflicting global pulls into independent local ones and guiding individuals around a single elite per subpopulation, thus avoiding interference, pseudo-attractors, and stagnation while preserving diversity through parallel front coverage. Information entropy further regulates this process: when low entropy reveals distribution imbalance and oscillation risk, the algorithm drives individuals toward sparse, higher-entropy regions, adaptively forming directionally convergent subgroups that escape pseudo-attractors and sustain both convergence efficiency and global exploration.

To overcome late-stage oscillation and stagnation caused by conflicting forces among multiple non-dominated solutions, we propose TCEG-APO: a Tri-Cluster Entropy-Guided Artificial Physics Optimization framework. The population is dynamically partitioned (elite, moderate, exploration) using non-domination ranking and IGD. Elite particles interact only within their cluster (shielded force) for fine Pareto exploitation; moderate particles invoke elite attraction when degradation emerges; exploration particles follow a triple-elite, inverse-distance weighted guidance to avoid ineffective long-range pulls and preserve diversity. Information entropy quantifies intra-cluster distribution, adaptively strengthens repulsion in sparse regions, and reweights archive performance indicators. This integrated hierarchical–entropy control sustains a convergence–diversity balance in high-dimensional objective spaces.

Section 2 briefly introduces the standard Artificial Physics Optimization (APO) algorithm and related research. Section 3 elaborates on the proposed TCEG-APO algorithm. Section 4 presents extensive comparative experimental results of TCEG-APO against various optimization algorithms on the DTLZ test suite. Finally, Sect. 5 concludes this study and outlines future work.

2 Related Work

This paper aims to propose an efficient Artificial Physics Optimization (APO) algorithm for handling many-objective optimization problems (MaOPs). To clearly elaborate on the proposed method, this section first introduces the fundamental APO algorithm, the multi-level learning framework, and the definition of information entropy.

2.1 Framework of the Artificial Physics Optimization Algorithm

In APO with population size n, each particle i ($i = 1, \ldots, n$) has:

- Mass: $m_i = \exp\left(\frac{f(\mathbf{x}_{\text{best}}) - f(\mathbf{x}_i)}{f(\mathbf{x}_{\text{worst}}) - f(\mathbf{x}_{\text{best}})}\right)$
- Position $\mathbf{x}_i$ and velocity $\boldsymbol{\nu}_i$ in each dimension

The net force on particle i ($i \neq$ best) in dimension k is:

$$F_{i,k} = \sum_{\substack{j=1 \\ j \neq i}}^{n} F_{ij,k} \quad \text{where} \quad F_{ij,k} = Gm_i m_j (x_{j,k} - x_{i,k}) \cdot \begin{cases} 1 & f(\mathbf{x}_i) \geq f(\mathbf{x}_j) \\ -1 & f(\mathbf{x}_i) < f(\mathbf{x}_j) \end{cases} \quad (1)$$

with gravitational constant $G = 10$. Velocity and position update as:

$$\nu_{i,k}(t+1) = w\nu_{i,k}(t) + \lambda F_{i,k}/m_i \quad (2)$$

$$x_{i,k}(t+1) = x_{i,k}(t) + \nu_{i,k}(t+1) \quad (3)$$

where $w \in [0.4, 0.9]$ decreases linearly, $\lambda \sim U(0,1)$, and bounded by:

$$\mathbf{x}_i \in [\mathbf{X}_{\min}, \mathbf{X}_{\max}], \quad \boldsymbol{\nu}_i \in [\boldsymbol{V}_{\min}, \boldsymbol{V}_{\max}]$$

2.2 Information Entropy

"Entropy" was originally introduced as a concept in thermodynamics. In 1948, Claude Shannon proposed information entropy to measure the uncertainty of a random variable [6] This theory holds that the greater the uncertainty of an event, the more information is obtained from the occurrence of that event.

After its establishment in the field of information theory, entropy has found numerous applications across various domains, each with its own interpretations and definitions. Information entropy can be applied in decision-making involving multiple evaluation attributes. The information entropy of an attribute reflects its degree of disorder: the smaller the information entropy of an attribute, the less the variation in values for that attribute among different individuals. Conversely, greater variation in values for that attribute among different individuals implies a higher degree of disorder for that attribute. The formula for calculating attribute entropy is as follows:

Assume a decision-making scenario involving N individuals and M evaluation attributes. The entropy for the k-th evaluation attribute across all individuals is calculated as follows:

$$E_k = -\frac{1}{\ln N} \sum_{j=1}^{N} p_{jk} \ln p_{jk} \quad (4)$$

$$p_{jk} = \frac{f_{jk}}{\sum_{j=1}^{N} f_{jk}} \quad (5)$$

$$\text{fn}_k(x_j) = \frac{f_k(x_j) - \min_{1 \le i \le N} f_k(x_i)}{\max_{1 \le i \le N} f_k(x_i) - \min_{1 \le i \le N} f_k(x_i)} \tag{6}$$

Here, f_{jk} represents the value of the k-th attribute for the j-th individual. p_{jk} is the normalized value, representing the proportion of the k-th attribute value of the j-th individual relative to the total sum for that attribute. Therefore, a smaller E_k indicates that the system is more concentrated. Conversely, a larger E_k indicates that the system is in a state of uniform distribution and carries a greater amount of information.

3 Approach

In this section, we propose a TCEG-APO algorithm, designed to effectively address the issue of instability during the later stages of iteration. This problem is characterized by the Inverted Generational Distance (IGD) metric fluctuating slightly around a sub-optimal value, which indicates that the particles are unable to escape the attraction of sub-optimal solutions or improve their convergence precision. Based on the characteristics of the algorithm, we have constructed the framework for TCEG-APO (Algorithm 1) and incorporated the following key designs: 1. introducing entropy to calculate the mass of individuals; 2. designing inter- and intra-cluster attraction-repulsion rules; and 3. enhancing the traditional Artificial Physics Optimization (APO) algorithm by establishing two archive sets, EFA and LEA.

3.1 The TCEG-APO Framework

In this study, the proposed TCEG-APO method differs significantly from MOAPO. The main distinctions include: (1) A new method for calculating particle mass and a new force strategy are proposed within TCEG-APO (Algorithm 2); (2) Two distinct search strategies, EFA and LEA, are designed (Algorithm 3 and Algorithm 4).

3.2 Entropy-Based Individual Mass Calculation

During the search phase, the mass m_i of each particle is calculated as:

$$m_i = \exp\left(\frac{\text{sortNo}}{N}\right)(1 + \alpha \cdot (-\log \rho_i)) \tag{7}$$

where $\alpha = 0.5$, sortNo is the non-dominated sorting rank, N is the population size, and $\rho_i = 1/(\bar{d}_i + \epsilon)$ with $\epsilon > 0$ preventing division by zero. The average distance $\bar{d}_i$ to k nearest neighbors is defined as:

$$\bar{d}_i = \frac{1}{k} \sum_{j \in \mathcal{N}_k(i)} \|\mathbf{f}_i - \mathbf{f}_j\|_2, \quad k = 5 + M \tag{8}$$

Algorithm 1. TCEG-APO framework

Initialize population;
Generate EFA,LEA;
while $FE < maxFE$ **do**
    Entropy-based Individual Mass Calculation according to Algorithm 2;
    Calculate Cluster individual force according to Algorithm 2;
    EFA update based on entropy weight method to Algorithm 3;
    Update the LEA according to Algorithm 4;
    Fusion and Update of the EFA and LEA Archives;
end while
return LEA

where M is the number of objectives, $\mathcal{N}_k(i)$ denotes the k-nearest neighbors in objective space, and the Euclidean distance between particles i and j is calculated using their M-dimensional objective vectors $\mathbf{f}_i = (f_{i1}, \ldots, f_{iM})$.

The quality measure m_i balances convergence and diversity by integrating a convergence term $\exp\left(\frac{\text{sortNo}}{N}\right)$ with a diversity term $(1 + \alpha \cdot (-\log \rho_i))$. Here, sortNO ranks individuals based on non-dominated rank and local entropy: superior rank and higher entropy yield smaller sortNO values, thus larger m_i. This mechanism highlights two key individuals—low-rank (driving convergence) and high-entropy (enhancing exploration)—ensuring faster convergence while preserving diversity. In dense regions, reduced inter-particle distance increases ρ_i, lowering entropy and indicating poor diversity, which risks local optima and Pareto front gaps. In sparse regions, larger distances reduce density and increase entropy, offering escape routes from local traps and enabling exploration of uncovered Pareto regions. Thus, entropy-guided ranking synergistically prevents premature convergence and strengthens front coverage.

3.3 Hierarchical Force Mechanism

3.3.1 Hierarchical Layering via Dynamic Clustering This paper employs K-means clustering on the particles based on their objective function values. For a given population of size N, the number of clusters, k, is dynamically determined by the formula: $k = \left\lceil \sqrt{\frac{N}{2}} \right\rceil$, where $\lceil \ldots \rceil$ denotes the ceiling function. This design offers two advantages: 1) the number of clusters increases as N grows, and 2) a maximum of 15 clusters is imposed to limit computational complexity.

K-means is performed on all objective vectors, and clusters are ranked in ascending IGD relative to the current Pareto front. They are then partitioned into three force-governed layers: (1) Elite (top 33%, closest to the PF, dominating convergence), (2) Intermediate (33–66%, retaining improvement potential), and (3) Exploration (remaining clusters, far from the PF, sustaining diversity). Each layer employs distinct attraction–repulsion rules to coordinate precise convergence with broad front coverage.

3.3.2 Inter-cluster Force Mechanism Elite particles are fully shielded to conduct stable, fine-grained Pareto exploitation. Intermediate particles remain isolated only while their IGD decreases; if it stalls or worsens they accept attraction from the nearest elite cluster. Exploration particles receive distance-weighted elite guidance that favors nearby high-quality sources. This tri-layer force design balances precise exploitation, adaptive convergence pressure, and controlled exploration, improving Pareto front coverage.

Particles within elite clusters interact exclusively with same-cluster particles, preserving solution quality. The inter-particle force follows:

$$\boldsymbol{F}_{ij} = Gm_i m_j(\boldsymbol{x}_j - \boldsymbol{x}_i)\begin{cases}\gamma_{\text{entropy}}\eta_a & m_i < m_j \\ -(1-\gamma_{\text{entropy}})\eta_r & m_i > m_j\end{cases} \tag{9}$$

where $\gamma_{\text{entropy}} = 1 - E^i_{\text{local}}/E_{\text{global}}$ with adjustment factors $\eta_a = 1.0$, $\eta_r = 0.5$. Global entropy is calculated using B solution bins:

$$E_{\text{global}} = -\sum_{k=1}^{B} p_k \log_2 p_k \tag{10}$$

where B represents the number of discretized intervals in the objective space of the population, and p_k is the probability of an individual falling into the k-th interval. This value reflects the overall diversity level of the population.

Local Information Entropy Calculation: For each particle i, the information entropy within its k-neighborhood (k=15) is calculated as:

$$E^i_{\text{local}} = \sum_{b=1}^{B'} q_b \log_2 q_b \tag{11}$$

where B' is the adaptively adjusted number of objective-space bins within a cluster, and q_b is the particle probability in bin b. Global entropy models the population's macroscopic distribution, while local entropy detects neighborhood aggregation, forming a dual macro-guidance + micro-regulation control. Early in optimization, local $\approx$ global entropy, so repulsion is strengthened and attraction weakened, helping elite particles escape suboptimal zones and driving broad exploration to avoid premature convergence. Later, as the population nears the Pareto front, local entropy falls well below global entropy; repulsion is damped, attraction concentrates search around elite solutions, and fine-grained exploitation improves distribution uniformity and convergence accuracy.

For particles in an intermediate cluster, the intra-cluster forces are identical to those in an elite cluster. To calculate the inter-cluster forces, it is first necessary to obtain the true Pareto front:

$$\mathcal{P}_f = \{x \mid \text{FrontNo}(x) = 1\} \tag{12}$$

Then, the cluster's Inverted Generational Distance (IGD) is calculated:

$$\mathrm{IGD}_c = \frac{1}{|\mathcal{G}_c|}\sum_{i\in\mathcal{G}_c}\min_{j\in\mathcal{P}_f j\neq i}\|f(i)-f(j)\|_2 \text{if } \mathcal{P}_f \neq \emptyset \text{ and } \exists j \neq i \tag{13}$$

Next, the IGD trend is quantified:

$$T_c^{(t)} = \begin{cases} -1 & \mathrm{IGD}_c^{(t-2)} > \mathrm{IGD}_c^{(t-1)} > \mathrm{IGD}_c^{(t)} \\ 1 & \mathrm{IGD}_c^{(t-2)} < \mathrm{IGD}_c^{(t-1)} < \mathrm{IGD}_c^{(t)} \\ 0 & \text{otherwise} \end{cases} \tag{14}$$

The basic rule is based on the IGD trend feedback, $T_c \in$ {-1, 0, 1}, where -1 signifies continuous improvement, 0 indicates fluctuation, and 1 indicates continuous deterioration. If the IGD shows continuous improvement, the cluster is not guided by the elite cluster. Conversely, if the IGD continuously deteriorates, the following elite guidance strategy is applied:

$$T_c \geq 0 \Rightarrow F_{i,\text{elite}} = G \cdot m_i m_j (x_j - x_i)\gamma_{\text{entropy}}\eta_{\text{mod-elite}} \cdot 1.5 \tag{15}$$

where $\eta_{exp-elite} = 1.2$ is the inter-cluster adjustment factor, and 1.5 is the elite guidance enhancement coefficient.

For particles in an exploration cluster, the intra-cluster interaction is the same as described above. The elite traction strategy is defined as:

$$F_i^{\text{total}} = \sum_{k=1}^{3}\left[\frac{G \cdot m_i m_k (x_k - x_i)\gamma_{\text{entropy}}\eta_{\text{exp-elite}} \cdot 2.0}{1+0.5k}\right] \tag{16}$$

These particles are subjected to a combined traction force from the three nearest elite particles. The distance decay factor, 1/(1+0.5k), prevents ineffective traction from distant particles, while a strong traction factor of 2.0 is used to accelerate the convergence of these exploration particles towards the elite particles. The process is shown in Algorithm 2.

3.4 EFA Archive Set Update Based on the Entropy Weight Method

The update mechanism for the EFA set is primarily based on the convergence performance of solutions F(x), with the objective of preserving and maintaining solutions that exhibit good convergence. By dynamically adjusting the members of the set, this mechanism ensures that the EFA consistently contains high-quality solutions with strong convergence performance, thereby guiding the population to converge towards more optimal solutions.

Algorithm 2. Particle swarm movement

Require: Population
Ensure: Offspring
Calculate the individual's Mass by Eql 7;
for each individual i in the Population **do**
    for each individual j in the Population, $j \neq i$ **do**
        Calculate the force on i between clusters based on the cluster type
    end for
end for
Update the velocity of individual i by Eql 2:
Update the position of individual i by Eql 3:
Generate offspring based on the updated positions of the individuals;
return Offspring

After normalizing objective values $\widehat{f}_k^{(i)}$, the dispersion weights are computed as:

$$P_{ik} = \frac{\widehat{f}_k^{(i)}}{\sum_{i=1}^{N} \widehat{f}_k^{(i)}}, \quad e_k = -\frac{1}{\ln N} \sum_{i=1}^{N} \delta(P_{ik}) P_{ik} \ln P_{ik}$$
$$\delta(P_{ik}) = \begin{cases} 1 & P_{ik} > 0 \\ 0 & P_{ik} = 0 \end{cases}, \quad w_k = \frac{e_k}{\sum_{j=1}^{M} e_j} \tag{17}$$

where $e_k \in [0, 1]$ measures objective k's dispersion (higher = more spread). The particle force aggregates weighted improvements:

$$F(i) = \sum_{\substack{j=1 \\ j \neq i}}^{N} \sum_{k=1}^{M} w_k \cdot \max\left(0, \widehat{f}_k^{(j)} - \widehat{f}_k^{(i)}\right) \tag{18}$$

In the TCEG-APO algorithm, the EFA maintains a fixed capacity. The update process begins by incorporating the solutions from the current population into the EFA and then calculates the convergence performance of each solution using Eq. (18) as shown in Algorithm 3.

Algorithm 3. UpdateEFA

Require: Population
$EFA = EFA \cup Population$
Calculate the fitness value of each solution in EFA by Eq. (18);
while $|EFA| > N$ **do**
    Delete the worst solution X_w from EFA;
end while

3.5 LEA Archive Set Update

The solution selection mechanism for the LEA set aims to maintain population diversity and ensure comprehensive coverage of the Pareto front by the solution set. This mechanism prevents premature convergence to local optima by selecting diverse solutions in the objective space, while simultaneously balancing solution distribution to avoid excessive differentiation. The process is shown in Algorithm 4

Algorithm 4. UpdateLEA

Require: Population
$LEA^* \leftarrow LEA \cup$ Population;
Move non-dominated solutions to LEA from LEA^*;
if $|LEA| < N$ **then return** ;
end if
Clear LEA;
Generate ideal points from each dimension of all points;
Find the solutions that are farthest and closest to the ideal point, and move them to LEA from LEA^*;
while $|LEA| < N$ **do**
    for each solution X_i in LEA^* **do**
        $S(X_i) = L_p\text{-norm}\,(X_i, X_{\text{farthest}}) + L_p\text{-norm}\,(X_i, X_{\text{closest}})$;
    end for
    Choose the solution X_i with the maximal $S(X_i)$, and move X_i from LEA^* to LEA;
end while

4 Experiments Settings

In this section, we conduct a comprehensive evaluation and analysis of the performance of the TCEG-APO method, comparing it with several representative methods to demonstrate its superiority.

4.1 Test Problems and Parameter Settings

To validate the performance of TCEG-APO on many-objective optimization problems (MaOPs), the experiments utilized two well-known benchmark test suites, including DTLZ [7]. The DTLZ suite comprises seven test problems. For each problem, the number of objectives, M, was set to 3, 5, 8, and 15, respectively, resulting in a total of 28 test instances. In the experiments, TCEG-APO was compared against six well-known many-objective evolutionary algorithms (MaOEAs). The algorithms involved are listed as follows: IBEA [8], MPSOD [9], NMPSO [10], hpaEA [11], PeEA [12], MOEA/D-UR [13]. To ensure a fair comparison, all algorithms used the same population size (N) and maximum number of function evaluations (MaxFEs). For M = 3, 5, 8, and 15, N was set to 120, 150, 180, and 240, respectively. For the other parameters of each algorithm, we adopted the settings recommended in their original literature. The computational environment was a 13th Gen Intel(R) Core(TM) i7-13700H 2.40 GHz processor with 16.0 GB of RAM, running Windows 11 and MATLAB R2023b. The proposed TCEG-APO was implemented on the PlatEMO platform [14].

To compare the algorithms, two popular performance metrics were employed: the Inverted Generational Distance (IGD) [15] and the Hypervolume (HV) [16]. Both metrics are capable of evaluating the convergence and diversity of the obtained Pareto front (PF). Each algorithm was run independently 30 times on each test instance, and the mean IGD and HV values were recorded. Additionally, the symbols "+/-/=" are used to summarize the overall performance. A "−" indicates that TCEG-APO outperformed the comparison algorithm on an instance. A "+" signifies that the comparison algorithm performed better than the proposed TCEG-APO. An "=" denotes that TCEG-APO and the comparison algorithm achieved similar performance.

4.2 Comparative Experiments

This experiment compared the performance of seven algorithms on the DTLZ test problem set, which includes 28 test cases with four different numbers of objectives. Table 1 presents the average IGD values obtained by TCEG-APO and the other six MaOEAs on the DTLZ benchmarks, with the best results highlighted in bold. The results show that TCEG-APO outperformed IBEA on 22 test instances. Specifically, TCEG-APO achieved better results on DTLZ1, DTLZ3, DTLZ4, and DTLZ7. TCEG-APO performed worse than IBEA on only 6 test instances. TCEG-APO performed worse than IBEA on only 6 test instances. Furthermore, TCEG-APO surpassed MPSOD on 20 test instances. TCEG-APO performed worse than MPSOD on only 8 test instances. TCEG-APO outperformed NMPSO on 23 test instances. It performed worse than NMPSO on only 5 test instances. TCEG-APO surpassed hpaEA on 25 test instances. It performed worse than hpaEA on only 3 test instances. TCEG-APO outperformed PeEA on 20 test instances. It performed worse than PeEA on only 8 test instances. TCEG-APO surpassed MOEA/D-UR on 18 test instances. It performed worse than MOEA/D-UR on only 9 test instances, with similar performance on 1 instance.

The results indicate that TCEG-APO outperformed IBEA on 16 test instances. TCEG-APO performed worse than IBEA on 11 test instances, with similar results on the remaining 1 instance. TCEG-APO outperformed MPSOD on 21 test instances, performing worse than MPSOD on only 7 test instances. TCEG-APO surpassed NMPSO on 18 test instances, performing worse than NMPSO on 10 test instances. TCEG-APO outperformed hpaEA on 19 test instances. TCEG-APO surpassed PeEA on 17 test instances, performing worse than PeEA on only 11 test instances. TCEG-APO outperformed MOEA/D-UR on 15 test instances, performing worse than MOEA/D-UR on only 13 test instances (Table 2).

Table 1. IGD values of the seven different algorithms on the DTLZ benchmarks

Problem	M	IBEA	MPSOD	NMPSO	hpaEA	PeEA	MOEADUR	TCEG-APO
DTLZ1	3	2.1696e-1 -	5.5907e-1 -	2.0302e-2 -	1.8728e-2 -	2.2494e-2 -	1.9621e-2 -	1.8544e-2
	5	2.0843e-1 -	1.0624e+0 -	6.4288e-2 -	6.0625e-2 -	6.1945e-2 -	5.8778e-2 -	5.8545e-2
	8	2.0848e-1 -	1.0780e+0 -	1.0752e-1 -	1.1892e-1 -	1.0621e-1 -	1.0558e-1 -	9.9532e-2
	15	2.9954e-1 -	7.4050e-1 -	2.2225e-1 -	1.5844e-1 -	1.3694e-1 +	1.6877e-1 -	1.4724e-1
DTLZ2	3	6.9834e-2 -	4.7432e-2 +	7.7416e-2 -	4.5572e-2 +	5.4733e-2 -	4.8448e-2 +	5.2856e-2
	5	2.0885e-1 -	1.9507e-1 -	2.0542e-1 -	1.8500e-1 -	1.9518e-1 -	1.8111e-1 +	1.9493e-1
	8	3.5429e-1 +	3.5272e-1 -	3.4685e-1 +	3.1932e-1 +	3.6774e-1 -	3.4432e-1 -	3.2110e-1
	15	5.5289e-1 +	5.2621e-1 +	6.2220e-1 -	6.6460e-1 -	5.7197e-1 -	5.8771e-1 -	5.4126e-1
DTLZ3	3	4.8285e-1 -	7.4399e+1 -	8.4466e+0 -	6.1568e-1 -	7.7013e-2 -	6.3058e-2 -	5.3776e-2
	5	6.0109e-1 -	5.8691e+1 -	6.7582e+0 -	2.5768e-1 -	2.0664e-1 -	2.1151e-1 -	1.9712e-1
	8	6.6863e-1 -	4.0168e+1 -	1.0857e+0 -	9.7967e+0 -	3.9910e-1 -	3.7478e-1 -	3.6232e-1
	15	8.0545e-1 +	4.6161e+1 -	6.2142e+1 -	2.8735e+1 -	7.8607e-1 +	6.8669e-1 +	8.5845e-1
DTLZ4	3	7.0138e-2 -	4.7532e-2 +	7.0807e-2 -	4.5804e-2 +	5.2820e-2 +	5.8516e-2 -	5.4829e-2
	5	4.2390e-1 -	1.9814e-1 -	2.0733e-1 -	4.2955e-1 -	1.9347e-1 -	1.9684e-1 -	1.9154e-1
	8	3.6272e-1 -	3.4479e-1 +	3.5289e-1 +	4.4053e-1 -	3.5355e-1 -	3.5180e-1 =	3.5171e-1
	15	5.6172e-1 -	5.6975e-1 -	5.4043e-1 -	5.4745e-1 -	5.3624e-1 -	5.6141e-1 -	5.3588e-1
DTLZ5	3	1.4263e-2 -	3.2498e-2 -	1.3425e-2 -	4.0806e-3 -	2.7925e-2 -	5.6225e-3 -	2.8488e-3
	5	7.5031e-2 -	9.6458e-2 -	3.8650e-2 +	1.7382e-1 -	5.3256e-2 -	1.4980e-2 +	7.3728e-2
	8	5.7781e-2 +	1.3994e-1 +	7.5482e-1 -	3.0965e-1 -	7.3416e-2 +	1.9409e-2 +	1.6572e-1
	15	1.5137e-1 +	1.4659e-1 +	7.5054e-1 -	4.8773e-1 -	2.3931e-1 +	1.3166e-2 +	4.6537e-1
DTLZ6	3	3.0307e-2 -	2.6334e-2 -	1.3086e-2 -	3.7574e-3 -	2.6952e-2 -	5.7454e-3 -	2.8140e-3
	5	8.7019e-2 -	8.5729e-2 -	3.4999e-2 +	4.3897e-1 -	5.9088e-2 +	1.7913e-2 +	8.4950e-2
	8	3.4214e-1 -	1.1941e-1 +	7.1672e-1 -	8.2187e-1 -	8.1312e-2 +	1.8849e-2 +	1.5990e-1
	15	3.9869e-1 +	1.3545e-1 +	7.4209e-1 +	2.5093e+0 -	1.3332e-1 +	1.4920e-2 +	9.8814e-1
DTLZ7	3	6.6768e-2 -	1.3764e-1 -	5.7899e-2 -	8.3766e-2 -	8.1040e-2 -	9.0188e-2 -	3.7996e-2
	5	2.7662e-1 -	7.1177e-1 -	2.5481e-1 -	5.5776e-1 -	2.6447e-1 -	4.1630e-1 -	2.4698e-1
	8	6.9105e-1 -	2.2217e+0 -	6.4689e-1 -	6.7727e-1 -	6.9170e-1 -	1.2312e+0 -	6.0803e-1
	15	2.3538e+0 -	2.6881e+0 -	2.3328e+0 -	3.4414e+0 -	1.5202e+0 -	2.9747e+0 -	1.5153e+0
+/-/=		6/22/0	8/20/0	5/23/0	3/25/0	8/20/0	9/18/1	

TCEG-APO demonstrates strong performance in the majority of test cases, exhibiting a significant advantage on complex multi-modal problems such as DTLZ1, DTLZ3 and DTLZ7. Its dynamic hierarchical architecture, which shields elite clusters from interference and applies adaptive forces to intermediate clusters, effectively maintains the search direction for optimal solutions in high-dimensional spaces, thereby overcoming the late-stage oscillation issue of the original APO. However, the algorithm shows notable limitations when addressing challenges in extremely high dimensions. This is primarily attributed to the "curse of dimensionality," which constrains the effectiveness of clustering, and the diminished regulatory precision of the information entropy mechanism in sparse, high-dimensional spaces. Consequently, key directions for future improvement lie in optimizing the clustering partition and further exploring the potential of the information entropy mechanism. Figure 1 illustrates the parallel coordinate plots of TCEG-APO and other MaOEAs on DTLZ7 with 8 objectives. The visu-

Table 2. HV values of the seven different algorithms on the DTLZ benchmarks

Problem	M	IBEA	MPSOD	NMPSO	hpaEA	PeEA	MOEADUR	TCEG-APO
DTLZ1	3	3.4807e-1 -	2.6965e-3 -	8.3766e-1 -	8.3812e-1 -	8.3390e-1 -	8.4158e-1 -	8.4326e-1
	5	7.1968e-1 -	0.0000e+0 -	9.5915e-1 -	9.7620e-1 +	9.5735e-1 -	9.7132e-1 +	9.6968e-1
	8	9.1493e-1 -	0.0000e+0 -	9.9062e-1 -	9.9002e-1 -	9.9347e-1 +	9.9145e-1 -	9.9217e-1
	15	9.4356e-1 -	1.7656e-5 -	9.8580e-1 -	9.9309e-1 -	9.9996e-1 +	9.8880e-1 -	9.9584e-1
DTLZ2	3	5.6266e-1 -	5.6227e-1 -	5.6212e-1 -	5.6639e-1 +	5.5953e-1 -	5.5773e-1 -	5.6491e-1
	5	8.0051e-1 +	7.8862e-1 +	8.0208e-1 +	8.0111e-1 +	7.5579e-1 -	7.8272e-1 -	7.8281e-1
	8	9.3789e-1 +	9.0958e-1 +	9.3115e-1 +	9.2178e-1 +	9.0798e-1 +	9.0623e-1 +	8.0937e-1
	15	9.9291e-1 +	9.8267e-1 +	9.7550e-1 +	8.9630e-1 +	9.9284e-1 +	9.3974e-1 +	8.1121e-1
DTLZ3	3	2.2585e-1 -	0.0000e+0 -	0.0000e+0 -	7.8464e-2 -	4.9951e-1 -	5.3845e-1 -	5.6190e-1
	5	3.6567e-1 -	0.0000e+0 -	0.0000e+0 -	6.0999e-1 -	7.6524e-1 -	7.5909e-1 -	7.7176e-1
	8	5.2910e-1 -	0.0000e+0 -	1.9108e-2 -	0.0000e+0 -	8.3462e-1 -	8.8635e-1 -	8.8848e-1
	15	7.4005e-1 +	0.0000e+0 -	0.0000e+0 -	0.0000e+0 -	6.9495e-1 +	9.1149e-1 +	3.8507e-1
DTLZ4	3	5.5670e-1 -	5.6166e-1 -	5.5670e-1 -	5.6631e-1 +	5.6164e-1 -	5.5294e-1 -	5.6472e-1
	5	7.2769e-1 =	7.8423e-1 +	8.0244e-1 +	6.8574e-1 -	7.8319e-1 +	7.8788e-1 +	7.2783e-1
	8	9.3923e-1 +	9.2315e-1 +	9.2999e-1 +	8.8421e-1 +	9.0947e-1 +	9.1960e-1 +	7.9738e-1
	15	9.9324e-1 +	9.9057e-1 +	9.9245e-1 +	9.7140e-1 +	9.9187e-1 +	9.8146e-1 +	8.0484e-1
DTLZ5	3	1.9904e-1 -	1.7039e-1 -	1.9702e-1 -	2.0026e-1 -	1.8576e-1 -	1.9877e-1 -	2.0144e-1
	5	1.1226e-1 -	1.0416e-1 -	1.1289e-1 -	1.0499e-2 -	1.1258e-1 -	1.2780e-1 +	1.1453e-1
	8	9.6013e-2 +	9.0732e-2 -	6.3878e-2 -	0.0000e+0 -	9.6375e-2 +	1.0501e-1 +	9.3552e-2
	15	9.0575e-2 -	7.0241e-2 -	7.2355e-2 -	0.0000e+0 -	9.0012e-2 -	9.5385e-2 +	9.0904e-2
DTLZ6	3	1.9581e-1 -	1.8614e-1 -	1.9691e-1 -	2.0051e-1 -	1.8467e-1 -	1.9910e-1 -	2.0157e-1
	5	1.1365e-1 +	1.1383e-2 -	1.1864e-1 +	1.3509e-4 -	1.1266e-1 +	1.2796e-1 +	6.4144e-2
	8	9.1159e-2 +	9.5400e-2 +	9.0909e-2 +	0.0000e+0 -	9.4706e-2 +	1.0487e-1 +	2.0264e-2
	15	9.0909e-2 -	9.1541e-2 -	9.0909e-2 -	0.0000e+0 -	9.0656e-2 -	9.5022e-2 +	9.1914e-2
DTLZ7	3	2.7768e-1 -	2.3457e-1 -	2.7909e-1 -	2.7028e-1 -	2.7283e-1 -	2.7100e-1 -	2.8487e-1
	5	2.5584e-1 -	1.2163e-2 -	2.6903e-1 -	2.3844e-1 -	2.4482e-1 -	2.1353e-1 -	2.7184e-1
	8	2.2009e-1 +	4.1573e-4 -	2.2774e-1 +	1.9219e-1 +	1.4428e-1 -	1.5768e-1 -	1.6374e-1
	15	1.4588e-1 +	9.4165e-9 -	1.6111e-1 +	3.2589e-2 -	7.8472e-2 -	1.9149e-2 -	9.5060e-2
+/-/=		11/16/1	7/21/0	10/18/0	9/19/0	11/17/0	13/15/0	

alization reveals that TCEG-APO achieve superior convergence and diversity, whereas IBEA and MPSOD exhibit limited diversity in their solutions.

4.3 Computational Complexity Analysis

A large number of objectives can increase the computational complexity of a many-objective optimization algorithm. In this section, we analyze the computational complexity of TCEG-APO. For TCEG-APO, its complexity is primarily determined by the clustering operation. The complexity of clustering is $O(N^2)$. The complexity of calculating information entropy is influenced by the population size, which is $O(N^2 \log N)$. Therefore, the overall complexity of TCEG-APO is $O(MN^2)$.

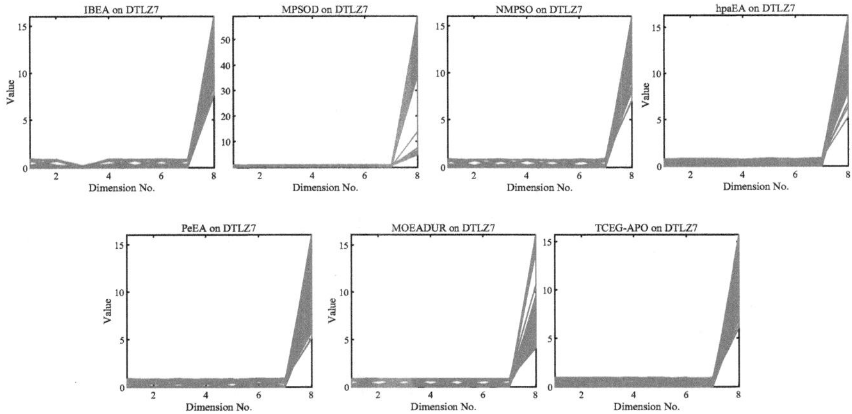

Fig. 1. The IGD values of seven different algorithms in the DTLZ benchmark test.

5 Conclusion

In conclusion, to address the issue of the many-objective Artificial Physics Optimization (APO) algorithm being prone to oscillating around sub-optimal solutions and exhibiting unstable convergence during its later iterations, this study proposes an improved algorithm, TCEG-APO, which integrates a clustering mechanism with information entropy theory. The core innovations of this algorithm are: 1) The construction of a three-layer dynamic hierarchical architecture—comprising elite, intermediate, and exploration clusters—based on K-means clustering and the IGD metric. This architecture effectively balances global exploration and local exploitation by shielding elite clusters from external forces for stable guidance, dynamically activating intermediate clusters under elite guidance, and enhancing diversity in exploration clusters through distance-weighting. 2) The introduction of an information entropy mechanism to dynamically adjust the interaction forces within each cluster, which precisely prevents premature population aggregation and stagnation, thereby achieving fine-tuned regulation of convergence and diversity in high-dimensional objective spaces. Systematic experiments on the DTLZ benchmark suite, covering 3, 5, 8, and even 15 objectives, demonstrate that the TCEG-APO algorithm significantly overcomes the late-stage instability of the original APO, and its overall performance is superior to that of several state-of-the-art many-objective evolutionary algorithms (MaOEAs). The experimental results fully validate the effectiveness of the proposed hierarchical clustering framework and information entropy regulation mechanism in enhancing the performance of many-objective optimization algorithms.

References

1. Xie, L.P., Zeng, J.C.: A global optimization based on physicomimetics framework. In: Proceedings of the First ACM/SIGEVO Summit on Genetic and Evolutionary Computation, pp. 609–616 (2009)
2. Wang, Y., Zeng, J., Tan, Y.: An artificial physics optimization algorithm for multi-objective problems based on virtual force sorting proceedings. In: Panigrahi, B.K., Das, S., Suganthan, P.N., Dash, S.S. (eds.) SEMCCO 2010. LNCS, vol. 6466, pp. 615–622. Springer, Heidelberg (2010). https://doi.org/10.1007/978-3-642-17563-3_72
3. Sun, B., Guo, N., Li, Z., et al.: Artificial physics constrained multi-objective optimization algorithm based on R2 indicator. J. Ordnance Equipment Eng. **43**(08), 186–192 (2022)
4. Sun, B., Zhang, L., Li, Z., et al.: Artificial physics multi-objective optimization algorithm based on partition-guided population evolution. Appl. Res. Comput. **40**(5), 1381–1388, 1409 (2023)
5. Sun, B., Zhang, L., Li, Z., Fan, K., Jin, Q., Guo, J.: An improved multi-objective artificial physics optimization algorithm based on multi-strategy fusion. IEEE Access **10**, 108736–108748 (2022)
6. Xue, Y., Sun, Y., Zhou, J., et al.: Multiattribute decision-making in wargames leveraging the entropy-weight method in conjunction with deep reinforcement learning. IEEE Trans. Games **16**(1), 151–161 (2023)
7. Deb, K., Thiele, L., Laumanns, M., et al.: Scalable multi-objective optimization test problems. In: Proceedings of the 2002 Congress on Evolutionary Computation. CEC 2002 (Cat. No. 02TH8600), vol. 1, pp. 825–830. IEEE (2002)
8. Zitzler, E., Künzli, S.: Indicator-based selection in multiobjective search. In: Yao, X., et al. (eds.) PPSN 2004. LNCS, vol. 3242, pp. 832–842. Springer, Heidelberg (2004). https://doi.org/10.1007/978-3-540-30217-9_84
9. Dai, C., Wang, Y., Ye, M.: A new multi-objective particle swarm optimization algorithm based on decomposition. Inf. Sci. **325**, 541–557 (2015)
10. Lin, Q., et al.: Particle swarm optimization with a balanceable fitness estimation for many-objective optimization problems. IEEE Trans. Evol. Comput. **22**(1), 32–46 (2018)
11. Chen, H., Tian, Y., Pedrycz, W., Wu, G., Wang, R., Wang, L.: Hyperplane assisted evolutionary algorithm for many-objective optimization problems. IEEE Trans. Cybern. **50**(7), 3367–3380 (2020)
12. Li, L., Yen, G.G., Sahoo, A., Chang, L., Gu, T.: On the estimation of pareto front and dimensional similarity in many-objective evolutionary algorithm. Inf. Sci. **563**, 375–400 (2021)
13. de Farias, L.R.C., Araújo, A.F.R.: A decomposition-based many-objective evolutionary algorithm updating weights when required. Swarm Evol. Comput. **68**, 100980 (2022)
14. Cheng, R., et al.: A benchmark test suite for evolutionary many-objective optimization. Complex Intell. Syst. **3**(1), 67–81 (2017). https://doi.org/10.1007/s40747-017-0039-7
15. Coello, C.A.C., Cortés, N.C.: Solving multiobjective optimization problems using an artificial immune system. Genet. Program Evolvable Mach. **6**, 163–190 (2005)
16. Zitzler, E., Thiele, L.: Multiobjective evolutionary algorithms: a comparative case study and the strength Pareto approach. IEEE Trans. Evol. Comput. **3**(4), 257–271 (1999)

Optimal Regulation of Green Heating Systems in Parks Considering Economy and Thermal Comfort

Yuanyuan Dai and Jianyan Tian(✉)

College of Electrical and Power Engineering, Taiyuan University of Technology, Taiyuan 030024, China
tut_tianjy@163.com

Abstract. In view of the problems existing in the traditional rule-based control methods in the control of green heating systems with heat pump units as the main equipment, such as high operating costs, poor user thermal experience, and adverse impact on equipment service life, this study proposes a multi-objective optimal regulation method based on the NSGA-II algorithm. The method takes economy and thermal comfort as the optimization goals, considers the influence of different outdoor temperatures on user-side heat load demand, building heat loss and heat pump energy efficiency, and dynamically adjusts the output of heat pump equipment and the opening degree of user-side valves under changing outdoor environments, thereby meeting the requirements of different index weights set by users and ensuring the economical and efficient operation of the system. Experiment results show that: the control method proposed in this paper can dynamically adjust the optimal output strategy of the equipment according to the different requirements of the user; compared with the traditional control method, under the same comfort weight requirements, the optimization method reduces the total power consumption of the heat pump unit by 5.85% while the thermal comfort indexes deviate from the optimal value by 19.85%. In addition, the frequency and magnitude of power fluctuations of the heat pump are significantly reduced, which also improves the safety and lifetime of the equipment. The optimal regulation strategy proposed in this paper has better economy, thermal comfort and stability, and can provide an optimized reference for the control of similar green heating systems.

Keywords: Heating systems · Optimized Control · Air source heat pumps · Comfort · Economy · NSGA-II (Non-dominated Sorting Genetic Algorithm II) · Pareto frontiers

1 Introduction

In the context of accelerating global sustainable development, buildings are major energy consumers, and their heating systems constitute a primary energy-consuming segment. Establishing a green and efficient heating system has become a core approach to energy conservation and consumption reduction in buildings. The previous studies have shown

H. Ma et al. (Eds.): IWACIII 2025, CCIS 2780, pp. 409–430, 2026.
https://doi.org/10.1007/978-981-95-6730-0_29

that air source heat pump unit has significant advantages in heating, which can efficiently transfer the heat in the air to the room, compared with the traditional heating method the energy utilization rate is greatly improved [1]. At the same time, air source heat pump units do not produce combustion exhaust during operation, which is important for improving air quality and reducing carbon emissions [2]. For building heating systems based on air-source heat pumps, it is necessary to balance efficiency improvement, cost reduction, and indoor comfort guarantee [3, 4]. Therefore, conducting research on multi-objective optimal regulation for the synergy of economy, comfort, and low carbon holds significant theoretical and practical value.

Traditional heating control is dominated by on-off control and PID control. On-off control starts or stops equipment based on thresholds, but it tends to cause significant room temperature fluctuations, low comfort levels, and frequent equipment wear [5]. PID control achieves precise regulation by dynamically tracking temperature deviations, but it is affected by factors such as differences in building structures and randomness of user energy consumption, making it difficult to establish accurate models and leading to a significant reduction in control accuracy under complex working conditions [6, 7]. To address these issues, modern control technologies have been introduced into the control of heating systems. Fuzzy control handles the nonlinear characteristics of the system by fuzzifying inputs such as temperature and humidity and making decisions based on an expert experience rule base [8, 9]. However, the rules of fuzzy controllers rely on specific experiences, making cross-scenario transplantation difficult [10]. Neural network control achieves precise scheduling by training models to predict heat loads through massive historical data (including indoor and outdoor temperatures, energy consumption time series, etc.) [11]. Nevertheless, it faces bottlenecks such as high data collection costs, large computing power requirements for training, and poor model interpretability. Moreover, it has not deeply integrated the collaborative optimization of economic and comfort goals [12]. In summary, these control methods are unable to meet the demand for improving the comprehensive performance of the system, and there is an urgent need to establish a multi-objective optimal regulation system that takes into account multiple requirements such as system operational economy and indoor comfort.

Multi-objective optimization aims to solve a set of optimal solutions that balance trade-offs among objectives. Genetic algorithms (GA) simulate natural evolution through selection, crossover, and mutation, have found extensive applications in heating system control. For example, Reference [13] utilized GA to balance energy costs and carbon dioxide emissions, and this approach can identify some potentially optimal solutions. However, GA often suffers from slow convergence in later stages, computational redundancy, and intermittent local optima, which undermine overall economic efficiency. Pombeiro et al. [14] optimized the Heating, Ventilation, and Air Conditioning (HVAC) system with comfort and economic efficiency as objectives, but the results indicate that GA requires a higher level of expertise for fine-tuning and is constrained by more limited development environments, thus limiting its practical applicability. Particle swarm optimization (PSO), which simulates bird flocking behavior, dynamically updates particle trajectories based on individual and group optima. Ma et al. [15] applied PSO to address uncertainties in heating user behavior, showing initial success in enhancing room temperature stability and comfort. However, under extreme weather conditions with sudden heat demand changes, particles tend to converge prematurely to local comfort optima. Haniff et al. [16] applied PSO for air conditioning system operational optimization. While achieving faster convergence, the algorithm struggled in later stages of the heating season due to interference from equipment aging and other factors, leading to search stagnation and an inability to balance the trade-off between comfort and cost. By comparison, the NSGA-II (Non-dominated Sorting Genetic Algorithm II) algorithm, as an outstanding representative of multi-objective evolutionary algorithms, features efficient non-dominated sorting and elite retention strategies [17, 18]. It can obtain multiple Pareto-optimal solutions in a single run, offering new perspectives for multi-objective optimization and scheduling in heating systems [19]. Garces-Jimenez et al. [20] used the NSGA-II algorithm to optimize operational parameters such as supply water temperature and flow rate in HVAC system, achieving efficient and stable regional heating network operations, and improving energy efficiency and economic performance of the heating system.

Based on the above analysis, this study focuses on a green heating system composed of multiple air source heat pumps. It fully considers factors such as user-side heat load demand, building heat loss, and the efficiency of air source heat pumps under different operating conditions. With system economic efficiency and thermal comfort as the optimization objectives, the NSGA-II algorithm is employed for multi-objective optimization scheduling. This approach dynamically adjusts the output of heat pump equipment and the valve openings on the user side under fluctuating outdoor environments, ensuring the system operates economically and efficiently. Case analysis further validates the feasibility and effectiveness of the proposed optimization and control method.

2 Modeling of Green Heating System

2.1 System Structure

This study takes the green heating system with air source heat pump as the main heat source, which has been put into operation in a business park in Shanxi, as shown in Fig. 1 for the system structure. The system mainly consists of three air source heat pumps (ASHPs), one water tank (WT), three types of heating buildings with different energy-using characteristics, and electric regulating valves (ERVs) used to independently control the inlet water flow of the three types of users, etc. Among them, ASHP absorbs low-grade heat energy in the air from the evaporator side, compresses and raises the temperature through the compressor, and then releases heat to the circulating water of the heating system on the condenser side; the heating equipment on the user side of the 3 types of different energy-using characteristics are mainly hot water floor radiant heating or cast-iron radiators and other equipment; ERV plays an important role of adjusting the flow rate of the heat medium and then controlling the heat distribution in the heating system, and its reasonable adjustment of the opening is important for meeting the demand of the heating system. The reasonable adjustment of its opening degree has a key impact on meeting the user's heat demand and realizing the overall optimization of the system.

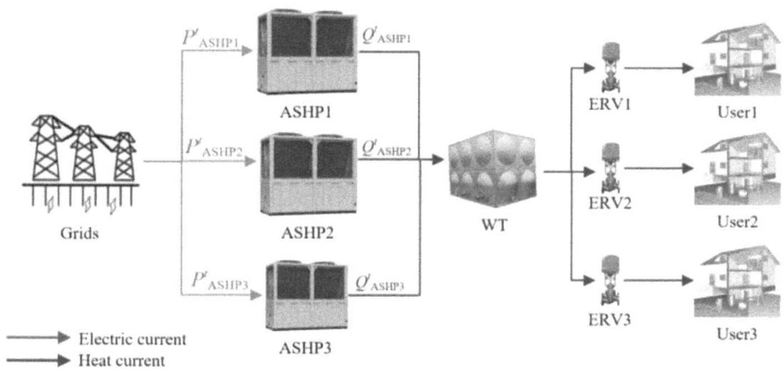

Fig. 1. Structural Diagram of the Green District Heating System

2.2 System Model

Accurate modeling of user heat demands, ASHPs, and ERVs, which are the core components of the heating system, is essential for achieving optimal regulation.

Model of user-side building heat loss

In the heating system, the user's heat load usually consists of two parts: the base heat load and the heat loss from the building. That is, the heating heat demand of the user is:

$$Q_{demand,k}(t) = Q_{\min,k} + Q_{loss,k} \tag{1}$$

where $Q_{\min,\,k}$ is the basic heat load of the k-th building, kW; $Q_{loss,\,k}$ is the heat loss of the k-th building, kW.

The basic heat load is the user's heat demand when considering economic optimization, which can be set as a fixed value by the user according to the characteristics of the building and the user's use of energy. That is:

$$Q_{\min,k}(t) = Q_{\text{base},k} \cdot S_{k,load} \tag{2}$$

where $Q_{\text{base},\,k}$ is the basic heat load per unit area corresponding to the k-th building, kW; $S_{k,\,load}$ is the heating area corresponding to the k-th building, m^2.

Building heat loss is the heat interaction with the outdoor and indoor, which can be expressed as [21]:

$$Q_{loss,k}(t) = S_{wall,k} \cdot u_{wall,k} \cdot \left(T_a(t) - T_{in,k}(t)\right)/3600 \tag{3}$$

where $Q_{loss,\,k}(t)$ is the k-th building in moment t of heat loss, kW; $S_{wall,\,k}$ is the k-th building and the outdoor environment for heat interaction of the building envelope area, m^2; $u_{wall,\,k}$ is the k-th building heat loss coefficient, J/(m^2·°C); $T_a(t)$ is the outdoor ambient temperature, and $T_{in,\,k}(t)$ is the indoor temperature of the k-th building.

Model of ASHP

The heat energy produced by an ASHP is:

$$Q_{ASHP,j}(t) = COP_{ASHP,j}(t) \cdot P_{ASHP,j}(t) \tag{4}$$

where $Q_{\text{ASHP},j}$ is the heating power of the j-th ASHP, kW; $P_{\text{ASHP},j}$ is the input electric power of the j-th ASHP, kW; $COP_{\text{ASHP},j}$ is the coefficient of performance (COP) of the ASHP, which is affected by a variety of factors, mainly including the ambient temperature and the outlet water temperature setpoint.

The operating power constraints of the ASHP are:

$$P_{ASHP,j}^{\min}(t) \le P_{ASHP,j}(t) \le P_{ASHP,j}^{\max}(t) \tag{5}$$

$$Q_{ASHP,j}^{\min}(t) \le Q_{ASHP,j}(t) \le Q_{ASHP,j}^{\max}(t) \tag{6}$$

where $P_{ASHP,j}$ and $Q_{ASHP,j}$ are the input electric power and the output thermal power of the ASHP, kW; $P_{ASHP,j}^{\min}$ and $P_{ASHP,j}^{\max}$ are the minimum and maximum electric power of the ASHP, kW; $Q_{ASHP,j}^{\min}$ and $Q_{ASHP,j}^{\max}$ are the minimum and maximum thermal power of the ASHP, kW, respectively.

Based on the system structure, the heat production of the air source heat pump should be equal to the total heat demand of the system. Thus, the following power balance constraints should be satisfied:

$$\sum_{j=1}^{J} Q_{ASHP,j}(t) = \sum_{k=1}^{K} Q_{demand,k}(t) \tag{7}$$

Model of ERVs

The flow rate of the ERV determines the thermal energy supplied to the user, and the model of the valve can be expressed as:

$$Q_{ERV,k}(t) = Q_{demand,k}(t) \tag{8}$$

$$Q_{ERV,k}(t) = C \cdot m_{ERV,k}(t) \cdot \left(T_{ERV,k}^{\text{out}}(t) - T_{ERV,k}^{\text{in}}(t)\right) \tag{9}$$

where $Q_{ERV,k}$ is the heat supply corresponding to the outlet of the k-th valve, kW; C is the specific heat capacity of water, 4.2×10^3 J/(kg·°C); $m_{ERV,k}$ is the flow rate corresponding to the opening of the k-th valve, kg/h; $T_{ERV,k}^{\text{in}}$ and $T_{ERV,k}^{\text{out}}$ are the temperature of the inlet and return water of the k-th valve, °C. Eq. (8) constrains the system to meet the heat demand of the user at each moment by controlling the valves, and Eq. (9) is used to model the relationship between the valve flow and heat. The valve used in this paper is of linear flow characteristics, and the flow rate is proportional to the opening:

$$m_{ERV,k}(t) = A_k \cdot CV_k(t) \tag{10}$$

where A_k is the linearity coefficient between valve opening and flow rate; CV_k is the opening of the k-th valve, %. And the valve opening at any moment should satisfy the following constraint:

$$CV_k^{\min} \le CV_k(t) \le CV_k^{\max} \tag{11}$$

where $CV_k^{\min}$ is the minimum valve opening to satisfy the basic thermal cycle of the system, %; $CV_k^{\max}$ is the maximum valve opening limit, %.

3 Multi-objective Optimal Regulation Model for Heating Systems

3.1 Optimizing Objective

Thermal Comfort Indicator

In heating system, comfort is one of the key indicators to measure its service quality, which is related to user's actual feeling [23]. Fanger [22] divided the thermal comfort into {+3 hot, +2 warm, +1 warmer, 0 moderate, −1 cooler, −2 cooler, −3 colder} seven grades that can represent the human body's thermal sensation and proposes to use the predicted mean vote index (PMV) to characterize the average value of the group's vote on the seven grades, which is calculated by the following equation [24]:

$$\begin{aligned} I_{PMV} = & \left(0.303e^{-0.036M} + 0.0275\right)\Big\{(M - W) \\ & - 3.05 \times 10^{-3}[5733 - 6.99(M - W) - P_a] \\ & - 0.42[(M - W) - 58.15] - 1.7 \times 10^{-5}M(5867 - P_a) \\ & - 0.0014M(34 - T_a) - 3.96 \times 10^{-8} f_{c1} h_c (T_{c1} - T_a)\Big\} \\ & - (T_r + 273)^4 \end{aligned} \tag{12}$$

where M is the metabolic rate of the human body; W is the human mechanical efficiency; P_a is the partial pressure of water vapor in the ambient air; T_a is the ambient air temperature; f_{cl} is the area coefficient of the garment; T_{cl} is the average skin temperature of clothed human body; T_r is the mean radiant temperature; and h_c is the convective heat transfer coefficient.

Considering that the values of the coefficients in Eq. (12) corresponding to different ages, genders, and activity levels of people in the same area are not the same, and this paper is concerned with the effect of indoor temperature on the comfort level. Therefore, use the model of the relationship between the indoor temperatures and the IPMV when other factors are at a comfortable level in literature [25]:

$$I_{PMV}(t) = \begin{cases} 0.3895 \cdot \left(T_{in}^{set}(t) - T_{in}^{set,0}\right), & T_{in}^{set}(t) \geq T_{in}^{set,0} \\ 0.4065 \cdot \left(-T_{in}^{set}(t) + T_{in}^{set,0}\right), & T_{in}^{set}(t) < T_{in}^{set,0} \end{cases} \tag{13}$$

where $T_{in}^{set,0}$ is the indoor temperature setting value corresponding to the best comfort, °C. I_{PMV} is based on the theory of human thermal comfort, and the size of the indoor temperature setting value directly determines the value of the comfort indicator I_{PMV}. When the indoor temperature is $T_{in}^{set,0}$, the I_{PMV} is 0, indicating that the user comfort is the highest at this time, and when the indoor temperature setting value deviates from $T_{in}^{set,0}$, the I_{PMV} increases, indicating that the user's thermal experience becomes worse.

Both international standard [26] and Chinese national standard [21] stipulate that for areas where people stay for a long time, the I_{PMV} ranging from −0.5 to 0.5 indicates a state of thermal comfort for the human body. According to the calculation using Eq. (13), the corresponding indoor temperature is 24.8 ~ 27.3 °C.

Economic indicators

The economic indicator of the system is mainly the electricity cost during the operation of the air-source heat pump unit, which can be expressed as:

$$I_{eco}(t) = \sum_{j=1}^{J} P_{ASHP,j}(t) \tag{14}$$

where J is the number of air source heat pumps; $P_{ASHP,\,j}(t)$is the input electric power of the j-th ASHP at moment t, kW.

Optimization objective

The purpose of this paper is to optimize the control of the system by considering the economy and comfort, and the objective function model of the optimization control is:

$$\min f = \min\{I_{PMV}(t), I_{eco}(t)\} \tag{15}$$

3.2 Optimization Models

The mathematical model of system optimal regulation constructed in this paper is a mixed integer linear programming problem, whose standard model form can be written as:

$$\begin{aligned} &\min f = \min(I_{PMV}, I_{eco}) \\ &s.t. \begin{cases} h_i(x) = 0, i = 1, 2, \cdots, m \\ g_j(x) \le 0, j = 1, 2, \cdots, n \\ x_{\min} \le x_k \le x_{\max}, k = 1, 2, \cdots, l \end{cases} \end{aligned} \tag{16}$$

where x is the optimization variable; $h_i(x) = 0$ is the equation constraint, mainly representing Eqs. (1) – (5) and Eqs. (8) – (11); $g_j(x) \le 0$ is the inequality constraint, mainly representing Eqs. (6) and (7) and Eq. (12); $x_{\max}$ and $x_{\min}$ are the upper and lower limits of the optimization variable. In the optimal regulation model constructed in this paper, the optimization variables are the output value of the air source heat pump unit and the opening degree of the electric control valve.

4 Optimal Regulation Model Based on NSGA-II

In the proposed optimal regulation model, the two objectives are directly affected by the outdoor ambient temperature, which is manifested as follows: when the outdoor temperature is low, the heat loss of the user side of the building increases, in order to ensure comfort, the heat pump needs to be increased in the heat production, and at this time, the energy efficiency of the air-source heat pump is relatively low, and the heat pump's power consumption is increased, the economy deteriorates. The two optimization objectives constrain each other, and a suitable multi-objective optimization algorithm is the key to achieve multi-objective optimal scheduling. NSGA-II is a multi-objective optimization algorithm, which aiming at solving the optimization problems with multiple conflicting objective functions, and then efficiently finding the Pareto optimal solution set for multi-objective optimization problems.This paper adopts NSGA-II algorithm to carry out the optimization control for the heating system with comprehensive consideration of economy and comfort, as shown in Fig. 2, which is the general block diagram of the optimal regulation algorithm for the heating system based on NSGA-II.

4.1 NSGA-II Algorithm Fundamentals

The NSGA-II algorithm is based on the idea of evolutionary algorithms, the core of lies in the non-dominated sorting and crowing distance calculation.

Non-dominated Sorting

In multi-objective optimization, individuals in the population are potential solutions that need to be evaluated through multi-objective functions. The conditions for individual i to dominate j are:

$$\text{① For all } m, f_m(i) \leq f_m(j) \tag{17}$$

$$\text{② There exists at least one } m, \text{ made } f_m(i) < f_m(j) \tag{18}$$

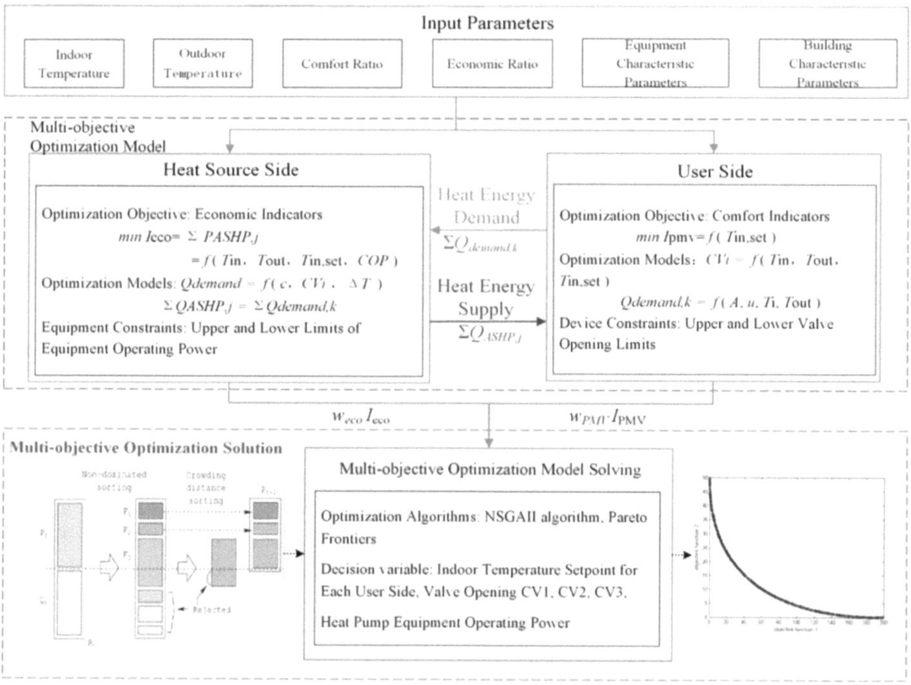

Fig. 2. Multi-Objective optimization regulation of the heating system based on NSGA-II

A solution that is not dominated by other individuals is a non-dominated solution. Through the above comparison, NSGA-II divides individuals into different non-dominated fronts: first, all non-dominated individuals are selected to form rank 1, then non-dominated individuals are selected from the remaining individuals to form rank 2, and so on, giving priority to retaining solutions with excellent performance in multi-objective trade-offs.

Crowding Distance Calculation For a set I of individuals at the same frontier, the calculation of the crowding distance d_k of individual $k \in I$ is as follows:

Step1: For each objective function f_m, sort I in ascending order into ${I_m}^{sort}$.

Step2: Calculate the crowding distance of boundary individuals ($i = 1$ and $i = |I|$):

$$d_{I_m^{sort}(1)} = d_{I_m^{sort}(|I|)} = \infty \tag{19}$$

where $|I|$ denotes the number of elements of set I.

Step3: For the intermediate individuals $i = 2,\ldots,|I|\text{-}1$, the congestion distance is calculated as:

$$d_{I_m^{sort}(i)} = \frac{f_m\left(I_m^{sort}(i+1)\right) - f_m\left(I_m^{sort}(i-1)\right)}{f_m^{\max} - f_m^{\min}} \tag{20}$$

where $f_m^{\max}$ and $f_m^{\min}$ are the maximum and minimum values of the objective function f_m in the set of individuals I.

Step4: The final crowding distance of individual k is the sum of the crowding distances in each objective function:

$$d_k = \sum_{m=1}^{M} d_{k,m} \tag{21}$$

4.2 Optimization Solution for Heating System Based on NSGA-II Algorithm

The steps of the multi-objective optimal regulation algorithm for heating system based on NSGA-II algorithm are as follows:

Population Initialization.

Firstly, randomly generate an initial population P_0 with a population size of N_{pop}. For each individual, randomly generate the air source heat pump output power $Q_{\text{ASHP,1}}$, $Q_{\text{ASHP,2}}$, $Q_{\text{ASHP,3}}$, and the user-side valve openings vectors CV_1, CV_2, CV_3 within a range of values of the decision variables. For example, the air source heat pump output power randomly generate a value within $[Q_{ASHP,j}^{\min}, Q_{ASHP,j}^{\max}]$; for each valve opening randomly generated a value within $[CV_k^{\min}, CV_k^{\max}]$.

Then, calculate the objective function values I_{pmv} and I_{eco} for each individual in the initial population and the degree to which the constraints are satisfied.

Non-dominated Ordering and Congestion Distance Calculation

Firstly, Eqs. (17) and (18) are used to perform a non-dominated ordering of the current population, dividing individuals into different non-dominated fronts F_1, F_2, and so on.

Then, for each non-dominated front, Eqs. (19) – (21) are used to compute the individual congestion distance.

Selection Operation

The binary tournament selection method is adopted to select parent individuals from the population.

Crossover Operation

For the selected parent individuals, the Simulated Binary Crossover (SBX) method is used.

Mutation Operation

For the offspring generated by crossover, the mutation operation is performed with a certain mutation probability p_m. For example, for ASHP output, a small random perturbation is performed within its value range, such as $Q_{ASHP,1}^{new} = Q_{ASHP,1}^{old} + \Delta Q_{ASHP,1}$, where $\Delta Q_{ASHP,1}$ is a random value within a certain range.

Mutation Offspring and Population Update Evaluation

Evaluate the objective functions and constraint satisfaction of the offspring. Merge the parent and offspring populations (with a size of $2N_{pop}$). After sorting and crowding distance calculation, select the first N_{pop} individuals as the next-generation population P_{next}.

Termination Condition Judgment

Set the maximum number of iterations G_{max}. If it is reached, terminate the algorithm and obtain the Pareto optimal solution set.

4.3 Optimization and Regulation Strategy

The NSGA-II algorithm is applied under the same outdoor temperature conditions, and the resulting Pareto front solution set is shown in Fig. 3, with the comfort index plotted on the horizontal axis and the economic index on the vertical axis. Different solutions distributed on the frontier curve reflect different trade-offs between comfort and economy: the solution near the horizontal axis focuses on comfort, with high and stable indoor temperature settings, but with relatively increased energy costs; the solution near the vertical axis pays more attention to economy and minimizes costs by appropriately lowering the temperature settings and finely regulating the heat pump and valves. Users can select the appropriate solution on the Pareto frontier according to their actual needs to achieve personalized optimal regulation.

The specific regulation ideas are: to meet user comfort demands, the system sorts the Pareto front solution set based on the computed comfort index values. The minimum and maximum comfort index values within the solution set are denoted as $I_{pmv}^{\min}$ and $I_{pmv}^{\max}$. When the user specifies a comfort weight ratio of $\frac{w_{pmv}}{I_{pmv}^{\min}}$, it is converted to a corresponding comfort setting value using Eq. (22). The solution on the Pareto front

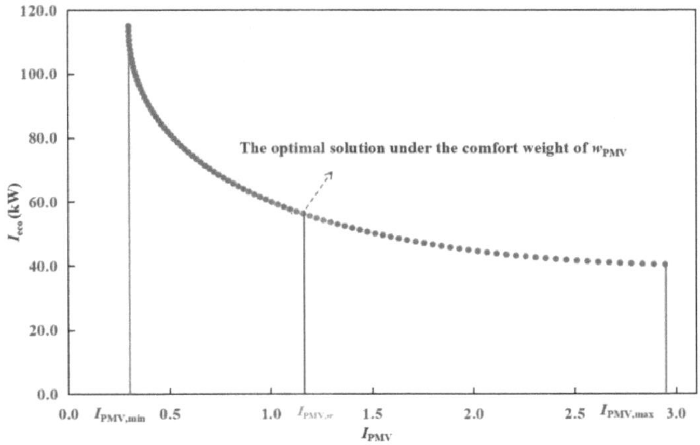

Fig. 3. Optimization Pareto frontier

corresponding to this value is selected as the optimal solution set under the given outdoor temperature conditions, which satisfies the user's comfort weight requirements. This process identifies the optimal combination of parameters, including indoor temperature settings, pump power output, and valve openings. It effectively optimizes the scheduling of the air source heat pump heating system, thereby enhancing the overall performance of the heating system.

$$I_{\mathrm{PMV},w} = I_{\mathrm{PMV}}^{\min} + (1 - w_{\mathrm{PMV}}) \cdot \left(I_{\mathrm{PMV}}^{\min} - I_{PMV}^{\max} \right) \tag{22}$$

Furthermore, at different times of each scheduling day, the system performs multi-objective optimization and selects regulation solutions based on the outdoor temperature at that specific time using the aforementioned method. This enables adaptive optimization and adjustment of the system at different moments, effectively avoiding fluctuations in indoor comfort caused by changes in the outdoor environment.

5 Case Studies

This paper based on Python's Anaconda environment and PyTorch 2.0 is used as the core to construct the system model and call the Cplex solver to deal with the system optimization regulation problem. Meanwhile, NSGA-II optimization algorithm is implemented using Python language combined with Numpy 1.24.3 library. The parameter settings of the NSGA-II algorithm comprehensively consider computational efficiency and optimization effects: the population size is set to 100, which is determined by balancing computational efficiency and solution diversity within the range of 50–200, this value can not only expand the solution space and reduce the risk of local optima but also take computational costs into account; the crossover probability of 0.8 is within the conventional range of 0.6–0.9, which is conducive to offspring inheriting excellent gene combinations from parents, accelerating the approach to the Pareto front while maintaining solution stability; the mutation probability of 0.05 can ensure the algorithm escapes local optima without excessive randomness; the maximum number of iterations of 200 is determined according to the problem complexity and computational resources to ensure sufficient convergence to the Pareto front. The outdoor temperature is obtained by the optimized regulation system through a crawler algorithm sourced from web-based open-source data and a combination of temperature sensor dynamics. This paper selected Yuci District, Shanxi Province, December 19, 2024, as a case study, the Fig. 4 shows a day of the outdoor temperature data.

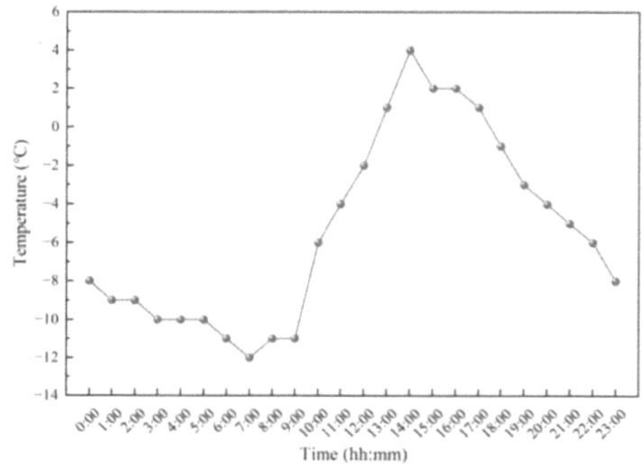

Fig. 4. Typical day outdoor temperature data

5.1 The Pareto Frontiers at Different Outdoor Temperatures

Multi-objective optimization is solved for the heating system at different outdoor ambient temperatures, as shown in Fig. 5. It gives the results of the pareto frontiers from the NSGA-II algorithm solved in Sect. 4 at outdoor temperatures of 5 °C, 0 °C, −8 °C, and − 15 °C, respectively.

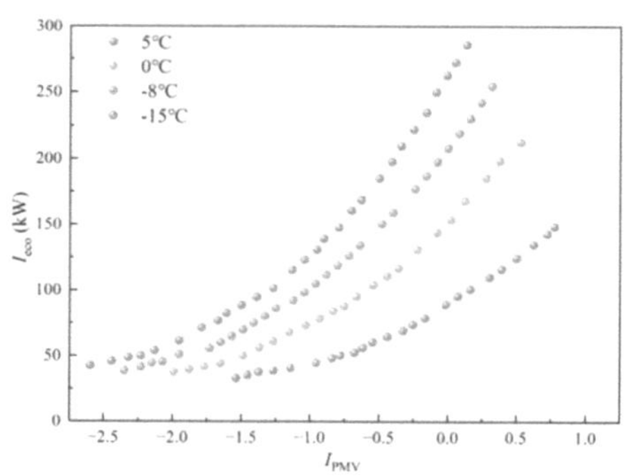

Fig. 5. Pareto frontiers at different outdoor temperatures

By analyzing the data in the graph, the following conclusions can be drawn:

(1) There is a negative correlation between heat pump power consumption and thermal comfort. At a certain outdoor temperature, as the thermal comfort requirements increase, the electricity consumption of ASHP increase. Meanwhile, this increase is not linear, which is manifested as follows: in the range of low thermal comfort, the electricity consumption of the heat pump increases relatively slowly; when the thermal comfort approaches a high level, the growth rate of electricity consumption accelerates. This is because the higher the comfort requirement, the greater the heat transfer resistance the heat pump needs to overcome or the higher the heating capacity limit it needs to reach, thus consuming more electricity.
(2) From the overall trend, at the same level of indoor thermal comfort, as the outdoor temperature decreases the electric power used by the heat pump increases significantly. For example, when the indoor thermal comfort indicator is 0, the heat pump power consumption is about 90 kW for 5 °C, 150 kW for 0 °C, and approach to 270 kW for −15 °C. This indicates that more power is needed to maintain the same level of indoor comfort in cold weather.
(3) Regarding the shapes of the Pareto front curves at different temperatures, the curve corresponding to 5 °C is relatively gentle, while the slopes of the curves at −8 °C and − 15 °C increase significantly. This means that in extremely cold environments, to further improve indoor thermal comfort, the power consumption of the heat pump will increase sharply, indicating that the cost of maintaining high indoor thermal comfort in low-temperature environments is relatively higher. Additionally, observing the starting points of the four curves in the low indoor thermal comfort region reveals that as outdoor temperature decreases, even maintaining a low level of indoor thermal comfort requires air source heat pump systems to consume more electricity to meet basic heating demands.

Overall, the outdoor temperature plays a decisive influence on the relationship between indoor thermal comfort and heat pump power consumption. That is the reason why this paper solves the optimal set of solutions separately according to different outdoor temperatures, so as to develop an accurate equipment regulation program.

5.2 Optimization Results with Different Comfort Weights

Results of Equipment Scheduling

Based on the Pareto frontiers at different outdoor temperatures shown in Fig. 6, the comfort weights are set to 20%, 50% and 80%. And the optimal solution that satisfies the comfort weight requirement is selected in each set of pareto-optimal solution sets using the strategy designed in Sect. 3.3. The Figs. 6, 7, and 8 show the regulation of the operating frequency of the three air-source heat pumps and the opening of three user-side regulating valves.

By comparing the working power of heat pump units and the valve openings under three different comfort weight settings shown in Figs. 6, 7, and 8. The following observations can be made: As the comfort weight increases, the indoor temperature setpoint for users gradually rises. To ensure that user-side comfort meets the requirements, the valve openings for all three buildings are adjusted upwards. For instance, considering the heat demand of Building 1 at 8:00 AM, the valve openings under the three comfort weight settings are 64%, 76%, and 83%, respectively. The increase in valve openings leads to a higher water flow rate in the heating network, which in turn raises the total heating load of the three buildings. To maintain supply-demand balance, the working power of the heat pump units is dynamically adjusted. Using the heat demand of Building 1 at 8:00 AM as an example, the total heat pump output under comfort weights of 20%, 50%, and 80% is 191 kW, 208 kW, and 286 kW, respectively. Hence, the proposed optimal regulation method can dynamically control system equipment in real-time to meet user heating demands under varying outdoor temperatures and comfort weight settings, demonstrating excellent adaptability.

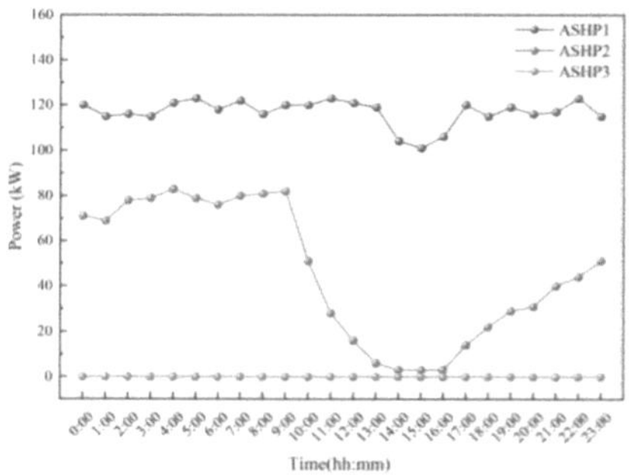

(a) Air Source Heat Pump Output

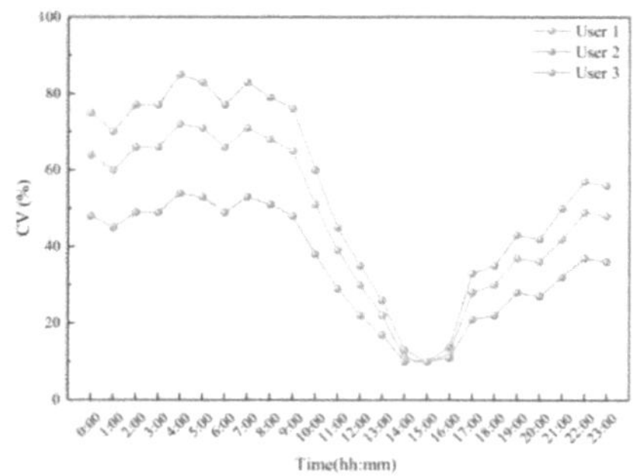

(b) Valve Opening

Fig. 6. Results of device regulation under 20% comfort weighting

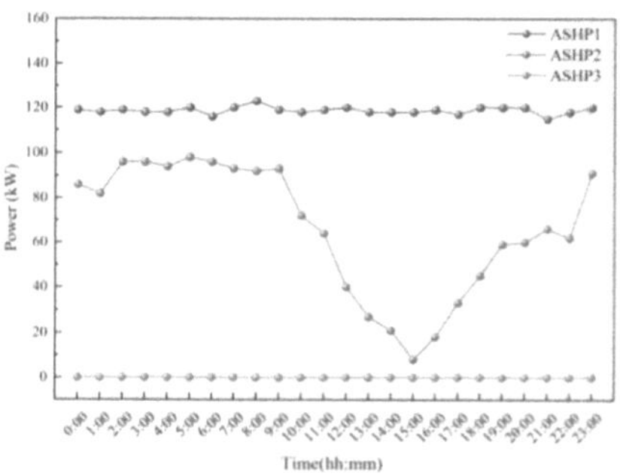

(a) Air source heat pump outputs

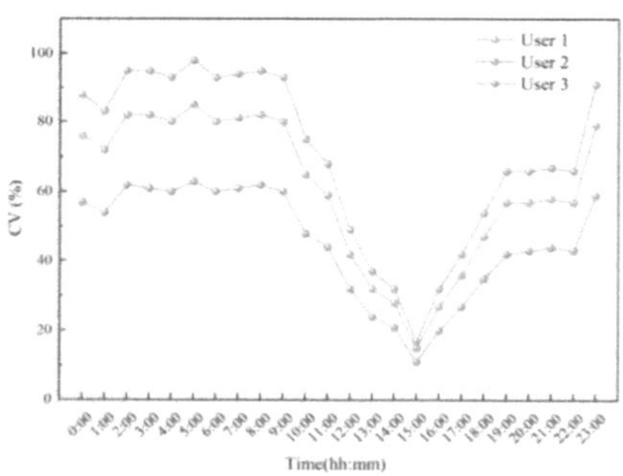

(b) Valve opening

Fig. 7. Results of device regulation under 50% comfort weighting

From the working power of the air source heat pumps shown in Fig. 6(a), Fig. 7(a), and Fig. 8(a), it can be observed that the system dynamically adjusts the output of the three pumps to ensure the balance of heat supply and demand. In this study, a stepwise recursive control strategy for multiple air source heat pumps was adopted to achieve efficient and stable heating for the user side. Specifically, the operation of the three air source heat pumps always follows the principle of prioritizing the activation of the first

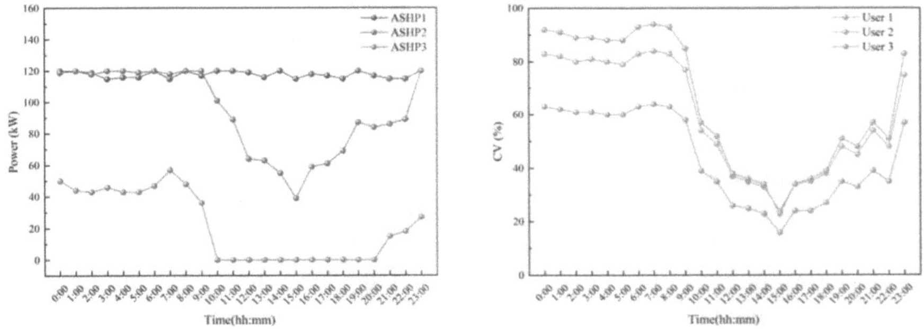

(a) Air source heat pump outputs (b) Valve opening

Fig. 8. Results of device regulation under 80% comfort weighting

pump. When monitoring the heat load continues to rise, only one heat pump has been difficult to maintain the preset room temperature, and its output is close to or reached the limit of the current operating conditions, the control system automatically triggers the second pump to start, in conjunction with the first pump to meet the demand for heat. Furthermore, when the first two pumps are still unable to meet the demand for heat supply, the third pump will then be put into operation. This step-by-step recursive control method on the one hand effectively reduces the frequency of heat pump start and stop and energy consumption, extends the life of the equipment. On the other hand, by controlling the heat pump with the heat load increases or decreases when starting or stopping one by one, to ensure the smooth output of heat, reduce the fluctuation of water temperature, to better maintain a constant temperature indoor comfort.

From the valve openings in Fig. 6(b), Fig. 7(b), and Fig. 8(b), it can be seen that the valve openings on the same user side are not kept constant at the same comfort setpoint, they are dynamically adjusted in response to changes in outdoor temperature. This is because when the outdoor temperature decreases (e.g., 16:00–23:00 in the figure), due to the increase in the temperature interaction between the building and the outdoors (i.e., the building heat loss increases), the system dynamically adjusts the valve opening to ensure that the indoor temperature is maintained near the indoor temperature setpoint corresponding to the set comfort level. Similarly, during the 7:00–14:00 period, when the outdoor temperature increases, the valve opening is reduced to prevent overheating in the room due to heat supply exceeding the actual heat demand. In the 14:00–16:00 period, the outdoor temperature is almost constant and reaches the maximum temperature of the day. At that time the valve opening is adjusted to the minimum opening, thus maintaining the minimum flow requirements in the heating network to ensure that various types of equipment are not damaged by freezing. It can be seen that the dynamic adjustment of the valve opening according to the outdoor temperature could make the system better meet the user's comfort requirements.

In addition, it can be seen from the valve opening curves of different user sides in Fig. 7(b), Fig. 8(b), and Fig. 9(b) that the valve openings of different user sides are not the same at the same time and under the same comfort weight. This is because the heating building characteristics of different user sides are different, and accordingly, the set values of the most comfortable indoor temperature on the user side and the heat loss of the buildings are also different. As a result, different buildings have different heat demands, so it is necessary to adjust the valve openings for different buildings respectively.

System Optimization Objectives

Under the regulation of equipment output and opening as shown in Figs. 6, 7 and 8, the electricity consumption of the heat pump units and the indoor thermal comfort of the users are shown in Fig. 9, and the corresponding total daily electricity consumption of the three heat pump units of the system and the average values of the indoor comfort indexes of the users are shown in Table 1.

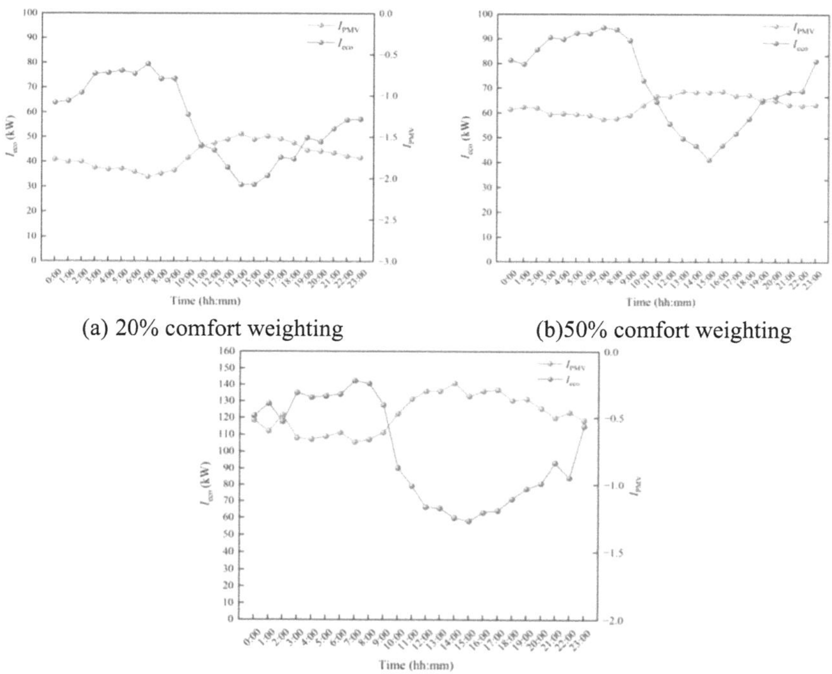

(a) 20% comfort weighting (b)50% comfort weighting

(c) 80% comfort weighting

Fig. 9. Variation curves of system optimization objectives under different weights

As shown in Fig. 9, the I_{PMV} indexes at each scheduling moment of the day are not constant values but fluctuate in a certain range under the same comfort weight setting values. This is because the proposed optimization algorithm optimizes and regulates the equipment after comprehensively considering the effects of outdoor temperature on the heat load demand on the user side, the building heat loss, and the energy efficiency of the air-source heat pumps under different operating conditions, and the I_{pmv} for different outdoor temperatures under the same comfort weights is not the same. This is also derived from the Pareto frontiers in Fig. 5 for different temperatures combined with

Table 1. Values of electricity consumption and indoor comfort of the unit

Comfort Weighting	Total Daily Electricity Consumption of Heat Pump Units (kWh)	Comfort Level Mean
20%	1337.06	−1.72
50%	1727.45	−1.09
80%	2380.79	−0.41

the optimized regulation strategy in Sect. 4.3. Taking Fig. 9(b) as an example, at the 07:00, the outdoor temperature is the lowest, and the COP of the air source heat pump is also the lowest at this time, which results in the highest power consumption and the lowest economy of the heat pump unit in generating the heat required by the user. The regulation algorithm needs to appropriately reduce the value of the comfort index in order to ensure that the weights of economy and comfort are both 50%. Therefore, the system uses the most electricity and the I_{pmv} indicator value is the lowest at 07:00; the outdoor temperature is the highest, the energy efficiency of the ASHP is the highest at 14:00. At that time, the heat loss of the building is the smallest, so the ASHP uses the least amount of electricity at this time, and the corresponding I_{pmv} indicator value reaches the maximum value of that day.

By comparing Fig. 9(a, b, and c), and also combining the data in Table 1, the indoor thermal comfort of the users and the energy consumption of the heat pump unit show significant differences under different comfort weights. When the comfort weight is 20%, the I_{pmv} index fluctuates between -1.95 and -1.45 during a day, the indoor thermal environment is inclined to be uncomfortable, and the electric power used by the heat pump unit fluctuates between 31.1 and 82.12 kW, and the energy consumption is relatively low. When the weighting is increased to 50%, the fluctuation range of I_{pmv} indicator becomes -1.26 ~ -0.93, and the stability of indoor thermal environment is improved, and the comfort is also gradually improved. Accordingly, the energy consumption of the heat pump unit also rises and compared with the total daily power consumption of the system when the comfort weight is 20%, it increases by 29.20% when the weight is 50%. The I_{pmv} is further improved at a comfort weight of 80%, with an average daily I_{pmv} of −0.49, which is close to the ideal comfort state, but the electric power used by the heat pump unit rises sharply. Compared with the total daily power consumption of the heat pump unit at 20% and 50%, it increased by 84.17% and 42.55% at 80% weighting, and the energy consumption increased significantly. Hence, with the increase of the comfort index percentage, the user's indoor thermal comfort is gradually optimized, which is more in line with the human body's thermal comfort needs. But inevitably leads to a significant increase in the energy consumption of the heat pump unit, which reveals the intrinsic correlation between comfort and economy in this system.

5.3 Performance Analysis

Comparative analysis of the comfort and economy of the system based on the proposed NSGA-II algorithm optimal regulation and the traditional control method with the system now in operation.

The control mode of the system now put into operation is a strategy designed by the system operators after several tests to comprehensively balance the comfort and economy. Its main features are the use of time-triggered way to adjust the valve opening, that is, in the 00:00–07:00 and 20:00–23:00 h when the outdoor temperature is low, the valve opening to 90%–100%. At 12:00–15:00 when the outdoor temperature is higher, the valve opening is adjusted to 10%–30%. Other periods of time are also according to the law of change of temperature at different times set to different fixed openings. This valve opening control does not consider the impact of different moments of outdoor temperature. In the air source heat pump control, is based on the overall level of outdoor temperature, set the air source heat pump water temperature. For example, when the average outdoor temperature is $-5 \sim 5$ °C, the water temperature is set to 45 °C, when the average outdoor temperature is $-10 \sim -15$ °C, the water temperature is set to 50 °C, and when the average outdoor temperature is lower than -15 °C, the water temperature is set to 55 °C, and by monitoring the water temperature of the water tank in order to control the start and stop of the heat pump. As shown in Fig. 10, the actual power consumption of the heat pump unit, and the trend of thermal comfort corresponding to the actual indoor temperature of the user in the typical day shown in Fig. 4 under the control strategy.

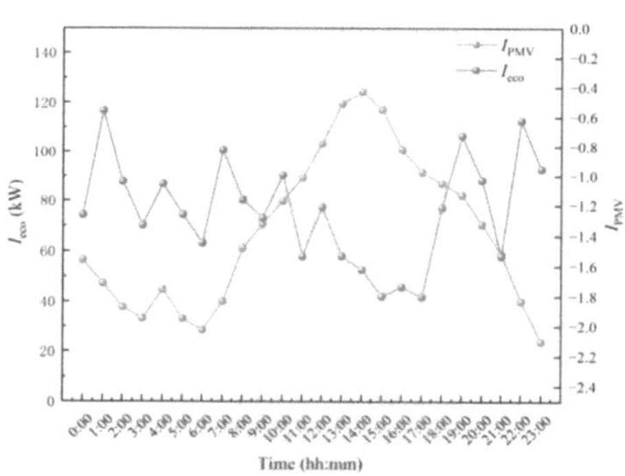

Fig. 10. System performance of conventional control method

The total power consumption of the heat pump on that day under the action of the control strategy currently used by the system is 1828.53 kW, and the average daily I_{PMV} index is -1.36. Comparing the index values under the 50% comfort weight in Table 1, it can be found that although the total electricity consumption of the traditional control method is 5.85% higher than that of the optimal regulation method proposed in this paper, its user-side comfort metrics are lower than those under the optimal control method. Further, comparing and analyzing the data in Fig. 10 with the data in Fig. 9(b), the following results can be obtained:

(1) From the point of view of the power consumption of the heat pump equipment, the power consumption curve of the heat pump in Fig. 9(b) is relatively smooth, and there is no substantial power fluctuation, indicating that the equipment is more stable in the operation process. On the other hand, the power consumption curve of the heat pump in Fig. 10 shows frequent high and low power switching, which means that the equipment needs to be constantly converted between high power and low power states. Such frequent power changes have a negative impact on the service life of the heat pump equipment. Because the equipment will be subjected to greater mechanical and thermal stresses during the high and low power switching process, which may lead to premature wear and tear and failure of the equipment components during long-term operation.
(2) In terms of thermal comfort indicators, the IPMV values in Fig. 9(b) are between [−1.26, −0.93], and their fluctuation ranges are relatively small, which indicates that the indoor thermal environment is relatively stable, and it can provide users with a more continuous and comfortable environmental experience. On the contrary, the IPMV values in Fig. 10 fluctuate in the range of [−0.43, −2.10], and this large fluctuation indicates that the indoor thermal environment is unstable, which may cause users to feel hot and cold, greatly affecting the indoor thermal comfort.
(3) In terms of the correlation between heat pump power consumption and thermal comfort indicators, it can be found that the relationship between those two in Fig. 9(b) is more coordinated, that is, changes in power consumption can maintain thermal comfort more smoothly. In Fig. 10, even though the large fluctuations in heat pump electricity consumption, the thermal comfort indicators are not well maintained, which means that the traditional control method in the heat source measurement (control of heat pump output) and the user side (control of valve opening) control there are still some problems.

In summary, the optimal control method considering economy and comfort proposed in this paper shows certain advantages in terms of indoor thermal comfort, energy utilization efficiency, and equipment lifetime. The traditional control method has problems such as equipment fragility, poor indoor thermal comfort, and poor energy control, which need to be further optimized and adjusted.

6 Conclusion

Against the backdrop of the growing demand for green development, and to address issues such as high operating costs and poor thermal experience for users associated with traditional heating control methods, this paper proposes an optimized control method based on the NSGA-II algorithm. With economy and comfort as the objectives, the method dynamically adjusts the output of heat pump equipment and the opening of user-side valves under different operating conditions by comprehensively considering the heat load demands of different user sides, building heat loss, and the dynamic changes in the energy efficiency of air-source heat pumps during actual operation. Experimental verification shows that:

(1) The proposed method can solve the optimal solution set and formulate a regulation strategy under different outdoor temperatures, enabling more real-time and accurate regulation of the output status of heating equipment, thereby making the system operate more rationally and precisely.
(2) Based on different building characteristics and user requirements, the valve opening degrees of each heat user are controlled separately. This ensures that each heat user connected to the same heat source can accurately obtain thermal energy on demand, thus alleviating the problem of uneven thermal comfort in traditional heating systems.
(3) The proposed optimized scheduling method demonstrates better economic efficiency and comfort. Compared with the traditional control method, the proposed method can reduce the total electricity consumption by 5.85%, while decreasing the deviation between the thermal comfort index and the optimal comfort value "0" by 19.85%.
(4) The proposed optimization method features better control stability. Compared with the traditional control method, the proposed method significantly reduces the fluctuation frequency and amplitude of the heat pump's power consumption; meanwhile, it narrows the fluctuation range of the thermal comfort index by 80.24%.

However, there are still some limitations in the research process of this paper. For instance, the modeling does not fully take into account the dynamic impact of complex working conditions (such as frost formation and defrosting of air-source heat pumps) on system performance. Future research can further integrate intelligent monitoring technology to correct model parameters in real time, thereby realizing more intelligent and efficient control of the heating system.

Acknowledgments. This research is supported by Fundamental Research Program of Shanxi Province of China (grant number: 202303021221026).

Disclosure of Interests The authors declare that they have no known competing financial interests or personal relationships that could have appeared to influence the work reported in this paper.

References

1. Dai, B., Qi, H., Dou, W., et al.: Life cycle energy, emissions and cost evaluation of CO2 air source heat pump system to replace traditional heating methods for residential heating in China: system configurations. Energy Convers. Manag. **218**, 112954 (2020)
2. Zhang, Q., Zhang, L., Nie, J., et al.: Techno-economic analysis of air source heat pump applied for space heating in northern China. Appl. Energy. **207**, 533–542 (2017)
3. Afroz, Z., Shafiullah, G.M., Urmee, T., et al.: Modeling techniques used in building HVAC control systems: a review. Renew. Sust. Energ. Rev. **83**, 64–84 (2018)
4. Michailidis, P., Michailidis, I., Gkelios, S., et al.: Artificial neural network applications for energy management in buildings: current trends and future directions. Energies. **17**(3), 570 (2024)
5. Ulpiani, G., Borgognoni, M., Romagnoli, A., et al.: Comparing the performance of on/off, PID and fuzzy controllers applied to the heating system of an energy-efficient building. Energ. Buildings. **116**, 1–17 (2016)
6. Rahmadini, V.F., Ma'arif, A., Abu, N.S.: Design of water heater temperature control system using PID control. Control Syst. Optim. Lett. **1**(2), 111–117 (2023)

7. Imal, M.: Design and implementation of energy efficiency in HVAC systems based on robust PID control for industrial applications. J Sens. **2015**(1), 954159 (2015)
8. Kang, C.S., Hyun, C.H., Park, M.: Fuzzy logic-based advanced on–off control for thermal comfort in residential buildings. Appl. Energy. **155**, 270–283 (2015)
9. Abdrakhmanov, R., Berkimbayev, K., Seitmuratov, A., et al.: Intelligent fuzzy-PID temperature control system for ensuring comfortable microclimate in an intelligent building. Int. J. Adv. Comput. Sci. Appl. **15**(3) (2024)
10. Belman-Flores, J.M., Rodríguez-Valderrama, D.A., Ledesma, S., et al.: A review on applications of fuzzy logic control for refrigeration systems. Appl. Sci. **12**(3), 1302 (2022)
11. Pezeshki, Z., Mazinani, S.M.: Comparison of artificial neural networks, fuzzy logic and neuro fuzzy for predicting optimization of building thermal consumption: a survey. Artif. Intell. Rev. **52**(1), 495–525 (2019)
12. Ahn, J., Chung, D.H., Cho, S.: Performance analysis of space heating smart control models for energy and control effectiveness in five different climate zones. Build. Environ. **115**, 316–331 (2017)
13. Schellenberg, C., Lohan, J., Dimache, L.: Operational optimisation of a heat pump system with sensible thermal energy storage using genetic algorithm. Therm. Sci. **22**(5), 2189–2202 (2018)
14. Pombeiro, H., Machado, M.J., et al.: Dynamic programming and genetic algorithms to control an HVAC system: maximizing thermal comfort and minimizing cost with PV production and storage. Sustain. Cities Soc. **34**, 228–238 (2017)
15. Ma, K., Hu, S., Hu, G., et al.: Energy management considering unknown dynamics based on extremum seeking control and particle swarm optimization. IEEE Trans. Control Syst. Technol. **28**(4), 1560–1568 (2019)
16. Haniff, M.F., Selamat, H., Khamis, N., et al.: Optimized scheduling for an air-conditioning system based on indoor thermal comfort using the multi-objective improved global particle swarm optimization. Energ. Effic. **12**(5), 1183–1201 (2019)
17. Deb, K., Pratap, A., Agarwal, S., et al.: A fast and elitist multiobjective genetic algorithm: NSGA-II. IEEE Trans. Evol. Comput. **6**(2), 182–197 (2002)
18. Coello, C.A.C., Lamont, G.B., Van Veldhuizen, D.A.: Evolutionary Algorithms for Solving Multi-Objective Problems. Springer Science & Business Media (2007)
19. Niyomubyeyi, O., Sicuaio, T.E., Díaz González, J.I., et al.: A comparative study of four meta-heuristic algorithms, AMOSA, MOABC, MSPSO, and NSGA-II for evacuation planning. Algorithms. **13**(1), 16 (2020)
20. Garces-Jimenez, A., Gomez-Pulido, J.M., Gallego-Salvador, N., et al.: Genetic and swarm algorithms for optimizing the control of building HVAC systems using real data: a comparative study. Mathematics. **18**(9), 2181 (2021)
21. Ministry of Housing and Urban-Rural Development of the People's Republic of China: Design Code for Heating Ventilation and Air Conditioning of Civil Buildings: GB 50736–2012. China Architecture&Building Press, Beijing (2012)
22. Yijie, L., Heqing, L., Liying, G., Tianyu, L.: Analysis of the PMV thermal comfort model and its index. Miner. Eng. Res. **31**(02), 70–75 (2016)
23. Fanger, P.O.: Thermal comfort: analysis and applications in environmental engineering. In: Thermal Comfort Analysis & Applications in Environmental. McGraw-Hill (1972)
24. International Organization for Standardization: Moderate Thermal Environments-Determination of the PMV and PPD Indices and Specification of the Conditions for Thermal Comfort: ISO 7730-2005, pp. 1–5, Switzerland, Geneva (2005)

25. Jingying, D.O.N.G., Bo, Z.H.O.U., Jinyuan, L.I.U., et al.: Joint dispatching model of virtual power plant considering user comfort. Elec. Pow. Constr. **40**(1), 19–26 (2019)
26. Ergonomics of the thermal environment - analytical determination and interpretation of thermal comfort using calculation of the PMV and PPD indices and local thermal comfort criteria (ISO 7730:2005); English version of DIN EN ISO 7730 (2006)

Improved Kepler Optimization Algorithm Based on Hybrid Strategies

Hongge Guo, Pengpeng Su(✉), and Zinan Liang

Taiyuan University of Science and Technology, Taiyuan 030024, China
hgguo@tyust.edu.cn, {supengpeng123,lzn}@stu.tyust.edu.cn

Abstract. Aiming at the defect that the Kepler Optimization Algorithm (KOA) is prone to falling into local optima in high-dimensional spaces, this paper proposes an improved KOA (IKOA) based on hybrid strategies. Firstly, the adaptive Levy flight strategy is introduced into the position formula of the distance between planets and the sun. Secondly, the adaptive Brownian motion fused with the golden sine algorithm is incorporated into the formula for updating planetary positions. Then, the dynamic random centroid opposition-based learning strategy is used to replace the original elite strategy. Finally, a simulation comparison is conducted between the IKOA and the original KOA on 12 test functions from the CEC2022 test suite. The simulation results show that IKOA has stronger convergence speed and global optimal optimization performance.

Keywords: Metaheuristic · Kepler Optimization Algorithm · Levy flight

1 Introduction

In the complex landscape of problem-solving in the real world, optimization stands as a critical endeavor, aiming to identify the most favorable solutions within predefined constraints across diverse domains such as engineering design, resource allocation, and artificial intelligence. While traditional optimization techniques are effective for certain well-structured problems, they often struggle to cope with the inherent complexity, non-linearity, and multimodality exhibited by many contemporary challenges. These limitations have spurred the development and advancement of meta-heuristic algorithms—a class of adaptive, high-level procedures that guide underlying heuristic methods to efficiently explore solution spaces [1].

Meta-heuristic algorithms can be classified from different dimensions. From the perspective of the operation mode of solutions in the optimization process, they are first divided into two types: the single-solution iterative improvement type that focuses on a single solution and gradually optimizes it through iteration, such as Simulated Annealing [2] and the Taboo Method [3], and the multi-solution parallel exploration type that processes a set of solutions simultaneously and explores potential optimal solutions in parallel, such as Particle Swarm Optimization [4], Duck Swarm Algorithm [5], Sand Cat Swarm Optimization Algorithm [6], Artificial Locust Swarm Optimization Algorithm [7], and the Bee Colony Method [8]. On this basis, according to the inspiration

H. Ma et al. (Eds.): IWACIII 2025, CCIS 2780, pp. 431–451, 2026.
https://doi.org/10.1007/978-981-95-6730-0_30

for algorithm design, they can be further subdivided into evolutionary algorithms that simulate biological evolution mechanisms, such as the Ebola Optimization Search Algorithm [9] and genetic algorithms [10], swarm-intelligence algorithms that draw on the social behavior patterns of animals and insects, such as the Butterfly Optimization Algorithm [11], Whale Optimization Algorithm [12], Zebra Optimization Algorithm [13], Ant Colony Optimization [14], Bat Algorithm [15], and Cuckoo Search [16], human-behavior-based algorithms that imitate the interaction logic of human society, such as the Hiking Optimization Algorithm [17] and Running City Game Optimizer [18] and physics-law-based algorithms that are modeled with reference to physical laws, such as the Gravitational Search Algorithm [19] and Kepler Optimization Algorithm [20].

However, despite the increasing emergence of meta-heuristic methods over the years, we are still confronted with numerous challenges, such as high computational complexity of algorithms, proneness to premature convergence, sensitivity to parameter settings, and limited scalability. These challenges, precisely, provide unique development opportunities for the optimization and upgrading of swarm intelligence algorithms [21]. This has aroused strong interest among many scholars in improving classical algorithms. Zhang H et al. [22] utilized the golden sine algorithm to improve the position update method of sparrow discoverers, thereby enhancing the global optimization speed of SSA. Gaussian perturbation was added to the position update of sparrow followers to prevent the algorithm from falling into local optima. Yu XW et al. [23] employed the good point set method to initialize the population to improve population richness, and introduced the beetle antennae search mechanism to avoid the algorithm getting trapped in local extrema, thus improving the grey wolf optimization algorithm. Long Wen et al. [24] used Logistic chaotic mapping for population initialization to lay a foundation for the global search of the algorithm, and utilized the elite opposition-based learning strategy to coordinate the exploration and exploitation capabilities of the algorithm, so as to improve the whale optimization algorithm. Li G et al. [25] introduced an adaptive weight factor to break the limitation of fixed weights on the optimization performance of the algorithm, and incorporated operators from the moth search algorithm to enhance the exploration ability of the algorithm, and proposed an improved MOES/DMS algorithm. And so on.

Kepler Optimization Algorithm (KOA) [20]. Its inspiration is based on Kepler's three laws of planetary motion, where planets orbit around the Sun. This algorithm simulates the way planets orbit the Sun to conduct searches within the solution space. Each planet is regarded as a candidate solution, and the Sun is treated as the global optimal solution of the current iteration. Through continuous iterations, planets are positioned at different locations to find the optimal solution. When an individual with a fitness value superior to the current global optimal solution emerges during the search process, this individual will replace the original current optimal solution. Such a cycle repeats, and the algorithm continuously performs iterative optimization. After undergoing multiple rounds of iterative computations, it converges to the global optimal solution.

However, despite the promising prospects of KOA, it may face problems such as being prone to falling into local optimum traps or having a slow convergence speed. To overcome these obstacles, this paper proposes an improved KOA based on a hybrid strategy. First, the adaptive Brownian motion and the golden sine algorithm [26] are integrated, and this integrated strategy is introduced into KOA to update the position formula of planets. Second, the adaptive Levy flight strategy [27] is introduced into KOA to update the position update formula of the distance between the planet and the Sun. Moreover, to enhance the population diversity and improve the convergence performance of the algorithm, a dynamic random centroid opposition - based learning strategy [28] is proposed to replace the original elite strategy. Finally, 12 test functions in CEC2022 are used for simulation comparison and verification. The simulation results show that IKOA has a stronger convergence speed and global optimal search performance.

2 The Original Kepler Optimization Algorithm

Kepler Optimization Algorithm (KOA) [20]. Its inspiration is based on Kepler's three laws of planetary motion, where planets orbit around the Sun. These laws are expressed as follows:

Kepler's First Law: All planets orbit the Sun along elliptical paths, with the Sun located at one of the foci of the ellipse. The elliptical equation of a planet's orbit can be expressed in polar coordinates as:

$$r = \frac{P}{1 + e\cos\theta} \tag{1}$$

where, r represents the distance from the planet to the Sun, θ is the angle between the line connecting the planet and the Sun and the major axis of the ellipse, P is the semi-latus rectum of the ellipse, and e denotes the eccentricity of the elliptical orbit, with a value ranging from 0 to 1. When $\theta = 0$, it indicates that the shape of the planet's orbit is circular.

Kepler's Second Law: The line segment joining a planet and the Sun sweeps out equal areas in equal intervals of time. Through this law, it is proven that the planet has different velocities at different positions on the elliptical orbit. When the planet is far from the Sun, its velocity decreases, and when it is close to the Sun, its velocity increases. The formula for the velocity of a planet orbiting the Sun is given by:

$$V = \left[\mu(M_S + m)\left(\frac{2}{R} - \frac{1}{a}\right)\right]^{\frac{1}{2}} \tag{2}$$

where, V represents the velocity of the planet orbiting the Sun, μ represents the gravitational constant, M_S represents the mass of the Sun, m represents the mass of the planet, R represents the distance between the Sun and the planet, and a represents the semi-major axis of the ellipse.

Kepler's Third Law: The square of the orbital period of each planet revolving around the Sun is proportional to the cube of the semi-major axis of its elliptical orbit. The expression representing the celestial orbital period T of a planet is given as follows:

$$T^2 = ka^3 \tag{3}$$

where, the proportionality coefficient k can be expressed as:

$$k = \frac{4\pi^2}{\mu(M_S + m)} \tag{4}$$

According to Kepler's three laws, there are four relatively key factors influencing the motion of a planet orbiting the Sun, including the mass m of the planet, its position x in the orbital path, its orbital velocity v, and the mutual gravitational force F between the Sun and the planet. These elements exhibit close correlations and interactions with one another and together form the core supporting components of the mathematical model of the Kepler Optimization Algorithm (KOA). The flow of the Kepler Algorithm is as follows:

Step 1: Initialization. All planets and the orbits revolving around the Sun are used to represent the solution space of the algorithm. Planets represent the candidate solutions to the problem, and the Sun represents the current optimal solution in terms of fitness. N represents the number of populations, and D represents the dimension. N planets are randomly distributed to arbitrary positions X in the D-dimensional orbit. X can be expressed as:

$$X_i^j = X_{i,low}^j + \text{rand}_{[0,1]} \times \left(X_{i,up}^j - X_{i,low}^j\right), \begin{cases} i = 1, 2, \ldots, N. \\ j = 1, 2, \ldots, D. \end{cases} \tag{5}$$

where, X_i represents the position of the i-th planet in the solution space, i ranges from 1 to N, j denotes the dimension of the planet and rand represents a random value taken between 0 and 1. Among them, $X_{i,up}^j$ and $X_{i,low}^j$ denote the upper bound and lower bound, respectively. The orbital period T of the i-th planet is randomly selected according to the normal distribution, and the expression for T is as follows:

$$T_i = |r| i = 1, \ldots, N \tag{6}$$

where, r represents a random number following a normal distribution.

The eccentricity e of the orbit of the i-th planet is randomly assigned a value between 0 and 1 and can be expressed by the following formula:

$$e_i = \text{rand}_{[0,1]}, i = 1, \ldots, N \tag{7}$$

Step 2: Define the gravitational force F. The Sun is the most massive object in the solar system. Relying on its gravitational force, it occupies a dominant position and plays a decisive role in the motion states of all planets within the solar system. Planets orbit around the Sun, and the fundamental reason lies in being constrained by the Sun's gravitational force, thus being maintained in specific orbits. The attractive force between the Sun and a planet is given by the universal gravitation formula as

$$F = e_i \times \mu(t) \times \frac{\overline{M}_S \times \overline{m}_i}{\overline{R}_i^2 + \varepsilon} + r_1 \tag{8}$$

where, e_i represents the eccentricity of planet i, μ(t) denotes the universal gravitational constant, $\overline{M}_S$ and $\overline{m}_i$ represent the normalized values of M_S and m_i. Their role is to make the data distribution more uniform and stable, avoiding the impact of excessively large or small values on the stability of the algorithm. They represent the masses of the Sun and each planet, respectively. ε is a very small value, preventing the denominator from being zero and violating mathematical operation rules. $\overline{R}_i$ represents the normalized value of the Euclidean distance between the Sun and the planet, and r_1 represents a randomly generated value within the range of 0–1. The Euclidean distance can be expressed by the following formula:

$$R_i(t) = \|X_S(t) - X_i(t)\|_2 = \sqrt{\sum_{j=1}^{D} \left(X_{Sj}(t) - X_{ij}(t)\right)^2} \quad (9)$$

where, R_i represents the Euclidean distance between the Sun X_S and the planet X_i across all dimensions. At time t, the masses of the Sun and each planet are calculated through fitness evaluation to account for the minimization problem:

$$M_S = r_2 \frac{fit_s(t) - worst(t)}{\sum_{k=1}^{N} (fit_k(t) - worst(t))} \quad (10)$$

$$m_i = \frac{fit_s(t) - worst(t)}{\sum_{k=1}^{N} (fit_k(t) - worst(t))} \quad (11)$$

where, r_2 represents a random number ranging from 0 to 1. At iteration time t, the fitness $fit_s(t)$ of the Sun (the current optimal solution) is equal to the minimum value among the fitness values $fit_k(k)$ of all N planetary individuals in the population. can be expressed as:

$$fit_s(t) = best(t) = \min_{k \in 1,2,\ldots N} fit_k(k) \quad (12)$$

At iteration time t, the maximum value of the fitness $fit_k(k)$ among all planetary individuals in the population is the "worst fitness $worst(t)$" of the current iteration step, which can be expressed as:

$$worst(t) = \max_{k \in 1,2,\ldots N} fit_k(k) \quad (13)$$

In Eq. 8, $\mu(t)$ is a function that decreases exponentially with time t and is used to control the search precision. Its expression is as follows:

$$\mu(t) = \mu_0 \times \exp\left(-\lambda \frac{t}{T_{\max}}\right) \quad (14)$$

where, λ represents a constant, μ_0 represents the initial value, and t and $T_{\max}$ represent the current number of iterations and the maximum number of iterations, respectively.

Step 3: Calculate the orbital speed of the planet. According to Kepler's Second Law, the gravitational force acting on a planet varies at different positions in its orbit, so the orbital speed of the planet also differs. When a planet moves away from the Sun, the gravitational force it experiences decreases, and its speed will slow down. When it approaches the Sun, the gravitational force on the planet increases, and its speed will increase accordingly. For the distance judgment formula $R_{i-norm}(t) \leq 0.5$, if it is satisfied, the velocity update formula adopts Eq. 15.

$$\begin{aligned} V_i(t) = \psi \times \left(2r_4 \vec{X}_i - \vec{X}_b\right) + \ddot{\psi} \times \left(\vec{X_a} - \vec{X_b}\right) + (1 - R_{i-\mathrm{norm}}(t)) \\ \times \xi \times \vec{U_1} \times \vec{r_5} \times \left(\vec{X}_{i,up} - \vec{X}_{i,low}\right) \end{aligned} \quad (15)$$

If $R_{i-norm}(t) \leq 0.5$ is not satisfied, it is represented by Formula 16.

$$V_i(t) = r_4 \times \varphi \times \left(\overrightarrow{X_a} - \overrightarrow{X}_i\right) + (1 - R_{i-\text{norm}}(t)) \times \varphi \times U_2 \times \overrightarrow{r_5} \times \left(r_3 \overrightarrow{X}_{i,up} - \overrightarrow{X}_{i,low}\right) \tag{16}$$

where:

$$\psi = \overrightarrow{U} \times \sigma \times \varphi \tag{17}$$

$$\varphi = \left[\mu(t) \times (M_S + m_i) \left| \frac{2}{R_i(t) + \varepsilon} - \frac{1}{a_{i(t)} + \varepsilon} \right| \right] \tag{18}$$

$$\sigma = (r_3 \times (1 - r_4) + r_4) \tag{19}$$

$$\overrightarrow{U} = \begin{cases} 0, \overrightarrow{r}_5 \leq \overrightarrow{r}_6 \\ 1, \ else \end{cases} \tag{20}$$

$$\xi = \begin{cases} 1, \overrightarrow{r}_4 \leq 0.5 \\ -1, \ else \end{cases} \tag{21}$$

$$\ddot{\psi} = \left(1 - \overrightarrow{U}\right) \times \sigma \times \varphi \tag{22}$$

$$\overrightarrow{\sigma} = \left(r_3 \times \left(1 - \overrightarrow{r}_5\right) + \overrightarrow{r}_5\right) \tag{23}$$

$$\overrightarrow{U}_1 = \begin{cases} 0, \overrightarrow{r}_5 \leq r_4 \\ 1, \ else \end{cases} \tag{24}$$

$$U_2 = \begin{cases} 0, r_3 \leq r_4 \\ 1, \ else \end{cases} \tag{25}$$

where, $\overrightarrow{V}_i(t)$ denotes the velocity of planet i at time t, and $\overrightarrow{X}_i(t)$ denotes the position of planet i at time t. r_3 and r_4 represent numbers randomly generated within the interval from 0 to 1, while $\overrightarrow{r}_5$ and $\overrightarrow{r}_6$ represent two vectors randomly generated within the range of 0 to 1. $\overrightarrow{X_a}$ and $\overrightarrow{X_b}$ are solutions selected from the overall set of planets. μ(t) stands for the gravitational constant, ε represents an extremely small value, and M_S and m_i denote the masses of the Sun and the planet, respectively. $R_i(t)$ represents the distance between the planet and the Sun at time t, where $a_i(t)$, which denotes the semi - major axis of the orbit of planet i at time t and is derived from Kepler's Third Law, plays a crucial role in accurately determining the planet's orbital dynamics and its gravitational effects on other celestial bodies within the broader solar system, and its expression is as follows:

$$a_i(t) = r_3 \times \left[T_i \times \frac{\mu(t) \times (M_S + m_i)}{4\pi^2}\right]^{\frac{1}{3}} \tag{26}$$

where, T_i represents the period of planet i. During the iteration process of the algorithm, the semi-major axis of the planet will gradually decrease, and its solution will move toward the possible global optimal solution. In Eqs. 15 and 16 $R_{i\text{-}norm}(t)$ represents the normalization of the Euclidean distance between X_S and X_i, and its definition is shown in:

$$R_{i-\text{norm}}(t) = \frac{R_i(t) - \min(R(t))}{\max(R(t)) - \min(R(t))} \tag{27}$$

If the value of $R_{i\text{-}norm}(t)$ is less than or equal to 0.5, it indicates that the distance between the planet and the Sun is close. At the same time, the orbital speed of the planet is increased to prevent drift caused by the enormous gravitational force of the Sun. If the value is greater than 0.5, the planet will decrease its orbital speed.

Step 4: Update the position of the planet. In the solar system, planets move periodically around the Sun in accordance with Kepler's laws. That is, within a specific time period, they first gradually approach the Sun and then gradually move away from the Sun along their orbits. Based on this motion characteristic, it provides the source of inspiration for the Kepler Optimization Algorithm (KOA). KOA designs the exploration and exploitation stages of the planet by simulating these two modes of planetary motion.

When a planet moves away from the Sun, the exploration stage is utilized. The KOA simulates the motion trajectory of a celestial body gradually moving away from the Sun. As the distance between the planet and the Sun continuously increases, the algorithm can comprehensively explore the entire search area. In the exploitation stage, the KOA simulates the motion state of a celestial body approaching the Sun. When an object approaches the Sun, the gravitational force intensifies, and the algorithm will deeply explore and accurately utilize the solutions in the area near the Sun to find new optimization points near the optimal solution. Then, the position update formula of the planet can be expressed as:

$$\begin{aligned} \overrightarrow{X_i}(t+1) &= \overrightarrow{X}_i(t) + \xi \times \overrightarrow{V}_i(t) \\ &+ \left(F_{gi}(t) + |r|\right) \times \overrightarrow{U} \times \left(\overrightarrow{X}_S(t) - \overrightarrow{X}_i(t)\right) \end{aligned} \tag{28}$$

where, $\overrightarrow{X}_i(t+1)$ represents the position of planet i at time $t+1$. $\overrightarrow{X}_i(t)$ represents the position of planet i at time t. Among them, $\overrightarrow{V}_i(t)$ represents the velocity required for the position update of planet i, ξ represents the search direction of the planet, and $F_{gi}(t)$ represents the gravitational force between the planet and the Sun.

Step 5: Update the distance to the Sun. To further optimize the planet exploration and exploitation operators, KOA simulates the typical behavior pattern of the natural change in the distance between the Sun and planets over time. When a planet approaches the Sun, KOA focuses on optimizing the exploitation operator; when a planet moves away from the Sun, KOA focuses on optimizing the exploration operator. The above rules depend on the dynamic change of the adjustment parameter h with the current time. When the value of h is large, the detection operator is activated to increase the distance between the planet's orbit and the Sun, prompting the algorithm to explore in a wider search space. Conversely, when the value of h is small, if the distance between the Sun and the planet is relatively close, the exploitation operator takes effect to conduct in-depth excavation of the area around the discovered optimal solution. The planet position update formula based on this idea is as follows:

$$\begin{aligned}\overrightarrow{X}_i(t+1) = \overrightarrow{X}_i(t) \times \overrightarrow{U}_1 + \left(1 - \overrightarrow{U}_1\right) \times \left[\frac{\overrightarrow{X}_i(t)+\overrightarrow{X}_S+\overrightarrow{X}_a(t)}{3.0}\right. \\ \left. +h \times \left(\frac{\overrightarrow{X}_i(t)+\overrightarrow{X}_S+\overrightarrow{X}_a(t)}{3.0} - \overrightarrow{X}_b(t)\right)\right]\end{aligned} \tag{29}$$

where h represents the adaptive factor for controlling the distance between the Sun and the planet at time t, and its expression is as follows:

$$h = \frac{1}{e^{\eta r}} \tag{30}$$

where, r is a number randomly generated based on a normal distribution, and η is a factor that decreases linearly from 1 to −2, with its expression as follows:

$$\eta = (a_2 - 1) \times r_4 + 1 \tag{31}$$

where, a_2 represents the cycle control parameter that gradually decreases from −1 to −2 for T cycles throughout the optimization process. Its expression can be represented by the following formula:

$$a_2 = -1 - 1 \times \left(\frac{t\% \frac{T_{\max}}{T}}{\frac{T_{\max}}{T}}\right) \tag{32}$$

Step 6: Elite Strategy. The elite strategy is the key for KOA to ensure convergence and solution quality. In each iteration, the algorithm evaluates the fitness of all candidate solutions according to the objective function and selects several of the optimal ones as elite individuals to retain. In subsequent iterations, the elite individuals do not participate in the ordinary position update. Instead, when updating the positions of other planets, weights are set to make them approach the elite individuals, guiding the population to evolve towards better solutions. This strategy helps the algorithm converge quickly, prevents the loss of high- quality solutions, and improves the reliability and stability of the algorithm in complex optimization problems. Its elite strategy formula is expressed as follows:

$$\overrightarrow{X}_{i,new}(t+1) = \begin{cases} \overrightarrow{X}_i(t+1), \, if \; f\left(\overrightarrow{X}_i(t+1)\right) \le f\left(\overrightarrow{X}_i(t)\right) \\ \overrightarrow{X}_i(t), \;\; else \end{cases} \tag{33}$$

The flow chart of the KOA is shown in Fig. 1.

The flowchart illustrates the process of a Kepler optimization algorithm related to celestial bodies. It starts with the initialization of a set of celestial bodies, including their random positions, orbital eccentricities, and orbital periods. Subsequently, the fitness values of the initial population are evaluated to determine the global optimal solution.

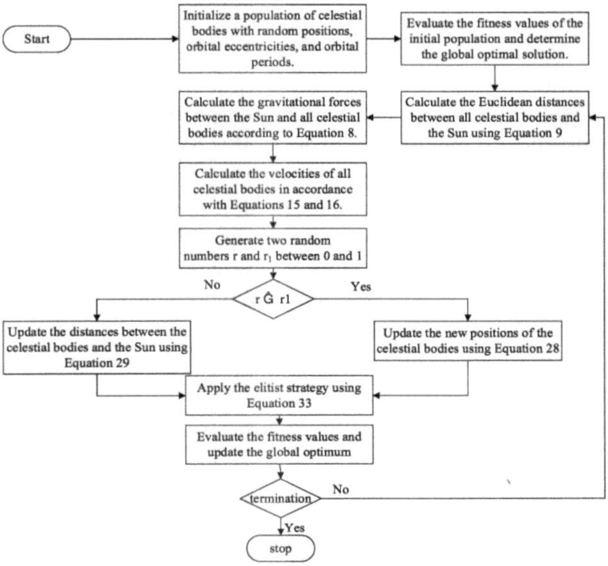

Fig. 1. Flowchart of the KOA

3 The Improved Kepler Optimization Algorithm Adopts a Hybrid Strategy

In the challenging research field of optimization algorithms, the Kepler Optimization Algorithm (KOA), as an emerging intelligent optimization algorithm based on the principles of celestial mechanics, demonstrates non - negligible theoretical research value and broad practical application potential by virtue of its unique solution However, like many other meta-heuristic algorithms, KOA still has some key issues to be improved in practical applications, which also constitute an important basis for research on its improvement. The original KOA often converges slowly when solving minimum value problems, resulting in the need for a large number of iterations to find a better solution. Meanwhile, KOA is prone to falling into local optimal traps during the search process. Therefore, this paper proposes three improvements to the original KOA algorithm to optimize the above-mentioned problems.

3.1 A Planet Position Update Method Based on Adaptive Brownian Motion and Golden Sine Algorithm

Integrate the adaptive Brownian motion and the golden sine algorithm [26], and introduce this fusion strategy into the position update formula of the planets in the KOA. The core principle of this strategy lies in using the periodic fluctuation characteristics of the sine function within the unit circle to simulate the fine - scanning mechanism of the solution space. The undulating changes of the sine wave enable the search area to be extensively expanded, and its waveform characteristics guide the algorithm to explore the potential solution space in different dimensions. When combined with the golden section ratio, the algorithm can intelligently shrink the search range according to the established rules, gradually approach the global optimal solution, and thus significantly improve the pertinence and efficiency of the search process.

However, in practical application scenarios, this algorithm tends to get trapped in local optimal solutions in the later stages of the search. To effectively address this issue, this paper introduces an adaptive Brownian motion mechanism. As a classical model describing the irregular motion of microscopic particles, Brownian motion, with its inherent randomness and uncertainty, injects new vitality into the search process. By dynamically adjusting the step size and direction parameters of Brownian motion, the algorithm can achieve leapfrog exploration outside the local optimal regions, thereby effectively evading the trap of local optima and further enhancing the optimization accuracy and reliability.

In the position $\overrightarrow{X}_{i+1}$ of the planet in Eq. 28, the introduction of the following position update formula can be expressed as:

$$\begin{aligned}\overrightarrow{X}_{i+2} = RB \times \overrightarrow{X}_{i+1} + \overrightarrow{X}_{i+1} \times |\sin(r_1)| - r_2 \\ \times \sin(r_1) \times \left|G_1 \times \overrightarrow{X}_S - G_2 \times \overrightarrow{X}_{i+1}\right|\end{aligned} \tag{34}$$

Where, r_1 and r_2 are random numbers between 0 and 1. G_1 and G_2 are golden-section coefficients, RB is the adaptive Brownian motion convergence factor and its expression is as follows:

$$G_1 = -\pi \times \left(1 - 0.5 \times \left(\sqrt{5} - 1\right)\right) + \pi \times 0.5 \times \left(\sqrt{5} - 1\right) \tag{35}$$

$$G_2 = -\pi \times \left(0.5 \times \left(\sqrt{5} - 1\right)\right) + \pi \times \left(1 - 0.5 \times \left(\sqrt{5} - 1\right)\right) \tag{36}$$

$$RB = \left(1 - \frac{t}{T_{max}}\right) \times \text{rand}(1, D) \tag{37}$$

t represents the current iteration number, $T_{\max}$ represents the maximum number of iterations, *rand* represents a random value, and D represents the dimension.

3.2 A Position Update Strategy for the Distance Between Planets and the Sun Based on the Adaptive Lévy Flight Strategy.

To balance exploration and exploitation in the search process, this paper introduces the adaptive Lévy flight strategy [27] into the position update formula of KOA for updating the distance between planets and the Sun. Lévy flight is a random walk model based on a heavy - tailed stable distribution, where the step size distribution is highly concentrated in the core region to ensure search stability, and the tail shows a slow decay characteristic to enable long-distance jumps, thus forming a dynamic balance between short-step fine-grained search and long-step global exploration. In heuristic algorithms, this strategy significantly expands the search range through long steps, effectively avoiding local optima and maintaining population diversity, and can quickly locate high-quality solution regions in the early stage of the search. In the later stage, short steps enhance the local exploitation capability, improving solution accuracy and accelerating convergence through refined search. This bimodal search mechanism coordinates short-distance exploitation and long-distance exploration probabilistically, overcoming the contradiction between exploitation and exploration in traditional algorithms, and is particularly suitable for complex multimodal optimization problems. The non-Gaussian random perturbations generated by its heavy-tailed characteristics can not only break through the limitations of the search space but also maintain the ergodicity of the iteration process, providing the algorithm with theoretically provable space coverage capability and serving as a key technical method to improve optimization performance.

To further enhance the performance of the Lévy flight strategy, this paper proposes an adaptive Lévy flight strategy. The core innovation of this strategy lies in the introduction of a dynamically varying factor, which can adaptively adjust the weight of the step size distribution according to different search stages of the algorithm. Specifically, in the early stage of the search, the varying factor tends to increase the probability of generating long steps to strengthen the global exploration capability; in the later stage of the search, the varying factor gradually reduces the weight of long steps while increasing the probability of generating short steps, thereby achieving fine-grained search in local regions. This adaptive mechanism can not only effectively improve the global search efficiency of the algorithm but also accelerate the convergence process while maintaining population diversity. The new position update formula is introduced into the position update formula $\overrightarrow{X}_{i+1}$ for the distance between the planet and the Sun in Eq. 29 as follows:

$$\overrightarrow{X}_{i+2} = X_s + R_L \times \left(\overrightarrow{X}_B - \overrightarrow{X}_{i+1}\right) \tag{38}$$

where, R_L represents the Lévy flight step length, and $\overrightarrow{X}_B$ represents the individual at a position near $\overrightarrow{X}_{i+1}$. The purpose is to avoid random directions generated by Lévy flight, which would lead to blind search. R_L can be expressed as:

$$R_L = w \times Levy(D) \tag{39}$$

where, w represents the adaptive factor, and $levy(D)$ represents the Lévy flight step length. Their expressions are respectively as follows:

$$w = \left(\frac{T_{max} - t}{T_{\max}}\right)^{\frac{t}{0.5 \times T_{\max}}} \tag{40}$$

$$\left[Levy(D) = \frac{u_1 \times \tau}{|u_2|^{\frac{1}{\eta}}}\right] \tag{41}$$

$$\tau = \left(\frac{\Gamma(1+\eta) \times \sin\left(\frac{\pi\eta}{2}\right)}{\Gamma\left(\frac{1+\eta}{2}\right) \times \eta \times 2^{\left(\frac{\eta-1}{2}\right)}}\right)^{\frac{1}{\eta}} \tag{42}$$

where, $T_{\max}$ is the maximum number of iterations in the search process, t is the current number of search iterations, u_1 and u_2 follow the standard normal distribution, the adjustment factor η is a parameter of the Lévy distribution with a fixed constant value of 1.5, and Γ is the gamma function.

3.3 An Elite Preservation Strategy Based on Dynamic Random Centroid Opposition-Based Learning

To enhance population diversity and improve the convergence performance of the algorithm, this paper proposes a dynamic random centroid opposition-based learning strategy [28]. Different from traditional centroid opposition-based learning strategies, dynamic centroid opposition-based learning randomly selects K individuals from the population in each iteration to calculate the centroid, rather than relying on the top-ranked K individuals with a fixed order. This random selection enhances population diversity, expands the global search scope, and can also increase the probability of finding better solutions. The centroid, serving as a guidance vector, will update each individual in the population, making it move a certain step length toward the centroid. After that, the updated individuals are compared with the individuals at corresponding positions in the original population, and better individuals are selected to enter the next generation based on fitness. The formula for the new elite guidance strategy is as follows:

$$\overrightarrow{X}_{i,new}(t+1) = \begin{cases} \overrightarrow{X}_i(t+1), \text{ if } f(X_i) \le f(X_i^*) \\ \overrightarrow{X_i^*}, \ else \end{cases} \tag{43}$$

$$\overrightarrow{X_i^*} = 2 \times C - \overrightarrow{X}_{Rand,i}, i = 1, 2, \ldots, N \tag{44}$$

$$C = \frac{\sum_{i=1}^{K} X_{Rand,i}}{K} \tag{45}$$

where, $\overrightarrow{X}_{Rand,i}$ represents the positions of randomly selected K individuals in the population. C denotes the centroid of the elite population. $\overrightarrow{X}_i^*$ represents the position of the new individual obtained by updating the individual in the original population based on the centroid individual.

The flowchart in Fig. 2 presents the improved Kepler optimization algorithm. It begins with initializing a population of celestial bodies, setting their random positions, orbital eccentricities, and orbital periods. Then, the fitness values of the initial population are evaluated to determine the global optimal solution. Next, the Euclidean distances between all celestial bodies and the Sun are calculated using Eq. 9, and the gravitational forces between the Sun and all celestial bodies are computed according to Eq. 8. After that, the velocities of all celestial bodies are calculated in line with Eqs. 15 and 16. the new positions of the celestial bodies are updated using Eq. 34; otherwise, the distances between the celestial bodies and the Sun are updated using Eq. 38. Subsequently, the elitist strategy is applied using Eq. 43. Then, the fitness values are evaluated again, and the global optimum is updated. Finally, a check for the termination condition is performed. If the condition is not met, the process loops back to calculating the Euclidean distances between all celestial bodies and the Sun; if met, the process stops.

The flow chart of the IKOA is shown in Fig. 2.

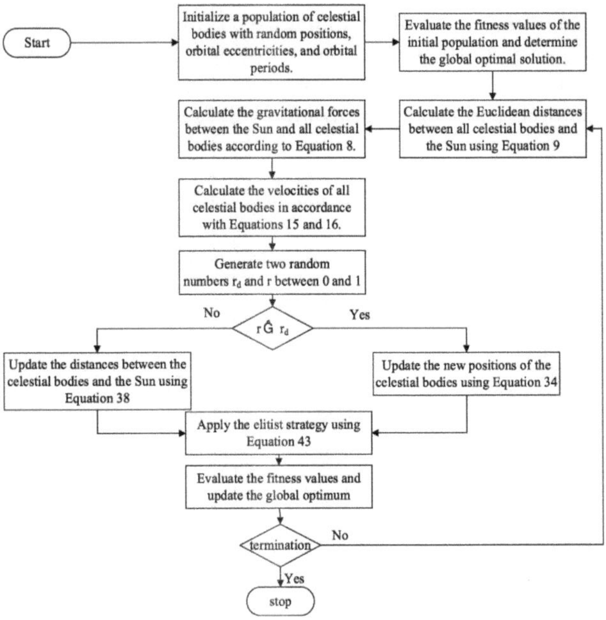

Fig. 2. Flowchart of the IKOA

4 Experimental Verification

In this paper, the CEC2022 [29] function test suite is employed to verify the improvement effect of the algorithm. CEC2022 is a single-objective optimization benchmark test suite launched by the IEEE International Conference on Evolutionary Computation in 2022, which is mainly used to evaluate the performance of various optimization algorithms, especially in dealing with complex optimization problems. Specifically, all functions in this suite are defined as minimization problems, including 12 benchmark functions classified into four categories: unimodal function F1, multimodal functions F2-F5, hybrid functions F6-F8, and composite functions F9-F12.

Among them, unimodal functions, which contain only one global optimal solution and no local extremum, are used to test the convergence speed and local search ability of algorithms. Multimodal functions, which involve multiple local extremum points, are employed to evaluate the ability of algorithms to escape from local optima. Hybrid functions divide multiple basic functions into different sub - components and introduce asymmetry and non-separability by means of rotation matrices and displacement operations to destroy variable independence, thereby increasing the difficulty of optimization. Their purpose is to examine the ability of algorithms to handle variable coupling and nonlinear problems. Composite functions simulate the complex optimization environment with multi-objectives and multi-constraints in practical engineering by weighted superposition of multiple hybrid functions or basic functions, along with the addition of bias values and dynamic weights. Their goal is to test the adaptability of algorithms in scenarios with dynamic weights and complex superposition.

The parameter settings of the algorithm are as follows: for 2-dimensional test functions, the population size is 10 and the number of iterations is 500; for 10-dimensional test functions, the population size is set to 30 and the maximum number of iterations is 1000. The test functions and ideal optimization values are shown in Table 1.

Table 1. Test functions and their ideal optimal values

Test function	Ideal optimization value
F1	300
F2	400
F3	600
F4	800
F5	900
F6	1800
F7	2000
F8	2200
F9	2300
F10	2400
F11	2600
F12	2700

Table 1 presents the 12 test functions F1-F12 from CEC2022, along with the corresponding target optimal values for each function, which are used to evaluate the performance of optimization algorithms.

To verify the effectiveness of IKOA, we solved 9 test functions F1-F5 and F9-F12 listed in Table 1, and the 2-dimensional optimization iteration results are shown in Fig. 3.

As shown in Fig. 3, subfigures (a-i) represent the iterative convergence graphs of F1-F5 and F9-F12, respectively, where the horizontal axis denotes the number of iterations and the vertical axis represents the target optimization value. For 2-dimensional optimization problems, the convergence curves of the 9 test functions reveal that IKOA exhibits a faster convergence speed. In most test functions, namely F1, F2, F3, F4, F5, F11, and F12, IKOA can quickly approach or reach its final convergence value within 50–150 iterations. In contrast, the convergence process of KOA is relatively slow, usually requiring more iterations (approximately 250–350 iterations) to stabilize, and its descending process often shows an obvious stepwise pattern, which indicates that KOA has certain difficulties in escaping local optima or balancing exploration and exploitation.

In terms of optimization accuracy, IKOA achieves better objective function values in all 9 tested functions, demonstrating that IKOA can find solutions of higher quality than KOA. Particularly in functions F2, F3, F10, and F12, the solutions of IKOA are numerically significantly superior to those of KOA. The 9 iterative convergence graphs in Fig. 3 intuitively prove that the IKOA algorithm has obvious advantages over the KOA algorithm in handling 2 - dimensional test functions. IKOA not only converges faster, being able to find high-quality solutions within fewer iterations, but also achieves higher final optimization accuracy. This indicates that the improvement effect is significant in low- dimensional scenarios.

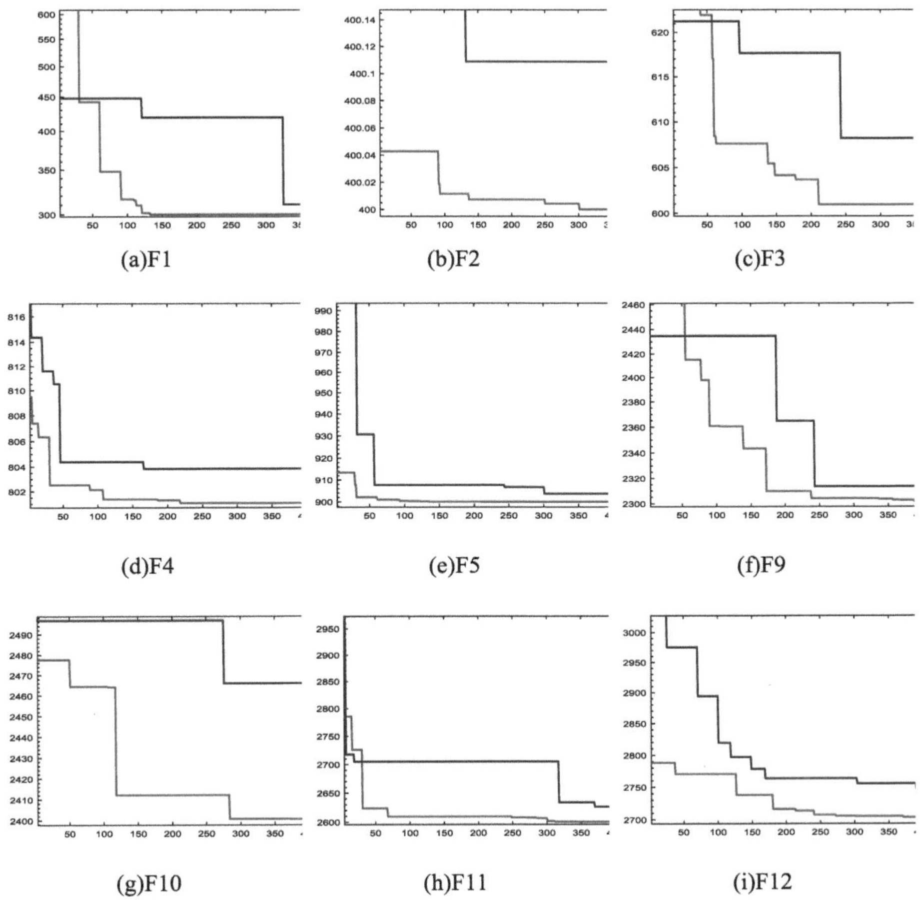

Fig. 3. Comparison of iterative curves for 2-dimensional functions

To better evaluate the comprehensive performance of the algorithms, we used KOA and IKOA respectively to solve the 9 test functions (F1-F5 and F9-F12) listed in Table 1. Meanwhile, since the optimal value can measure the convergence accuracy of the algorithms, the standard deviation can assess the stability of the algorithms, and the average running time can gauge the computational efficiency of the algorithms, we ran each algorithm independently 30 times to compare their mean values, standard deviations, and running times. The experimental results are presented in Table 2.

The data in Table 2 presents the performance metrics of KOA and IKOA, including the running mean, standard deviation, and average running time. A comparison of their performance metrics reveals that for the test functions F1-F5 and F9-F12, the mean values of IKOA are all superior to those of KOA. Although the differences are slight for individual functions, IKOA generally yields better solutions for 2-dimensional problems. For all test functions listed in Table 2, the standard deviation of IKOA is significantly lower than that of KOA. Particularly in F1, F4, F11, and F12, the standard deviation of IKOA is several orders of magnitude smaller than that of KOA. This indicates that IKOA has higher stability, being able to consistently converge to similar high-quality solutions across different runs, whereas the results of KOA are more volatile compared to IKOA. In terms of running time, for the 9 test functions in Table 2, the running time of IKOA is not significantly different from that of KOA, which suggests that the improved algorithm (IKOA) does not compromise computational efficiency.

Table 2. Iteration results of the 2-dimensional test function algorithm

Function	Algorithm	Average optimal value	Standard deviation	Average running time/s
F1	KOA	302.61	3.5824	0.0875
F1	IKOA	300.04	0.0017	0.0967
F2	KOA	401.54	2.3145	0.0742
F2	IKOA	400.18	0.0541	0.0724
F3	KOA	601.59	0.9231	0.0863
F3	IKOA	600.08	0.0698	0.0648
F4	KOA	804.85	4.2574	0.0142
F4	IKOA	800.75	0.0751	0.0174
F5	KOA	901.41	0.8512	0.0545
F5	IKOA	900.21	0.0841	0.0478
F9	KOA	2302.21	2.1521	0.0785
F9	IKOA	2300.14	0.4151	0.0854
F10	KOA	2402.86	1.5784	0.0698
F10	IKOA	2400.54	0.4875	0.0578
F11	KOA	2610.64	6.7641	0.0853
F11	IKOA	2600.01	0.0264	0.0777
F12	KOA	2707.25	6.5874	0.0745
F12	IKOA	2700.12	0.0741	0.0885

To more comprehensively evaluate the effect of the improvement, we solved all 12 test functions listed in Table 1 in the 10-dimensional space, and the iterative optimization curves are shown in Fig. 4.

Based on the 12 sets of convergence curve comparison graphs shown in Fig. 4, it can be observed that the convergence speed of IKOA is superior to that of KOA. In the early stages of iteration, IKOA can reduce the objective function value more rapidly, especially

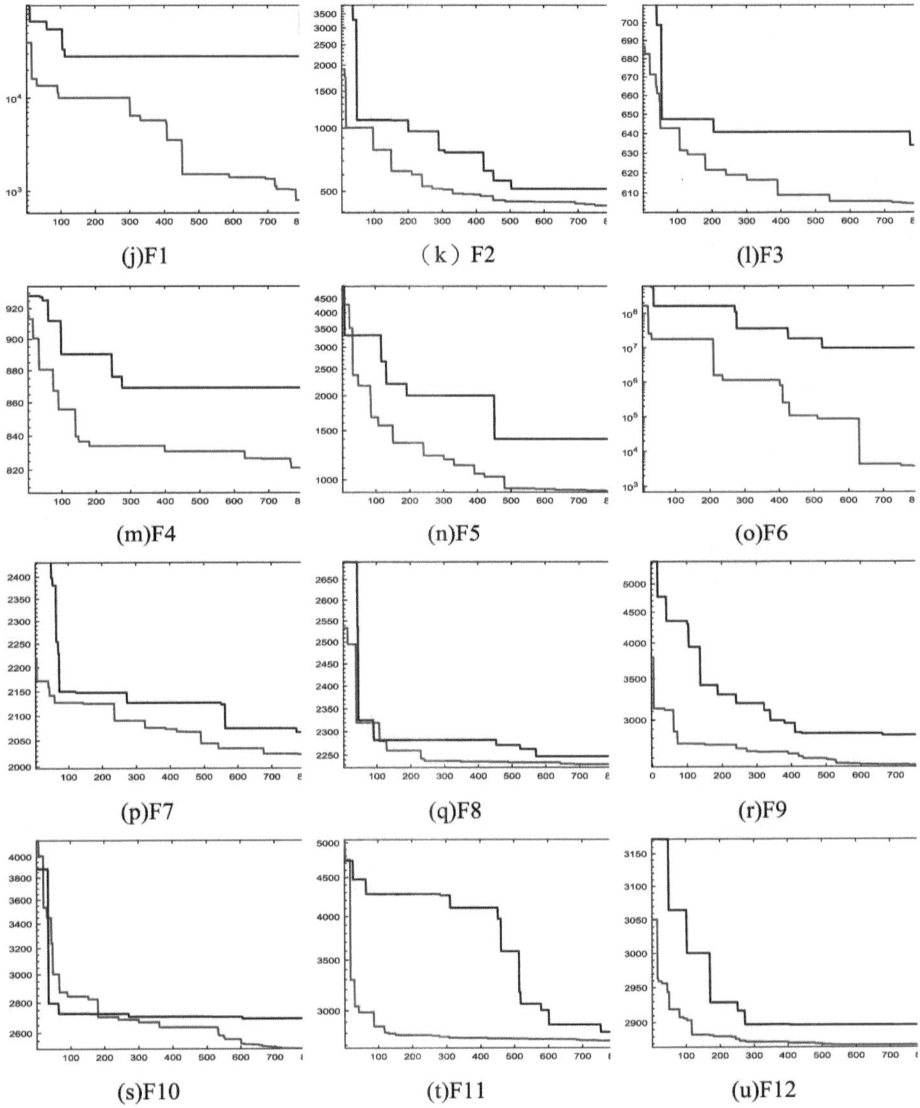

Fig. 4. Comparison of iteration curves for 10-dimensional functions

for functions F1 and F11, where the initial decline rate of IKOA is particularly fast. In contrast, the convergence process of KOA is generally slower, and it exhibits obvious slow stepwise decline or long-term stagnation in multiple functions such as F1, F2, F3, F4, F5, F7, and F10, indicating that KOA is more prone to falling into local optima in the 10-dimensional space. In terms of optimization performance, at the end of 1000 iterations, IKOA converges to better objective function values in all 12 test functions. For functions F1, F3, F4, F5, F6, F7, and F10, the quality of solutions obtained by IKOA is significantly superior to that of KOA, with a large difference in the final fitness values between the two. This indicates that the global search capability and final optimization accuracy of IKOA in high-dimensional spaces have also been significantly improved. In summary, IKOA still maintains a fast convergence speed and high optimization accuracy in the 10-dimensional space, demonstrating better robustness and the ability to handle high-dimensional optimization problems.

Similarly, for a more intuitive analysis, we calculated the average optimal values, standard deviations, and average running times (in seconds) of the 12 test functions listed in Table 1 under the 10-dimensional setting, and the results are presented in Table 3.

By analyzing Table 3, we can intuitively observe the convergence performance of KOA and IKOA on the 10-dimensional test functions F1-F12. Among the mean values of the 12 10-dimensional test functions, the average optimal values of IKOA are all lower than those of KOA, with particularly significant advantages in functions F1, F5, F6, F9, F11, and F12. This indicates that IKOA has stronger global search capability and higher optimization accuracy in the 10-dimensional space. In terms of standard deviation, for all 12 10-dimensional test functions, the standard deviation of IKOA is significantly lower than that of KOA. In functions F1, F3, F4, F6, F7, F10, and F11, the standard deviation of IKOA is one or more orders of magnitude smaller than that of KOA, which demonstrates that the results of IKOA are stable and reliable, and similar high-quality solutions can be obtained in multiple runs. In terms of computational efficiency, although IKOA does not achieve time advantages in all functions and takes slightly more time in some functions, considering the significant improvements made by IKOA in optimization accuracy and stability, such minor fluctuations in running time or slight increases in some cases are generally acceptable.

Overall, the data in Table 3 can prove that the comprehensive performance of IKOA is significantly superior to the original KOA when dealing with 10-dimensional optimization problems.

Table 3. Iteration results of the 10-dimensional test function algorithm

Function	Algorithm	Average optimal value	Standard deviation	Average running time/s
F1	KOA	12884	640	0.1541
F1	IKOA	305.71	2.84	0.1210
F2	KOA	512.54	9.03	0.1752
F2	IKOA	410.86	3.82	0.1530
F3	KOA	642.12	15.67	0.1775
F3	IKOA	607.52	2.14	0.1452
F4	KOA	867.14	16.74	0.1475
F4	IKOA	804.56	1.58	0.1254
F5	KOA	1486.54	7.88	0.2131
F5	IKOA	912.21	5.21	0.1485
F6	KOA	4014.24	158.26	0.0957
F6	IKOA	1804.62	1.85	0.1452
F7	KOA	2087.58	52.27	0.3214
F7	IKOA	2011.32	4.55	0.4234
F8	KOA	2287.21	8.54	0.1475
F8	IKOA	2206.27	5.21	0.1236
F9	KOA	2645.12	6.26	0.1451
F9	IKOA	2304.12	4.14	0.2354
F10	KOA	2520.32	32.32	0.1054
F10	IKOA	2402.24	2.55	0.1595
F11	KOA	2733.26	19.24	0.1425
F11	IKOA	2608.25	5.64	0.2451
F12	KOA	2894.74	15.93	0.3214
F12	IKOA	2721.74	8.21	0.3654

5 Conclusion

This paper first elaborates on the basic principles, mathematical models, and core processes of the Kepler Optimization Algorithm (KOA). Drawing inspiration from Kepler's three laws of planetary motion, this algorithm simulates the process of planets orbiting the sun for optimization purposes, where planets are regarded as candidate solutions and the sun as the current optimal solution. Based on an in-depth analysis of the KOA's principles, it is found that the original KOA may have shortcomings such as slow convergence speed and a tendency to fall into local optima when solving complex optimization problems. To overcome these defects, this paper proposes the Improved Kepler Optimization Algorithm (IKOA), which integrates three strategies. Firstly, a golden sine algorithm combined with adaptive Brownian motion is introduced to update the positions of planets, aiming to balance the global exploration and local exploitation capabilities. Secondly, an adaptive Lévy flight strategy is incorporated into the position update related to the distance between planets and the sun, so as to enhance population diversity and improve the ability to jump out of local optima. Thirdly, a dynamic random centroid opposition-based learning strategy is adopted to replace the original elitism retention strategy, further accelerating convergence and improving the quality of solutions.

Finally, to verify the performance of IKOA, 12 benchmark functions from the CEC2022 standard test function set are selected, and comprehensive simulation comparison experiments between IKOA and the original KOA are conducted under different dimensions. The experimental results show that IKOA outperforms the original KOA in terms of convergence speed, solution accuracy, and stability, which proves the effectiveness of the proposed improvement strategies in this paper.

References

1. Kaveh, M., Mesgari, M.S.: Application of meta-heuristic algorithms for training neural networks and deep learning architectures: a comprehensive review. Neural Process Lett. **55**(4), 4519–4622 (2023)
2. Kirkpatrick, S., Gelatt Jr., C.D., Vecchi, M.P.: Optimization by simulated annealing. science. **220**(4598), 671–680 (1983)
3. Prajapati, V.K., Jain, M., Chouhan, L.: Tabu Search Algorithm (TSA): a Comprehensive Survey. In: 2020 3rd International Conference on Emerging Technologies in Computer Engineering: Machine Learning and Internet of Things (ICETCE). IEEE (2020)
4. Shami, T.M., et al.: Particle swarm optimization: a comprehensive survey. IEEE Access. **10**, 10031–10061 (2022)
5. Zhang, M., Wen, G.: Duck swarm algorithm: theory, numerical optimization, and applications. Clust. Comput. **27**(5), 6441–6469 (2024)
6. Li, Y., Yu, Q., Du, Z.: Sand cat swarm optimization algorithm and its application integrating elite decentralization and crossbar strategy. Sci. Rep. **14**(1), 8927 (2024)
7. Kesemen, O., et al.: Artificial locust swarm optimization algorithm: Kesemen, O et al. Soft. Comput. **27**(9), 5663–5701 (2023)
8. Bencherqui, A., et al.: Optimal reconstruction and compression of signals and images by Hahn moments and artificial bee Colony (ABC) algorithm. Multimed. Tools Appl. **81**(21), 29753–29783 (2022)
9. Oyelade, O.N., et al.: Ebola optimization search algorithm: a new nature-inspired metaheuristic optimization algorithm. IEEE Access. **10**, 16150–16177 (2022)
10. Sohail, A.: Genetic algorithms in the fields of artificial intelligence and data sciences. Ann. Data Sci. **10**(4), 1007–1018 (2023)
11. Arora, S., Singh, S.: Butterfly optimization algorithm: a novel approach for global optimization. Soft. Comput. **23**(3), 715–734 (2019)
12. Mirjalili, S., Lewis, A.: The whale optimization algorithm. Adv. Eng. Softw. **95**, 51–67 (2016)
13. Trojovská, E., Dehghani, M., Trojovský, P.: Zebra optimization algorithm: a new bio-inspired optimization algorithm for solving optimization algorithm. IEEE Access. **10**, 49445–49473 (2022)
14. Dorigo, M., Birattari, M., Stutzle, T.: Ant colony optimization. IEEE Comput. Intell. Mag. **1**(4), 28–39 (2007)

15. Yang, X.S.: A new metaheuristic bat-inspired algorithm. In: Nature inspired cooperative strategies for optimization (NICSO 2010), pp. 65–74. Springer Berlin Heidelberg, Berlin, Heidelberg (2010)
16. Gandomi, A.H., Yang, X.S., Alavi, A.H.: Cuckoo search algorithm: a metaheuristic approach to solve structural optimization problems. Eng. Comput. **29**(1), 17–35 (2013)
17. Oladejo, S.O., Ekwe, S.O., Mirjalili, S.: The hiking optimization algorithm: a novel human-based metaheuristic approach. Knowl.-Based Syst. **296**, 111880 (2024)
18. Ma, B., et al.: Running city game optimizer: a game-based metaheuristic optimization algorithm for global optimization. J. Comput. Des. Eng. **10**(1), 65–107 (2023)
19. Rashedi, E., Nezamabadi-Pour, H., Saryazdi, S.: A gravitational search algorithm. Inf. Sci. **179**(13), 2232–2248 (2009)
20. Abdel-Basset, M., et al.: Kepler optimization algorithm: a new metaheuristic algorithm inspired by Kepler's laws of planetary motion. Knowl.-Based Syst. **268**, 110454 (2023)
21. Chakraborty, A., Kar, A.K.: Swarm intelligence: a review of algorithms. In: Nature-inspired computing and optimization, pp. 475–494 (2017)
22. Zhang, H., et al.: A multi-strategy improved sparrow search algorithm for solving the node localization problem in heterogeneous wireless sensor networks. Appl. Sci. **12**(10), 5080 (2022)
23. Yu, X. W, et al.: WSN node location based on beetle antennae search to improve the gray wolf algorithm. Wirel. Netw 28(2), 539–549 (2022), Wirel. Networks
24. Long, W., Cai, S.H., Jiao, J.J.: Hybrid grey wolf optimization algorithm for high-dimensional optimization. Control Decis. **31**(11), 1991–1997 (2016)
25. Li, G., Wang, G.G., Xiao, R.B.: A novel adaptive weight algorithm based on decomposition and two-part update strategy for many-objective optimization. Inf. Sci. **615**, 323–347 (2022)
26. Tanyildizi, E., Demir, G.: Golden sine algorithm: a novel math-inspired algorithm. Adv. Electr. Comput. Eng. **17**(2), 71–78 (2017)
27. Kamaruzaman, A.F., et al.: Levy flight algorithm for optimization problems-a literature review. Appl. Mech. Mater. **421**, 496–501 (2013)
28. Si, T., et al.: Pcobl: a novel opposition-based learning strategy to improve metaheuristics exploration and exploitation for solving global optimization problems. IEEE Access. **11**, 46413–46440 (2023)
29. Luo, W., et al.: Benchmark functions for CEC 2022 competition on seeking multiple optima in dynamic environments. arxiv preprint arxiv 2201(00523). (2022)

Application of Multi-strategy Harris Hawks Optimization Algorithm in Path Planning

Yuyan Zhang, Chunmei Zhang(✉), Fanzhu Hao, and Long Ding

Taiyuan University of Science and Technology, Taiyuan 0330024, China
{s202315110514,s202315110490,s202215110494}@stu.tyust.edu.cn,
zhangchunmei@tyust.edu.cn

Abstract. In complex environments, traditional path planning algorithms are prone to issues such as local optima. To address this, we propose a Multi-Strategy Harris Hawks Optimization (MSHHO) algorithm. First, the Harris Hawks population is initialized using chaotic mapping to enhance diversity. Second, criss-cross optimization strategy is introduced to improve the algorithm's global search capability, accelerate convergence, and escape local optima. Finally, a convergence factor update strategy is proposed to dynamically adjust the prey's escape energy, balancing global exploration and local exploitation. To verify the algorithm's effectiveness, the proposed MSHHO is comprehensively tested against five other metaheuristic optimization algorithms in dual-grid map pathfinding experiments. Experimental results demonstrate that MSHHO achieves superior performance and shorter optimal paths compared to competing methods.

Keywords: Harris hawk optimization algorithm · Vertical and horizontal crossover strategy · Chaotic mapping · Convergence factor

1 Introduction

In the field of intelligent navigation and mobile robotics, path planning plays a crucial role in enabling mobile robots to find safe and efficient routes [1, 2]. Traditional path planning algorithms, such as the Dijkstra algorithm [3], the A* algorithm [4] and the RRT algorithm [5] etc., have been developed for both global and local path planning based on their respective applicability. With the increasing complexity of environments and higher real-time requirements, traditional algorithms have shown limitations in adaptability and efficiency, making metaheuristic optimization algorithms a more promising alternative due to their superior performance.

Metaheuristic optimization algorithms, primarily inspired by biological behaviors in nature or physical phenomena [6], represent a class of high-level heuristic strategies widely applied in complex optimization problems. Based on their structural characteristics and the nature of the inspirations they emulate, these algorithms are generally categorized into three main groups: evolution-based, physics-based, and swarm-based methods. Evolution-based metaheuristic algorithms, such as Genetic Algorithms (GA)

H. Ma et al. (Eds.): IWACIII 2025, CCIS 2780, pp. 452–468, 2026.
https://doi.org/10.1007/978-981-95-6730-0_31

[7], Differential Evolution (DE) [8], et al., mimic natural evolutionary processes—namely selection, mutation, and recombination—to effectively explore global optima within complex search spaces. Physics-based metaheuristics are designed by modeling physical phenomena. Representative examples include Simulated Annealing (SA) [9], the Gravitational Search Algorithm (GSA) [10], and others, which leverage analogies to thermodynamics, gravity, and related principles to guide the search process. Swarm-based algorithms emulate the collective behavior of social organisms. Algorithms such as the Artificial Bee Colony (ABC) algorithm [11], the Firefly Algorithm (FA) [12], etc., utilize a population of interacting agents that collaboratively evolve toward high-quality solutions through information sharing and distributed search dynamics.

The standard Harris Hawks Optimization (HHO) algorithm [13], proposed in 2019, is a swarm-based metaheuristic algorithm inspired by the cooperative hunting behavior of Harris hawks. Due to its outstanding exploration and exploitation capabilities, HHO has been widely applied across various domains. For instance, to prevent disease progression and enable early treatment of Alzheimer's disease (AD), Zhang et al. [14] utilized an enhanced version of the HHO algorithm to successfully identify relevant features and improve classification performance in AD diagnosis. Jiang et al. [15] developed a coupled model based on HHO for accurately predicting water demand, providing decision-makers with effective tools for regional water resource allocation and promoting sustainable water usage. Xiao et al. [16] employed an improved HHO algorithm in the field of visual target tracking. In recent years, many researchers have focused on enhancing the performance of the HHO algorithm. For example, Ouyang et al. [17] proposed improvements by modifying the intermittent energy factor, decay vector, and flight step length using a Cardano-based function, thereby enhancing local search ability and computational accuracy. Juan et al. [18] introduced enhancements involving circle mapping, a revised escape energy formula, an improved exploitation phase, and hybrid mutation mechanisms to improve the optimization performance of HHO. Additionally, Li et al. [19] proposed exploration strategies based on logarithmic spirals and opposition-based learning to further boost the global search capability of the algorithm.

This paper proposes a multi-strategy Harris Hawks Optimization (MSHHO) algorithm. First, the initial population of Harris hawks is generated using one of three chaotic maps—Tent, Sine, or Cubic—randomly selected via a random number generator. This hybrid chaotic initialization produces a more uniformly distributed population. Second, crisscross optimization strategy is introduced to address the premature convergence issue of the standard HHO, thereby enhancing global search capability, improving convergence speed. Finally, a novel convergence factor updating mechanism is designed to recalculate the prey's escape energy, effectively balancing the global exploration and local exploitation phases and improving the algorithm's precision. Experimental results in a dual-grid map demonstrate that, compared with the standard HHO and four other metaheuristic algorithms, MSHHO exhibits strong competitiveness in numerical optimization.

The remainder of this paper is organized as follows: Sect. 2 reviews the fundamental principles of the standard HHO; Sect. 3 describes the proposed MSHHO algorithm in detail; Sect. 4 discusses the application of MSHHO to path planning in a dual-grid map; and Sect. 5 concludes the paper and outlines directions for future research.

2 Background of Harris Hawks Optimization (HHO)

In 2019, Heidari was inspired by the cooperative hunting behavior of Harris's hawks and proposed the population-based Harris Hawks Optimization (HHO) algorithm [13]. The hawks cooperate as a group, launching attacks from different directions, surrounding the escaping prey simultaneously once it is targeted, and eventually achieving a successful capture. Moreover, Harris's hawks are capable of exhibiting various chasing styles in response to dynamic environmental conditions and the evasive behavior of the prey.

The standard HHO consists of two main phases: the exploration phase and the exploitation phase. The exploration phase includes two different position update strategies, while the exploitation phase comprises four position update strategies. Figure 1 illustrates the dynamic stages of the standard HHO algorithm.

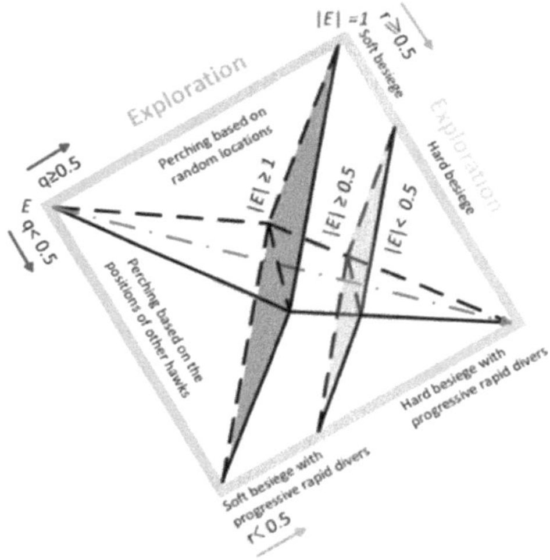

Fig. 1. The different stages in the standard Harris Hawks Optimization (HHO) algorithm

2.1 Exploration Phases

As shown in Fig. 1, the standard HHO algorithm employs two distinct search strategies during the exploration phase. Assuming each strategy has an equal probability q, when $q \geq 0.5$, the Harris hawks perform a random exploration based on the position information of family members and their own positions. In contrast, when $q < 0.5$, the hawks perch randomly within the search space to explore potential target regions. The corresponding equations are as follows:

$$X(t+1) = \begin{cases} X_{rand}(t) - r_1 |X_{rand}(t) - 2r_2 X(t)| & if\ q \geq 0.5 \\ X_{rabbit}(t) - X_m(t) - r_3[LB + r_4(UB - LB)] & if\ q < 0.5 \end{cases} \tag{1}$$

Where $X(t+1)$ represents the position vector of a hawk at iteration $t+1$, and $X(t)$ is its current position vector. $X_{rand}(t)$ denotes the position of a randomly selected hawk at iteration t, while $X_{rabbit}(t)$ is the position of the prey (rabbit). The parameters q, r_1, r_2, r_3, and r_4 are random numbers in the range (0,1) and are updated in each iteration. The vectors UB and LB represent the upper and lower bounds of the search space, respectively. $X_m(t)$ refers to the mean position of the current population, which is calculated as follows:

$$X_m(t) = \frac{1}{N}\sum\nolimits_{i=1}^{N} X_i(t) \tag{2}$$

Where $X_i(t)$ represents the position of each Harris hawk at the t-th iteration, and N denotes the population size of the Harris hawks.

2.2 Transition Phase

As shown in Fig. 1, during the exploration phase, the Harris hawk population primarily adopts strategies to search for and locate the prey. In the exploitation phase, the population employs multiple strategies to capture the prey. The standard HHO algorithm is capable of switching among different exploitation behaviors during the transition from exploration to exploitation, based on the prey's escape energy. To simulate the energy consumption during the prey's escape process, the following model is established:

$$E = 2E_0\left(1 - \frac{t}{T}\right) \tag{3}$$

Where E represents the escape energy of the prey, and T is the maximum number of iterations. E_0 represents the initial energy, which is randomly generated within the range (−1,1) at each iteration in the standard HHO algorithm. The escape energy E gradually decreases during the iteration process. When $|E| \geq 1$, the exploration phase is executed; when $|E| < 1$, the exploitation phase is carried out.

2.3 Exploitation Phase

Assuming r is the escape probability of the prey, the prey is considered to have successfully escaped when $r \geq 0.5$, and to have failed to escape when $r < 0.5$. Based on the prey's escape behavior and the pursuit strategies of the Harris hawks, the standard HHO algorithm defines four distinct attack strategies.

Soft Besiege. The prey still possesses sufficient escape energy, but the Harris hawks adopt a soft besiege strategy that gradually depletes the prey's energy. Once the prey becomes exhausted, the hawks launch a sudden attack. In this case, $|E| \geq 0.5$ and $r \geq 0.5$. This strategy is represented by the following equation:

$$X(t+1) = \Delta X(t) - E|JX_{rabbit}(t) - X(t)| \tag{4}$$

$$\Delta X(t) = X_{rabbit}(t) - X(t) \tag{5}$$

$$J = 2(1 - r_5) \tag{6}$$

Where $\Delta X(t)$ represents the difference between the prey's position vector and the current position of the Harris hawk at iteration t, and $X(t)$ denotes the position vector of the Harris hawk at iteration t. Accordingly, $X(t+1)$ is the position vector of the Harris hawk in the next iteration. $X_{rabbit}(t)$ indicates the best position vector of the prey at iteration t, and J is the random jump strength of the prey, used to simulate its leaping behavior.

Hard Besiege. At this stage, the prey's escape energy is nearly depleted, and the Harris hawks launch a direct sudden attack. In this case, $|E| < 0.5$ and $r \geq 0.5$. This strategy is described by the following equation:

$$X(t+1) = X_{rabbit}(t) - E|\Delta X(t)| \tag{7}$$

Where E is obtained from the above Eq. (3).

Soft Besiege with Progressive Rapid Dives. In this phase, the standard HHO algorithm incorporates the concept of Lévy flight. The prey possesses sufficient escape energy to evade the Harris hawks' encirclement. The hawks adopt a gradual, rapid-diving soft besiege strategy, while the prey employs various deceptive maneuvers to escape the attack. In this case, $|E| \geq 0.5$ and $r < 0.5$. The specific approach is described by the following equation:

$$Y = X_{rabbit}(t) - E|JX_{rabbit}(t) - X(t)| \tag{8}$$

$$Z = Y + S \times LF(D) \tag{9}$$

$$LF(x) = 0.01 \times \frac{u \times \sigma}{|v|^{\frac{1}{\beta}}}, \sigma = \left(\frac{\Gamma(1+\beta) \times \sin\left(\frac{\pi\beta}{2}\right)}{\Gamma\left(\frac{1+\beta}{2}\right) \times \beta \times 2^{\left(\frac{\beta-1}{2}\right)}} \right)^{\frac{1}{\beta}} \tag{10}$$

Where D represents the dimensionality of the search space, S is a random vector of size $1\times D$, and LF represents the Lévy flight function. u and v are random numbers within the range (0,1), and β is a constant set to 1.5.

Therefore, the final update strategy for this phase is summarized as follows:

$$X(t+1) = \begin{cases} Y & if\ F(Y) < F(X(t)) \\ Z & if\ F(Z) < F(X(t)) \end{cases} \tag{11}$$

Hard Besiege with Progressive Rapid Dives. The prey lacks sufficient escape energy, prompting the Harris hawks to reduce the distance between the average position of the population and the prey, forming an encirclement before launching an assault. In this case, $|E| < 0.5$ and $r < 0.5$. The corresponding mathematical model is given as follows:

$$Y = X_{rabbit}(t) - E|JX_{rabbit}(t) - X_m(t)| \tag{12}$$

$$Z = Y + S \times LF(D) \tag{13}$$

$$X(t+1) = \begin{cases} Y & if\ F(Y) < F(X(t)) \\ Z & if\ F(Z) < F(X(t)) \end{cases} \tag{14}$$

Where $X_m(t)$ represents the average position of the Harris hawk population at iteration t, as defined in Eq. (2) above.

3 Method of Improvement

3.1 Chaotic Mapping Initialization

Compared with random initialization, chaotic mapping possesses strong ergodicity. Common chaotic maps include the following types [20, 21]: Logistic, PWLCM, Singer, Sine, Gaussian, Tent, Bernoulli, Chebyshev, Circle, Cubic, Sinusoidal, and ICMIC. Due to their superior properties, chaotic maps have been widely applied to swarm intelligence optimization algorithms by various researchers, such as the chaotic fruit fly optimization algorithm [22], chaotic bat algorithm [23], chaotic accelerated particle swarm optimization [24], chaotic firefly algorithm [25], chaotic artificial bee colony algorithm [26], and chaotic cuckoo search algorithm [27].

The Tent map exhibits an approximately uniform distribution and good ergodicity; the Sine map introduces nonlinear perturbations, enabling the search to escape locally concentrated regions; the Cubic map possesses strong chaotic behavior, allowing for rapid dispersion. In this section, chaotic maps are used to initialize the search agents, which helps generate a more uniformly distributed initial population at a faster rate. The comparison diagram of the initial population distribution is shown in Fig. 2. The corresponding model is defined as follows:

$$\begin{cases} Tent: T_{k+1} = \begin{cases} T_k/\gamma, & 0 < T_k \leq \gamma \\ (1-T_k)/(1-\gamma), & \gamma < T_k \leq 1 \end{cases}, & 0 < p \leq \frac{1}{3} \\ Cubic: C_{k+1} = \rho C_k\left(1 - C_k^2\right), & \frac{1}{3} < p \leq \frac{2}{3} \\ Sine: S_{k+1} = \frac{a}{4}\sin(\pi S_k), & \frac{2}{3} < p \leq 1 \end{cases} \tag{15}$$

For chaotic maps, appropriate parameter settings can significantly enhance their performance. Several existing studies [20, 28] have experimentally verified the feasibility and stability of certain parameter values in practical applications. Therefore, this paper adopts these validated parameter settings, as summarized in the table below (Table 1).

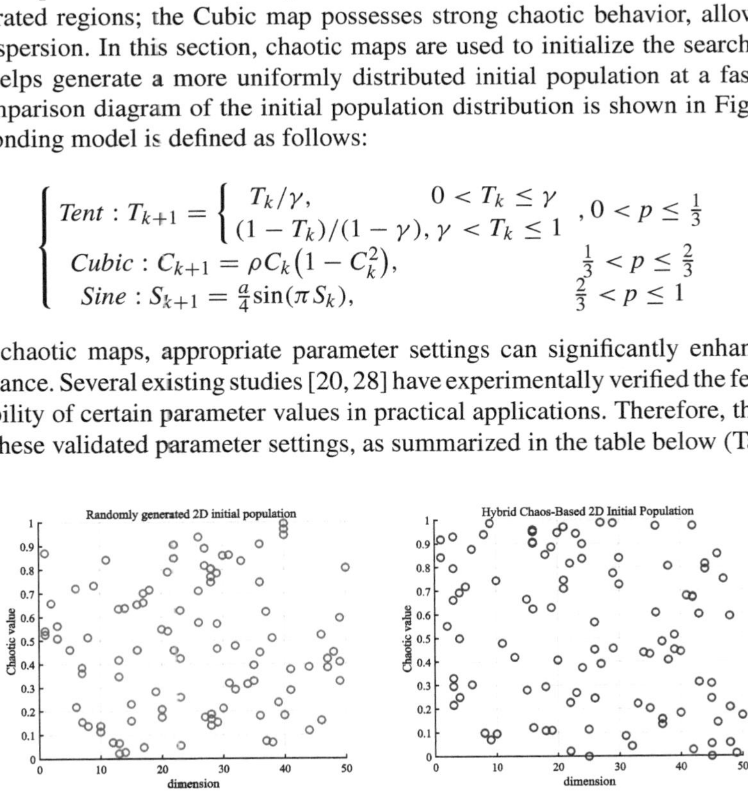

Fig. 2. Comparison of initial population distribution

Table 1. Parameter values

γ	T_0	ρ	C_0	a	S_0
0.4	0.154	2.59	0.242	4	0.152

3.2 Crisscross Optimization Strategy

Crisscross optimization algorithm is a population-based optimization algorithm [29]. This strategy enhances both the global exploration and local exploitation capabilities of the algorithm by simulating two orthogonal search mechanisms—horizontal crossover and vertical crossover—thereby preventing premature convergence. The horizontal crossover simulates information exchange among individuals along the same dimension, similar to the crossover operation in genetic algorithms, which strengthens local exploitation. In contrast, the vertical crossover simulates information exchange across different dimensions, breaking inter-dimensional correlations and enhancing global exploration.

Let the optimization problem be defined in a *D*-dimensional search space, with a maximum number of iterations set to T, and a population of N individuals initialized $X(X_1, X_2, \ldots, X_N), X_i(x_{i,1}, x_{i,2}, \ldots, x_{i,D})$.

The initial values in each dimension are generated according to a uniform distribution:

$$x_{i,j} = LB_j + \text{rand}(0, 1) \times \left(UB_j - LB_j\right), j = 1, 2, \ldots, D \tag{16}$$

Where LB_j and UB_j represent the lower and upper bounds of the j-th dimension, respectively, and $rand(0, 1)$ is a random number uniformly distributed within the interval [0,1].

When entering the global exploration phase, the algorithm prioritizes vertical crossover to enhance population diversity. For $P_{VC} \cdot N$ individuals randomly selected from the population (where P_{VC} denotes the vertical crossover probability), two different dimensions k and g are randomly chosen for each selected individual, and the update is performed according to the following equation:

$$x_{\text{i,j}}^{\text{new}} = \begin{cases} x_{\text{i,k}}^{\text{new}} = r \cdot x_{i,k} + (1 - r) \cdot x_{i,g} \\ x_{\text{i,g}}^{\text{new}} = r \cdot x_{i,g} + (1 - r) \cdot x_{i,k} \end{cases} \tag{17}$$

Where $r \in [0, 1]$ is a uniformly distributed random number. If the updated dimensional values exceed the predefined boundaries after crossover, a reflection boundary mechanism is applied for correction, as shown in the following equation:

$$x_{\text{i,j}}^{\text{new}} = \begin{cases} 2LB_j - x_{\text{i,j}}^{\text{new}} \text{ if } x_{i,j}^{new} \leq \text{LB}_j \\ 2UB_j - x_{\text{i,j}}^{\text{new}} \text{ if } x_{i,j}^{new} \leq \text{UB}_j \end{cases} \tag{18}$$

When $|E| < 1$, the algorithm enters the exploitation phase. Based on the value of E and a random number r, one of the following four predation strategies is selected, with horizontal crossover embedded to enhance local search precision.

Soft Besiege The soft besiege strategy generates a temporary encircling solution based on the relative distance between the prey's position and the current position of the individual, as defined by the following equation:

$$X^{temp} = X_{rabbit} - E \cdot |JX_{rabbit} - X_i| \tag{19}$$

In the equation, $X_{rabbit} \in R^D$ represents the global optimal solution, and $X_i \in R^D$ represents the current position of the individual. This formulation employs the energy coefficient E to dynamically adjust the encircling radius, enabling a smooth transition from coarse-grained search to fine-grained exploitation.

To enhance local search efficiency and avoid premature convergence, the algorithm introduces a horizontal crossover operation. Specifically, for the current individual X_i, the individual X_j with the smallest Euclidean distance within the population is selected:

$$D_j = \arg \min_{k \neq i} \|X_i - X_j\|_2 \tag{20}$$

After generating a new solution through the predation strategy, a same-dimension crossover is performed between X_i^{new} and its neighboring individual X_j with a horizontal crossover probability P_{HC} , as defined below:

$$X_i^{new} = r_2 \cdot X^{\text{temp}} + (1 - r_2) \cdot X_j + c \cdot \left(X^{temp} - X_j\right) \tag{21}$$

$$x_{i,k}^{\text{new}} = r_2 \cdot x_{i,k}^{\text{temp}} + (1 - r_2) \cdot x_{j,k} + c \cdot \left(x_{i,k}^{temp} - x_{j,k}\right) \tag{22}$$

In the equation, r_2 is a random weight, and c is a perturbation factor.

The final position of the individual is determined based on fitness comparison.

$$X_i^{\text{new}} = \begin{cases} X_i^{new} & f\left(X_i^{new}\right) \leq f\left(X^{temp}\right) \\ X^{\text{temp}} & \text{otherwise} \end{cases} \tag{23}$$

If the new solution is better than the current global best, the prey's position is updated accordingly.

$$X_{\text{rabbit}} = \begin{cases} X_i^{\text{new}} & f\left(X_i^{new}\right) \leq f\left(X_{rabbit}\right) \\ X_{\text{rabbit}} & otherwise \end{cases} \tag{24}$$

Hard Besiege. The hard besiege strategy approaches the prey's position directly through linear contraction, as defined by the following equation:

$$X_i^{new} = X_{\text{rabbit}} - E \cdot |X_{\text{rabbit}} - X_i| \tag{25}$$

To avoid local optima, random perturbation is added to the solution with a probability $P_{HC} = 0.3$, as shown below:

$$x_{i,k}^{new} = x_{i,k}^{new} + c \cdot \left(x_{rabbit,k} - x_{i,k}^{new}\right) \cdot N(0, 1) \tag{26}$$

In the equation, $N(0, 1)$ is a random number following a standard normal distribution. A greedy selection rule is also applied to retain the solution with better fitness as the new position of the individual. If the new solution outperforms the current global best, the prey's position is updated according to Eq. (24).

$$X_i^{new} = \begin{cases} X_i^{new} \ \mathrm{f}\left(X_i^{new}\right) \le f(X_i) \\ X_i \quad \text{otherwise} \end{cases} \tag{27}$$

Soft Besiege with Progressive Rapid Dives. Based on the relative distance between the prey position and the current individual, an initial candidate solution is generated, as shown in Eq. (28). A long-distance jump candidate solution is further generated by introducing Lévy flight, as shown in Eq. (29). Horizontal crossover operations (as in the soft besiege strategy) are then performed on both Y and Z , producing the optimized solutions Y^{HC} and Z^{HC}.

$$Y^{HC} = r_2 \cdot Y + (1 - r_2) \cdot X_j + c \cdot \left(Y - X_j\right) \tag{28}$$

$$Z^{\mathrm{HC}} = r_2 \cdot Z + (1 - r_2) \cdot X_j + c \cdot \left(Z - X_j\right) \tag{29}$$

The solution with better fitness is selected as the final candidate.

$$X^{opt} = \arg \min_{X \in \left\{Y^{\mathrm{HC}}, Z^{\mathrm{HC}}\right\}} f(X) \tag{30}$$

The better solution is retained as the new position of the individual. Similarly, the prey position is updated according to Eq. (24).

$$X_i^{new} = \begin{cases} X^{opt} \ \mathrm{f}\left(X^{opt}\right) \le f(X_i) \\ X_i \quad \text{otherwise} \end{cases} \tag{31}$$

Hard Besiege with Progressive Rapid Dives. A random search guided by the population's mean position is introduced near the global optimum, as shown below.

$$X_i^{new} = X_{rabbit} - E \cdot \left|JX_{rabbit} - X_{avg}\right| \tag{32}$$

$$X_{avg} = \frac{1}{N}\sum\nolimits_{i=1}^{N} X_i \tag{33}$$

In the equation, X_{avg} represents the mean position of the population.

A random subset of $d = [0.2D]$ dimensions is selected and exchanged with the corresponding dimensions of the historical best solution, as shown below.

$$x_{\mathrm{i,k}}^{\mathrm{new}} = \begin{cases} x_{rabbit,k} \ \text{if k} \in \text{RandomSubset}(1 : D, d) \\ x_{\mathrm{i,k}}^{\mathrm{new}} \quad \text{otherwise} \end{cases} \tag{34}$$

The better solution is retained as the new position of the individual. The prey position is updated according to Eq. (27).

3.3 Convergence Factor Update Strategy

This paper adopts a nonlinear energy convergence factor adjustment strategy based on an exponential function to optimize the escape energy mechanism. The strategy uses the absolute value of escape energy as the convergence factor and achieves a smooth transition in the algorithm's search behavior through dynamic adjustment. In the early stages of iteration, a larger escape energy value effectively reduces the risk of premature convergence to local optima. In the later stages, a smaller escape energy value significantly improves the convergence accuracy. Compared to the standard HHO algorithm, this improved strategy better balances global exploration and local exploitation, enabling the algorithm to adaptively adjust its search intensity at different stages of the optimization process (Fig. 3).

$$E = 2E_0\left(1 - e^{1-\frac{t}{T}}\right) \tag{35}$$

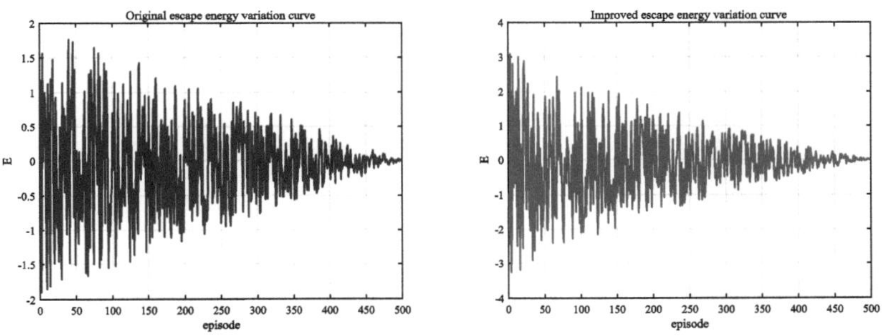

Fig. 3. Original and improved escape energy variation curves

3.4 Pseudocode of MSHHO

The pseudocode of the improved MSHHO algorithm is presented in Algorithm 1.

Algorithm 1: The pseudo-code of the MSHHO

1: **Inputs**: The population size N and maximum number of iterations T
2: **Outputs**: The location of the rabbit(X_{rabbit}) and its fitness value
3: Initialize Harris hawks population X_i(i=1,2,...,N) using Eq.(15)
4: Calculate the fitness values of hawks
5: Set X_{rabbit} as the location of rabbit (best location)
while (t ≤ T)
  For(each Harris hawks position(X_i))
  Update E using Eq.(35)
    if (|E|≥ 1) then
      Update the position vector using Eq. (17)(18)
    Else if (|E|< 1) then
      if (r ≥0.5 and |E|≥ 0.5) then
        Update the position vector using Eq. (24)
      Else if (r ≥0.5 and |E|< 0.5) then
        Update the position vector using Eq. (27)
      Else if (r < 0.5 and |E|≥ 0.5) then
        Update the position vector using Eq. (31)
      Else if (r < 0.5 and |E|< 0.5) then
        Update the position vector using Eq. (27)
      End if
    End if
End for
End while
Return X_{rabbit}

4 Experiment on Path Planning

4.1 2D Environment Model

Because of its simplicity and efficiency, as well as its strong adaptability to obstacles and significant reduction in environmental modeling complexity, the grid method has become a widely used technique for environmental map modeling in path planning [30]. In this study, we adopt the grid method to construct the working environment model, where white cells represent traversable space and black cells indicate obstacle regions. The starting point is marked by a blue cell, and the goal point is represented by a red cell, as shown in Fig. 4.

In this paper, the path search adopts the eight-neighborhood representation, where the agent can move from the current grid cell to its neighboring cells D1–D8, corresponding to the eight directions: up, down, left, right, upper left, upper right, lower left, and lower right. Compared with the four-neighborhood (up, down, left, and right only), this approach offers greater flexibility and higher path freedom. As illustrated in Fig. 4, the center of each grid cell is defined as the movement position, and arrows indicate the possible directions of movement.

In the path planning process, the algorithm represents the robot's movement on a two-dimensional grid map by continuously updating the position coordinates of the population. In this study, a constrained fitness function is constructed to evaluate the feasibility and quality of the generated paths. This function retains solutions that satisfy the constraints and eliminates those that do not. The algorithm ensures that each node (x_i, y_i), which does not cross any obstacles in the path, adheres to the basic constraint of staying within the map boundaries, while seeking the optimal path from the starting point to the goal. The robot's position at the next time step should satisfy $\mathrm{x}_{t+1} > \mathrm{x}_t$ or $\mathrm{y}_{t+1} > \mathrm{y}_t$.

$$F_{fit} = \sum_{i=1}^{m} \sqrt{(\mathrm{x}_{i+1} - \mathrm{x}_i)^2 + (\mathrm{y}_{i+1} - \mathrm{y}_i)^2} \tag{36}$$

$$\begin{cases} lb_x \leq x_i \leq ub_x \\ lb_y \leq y_i \leq ub_y \end{cases} \tag{37}$$

Where lb_x and ub_x are the lower and upper limits of the horizontal boundary, and lb_y and ub_y are the lower and upper limits of the vertical boundary. F_{fit} is the fitness value representing the total path length. The coordinates (x_i, y_i) and (x_{i+1}, y_{i+1}) denote two adjacent nodes on the path, with m representing the total number of nodes in the path.

To achieve path optimization, the planning objective is to minimize the fitness based on path length. By minimizing the path length, the algorithm seeks to find the optimal path that satisfies all constraints.

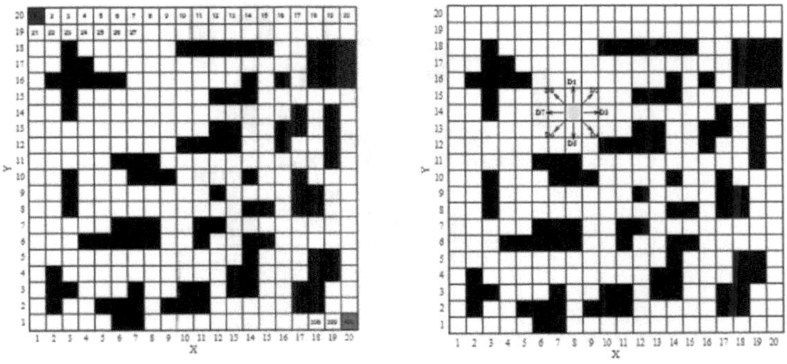

Fig. 4. Two-dimensional grid map and movement Direction Diagram

4.2 Parameter Settings for 2D Experiments

In this paper, the proposed MSHHO algorithm is compared with the original HHO, Ant Colony Optimization (ACO), Gorilla Troops Optimization (GTO), Sea-Horse Optimizer (SHO), and Particle Swarm Optimization (PSO) under the same parameter settings through path planning experiments. These six algorithms are tested on both a 20 × 20 and a 40 × 40 grid map, as shown in Fig. 5.

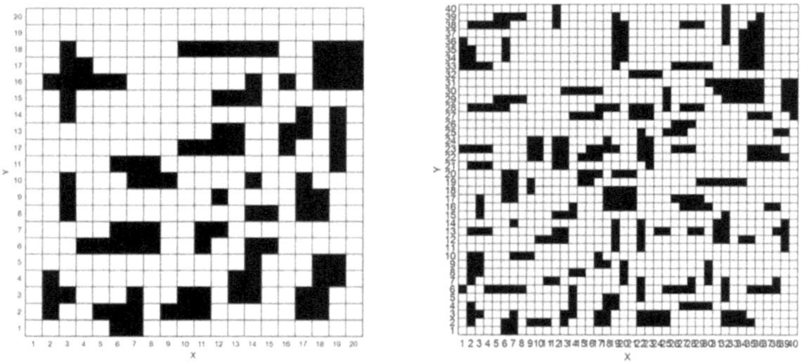

Fig. 5. Two-dimensional grid map of 20x20 and 40x40

Each algorithm is independently executed 20 times with 100 iterations on 20x20 grid map and executed 20 times with 500 iterations on 40x40 grid map. The trajectory routes of each algorithm are shown in Figs. 6 and 7, and Convergence curve of the path length is shown in Fig. 8. The corresponding numerical results are summarized in Table 2, where the best values are highlighted in bold.

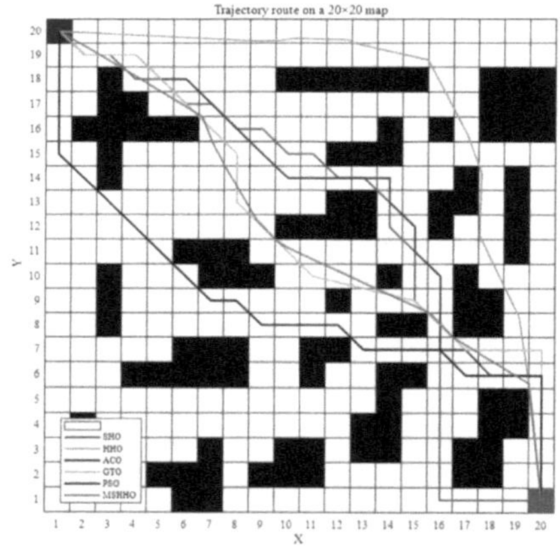

Fig. 6. Trajectory route on a 20 × 20 map

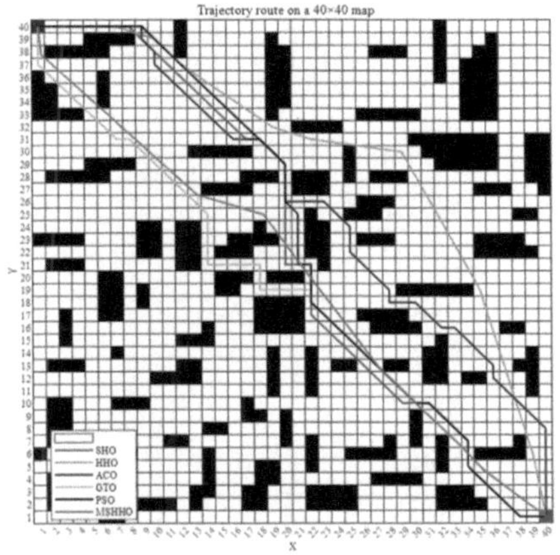

Fig. 7. Trajectory route on a 40 × 40 map

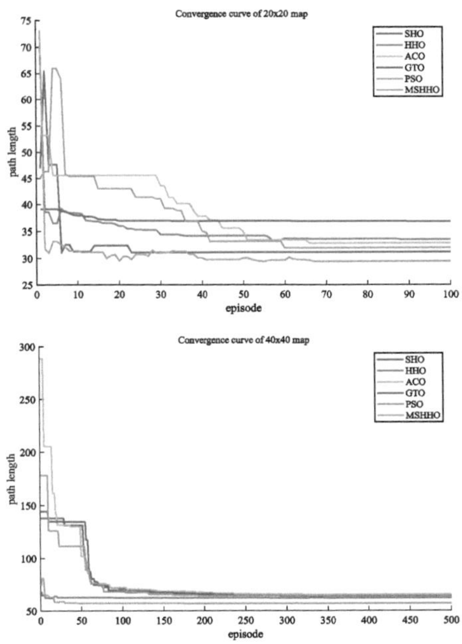

Fig. 8. Convergence curve of grid maps

Table 2. Comparison of path planning experimental analysis results

map	algorithm	Opt	Avg
20x20	MSHHO	**29.28**	30.38
	HHO	33.10	34.90
	ACO	32.37	37.75
	GTO	39.14	37.08
	SHO	29.95	32.24
	PSO	31.89	36.90
40x40	MSHHO	**56.51**	57.30
	HHO	62.07	62.38
	ACO	65.11	77.14
	GTO	63.94	73.92
	SHO	64.77	74.06
	PSO	64.18	72.17

4.3 Results and Analysis of 2D Experiments

Based on the data presented in Fig. 7, Fig. 8, and Table 2, the following conclusions can be drawn. On the 20 × 20 grid map, compared to MSHHO, the improvements in optimal path length are 11.54% (HHO), 9.56% (ACO), 25.20% (GTO), 2.24% (SHO), and 8.18% (PSO), respectively. In terms of average path length, the improvements are 12.96% (HHO), 19.53% (ACO), 18.05% (GTO), 5.77% (SHO), and 17.66% (PSO). On the 40 × 40 grid map, MSHHO achieves improvements in optimal path length of 8.97% (HHO), 13.23% (ACO), 11.63% (GTO), 12.76% (SHO), and 11.96% (PSO). For the average path length, the corresponding improvements are 8.14%, 25.72%, 22.47%, 22.59%, and 20.57%, respectively.

In all other submaps, the curve of MSHHO lies below the curves of other algorithms, indicating that MSHHO can find shorter paths in the above-mentioned maps. On the 20 × 20 and 40 × 40 grid maps, the paths generated by MSHHO are shorter and smoother. Experimental results in a dual-grid map demonstrate that, compared with the standard HHO and four other metaheuristic algorithms, MSHHO exhibits strong competitiveness in numerical optimization.

5 Conclusions and Future Works

To address the shortcomings of the standard Harris Hawks Optimization (HHO) algorithm, this study proposes several enhancements. First, a hybrid chaotic initialization strategy is employed during the population initialization phase to increase population diversity and ensure a more uniform distribution of Harris hawks, thereby improving the likelihood of successful hunting. Second, crisscross optimization strategy is introduced in the exploration and exploitation phases, which not only enhances the algorithm's global search capability and accelerates convergence but also helps to escape local optima. Third, a convergence factor update strategy is adopted to recalculate the escaping energy of the prey, enabling a better balance between global exploration and local exploitation. This allows the algorithm to adaptively adjust its search intensity across different search stages. The proposed MSHHO algorithm is further validated through optimization experiments on a binary grid map.

In future work, we aim to extend the MSHHO algorithm to solve multi-objective and high-dimensional optimization problems, offering more practical solutions for path planning tasks.

Acknowledgments. This study was funded by the Natural Science Foundation of Shanxi Province (grant number 202203021211198).

Disclosure of Interests The authors have no competing interests to declare that are relevant to the content of this article.

References

1. Chen, P.F., Huang, Y.M., Papadimitriou, E., Mou, J.M., Gelder, P.V.: Global path planning for autonomous ship: a hybrid approach of fast Marching Square and velocity obstacles methods. Ocean Eng. **214**, 107793 (2020)
2. Zheng, A.R., Liu, T.C., Song, Y.P., Wang, X.S., Liu, L.L., Qi, L.: Minimum dose path planning method for virtual nuclear facilities based on navigation mesh. Ann. Nucl. Energy. **195**, 110104 (2024)
3. Zhao, Y.: Dynamic path planning analysis of warehouse handling robot. J. Sensors. **2023** (2022)
4. Hart, P.E., Nilsson, N.J., Raphael, B.: A formal basis for the heuristic determination of minimum cost paths. IEEE Trans. Syst. Sci. Cybern. **4**, 100–107 (1968)
5. Gu, Q.Y., Zhen, R., Liu, J.L., Li, C.: An improved RRT algorithm based on prior AIS information and DP compression for ship path planning. Ocean Eng. **279**, 114595 (2023)
6. Aljarah, I., Mafarja, M., Heidari, A.A., Faris, H., Mirjalili, S.: Multi-verse optimizer: theory, literature review, and application in data clustering. Nat.-Inspired Optim. **811**, 123–141 (2020)
7. Chambers, L.D.: Practical Handbook of Genetic Algorithms: Complex Coding Systems, 3rd edn. CRC Press (1998)
8. Zhang, C.M., Chen, J., Xin, B.: Distributed memetic differential evolution with the synergy of Lamarckian and Baldwinian learning. Appl. Soft Comput. **13**, 2947–2959 (2013)
9. Ingber, L.: Simulated annealing: practice versus theory. Math. Comput. Model. **18**, 29–57 (1993)
10. Rashedi, E., Nezamabadi-Pour, H., Saryazdi, S.: GSA: a gravitational search algorithm. Inf. Sci. **179**, 2232–2248 (2009)
11. Karaboga, D., Basturk, B.: A powerful and efficient algorithm for numerical function optimization: artificial bee colony (ABC) algorithm. J. Glob. Optim. **39**, 459–471 (2007)
12. Yang, Y.S.: Multiobjective firefly algorithm for continuous optimization. Eng. Comput. **29**, 175–184 (2013)
13. Heidari, A.A., Mirjalili, S., Faris, H., Aljarah, I., Mafarja, M., Chen, H.: Harris hawks optimization: algorithm and applications. Future Gener. Comput. Syst. **97**, 849–872 (2019)

14. Zhang, Q., Sheng, J.H., Zhang, Q., Wang, L.Y., Yang, Z., Xin, Y.: Enhanced Harris hawks optimization-based fuzzy k-nearest neighbor algorithm for diagnosis of Alzheimer's disease. Comput. Biol. Med. **165**, 107392 (2023)
15. Jiang, Q.X., Guo, W.P., Wang, Z.L., Wu, Y.X., Zhao, Y.Z., Tao, M.Y., et al.: Forecasting regional water demand using multi-fidelity data and Harris hawks optimization of generalized regression neural network models – a case study of heilongjiang province, china. Sci. Total Environ. **634**, 131084 (2024)
16. Xiao, Y.Q., Wu, Y.J.: Robust visual tracking via modified Harris hawks optimization. Image Vis. Comput. **144**, 104959 (2024)
17. Ouyang, C.T., Liao, C., Zhu, D.L., Zheng, Y.Y., Zhou, C.J., Li, T.Y.: Integrated improved Harris hawks optimization for global and engineering optimization. Sci. Rep. **14**, 10100 (2024)
18. Nan, J., Wang, J., Wu, H., Li, K.: Optimized extreme learning machine by an improved Harris hawks optimization algorithm for mine fire flame recognition. Min. Metall. Explor. **40**, 367–388 (2023)
19. Li, C.Y., Li, J., Chen, H.L., Jin, M., Ren, H.: Enhanced Harris hawks optimization with multi-strategy for global optimization tasks. Expert Syst. Appl. **185**, 115499 (2021)
20. Yu, Y., Gao, S.C., Cheng, S., Wang, Y.R., Song, S.Y., Yuan, F.G.: CBSO: a memetic brain storm optimization with chaotic local search. Memet. Comput. **10**, 353–367 (2018)
21. Yang, L., Gao, S.C., Yang, H.C., Cai, Z.H., Lei, Z.Y., Todo, Y.: Adaptive chaotic spherical evolution algorithm. Memet. Comput. **13**, 383–411 (2021)
22. Mitić, M., Vuković, N., Petrović, M., Miljković, Z.: Chaotic fruit fly optimization algorithm. Knowl.-Based Syst. **89**, 446–458 (2015)
23. Khenissi, I., Guesmi, T., Alshammari, B.M., Alqunun, K., Almalaq, A., Alturkietl, M., et al.: A hybrid chaotic bat algorithm for optimal placement and sizing of DG units in radial distribution networks. Energy Rep. **12**, 1723–1741 (2024)
24. Kang, L.L., Lai, Y., Wang, J., Cao, W.L.: A pacesetter-Lévy multi-objective particle swarm optimization with Arnold chaotic map and opposition-based learning. Inf. Sci. **678**, 121048 (2024)
25. Gümüşçü, A., Kaya, S., Tenekeci, M.E., Karacizmeli, İ.H., Aydilek, İ.B.: The impact of local search strategies on chaotic hybrid firefly particle swarm optimization algorithm in flow-shop scheduling. J. King Saud Univ. Comput. Inf. Sci. **34**, 6432–6440 (2022)
26. Ayan, K., Kılıç, U., Baraklı, B.: Chaotic artificial bee colony algorithm based solution of security and transient stability constrained optimal power flow. Int. J. Electr. Power Energy Syst. **64**, 136–147 (2015)
27. Yin, Z.D., Wang, L., Zhang, Y.J., Gao, Y.: Parameter identification of DC arc models using chaotic quantum cuckoo search. Appl. Soft Comput. **108**, 107451 (2021)
28. Gao, S.C., Yu, Y., Wang, Y.R., Wang, J.H., Cheng, J.J., Zhou, M.C.: Chaotic local search-based differential evolution algorithms for optimization. IEEE Trans. Syst. Man Cybern. Syst. **51**, 3954–3967 (2019)
29. Meng, A.B., Chen, Y.C., Yin, H., Chen, S.Z.: Crisscross optimization algorithm and its application. Knowl.-Based Syst. **67**, 218–229 (2014)
30. Gong, S., Chen, R., Xie, Z., Li, X.: Major approaches to robot environmental modeling. In: Proceedings of the 6th International Symposium on Computer and Information Processing Technology (ISCIPT), pp. 218–222 (2021)

Three-dimensional Path Planning for Multiple UCAVs Based on the MP-GWO Algorithm

Chenghao Li, Chunmei Zhang(✉), Fanzhu Hao, and Zhenyang Zhao

Taiyuan University of Science and Technology, Taiyuan 030024, People's Republic of China
{s202415210594,s202315110490,s202315210615}@stu.tyust.edu.cn,
zhangchunmei@tyust.edu.cn

Abstract. With the extensive use of drones in both military and civilian fields, multi-drone collaborative trajectory planning has become a research hotspot. This paper proposes a multi-drone three-dimensional trajectory planning method based on a multi-population gray wolf optimization algorithm (MP-GWO), aimed at addressing multi-objective optimization problems in complex environments. Building on the GWO algorithm, this study first introduces a multi-population collaborative mechanism, dividing the search space into multiple sub-populations through clustering for separate optimization, thereby enhancing global search capability. Furthermore, based on a greedy approach and random initialization method, the average fitness value of the initial population is improved. Finally, the convergence factor in the GWO algorithm is adjusted from linear to a nonlinear control parameter; the adjusted convergence factor decreases rapidly in early iterations to enhance global search capability, avoiding local optima, and subsequently decreases gradually in later iterations to increase local search capability. Experimental results indicate that compared with the standard GWO algorithm, the MP-GWO algorithm demonstrates significant improvements in both convergence speed and trajectory quality, effectively planning three-dimensional collaborative trajectories that meet constraint conditions.

Keywords: Drones · Gray Wolf Optimization Algorithm · Multi-objective Optimization · Trajectory Planning

1 Introduction

In recent years, with the rapid development of autonomous aerial combat theory and cooperative control technologies for unmanned aerial vehicles, Unmanned Combat Aerial Vehicles (UCAVs) have attracted widespread attention from military powers around the world [1] due to their capability to perform crucial missions in remote and high-risk environments. Trajectory planning is an important component of UCAV mission management systems; a rational trajectory planning strategy can effectively enhance the survival probability of UCAVs. Additionally, the cooperative path planning of multiple UCAVs is one of the key technologies for UCAV joint operations.

At present, some valuable research results have been achieved in the research on multi-UAV trajectory planning at home and abroad. Fan Zhengzheng et al. [2] proposed a

H. Ma et al. (Eds.): IWACIII 2025, CCIS 2780, pp. 469–483, 2026.
https://doi.org/10.1007/978-981-95-6730-0_32

multi-UAV real-time strike track planning method based on A* algorithm in order to meet the real-time requirements of multi-target and multi-directional strike track planning in complex battlefield environments. Xu Wenyu et al. [3] proposed a local track planning method that integrates pigeon flock optimization algorithm and artificial potential field method in order to solve the problem of track re-planning in the multi-UAV cooperative mission environment. In order to solve the problem of time coordination, Zhou Lingzi proposed a multi-UAV collaborative formation assembly method based on pigeon flock optimization algorithm [4]. Wang Jinjin et al. proposed a consistent formation assembly control algorithm based on Dubins curve to solve the problem of rapid assembly of UAV formations [5].

However, with the increase of complexity, the GWO algorithm will have problems such as slow convergence speed and poor optimization effect [6], so some improved GWO algorithms have been proposed to improve the above problems to a certain extent. Yin Lingyi et al. [7] proposed an improved gray wolf algorithm for hypercube sampling initialization population and dynamic weighting, which made the algorithm obtain better solution accuracy. Liu Ningning [8] balanced the global and local search capabilities by improving the convergence factor, and improved the stability of the algorithm. Xu Le et al. [9] proposed to use the reverse learning strategy to initialize the individual, and design the convergence factor to optimize the individual position, so that the algorithm has a good convergence speed. Yang et al. [10] proposed a variety of swarm chaotic gray wolf algorithms, introduced a variety of swarm ideas and chaotic search strategies, and effectively solved the problem that the gray wolf algorithm is prone to fall into local optimality in the process of solving multi-UAV trajectory planning. Although the aforementioned studies have improved the optimization capability of the Grey Wolf Optimization (GWO) algorithm to some extent, most of them focus on addressing the trajectory planning of a single unmanned aerial vehicle (UAV) in two-dimensional or three-dimensional environments. Furthermore, when the GWO algorithm is employed for multi-Unmanned Combat Aerial Vehicle (UCAV) three-dimensional collaborative trajectory planning, issues such as reduced population diversity, susceptibility to local optima, slow convergence rates, low solution accuracy, and weak collaborative capability arise [11]. Consequently, multi-UCAV collaborative trajectory planning has gradually become a mainstream trend in development [12], and there is a scarcity of research outcomes regarding three-dimensional collaborative trajectory planning for multiple UCAVs based on the GWO algorithm.

This article addresses the three-dimensional trajectory planning issue for multiple UCAVs, considering various constraints such as fuel cost, altitude constraints, threat cost, time constraints, and spatial constraints. It establishes an objective function and employs the MP-GWO algorithm for trajectory planning of multiple UCAVs. Finally, simulations are conducted under the three-dimensional cooperative trajectory planning model for multiple UCAVs, with results indicating that the proposed algorithm demonstrates better cooperativeness compared to the GWO, PSO, and EA algorithms, significantly improving the solution accuracy and convergence speed.

2 Problem Description and Model Establishment

2.1 Description of the Cooperative Trajectory Planning Problem

Collaborative constraint refers to the fact that in order to ensure that the track meets the constraints of the single aircraft track, each UAV in the entire UCAV formation can successfully complete the task. Let be the set of Unmanned Combat Aerial Vehicles (UCAVs) assigned to execute the mission, $M = \{M_i, i = 1, 2, \cdots, N\}$ is the collection of enemy threats. For a single UCAV, the path from the starting point to the target point consists of a series of waypoints. By connecting all the waypoints between the starting point and the target point according to established rules, one obtains the flight path. Let the maximum range of the V_i be $L_{max}^{(i)}$, the maximum angle of rotation is $\theta_{max}^{(i)}$, and the speed range is $\left[v_{\min}^{(i)}, v_{\max}^{(i)}\right]$. Therefore, the track node of the V_i can be represented by Eq. 1:

$$S_i \overset{\Gamma(q)}{\rightarrow} P_{i-1} \cdot P_{i-(n-1)} \overset{\Gamma(q)}{\rightarrow} G_i \tag{1}$$

where S_i is the starting point of the task, G_i is the target point, $P_{i-k}(k = 1, 2, \cdots, n-1)$ is the track node, and $\Gamma(q)$ represents the constraint condition, that is, the q trajectory constraint parameter.

2.2 Establish a Fitness Function Model

The evaluation of the advantages and disadvantages of drone flight paths is composed of multiple indicators. Currently, the evaluation metrics for a single drone mainly consist of fuel consumption, flight altitude, and threat cost. Based on the description of path planning issues above, the fitness function for evaluating multiple UCAV flight paths is primarily built upon the metrics for a single drone while incorporating the temporal and spatial coordination of multiple machines, with a main focus on time and spatial costs. Therefore, the comprehensive model for the cooperative flight path evaluation function is as follows:

$$J = \sum_{i=1}^{m} (w_1 f_{Oi} + w_2 f_{Hi} + w_3 f_{Ti} + w_4 f_{Mi} + w_5 f_{Ci}) \tag{2}$$

Cost of Fuel Consumption:

$f_{Oi} = \frac{L_i}{L_{max}}$,

$$L_i = \sqrt{\left[(x_{k+1} - x_k)^2 + (y_{k+1} - y_k)^2 + (z_{k+1} - z_k)^2\right]} \tag{3}$$

Cost of Flight Altitude:

$$f_{Hi} = \begin{cases} z_k - H_{max}, & z_k > H_{max} \\ 0, & H_{min} \le z_k \le H_{max} \\ H_{min} - z_k, & z_k < H_{min} \end{cases} \tag{4}$$

Cost of Threatened Areas:

$$f_{Ti} = \begin{cases} 0, & d_r > d_t \\ 1, & d_r \le d_t \end{cases}$$

$$d_r = \sqrt{\left[(x_k - E_{xt})^2 + \left(y_k - E_{yt}\right)^2 + (z_k - E_{zt})^2\right]} \quad (5)$$

Cost of Time Coordination:

$$f_{Mi} = \begin{cases} 0 & , t_{min}^i \leqslant t_c \leqslant t_{max}^i \\ |t_i - t_c| , & others \end{cases} \quad (6)$$

Cost of Spatial Collaboration:

$$f_{Ci} = \begin{cases} 1 & , d_{i-j} < d_s \\ 0 & , d_{i-j} \geqslant d_s \end{cases} \quad (7)$$

Total Constraints of the Model:

$$s.t. \begin{cases} t_{min}^i \leqslant t_c \leqslant t_{max}^i \\ \omega_1 + \omega_2 + \omega_3 + \omega_4 + \omega_5 = 1 \\ L_{min} \le L_i \le L_{max} \\ H_{min} \le z_k \le H_{max} \\ d_r > d_t \\ 0.3Ma \le v_i \le 0.7Ma \\ \Phi_{max} \le 60^\circ \\ \theta_{max} \le 45^\circ \\ d_{i-j} \geqslant d_s \\ t_c \le 160 \\ j \le i - 1 \end{cases} \quad (8)$$

where m is the number of UCAVs in the objective function (2), and f_{Oi} 、 f_{Hi} 、 f_{Ti} 、 f_{Mi} and f_{Ci} represent the fuel, altitude, threat, time and collision cost of the i-th UCAV, respectively. ω_i is the weight of each cost; In Eq. (3), L_i is the track length of the i-th UCAV, x_k and y_k are the horizontal and longitudinal coordinates of the kth waypoint, respectively, and z_k is the altitude of the kth waypoint. In Eq. (4), H_{min} and H_{max} represent the maximum and minimum altitudes of UCAV flight, respectively. In Eq. (5), d_r is the distance between the UCAV and the threat region, and E_{xt}, E_{yt}, and E_{zt} are the coordinates of the center of the tth threat region. Equation (6) sets the command time t_c for the UCAV to arrive at its destination, solving for the actual arrival time t_i. By assessing the relationship between the two times, we can establish the time cost function. If $t_{min}^i \leqslant t_c \leqslant t_{max}^i$, the ith UCAV can complete the mission within the designated time; otherwise, it cannot, resulting in a lack of coordination in the formation. Equation (7) calculates the collision cost by first determining the current time of the ith UCAV at each waypoint and the distances to the previous *i*-1 UCAVs. If this distance is less than the minimum safe flying distance, a collision is recorded. This process is repeated for all waypoints along the flight path to total the number of collisions for the UCAV's trajectory. Here, d_{i-j} represents the minimum distance between the ith UCAV and the jth UCAV's waypoints during flight, and d_s denotes the minimum safe flying distance between UCAVs, where $j \leq i-1$ indicates that the ith UCAV is being compared to the preceding i-1 UCAVs. Constraint (8) ensures the rationality and feasibility of the final path.

3 Improvements to the Grey Wolf Optimization Algorithm

The Gray Wolf Algorithm (GWO) has been widely used in the field of planning, and the results obtained are relatively good. Although the standard gray wolf optimization algorithm has the advantages of fewer control parameters and avoids falling into local optimal to a certain extent, the gray wolf optimization algorithm also has certain defects in solving the three-dimensional trajectory planning problem. When the gray wolf optimization algorithm initializes the individuals randomly, the generated waypoints are disorderly, resulting in low individual fitness. Because the distance control parameter a is linearly decreasing, the development capacity is insufficient [13]. In order to better solve the problem of instability in the multi-machine cooperative track planning solution, the MP-GWO algorithm is improved on the basis of the original GWO algorithm and combined with the idea of multiple groups [14]. The algorithm flow diagram is shown in Fig. 1.

3.1 Minimum Value Clustering Strategy

The GWO algorithm mostly employs a single population model for searching, and all search processes are confined within one population. If the individuals tend to converge, the algorithm is prone to getting trapped in local optima. Therefore, in this article, in collaborative track planning under multi-index conditions, the initial track population of UCAV is first subjected to clustering analysis. Multiple indicators are considered separately, and the track population is evaluated accordingly. Then, good genes from the sub-populations are selected, and the remaining individuals are formed into a new reserved population to store the good individuals from each sub-population.

3.2 Hybrid Initialization Method

The GWO algorithm employs a random generation method to initialize the population [15]; However, this method demonstrates an uneven distribution throughout the entire space. As a result, the waypoints generated by the GWO algorithm appear chaotic and disorganized, making it difficult to maintain the diversity of the population, which can also lead to lower fitness levels among individuals. Therefore, a greedy approach has been incorporated into the initialization, namely: The selection interval for the individual waypoints uses the coordinates of the previous waypoint as the lower boundary (the first waypoint uses the starting point as the lower boundary) and the coordinates of the target point as the upper boundary, ensuring that the distribution of the initialized individual waypoints approximately aligns with the direction of the trajectory (the starting point pointing towards the target point). The random selection of waypoints follows a uniform distribution within the specified interval. The initialization method based on the greedy principle is:

$$x_{i+1} \sim U(x_i, e_x), y_{i+1} \sim U(y_i, e_y) \tag{9}$$

where x_i, x_{i+1} is the abscissa of the coordinates of two adjacent waypoints in an individual; y_i, y_{i+1} is the ordinate of two adjacent waypoint coordinates in the individual; e_x, e_y is the abscissa and ordinate of the target point coordinates; U is a uniform distribution function of waypoint coordinates.The initialization method based on greedy thought improves the average fitness of individuals in the population to a certain extent, but may also reduce the diversity of the population, so the mutation strategy is added, and a hybrid initialization method is formed through the control of the mutation threshold, that is, when the mutation parameter σ is less than the threshold value, the initialization method based on the idea of greed is used for the initialization of the individual. Instead, the instance uses the original random initialization method.

Fig. 1. Flowchart of the MP-GWO Algorithm

3.3 Improved Design of Convergence Factors

In the standard Grey Wolf Optimization (GWO) algorithm, the convergence factor a decreases linearly from 2 to 0, which is used to control the exploration and exploitation capabilities of the algorithm [16]. However, the linear decrease strategy may lead to issues in practical applications, such as converging too quickly and getting trapped in local optima, or converging too slowly and affecting the efficiency of the algorithm. To address this, this paper proposes *a* nonlinear periodic adjustment of the convergence factor a, introducing a cosine function, so that the convergence factor exhibits periodic fluctuations with iterations. The specific design is as follows:

$$a = a_{ini} \cdot \left(\frac{1+\cos(\pi \cdot \frac{t}{T})}{2} \right) \tag{10}$$

Among them, the initial value a_{ini} is 2, t is the current iteration count, and T is the maximum number of iterations. This convergence factor fluctuates through the cosine function, allowing the algorithm to periodically enhance its exploratory capability during the iterative process, thereby avoiding premature convergence.

3.4 Pseudocode of MP-GWO

The overall pseudocode process of the MP-GWO algorithm is as follows.

Algorithm 1: MP-GWO(Multi-Population Grey Wolf Optimization)

```
Inputs: The population size N and maximum number of iterations T
Outputs: The location of the alpha wolf (Xα) and its fitness value
Initialize the gray wolf population X_i(i = 1,2,···,N) using the hybrid initiali zation method(Eq. (9))
Evaluate the fitness of each wolf using Eq. (2)
Cluster the population into KK subpopulations using minimum value clustering strategy
Set X_α,X_β,X_δ  as the top three best solutions
while t≤T do
   for each subpopulation k=1 to K do
      Update convergence factor aa using Eq. (10)
         for each wolf i in subpopulation k do
            Update position based on α,β,δ leaders
            Apply mutation if σ< threshold
            Evaluate new fitness
         end for
   end for
   Update X_α,X_β,X_δ from all subpopulations
   Merge and reassign subpopulations periodically
end while
return X_α
```

4 Simulation Experiments and Result Analysis

4.1 Simulation Environment Settings

The experiment assumes that 3 UCAVs attack 3 ground targets, and the starting coordinates and 3 target coordinates of 3 UCAVs are shown in Table 1.

Table 1. UCAV coordinates

Drone	UCAV1	UCAV2	UCAV3
Starting point	(0,0)	(0,100)	(300,0)
Target point	(875,875)	(800,875)	(875,800)

According to the space kinematic constraints [17], the maximum total length of the UCAV L_{max} is 1.5 times of the straight-line distance between the starting point and the target point, the minimum track length is l_{min} =2 km, the minimum flight altitude is H_{min} =20 m, the range of flight speed is [0.3*Ma*, 0.7Ma], the maximum declination angle $\Phi = 60°$, and the maximum inclination angle $\theta = 45°$. The command arrival time of UCAV to the target destination t_c =160 s; The safety distance between each UCAV is d_s =25 km; The specific parameters of the threat are shown in the simulation diagram. According to the combat environment, the fuel cost weight ω_1, altitude cost ω_2, threat cost ω_3, time cost ω_4 and collision cost ω_5 of the track planning evaluation function are 0.05, 0.05, 0.1, 0.7 and 0.1, respectively.

Table 2. Parameter Settings

Population size	100
Maximum number of iterations	500
Number of subgroups	5
a_{ini}, a_{fin}	2, 0

This paper uses Matlab R2024a for simulation, running memory size of 16.0GB, processor 11th generation Intel Core i5-11400H.The algorithm parameter settings are shown in Table 2.

4.2 Analysis of the Comparative Experiments on Improving Convergence Factors

In order to validate the effectiveness of the improved convergence factor on the algorithmic search, this paper conducts comparative experiments between the proposed convergence factor and other types of convergence factors [18], which are listed as follows:

$$\begin{cases} a_1 = 2 - 2 \cdot \frac{t}{T} \\ a_2 = 2 \cdot \cos\left(\frac{\pi}{2} \cdot \frac{t}{T}\right) \\ a_3 = \left(a_{ini} - a_{fin}\right) \cdot \cos\left(\frac{t}{T}\right) \end{cases} \tag{11}$$

Among them, the final value a_{fin} is 0, as shown in the comparison experimental graph below:

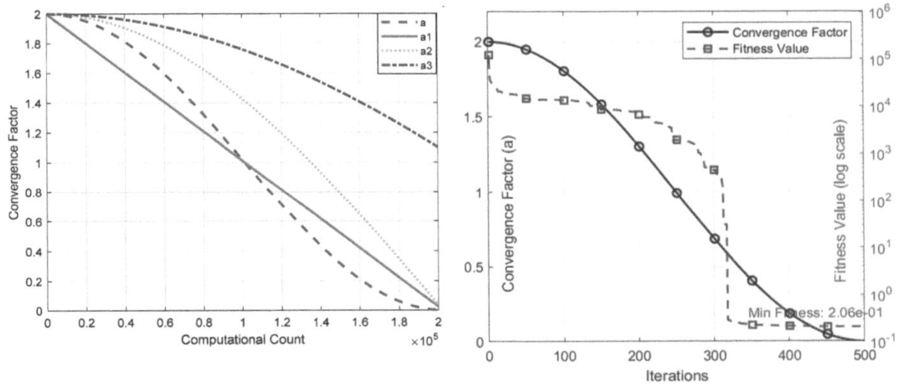

Fig. 2. Simulation results of the convergence factor.

The figure illustrates that the convergence factor of the GWO algorithm (a_1) decreases at a uniform rate, leading to an inadequate balance between global and local search during the search process. Both a_2 and a_3 maintain relatively high values in the early iterations of the algorithm; however, a_2 tends to miss the optimal solution due to a large step size in the later stages, which can lead to local optima [19], while a_3 ceases to search during later iterations. In contrast, the improved convergence factor a proposed in this paper demonstrates a gradual decrease in the early iterations, ensuring that the convergence factor maintains a relatively high value over a certain period. This promotes a larger A value, thus allowing the algorithm to retain strong global search capabilities in the initial phase. As the iteration progresses into the later stages, the convergence factor gradually decreases to a lower value, allowing the algorithm to perform local exploration and preventing the step size from becoming too large and missing the optimal solution. Therefore, the improved convergence factor a presented in this paper effectively balances the exploratory and exploitative capabilities of the algorithm.

4.3 Multiple UCAV Path Planning Simulation Experiments

The MP-GWO algorithm is used for track planning based on the above simulation environment, and the number of runs of the two algorithms is set to be 30 times. The simulation results are shown in Fig. 3 and Table 3.

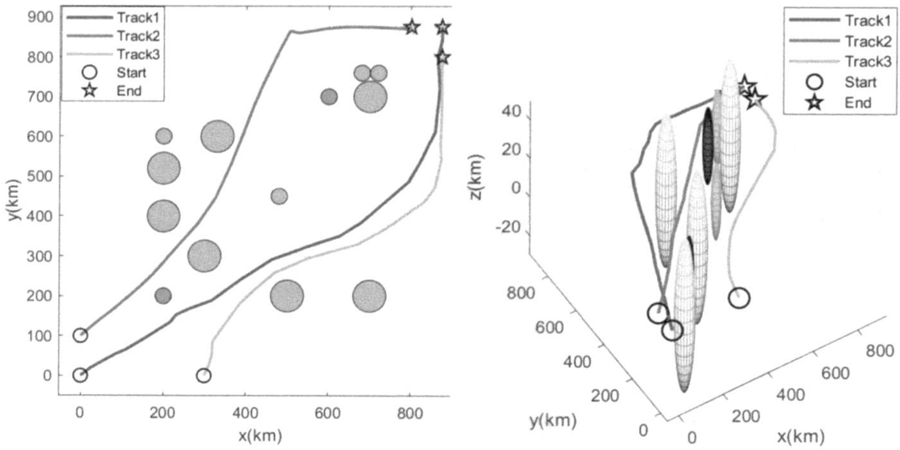

(a)2D (b)3D

Fig. 3. MP-GWO Track Planning

Figure 3 are the two-dimensional and three-dimensional plane search process diagrams of the three UCAV tracks based on the MP-GWO algorithm, respectively, in which the black area represents the meteorological threat area, and the blue area represents the missile, artillery, radar and other threat areas. As can be seen from Fig. 2, the three UCAVs can effectively avoid the threat area and reach the target destination.

Table 3 shows the time, velocity, distance and collision simulation results of each UCAV in the track planning, from which it can be seen that the error between the flight time and the command time of each planned track is within 1%, and there is no collision between each UCAV.

Table 3. UCAV simulation results

Simulation results	UCAV1	UCAV2	UCAV3
Flight time/s	9134.39	7246.80	4939.46
Flight range / km	1310.08	1351.72	1083.29
Flight speed/Ma	0.422	0.549	0.645
Number of collisions	0	0	0

In order to verify the performance of the improved MP-GWO algorithm, a comparative experiment was carried out. The original PSO algorithm, GWO algorithm and EA algorithm are used for three-dimensional trajectory planning based on the above environment [21, 22], and the optimization time consumption, the optimal function value and the number of iterations to achieve the optimal are compared and analyzed. The simulation results are shown in Table 4:

Table 4. Simulation results of different algorithms

Algorithm	Optimize Time	Convergence Value	Average flight distance
MP-GWO	92	0.19	1185.24
GWO	105	0.23	1196.97
PSO	155	0.35	1687.94
EA	104	0.21	3962.14

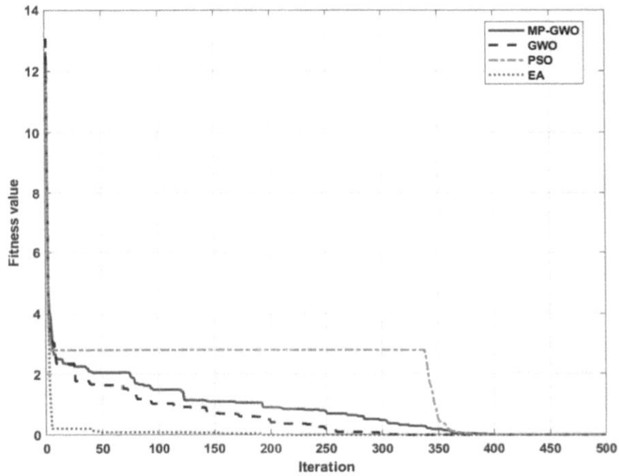

Fig. 4. Comparison of convergence curves of different algorithms

Figure 4 shows the convergence curves obtained by different algorithms, and as can be seen from Table 4 and Fig. 2, the MP-GWO algorithm has a faster convergence speed and better convergence accuracy than other algorithms, and can reach the optimal function value with an appropriate number of iterations.

4.4 Simulation Experiments in Complex Environments

In order to verify the stability of the MP-GWO algorithm in the track planning problem, a more realistic actual situation is simulated by adding new threats [23], and the parameters of the new threats are directly marked on the simulation map, and the simulation results are shown in Figs. 5 and 6.

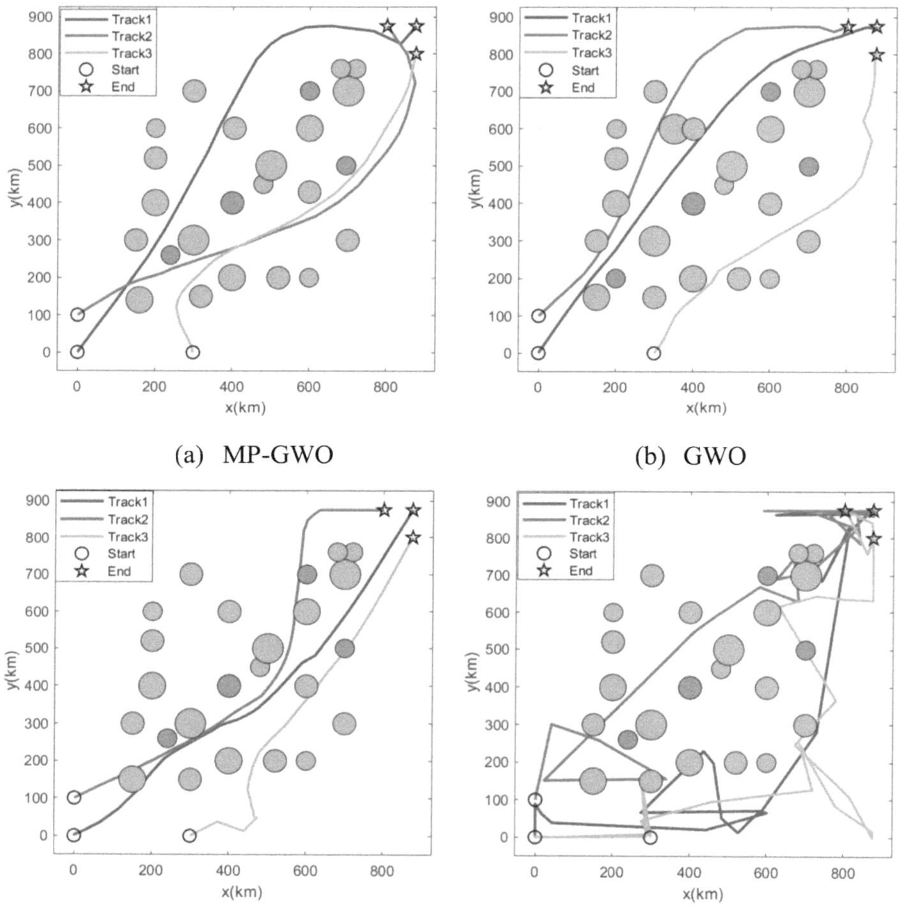

(a) MP-GWO (b) GWO

(c) PSO (d) EA

Fig. 5. 2D planar path planning

From the 2D and 3D path planning diagrams in Figs. 5 and 6, it can be observed that the trajectory of the MP-GWO algorithm is the safest, and the planned path is smoother, with significantly fewer turning points compared to the other three algorithms, UCAV can still search for low-cost tracks and effectively avoid various threat areas.

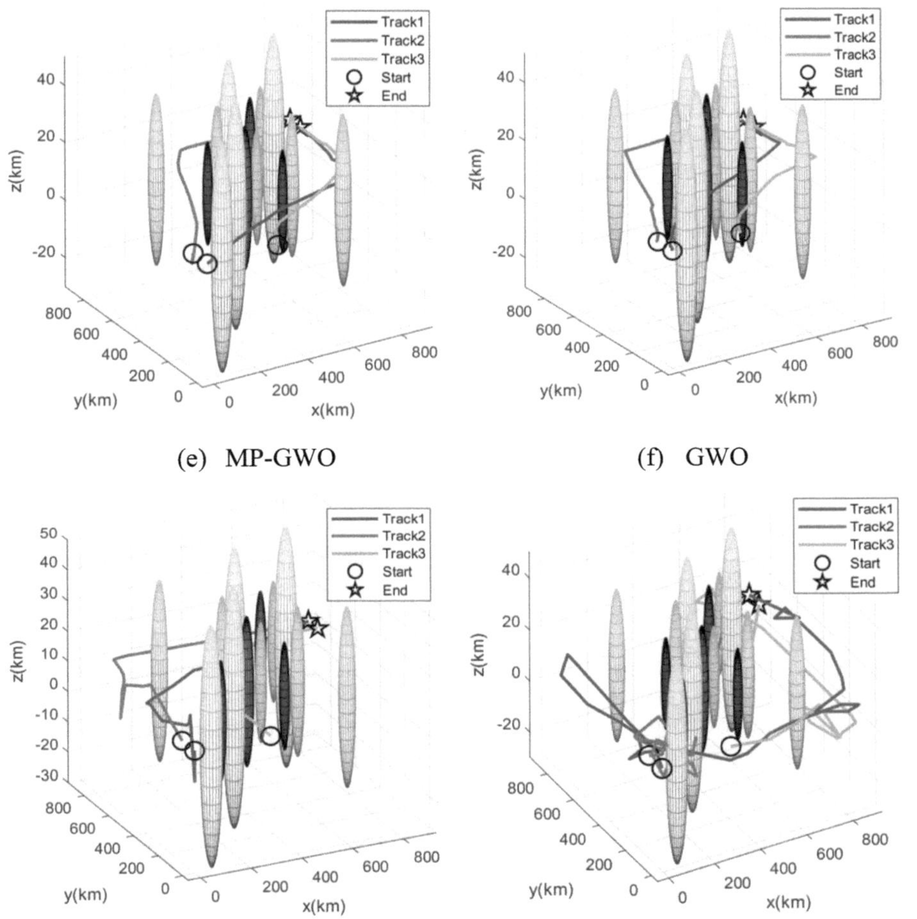

(e) MP-GWO (f) GWO

(g) PSO (h) EA

Fig. 6. 3D planar path planning

5 Conclusion

This paper addresses the collaborative trajectory planning of multiple unmanned combat aerial vehicles (UCAVs) in complex environments, proposing a method based on the Multi-swarm Gray Wolf Optimization (MP-GWO) algorithm for three-dimensional trajectory planning. By introducing multi-swarm coordination mechanisms, a hybrid initialization strategy, and improvements to the nonlinear convergence factor, the algorithm's global search capability and convergence speed have been significantly enhanced. Simulation experiments show that the MP-GWO algorithm outperforms traditional PSO, GWO, and EA algorithms in optimization time, convergence accuracy, and trajectory smoothness, effectively planning collaborative trajectories that meet spatiotemporal constraints while ensuring good obstacle avoidance performance. In complex dynamic threat environments, the algorithm continues to demonstrate strong stability and adaptability, validating its practical application value. Future work can further optimize the algorithm's real-time performance and integrate other intelligent optimization methods to address more complex scenarios.

Acknowledgments. This study was funded by the Natural Science Foundation of Shanxi Province (grant number 202203021211198).

Disclosure of Interests The authors have no competing interests to declare that are relevant to the content of this article.

References

1. Wang, G., Guo, L., Duan, H., Liu, L., Wang, H.: A modified firefly algorithm for UCAV path planning. Int. J. Hybrid Inf. Technol. **5**(3), 123–144 (2012)
2. Fan, Z., Wang, Z., Ge, J.: Multi-UAV real-time strike path planning based on A* algorithm. Tact. Missile Technol. **5**, 94–101+112 (2021)
3. Xu, W., Ao, H., Liu, Y.: Local trajectory re-planning for multiple unmanned aerial vehicles based on pigeon flock optimization algorithm. Tact. Missile Technol. **1**, 46–52 (2022)
4. Zhou, L.: Multi-UAV Cooperative Trajectory Planning Based on Pigeon Swarm Optimization Algorithm. Nanjing University of Aeronautics and Astronautics (2021)
5. Wang, J., Qi, S., Zhong, H., et al.: Uniform formation gathering control based on Dubins curves. Comput. Simul. **38**(7), 40–45 (2021)
6. Han, T., Tang, A., Zhou, H., et al.: A multi-UAV cooperative trajectory planning method based on the LASSA algorithm. J. Syst. Eng. Electron. **44**(1), 233–241 (2022)
7. Yin, L., Xiang, F.: Three-dimensional path planning based on improved grey wolf optimization algorithm. Telev. Technol. **46**(1), 35–40 (2022)
8. Liu, N.: Path Planning for Mobile Robots Based on an Improved Grey Wolf Optimization Algorithm. Xinjiang University (2019)
9. Xu, L., Zhao, W.: Path planning of unmanned aerial vehicles based on a novel grey wolf optimization algorithm. Electron. Meas. Technol. **45**(5), 55–61 (2022)
10. Yang, L., Guo, J., Liu, Y.: Three-dimensional UAV co-operative path planning based on the MP-CGWO algorithm. Int. J. Innov. Comput. Inf. Control. **16**(3), 991–1006 (2020)
11. Dai, L., Cao, Q., Xia, Y., et al.: Distributed MPC for formation of multi-agent systems with collision avoidance and obstacle avoidance. J. Franklin Inst. **354**(4), 2068–2085 (2017)
12. Wenchao, Dong, Xie, W., et al.: A three-dimensional cooperative trajectory planning method for multiple unmanned aerial vehicles based on CEA-GA. J. Beihang Univ., 1-19 (2022).
13. Jia, D., Vagners, J.: Parallel evolutionary algorithms for UAV path planning. AIAA 1st Intelligent Systems Technical Conference, 6230 (2004)
14. Panda, M., Das, B., Pati, B.: Global path planning for multiple AUVs using GWO. Arch. Control Sci. **30**(1), 77–100 (2020)
15. Dewangan, R.K., Shukla, A., Godfrey, W.W.: Three dimensional path planning using Grey wolf optimizer for UAVs. Appl. Intell. **49**, 2201–2217 (2019)
16. Cheng, X., Li, J., Zheng, C., et al.: An improved PSO-GWO algorithm with chaos and adaptive inertial weight for robot path planning. Front. Neurorobot. **15**, 770361 (2021)

17. Gul, F., Rahiman, W., Alhady, S.S.N., et al.: Meta-heuristic approach for solving multi-objective path planning for autonomous guided robot using PSO-GWO optimization algorithm with evolutionary programming. J. Ambient. Intell. Humaniz. Comput. **12**(7), 7873–7890 (2021)
18. Yu, X., Jiang, N., Wang, X., et al.: A hybrid algorithm based on grey wolf optimizer and differential evolution for UAV path planning. Expert Syst. Appl. **215**, 119327 (2023)
19. Luo, Y., Qin, Q., Hu, Z., et al.: Path planning for unmanned delivery robots based on EWB-GWO algorithm. Sensors. **23**(4), 1867 (2023)
20. Kumar, R., Singh, L., Tiwari, R.: Path planning for the autonomous robots using modified grey wolf optimization approach. J. Intell. Fuzzy Syst. **40**(5), 9453–9470 (2021)
21. Zhang, W., Zhang, S., Wu, F., et al.: Path planning of UAV based on improved adaptive grey wolf optimization algorithm. IEEE Access. **9**, 89400–89411 (2021)
22. Kiani, F., Seyyedabbasi, A., Aliyev, R., et al.: 3D path planning method for multi-UAVs inspired by grey wolf algorithms. J. Internet Technol. **22**(4), 743–755 (2021)
23. Ding, Q., Xu, X.: Improved GWO Algorithm for UAV Path Planning on Crop Pest Monitoring. Int. J. Interact. Multimed. Artif. Intell. **7**, 30–39 (2022)

MRLLM: Multimodal Knowledge and Feedback Based Refinement Assist for Robotic Arm Operations Using Large Language Model Reasoning

Zhibin Yang[1] and Zhi Zheng[1,2](✉)

[1] College of Computer and Cyber Security, Fujian Normal University, Fuzhou 350117, China
[2] College of Control Science and Engineering, Zhejiang University, Hangzhou 310027, China
zhengz@fjnu.edu.cn

Abstract. The Large Language Model (LLM) has been proven to be capable of performing advanced planning for long-term robot tasks, but existing methods still have shortcomings in solving the illusion problem that is answered by large models. Therefore, how to provide task related knowledge to large models and help them make correct decisions has become a hot research topic. Based on the above issues, we propose a large model planning framework called MRLLM (Mutilmodal and Reflection LLM) based on multimodal knowledge and reflective self refinement. This framework provides knowledge sources and enhances reasoning for LLM through multimodal knowledge regularization and error feedback self refinement iteration, and constrains LLM output through template prompts, which to some extent alleviates the illusion problem of LLM. The experiment shows that our method has better performance compared to traditional methods in the classical planning domain of Blockworld and Housework.

Keywords: Multimodal · LLM · Reflection · Long Sequence Task Planning

1 Introduction

Task Planning (TP) is an action plan generated for tasks in complex environments. Traditional methods convert problems into symbolic languages (such as PDDL [1]: Planning Domain Definition Language) and generate high-level symbolic plans to determine the abstract action sequence required to complete from

The work is supported by the National Key Research and Development Program of China (No.2022YFB4201603), the "Pioneer" and "Leading Goose" R&D Program of Zhejiang (No.2024SJCZX0005), and the Natural Science Foundation of Fujian Province (No.2024H0012).

H. Ma et al. (Eds.): IWACIII 2025, CCIS 2780, pp. 484–498, 2026.
https://doi.org/10.1007/978-981-95-6730-0_33

the initial state to the target state. Recently, natural language models represented by chatGPT [14] have shown excellent reasoning and dialogue generation abilities. Meanwhile, the use of TAMP's PDDL to describe task specifications has been proven to be highly successful.

The intuitiveness of natural language for task description paradoxically highlights a key shortcoming of PDDL—the impracticality of manual, real-time encoding of world dynamics, which generates complex syntax and hinders usability [3]. This very challenge is being addressed by burgeoning research in pre-trained LLMs (e.g., GPT-4 [14]), whose proficiency in natural language tasks is rapidly advancing their application in task planning [20] and TAMP [4,16].

In the domain of task planning (TP), approaches leveraging large language models (LLMs) frequently employ pre-trained models like GPT-4 to interpret natural language task descriptions [22]. These models are tasked with generating symbolic plans and performing fault inference, where the quality of contextual prompts significantly influences performance [6]. Inadequate prompting, however, can substantially impair planning and reasoning capabilities. To address this, recent research has explored various prompt engineering strategies. For instance, Wang et al. [22] introduced template-based prompts that use predefined structures with placeholders to express spatial and geometric relations. Similarly, Ding et al. [5] employed separate templates for scene description and environment state representation. Another line of work, such as Li et al. [9], enables interactive task planning by combining user inputs with stored task guidelines to dynamically construct prompts. Despite these advances, many prompt generation techniques still depend on rigid templates, which can produce repetitive and contextually inflexible prompts. Such formulations often fail to adapt to dynamic environmental conditions, limiting their applicability in novel or unforeseen scenarios. Moreover, the simplification inherent in template-based approaches may overlook nuanced task contexts, potentially yielding incomplete or ambiguous guidance that misdirects the planning process. Given the critical role of well-designed prompts in enhancing LLM efficacy [8], there is a growing need for more adaptive and nuanced prompt engineering methods to enable accurate and reliable task planning.

In some studies, it has been found that providing diversified information in prompt templates can improve the reasoning ability of LLM, such as scene images [12], network knowledge [15], or ontology knowledge [4]. Through these diverse information and problem constraints, LLM can improve its accuracy in solving new problems. Using network knowledge combined with expert data to solve QA problems in [16].

To overcome the limitations discussed above, this paper introduces a prompt optimization approach that integrates multiple knowledge sources through knowledge based reasoning, enhancing and refining user-provided prompts within specific task contexts. Building upon the existing ISR-LLM framework [25], the proposed method incorporates a novel prompt tuning strategy. While prior work, including that of Liu et al. [10], has explored the incorporation of multimodal knowledge into LLM-driven task planning, such approaches often face challenges

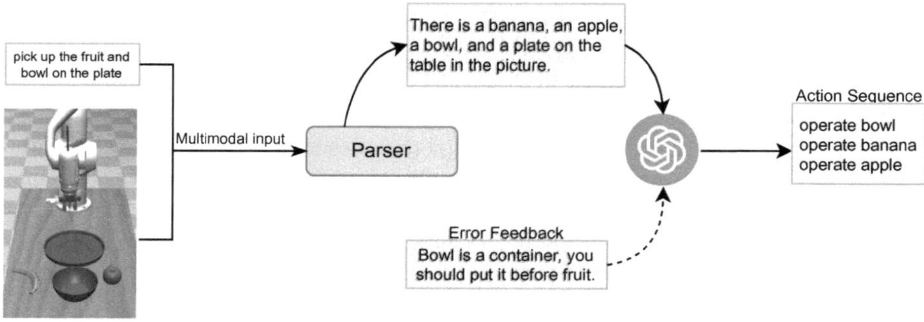

Fig. 1. The Framework of MRLLM. Using the method described in this article to optimize LLM prompts, the parser analyzes multimodal data (such as text and images) and converts it into prompt words for input to LLM. Error feedback is added to the self refinement process, and LLM self reflects and adjusts its output strategy to generate correct and executable action sequences.

related to multimodal data alignment and computational expense—leading to limited practical gains, as noted by Xu et al. [24]. In contrast, our method employs knowledge-guided reasoning for prompt refinement, with the dual objective of resolving real-world constraints and optimizing action sequences. By formalizing acquired multimodal knowledge rules as structured prompts, our approach ensures that the resulting plans are both executable and semantically consistent.

To illustrate, we examine a scenario from Din et al. [4] (Fig. 1), in which fruits and bowls are arranged on a plate according to scene perception and user directives. A typical symbolic planner might output an action sequence such as: "pick up bananas and apples and put them on the plate, then pick up bowls and place them on the plate." While syntactically valid, this sequence is semantically inconsistent—it could lead to bowls being positioned on top of the fruits. Our prompt tuning approach addresses this issue by incorporating a validator and domain-specific knowledge. When an error is detected, the system retrieves relevant knowledge (e.g., "bowls are containers and should be handled before apples and bananas" [4]) to refine the prompt. This enriched input guides the LLM to produce plans that are both semantically sound and logically ordered. The main contribution of this work is MRLLM, a multimodal knowledge-aware prompt optimization framework for task planning, designed to help LLMs generate rule-compliant and contextually appropriate plans. Building on this framework, the key contributions include:

- Automated prompt tuning with task-context reasoning: The method extracts commonsense and procedural knowledge from environmental inputs and problem statements to guide the LLM toward generating plans that are semantically coherent and logically sequenced.

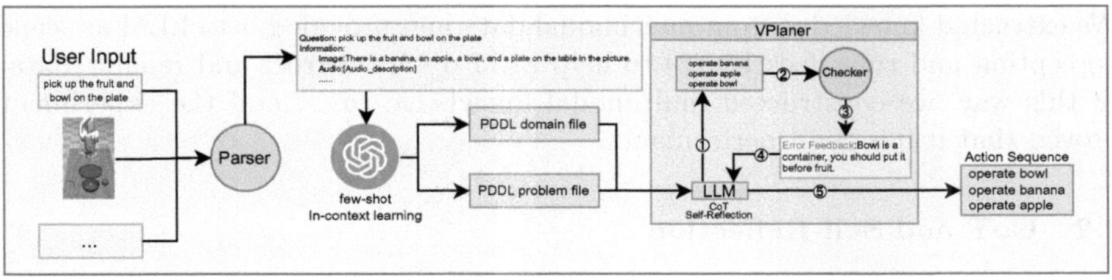

Fig. 2. The MRLLM framework. Converts multimodal data into knowledge rules, which are then translated into PDDL by LLM. Reflection planning is performed using the reflection planner VPlanner, and error feedback is provided after verification by the validator until the correct solution is evaluated (as shown in the Step ① ~⑤). The validator will collect relevant information through information retrieval for result verification.

- Knowledge-enhanced environment state representation: We leverage multimodal knowledge to construct rich, context-aware environment descriptions, strengthening the prompt tuning process for more reliable task planning.
- Integration of an iterative self-refinement mechanism: By incorporating chain-of-thought reasoning with error feedback, the approach progressively improves the robustness and accuracy of LLM-based planning. Experimental validation across two distinct planning problems demonstrates the effectiveness of the proposed method.

In Sect. 2, we present related work to position our research within the existing literature. Section 3 details our proposed methodology, including key innovations and implementation. To validate the effectiveness of our approach, Sect. 4 provides comprehensive experimental results with comparative analysis. Finally, Sect. 5 concludes the paper by summarizing contributions and discussing future directions.

2 Related Work

2.1 Multimodal Knowledge Construction

Multimodal knowledge refers to the combination of different types of data, such as text, images, and audio, to obtain more comprehensive and accurate knowledge. The traditional LLM uses natural language as input. Although it can handle QA problems with good results, in the face of tasks requiring environmental awareness (such as robot task planning, real-time question and answer), the scheme given by LLM is usually difficult to achieve the goal. We found that by combining natural language with other types of data (such as images), we can provide more accurate hints for LLM. Compared with previous work, we found that directly using multimodal data as input to LLM can not make the model fully obtain the knowledge implied in these data, which may lead to the omission or loss of some key information in the subsequent problem-solving reasoning.

We extracted knowledge from multimodal data and provided it to LLM as scene perception and rules (see Fig. 2) to help LLM obtain correct and reliable data. In this way, we constructed multimodal knowledge rules, and the experiment proved that it has good performance.

2.2 CoT and Self-Reflection

Chain of thought (CoT) is a hint strategy proposed by Jason Wei (2022) [23] to improve the performance of LLM in complex reasoning tasks, such as arithmetic reasoning, common sense reasoning and symbolic reasoning. CoT does not simply use input and output pairs to construct prompts as in context learning, but combines intermediate reasoning steps, which can introduce the final output into the prompts. In the prompt words, it is common to require LLM to use step-by-step generation for reasoning. This helps LLM to think explicitly when reasoning, and improves the feasibility of generating planning schemes.

Reflection refers to the ability of a model to evaluate and correct its output in the process of generating text. Reflection is not only a simple process, but also involves the understanding, analysis and self-improvement of the model on its own decision-making. By introducing the reflection mechanism, the model can better adapt to complex tasks, so as to improve the quality and consistency of its generated content. In recent years, researchers have made outstanding achievements in reflection technology. In 2023, Shinn [19] proposed the"reflective generation" method, which emphasized how to improve the quality of generated content through self-examination. Their research shows that the model using reflective mechanism performs better in Generative tasks and can generate more coherent and consistent sentences. Chen et al. (2023) [2] proposed a feedback generation framework based on self-assessment. In this framework, the model will continuously analyze its own output during the generation process, and improve the quality of the next generation by adjusting parameters, which makes the model more resilient when dealing with complex naturallanguageprocessing tasks. Renze(2024) [17] proposed a framework of "reflection generation" by using the reflection of the dialogue generation model. The model adjusts the generation strategy through the reflection of the historical dialogue, so as to better understand the user intention.

In this paper, we propose to combine the iterative self reflection [11] and thinking chain into the task planning of LLM, guide LLM to reflect on reasoning through specific prompts in Sect. 2.3, generate solutions iteratively and evaluate them. The task planning scheme generated in this way shows better stability and success rate in the experiment.

2.3 Error Feedback

Error feedback refers to responding to the error information of the superior module when the current solution is executed incorrectly, and helping the control module analyze the error and re plan according to the error information. Mei [13] pointed out in ReplanVLM that repeated query of LLM can realize the verification of solutions, but in some scenarios, LLM's perception of the environment and understanding of problems are limited, which may lead to misjudgment in the verification process. LLM-SAP [21] proposed to realize error feedback through mutual guidance of LLM, and to correct plan errors through mutual criticism of multi-agent. In the above methods, it is emphasized that LLM should reflect and correct errors by itself. However, without external guidance, LLM is likely to misjudge. Therefore, we propose to provide external knowledge to help LLM check to solve the misjudgment caused by insufficient knowledge.

3 Methods

In order to ensure the reliability of the LLM generated solution, we design a modulator to convert multimodal data into knowledge rules, and combine with the chain of thought (CoT) and reflection mechanism, design a self refinement iteration method, and introduce an error feedback mechanism in the verification process to check the errors through LLM's self verification, and use it as a correction rule to guide LLM to self refine the existing solution.

3.1 Multimodal Knowledge Rule Construction

In this article, the problem is defined as

$$P = (s_{init}, K, S, T, A, G), \tag{1}$$

where:

- $s_{\text{init}} \in S$ represents the initial state.
- k stands for multimodal knowledge rules.
- S represents all possible states.
- $T : S_n \times (A|K) \rightarrow S_{n+1}$ represents the state transition function.
- A means all possible actions.

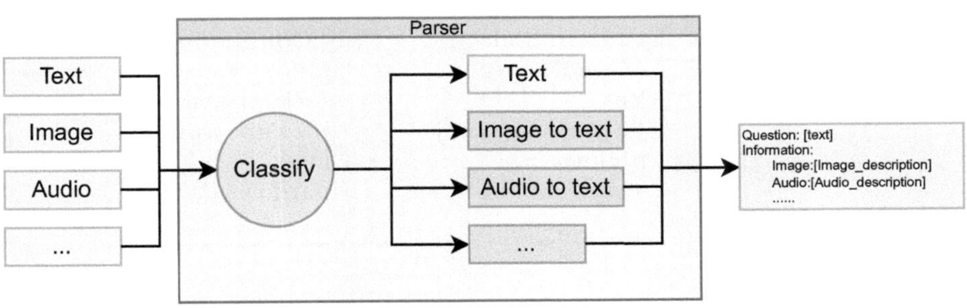

Fig. 3. Design of parser. We collect mutilmodal data for environmental perception, including text, environmental image and others. Specially, in the homework, audio can be important information, we collect it and in our experience.

– $G \in S$ indicates the target status.

In order to realize the input of different types of data as a large model, we designed a parser (Fig. 3). According to the data type, the parser will classify the data and direct it to different parsing modules for semantic analysis and rule writing. Through this parsing method, extensible and plug and play data mediation [15] can be realized to convert multimodal data into rules in a unified format and provide them to LLM for problem analysis.

3.2 Self-Refinement Iteration Combined with CoT

In order to make the solution generated by LLM reliable, we introduce CoT into the prompt template and realize the extraction of the solution by outputting the normalized constraint [7]. The prompt template we used is shown in the Table 1.

Table 1. Prompt structure and Few-shot for VPlaner in two scenarios.

System Prompt	Few Shot in Blockworld	Few Shot in Housework
You are a confident assistant that ... Output format: """ {Analysis} If solution exists, evaluate effective-ness. {Conclusion} "Yes" or "No" {Final Answer} Proposed solution """	Question: PDDL domain: ... action pickup ... action putdown ... action stack ... action unstack ... PDDL problem: init: ... goal: ... LLM Response: {Analysis} here is no solution, I will analysis the problem ... {Conclusion} Yes {Final Answer} pickup b2 stack b2 b1 ...	Question: PDDL domain: ... action pickup... action move... action putdown... PDDL problem: init: ... goal: ... LLM Response: {Analysis} here is no solution, I will analysis the problem ... {Conclusion} Yes {Final Answer} pickup apple move putdown ...

On the verification of the solution, we designed a self refining iterative planner combined with CoT, and the Algorithm 1 is a specific description of the method.

Algorithm 1. Refinement Planner

Input: q ▷ q: user question
Initialize:
$pddl \leftarrow$ TRANSLATOR(q), $feedback \leftarrow \varnothing$ ▷ $pddl$: nature language to PDDL
LLM.Init()
$action, res \leftarrow$ LLM.plan($pddl$) ▷ $action$: action sequence
▷ res:LLM refienment answer.
for $i \leftarrow 1$ **to** MaxRefine **do**
  $feedback, res \leftarrow$ LLM.check($action$)
  if res = "Yes" **then** ▷ LLM refine successfully
    Break
  end if
  $action \leftarrow$ LLM.plan($action, feedback$)
end for
OutPut: $action$

3.3 Implement a Verifier with LLM Self-Evaluation

For the verifier described in Sect. 2.2, we adopt the following design ideas:

LLM Self Certification. After LLM analyzes the problem and gives a solution, the verifier will test the current solution to verify its feasibility. If the verifier detects a condition conflict or $g \neq G$ (g represents the target status of the current plan) during the verification process, it will give error feedback, analyze the possible errors and give suggestions for modification.

Information Integration. If LLM has a deviation in understanding the current problem, it will give wrong verification information and irrelevant error feedback. To solve this problem, we allow LLM to use the tool [18] to query online rules, and integrate relevant knowledge to verify and analyze the problem and give error feedback.

Based on the above methods, we designed a validator (Fig. 2) for implementing LLM evaluation. In order to provide a certain evaluation ability during the initialization of the validator, we provide few samples for context learning and constrained the LLM validation output. Based on this processing method, we have simplified the extraction of LLM generation planning and can demonstrate that LLM can acquire certain skills about prompt samples under appropriate prompt learning.

4 Experimental Results

To prove the reliability and feasibility of our method, we tested it on the planning problem blockworld and the housework problem [4]. See Sect. 4.1 for the description of the problem. Blockworld simulates the scene in the simulator MUJOCO and obtains the perception information, which is planned in GPT-4.

4.1 Problem Description

BlockWorld. Blockworld is a classical planning problem in the field of artificial intelligence, which is used to study symbolic logic, state space search and automatic planning algorithm. The reasoning problem of stacking n wooden blocks is considered. In the problem planning, n randomly placed wooden blocks are stacked according to the target sequence. The blockworld rules are shown in Table 2.

Table 2. Scenario description of BlockWorld domain

Element	BlockWorld
Initial State	– n blocks , placed randomly on the table. – No stacking or Random Stacking.
Goal State	– Stacking Sequence (From Bottom to Top) – as shown in Figure 4: 1 → 2 → 4 → 3
Actions	1. Pickup(a): pick up block a from hand. 2. Putdown(a): put down block a from hand. 3. Unstack(a,b): unstack a from b. 4. Stack(a,b): stack a on b.
Constraint	– Only 1 block at a time – The top of the target block must be empty when stacking. – Stack in strict order

According to the planning requirements, we selecte three tasks for the experiment. The task description is as follows Fig. 4, in which:

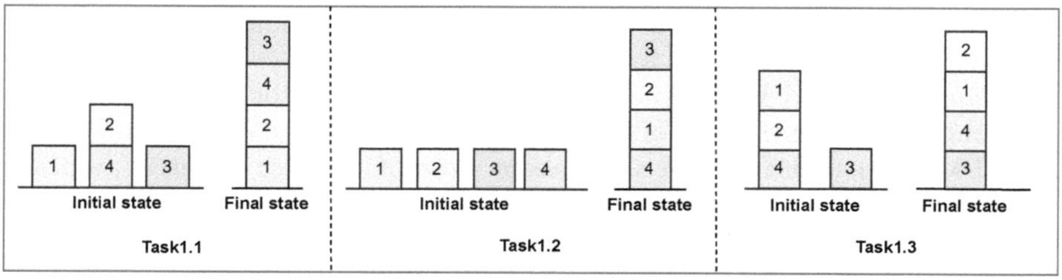

Fig. 4. Blockworld specific tasks used in this article.

Housework. The Housework problem is the impact of Din [4] on the planning results under different problem descriptions of LLM (Fig. 1). In the original problem, the author discussed that when a simple task description is given without a description of the properties of the operation object, the correct object state but unreasonable stacking may occur. In this paper, we select some of them for testing (Table 3), and we set the target state "fruit in the bowl and bowl on the plate". Each experiment is repeated for 30 times. Here is a correct simulation in Fig. 5. In our experiment, We find that when ontology priority or operation constraints are not given during description, the following error conditions will occur: pick up all fruit in the plate and put bowl on it.

In the scene, in order to confirm the sequence of planning, we simplified the operation. We define the action: pickup, move and putdown as complete actions, they can perform object grabbing - moving to a specified position and put down it.

Table 3. Housework Problem in our experiments.

Task	Problem	Goal State
Task2.1	put banana, apple and bowl on plate.	bowl on the plate and fruit in the bowl.
Task2.2	put apple, banana and bowl on plate.	
Task2.3	clear the table, put all things on plate.	

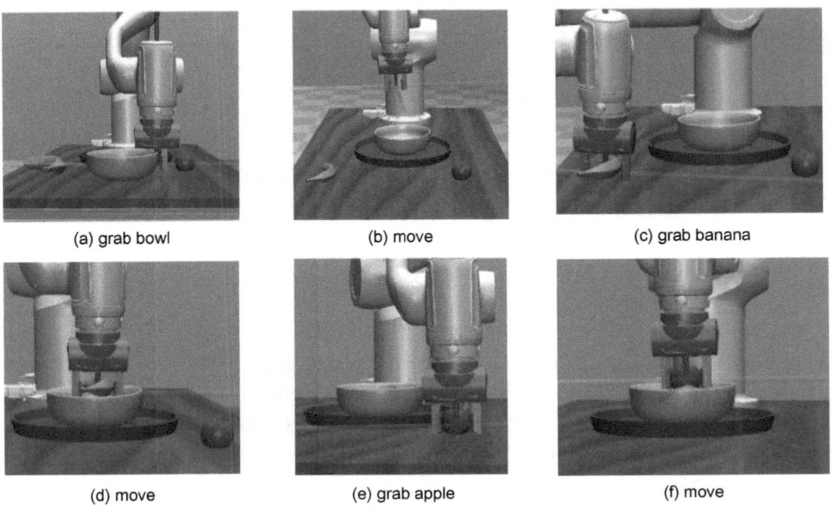

Fig. 5. Visualization Task 2.3. We perform the Task 2.3: clear the table, put all things on plate. The task 2.1, task 2.2 and task 2.3 are the same task, only different in task description.

4.2 Evaluation Parameters

According to the content of the experiment, we developed the following evaluation indicators:

- Data Resolution rate (DR): This indicator evaluates the success rate of converting multimodal data into rules. We calculate this indicator based on the number of successful object state parsing in the visual image, OS, such as the state of apples, bananas and plates in the figure. We mark the object state in the scene for evaluation. If the object state obtained by the parser is consistent with the facts, the statistics are as follows:

$$DR = \frac{correct\ object\ states\ number\ by\ parser\ analyzed}{all\ object\ states\ number}. \tag{2}$$

- Plan Success rate (PS): This indicator represents the feasibility of the action sequence planned by LLM. We count the reasonable actions in the action sequence planned each time. Here we design an external verifier for evaluation, and calculate the feasibility of planning through action state transition. The calculation formula is as follows:

$$PS = \frac{success\ actions}{total\ actions}. \tag{3}$$

- Accuracy (AC): This indicator evaluates the success rate of planning implementation. Unlike PS, we have simulated the action sequence, which may have some potential conditions not taken into account, which can more truly reflect the feasibility of planning. The calculation formula is as follows:

$$AC = \frac{success\ executable\ actions}{total\ actions}. \tag{4}$$

In two different problems, we test and get the following Table 5 and Fig. 6 for blockworld and Table 6 and Fig. 7 for housework.

From Table 5 and Fig. 6, it can be seen that compared to the baseline, our method has better performance in long-term task planning. Under the same task, the PS and AC of planning have increased by up to 13% and 31%, respectively. Moreover, under the premise of successful planning, our method can generate solutions with shorter planning paths. From Table 6 and Fig. 7, it can be observed that in terms of factual cognition, our method has a more ambiguous task description, such as Task 2.3, with higher PS and AC than the baseline, increasing by 24% and 18% respectively. In two scenarios and six tasks, our method exceeded the performance of the extreme method on four tasks. Although our task execution success rates were low on tasks 2.1 and 2.2, using LLM and multimodal data analysis methods does not require building an ontology knowledge base (Table 4), greatly saving manual operations. In comparison with ASL, our method can achieve planning lengths similar to the extreme method, further proving that our method has better generalization than the baseline.

Table 4. The difference between our method and ontology-driven. Error correction is realized through self checking of LLM, without the need to build ontology knowledge rules in advance.

Ontology-driven [4]	MRLLM
Rule1 ⊑ ActionPriority ⊓∃ hasAction.Put ⊓∃ hasObjectType.Crockery ⊓∃ hasPriority.1 Rule2 ⊑ ActionPriority ⊓∃ hasAction.Put ⊓∃ hasObjectType.FoodItem ⊓∃ hasPriority.2	[LLM PLAN] operate banana operate apple operate bowl [Checker Feedback] Bowl is a container, you should put it before fruit.

In order to test that our method can generate better schemes, we designed the evaluation index ASL (average sequence length). The calculation formula is as follows:

$$ASL = \frac{\sum_{i=1}^{n} L_i}{n}, \tag{5}$$

where:

- n represents the number of repeated tests for a single task. For example, n=30 represents 30 times of task execution.
- L_i represents the length of the action sequence generated by the ith task. The actions in the action sequence are motion primitives. In our method, we define the PDDL actions as motion primitives (Table 1).

Table 5. Comparative analysis of MRLLM performance in blockworld. The arrow represents the optimization direction of this indicator. ↑ repersents that the higher the indicator, the better. ↓ repersents that the lower the indicator, the better. The bolded part represents the best data.

Method	Task	DR↑	PS ↑	AC ↑	ASL ↓
ISR-LLM [25]	task1.1	/	48%	45%	6.8
	task1.2	/	51%	48%	6.5
	task1.3	/	45%	39%	12.2
Ontology-driven [4]	task1.1	/	15%	21%	12
	task1.2	/	21%	24%	15.4
	task1.3	/	18%	12%	13.3
MRLLM(Ours)	task1.1	87%	**51%**	**51%**	**6.4**
	task1.2	90%	**62%**	**60%**	**6.2**
	task1.3	87%	**68%**	**70%**	**11**

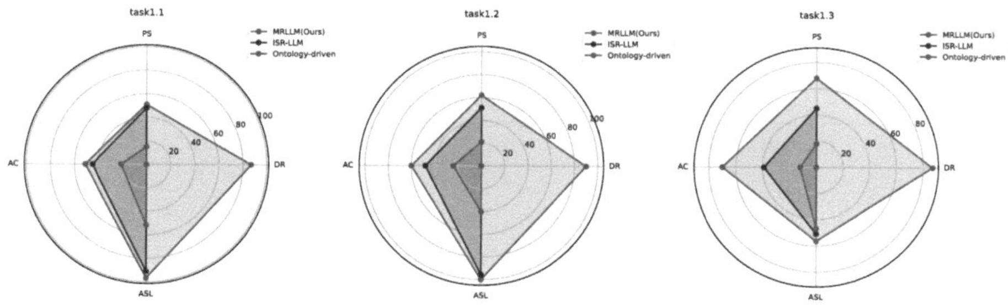

Fig. 6. Radar chart of experimental results for the blockworld. In order to visually demonstrate ASL, we have inverted it so that the larger the coverage area, the better the ASL.

The execution success rate ASL of action sequences generated by multiple tasks is tested on blockworld and housework. We conduct 20 correct experiments for each task. The results showed that Table 6 compared with the base-

Table 6. Comparative analysis of MRLLM performance in housework. The meaning of arrow is the same as that in Table 5. The bolded part represents the best data.

Method	Task	DR↑	PS ↑	AC ↑	ASL ↓
ISR-LLM [25]	task2.1	/	39%	42%	4.3
	task2.2	/	45%	42%	4.1
	task2.3	/	48%	45%	4.7
Ontology-driven [4]	task2.1	/	**100%**	**92%**	3.5
	task2.2	/	**100%**	**95%**	3.7
	task2.3	/	36%	30%	6.5
MRLLM(Ours)	task2.1	100%	90%	87%	**3.4**
	task2.2	100%	84%	81%	**3.7**
	task2.3	100%	**60%**	**48%**	**3.4**

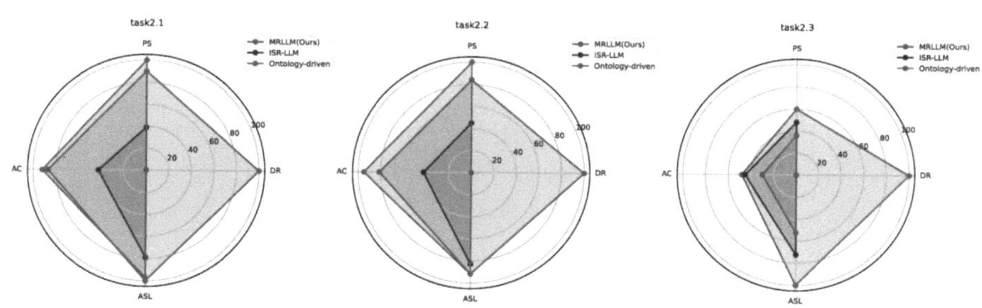

Fig. 7. Radar chart of experimental results for the housework. In order to visually demonstrate ASL, we have inverted it so that the larger the coverage area, the better the ASL.

line method in reasoning, our method can generate correct action sequences with fewer average steps. In factual cognition, MRLLM can achieve ASL as good as Ontology-driven method.

5 Conclusion

This paper discusses the method of enhancing LLM reasoning and optimizing with multimodal data, and proves the effectiveness of the method through experiments. However, although we can conduct reasoning proof through the displayed guided LLM, due to the characteristics of LLM design, we can not guarantee to achieve complete correctness. Therefore, in the subsequent work, we will consider designing and implementing the verifier and error feedback mechanism in a more reliable and reasonable way.

References

1. Aeronautiques, C., Howe, A., Knoblock, C., et al.: Pddl| the planning domain definition language. Tech. Rep, Technical Report (1998)
2. Chen, X., Lin, M., Schärli, N., Zhou, D.: Teaching large language models to self-debug. CoRR abs/2304.05128 (2023). https://doi.org/10.48550/ARXIV.2304.05128
3. Chen, Y., Arkin, J., Dawson, C., et al.: Autotamp: autoregressive task and motion planning with LLMs as translators and checkers. In: 2024 IEEE International Conference on Robotics and Automation (ICRA), pp. 6695–6702 (2024). https://doi.org/10.1109/ICRA57147.2024.10611163
4. Din, M.U., Rosell, J., Akramet, et al.: Ontology-driven prompt tuning for LLM-based task and motion planning. CoRR (2024). https://doi.org/10.48550/ARXIV.2412.07493
5. Ding, Y., Zhang, X., Paxton, C., Zhang, S.: Task and motion planning with large language models for object rearrangement. In: 2023 IEEE/RSJ International Conference on Intelligent Robots and Systems (IROS), pp. 2086–2092 (2023). https://doi.org/10.1109/IROS55552.2023.10342169
6. Dong, Q., Li, L., Dai, D., et al.: A survey on in-context learning. In: Proceedings of the 2024 Conference on Empirical Methods in Natural Language Processing(EMNLP), pp. 1107–1128 (2024). https://doi.org/10.18653/V1/2024.EMNLP-MAIN.64
7. Hao, Y., Sun, Y., Dong, L., et al.: Structured prompting: scaling in-context learning to 1, 000 examples. CoRR abs/2212.06713 (2022). https://doi.org/10.48550/ARXIV.2212.06713
8. Jin, Y., Li, D., A, Y., et al.: Robotgpt: robot manipulation learning from chatgpt. IEEE Robot. Autom. Lett. **9**(3) (2024). https://doi.org/10.1109/LRA.2024.3357432
9. Li, B., Wu, P., Abbeel, P., Malik, J.: Interactive task planning with language models. Trans. Mach. Learn. Res. (2025). https://openreview.net/forum?id=VmfWywWuYQ
10. Liu, H., Zhu, Y., Kato, K., et al.: Enhancing the LLM-based robot manipulation through human-robot collaboration. IEEE Robot. Automation Lett. **9**, 6904–6911 (2024). https://doi.org/10.1109/LRA.2024.3415931

11. Madaan, A., Tandon, N., Guptaet, P., et al.: Self-refine: iterative refinement with self-feedback. CoRR abs/2303.17651 (2023). https://doi.org/10.48550/ARXIV.2303.17651
12. McKinzie, B., Gan, Z., Fauconnier, J., et al.: MM1: methods, analysis and insights from multimodal LLM pre-training. In: Computer Vision - ECCV 2024 - 18th European Conference. vol. 15087, pp. 304–323 (2024). https://doi.org/10.1007/978-3-031-73397-0_18
13. Mei, A., Zhu, G.N., Zhang, H., Gan, Z.: Replanvlm: replanning robotic tasks with visual language models. IEEE Robot. Automation Lett. **9** (2024).https://doi.org/10.1109/LRA.2024.3471457
14. OpenAI: GPT-4 technical report. CoRR abs/2303.08774 (2023). https://doi.org/10.48550/ARXIV.2303.08774
15. Pan, Z., Luo, H., Li, M., et al.: Chain-of-action: faithful and multimodal question answering through large language models. In: The Thirteenth International Conference on Learning Representations (2025). https://openreview.net/forum?id=1BdPHbuimc
16. Qian, H., Bai, C., Zhang, J., et al.: Discriminator-guided embodied planning for LLM agent. In: The Thirteenth International Conference on Learning Representations (2025). https://openreview.net/forum?id=TjP1d8PP8l
17. Renze, M., Guven, E.: Self-reflection in LLM agents: effects on problem-solving performance. CoRR abs/2405.06682 (2024). https://doi.org/10.48550/ARXIV.2405.06682
18. Schick, T., Dwivedi-Yu, J., Dessì, R., et al.: Toolformer: language models can teach themselves to use tools. CoRR abs/2302.04761 (2023). https://doi.org/10.48550/ARXIV.2302.04761
19. Shinn, N., Cassano, F., Gopinathet, A., et al.: Reflexion: language agents with verbal reinforcement learning. In: Annual Conference on Neural Information Processing Systems 2023(NeurIPS) (2023). http://papers.nips.cc/paper_files/paper/2023/hash/1b44b878bb782e6954cd888628510e90-Abstract-Conference.html
20. Singh, I., Blukis, V., Mousavian, A., et al.: Progprompt: generating situated robot task plans using large language models. CoRR (2022). https://doi.org/10.48550/ARXIV.2209.11302
21. Wang, L., Zhong, H.: LLM-sap: large language models situational awareness-based planning, pp. 1–6 (2024). https://doi.org/10.1109/ICMEW63481.2024.10645429
22. Wang, S., Han, M., Jiao, Z., et al.: LLM3: large language model-based task and motion planning with motion failure reasoning. CoRR abs/2403.11552 (2024). https://doi.org/10.48550/ARXIV.2403.11552
23. Wei, J., Wang, X., Schuurmans, D., et al.: Chain-of-thought prompting elicits reasoning in large language models. In: Advances in Neural Information Processing Systems 35: Annual Conference on Neural Information Processing Systems 2022(NeurIPS) (2022). http://papers.nips.cc/paper_files/paper/2022/hash/9d5609613524ecf4f15af0f7b31abca4-Abstract-Conference.html
24. Xu, M., Yin, W., Cai, D., et al.: A survey of resource-efficient LLM and multimodal foundation models. CoRR abs/2401.08092 (2024). https://doi.org/10.48550/ARXIV.2401.08092
25. Zhou, Z., Song, J., Yao, K., et al.: ISR-LLM: Iterative self-refined large language model for long-horizon sequential task planning. In: 2024 IEEE International Conference on Robotics and Automation (ICRA), pp. 2081–2088 (2024). https://doi.org/10.1109/ICRA57147.2024.10610065

DBSCAN Clustering and Angle-Correlation Driven Bi-Population Constrained Multi-Objective Artificial Physical Optimization

Kaipeng Song[1] and Liping Xie[1,2(✉)]

[1] School of Computer Science and Technology, Taiyuan University of Science and Technology, Taiyuan 030024, China
S202320110978@stu.tyust.edu.cn

[2] Shanxi Key Laboratory of Complex Systems and Computational Intelligence, Taiyuan University of Science and Technology, Taiyuan, Shanxi 030024, China
xieliping@tyust.edu.cn

Abstract. Constrained multi-objective optimization problems (CMOPs) involve both conflicting objective functions and various complex constraints. Artificial Physical Optimization is susceptible to local optimality and it is challenging to identify a solution that simultaneously achieves constrained feasibility and objective optimization. To address this problem, we propose a two-population algorithm called BiACMOAPO. In BiACMOAPO, the main population is made to search for the optimal Pareto front that satisfies the constraints within the identified feasible region by clustering and angle-based screening and by calculating the correlation between individuals using the APO force rule to attract the good particles to the poor ones. The auxiliary population enables particles to traverse the infeasible region by initially disregarding the constraints entirely, and subsequently addressing the constraints through separate cases, and guides the individuals of the main population toward the constrained optimal frontier. For the complex situation of multiple constrained regions, a fitness evaluation model based on DBSCAN clustering and angle is introduced, which divides the space into multiple regions through DBSCAN clustering disassembly and adjusts the individual fitness values by using the angle information to effectively maintain the search diversity and to prevent different feasible regions from interfering with the current region. Finally, the promising particles searched are deposited into the archive set by the update strategy of clustering and angle. In the experiments, the MW series and DTLZ series are selected as the benchmark test functions to compare the proposed algorithm with some state-of-the-art algorithms, and the experimental results suggest that the BiACMOAPO algorithm can obtain quite competitive performance in comparison to other algorithms on both IGD and HV test problems.

H. Ma et al. (Eds.): IWACIII 2025, CCIS 2780, pp. 499–515, 2026.
https://doi.org/10.1007/978-981-95-6730-0_34

Keywords: artificial physical optimization · Angle-based selection · constrained multi-objective optimization · DBSCAN clustering

1 Introduction

Constrained multi-objective optimization problems (CMOPs) are widely existed in the real world. such as Vehicle Path Planning [1], Concrete Structure Optimization [2], Energy Management Optimization [3].A CMOP can be defined as:

$$\begin{aligned} \min\ & F(x) = \big(f_1(x), f_2(x), \ldots, f_m(x)\big) \in \mathbb{R}^m \\ \text{s.t.}\ & x \in \Omega \\ & g_j(x) \leq 0, \quad j = 1, \ldots, s \\ & h_j(x) = 0, \quad j = s+1, \ldots, w \\ & x = (x_1, x_2, \ldots, x_n) \in \mathbb{R}^n \end{aligned} \tag{1}$$

where $F(x)$ is a vector of m objective functions, $x = (x_1, x_1, ..., x_n)$ is a solution consisting of n variables, $g_{\mathrm{j}}(x)$ are p inequality constraints, $h_{\mathrm{j}}(x)$ are q equality constraints. The problem to be solved by the constrained multi-objective optimization algorithm is to reduce each objective $f(x)$ as much as possible while satisfying all the inequality constraints $g_{\mathrm{j}}(x)$ and the equation constraints $h_{\mathrm{j}}(x)$.

In the constrained multi-objective optimization algorithm,the degree of constraint violation is employed to calculate the violation value for the i-th constraint, and the formula can be define as:

$$c_j(x) = \begin{cases} \max\big(0, g_j(x)\big), & j \leq s, \\ \max\big(0, |h_j(x)| - \epsilon\big), & \text{otherwise.} \end{cases} \tag{2}$$

where ϵ is a very small positive value, due to the precision of floating-point calculations, relaxing the equation constraints is necessary. For the degree of violation on all constraints (referred as: the overall constraint violation, CV), and the formula can be define as:

$$\mathrm{CV}(x) = \sum_{j=1}^{w} c_j(x) \tag{3}$$

When $CV(x) = 0$, x is feasible. The feasible region $\mathbb{F}$ can be defined as $\mathbb{F} = \{x \in \mathbb{R}^n | CV(x) = 0\}$. In a multi-objective optimization algorithm, assume that there are two decision variables x and y. if $x, y \in \mathbb{R}_n, \forall i \in \{1, 2, ..., m\}, f_i(x) \leq \mathrm{f_i}(y)$ and $\exists i \in \{1, 2, ..., m\}, f_i(x) < \mathrm{f_i}(y)$, is considered x Pareto domination y (denote as $x \prec y$). If not exist other solution dominates x^*, x^* considered as Pareto optimal solution. The set of all Pareto optimal solutions is called the Pareto set(PS) and the mirror image in the objective space is called the Pareto front(PF).

In constrained multi-objective optimization algorithms, if the constraints are completely ignored, the Pareto optimal frontier that is found is often referred to as the Utopia Pareto Front (UPF), whereas if it is the Pareto optimal frontier within the feasible region, it is usually referred to as the Constrained Pareto Front (*CPF*). The ultimate goal of the algorithm is to find a well-distributed Pareto optimal solution in the feasible region (finding a solution in the CPF region).

However achieving this goal in constrained multi-objective optimization problems is not an easy task. To address these difficulties, in this paper, we propose a new two-population co-evolutionary algorithm called BiACMOAPO.The contribution of this work are as follows:

1) During initial solution generation, traditional random initialization can lead to issues such as local clustering or sparse distribution of samples in the decision space due to random fluctuations. Some regions may suffer from excessive sample concentration, while others may lack sufficient sample coverage, making it difficult to uncover potential high-quality solutions. To address this issue, introduces the Latin hypercube sampling strategy. By performing hierarchical processing on each dimension of the decision space, each dimension is divided into an equal number of sub intervals, and sampling points are randomly selected within each sub interval. This ensures that the initial samples maintain a uniformly distributed characteristic across each dimension of the entire decision space.

2) The current algorithm has limitations in comprehensively exploring all feasible regions when dealing with problems involving multiple feasible regions. Specifically, the population interaction strategy in the existing two-population exploration mechanism often fails to fully leverage the information advantages of each population when faced with scenarios involving multiple feasible regions, resulting in the algorithm being unable to perform a complete search of all feasible regions. To address this core issue, this paper proposes a dual-population collaborative search strategy: by having the auxiliary population perform directed exploration of UPF regions, it provides location guidance for the main population in feasible regions; simultaneously, clustering methods are introduced to prompt the main population to conduct high-precision searches within already discovered feasible regions, thereby enhancing the algorithm's global exploration capability and local optimization accuracy for scenarios with multiple feasible regions.

3) Crowding distance characterizes density features by calculating the distance differences between an individual and its directly adjacent individuals across various target dimensions. This local assessment mechanism struggles to capture the population's distribution patterns across the entire population. Angular information can effectively characterize the relative positional differences of individuals in space. When the population exhibits uneven distribution, changes in angle can sensitively capture the boundaries between dense and sparse regions, enabling the algorithm to selectively retain individuals in sparse regions. This ensures that diversity in the solution is maintained even in scenarios with uneven distribution.

In the remainder of this paper, we first present the background as well as related work in Sect. 2. Then the design ideas and technical of the BiACMOAPO algorithm are described in detail in Sect. 3. Afterwards, we design experiments to compare some excellent algorithms in Sect. 4. Finally, summarizes the article and future work is outlined in Sect. 5.

2 Related Work

2.1 Constraint Handle Techniques

Currently, many constraint handling techniques have been proposed, including constraint dominance techniques, ε-constraint handling methods, penalty function methods, etc. This section will provide a detailed introduction to constraint dominance criteria and ε-constraint handling methods.

Constrained Dominance Principle (CDP): Deb et al. [4] first proposed the constrained dominance principle in NSGA-II. The idea is as follows: given two solutions x and y. If two solutions violate constraint $CV(x) \geq 0$, then the constraint violation that is smaller dominates the constraint violation that is larger. If the violation degrees of the constraints for both solutions are $CV(x) = 0$, then the dominance relationship is determined based on the objective values of the two solutions. If the following formula is satisfied, it indicates that $x \prec y$.

$$\begin{array}{l} \forall \mathrm{i} \in \{1, ..., m\}\ f_i(x) \leq f_i(y) \\ \exists \mathrm{i} \in \{1, ..., m\}\ f_i(x) < f_i(y) \end{array} \tag{4}$$

The CDP method emphasizes feasibility and is therefore widely used in CMOP problems. However, due to the complexity of constraints, such as discontinuous feasible regions and feasible regions with inner and outer layers, this method of ignoring infeasible solutions can easily cause the algorithm to get stuck in a local optimum or prone to premature.

ε-constraint handling method: Takahama et al. [5]. proposed a ε-constraint processing method.This method sets a relaxation factor ε. When the constraint violation degree of individuals in a population is less than ε, they are also considered to satisfy the constraint conditions. By retaining some feasible solutions with good fitness but violating constraints in the next generation. Therefore, this method enables the algorithm to escape from local infeasible regions while helping to find narrow feasible regions. Given two solutions x and y in the population, if any of the following conditions are met, x is considered to constraint dominate y (referred as: $x \prec_{\varepsilon} y$).

$$\begin{cases} CV(x) > \varepsilon, \quad CV(y) > \varepsilon, \quad CV(x) < CV(y) \\ CV(x) \leq \varepsilon, \quad CV(y) \leq \varepsilon, \quad x \prec y \end{cases} \quad \forall x \neq y \tag{5}$$

Fan et al. designed an improved ε update strategy in PPS [6]. Constraint processing is divided into two stages, namely the push stage and the pull stage. The update formula is as follows:

$$\varepsilon(k) = \begin{cases} (1-\tau)\,\varepsilon(k-1), & rf < \alpha \\ \varepsilon(0)\,(1-\frac{k}{T_c})^{cp}, & rf \geq \alpha \end{cases} \tag{6}$$

2.2 Artificial Physical Optimization

Artificial Physical Optimization (APO), first proposed by Spears et al. [7]. Then Xie et al. [8] introduced it into optimization algorithms. The APO algorithm is one of the heuristic algorithms inspired by Newton's second law. The main idea of this algorithm is that each individual in the population adjusts its movement based on the combined forces exerted by other individuals. By adjusting the attraction and repulsion of each individual toward other particles, as well as the mass of each individual, superior individuals attract inferior individuals, while inferior individuals repel superior individuals. Ultimately, the position of each individual is updated, causing each individual in the population to move toward the Pareto optimal direction. In the APO algorithm, the quality function formula for the i-th individual is calculated as follows:

$$m_i = e^{\frac{f(x_{best})-f(x_i)}{f(x_{worst})-f(x_{best})}}, m_i \in (0,1] \tag{7}$$

The formula for the force exerted on each individual by other individuals is as follows:

$$F_{ij,k} = \begin{cases} Gm_i m_j (x_{j,k} - x_{i,k}), & f(x_i) > f(x_j) \\ -Gm_i m_j (x_{j,k} - x_{i,k}), & f(x_i) \leq f(x_j) \end{cases} \forall i \neq j \text{ and } i \neq best \tag{8}$$

$$F_{i,k} = \sum_{\substack{i=1 \\ i \neq j}}^{n} F_{ij,k}, \forall i \neq best \tag{9}$$

The speed and position of individual x_i are updated using the following formula:

$$\begin{cases} v_{i,k}(t+1) = \omega v_{i,k}(t) + \lambda F_{i,k}/m_i \\ x_{i,k}(t+1) = x_{i,k}(t) + v_{i,k}(t+1) \end{cases} \tag{10}$$

Wang et al. [9] proposed a rank-based multi-objective artificial physical algorithm (RMOAPO). Applying artificial physical optimization to multi-objective optimization algorithms, this algorithm calculates the quality of individuals by finding their individual sequence values through non-dominant sorting and crowding. The specific calculation formula is as follows.

$$rank = \frac{\left(\sum_{i=1}^{n} (\max |E_i|)^2\right)^{\frac{1}{2}}}{archivesize - 1} \tag{11}$$

$$m_i = \exp\left(\frac{-rank_i}{N}\right) \tag{12}$$

Based on the RMOAPO algorithm, Wang [10] proposed the VDCMOAPO algorithm, which applies mimetic physics to constrained multi-objective optimization algorithms. By redesigning the mass of individuals, it distinguishes between individuals in the feasible region and those in the infeasible region. Set the quality value range to $(0, 2]$, where the quality range for the infeasible region is set to $(0, 1]$ and the quality range for the feasible region is set to $(1, 2]$. The formula is as follows.

$$m_i = \begin{cases} 1 + e^{\frac{f(x_{\text{best}}) - f(x_i)}{f(x_{\text{worst}}) - f(x_{\text{best}})}}, & x_i \in \text{feasible region} \\ e^{\frac{-\varphi(x_i)}{\varphi(x)}}, & x_i \in \text{infeasible region} \end{cases} \tag{13}$$

3 Proposed Algorithm

3.1 Framework of BiACMOAPO

This paper attempts to solve CMOPs problems through two populations and proposes a new algorithm called Bi-Population Angle correlation multi-objective artificial physical optimization (BiACMOAPO). Next, we will introduce the algorithm design in detail.

Algorithm 1. shows the main framework of BiACMOAPO. During initialization, a Latin hypercube sampling strategy was adopted. This method uses stratified sampling to ensure that sample points are evenly distributed across each dimension. This allows the algorithm to better cover the target space in the initial stage, enabling it to obtain more valuable information and converge more quickly to the Pareto front. The Archive is initially set to empty. During the search process, the main population adopts a artificial physical movement strategy and calculates fitness values using new variants derived from the SPEA2 fitness calculation, thereby guiding individuals in the population toward the CPF direction. Meanwhile, when the auxiliary population identifies a new feasible region, individuals with lower fitness values in the main population are transferred to the new feasible region. The auxiliary population uses an arithmetic crossover operator to calculate new adaptive fitness values, dividing the particles in the population into two parts, x_f and x_l, which then randomly learn from each other to generate new individuals. Finally, the current round optimal individuals are saved to the archive set through non-dominated sorting and angle-based update strategies. In short, BiACMOAPO consists of three key components:1) Update main population, 2) Update auxiliary population, 3) Generate archive set.

3.2 Angle-Correlation Individual Evaluation

In order to enable the main population and auxiliary population to efficiently perform their respective functions, we redesigned two methods for calculating fitness rules, which are applied to the main population and auxiliary population, respectively.

The main purpose of the main population is to perform a fine search within the feasible region. At the beginning, all individuals are outside the feasible region, so the main population hopes that individuals with fewer constraint

Algorithm 1. Framework of BiACMOAPO

```
Require: a CMOP and the population size N, termination condition T;
Ensure: Archive(final output);
Initialize k = 0, Archive = φ
Initialize MainPopulation(MP) by Latin Hypercube Sampling
Initialize AuxiliaryPopulation(AP) by Latin Hypercube Sampling
Update Archive from (MP ∪ AP) by angle-based fitness selection
set k = 0
while k < T do
    Update MP and ε(k) driven by modified APO force rules with
  angle-correction and DBSCAN clustering
    Update AP from AP and some MP by GA
    Update MP by angle-correction attraction from AP
    Select MP by angle-based fitness function with ε(k) from MP
    Select AP by angle-based fitness function from AP
    Update Archive from (Archive ∪ MP ∪ AP) by angle-based fitness
  selection
    k ← k + 1
end while
return Archive
```

violations will dominate those with more constraint violations. When two individuals are both feasible, they should be sorted according to the non-domination relationship and the density of the individuals.

If two individuals x and y exist, when $x \prec y$, the fitness of x is higher than that of y. If x and y do not dominate each other, they are ranked according to their density. If the density of x is lower than that of y, the fitness of x is greater than that of y.

In order to better explore feasible regions, the main population uses the ε-constraint method strategy. In the initial stage, the proportion of feasible solutions is relatively small, so the tightening speed of ε is slow, making full use of the information in the infeasible region. In the later stage, most individuals have entered the feasible region constrained by ε, so the tightening speed of ε is accelerated to drive the individuals in the population toward the true feasible region. The following formula shows the rules for updating the ε constraint.

$$\varepsilon(t) = \begin{cases} \varepsilon(0) \times e^{-\gamma \times t}, & rf < \alpha \text{ and t} < T_c \\ CV(t)_{\max} \times \left(1 - \frac{t}{T_c}\right)^{cp}, & rf \geq \alpha \text{ and t} < T_c \\ 0, & t \geq T_c \end{cases} \tag{14}$$

where $\varepsilon(0)$ refers to the maximum overall constraint value of particles in the population in the 0th generation. γ is the parameter controlling the relaxation of constraints. rf is the proportion of individuals with a total constraint violation degree of 0. α controls the search preference between feasible and infeasible regions. Tc is the number of boundary iterations. When t is less than Tc, the ε

constraint condition gradually tightens. When t is greater than or equal to Tc, it is equivalent to ε being 0, and the constraint violation degree is strictly checked.

Therefore, the fitness of the main population was designed with reference to the SPEA2 algorithm [11] and improved. The calculation formula is as follows:

$$fit\left(x_i\right) = R\left(x_i\right) + D\left(x_i\right), x_i \in MP \tag{15}$$

where $R(x_i)$ represents the number of dominant individuals, and $D(x_i)$ represents the degree of crowding at the location of that individual. The formula for $R(x_i)$ is as follows:

$$R\left(x_i\right) = \sum_{\substack{j=1 \\ i \neq j}}^{N} r\left(x_i\right) \tag{16}$$

where $r\left(x_i\right)$ indicates whether it is dominated. If it is dominated, the value is 1; otherwise, the value is 0. The formula is:

$$r\left(x_i\right) = \begin{cases} 1, x_j \prec x_i \\ 0, \text{others} \end{cases}, x_i, x_j \in MP \tag{17}$$

$D\left(x_i\right)$ is used to estimate the crowding degree of x_i.An angle-based calculation method is used to compare the angle sizes between each individual and others, thereby obtaining the crowding degree of each individual.The formula is:

$$D\left(x_j\right) = \min\left(\theta_{x_i, x_k}\right), x_j, x_k \in MP, j \neq k \tag{18}$$

where θ_{v_i, v_k} is the angle value between the two individuals x_j and x_k. To calculate the angle value,it has to find the ideal point, $Z^{\min} = \left(z_1^{\min}, z_2^{\min}, ..., z_m^{\min}\right)$ and $Z^{\max} = \left(z_1^{\max}, z_2^{\max}, ..., z_m^{\max}\right)$. where $Z_i^{\min}$ and $Z_i^{\max}$ represent the minimum and maximum values of the i-th objective among all solutions, respectively. Then, the objective vector of the j-th solution can be normalized as follows: $F^{'}\left(x_j\right) = \left(f_1^{'}\left(x_j\right), f_2^{'}\left(x_j\right), ..., f_m^{'}\left(x_j\right)\right)$, where the formula for each dimension is as follows:

$$f_i^{'}\left(x_j\right) = \frac{z_i^{\max} - f_i\left(x_j\right)}{z_i^{\max} - z_i^{\min} + eps} \tag{19}$$

where $f_i\left(x_j\right)$ is the original vector value of individual x_i in the j-th dimension, $f_i^{'}\left(x_j\right)$ is the normalized vector value of individual x_i in the j-th dimension, eps is a very small positive number whose purpose is to avoid possible division by zero. The magnitude of eps is much smaller than the numerical value of the vector, so it does not affect the calculation accuracy of normalization. Then, the vector angle between the two solutions v_j and v_k is calculated using the following formula:

$$\theta_{x_j, x_k} = \arccos\left|\frac{F^{'}\left(x_j\right) \bullet F^{'}\left(x_k\right)}{\left\|F^{'}\left(x_j\right)\right\| \times \left\|F^{'}\left(x_k\right)\right\|}\right| \tag{20}$$

where $F^{'}(x_j) \bullet F^{'}(x_k)$ represents the inner product of $F^{'}(x_j)$ and $F^{'}(x_k)$. $\|\cdot\|$ denotes the L2 norm of a vector. In Fig. 1, assume that A, B, C, and D are four points in the solution space. So we can find that A and D are in different search directions, so $\theta_{A,D}$ is very large, while B and C are in similar directions, so $\theta_{\mathrm{B,C}}$ is relatively small.

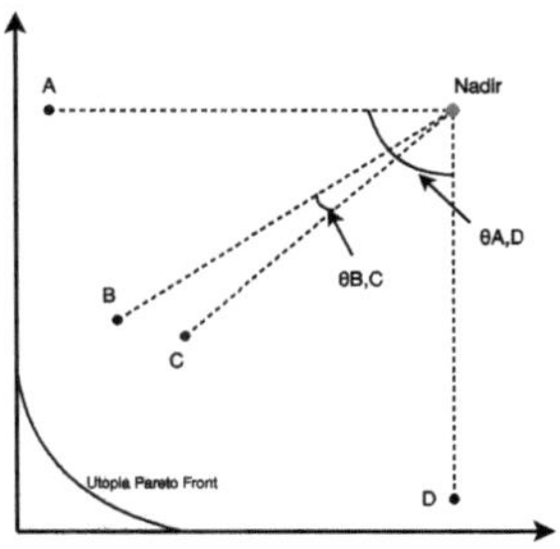

Fig. 1. Illustration of the vector angle with two objective functions.

The mass of each individual in the population is evaluated using the fitness function, as shown in the following formula.

$$m_i = \begin{cases} 1 + e^{\frac{-sort(x_i)}{N}}, & CV(x_i) = 0 \\ \frac{CV(x_i) - \min(CV)}{\max(CV) - \min(CV) + eps}, & CV(x_i) > 0 \end{cases}, x_i \in MP \tag{21}$$

where N is the population size, $sort(x_i)$ is the ranking value of individuals based on fitness, and $CV(x_i)$ is the overall constraint violation degree of x_i. Through this quality function, the quality of infeasible solutions can be limited to the range $(0, 1]$, while the quality of feasible solutions can be limited to the range $(1, 2]$, so that the quality of feasible solutions is always greater than that of infeasible solutions.

To efficiently explore the feasible region, all individuals in the auxiliary population initially move toward the UPF. At this stage, convergence is the primary consideration, and constraint violation severity is completely ignored. The fitness value should focus on feasibility. Once the dominance hierarchy within the auxiliary population has only one layer, begins to determine whether the UPF and CPF overlap. If they do not overlap at all, begins to prioritize constraint violation,causing individuals in the auxiliary population to move toward reducing constraint violation.

Therefore, the fitness rules for the auxiliary population differ from those for the main population. The formula is as follows.

$$fit(x_i) = R(x_i) + D(x_i) + fp \times CVrank(x_i), x_i \in AP \tag{22}$$

where, fp refers to whether constraints are considered, with $fp = 0$ indicating that constraints are completely ignored and fp=1 indicating that constraints are considered.The CV_{rank} metric denotes the ranking of constraint violation degrees, with a higher degree of constraint violation corresponding to a higher ranking. In the initial stage of the algorithm, the primary task of the auxiliary population is to rapidly approach the Pareto region with good convergence properties. To fully explore the UPF region, the fitness evaluation completely ignores the degree of constraint violation at this stage. However, in the later stages, if the UPF and CPF regions are completely non-overlapping, to better guide the particles in the main population, fp is set to 1, and the particles in the auxiliary population begin to move toward the UPF direction. At this point, the size of the CV value is considered when calculating the number of dominant individuals.The formula for R_i is the same as (Eq. 16), while the formula for $r(x_i)$ is as follows:

$$r(x_i) = \begin{cases} 1, \ x_j \prec x_i \text{ and } CV(x_j) < CV(x_i) \\ 0, \text{ others} \end{cases}, x_i \in AP \tag{23}$$

3.3 Main Population Based on Angle-Correlation and DBSCAN Clustering

Main population searches for the feasible region of the constrained optimal frontier, efficient force design can significantly improve the algorithm's performance. However, in current heuristic algorithms, the forces acting on individuals originate from global forces. Typically, many forces cancel each other out or interfere with one another, which can easily affect the algorithm's movement. Additionally, due to the constraints imposed by the conditions, there are often many feasible regions. Global forces can cause the algorithm to easily exit local feasible regions, causing it to lose the feasible regions it has already explored.

This study presents a redesigned framework for force rules. Based on the traditional APO algorithm, it incorporates a novel strategy that integrates angle-based mechanisms and DBSCAN clustering. Initially,the algorithm has not reach the feasible region. At this stage, an angle correlation strategy is adopted to compare the angle value differences between two particles and dynamically adjust the magnitude of the force exerted by each particle on other particles. The correlation weight formula (see Eq. 24), Through the correlation weighting, nearby particles can influence the current particle, while distant particles do not influence the current particle.

$$Co(x_{i,j}) = \exp\left(\frac{-AngleRank\,(x_{i,j})}{0.3 \times size\,(AngleRank\,(x_{i,j}))}\right)^{-3} \tag{24}$$

where AngleRank is defined as the ranking of the angle distance between vectors x_i and x_j, When the angle is small, AngleRank is small; conversely, when the angle is large, AngleRank is large.

In the later stages, when particles find multiple feasible regions, DBSCAN clustering is used. Specifically, by binding particles to clusters, particles are confined to an influence domain. Particles within the same influence domain can interact with each other, while particles from other influence domains cannot affect the current influence domain. Specifically, DBSCAN defines clusters and noise based on density. First, two parameters, δ and $MinPts$, are initialized.

Where ε is the neighborhood radius, which defines the size of the neighborhood range of a point. MinPts is the minimum number of points, which defines the minimum number of points contained in the core point neighborhood. Next, find the core points. For each point x_i in the population, calculate the number of points within its domain $|N_{\in}(x_{\mathrm{i}})|$. If $|N_{\in}(x_i)| \geq MinPts$, mark it as a core point x_i. The formula is as follows.

$$N_{\in}(x_i) = \{x_j \in P|\, \mathrm{dist}(x_i, x_j) \leq \delta\}\, , i \neq j \tag{25}$$

Then, starting from any not visited core point x_i, mark it as visited and add all points within its δ-neighborhood to the current cluster C. For these newly added points x_j, if x_j is also a core point (denote as CP), continue to expand the points within ε-neighborhood of x_j until no new points can be added. The formula is as follows:

$$\begin{cases} x_i \in CP, \, |N_{\in}(x_{\mathrm{i}})| \geq \mathrm{MinPts} \\ x_i \notin CP, \text{ others} \end{cases} \tag{26}$$

Finally, for points that are not marked as clusters, if they are within the ε domain of a cluster, they are marked as boundary points and added to the corresponding cluster; otherwise, they are marked as noise points. Repeat the above steps until all points have been visited. At this point, the individuals in the population will be divided into multiple clusters $C = \{c_1, c_2, ..., c_r\}$. Next, find the center of each cluster. To more accurately reflect the characteristics of the cluster, select all core points within the cluster and calculate their arithmetic mean (referred as center of the Core points).

$$F_{ij,k}^{MP} = \begin{cases} G \cdot Co(x_{ij}) m_i m_j (x_{j,k} - x_{i,k}), & t < \rho T, \ x_j \prec x_i \\ -G \cdot Co(x_{ij}) m_i m_j (x_{j,k} - x_{i,k}), & t < \rho T, \ x_i \prec x_j \\ G \cdot m_i m_j (x_{j,k} - x_{i,k}), & t > \rho T, \ x_i, x_j \in c_r \\ -G \cdot m_i m_j (x_{j,k} - x_{i,k}), & t > \rho T, \ x_i, x_j \in c_r \\ 0, & \text{otherwise} \end{cases}, \ \begin{matrix} x_i, x_j \in MP, \\ \forall i \neq j \ , \ i \neq best \end{matrix} \tag{27}$$

where ρ is the coefficient determining the two stages, when $t < \rho T$, the population moves through angle correlation, and when $t > \rho T$, the population moves through DBSCAN. In the second stage, when x_i and x_j belong to the same cluster c_r, there is an interaction between individuals, and when x_i and x_j do not belong to the same cluster, there is no interaction between individuals.

3.4 Bi-Population Interaction Strategy Based on Angle Selection

The existing bi-population algorithm fails to fully utilize each other's information resources, resulting in poor collaboration. To address this issue,proposes a new bi-population interaction mechanism that fully utilizes and integrates the information of both sides, enabling the algorithm to significantly improve population diversity while ensuring convergence performance.

The update of the auxiliary population is based on arithmetic crossover operations. First, 10% of individuals from the main population and the auxiliary population are selected through a competition according to the fitness rules to choose the more superior particles.Help enhance the diversity of auxiliary populations and prevent premature of auxiliary populations.

The main population calculates the angle relationships between individuals in the auxiliary population and selects the 10% of auxiliary population individuals with the closest angles to the main population individuals. From these, it selects the individuals with the optimal fitness. By leveraging the attraction of these selected individuals, it provides the main population individuals with a guiding force toward the Pareto frontier, driving the main population toward the optimal solution region. The update formula for the kth dimension is as follows:

$$F_{ij,k}^{AP} = \begin{cases} Gm_i m_j \left(x_{j,k} - x_{i,k}\right), x_j \prec x_i \\ 0, \qquad\qquad\qquad\qquad others \end{cases}, x_i \in MP, x_j \in AP \tag{28}$$

In the main population (MP), the velocity and position of each individual are updated using the following formula:

$$v_{i,k}\left(t+1\right) = r_0 v_{i,k}\left(t\right) + \lambda F_{i,k}/m_i + r_1\left(v_{j,k}\left(t\right) - v_{i,k}\left(t\right)\right), x_i \in MP, x_j \in AP \tag{29}$$

$$x_{i,k}\left(t+1\right) = x_{i,k}\left(t\right) + v_{i,k}\left(t+1\right), x_i \in MP \tag{30}$$

where r_0 and r_1 are random numbers between 0 to 1, λ is the weight of the original speed, and v_j is the individual selected from the auxiliary population to attract individuals from the main population.

Subsequently, the force acting on individuals within the main population and exerted by the auxiliary population is calculated see in Eq. 31. Through this strategy,movements of one individual are not affected by other individuals.

$$F_{i,k} = \sum_{\substack{i=1 \\ i\neq j}}^{n} F_{ij,k}^{MP} + F_{im,k}^{AP}\ \forall i \neq best \tag{31}$$

The algorithm flow of the main population is given in Algorithm 2.

Algorithm 2. Updating of the Main Population

Require: MainPopulation MP, AuxiliaryPopulation AP, constraint relaxation $factor\varepsilon(k)$
Ensure: MainPopulation $MP^{'}$
Initialize $AP^{'} = \phi$
Calculate CV using Equation(3)
Calculate MP fitness using Equation(15)
Calculate MP mass using Equation(21) where ε is used
if stage = 1 **then**
    Calculate the angle between MP using Equation(19,20)
    Calculate MP Force using Equation(27)
else if stage = 2 **then**
    DBSCAN clustering on MP using Equation(25,26)
    Calculate MP Force using Equation(27)
end if
Calculate Cross Angle between MP and AP using Equation(19,20)
Calculate Attraction force by AP using Equation(28)
Calculate all Force using Equation(32)
$MP^{'} \leftarrow$ Calculate Position using Equation(29,30)
return $MP^{'}$

4 Experimental Research

4.1 Experimental Setup

In order to test the performance of the proposed algorithm BiACMOAPO, this algorithm will be compared with PPS [6], ToP [12], TiGE2 [13], CAEAD [14], and DPPPS [15] algorithms on the MW [16] ,C-DTLZ and DC-DTLZ [17,18] benchmark test set by IGD [20] and HV [21] metrics.

In the comparison algorithms, the parameter settings are consistent with the original text. In the algorithm of this paper, the gravitational G is set to 0.02, the feasible solution ratio α is set to 0.9, the constraint relaxation speed cp is set to 2, Tc is set to N*90%, and minPts is set to 4. The population size for all algorithms is set to 100. The maximum evaluation count on the MW test set is set to $3x10^4$, and others is set to $1x10^5$. All algorithms are run independently 30 times. All experiments will be conducted on the PlatEMO [19] platform.

4.2 Experimental Results

In order to validate the effectiveness of the algorithms in solving constrained multi-objective problems, we conducted comparative experiments between BiACMOAPO and PPS [6], ToP [12], TiGE2 [13], CAEAD [14], and DPPPS [15]. All the experiments were subjected to Wilcoxon rank sum test at 0.05 level of significance to compare the statistical significance level of the experimental results of the two algorithms. Where "+", "-" and "≈" represent that the compared algorithms are better than, worse than and similar to BiACMOAPO.

Table 1 shows the results of the IGD metrics for the six algorithms on the MW test problem. In this series of tests, BiACMOAPO outperformed PPS, Top, TiGE2, CAEAD, and DPPPS in the IGD metric by 12, 8, 13, 12, and 6 instances, respectively. Although BiACMOAPO slightly behind DPPPS in the IGD metric, it achieved superior IGD values in most test problems. Table 2 shows the results of the HV metric for the six algorithms on the MW test problems. In this series of tests, BiACMOAPO outperformed PPS, ToP, TiGE2, CAEAD, and DPPPS in the HV metric for 13, 8, 13, 13, and 7 problems, respectively. Table 3 shows the IGD metric results for the six algorithms on the DTLZ series of test problems, and Table 4 shows the HV metric results for the six algorithms on the DTLZ series of test problems. The results show that BiACMOAPO has better IGD and HV values than other algorithms on the DTLZ series of test problems.

Table 1. Performance Compared on MW Problems with Respect to the IGD Values. The Best Average IGD Among the Algorithms is Highlighted in Gray

Problem	M	PPS	ToP	TiGE2	CAEAD	DPPPS	BiACMOAPO
MW1	2	4.5284×10^{-2} (1.17×10^{-1}) −	NaN (NaN)	8.5955×10^{-2} (8.59×10^{-2}) −	1.8517×10^{-1} (2.76×10^{-1}) −	2.8013×10^{-3} (4.20×10^{-3}) −	1.6617×10^{-3} (2.36×10^{-5})
MW2	2	1.4280×10^{-1} (8.86×10^{-2}) −	3.0873×10^{-1} (1.77×10^{-1}) −	3.5554×10^{-2} (1.34×10^{-2}) −	9.3088×10^{-2} (4.16×10^{-2}) −	1.7009×10^{-2} (6.48×10^{-3}) +	3.2042×10^{-2} (2.14×10^{-2})
MW3	2	6.2038×10^{-3} (6.07×10^{-4}) −	6.2823×10^{-1} (3.03×10^{-1}) −	2.7649×10^{-2} (1.92×10^{-2}) −	6.4213×10^{-3} (3.20×10^{-4}) −	5.2499×10^{-3} (2.52×10^{-4}) +	5.6253×10^{-3} (3.35×10^{-4})
MW4	3	6.4698×10^{-2} (4.22×10^{-3}) −	NaN (NaN)	8.1419×10^{-2} (1.63×10^{-2}) −	1.4420×10^{-1} (9.31×10^{-2}) −	4.5841×10^{-2} (1.15×10^{-3}) −	4.1492×10^{-2} (3.57×10^{-4})
MW5	2	3.8623×10^{-1} (3.54×10^{-1}) −	NaN (NaN)	6.7600×10^{-2} (3.07×10^{-2}) −	4.1441×10^{-1} (3.33×10^{-1}) −	5.5742×10^{-2} (1.83×10^{-1}) −	5.0639×10^{-3} (1.39×10^{-3})
MW6	2	6.5138×10^{-1} (3.62×10^{-1}) −	1.0036×10^{0} (2.44×10^{-1}) −	3.0881×10^{-1} (3.15×10^{-1}) −	4.6220×10^{-1} (1.93×10^{-1}) −	1.0866×10^{-2} (5.82×10^{-3}) +	4.1012×10^{-2} (8.23×10^{-2})
MW7	2	5.7297×10^{-3} (5.22×10^{-4}) −	3.6661×10^{-1} (2.38×10^{-1}) −	4.6377×10^{-2} (1.68×10^{-2}) −	6.5815×10^{-3} (6.94×10^{-4}) −	4.8217×10^{-3} (3.49×10^{-4}) +	6.4059×10^{-3} (1.34×10^{-3})
MW8	3	2.3174×10^{-1} (2.13×10^{-1}) −	8.4968×10^{-1} (2.43×10^{-1}) −	1.0070×10^{-1} (7.85×10^{-3}) −	1.1942×10^{-1} (4.12×10^{-2}) −	4.7546×10^{-2} (2.18×10^{-3}) =	4.7884×10^{-2} (6.10×10^{-3})
MW9	2	3.8785×10^{-1} (3.40×10^{-1}) −	NaN (NaN)	1.1521×10^{-1} (2.05×10^{-1}) −	5.4522×10^{-1} (2.72×10^{-1}) −	3.1552×10^{-2} (1.29×10^{-1}) −	5.2674×10^{-3} (3.03×10^{-4})
MW10	2	4.0684×10^{-1} (2.35×10^{-1}) −	NaN (NaN)	8.0252×10^{-2} (1.03×10^{-1}) =	4.9034×10^{-1} (2.07×10^{-1}) −	1.8946×10^{-2} (2.37×10^{-2}) +	5.9299×10^{-2} (4.00×10^{-2})
MW11	2	7.2904×10^{-3} (3.49×10^{-4}) −	8.0966×10^{-1} (1.48×10^{-1}) −	4.1405×10^{-2} (8.18×10^{-3}) −	6.8316×10^{-3} (2.64×10^{-4}) −	6.1259×10^{-3} (1.57×10^{-4}) +	1.3508×10^{-2} (6.92×10^{-3})
MW12	2	2.1426×10^{-1} (2.77×10^{-1}) −	NaN (NaN)	1.3064×10^{-1} (2.49×10^{-1}) −	4.1427×10^{-1} (3.46×10^{-1}) −	3.2413×10^{-2} (1.40×10^{-1}) −	2.7428×10^{-2} (1.21×10^{-1})
MW13	2	4.5063×10^{-1} (3.62×10^{-1}) −	9.5251×10^{-1} (6.04×10^{-1}) −	4.3039×10^{-1} (1.73×10^{-1}) −	2.7352×10^{-1} (1.27×10^{-1}) −	2.4425×10^{-2} (1.62×10^{-2}) +	7.1994×10^{-2} (3.59×10^{-2})
MW14	3	1.8277×10^{-1} (5.07×10^{-2}) −	4.5408×10^{-1} (5.51×10^{-1}) −	1.6091×10^{-1} (1.09×10^{-2}) −	3.0897×10^{-1} (7.60×10^{-2}) −	1.3249×10^{-1} (5.14×10^{-2}) −	9.7235×10^{-2} (1.29×10^{-3})
+/-/=		1/12/1	0/8/0	0/13/1	1/12/1	7/6/1	

Table 2. Performance Compared on MW Problems with Respect to the HV Values. The Best Average HV Among the Algorithms Highlighted in Gray

Problem	M	PPS	ToP	TiGE2	CAEAD	DPPPS	BiACMOAPO
MW1	2	4.4030×10^{-1} (9.90×10^{-2}) −	NaN (NaN)	4.0743×10^{-1} (6.67×10^{-2}) −	3.3619×10^{-1} (1.89×10^{-1}) −	4.8756×10^{-1} (5.71×10^{-3}) −	4.8973×10^{-1} (1.10×10^{-4})
MW2	2	3.9959×10^{-1} (9.58×10^{-2}) −	2.6129×10^{-1} (1.38×10^{-1}) −	5.3254×10^{-1} (1.64×10^{-2}) −	4.5449×10^{-1} (5.14×10^{-2}) −	5.5916×10^{-1} (1.10×10^{-2}) +	5.3679×10^{-1} (2.99×10^{-2})
MW3	2	5.4282×10^{-1} (9.56×10^{-4}) =	1.2514×10^{-1} (1.25×10^{-1}) −	5.2599×10^{-1} (1.26×10^{-2}) −	5.4111×10^{-1} (5.54×10^{-4}) −	5.4333×10^{-1} (5.57×10^{-4}) =	5.4321×10^{-1} (5.59×10^{-4})
MW4	3	7.9750×10^{-1} (7.92×10^{-3}) −	NaN (NaN)	7.9390×10^{-1} (1.75×10^{-2}) −	6.8405×10^{-1} (1.32×10^{-1}) −	8.3547×10^{-1} (1.87×10^{-3}) −	8.3960×10^{-1} (7.58×10^{-4})
MW5	2	1.9289×10^{-1} (1.04×10^{-1}) −	NaN (NaN)	2.8203×10^{-1} (1.48×10^{-2}) −	1.4922×10^{-1} (8.62×10^{-2}) −	3.0233×10^{-1} (6.14×10^{-2}) −	3.2132×10^{-1} (9.75×10^{-4})
MW6	2	7.6403×10^{-2} (7.61×10^{-2}) −	2.5390×10^{-2} (3.09×10^{-2}) −	2.0785×10^{-1} (8.31×10^{-2}) −	1.1518×10^{-1} (6.64×10^{-2}) −	3.1546×10^{-1} (8.63×10^{-3}) +	2.9010×10^{-1} (3.30×10^{-2})
MW7	2	4.1122×10^{-1} (7.00×10^{-4}) −	2.2300×10^{-1} (8.97×10^{-2}) −	3.8928×10^{-1} (3.74×10^{-3}) −	4.0963×10^{-1} (7.54×10^{-4}) −	4.1147×10^{-1} (4.49×10^{-4}) −	4.1192×10^{-1} (4.92×10^{-4})
MW8	3	2.7008×10^{-1} (1.16×10^{-1}) −	7.8450×10^{-2} (9.20×10^{-2}) −	4.7392×10^{-1} (2.27×10^{-2}) −	3.8369×10^{-1} (6.97×10^{-2}) −	5.4008×10^{-1} (1.02×10^{-2}) +	5.2509×10^{-1} (1.72×10^{-2})
MW9	2	1.6756×10^{-1} (1.83×10^{-1}) −	NaN (NaN)	3.0954×10^{-1} (1.06×10^{-1}) −	7.7066×10^{-2} (1.40×10^{-1}) −	3.8012×10^{-1} (7.18×10^{-2}) −	3.9580×10^{-1} (2.25×10^{-3})
MW10	2	2.2266×10^{-1} (1.09×10^{-1}) −	NaN (NaN)	3.8762×10^{-1} (5.33×10^{-2}) =	1.8292×10^{-1} (9.36×10^{-2}) −	4.3621×10^{-1} (2.01×10^{-2}) +	4.0244×10^{-1} (2.77×10^{-2})
MW11	2	4.4723×10^{-1} (1.71×10^{-4}) +	2.5255×10^{-1} (2.65×10^{-2}) −	4.3116×10^{-1} (2.86×10^{-3}) −	4.4656×10^{-1} (3.57×10^{-4}) +	4.4706×10^{-1} (2.27×10^{-4}) +	4.4434×10^{-1} (1.73×10^{-3})
MW12	2	4.1190×10^{-1} (2.44×10^{-1}) −	NaN (NaN)	4.9629×10^{-1} (1.97×10^{-1}) −	2.7214×10^{-1} (2.70×10^{-1}) −	5.8200×10^{-1} (1.10×10^{-1}) −	5.8380×10^{-1} (1.06×10^{-1})
MW13	2	2.7329×10^{-1} (1.20×10^{-1}) −	2.0174×10^{-1} (1.31×10^{-1}) −	3.4023×10^{-1} (4.46×10^{-2}) −	3.2008×10^{-1} (7.03×10^{-2}) −	4.6743×10^{-1} (8.50×10^{-3}) +	4.4515×10^{-1} (1.83×10^{-2})
MW14	3	4.2966×10^{-1} (1.29×10^{-2}) −	3.5035×10^{-1} (1.61×10^{-1}) −	4.4821×10^{-1} (5.53×10^{-3}) −	3.8202×10^{-1} (3.06×10^{-2}) −	4.5865×10^{-1} (1.64×10^{-2}) −	4.7116×10^{-1} (1.68×10^{-3})
+/-/=		1/12/1	0/8/0	0/13/1	1/13/0	6/7/1	

Figure 2 shows that BiACMOAPO can find multiple feasible regions in multimodal testing problems and push particles in the population toward the optimal Pareto frontier region that satisfies the constraints. Other algorithms found the

Table 3. Performance Compared on DTLZ Problems with Respect to the IGD Values. The Best Average IGD Among the Algorithms is Highlighted in Gray

Problem	M	PPS	ToP	TiGE2	CAEAD	DPPPS	BiACMOAPO
C1_DTLZ1	3	$2.5849 \times 10^{-2} (7.34 \times 10^{-4})$ –	$NaN (NaN)$	$3.1337 \times 10^{-1} (1.28 \times 10^{-1})$ –	$2.3297 \times 10^{-2} (5.90 \times 10^{-4})$ –	$2.1515 \times 10^{-2} (2.75 \times 10^{-4})$ =	$2.1399 \times 10^{-2} (2.84 \times 10^{-4})$
C1_DTLZ3	3	$3.2565 \times 10^{0} (3.96 \times 10^{0})$ –	$1.0654 \times 10^{0} (2.40 \times 10^{0})$ –	$6.2099 \times 10^{0} (3.10 \times 10^{0})$ –	$1.9051 \times 10^{0} (2.88 \times 10^{0})$ –	$5.8497 \times 10^{-2} (8.96 \times 10^{-4})$ =	$5.8664 \times 10^{-2} (2.39 \times 10^{-3})$
C2_DTLZ2	3	$5.5964 \times 10^{-2} (1.82 \times 10^{-3})$ –	$6.2680 \times 10^{-2} (5.01 \times 10^{-3})$ –	$1.0518 \times 10^{-1} (1.61 \times 10^{-2})$ –	$5.3192 \times 10^{-2} (1.21 \times 10^{-3})$ –	$4.5803 \times 10^{-2} (5.82 \times 10^{-4})$ =	$4.5474 \times 10^{-2} (6.45 \times 10^{-4})$
C3_DTLZ4	3	$1.5708 \times 10^{-1} (8.61 \times 10^{-2})$ –	$1.5024 \times 10^{-1} (5.18 \times 10^{-3})$ –	$1.9306 \times 10^{-1} (1.43 \times 10^{-2})$ –	$1.1788 \times 10^{-1} (3.41 \times 10^{-3})$ –	$1.0823 \times 10^{-1} (2.93 \times 10^{-3})$ –	$1.0270 \times 10^{-1} (2.65 \times 10^{-3})$
DC1_DTLZ1	3	$3.0795 \times 10^{-2} (9.14 \times 10^{-3})$ –	$2.4089 \times 10^{-2} (6.36 \times 10^{-3})$ –	$5.3116 \times 10^{-1} (2.61 \times 10^{-1})$ –	$6.9361 \times 10^{-2} (7.89 \times 10^{-2})$ –	$1.2761 \times 10^{-2} (3.36 \times 10^{-4})$ –	$1.2294 \times 10^{-2} (1.20 \times 10^{-4})$
DC1_DTLZ3	3	$3.9479 \times 10^{-1} (2.90 \times 10^{-1})$ –	$1.0487 \times 10^{0} (1.53 \times 10^{0})$ –	$1.9400 \times 10^{0} (5.22 \times 10^{-1})$ –	$1.2246 \times 10^{0} (1.23 \times 10^{0})$ –	$3.6858 \times 10^{-2} (6.20 \times 10^{-4})$ –	$3.5713 \times 10^{-2} (7.20 \times 10^{-4})$
DC2_DTLZ1	3	$4.6979 \times 10^{-2} (4.97 \times 10^{-2})$ –	$NaN (NaN)$	$3.4931 \times 10^{-1} (7.29 \times 10^{-2})$ –	$7.0640 \times 10^{-2} (6.45 \times 10^{-2})$ –	$2.2769 \times 10^{-2} (6.75 \times 10^{-4})$ –	$2.1503 \times 10^{-2} (2.29 \times 10^{-4})$
DC2_DTLZ3	3	$5.1723 \times 10^{-1} (2.26 \times 10^{-1})$ –	$NaN (NaN)$	$6.9089 \times 10^{-1} (9.68 \times 10^{-3})$ –	$3.8489 \times 10^{-1} (2.51 \times 10^{-1})$ –	$1.6990 \times 10^{-1} (2.05 \times 10^{-1})$ –	$5.6489 \times 10^{-2} (5.70 \times 10^{-4})$
DC3_DTLZ1	3	$7.0453 \times 10^{-1} (1.19 \times 10^{0})$ –	$2.3773 \times 10^{0} (4.16 \times 10^{0})$ –	$1.4501 \times 10^{0} (1.03 \times 10^{0})$ –	$1.0360 \times 10^{0} (1.01 \times 10^{0})$ –	$7.6624 \times 10^{-3} (3.50 \times 10^{-4})$ –	$7.3684 \times 10^{-3} (1.12 \times 10^{-4})$
DC3_DTLZ3	3	$2.5250 \times 10^{0} (1.78 \times 10^{0})$ –	$6.8676 \times 10^{0} (4.21 \times 10^{0})$ –	$3.5166 \times 10^{0} (8.62 \times 10^{-1})$ –	$3.8378 \times 10^{0} (2.82 \times 10^{0})$ –	$2.6228 \times 10^{-1} (2.46 \times 10^{-1})$ –	$2.3438 \times 10^{-2} (1.19 \times 10^{-2})$
+/-/=		0/10/0	0/7/0	0/10/0	0/10/0	0/7/3	

Table 4. Performance Compared on DTLZ Problems with Respect to the HV Values. The Best Average HV Among the Algorithms is Highlighted in Gray

Problem	M	PPS	ToP	TiGE2	CAEAD	DPPPS	BiACMOAPO
C1_DTLZ1	3	$8.1805 \times 10^{-1} (4.49 \times 10^{-3})$ –	$NaN (NaN)$	$2.8684 \times 10^{-1} (2.36 \times 10^{-1})$ –	$8.1790 \times 10^{-1} (4.23 \times 10^{-3})$ –	$8.3354 \times 10^{-1} (3.59 \times 10^{-3})$ –	$8.3865 \times 10^{-1} (1.83 \times 10^{-3})$
C1_DTLZ3	3	$3.0851 \times 10^{-1} (2.56 \times 10^{-1})$ –	$3.1226 \times 10^{-1} (2.20 \times 10^{-1})$ –	$0.0000 \times 10^{0} (0.00 \times 10^{0})$ –	$1.3559 \times 10^{-1} (1.93 \times 10^{-1})$ –	$5.5247 \times 10^{-1} (3.40 \times 10^{-3})$ –	$5.5329 \times 10^{-1} (3.07 \times 10^{-3})$
C2_DTLZ2	3	$4.9775 \times 10^{-1} (3.67 \times 10^{-3})$ –	$4.6282 \times 10^{-1} (1.48 \times 10^{-2})$ –	$4.5192 \times 10^{-1} (1.44 \times 10^{-2})$ –	$4.8515 \times 10^{-1} (3.02 \times 10^{-3})$ –	$5.1294 \times 10^{-1} (1.26 \times 10^{-3})$ +	$5.1170 \times 10^{-1} (2.00 \times 10^{-3})$
C3_DTLZ4	3	$7.6228 \times 10^{-1} (2.50 \times 10^{-2})$ –	$7.5006 \times 10^{-1} (6.20 \times 10^{-3})$ –	$7.6632 \times 10^{-1} (4.09 \times 10^{-3})$ –	$7.7606 \times 10^{-1} (2.42 \times 10^{-3})$ –	$7.8352 \times 10^{-1} (1.49 \times 10^{-3})$ –	$7.8832 \times 10^{-1} (1.52 \times 10^{-3})$
DC1_DTLZ1	3	$5.7342 \times 10^{-1} (2.29 \times 10^{-2})$ –	$5.7416 \times 10^{-1} (2.36 \times 10^{-2})$ –	$7.4893 \times 10^{-2} (5.84 \times 10^{-2})$ –	$4.6282 \times 10^{-1} (1.83 \times 10^{-1})$ –	$6.2783 \times 10^{-1} (1.94 \times 10^{-3})$ –	$6.3068 \times 10^{-1} (7.36 \times 10^{-4})$
DC1_DTLZ3	3	$2.8396 \times 10^{-1} (1.22 \times 10^{-1})$ –	$1.8169 \times 10^{-1} (1.70 \times 10^{-1})$ –	$0.0000 \times 10^{0} (0.00 \times 10^{0})$ –	$1.0005 \times 10^{-1} (1.60 \times 10^{-1})$ –	$4.6953 \times 10^{-1} (1.65 \times 10^{-3})$ –	$4.7007 \times 10^{-1} (2.25 \times 10^{-3})$
DC2_DTLZ1	3	$7.5740 \times 10^{-1} (1.31 \times 10^{-1})$ –	$NaN (NaN)$	$2.2117 \times 10^{-1} (1.15 \times 10^{-1})$ –	$7.0658 \times 10^{-1} (1.67 \times 10^{-1})$ –	$8.3506 \times 10^{-1} (2.47 \times 10^{-3})$ –	$8.3920 \times 10^{-1} (5.67 \times 10^{-4})$
DC2_DTLZ3	3	$9.5186 \times 10^{-2} (1.94 \times 10^{-1})$ –	$NaN (NaN)$	$0.0000 \times 10^{0} (0.00 \times 10^{0})$ –	$2.0119 \times 10^{-1} (2.46 \times 10^{-1})$ –	$4.2542 \times 10^{-1} (2.16 \times 10^{-1})$ –	$5.5696 \times 10^{-1} (2.34 \times 10^{-3})$
DC3_DTLZ1	3	$1.3659 \times 10^{-1} (1.67 \times 10^{-1})$ –	$2.3459 \times 10^{-2} (7.02 \times 10^{-2})$ –	$1.0634 \times 10^{-2} (3.50 \times 10^{-2})$ –	$6.6351 \times 10^{-2} (1.62 \times 10^{-1})$ –	$5.3127 \times 10^{-1} (2.96 \times 10^{-3})$ –	$5.3242 \times 10^{-1} (1.49 \times 10^{-3})$
DC3_DTLZ3	3	$3.9329 \times 10^{-2} (1.07 \times 10^{-1})$ –	$8.7130 \times 10^{-3} (4.77 \times 10^{-2})$ –	$0.0000 \times 10^{0} (0.00 \times 10^{0})$ –	$1.3917 \times 10^{-2} (6.50 \times 10^{-2})$ –	$1.8334 \times 10^{-1} (1.67 \times 10^{-1})$ –	$3.5729 \times 10^{-1} (1.74 \times 10^{-2})$
+/-/=		0/10/0	0/7/0	0/10/0	0/10/0	1/6/3	

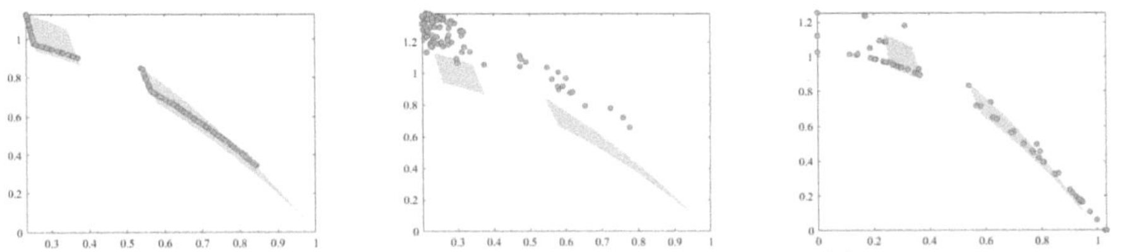

(a) PPS on MW10 (b) ToP on MW10 (c) TiGE2 on MW10

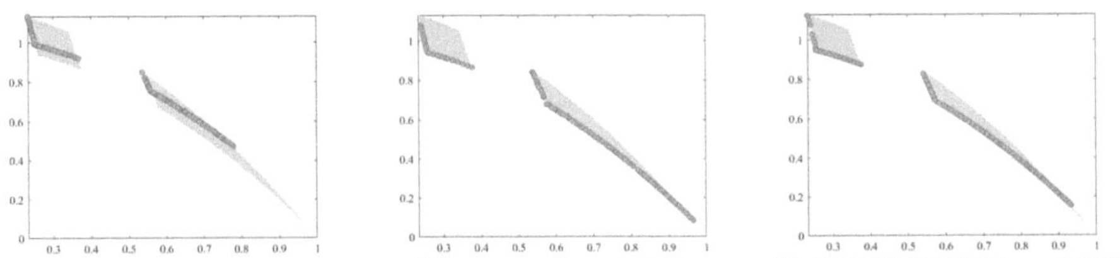

(d) CAEAD on MW10 (e) DPPPS on MW10 (f) BiACMOAPO on MW-10

Fig. 2. The non-dominated solutions achieved by each algorithm on MW10.

optimal region but did not push toward the optimal Pareto frontier that satisfies the constraints, while some algorithms were unable to find feasible regions.

Figure 3 shows that BiACMOAPO can cross large infeasible regions toward the optimal Pareto region that satisfies the constraints. Other algorithms find feasible regions but do not move toward the optimal Pareto region that satisfies the constraints. Some algorithms only find the outer feasible region.

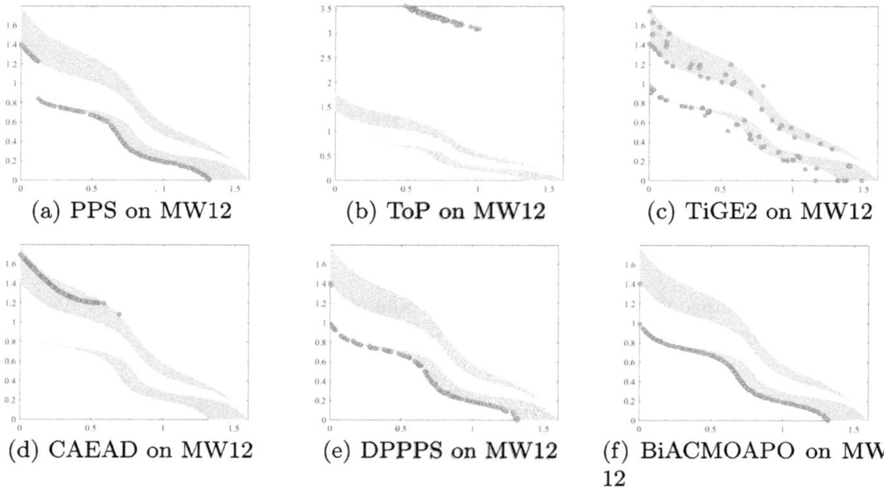

(a) PPS on MW12 (b) ToP on MW12 (c) TiGE2 on MW12
(d) CAEAD on MW12 (e) DPPPS on MW12 (f) BiACMOAPO on MW-12

Fig. 3. The non-dominated solutions achieved by each algorithm on MW12.

5 Conclusion

In this paper, a novel CMOEA called BiACMOAPO is proposed. The algorithm employs co-evolution of two population. The main population exploring the constrained Pareto front within the feasible region, while the auxiliary population ignore the infeasible region to explores new feasible domains and guides the main population towards the Constrained Pareto Front. To verify the effectiveness of the algorithm, the benchmark of MW and DTLZ series are selected and compares it with five state-of-the-art algorithms, the experiments show that this algorithm can efficiently find the Pareto optimal solution satisfying the constraints.

In future work, we will design new force updating rules to enable more efficient exploration of feasible regions between particles. And it also planned for real-world constrained multi-objective optimization.

References

1. Peng, C., Huang, X., Wu, Y., Kang, J.: Constrained multi-objective optimization for UAV-enabled mobile edge computing: offloading optimization and path planning. IEEE Wireless Commun. Lett. **11**(4), 861–865 (2022)
2. Afshari, H., Hare, W., Tesfamariam, S.: Constrained multi-objective optimization algorithms: review and comparison with application in reinforced concrete structures. Appl. Soft Comput. **83**, 105631 (2019)
3. Lv, Y., Li, K., Zhao, H., Lei, H.: A multi-stage constraint-handling multi-objective optimization method for resilient microgrid energy management. Appl. Sci. **14**(8), 3253 (2024)
4. Deb, K., Pratap, A., Agarwal, S., et al.: A fast and elitist multiobjective genetic algorithm: NSGA-II. IEEE Trans. Evol. Comput. **6**(2), 182–197 (2002)

5. Takahama, T., Sakai, S.: Constrained optimization by constrained particle swarm optimizer with-level control. Soft Comput. Transdis. Sci. Technol., 1019–1029 (2005)
6. Ma, Z., Wang, Y.: Push and pull search for solving constrained multi-objective optimization problems. Swarm Evol. Comput. **44**, 665–679 (2019)
7. Spears, W. M., Spears, D. F., Heil, R., Kerr, W., Hettiarachchi, S.: An overview of physicomimetics. Swarm Robot., 84–97 (2005)
8. Xie, L.P., Zeng, J.C.: A global optimization based on physicomimetics framework. In: Proceedings of the First ACM/SIGEVO Summit on Genetic and Evolutionary Computation, pp. 609–616 (2009)
9. Wang, Y., Zeng, J.: A multi-objective artificial physics optimization algorithm based on ranks of individuals. Soft. Comput. **17**(6), 939–952 (2013)
10. Wang, Y.: Research on multi-objective mimic physics optimization algorithm and its application. Lanzhou University of Technology (2011)
11. Zitzler, E., Laumanns, M., Thiele, L.: SPEA2: improving the strength pareto evolutionary algorithm for multiobjective optimization. evolutionary methods for design, optimization and control with applications to industrial problems. Proc. EUROGEN'2001 (2001)
12. Liu, Z.-Z., Wang, Y.: Handling constrained multiobjective optimization problems with constraints in both the decision and objective spaces. IEEE Trans. Evol. Comput. **23**(5), 870–884 (2019)
13. Zhou, Y., Zhu, M., Wang, J., Zhang, Z., Xiang, Y., Zhang, J.: Tri-Goal evolution framework for constrained many-objective optimization. IEEE Trans. Syst. Man Cybern. Syst. **50**(8), 3086–3099 (2020)
14. Zou, J., Sun, R., Yang, S., Zheng, J.: A dual-population algorithm based on alternative evolution and degeneration for solving constrained multi-objective optimization problems. Inf. Sci. **579**, 89–102 (2021)
15. Ming, F., Gong, W., Wang, L., Lu, C.: A tri-population based co-evolutionary framework for constrained multi-objective optimization problems. Swarm Evol. Comput. **70**, 101055 (2022)
16. Ma, Z., Wang, Y.: Evolutionary constrained multiobjective optimization: test suite construction and performance comparisons. IEEE Trans. Evol. Comput. **23**(6), 972–986 (2019)
17. Deb, K., Jain, H.: An evolutionary many-objective optimization algorithm using reference-point-based nondominated sorting approach, part I: solving problems with box constraints. IEEE Trans. Evol. Comput. **18**(4), 577–601 (2013)
18. Li, K., Chen, R., Fu, G., et al.: Two-archive evolutionary algorithm for constrained multiobjective optimization. IEEE Trans. Evol. Comput. **23**(2), 303–315 (2018)
19. Tian, Y., Cheng, R., Zhang, X., et al.: PlatEMO: a MATLAB platform for evolutionary multi-objective optimization. IEEE Comput. Intell. Mag. **12**(4), 73–87 (2017)
20. Zitzler, E., Thiele, L., Laumanns, M., et al.: Performance assessment of multiobjective optimizers: an analysis and review. IEEE Trans. Evol. Comput. **7**(2), 117–132 (2003)
21. Zitzler, E., Thiele, L.: Multiobjective evolutionary algorithms: a comparative case study and the strength pareto approach. IEEE Trans. Evol. Comput. **3**(4), 257–271 (1999)

Intelligent Decision-Making Technology Based on Brain-Inspired Spiking Neural Networks

Song Xu(✉), Wenjing Duan, Jie Li, and Wang Lina

National Laboratory of Aerospace Intelligent Control Technology, Beijing Aerospace Automatic Control Institute, Beijing, China
xusong618@163.com.cn

Abstract. Current intelligent agent decision control systems lack comprehensive self-assessment of operational status and effective decision-driving mechanisms, often resulting in limited adaptability to emergencies and unknown environments, as well as constrained task performance. To address the demands of intelligent agent decision-making, this paper proposes a novel brain-inspired decision-making method based on Spiking Neural Networks (SNN). First, a spiking encoding method for multi-source heterogeneous data is introduced to transform the agent's multi-source state data into spiking signals. Subsequently, leveraging the encoded multi-source state information, a brain-inspired decision control computational model based on SNN is established to achieve real-time, accurate, and stable state identification and decision-making for the agent. The proposed brain-inspired intelligent decision-making method provides a technical solution for developing next-generation intelligent agent decision systems, with the potential to accelerate the advancement of intelligent agents toward greater autonomy and intelligence.

Keywords: Brain-inspired decision-making · Spiking Neural Networks · Intelligent agent · Intelligent decision and control

1 Introduction

The theoretical study of intelligent autonomous decision-making technology has long attracted widespread attention from scholars worldwide and remains a hotspot in the field of decision control science. Its essence lies in integrating human-like intelligence into decision systems, rather than merely focusing on controller algorithms [1]. Intelligent agent decision-making enables the sharing and interoperability of state information, real-time assessment of current capabilities during operation, and automatic resource allocation and redundancy switching based on task requirements. This approach optimizes decision-making and autonomous control, equipping agents with the ability to execute dynamic tasks in complex environments while enhancing reliability. In recent years, leading international research institutions have significantly increased investments in intelligent autonomous decision-making, achieving a series of notable advancements.

H. Ma et al. (Eds.): IWACIII 2025, CCIS 2780, pp. 516–526, 2026.
https://doi.org/10.1007/978-981-95-6730-0_35

Expert system-based decision-making methods rely on domain-specific knowledge and experience to simulate human expert reasoning and judgment, addressing complex problems that typically require expert intervention [2]. For instance, [3] proposed an object-oriented rule-based knowledge construction method and inference mechanism, developing an intelligent safety control decision system. While this system flexibly processes diverse input information, its inference capability is limited, making it unsuitable for complex task scenarios. Due to the simplicity of expert system reasoning and inflexible diagnostic strategies, inference performance is often weak, and errors in the knowledge base may lead to incorrect decisions. Bayesian theory-based decision methods classify events based on their probabilities and credibility, offering modularity and transparency. These methods hold significant research value and broad application prospects in knowledge discovery from massive datasets. For example, [4] proposed a dynamic Bayesian network-based decision model for unmanned aerial vehicle (UAV) air combat, addressing real-time requirements and independence from kinematic models. This model improves UAV adaptability to battlefield environments and enhances the flexibility and effectiveness of maneuver decisions. However, many assumptions in Bayesian network learning are difficult to satisfy in practical applications, and the computational complexity escalates when solving complex problems. Neural network-based decision methods simulate the human brain's nervous system, endowing machines with perception, learning, and reasoning capabilities akin to human intelligence. Inspired by deep learning, [5] demonstrated the effectiveness of a hierarchical structure integrating convolutional neural networks into the decision-making process for indoor obstacle avoidance in mobile robots. By using raw depth images as input and control commands as output, the system accomplished model-free obstacle avoidance tasks.

In summary, current research on decision rule generation mechanisms is progressing toward greater intelligence and autonomy. However, existing intelligent agent decision-driving architectures face challenges in accurately identifying motion modes, online planning, and real-time decision-making. Establishing a theoretical framework for motion mode recognition and system operation analysis based on multi-source heterogeneous operational and environmental information is a highly challenging problem. Such a framework would enable accurate identification of dynamic transitions in agent motion modes and uncover decision rule generation mechanisms under complex tasks and abnormal states. Leveraging the brain-like working mechanism and low-power, high-performance characteristics of SNN [6], this paper designs a hybrid artificial neural network (ANN)-SNN model with an encoder-feature extractor-decoder hierarchical structure. This model directly processes multi-source heterogeneous flight data encoded as spiking signals, improving precision and adaptability.

2 Methodology

Spiking Neural Networks (SNN) hold the potential to achieve higher levels of intelligence while maintaining low power consumption, offering a more efficient computational model for intelligent tasks. Aligning with data characteristics and task requirements, this paper constructs a deep SNN model with an encoder-feature extractor-decoder hierarchical structure. The model incorporates a learnable spiking encoder inspired by traditional

ANN, enabling direct processing of analog input data. Furthermore, using error back-propagation across temporal and spatial dimensions, the deep SNN is trained end-to-end while preserving the dynamic properties of spiking neurons. The technical roadmap is illustrated in Fig. 1.

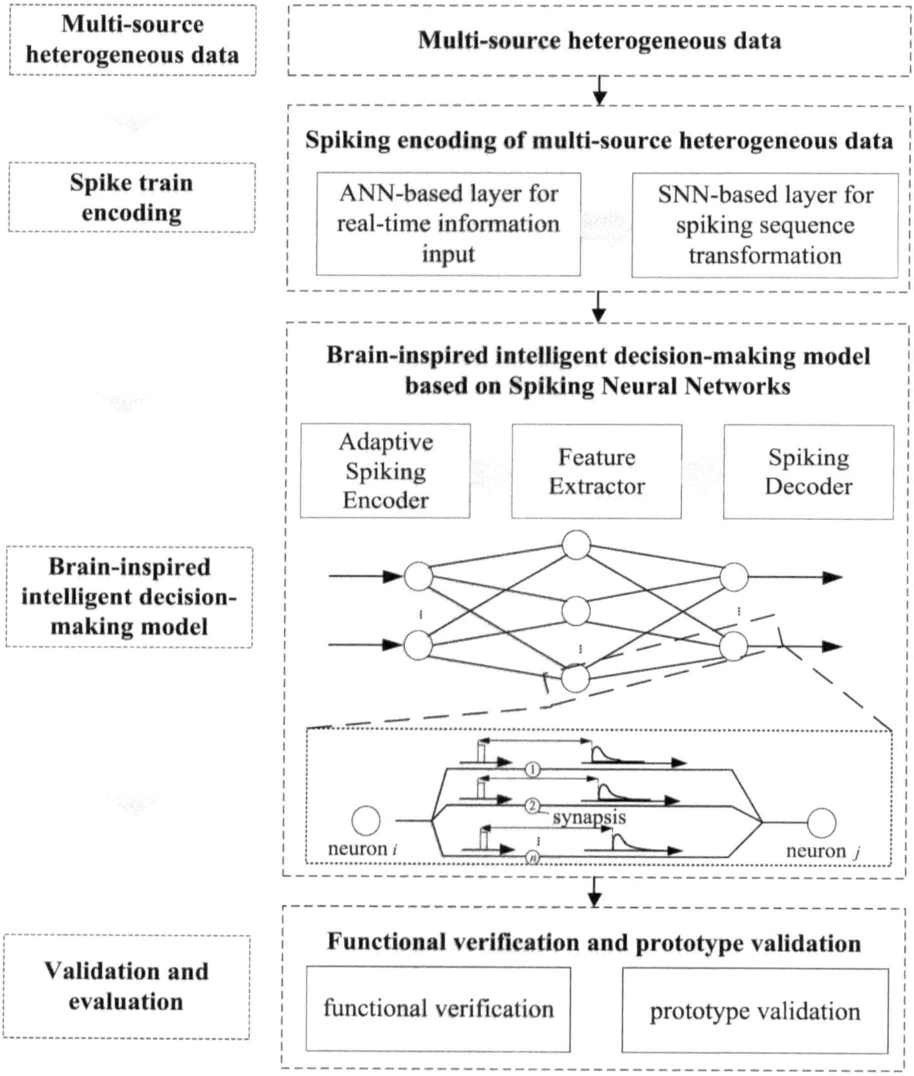

Fig. 1. Schematic diagram of the technical roadmap

2.1 Spiking Neuron Computational Model Design

Spike trains in response to external stimuli, necessitating the incorporation of spatiotemporal dynamics to express rich neural computational properties. Various spiking neuron models describe the dynamic changes in membrane potential and spike generation mechanisms at different levels of detail. The Leaky Integrate-and-Fire (LIF) model quantitatively studies neuronal spiking patterns through analytical expressions of membrane potential. Its mathematical simplicity and ability to capture key dynamics of biological neural networks have made it widely applicable in neuromorphic computing and brain science. Given its computational efficiency and neuro-dynamic properties, this paper adopts the LIF model as the fundamental computational unit of the SNN.

2.1.1 Spiking Neuron Computational Model Design

The LIF neuron is a concise phenomenological spiking neuron model that characterizes internal states through membrane potential dynamics and spiking mechanisms. As shown in Fig. 2(a), the LIF model simplifies neurons into nodes, representing them as parallel combinations of "leaky" resistors and capacitors.

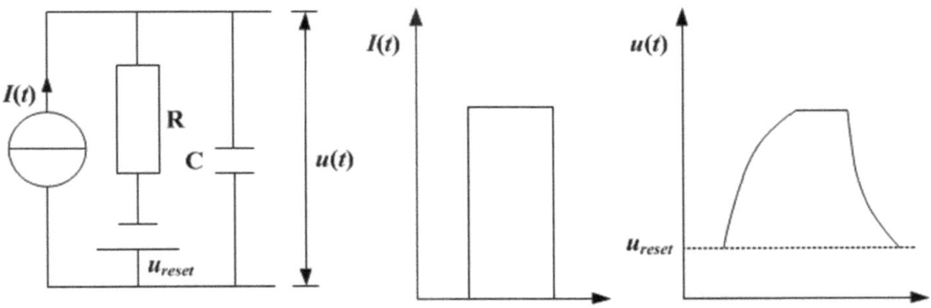

Fig. 2. Physical model of the LIF neuron: (a) Equivalent RC circuit; (b) Schematic diagram of membrane potential response.

Based on circuit principles, the input current $I(t)$ equals the sum of currents through the resistor R and capacitor C, i.e., $I(t) = I_R + I_C$. For the resistor current, Ohm's law yields $I_R = \frac{u-u_{reset}}{R}$, while the capacitor current can be expressed as $I_c = C\frac{du}{dt}$ based on current definitions. Rearranging these terms gives:

$$I(t) = \frac{u(t) - u_{reset}}{R} + C\frac{du}{dt} \tag{1}$$

Multiplying both sides by R and moving the $RC\frac{du}{dt}$ term to the left yields:

$$RC\frac{du}{dt} = -(u - u_{reset}) + RI(t) \tag{2}$$

Here, RC is the membrane time constant (τ_m), indicating the time required for the membrane potential to reach equilibrium. The standard LIF model's differential equation is:

$$\tau_m \frac{du}{dt} = -(u - u_{reset}) + RI(t) \tag{3}$$

$$\text{spike}\&u = u_{reset}, \quad u > \theta \tag{4}$$

where θ is the firing threshold. When the membrane potential reaches this threshold, the neuron generates an action potential. Eq. (3) introduces a leak current term, maintaining the membrane potential at rest in the absence of external stimuli, thereby explaining ion diffusion effects in biological neurons. Upon exceeding θ, the neuron emits a spike and resets its membrane potential to u_{reset}, In SNN, each presynaptic spike is modulated by synaptic weights to produce postsynaptic input currents $I(t)$. Thus, the input current from the j-th neuron can be expressed as:

$$I(t) = \sum\nolimits_j w_{ij} \sum\nolimits_f \alpha\left(t - t_j^{(f)}\right) \tag{5}$$

where w_{ij} is the synaptic weight, $t_j^{(f)}$ is the time of the f-th spike from neuron i, and $\alpha(\cdot)$ may take the simplest form of a Dirac function.

2.1.2 Spiking Neuron Computational Model Design

As described, the standard LIF neuron is a continuous-time dynamic system defined by differential equations, as shown in Eq. (6):

$$\tau_m \frac{du_i^{(l)}}{dt} = -\left(u_i^{(n)} - u_{reset}\right) + RI_i^{(n)} \tag{6}$$

Here, $u_i^{(n)}$ denotes the membrane potential of neuron ii in layer l, $I_i^{(n)}$ is the external input current, u_{reset} is the resting potential, τ_m is the membrane time constant, and R is the input resistance. Although Eqs. (1) to (2) detail the LIF model's dynamics, this section discretizes the LIF model for practical implementation. Assuming $R = 1$ without loss of generality and sampling step size $\Delta t > 0$, the discretized LIF neuron's input, membrane potential update, reset, and spiking dynamics are:

$$I_i^{(l)}[t+1] = \sum\nolimits_{j=1}^{(l-1)} w_{ij}^{(l)} o_j^{(l)}[t+1] \tag{7}$$

$$du_i^{(l)} = \frac{\Delta t}{\tau_{mem}}\left(u_{reset} - u_i^{(l)}[t]\right) + I_i^{(l)} \tag{8}$$

$$u_i^{(l)}[t+1] = \left(1 - o_i^{(l)}[t+1]\right) \cdot \left(u_i^{(l)}[t] + du_i^{(l)}[t] \cdot \Delta t\right) + o_i^{(l)}[t+1] \cdot u_{reset} \tag{9}$$

$$\mathrm{u}_i^{(l)}[\mathrm{t}+1] = \left(1 - \mathrm{o}_i^{(l)}[\mathrm{t}+1]\right) \bullet \left(\mathrm{u}_i^{(l)}[\mathrm{t}] + \mathrm{du}_i^{(l)}[\mathrm{t}] \bullet \Delta \mathrm{t}\right) + \mathrm{o}_i^{(l)}[\mathrm{t}+1] \bullet \mathrm{u}_{\mathrm{reset}}$$

$$o_i^{(l)}[t] = h\left(u_i^{(l)} - \theta\right) \tag{10}$$

where $w_{ij}^{(l)}$ represents synaptic strength, $o_i^{(l)}$ denotes output spike events, and $du_i^{(l)}$ is the membrane potential increment of neuron i in layer l over Δt. The step function $h(x)$, outputting 0 or 1, models the spiking process:

$$h(x) = \begin{cases} 1, & x \geq 0 \\ 0, & x < 0 \end{cases} \tag{11}$$

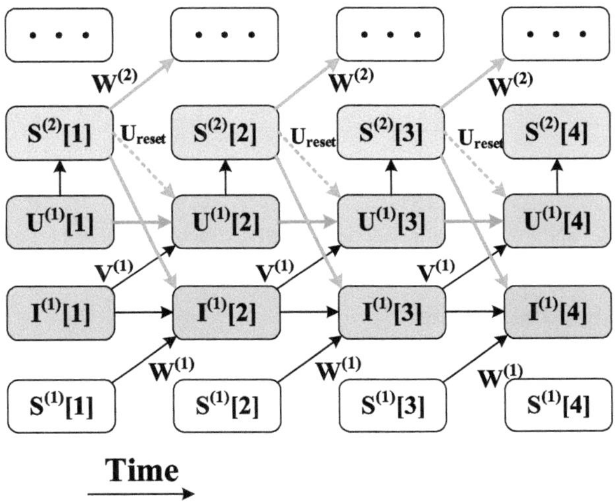

Fig. 3. Computational graph of the spiking neural network in discrete time

The LIF spiking neuron is a specialized recurrent neural network structure due to its temporal dimension. Its behavior can be described by three discrete equations: charging, resetting, and spiking. Fig. 3 illustrates the SNN's computational graph in discrete time, where inputs are external currents multiplied by membrane resistance, hidden states are membrane potentials, and outputs are spike trains. Input spikes propagate upward through layers, with synaptic currents decaying at each time step and influencing membrane potential changes. Threshold nonlinearities generate spike trains at each step, and the Markov property ensures current outputs depend only on current inputs and neuron states.

2.2 Brain-Inspired Decision Computational Network Design for Intelligent Agent Systems

The discretized LIF model can be simulated using mature deep learning frameworks, facilitating flexible design and implementation of deep SNNs for machine learning tasks. Compared to traditional deep neural networks, SNN-based machine learning involves three stages: spiking encoding, feature extraction, and decoding. This paper designs a hybrid ANN-SNN model with an encoder-feature extractor-decoder hierarchy, directly processing diverse input data to improve precision (Fig. 4).

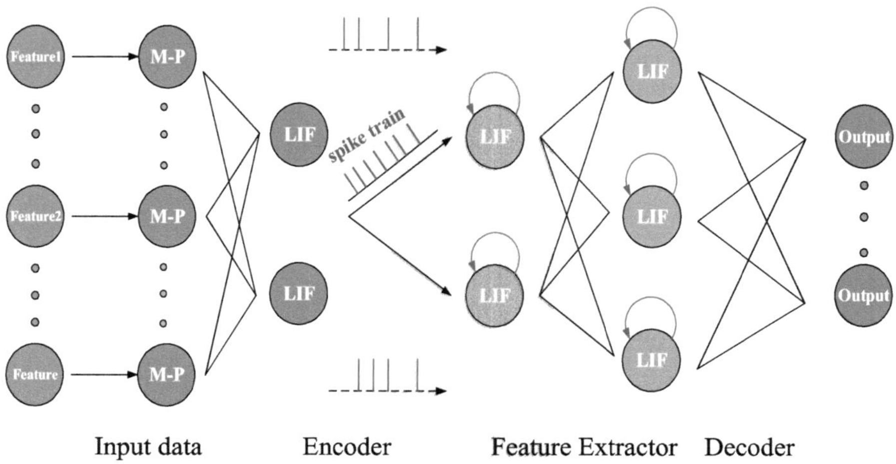

Fig. 4. SNN model with encoder-feature extractor-decoder hierarchy

2.2.1 Adaptive Spiking Encoder

SNNs receive spike trains as input, necessitating tailored encoding schemes for different data types. Independent encoding methods may lose original features and prolong simulation cycles, exacerbating gradient vanishing. To address this, an adaptive spiking encoder based on neural networks is proposed. Each neuron in this layer processes real-valued inputs through one or more ANN layers before generating spike trains via LIF neurons.

For heterogeneous data, hybrid ANN-SNN encoder layers employ task-specific ANN architectures to extract deeper structural information, while SNN layers encode data into spikes for binary representation. The ANN layer's flexible structure and learnable parameters optimize overall performance, yielding precise spatiotemporal patterns.

2.2.2 Feature Extractor

Once the initial data conversion into spike trains is completed, the construction of SNNs can conceptually follow the layered architecture prevalent in traditional deep networks. The feature extraction component typically comprises one or more layers of Leaky Integrate-and-Fire (LIF) neurons, which can be interconnected via full (dense), convolutional, or recurrent synaptic connections, depending on the specific data structure and task requirements. To manage computational complexity and control for potential overfitting, pooling operations, such as max-pooling or average-pooling, are employed to reduce the spatial dimensionality of the spiking activity. In the context of spiking convolutional layers, the process involves convolving the input spike trains with learnable weight kernels at every single time step. This convolution generates synaptic currents that subsequently integrate within postsynaptic neurons, gradually accumulating towards their firing threshold. Conversely, spiking average pooling utilizes fixed, non-trainable kernels to compute the mean of the membrane potentials within a receptive field. This approach effectively downsamples the feature maps while consciously preserving the underlying spatial topology of the information.

2.2.3 Spiking Decoder

The decoder translates output layer spikes into task-specific results. Fig. 5(a) shows firing rates indicating class responses, where the correct class neuron spikes most frequently. Fig. 5(b) illustrates membrane potential dynamics over simulation cycles.

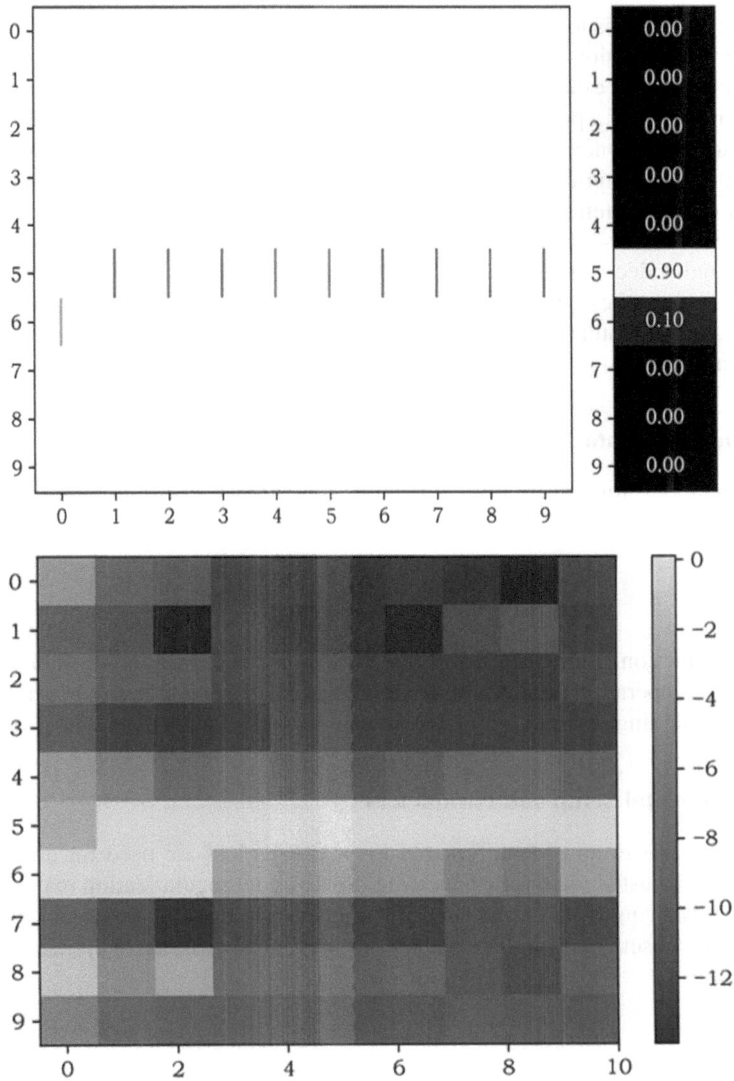

Fig. 5. Output layer neuron behavior: (a) Spike rates; (b) Membrane potentials

The proposed model processes input data through the encoder, extracts spatiotemporal features via the feature extractor, and decodes results for classification. Notably, parameters (e.g., convolutional, pooling, or fully connected layers) are shared across all simulation cycles.

3 System Identification and Decision Analysis for Agent State Data

This section is dedicated to the comprehensive validation and performance evaluation of the proposed Spiking Neural Network (SNN)-based methodology. The primary objective of this methodology is to achieve accurate and real-time agent state identification while ensuring robust decision-making capabilities under a spectrum of typical power system failure scenarios. To establish a rigorous and fair assessment framework, high-fidelity simulation models of a power grid are meticulously constructed. These models are used to generate corresponding, time-synchronized state data, with a particular emphasis on replicating challenging, complex, and multi-stage cascading failure events that test the limits of the system. Against this simulated backdrop, the feasibility and overall effectiveness of our specialized neural architecture—comprising an encoder, a feature extractor, and a decoder within the SNN framework—are systematically tested. The analysis focuses on key performance metrics, including identification accuracy, decision optimality, and computational efficiency, thereby providing a holistic verification of the proposed approach

3.1 Agent State Data Acquisition under Power Failures

Simulation models for power failures are built, with thrust transfer functions represented as:

$$\frac{P(s)}{P_c(s)} = k_e e^{-\tau_e s} \frac{w^2}{s^2 + 2w\varepsilon s + w^2} \tag{12}$$

Power reduction ratios (0%, 30%, 60%, 90%) are tested, with failures uniformly distributed across operation time. Multiple simulations with random noise generate motion trajectories, yielding velocity and position datasets for power failure states.

3.2 Experimental Setup and Parameters

The Adam optimizer and Mean Squared Error (MSE) loss are used for training, with learning rates adjusted via cosine annealing. Spike count regularization maintains sparsity, balancing firing rates across layers. Time-specific dropout randomly deactivates fixed neuron subsets per simulation cycle to prevent overfitting. Key parameters are listed in Table 1.

Table 1. Key parameters

Paremeter	Description	Value
Δt	Intregate time	1 ms
T	Simulation time	10 ~ 20 ms
θ	Threshold	1
τ_{mem}	membrane time constant	100
λ	Parameter	0.1
k	degree	1

3.3 State Identification and Decision-Making under Power Failures

The proposed SNN model is evaluated on labeled simulation data. Comparative results (Table 2) demonstrate higher accuracy than traditional machine learning methods and competitive performance against ANN, with lower computational overhead and suitability for neuromorphic hardware.

Table 2. Identification accuracy for the state of intelligent agent

Method	Accurancy
Decision tree	91.2%
ANN	95.1%
Proposed method	95%

4 Conclusions

Intelligence is the core direction for the future development of smart systems. To address the growing demands of complex tasks, this paper innovatively proposes a brain-inspired intelligent decision-making system based on Spiking Neural Networks (SNNs). This system, by simulating the neural information processing mechanisms of the brain, lays a key technological foundation for constructing next-generation agent decision-control systems. The proposed method aims to significantly enhance an agent's capabilities for precise state identification and decision-making in complex, dynamic environments, thereby effectively improving its information processing efficiency and environmental adaptability. Through the integration of insights from brain science and neuromorphic engineering into decision control, this study successfully establishes an efficient perception-decision-action closed-loop system. This advancement promotes a fundamental shift for intelligent agents from traditional rule-driven approaches to autonomous intelligent decision-making, injecting new vitality into theoretical and disciplinary innovation in related fields.

References

1. Wu, H., Chang, Y.: Brief introduction and research conception of intelligent control system. Space Control Technology and Application. **45**(4), 1 (2020)
2. Sun, X., Yan, X., Shang, Y., et al.: An improved rete algorithm based on shared degree model. Acta Automat. Sin. **43**(9), 1571–1579 (2017)
3. Li, Y., Zhao, G., Huang, X., Qiao, Y., Li, W.: Design of Object-Based Intelligent Decision-Making System for rocket safety control. Journal of Chongqing University (Natural Science Edition). **27**(12), 102–106 (2004)
4. Meng, G., Luo, Y., Liang, X., Xu, Y.: Air combat decision-making method based on dynamic Bayesian networks. Command, Control and Simulation. **39**(3), 49–54 (2017)
5. Tai, L., Li, S., Liu, M.: A deep-network solution towards model-less obstacle avoidance[C]//2016 IEEE/RSJ international conference on intelligent robots and systems (IROS). IEEE, 2759–2764 (2016)
6. Kaushalya Kumarasinghe, Nikola Kasabov, Denise Taylor: Deep learning and deep knowledge representation in spiking neural networks for brain-computer interfaces. Neural Netw. **121**, 169–185 (2020)
7. Midya, R., et al: Artificial neural network (ANN) to spiking neural network (SNN) converters based on diffusive memristors. Adv. Electron. Mater. **5**(9), 1900060 (2019)
8. Kyle D Julian, Mykel J Kochenderfer, Michael P Owen: Deep neural network compression for aircraft collision avoidance systems. J. Guid. Control. Dyn. **42**(3), 598–608 (2019)

Quantum-Hybrid Multi-objective Artificial Physics Optimization Algorithm

Xingyu Chen[1] and Liping Xie[2(✉)]

[1] School of Computer Science and Technology, Taiyuan University of Science and Technology, Taiyuan 030024, Shanxi, China
S202320110928@stu.tyust.edu.cn

[2] Shanxi Key Laboratory of Complex Systems and Computational Intelligence, Taiyuan University of Science and Technology, Taiyuan 030024, Shanxi, China
xieliping@tyust.edu.cn

Abstract. The Multi-Objective Artificial Physics Optimization algorithm faces significant challenges due to its population-dependent update mechanism. This dependency often restricts the exploration of the search space, potentially leading to premature convergence to local optimization and a reduction in solution diversity, thus limiting its effectiveness in complex multi-objective problems. To address these issues, this paper introduces a novel quantum computing-based MOAPO algorithm. By leveraging quantum parallelism and superposition, the proposed method aims for more effective exploration. Specifically, it employs quantum state initialization to establish a diverse initial population, then utilizes an angle-based quantum space rotation mechanism to refine solution updates and guide the search trajectory. This hybrid strategy integrates the global exploration of quantum computing with the local exploitation advantages of APO forces, effectively enhancing the algorithm's ability to escape local optimization and maintain solution diversity. Extensive comparisons against seven algorithms on twenty-one benchmark functions demonstrate that the proposed quantum-based MOAPO significantly improves both Pareto front diversity and convergence performance, indicating its potential for solving complex multi-objective optimization problems.

Keywords: Multi-objective Optimization · Artificial Physics Optimization · Quantum Computing

1 Introduction

Multi-objective Optimization Problems (MOPs), which aim to simultaneously optimize multiple conflicting objectives, are prevalent in scientific research and engineering practice. To solve such problems, various metaheuristic algorithms have been proposed, such as the Multi-objective Genetic Algorithm [1], the Multi-objective Particle Swarm Optimization [2], the Multi-objective Artificial Bee Colony [3], and the Multi-objective Grey Wolf Algorithm [4]. Among

H. Ma et al. (Eds.): IWACIII 2025, CCIS 2780, pp. 527–543, 2026.
https://doi.org/10.1007/978-981-95-6730-0_36

these, the Artificial Physics Optimization (APO) algorithm, proposed by Xie et al. [5], has garnered significant attention for its advantages, including fast convergence, high precision, and simple implementation. Subsequently, Wang et al. [6] extended APO to the multi-objective domain, proposing the Multi-objective Artificial Physics Optimization (MOAPO) algorithm. This algorithm aggregates objectives using random weights and utilizes APO's search mechanism to find the Pareto optimal set. However, when dealing with complex problems, MOAPO and its subsequent improvements, such as PEMOAPO by Sun [7], are still constrained by the inherent drawback of insufficient diversity and uniformity in the solution set. These inherent challenges limit the algorithm's performance when confronting high-dimensional and complex Pareto fronts.

Quantum computing (QC), founded on principles of quantum superposition, entanglement, interference, and collapse [8,9], offers a revolutionary paradigm for efficiently solving complex optimization problems through its inherent parallel processing capabilities. Researchers have enhanced metaheuristic algorithms by integrating quantum computing concepts. Early algorithms such as the Genetic Quantum Algorithm (GQA) [10] and the Quantum-inspired Evolutionary Algorithm (QEA) [11], embedded principles like qubits and quantum gates into classical evolutionary frameworks. Subsequent studies have leveraged quantum mechanisms to improve population diversity and balance search strategies. For instance, QHDBO [12] enhances population uniformity and diversity, QARO [13] improves global search efficiency, and QDGWO [14] uses quantum superposition to prevent premature convergence. DM-QPSO [15], building on QPSO [16], introduces a diversity migration strategy, while MOQPSO-DSCT [17] employs a dual search strategy and a circular transposon mechanism to enhance both solution convergence and distribution.

Despite significant advances in multi-objective optimization, balancing convergence and diversity remains a persistent challenge for many metaheuristic algorithms, including the Multi-objective Artificial Physics Optimization (MOAPO). While MOAPO's search mechanism promotes rapid convergence, it often leads to premature convergence to local optima due to a loss of population diversity. To address these limitations, this paper proposes a Quantum Computing-integrated Multi-objective Artificial Physics Optimization (QC-MO APO) algorithm.

The main contributions of this paper are as follows: (1) For initialization, it leverages the superposition and measurement collapse properties of qubits, combined with adaptive weight mapping, to generate a diverse initial population, thereby overcoming the limitations of traditional random initialization. (2) During the evolutionary process, an adaptive quantum rotation strategy is introduced. This strategy employs multi-layered weighting for precise qubit control and a dual-vector angle model to simultaneously enhance both the approximation of the Pareto front (convergence) and population diversity. (3) At the framework level, a diversity evaluation mechanism enables dynamic switching between the quantum and real-valued populations. This approach effectively balances exploration and exploitation, preventing the algorithm from converging prematurely.

The remainder of this paper is organized as follows: Sect. 2 briefly reviews relevant basic concepts. Section 3 elaborates on the proposed algorithm. Section 4 details the experimental setup and analyzes the results. Finally, Sect. 5 concludes the paper.

2 Related Work

2.1 Multi-objective Artificial Physics Optimization (MOAPO)

The Multi-objective Artificial Physics Optimization (MOAPO) algorithm [6] models solutions as particles whose movement toward Pareto-optimal solutions is driven by virtual forces determined by mass, which is assigned based on a rank ($rank_i$) derived from dominance and crowding distance, where a lower rank indicates a superior solution. The mass is calculated from this rank via Eq. (1).

$$m_i = \exp(\frac{-rank_i}{N}) \tag{1}$$

Herein, N denotes the population size. This rank-based quality function can effectively differentiate the quality among individuals, providing a crucial reference for the subsequent evolutionary and optimization process, and guiding the population to evolve towards more optimal solutions.

The force $F_{ij,k}$ exerted by individual j on individual i along the k-th dimension is initially defined as Eq. (2):

$$F_{ij,k} = \begin{cases} Gm_i m_j (x_{j,k} - x_{i,k}) & \text{if } m_i < m_j \\ -Gm_i m_j (x_{j,k} - x_{i,k}) & \text{if } m_i \geq m_j \\ \forall i \neq j \end{cases} \tag{2}$$

where G is the gravitational constant. For multi-objective problems, this formulation is adapted to incorporate an external archive set, modifying the force calculation as shown in Eq. (3):

$$F_{ij,k} = \begin{cases} Gm_i m_j (x_{j,k} - x_{i,k}), & j \in P(i) \text{ and } x_j \in DA(i) \\ -Gm_i m_j (x_{j,k} - x_{i,k}), & j \in P(i) \text{ and } x_j \in DA(i) \\ 0, & \text{otherwise} \end{cases} \tag{3}$$

Simultaneously, individual i updates its position $x_{i,k}$ in the k-th dimension according to Eq. (4) and (5), respectively.

$$x_{i,k}(t+1) = x_{i,k}(t) + \lambda \cdot F_{i,k}/m_i \tag{4}$$

$$x_{i,k}(t+1) = x_{i,k}(t) + \beta \cdot (x_{r1,k} - x_{r2,k}) + \gamma \cdot (x_{\text{best},k} - x_{i,k}) \tag{5}$$

$\omega \in [0, 1]$ is the inertia weight; λ, β, and γ are random numbers drawn from a uniform distribution on (0, 1). The terms $x_{r1,k}$ and $x_{r2,k}$ represent the positions of the two individuals closest to $x_{i,k}$ along the k-th dimension. Additionally, the position of each individual, x_i, is bounded by the search range $[x_{min}, x_{max}]$.

2.2 Quantum Representation

A quantum state represents any possible configuration of a quantum mechanical system. Qubits, as the fundamental information units in quantum computing, differ from classical bits by existing in superpositions of basis states denoted as $|0\rangle$ and $|1\rangle$, as formalized in Eq. (6).

$$|\psi\rangle = \alpha|0\rangle + \beta|1\rangle, |0\rangle = \begin{bmatrix}1\\0\end{bmatrix}, |1\rangle = \begin{bmatrix}0\\1\end{bmatrix} \tag{6}$$

In quantum mechanics, a qubit in superposition state (Eq. (6)) is characterized by complex probability amplitudes α and β. When measured in the computational basis, the qubit collapses to either the $|0\rangle$ state with probability $|\alpha|^2$ or the $|1\rangle$ state with probability $|\beta|^2$, satisfying the normalization condition $|\alpha|^2 + |\beta|^2 = 1$ [18]. This measurement-induced collapse to the corresponding eigenstate constitutes a fundamental operation for extracting classical information from quantum systems.

In quantum computational frameworks, information is encoded within multi-qubit registers where collections of qubits form a Quantum Population (QP). Each quantum individual $Q_i = (q_{i,1}, q_{i,2}, \dots, q_{i,D})$ encodes a candidate solution, with the quantum information representable through either probability amplitude mappings to classical real-valued solutions or angle-based representations that facilitate quantum domain operations. Each quantum unit q_{ij} within these individuals is formally defined as shown in Eq. (7).

$$q_{ij} = (q_{ij,1}, ..., q_{ij,k} ..., q_{ij,n})(1 \leq i \leq N, 1 \leq j \leq D, 1 \leq k \leq n) \tag{7}$$

Each quantum unit q_{ij} is composed of a string of qubits, the specific form of which is shown in Eq. (8) [14].

$$q = \begin{bmatrix}\cos(\theta_1)\ \cos(\theta_2)\ ...\ \cos(\theta_n)\\ \sin(\theta_1)\ \sin(\theta_2)\ ...\ \sin(\theta_n)\end{bmatrix} \tag{8}$$

In quantum computing, a qubit's state $(\cos(\theta_i), \sin(\theta_i))$ collapses upon measurement into a classical binary string (Eq. (9)). Specifically, each quantum individual in the QP generates a binary string composed of n bits of 0 or 1 after measurement, subsequently mapped to a D-dimensional real-valued vector $X_i = (x_{i1}, \dots, x_{iD})$ [10]. This process transforms the quantum representation into a classical, real-valued solution, thereby bridging quantum computing with classical optimization and facilitating individual evaluation within the solution space.

$$qb_{ij}(t) = b_{n-1}(t)b_{n-2}(t)...b_1(t)b_0(t) \tag{9}$$

In quantum representation, qubits are encoded using phase angles (θ_i) that enable coherent evolution without state collapse [19,20]. Through quantum gates, particularly rotation gates, these phase angles can be precisely modified, facilitating complex operations on quantum states. The θ_i directly converts to

$(\cos(\theta_i), \sin(\theta_i))$ via trigonometric functions, allowing computational flexibility based on specific requirements. For a system where n denotes the length of individual qubits, the corresponding phase vector (q_θ) [18] comprises n phase angles (θ_i) and can be formally expressed as Eq. (10).

$$q_\theta = [\theta_1, \theta_2, ..., \theta_n], \theta \in [0, \pi] \tag{10}$$

Quantum gates are categorized as single-qubit or multi-qubit gates based on their operational scope. Single-qubit gates manipulate individual qubits and include Pauli gates (X, Y, Z), the Hadamard gate, and various phase gates. Geometrically, these gates represent rotations on the Bloch sphere around the respective axes, implemented through exponentiated Pauli matrices. The X gate functions as a bit-flip operation, while the Z gate performs phase-flips. These rotational operations are mathematically expressed in Eqs. (11) and (12).

$$R_x(\theta) \equiv e^{-i\theta X/2} = \cos\frac{\theta}{2}I - i\sin\frac{\theta}{2}X = \begin{bmatrix} \cos\frac{\theta}{2} & -i\sin\frac{\theta}{2} \\ -i\sin\frac{\theta}{2} & \cos\frac{\theta}{2} \end{bmatrix} \tag{11}$$

$$R_z(\theta) \equiv e^{-i\theta Z/2} = \cos\frac{\theta}{2}I - i\sin\frac{\theta}{2}Z = \begin{bmatrix} e^{-i\theta/2} & 0 \\ 0 & e^{i\theta/2} \end{bmatrix} \tag{12}$$

Multi-qubit gates operate on two or more qubits. A crucial example is the Controlled-NOT (CNOT) gate, which is frequently used in algorithms such as QAOA [21]. The CNOT gate acts on a control and a target qubit: the state of the target qubit is flipped (equivalent to applying an X gate) if and only if the control qubit is in the state $(|1\rangle)$. Its matrix representation is as follows:

$$U_{CNOT} = \begin{bmatrix} 1\,0\,0\,0 \\ 0\,1\,0\,0 \\ 0\,0\,0\,1 \\ 0\,0\,1\,0 \end{bmatrix} \tag{13}$$

3 Proposed Method

3.1 Frame Work

Based on the quantum computing paradigm, the QC-MOAPO algorithm introduces innovative population initialization and dynamic update strategies. This method aims to significantly enhance population diversity and effectively avoids premature convergence to local optima by intelligently balancing global exploration and local convergence. The specific implementation process of the Algorithm 1 is as follows.

First, initialize a quantum population (QP) containing N quantum individuals. Subsequently, through a mapping mechanism, each quantum individual $Q_i(t)$ generates a corresponding real-valued individual $X_i(t)$, forming the initial real-valued population (DP). Simultaneously, establish an archive set (DA) for storing the optimal solutions in the real-valued space.

Algorithm 1. Quantum Population QP Initialization

Require: Population size N, objective space dimension M, number of qubits n, maximum iterations $T_{\max}$
Ensure: Optimal solution DA
Initialize quantum population Q with N random individuals and rotation coefficients//Algorithm 2
Establish real number archive DA
while $t < T_{\max}$ **do**
  Evaluate quantum individuals//Algorithm 3
  if quantum diversity > threshold **then**
    Update using the real number population
  else
    Update using the quantum population//Algorithm 5
    if quantum population diversity improves **then**
      Generate a new real number population for update//Algorithm 4
    end if
  end if
  Select the next generation population QP and DP through environmental selection
  Update archive DA
end while

Then in each iteration, the algorithm's core decision-making is based on whether the Quantum Diversity (QD) exceeds a predefined threshold, and different update strategies are adopted accordingly.

Path 1: High Quantum Diversity. When population diversity is above the threshold, the algorithm switches to the real-number space for exploitation. The quantum population is collapsed into a DP, which is then evolved using the classical Artificial Physics Optimization (MOAPO) method. During this process, the DA is updated with the newly generated elite solutions. Path 2: Low Quantum Diversity. Conversely, when diversity falls below the threshold, the algorithm operates within the quantum space to enhance exploration. New quantum individuals are generated via angle-based quantum rotation gates. Subsequently, the algorithm assesses whether the change in diversity necessitates collapsing the new quantum population into a real-valued one to update the archive. After completing the update via either path, the algorithm checks the termination condition. If satisfied, the optimal solution from the DA is output; otherwise, the next iteration begins.

The algorithm executes four core steps in each iteration: mapping (measuring quantum states to obtain classical solutions), evaluation (finding the Pareto front), adaptive adjustment (updating weights based on solution quality), and quantum update (evolving quantum states through rotation gates). Through cyclical execution of these steps, the entire quantum population is driven to evolve toward regions of optimal objective function values.

3.2 Population Initialization Based on Quantum Computation

In the QC-MOAPO algorithm, population initialization adopts a quantum co mputing-based approach, which differs from the initialization strategies of traditional algorithms like PLATEMO. The core of this method lies in utilizing quantum computing principles to represent and evolve the population of solutions.

First, initialization employs quantum computing principles to represent and evolve potential solutions. A population comprising N quantum individuals is constructed, each characterized by D dimensions where each dimension is encoded using n qubits. To ensure complete impartiality at the algorithm's inception, all fine-grained qubit weights $w_{i,k,j}$ are uniformly initialized to a constant value of 0.05, establishing an unbiased search space. Concurrently, an initial weight w_i is assigned to each individual i according to Eq. (14), maintaining population diversity while preserving the equitable treatment of all potential solutions.

$$w_i = \frac{1}{N} \tag{14}$$

This uniform weight setting ensures the impartiality of the initial population.

For each dimension k of a quantum individual i, its quantum state $|\theta_{i,k}\rangle$ is measured. To reduce the stochasticity of quantum measurement, this process is repeated T times. In a single measurement, a 20-bit binary string $b_{i,k}$ is first obtained and then converted into an integer according to Eq. (15). Subsequently, by averaging all integer values obtained from the T measurements, the classical corresponding value $s_{i,k}$ for that dimension is calculated (see Eq. (16)), which serves as a stable output.

$$s_{i,k} = s_{i,k} + \sum_{j=1}^{20} b_{i,k,j} \cdot 2^{20-j} \tag{15}$$

$$s_{i,k} = \frac{s_{i,k}}{T} \tag{16}$$

Furthermore, the expectation value of the Pauli-Z operator, $\langle Z \rangle_{i,k}$, is calculated for the quantum state of this dimension (see Eq. (17)). This value quantifies the degree to which the qubit is biased towards the $|0\rangle$ or $|1\rangle$ state.

$$\langle Z \rangle_{i,k} = \frac{1}{n} \sum_{j=1}^{n} (|\alpha_{i,k,j}|^2 - |\beta_{i,k,j}|^2) \tag{17}$$

Then, $s_{i,k}$ and $\langle Z \rangle_{i,k}$ are combined to calculate the final numerical value $q_{i,k}$ (Eq. (18)). Meanwhile, to increase population diversity or exploration capability, a small perturbation term $0.1 \cdot \langle Z \rangle_{i,k}$ is introduced. Subsequently, the numerical values of all dimensions are combined into a D-dimensional vector, serving as the solution of the quantum individual in the space. The vectors of all individuals constitute the quantum population QP. The Algorithm 2 presents an iterative quantum population-based optimization process.

$$\theta_{i,k} = \frac{s_{i,k}}{2^{20} - 1} + 0.1 \cdot \langle Z \rangle_{i,k} \tag{18}$$

Algorithm 2. Quantum Population QP Initialization

Require: Quantum population QP, number of objectives M
Ensure: Updated quantum population QP

Initialize weights (Eq. (14))
$Q \leftarrow$ empty population
for $i \leftarrow 1$ to N **do**
  for $k \leftarrow 1$ to D **do**
    Measure $|\beta_{i,k}\rangle$ T times to obtain $S_{i,k}$ (Eqs. (15)–(16))
    Calculate qubit expected value $Z_{i,k}$ (Eq. (17))
    Generate final value $\theta_{i,k}$ (Eq. (18))
  end for
  Add $[\theta_{i,1}, ..., \theta_{i,D}]$ to QP
end for

3.3 Quantum Population Diversity Assessment and Spatial Mapping

Following quantum population initialization, the algorithm evaluates population diversity using the Weighted Pareto Front Score (WPFS) indicator (Eq. (19)). This assessment determines whether to trigger a mapping operation, thereby balancing global exploration and local exploitation.

$$WPFS = |ND| + \alpha \cdot NPI + \beta \cdot \sum_{m=1}^{M} DSM \qquad 0 < \alpha, \beta < 1 \tag{19}$$

This comprehensive metric evaluates three critical aspects: the number of non-dominated solutions in objective space ($|ND|$), the aggregate objective function values of all non-dominated solutions (Eq. (20)), and the dimensional span of the non-dominated front (Eq. (21)). Through this multi-dimensional assessment framework, the algorithm scientifically determines optimal transition points between quantum and real-number updates, effectively balancing exploration and exploitation capabilities. This strategic approach prevents convergence to local optima while maintaining solution diversity, ultimately enhancing both search effectiveness and computational efficiency.

$$NPI = \frac{1}{1 + \frac{\sum\limits_{\theta \in ND} \sum_{m=1}^{M} f_m(\theta)}{f_{\max} - f_{\min}}} \tag{20}$$

In Eq. (20), ND represents the non-dominated solution set approximating the Pareto front, where θ denotes an individual solution, M indicates the total objective count, and $f_m(\theta)$ expresses the m-th objective value for solution θ. The parameters f_{max} and f_{min} correspond to the maximum and minimum objective values across the population. For minimization problems, superior non-dominated sets exhibit lower aggregate objective values. The formula quantifies this relationship by summing objective values across all solutions in ND, normalizing by the objective value range $(f_{max} - f_{min})$, and applying an inverse mapping that causes NPI to approach 1 as convergence improves. Consequently, higher NPI values indicate enhanced convergence performance.

$$DSF_m = \max_{\theta \in ND} f_m(\theta) - \min_{\theta \in ND} f_m(\theta) \tag{21}$$

Equation (21) calculates the objective value range (DSF_m) across all non-dominated solutions, providing a dimensional measure of the Pareto front's spread within each objective space, thereby quantifying the diversity of the solution set.

This method helps maintain population diversity during the optimization process, thereby avoiding premature convergence. The specific algorithm flow is shown in Algorithm 3.

Algorithm 3. The WPFS (Weighted Pareto Front Score) indicator

Require: Quantum population QP, maximum iterations T_{max}
Ensure: Next update method

```
while t < T_max/2 do
    Calculate non-dominated solutions in QP (denoted as F)
    Calculate WPFS by summing objectives of solutions in F
    if 1.5N < WPFS < 2N then
        if WPFS(t+1) − WPFS(t) > 0 then
            Use real number population for next update
        else
            Perform quantum iteration
        end if
    else
        Use real number population for next update
    end if
end while
```

In multi-qubit systems, quantum mechanical principles allow states to be represented as superpositions through complex probability amplitudes. Following the quantum superposition principle, a quantum state exists as a linear combination of all possible basis states. This research framework employs quantum individuals to represent superposition states, while binary individuals correspond to basis states. Prior to fitness evaluation, quantum measurement and re-preparation processes are essential for obtaining real-valued individuals X [15].

The measurement operation maps quantum individuals to real-valued individuals, thereby establishing a real-valued population (see Eq. (22)). Algorithm 4 details this transformation process.

$$f(x_{ij}^t) = \frac{1}{n}\sum_{k=1}^{n} p^i x_{ijk}^t \tag{22}$$

Algorithm 4. Quantum Population Mapping

Require: Quantum population QP
Ensure: Real number population DP

```
for i ← 1 to N do
    for j ← 1 to D do
        Extract probability amplitudes α and β from quantum individual where |α|^2 + |β|^2 = 1
        Perform T measurements on each qubit q_ijk (k = 1, ..., n) with probability |α|^2
        Calculate mapping values x_ij^t (t = 1, ..., T) using Eq. (22), where p^t is the state weight
        Set x_ij = (1/T) Σ_{t=1}^{T} f(x_ij^t)
    end for
    Construct real number individual x_i = [x_i1, x_i2, ..., x_iD]
end for
Return real number population DP = {x_1, x_2, ..., x_N}
Optional: Adjust α and β and repeat until convergence or maximum iterations reached
```

3.4 Quantum Individual Rotation Strategy Based on Angle

To address the convergence challenges posed by the balance between exploration and exploitation in multi-objective optimization problems, this paper proposes an angle-based quantum individual adaptive rotation update method(The steps of this algorithm are shown in Algorithm 5). This method achieves efficient exploration of the search space and precise approximation of Pareto optimal solutions through dynamic evaluation of the distribution characteristics and individual contributions of the quantum population. The main innovations of this algorithm include:

First, we designed an individual weight allocation mechanism based on Euclidean distance, which assigns higher weights to distant individuals with strong exploration capabilities by calculating the Euclidean distance $distance_i$ from each quantum individual Q_i to the ideal point (Eq. (23)). Furthermore, we introduced the approximate gradient contribution $c_{i,k,j}$ to adjust the more fine-grained weights $w_{i,k,j}$ (Eq. (24)), thereby achieving precise control over quantum bits and effectively improving the algorithm's fitting capability for complex Pareto fronts.

$$\text{distance}_i = \min_{i^*} \sqrt{\sum_{k=1}^{D} (\theta_{i,k} - \theta_{i^*,k})^2} \tag{23}$$

$$w_i = \frac{\text{distance}_i}{\sum\limits_{i=1}^{N} \text{distance}_i}, \qquad w_{i,k,j} = \frac{c_{i,k,j}}{\sum\limits_{j=1}^{20} c_{i,k,j}} \tag{24}$$

Algorithm 5. Quantum Individual Rotation Strategy

Require: Quantum population QP, n, M
Ensure: Updated quantum population QP
for $i \leftarrow 1$ to N **do**
    Compute Euclidean distance to ideal point (Eq. (23))
end for
Normalize distances and update weights w_i (Eq. (24))
for $i \leftarrow 1$ to N **do**
    for $k \leftarrow 1$ to D **do**
        for $j \leftarrow 1$ to 20 **do**
            Compute $c_{i,k,j}$ and update $w_{i,k,j}$
        end for
    end for
end for
Initialize: $w_g = 0.6$, $w_f = 0.4$
for $t \leftarrow 1$ to $T_{\max}$ **do**
    if $D_{pop} < D_{threshold}$ **then**
        Adjust weights (Eq. (25))
    end if
    for $i \leftarrow 1$ to N, $k \leftarrow 1$ to D, $j \leftarrow 1$ to 20 **do**
        Calculate rotation angle (Eqs. (27)–(29))
        Apply rotation (Eq. (30))
    end for
    Update $|\psi'_i\rangle = \bigotimes_{k=1}^{D} \bigotimes_{j=1}^{20} |\phi'_{i,k,j}\rangle$
end for
return updated QP

Second, we proposed a population diversity adaptive control mechanism that dynamically balances the global attraction weight w_g and the forcing weight w_f (Eq. (25)). By introducing a comparison between the population distribution index D_{pop} (Eq. (26)) where $f_i^t = [f_1(\theta_i^t), ..., f_M(\theta_i^t)] \in R^n$ and the threshold $D_{\text{threshold}}$, the algorithm can intelligently switch between exploitation and exploration strategies. When population clustering is high, exploration is increased to avoid local optima; when the distribution is overly dispersed, convergence is enhanced to accelerate optimization.

$$\begin{aligned} w_g &= w_g + \eta \cdot (1 - \frac{D_{\text{pop}}}{D_{\text{threshold}}}), \\ w_f &= 1 - w_g \end{aligned} \quad (25)$$

$$D_{pop} = \frac{2}{N(N-1)} \sum_{1 \leq i \leq j \leq N} \left\| f_j^t - f_i^t \right\|_2 \quad (26)$$

Finally, we constructed a dual quantum rotation angle update model (Eq. (27)) that simultaneously considers the θ_{global} direction toward Pareto optimal solutions and the θ_{force} direction toward the population average state (Eq. (28)), with comprehensive regulation through a multi-layer weight system (w_g, w_f, w_i, $w_{i,k,j}$).

$$\theta_{global} = \text{sign}(\alpha_{i,k,j}\beta_{i^*,k,j} - \beta_{i,k,j}\alpha_{i^*,k,j}) \quad (27)$$

$$\theta_{force} = \text{sign}(\alpha_{i,k,j}\beta_{\text{avg},k,j} - \beta_{i,k,j}\alpha_{\text{avg},k,j}) \quad (28)$$

$$\theta_{i,k,j} = 0.01\pi[w_g \cdot w_i \cdot w_{i,k,j} \cdot \theta_{\text{global}} + w_f \cdot \theta_{\text{force}} \quad (29)$$

$$|\varphi'_{i,k,j}\rangle = R(\theta_{i,k,j})|\varphi_{i,k,j}\rangle \quad (30)$$

This angle update strategy (Eq. (29)) ensures that the quantum rotation gate $R(\theta_{i,k,j})$ can efficiently update quantum bit states (Eq. (30)), improving both the convergence speed and solution quality of the algorithm while maintaining population diversity.

4 Experimental Setup and Results

4.1 Experimental Environment

The experimental hardware environment is AMD Ryzen 5 7530U with Radeon Graphics 2.00 GHz, 16 GB RAM. The experimental platform was Matlab R2023a.

To validate the performance of the proposed QCMOAPO algorithm, we compared it against seven mainstream algorithms: MOPSOCD, MMOPSO, CMOPSO, SMPSO, dMOPSO, MOEAD, and MOEADDQN. The experiments were conducted on the standard ZDT, DTLZ, and WFG test suites, which are widely recognized as benchmarks for evaluating the performance of multi-objective optimization algorithms.

The parameter settings for the compared algorithms were maintained consistent with their original publications. For all algorithms, the QP size was set to 100, the qubit encoding length is set to n = 20. The maximum number of evaluations was configured as 1×10^4 for the DTLZ test suite and 3×10^4 for other test suites. Each algorithm was executed independently for 30 runs. All experiments were conducted on the PlatEMO platform. We evaluated our algorithm using two standard metrics: Hypervolume (HV) and Inverted Generational Distance (IGD). [22,23].

4.2 Experimental Results

For all test problems, a population size of 100 was consistently employed. QC-MOAPO was benchmarked against seven other algorithms. Twenty-one test functions, namely ZDT1-ZDT4, ZDT6, DTLZ1-DTLZ6, and WFG1-WFG9, were selected from the ZDT, DTLZ, and WFG series. Following 30 independent runs, the statistical results for IGD and HV are summarized in Tables 1 and 2. In these tables, '+', '−', and '=' denote, respectively, a significantly better, worse, or similar performance of the comparative algorithm compared to QC-MOAPO. Bold and shaded entries indicate the best results achieved for each test function.

Table 1. IGD Performance of QC-MOAPO and Comparative Algorithms.

Problem	M	D	MOPSOCD	MMOPSO	CMOPSO	SMPSO	dMOPSO	MOEAD	MOEADDQN	QCMOAPO
DTLZ1	3	7	1.1405e+1 -	3.6958e-2 -	5.9388e-1 -	8.3443e-1 -	1.0062e+1 -	2.0569e-2 -	2.0566e-2 -	**2.0049E-02**
DTLZ2	3	12	6.6083e-2 -	7.1210e-2 -	6.1060e-2 -	7.3165e-2 -	1.6145e-1 -	6.8275e-2 =	5.6882e-2 -	**5.4518E-02**
DTLZ3	3	12	8.4925e+1 -	3.4917e-1 -	3.1635e+1 -	1.1299e+1 -	1.1534e+2 -	3.5537e+0 -	8.5925e+0 -	**5.3271E-02**
DTLZ4	3	12	1.0911e-1 -	1.0053e-1 -	1.2229e-1 -	5.2124e-1 -	4.9761e-1 -	6.0804e-1 -	1.1047e-1 -	**5.4727E-02**
DTLZ5	3	12	4.9950e-3 -	6.2609e-3 -	6.8577e-3 -	6.0380e-3 -	4.8259e-2 -	3.3479e-2 -	2.2765e-2 -	**4.4612E-03**
DTLZ6	3	12	4.8099e-3 -	6.9142e-3 -	7.0822e-2 -	6.3622e-1 -	3.3467e-2 -	3.3119e-2 -	2.2831e-2 -	**4.0612E-03**
ZDT1	2	30	4.0007e-3 -	4.8800e-3 -	4.1315e-3 -	5.0099e-3 -	1.3484e-2 -	5.2768e-3 -	3.9229e-3 =	**3.9059E-03**
ZDT2	2	30	4.0667e-3 -	5.2612e-3 -	4.0956e-3 -	5.0868e-3 -	7.3468e-2 -	1.4756e-2 -	**3.8334e-3** +	3.9253E-03
ZDT3	2	30	4.7292e-3 +	5.5624e-3 -	**4.6419e-3** +	5.2522e-3 -	1.3281e-2 -	1.8352e-2 -	1.0946e-2 -	4.8104E-03
ZDT4	2	10	3.7352e+0 -	1.0341e-2 +	9.5272e-2 =	7.4911e-1 -	6.6416e-3 +	1.2395e-2 +	**4.1252e-3** +	2.0742E-01
ZDT6	2	10	3.3814e-3 -	4.2441e-3 -	3.1050e-3 -	3.8536e-3 -	3.1272e-3 -	5.1114e-3 -	3.1039e-3 -	**3.0747E-03**
WFG1	3	12	2.0191e+0 -	2.8351e-1 =	1.2331e+0 -	1.4806e+0 -	1.4607e+0 -	**2.5722e-1** =	2.7399e-1 +	4.1727E-01
WFG2	3	12	2.4886e-1 -	2.3236e-1 -	1.9171e-1 -	2.3235e-1 -	3.4099e-1 -	2.3729e-1 -	2.6716e-1 -	**1.8021E-01**
WFG3	3	12	2.0403e-1 -	**1.0302e-1** +	1.6196e-1 -	1.3423e-1 =	3.5786e-1 -	1.6622e-1 -	1.3670e-1 +	1.4845E-01
WFG4	3	12	3.4255e-1 -	2.9322e-1 -	2.7270e-1 -	3.2350e-1 -	3.4075e-1 -	2.5601e-1 -	3.1084e-1 -	**2.2107E-01**
WFG5	3	12	2.7878e-1 -	2.8811e-1 -	2.6089e-1 -	2.9962e-1 -	3.0837e-1 -	2.4970e-1 -	2.9861e-1 -	**2.2369E-01**
WFG6	3	12	3.4577e-1 -	3.1842e-1 -	2.5064e-1 -	3.1611e-1 -	4.1412e-1 -	2.7643e-1 -	3.2547e-1 -	**2.3152E-01**
WFG7	3	12	2.7915e-1 -	2.8705e-1 -	2.4512e-1 -	2.9980e-1 -	3.9332e-1 -	3.0357e-1 -	2.8152e-1 -	**2.1918E-01**
WFG8	3	12	4.4239e-1 -	3.7529e-1 -	3.4561e-1 -	4.2963e-1 -	5.0374e-1 -	3.1072e-1 -	3.5816e-1 -	**3.0309E-01**
WFG9	3	12	3.4295e-1 -	2.8433e-1 -	2.2987e-1 -	2.8221e-1 -	2.8900e-1 -	2.7139e-1 -	3.3529e-1 -	**2.2235E-01**
+/-/=			1/19/0	2/17/1	1/18/1	0/19/1	1/19/0	1/17/2	4/15/1	

The IGD data presented in Table 1 indicates that the QC-MOAPO algorithm achieved the best IGD values on 15 test functions. It is worth noting that the proposed method demonstrates outstanding performance on the three-objective WFG1 9 and DTLZ1 6 problems, while its effectiveness on ZDT problems is relatively less significant. Compared to the original MOAPO algorithm, QC-MOAPO obtained superior IGD values across all 21 test functions, fully demonstrating the significant effectiveness of the improvement strategies proposed in this paper.

Table 2. HV Performance of QC-MOAPO and Comparative Algorithms.

Problem	M	D	MOPSOCD	MMOPSO	CMOPSO	SMPSO	dMOPSO	MOEAD	MOEADDQN	QCMOAPO
DTLZ1	3	7	0.0000e+0 -	7.9724e-1 -	4.1525e-1 -	2.5490e-1 -	0.0000e+0 -	8.4157e-1 -	8.4154e-1 -	**8.4295E-01**
DTLZ2	3	12	5.3363e-1 -	5.3002e-1 -	5.3845e-1 -	5.1855e-1 -	3.4935e-1 -	5.4057e-1 -	5.5542e-1 -	**5.5622E-01**
DTLZ3	3	12	0.0000e+0 -	4.1184e-1 -	0.0000e+0 -	6.2011e-2 -	1.8775e-3 -	3.6905e-1 -	3.7575e-1 -	**5.6203E-01**
DTLZ4	3	12	5.3456e-1 -	5.1472e-1 -	5.0248e-1 -	2.8361e-1 -	1.9667e-1 -	2.8228e-1 -	5.3707e-1 -	**5.5618E-01**
DTLZ5	3	12	1.9946e-1 =	1.9895e-1 -	1.9737e-1 -	1.9852e-1 -	1.4925e-1 -	1.8205e-1 -	1.9037e-1 -	**1.9949E-01**
DTLZ6	3	12	1.9953e-1 -	1.9925e-1 -	1.9322e-1 -	1.3286e-1 -	1.8218e-1 -	1.8105e-1 -	1.9049e-1 -	**2.0008E-01**
ZDT1	2	30	7.2028e-1 -	7.1927e-1 -	7.1947e-1 -	7.1881e-1 -	7.0575e-1 -	7.1745e-1 -	7.2010e-1 -	**7.2039E-01**
ZDT2	2	30	**4.4495e-1** +	4.4410e-1 -	4.4415e-1 -	4.4339e-1 -	3.9061e-1 -	4.3058e-1 -	4.4481e-1 +	4.4476E-01
ZDT3	2	30	5.9976e-1 +	5.9940e-1 -	5.9972e-1 +	5.9974e-1 +	5.9714e-1 -	**6.0509e-1** +	5.9796e-1 -	5.9966E-01
ZDT4	2	10	2.9861e-5 -	7.1220e-1 +	6.1213e-1 -	3.3591e-1 -	7.1471e-1 +	7.0688e-1 +	**7.1951e-1** +	6.8386E-01
ZDT6	2	10	3.8854e-1 -	3.8776e-1 -	3.8887e-1 -	3.8815e-1 -	3.8886e-1 -	3.8477e-1 -	3.8891e-1 -	**3.8892E-01**
WFG1	3	12	4.8323e-2 -	8.8731e-1 +	4.0660e-1 -	3.0608e-1 -	3.0531e-1 -	**8.9783e-1** +	8.9529e-1 +	7.8943E-01
WFG2	3	12	9.0415e-1 -	9.1405e-1 +	**9.2682e-1** +	9.0599e-1 -	7.9752e-1 -	9.0744e-1 -	8.7977e-1 -	9.1165E-01
WFG3	3	12	3.4136e-1 -	**3.9622e-1** +	3.5282e-1 -	3.7313e-1 +	2.7075e-1 -	3.5915e-1 +	3.7356e-1 +	3.5852E-01
WFG4	3	12	4.5945e-1 -	5.0468e-1 -	4.8664e-1 -	4.6507e-1 -	4.5292e-1 -	**5.3933e-1** +	5.1971e-1 =	5.29E-01
WFG5	3	12	4.8024e-1 -	4.8986e-1 -	4.7369e-1 -	4.6510e-1 -	4.5675e-1 -	4.9948e-1 -	4.7069e-1 -	**5.0964E-01**
WFG6	3	12	4.5391e-1 -	4.7634e-1 -	**5.1427e-1** =	4.7357e-1 -	3.8514e-1 -	4.8843e-1 -	4.4540e-1 -	5.1254E-01
WFG7	3	12	5.1023e-1 -	5.2255e-1 -	5.2071e-1 -	4.9430e-1 -	3.9487e-1 -	5.2382e-1 -	5.2549e-1 =	**5.3637E-01**
WFG8	3	12	3.8156e-1 -	4.3508e-1 -	4.2847e-1 -	3.8427e-1 -	3.2822e-1 -	**4.6033e-1** +	4.2591e-1 -	4.5069E-01
WFG9	3	12	4.2880e-1 -	5.0093e-1 -	**5.1295e-1** =	4.8690e-1 -	4.7965e-1 -	4.9140e-1 -	4.4766e-1 -	5.0977E-01
+/-/=			2/17/1	4/16/0	2/16/2	2/18/0	1/19/0	6/14/0	4/14/2	

Further analysis of the HV data in Table 2 reveals that QC-MOAPO exhibited the best performance on 10 test functions, yielding the highest HV values. It is worthy that the proposed method demonstrates excellent performance on WFG problems, while its effectiveness on other problems still requires improvement. Synthesizing both IGD and HV metrics, the experimental results indicate that the QC-MOAPO algorithm is capable of obtaining solution sets with stronger convergence and more uniform distribution on the vast majority of ZDT, DTLZ, and WFG series test problems, thereby exhibiting excellent comprehensive performance and optimization capability.

Figure 1 shows the Pareto front distributions of the QC-MOAPO algorithm and other comparative algorithms on the DTLZ7 test function from a single independent run. The results indicate that the solution set obtained by the QC-MOAPO algorithm not only provides a closer approximation to the true Pareto front but also exhibits a more uniform distribution in the objective space.

In conclusion, the collective experimental results, incorporating performance metrics such as IGD and HV along with the Pareto front distribution plots, consistently demonstrate that the proposed QC-MOAPO algorithm offers significant advantages in both convergence and diversity. This algorithm not only achieves optimal comprehensive performance but also fully validates its effectiveness.

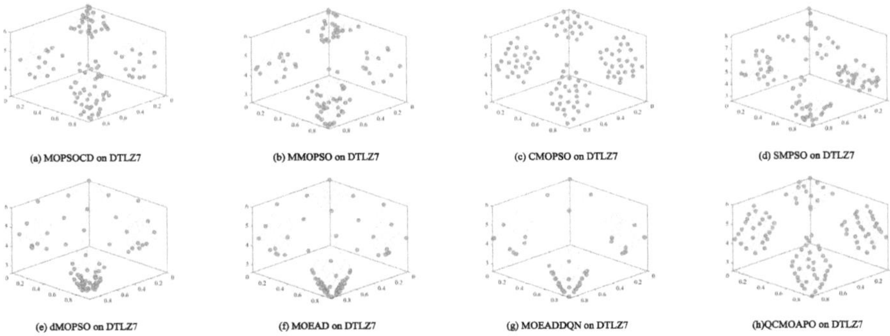

Fig. 1. Pareto fronts obtained by eight algorithms on the DTLZ7 test function.

5 Conclusion

To address the limitations of traditional multi-objective optimization algorithms (MOAPO), such as insufficient population diversity and weak global search capabilities, this paper proposes a multi-objective optimization algorithm integrated with quantum computing (QC-MOAPO). The core of this algorithm is the deep integration of quantum mechanisms to enhance its performance. Specifically, it utilizes qubit encoding and the principle of quantum superposition for a novel population initialization strategy, significantly improving the diversity of initial solutions. Furthermore, an individual update mechanism based on quantum measurement and rotation gates is constructed to strengthen the algorithm's global exploration capability. To balance exploration and exploitation, an adaptive co-evolutionary strategy between quantum and real-valued individuals is developed, which effectively prevents premature convergence and entrapment in local optima.

Comparative experiments on 21 standard test functions demonstrate that QC-MOAPO achieves superior comprehensive performance in both solution set diversity and convergence when compared to seven representative algorithms, thus validating the effectiveness of the proposed strategies.

Future research will proceed in two primary directions. The first involves generalizing the algorithm for application to practical engineering problems, such as AGV path planning in intelligent warehousing systems. The second will focus on enhancing the algorithmic framework by optimizing the evaluation and update mechanisms for quantum individuals, thereby addressing complex optimization challenges characterized by discrete or degenerate Pareto fronts.

References

1. Verma, S., Pant, M., Snasel, V.: A comprehensive review on NSGA-II for multi-objective combinatorial optimization problems. IEEE Access **9**, 57757–57791 (2021)

2. Cui, Y., Meng, X., Qiao, J.: A multi-objective particle swarm optimization algorithm based on two-archive mechanism. Appl. Soft Comput. **119**, 108532 (2022)
3. Luo, J., Liu, Q., Yang, Y.: An artificial bee colony algorithm for multi-objective optimisation. Appl. Soft Comput. **50**, 235–251 (2017)
4. Makhadmeh, S.N., Alomari, O.A., Mirjalili, S.: Recent advances in multi-objective grey wolf optimizer, its versions and applications. Neural Comput. Appl. **34**(22), 19723–19749 (2022)
5. An, Z., Xie, L.: Multi-state metaphor physics optimization algorithm. J. Taiyuan Univ. Sci. Technol. **40**(1), 19–25 (2019)
6. Wang, Y., Zeng, J.: Multi-objective optimization algorithm based on artificial physics optimization. Control Decis. **25**(7), 1040–1044 (2010)
7. Sun, B., Guo, N., Li, Z.: Quasi-physics constrained multi-objective optimization algorithm based on R2 indicator. J. Ordnance Equipment Eng. **43**(8), 186–192 (2022)
8. Benioff, P.: The computer as a physical system: a microscopic quantum mechanical Hamiltonian model of computers as represented by Turing machines. J. Stat. Phys. **22**(5), 563–591 (1980)
9. Feynman, R.P.: Simulating physics with computers. Int. J. Theor. Phys. **21**(6), 467–488 (1982)
10. Han, K.H., Kim, J.H.: Genetic quantum algorithm and its application to combinatorial optimization problems. In: International Congress on Evolutionary Computation (CEC 2000), San Diego, CA, USA, pp. 1354–1360 (2000)
11. Han, K.H., Kim, J.H.: Quantum-inspired evolutionary algorithm for a class of combinatorial optimization. IEEE Trans. Evol. Comput. **6**(6), 580–593 (2002)
12. Zhu, F., Li, G., Tang, H., et al.: Dung beetle optimization algorithm based on quantum computing and multi-strategy fusion for solving engineering problems. Expert Syst. Appl. **236**, 121219 (2024)
13. Alamir, N., Kamel, S., Hassan, M.H., et al.: An effective quantum artificial rabbits optimizer for energy management in microgrid considering demand response. Soft. Comput. **27**, 15741–15768 (2023)
14. Wang, Y., Wang, W.: Quantum-inspired differential evolution with grey wolf optimizer for 0–1 Knapsack problem. Mathematics **2**(5), 1233 (2021)
15. Gong, C., Zhou, N., Xia, S., Huang, S., et al.: Quantum particle swarm optimization algorithm based on diversity migration strategy. Future Gener. Comput. Syst. **157**(11), 445–458 (2024)
16. Liu, T., Jiao, L., Ma, W., Ma, J., Shang, R.: A new quantum-behaved particle swarm optimization based on cultural evolution mechanism for multi objective problems. Knowl.-Based Syst. **2018**(1), 8702820 (2016)
17. Han, F., Sun, Y., Ling, Q.: An improved multiobjective quantum-behaved particle swarm optimization based on double search strategy and circular transposon mechanism. Complexity **157**(11), 445–458 (2018)
18. Hota, A.R., Pat, A.: An adaptive quantum-inspired differential evolution algorithm for 0–1 knapsack problem. In: 2010 Second World Congress on Nature and Biologically Inspired Computing (NaBIC), pp. 301–306. IEEE, Fukuoka, Japan (2010)
19. Patrick, R.: Faster coherent quantum algorithms for phase, energy, and amplitude estimation. Quantum **5**, 566 (2021)
20. Zubairy, M.S.: Quantum Mechanics for Beginners. With Applications to Quantum Communication and Quantum Computing. Oxford University Press, Oxford (2020)
21. Farhi, E., Goldstone, J., Gutmann, S.: A quantum approximate optimization algorithm. Quantum Phys. **14**(13–14), 1220–1244 (2014)

22. Shang, K., Chen, W., Liao, W., et al.: HV-net: hypervolume approximation based on DeepSets. IEEE Trans. Evol. Comput. **27**(4), 1154–1160 (2022)
23. Kouka, N., BenSaid, F., Fdhila, R., et al.: A novel approach of many-objective particle swarm optimization with cooperative agents based on an inverted generational distance indicator. Inf. Sci. **623**, 220–241 (2023)

An Angle-Penalty Distance and Polarity-Division Guided Algorithm for Many-Objective Electric Field Optimization

Yue Zhang[1] and Liping Xie[1,2](✉)

[1] School of Computer Science and Technology, Taiyuan University of Science and Technology, Taiyuan 030024, China
s202320110964@stu.tyust.edu.cn

[2] Shanxi Key Laboratory of Complex Systems and Computational Intelligence, Taiyuan University of Science and Technology, Taiyuan 030024, Shanxi, China
xieliping@tyust.edu.cn

Abstract. Many-objective Optimization Problems (MaOPs) often involve more than three objectives. As the number of objectives increases, traditional algorithms face critical challenges: weakened selection pressure due to the explosion of non-dominated solutions, difficulty in balancing convergence and diversity and inefficient search in many objective spaces. This paper addresses these issues by proposing an angle-penalty distance and polarity-division guided algorithm for many-objective electric field optimization (PD-MaOEFO). Firstly, a charge function representing the quality of particles has constructed based on the angle-penalty distance(APD), such that the better the particle, the greater the charge it carries. Secondly, a particle polarity division strategy is introduced. Non-dominant solutions with smaller APD values are classified as protons, particles with smaller APD values in the solution set are classified as electrons, and particles with larger APD values in the solution set are classified as neutrons. Electrons move under the influence of proton Coulomb attraction and neighboring electron Coulomb repulsion, while neutrons move under the influence of proton nuclear force, the algorithm achieves a balance between development and exploration capabilities. Finally, a proton pool update strategy based on reference vectors and APD is introduced to achieve a balance between protons convergence and diversity. To validate the effectiveness of the proposed algorithm, we conducted comparative experiments with seven many-objective optimization algorithms, covering 13 many-objective optimization test problems. The experimental results demonstrate that the proposed algorithm can effectively solve many-objective optimization problems and exhibits good stability under varying numbers of objectives.

Keywords: Many-objective optimization · Reference vectors · Polarity-division · Angle-penalty distance · Electric Field Optimization

H. Ma et al. (Eds.): IWACIII 2025, CCIS 2780, pp. 544–560, 2026.
https://doi.org/10.1007/978-981-95-6730-0_37

1 Introduction

Formally, a Multi-Objective Optimization Problem (MOP) can be mathematically formulated as:

$$\begin{cases} \min F(X) = (f_1(X), \ldots, f_M(X))^T \\ X \in \Omega \end{cases} \tag{1}$$

where $X = (x_1, \ldots, x_D)^T$ is the decision vector; Ω is the decision space of the problem; D is the number of decision variables; $f_i(x)$ is the ith objective function and M is the number of objectives.

Many-objective Optimization Problems(MaOPs) refers to complex optimization problems with more than three objective function dimensions [1]. With the increase of objective dimensions, the solution set size grows exponentially. A significant increase in the probability of failure of the traditional Pareto dominance criterion. A surge in the number of non-dominant solutions can lead to weakened selection pressure and challenges in balancing convergence and diversity. Therefore, it is particularly necessary to design new Many-objective Optimization Problems algorithms.

In recent years, many heuristic optimization algorithms based on physics principles have emerged. For example, Artificial Physics Optimization Algorithm (APO) [2–5], Simulated Annealing Algorithm [6], Electromagnetic-like Mechanism Algorithm [7] and Central Force Algorithm [8]. Other algorithms based on the "electric" property include the Artificial Electric Field Algorithm (AEFA) [9] and the Electromagnetic Field Optimization Algorithm (EFO) [10,11]. Among them, AEFA is a global optimization algorithm based on Coulomb electrostatic force and Newton's laws of motion. This algorithm treats the individuals in the optimization problem as charged particles, and guides the search process through the attraction and repulsion forces between the particles. Chauhan et al. proposes a locally guided artificial electric field algorithm (L-AEFA) that dynamically determines Kbest by utilizing the positions of the historically best individual, the currently worst individual, and other individuals, thereby guiding movement more effectively [12]. Zhou et al. develops a multi-objective artificial electric field algorithm with a SARSA mechanism for automotive hybrid production line problems [13]. To date, the AEFA algorithm has been applied to multi-objective optimization, but only for two objectives. EFO guides the search by simulating the interactions of different electromagnetic fields, which provides a strong global search capability and convergence speed. In summary, physics-based algorithms possess powerful search capabilities and are well-suited for MaOPs.

Decomposition-based MOEA is one of the traditional MOEAs. It decomposes a multi-objective problem into multiple single-objective subproblems and guides the search by reference vectors. Although the traditional decomposition-based MOEA represented by MOEA/D [14] can handle MaOPs, it is observed from previous studies that its unstable performance and low search efficiency are not satisfactory. In recent years, another decomposition strategy has been proposed to decompose the objective space into a set of subspaces using reference vectors [15]. RVEA [16] is one of the most representative algorithms in this branch. There have been some improvement researches to address the inefficiency of RVEA. Qiu et al. [17] points out that the selection pressure exerted by RVEA on the convergence properties of solutions primarily stems from the length of the objective vector after translation when computing the APD. Therefore, an improved APD is proposed to enhance convergence. On the one hand, decomposition-based MOEA avoids additional challenges such as non-dominance and high computational cost while proving effective for uniformity comparisons. On the other hand, convergence issues can be addressed by integrating specific methods, ultimately achieving a balance between convergence and diversity.

In summary, physics-based optimization algorithms exhibit strong search capabilities. Decomposition-based MOEAs can effectively balance convergence and diversity. Therefore, we propose an angle-penalty distance and polarity-division guided algorithm for many-objective electric field optimization (PD-MaOEFO), which treats the decision space as an electric field, and the solution set as particles with different polarities that are in the electric field. This paper focuses on exploring how to balance convergence and diversity through reference vectors and APD, as well as designing more efficient particle force rules to enhance search efficiency within complex solution spaces. Specific contributions include:

(1) This paper constructs a charge function by APD, establishes a negative correlation mapping relationship between APD and charge, so as to better characterize the quality of the distribution of particles in the Pareto Front.
(2) Design the polarity division strategy of particles, regarding the non-dominated particles as protons and deposit them into the proton pool; regarding the particles in the solution set as electrons; after optimizing the electrons, screen the electrons with increased APD value to be converted into neutrons
(3) A two-stage action force rule is designed. In the early stage of the algorithm, multi-proton attraction and multi-electron repulsion force rules are designed for the electrons to enable them to explore a wider region in the decision space; and in the late stage of the algorithm, a specific proton from the proton pool is selected for each electron as a guide to speed up the convergence of the algorithm.
(4) A proton pool updating strategy based on reference vectors and APD is designed. By selecting the proton with the smallest APD for each reference vector to ensure that the protons outputted by the algorithm in each generation are all high-quality protons, thus realizing an effective balance between the convergence and diversity of the algorithm.

The rest of the paper is organized as follows. Section 2 describes the reference vectors, the APD and the associated physical principles. Section 3 presents the details of the PD-MaOEFO algorithm. Section 4 presents the results of comparative experiments between the PD-MaOEFO algorithm and seven other algorithms. Finally, Sect. 5 presents the summary and future work.

2 Related Work

2.1 Reference Vectors

The set of uniformly distributed reference vectors is the core tool of decomposition-based MOEA. First, uniformly distributed points are generated on the unit simplex. The number of generated points $N = \begin{pmatrix} P+M-1 \\ M-1 \end{pmatrix}$ is determined by the number of intervals P between consecutive points on each objective coordinate axis and the number of objectives M. After that, a set of reference vectors is obtained by connecting these reference points from the coordinate origin. Converting these obtained reference vectors into unit vectors facilitates the subsequent computation, so each reference vector is unitized beforehand, i.e., dividing each reference vector by its mode length [18]. The reference vector set is represented as shown in Eq. (2).

$$V = \{V_1, \ldots, V_j, \ldots, V_N\} \tag{2}$$

2.2 Angle-Penalty Distance(APD)

APD is a comprehensive evaluation metric proposed by the RVEA algorithm to balance convergence and diversity [16]. Its mathematical expression is shown in Eq. (3):

$$d_{i,j}(t) = (1 + P(\theta_{i,j}(t))) \times \|f_i'(t)\|_2 \tag{3}$$

where $\|f_i'(t)\|_2$ is the convergence criterion, which is expressed as the Euclidean distance from the ith solution to the ideal point. The mathematical expression of $f_i'(t)$ is shown in Eq. (4):

$$f_i'(t) = f_i(t) - Z_{\min}(t) \tag{4}$$

where $Z_{\min}(t) = \left(Z_{\min}^1(t), Z_{\min}^2(t), \ldots, Z_{\min}^M(t)\right)$ is the ideal point representing the minimum objective value on each target dimension.

$$P(\theta_{i,j}(t)) = M \times \left(\frac{t}{t_{\max}}\right)^{\alpha} \times \frac{\theta_{i,j}(t)}{\gamma_{V_j}(t)} \tag{5}$$

$$\gamma_{V_j}(t) = \min\{V_i, V_j\} \quad i \in \{1, \ldots, N\}, i \neq j \tag{6}$$

Equation (5) is a function of the diversity criterion. $\theta_{i,j}(t)$is the angle between the ith objective vector and the jth reference vector at time t. α is a parameter that controls the rate of change of the $P(\theta_{i,j}(t))$ function. $\gamma_{V_j}(t)$ is the minimum angle value between the reference vector V_j and other reference vectors in the current generation. Its mathematical expression is shown in Eq. (6).

2.3 Physical Principle

(1) **Coulomb's Law.**Its mathematical expression is shown in Eq. (7):

$$F_{ij} = K\frac{Q_i \times Q_j}{R^2} \tag{7}$$

where F_{ij} is the Coulomb force between the two particles; K is the Coulomb force constant; Q_i and Q_j are the charges carried by the ith and jth particles, respectively; and R is the distance between the two charged particles.

(2) **Proton-Neutron Nuclear Force.**It is expressed by the negative gradient of the "Tukawa potential". "Yukawa potential" is shown in Eq. (8):

$$V_{Yukawa}(r) = -g^2\frac{e^{-r/r_0}}{r} \tag{8}$$

where $V(r)$ is the value of the potential energy, g is the coupling constant, which controls the strength of the interaction, r is the distance between the neutron and the proton, and r_0 is the range of the "Yukawa potential".

(3) **Newton's Second Law.**Its mathematical expression is shown in Eq. (9):

$$a = \frac{F}{m} \tag{9}$$

where F is the combined force on the object; m is the mass of the object; and a is the acceleration of the object.

3 PD-MaOEFO Algorithm

3.1 Algorithmic Idea

The algorithm simulates the atomic structure based on the particle polarity division strategy, and divides the solution set into protons (non-dominant solutions with smaller APD values), electrons (well-adapted particles) and neutrons (poorly adapted particles). Electrons perform global search driven by proton Coulomb attraction and electron Coulomb repulsion, and multi-directional forces drive electrons to explore a wider solution space to enhance the diversity of particles, while neutrons jump to the vicinity of protons through the "nuclear force" to perform a localized and fine-grained search to enhance the quality of the solution set. When the number of protons reaches the maximum size of the proton pool, the best protons in each direction are filtered based on reference vectors and APD indexes to ensure the convergence and uniformity of the Pareto Front. The proton as the final output has a dual advantage: it represents the current optimal solution in multi-objective optimization, and its Coulomb repulsion property prevents aggregation of the solution set and maintains distribution uniformity. The algorithm combines global exploration (electron repulsion against precocity), local exploitation (neutron nucleus force optimization) and elite retention (dynamic screening of the proton pool), and achieves a balance between algorithmic exploration and exploitation through the principles of electromagnetism and nucleus action.

3.2 Particle Polarity Division

The algorithm marks non-dominated solutions as protons and stores them in the proton pool, thereby characterizing the high-quality features of elite solutions; the solution set particles are mapped to electrons, which constitute the movable search body. The electrons whose APD values become larger after screening and optimization are converted into neutrons. The particle charge function is designed based on APD. The algorithm uses this to evaluate the merits of the charged particles, i.e., the smaller the APD value of the particle, the larger the charge it carries, and the better adapted the charged particles are. The charge of the particle is calculated as shown in Eq. (10):

$$Q_{i,j}(t) = \exp\left(-\frac{(APD_{i,j}(t) - APD_{\min}(t))}{(APD_{\max}(t) - APD_{\min}(t))}\right) \tag{10}$$

where $APD_{i,j}(t)$ is the APD value of the ith proton or the ith electron on V_j at time t; $APD_{\max}(t)$ and $APD_{\min}(t)$ are the maximum and minimum values of the APD values of all protons or electrons on V_j, respectively.

For a group of electrons, the APD value of each electron is first calculated, then the maximum and minimum APD values in this group are compared, and finally the charge of the electrons is calculated by Eq. (10); for a group of protons, the algorithm first calculates the APD value of each proton on each reference vector, and for the jth reference vector V_j, the charge of the ith proton on the reference vector V_j is calculated according to Eq. (10).

3.3 Coulomb Force Drive Optimization

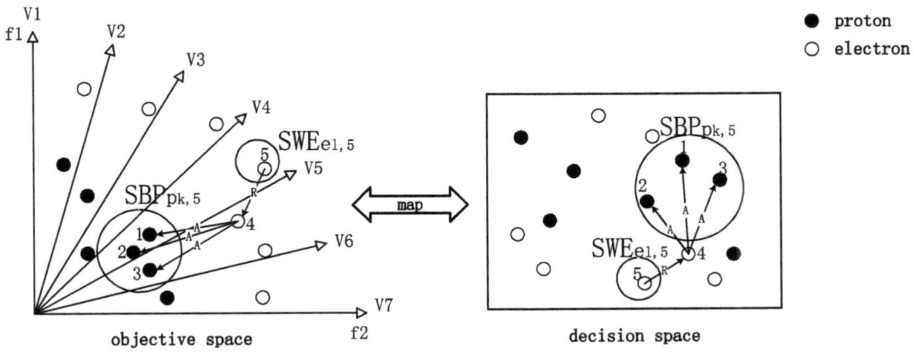

Fig. 1. Schematic representation of the forces on the electrons in the first stage.

First, for the ith electron (name it as e_i), its associated reference vector is V_j and its APD value is $APD_{e_{i,j}}(t)$. For the kth proton in the proton pool (name it as p_k), compute the APD value of p_k on V_j, $APD_{p_{k,j}}(t)$, and select the protons that satisfy $APD_{p_{k,j}}(t) < APD_{e_{i,j}}(t)$ to be deposited into the set $SBP_{p_{k,j}}$, which is denoted as the elite set of protons that are better on the vector V_j for the electron e_i as shown in Eq. (11):

$$SBP_{p_{k,j}} = \{p_k \mid APD_{p_{k,j}}(t) < APD_{e_{i,j}}(t), e_i \in Electron, p_k \in Proton\} \quad (11)$$

where $e_i \in Electron$ denotes that the ith electron belongs to the *Electron*; $p_k \in Proton$ denotes that the kth proton belongs to the *Proton*.

In the electron set "*Electron*", there may be an electron e_l associated with the same reference vector V_j as e_i. Calculate the APD value $APD_{e_{l,j}}(t)$ of e_l on V_j, and select the electrons that satisfy $APD_{e_{l,j}}(t) > APD_{e_{i,j}}(t)$ to be deposited in the set $SWE_{e_{l,j}}$, which is expressed as the set of electrons that are worse than electron e_i on vector V_j, as shown in Eq. (12):

$$SWE_{e_{i,j}} = \{e_l \mid APD_{e_{i,j}}(t) > APD_{e_{l,j}}(t), e_i, e_l \in Electron\} \quad (12)$$

where $e_l \in Electron$ denotes the e_l belongs to the *Electron*.

A two-stage force rule is applied to the electrons. The algorithm preliminarily provides the electrons with a broad search range in order to search for a more optimal region, where the e_i is subjected to the Coulomb attractive force of the protons in set $SBP_{p_{k,j}}$ and the Coulomb repulsive force of electrons in set $SWE_{e_{l,j}}$. The first stage of the electron force schematic is shown in Fig. 1. Where "A" represents the Coulomb attractive force, "R" represents the Coulomb repulsive force, for the electron 4, its associated reference vector is V_5, assuming that the APD value of protons 1,2,3 on V_5 is smaller than the APD value of electron 4, the electron 4 is subjected to the Coulomb attractive force of protons 1,2,3; at the same time, electron 5 is also associated with V_5 and the APD value of electron 5 is larger than that of electron 4, so electron 4 is subject to electron 5's Coulomb repulsion.

In the D-dimensional decision space, the e_i is subject to the Coulomb attractive force vector $F_{ik}^D(t)$ of the p_k as shown in Eq. (13):

$$F_{ik}^D(t) = K \times Q_{p_{k,j}}(t) \times Q_{e_{i,j}}(t) \times \frac{X_{p_k}(t) - X_{e_i}(t)}{[\text{norm}(X_{p_k}(t) - X_{e_i}(t))]^2}, p_k \in SBP_{p_{k,j}} \quad (13)$$

where K is the Coulomb force constant; $Q_{p_{k,j}}(t)$ and $Q_{e_{i,j}}(t)$ are the charges of the p_k on the vector V_j and the e_i on the vector V_j at the moment t, respectively; $X_{p_k}(t)$ and $X_{e_i}(t)$ are the positions of p_k and e_i at the moment t; and *norm*() is the parametric operator for solving the distance between p_k and e_i.

In the D-dimensional decision space, the e_i is subject to the Coulomb repulsive force vector $F_{il}^D(t)$ of the e_l as shown in Eq. (14):

$$F_{il}^D(t) = K \times Q_{e_{l,j}}(t) \times Q_{e_{i,j}}(t) \times \frac{X_{e_i}(t) - X_{e_l}(t)}{[\text{norm}(X_{e_i}(t) - X_{e_l}(t))]^2}, e_l \in SWE_{e_{l,j}} \quad (14)$$

where K is the Coulomb force constant; $Q_{e_{l,j}}(t)$ and $Q_{e_{i,j}}(t)$ are the charges of the e_l on the vector V_j and the e_i on the vector V_j at the moment t, respectively;

$X_{e_l}(t)$ and $X_{e_i}(t)$ are the positions of e_l and e_i at the moment t; and $norm()$ is the parametric operator for solving the distance between e_l and e_i.

Finally, the Coulomb repulsive and Coulomb attractive force vector on e_i are accumulated and the combined force vector $F_i^D(t)$ on it is shown in Eq. (15):

$$F_i^D(t) = \sum_{k \in SBP} F_{ik}^D(t) + \sum_{l \in SWE} F_{il}^D(t) \tag{15}$$

where SBP is the set of protons screened from the proton pool against e_i; and SWE is the set of electrons screened from the electron set against e_i.

At the late stage of the algorithm, in order to improve the convergence of the algorithm, the electron e_i is only subjected to the Coulomb attractive force of the proton with the smallest APD value in the $SBP_{p_{k,j}}$, which is the combined force exerted on the electron e_i. The force on the electron in the late stage of the algorithm is shown in Fig. 2. Where "A" represents the Coulomb attractive force, the reference vector associated with electron 2 is V_5, and the proton with the smallest APD value from the $SBP_{p_{k,j}}$ is screened as proton 1, so electron 2 is subject to the Coulomb attractive force of proton 1.

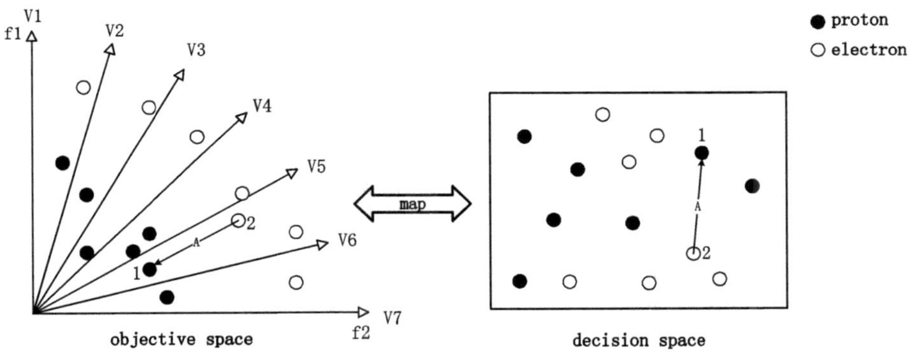

Fig. 2. Schematic of forces on electrons in the algorithm's late stages.

The combined force on the electrons is obtained, and then the acceleration is obtained by "Newton's second law", where the mass of all particles is "1" by default, so the combined force on the electrons is the acceleration. So the velocity is updated as shown in Eq. (16):

$$v_i(t+1) = r_1 \times v_i(t) + r_2 \times F_i^D(t) \tag{16}$$

where r_1 and r_2 are random numbers between 0 and 1, respectively.

The particle position is updated as shown in Eq. (17):

$$X_i(t+1) = X_i(t) + v_i(t+1) \tag{17}$$

The maximum velocity of the particle is shown in Eq. (18):

$$v_{\max} = (X_{\max} - X_{\min}) \times 0.5 \tag{18}$$

The velocity limit of the particle is shown in Eq. (19):

$$v_i(t+1) = \max(\min(v_i(t+1), v_{\max}), -v_{\max}) \tag{19}$$

The positional limits of the particles are shown in Eq. (20):

$$X_i(t+1) = \min(\max(X_i(t+1), X_{\min}), X_{\max}) \tag{20}$$

where $X_{\max}$ and $X_{\min}$ are the upper and lower bounds of the decision space.

3.4 Neutron State Strengthening Mechanism

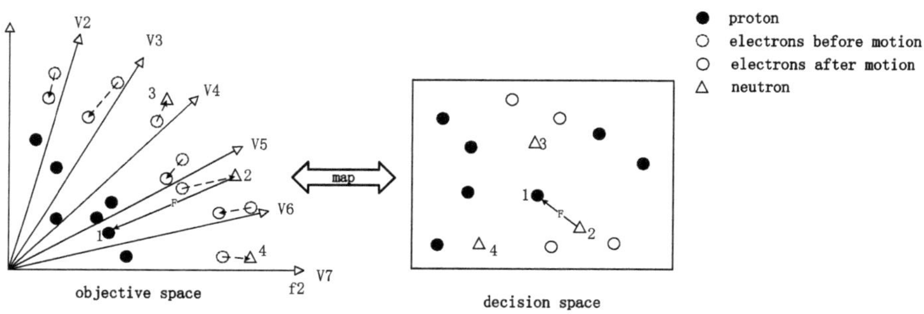

Fig. 3. Schematic diagram of neutron force.

This strategy simulates the nuclear force between protons and neutrons to drive neutron motion. Specifically, electrons with increased APD will be converted to neutrons through "Coulomb force-driven optimization", and local refinement will be performed through "proton-neutron nuclear force" to enhance the local search capability and achieve a balance between exploration and exploitation. The force on a neutron is shown in Fig. 3. Where "F" stands for the nuclear force. From Fig. 3, it can be seen that there are 2,3,4 converted from electrons to neutrons, taking neutron 2 as an example, the reference vector associated with it is V_5, and the proton 1 with the smallest APD value is obtained from the set $SBP_{p_{k,j}}$. The neutron 2 is subjected to the nuclear force of the proton 1.

The nuclear force is represented by the negative gradient of the "Yukawa potential", i.e., in the D-dimensional decision space, the ith neutron (denoted by n_i) is subjected to the nuclear force vector $F_{n_i}^D$ of the proton p as shown in Eq. (21):

$$F_{n_i}^D = g^2 \left(\frac{\exp\left(-\frac{r}{r_0}\right)}{r^2} + \frac{\exp\left(-\frac{r}{r_0}\right)}{(r \cdot r_0)} \right) \times \frac{(X_p(t) - X_{n_i}(t))}{r}, p \in SBP_{p_{k,j}} \tag{21}$$

where g is the coupling constant for controlling the strength of the nuclear force; r is the distance between the proton p with the smallest APD value in $SBP_{p_{k,j}}$ and the ith neutron n_i; $X_p(t)$ and $X_{n_i}(t)$ are the position of the proton p and the position of the n_i at the time t, respectively; r_0 is the range of action, which is set to be the minimum of the absolute values of the decision variables of the proton p, and then the decision variables of the neutrons in each dimension of the n_i are scaled by the scaling factor to $[-r_0, r_0]$ range, which facilitates the neutron n_i to do fine search within the circle of proton p with radius r_0.

3.5 Strategy for Updating Proton Pool

In the PD-MaOEFO algorithm, in order to update and retain the primes generated at each iteration, a reference vector and APD-based protons pool updating strategy for storing high-quality non-dominated solutions is proposed.The specific process is shown in Algorithm 1.

3.6 The Main Framework of the PD-MaOEFO Algorithm

Algorithm 2 shows the pseudo-code of the PD-MaOEFO algorithm.

4 Experiments

In this study, seven many-objective optimization algorithms are selected as benchmarks for comparison, including MPSO/D [19], NMPSO [20], RVEA [16], hpaEA [21], PeEA [22], MaOEA- IGD [23] and MaOEA-IT [24]. The experiments are based on the PlatEMO evolutionary multi-objective optimization platform, and systematic evaluation is carried out on 13 benchmark problems from two standard test suites, DTLZ and WFG. For each test problem, many-objective optimization is performed in the 3-, 5-, 8-, 10-, and 15-dimensional objective space, and 20 independent repetitions of the experiment are executed for each dimensional configuration to ensure the statistical reliability of the results.

4.1 Evaluation Indicators and Parameter Setting

The comprehensive evaluation indicators used are the hypervolume metric (HV) and the improved inverse generational distance (IGD^+), respectively.

The algorithm comparison parameters involve the solution set size and the maximum number of function evaluations, both of which vary according to the number of objectives M. The algorithm comparison parameters are set to the maximum number of function evaluations. The specific parameter settings are shown in Table 1.

Algorithm 1. Strategy for Updating Proton Pool

Input: Electron Set,*Electron*; Proton Pool,*Proton*; Reference Vector,V; Proton Pool Size,$nArc$
Output: Proton Pool,*Proton*

1: $Particle \leftarrow Proton \bigcup Electron$
2: $Proton \leftarrow$ Undominated particles in *Particle*
3: Calculate the APD value of *Proton*
4: **if** $|Proton| <= nArc$ **then**
5:     *return Proton*
6: **else**
7:     **for** $i = 1 : |V|$ **do**
8:         Get the prime index of the V_i association:$Current$
9:         **if** $Current == NULL$ **then**
10:             Get the V_i neighborhood and its associated prime indexes: as
11:             **if** $as == NULL$ **then**
12:                 Select the proton with the smallest APD value from the remaining unselected protons of *Proton*
13:             **else**
14:                 Select the proton with the smallest APD value from as
15:             **end if**
16:         **else**
17:             Select the proton with the smallest APD value from $Current$
18:         **end if**
19:     **end for**
20: **end if**

Algorithm 2. The main framework of the PD-MaOEFO algorithm

Input: Electron Set,*Electron*; Size,*N*; Proton Pool,*Proton*; Reference Vector,V; Maximum Number of Function Evaluations,$maxFEs$
Output: Proton Pool,*Proton*

1: Generate reference vectors V
2: initialization:$Electron$; $Proton = NULL$
3: The objective space is divided into multiple subregions by V and APD
4: Updating the proton pool (Algorithm 1)
5: **while** $FE < maxFEs$ **do**
6:     The APD of each electron and proton is calculated by Equation (3)
7:     Calculate the charge of each proton or electron on V_j by Equation (10)
8:     Coulomb Force Drive Optimization (3.3)
9:     Neutron State Strengthening Mechanism (3.4)
10:     **if** $APD(n) < APD(e)$ **then**
11:         Replacement of poorer electrons with newer neutrons: $e \leftarrow n$
12:     **end if**
13:     Get the superior $Electron(t+1)$ and repeat step 3
14:     Using $Proton \bigcup Electron(t+1)$ updating *Proton* (Algorithm 1)
15:     Evolutionary search for proton pool: S
16:     Using $Proton \bigcup S$ updating *Proton* (Algorithm 1)
17:     Reference Vector Adaptive
18: **end while**

Table 1. Parameter setting

Number of objectives (M)	Solution set size (N)	Maximum number of function evaluations ($MaxFEs$)
3	153	50000
5	210	110000
8	156	140000
10	275	275000
15	135	210000

4.2 Experimental Results and Analysis

(1) DTLZ experimental results and analysis

Table 2 summarizes the mean and standard deviation of HV obtained from 20 independent runs of each algorithm on DTLZ1-DTLZ4. From Table 2, it can be seen that PD-MaOEFO obtains 15 optimal results in the comparison of HV means and standard deviations for the 20 test problems. Specifically, PD-MaOEFO outperforms most of the competing algorithms on DTLZ3 and DTLZ4, indicating that the PD-MaOEFO algorithm has an advantage in dealing with test problems with nonlinear PFs with multiple poles and with nonlinear PFs with non-uniform solution distribution. The 15-objective DTLZ3 optimization results of each algorithm are shown in Fig. 4, in which both RVEA and PD-MaOEFO converge to the real Pareto front with no significant difference in effect, and the other algorithms do not converge to the real Pareto front, and there is still a gap compared with the first two algorithms. On the DTLZ1 test problem, although PD-MaOEFO only obtained 2 optimal HV results, it is not much different from the optimal algorithm RVEA in most dimensions.

Table 2. Comparison of HV mean and standard deviation of algorithms for DTLZ

Problem	M	D	hpaEA	MPSOD	MaOEAIT	PeEA	RVEA	NMPSO	MaOEAIGD	PD-MaOEFO
	3	7	8.4302e-1 (6.17e-4) -	1.1767e-1 (1.41e-1) -	4.5608e-1 (1.22e-1) -	8.2801e-1 (6.21e-3) -	8.4112e-1 (9.55e-4) -	8.2480e-1 (7.66e-2) -	4.2169e-1 (3.78e-1) -	8.4802e-1 (4.61e-3)
	5	9	9.7670e-1 (3.38e-4) -	2.9453e-1 (1.94e-1) -	1.3286e-1 (2.62e-1) -	9.5627e-1 (6.37e-3) -	9.7486e-1 (1.85e-4) -	9.6619e-1 (3.44e-3) -	5.1604e-1 (3.66e-1) -	9.7986e-1 (2.01e-4)
DTLZ1	8	12	9.7173e-1 (1.14e-1) =	9.3828e-1 (7.77e-2) -	1.5217e-6 (6.81e-6) -	9.9311e-1 (3.45e-3) -	9.9728e-1 (5.17e-5) +	9.6882e-1 (1.06e-2) -	4.6803e-1 (4.12e-1) -	9.9553e-1 (6.37e-3)
	10	14	9.9739e-1 (6.14e-3) =	8.8914e-1 (7.88e-2) -	0.0000e+0 (0.00e+0) -	9.9858e-1 (1.06e-3) -	9.9968e-1 (1.52e-5) +	9.8784e-1 (8.14e-3) -	9.8808e-1 (9.24e-3) -	9.9932e-1 (5.29e-4)
	15	19	9.9937e-1 (1.33e-3) +	9.8222e-1 (6.29e-3) -	3.5961e-3 (1.61e-2) -	9.9925e-1 (4.34e-4) =	9.9992e-1 (1.14e-5) +	1.1826e-1 (2.81e-1) -	4.5029e-1 (3.86e-1) -	9.8888e-1 (2.51e-2)
	3	12	5.6082e-1 (3.51e-4) -	5.5857e-1 (1.56e-4) -	1.0994e-1 (6.70e-2) -	5.5490e-1 (1.82e-3) -	5.5957e-1 (3.88e-5) -	5.6883e-1 (9.16e-4) -	5.5010e-1 (2.28e-4) -	5.6926e-1 (9.57e-4)
	5	14	8.0029e-1 (1.16e-3) +	7.9281e-1 (6.71e-4) -	1.0238e-1 (3.36e-2) -	7.7458e-1 (3.74e-3) -	7.9496e-1 (3.80e-4) -	8.0980e-1 (1.46e-3) +	8.1256e-1 (5.72e-4) +	7.9698e-1 (1.61e-3)
DTLZ2	8	17	9.1873e-1 (9.94e-3) -	9.1575e-1 (5.59e-4) -	1.5972e-1 (7.09e-2) -	9.0009e-1 (6.23e-3) -	9.1901e-1 (2.95e-4) -	9.2753e-1 (1.57e-3) -	9.1973e-1 (3.03e-3) -	9.3364e-1 (5.16e-4)
	10	19	9.4432e-1 (1.26e-2) -	9.6650e-1 (2.71e-4) -	8.0860e-2 (3.89e-2) -	9.4925e-1 (1.57e-2) -	9.6977e-1 (1.23e-4) -	9.6773e-1 (1.61e-3) -	9.6822e-1 (9.43e-4) -	9.7387e-1 (2.23e-3)
	15	24	9.2317e-1 (1.56e-2) -	9.8941e-1 (1.87e-4) -	6.6103e-2 (3.31e-2) -	9.8079e-1 (1.58e-2) -	9.9039e-1 (9.13e-4) -	8.5534e-1 (1.33e-1) -	5.8812e-1 (1.66e-1) -	9.9095e-1 (5.98e-4)
	3	12	5.4278e-1 (8.25e-3) -	0.0000e+0 (0.00e+0) -	3.0856e-2 (7.62e-2) -	5.4280e-1 (8.23e-3) -	4.7076e-1 (1.64e-1) -	0.0000e+0 (0.00e+0) -	0.0000e+0 (0.00e+0) -	5.5841e-1 (3.14e-2)
	5	14	7.5813e-1 (1.83e-2) -	0.0000e+0 (0.00e+0) -	0.0000e+0 (0.00e+0) -	7.6285e-1 (6.70e-3) -	7.8189e-1 (8.94e-3) -	7.1243e-1 (2.09e-1) =	0.0000e+0 (0.00e+0) -	8.0054e-1 (1.48e-3)
DTLZ3	8	17	0.0000e+0 (0.00e+0) -	0.0000e+0 (0.00e+0) -	0.0000e+0 (0.00e+0) -	8.9313e-1 (1.01e-2) -	9.1743e-1 (1.29e-3) -	8.2545e-1 (2.09e-1) -	0.0000e+0 (0.00e+0) -	9.3378e-1 (6.79e-4)
	10	19	0.0000e+0 (0.00e+0) -	0.0000e+0 (0.00e+0) -	0.0000e+0 (0.00e+0) -	9.4269e-1 (1.22e-2) -	9.6954e-1 (2.20e-4) -	8.5382e-1 (1.94e-1) -	0.0000e+0 (0.00e+0) -	9.7432e-1 (2.13e-3)
	15	24	0.0000e+0 (0.00e+0) -	0.0000e+0 (0.00e+0) -	0.0000e+0 (0.00e+0) -	9.8358e-1 (1.23e-2) -	9.9022e-1 (1.68e-3) -	1.0711e-2 (2.84e-2) -	3.3263e-3 (1.49e-2) -	9.9056e-1 (1.41e-3)
	3	12	3.9968e-1 (1.52e-1) -	5.5840e-1 (2.16e-4) -	9.1241e-2 (2.50e-2) -	5.5368e-1 (1.35e-3) -	5.5957e-1 (3.52e-5) -	5.5873e-1 (4.96e-2) =	4.5539e-1 (1.31e-1) -	5.6968e-1 (4.95e-4)
	5	14	6.6930e-1 (1.35e-1) -	7.9100e-1 (7.71e-4) -	1.3209e-1 (8.75e-2) -	7.4845e-1 (1.41e-1) -	7.9489e-1 (4.33e-4) -	8.1140e-1 (1.48e-3) +	7.9307e-1 (4.08e-2) -	8.0035e-1 (2.25e-3)
DTLZ4	8	17	9.0675e-1 (3.59e-2) -	9.1855e-1 (3.57e-4) -	9.3281e-2 (1.55e-1) -	8.9425e-1 (9.71e-3) -	9.1896e-1 (2.06e-4) -	8.6644e-1 (1.09e-1) -	9.1076e-1 (2.42e-2) -	9.3385e-1 (1.03e-3)
	10	19	9.6759e-1 (2.18e-3) -	9.7108e-1 (2.80e-4) -	1.1968e-1 (8.63e-2) -	9.4861e-1 (8.82e-3) -	9.6971e-1 (1.43e-4) -	9.6871e-1 (1.00e-2) -	9.7004e-1 (2.20e-3) -	9.7461e-1 (1.63e-3)
	15	24	9.8645e-1 (3.10e-3) -	9.9067e-1 (7.79e-5) -	3.9542e-2 (7.29e-2) -	9.7066e-1 (1.57e-2) -	9.8998e-1 (1.66e-3) -	3.9073e-1 (3.65e-1) -	9.8483e-1 (7.43e-3) -	9.9133e-1 (1.55e-4)
+/-/=			2/16/2	0/20/0	0/20/0	0/19/1	3/17/0	2/16/2	1/19/0	

Table 3 summarizes the IGD^{+}-means and standard deviations obtained for each algorithm for 20 independent runs on DTLZ1-DTLZ4. It can be seen that PD-MaOEFO obtains the 15 best results in the comparison of IGD^{+}-means and standard deviations for the 20 test problems. PD-MaOEFO obtains the optimal results for all objectives on the DTLZ3 test problem except for M=15 where the IGD^{+}-means and standard deviations are slightly inferior to PeEA; On the DTLZ4 test problem, the PD-MaOEFO algorithm achieves optimal results across all objectives, demonstrating superior performance in comparison to other algorithms. For the DTLZ1 and DTLZ2 test problems, although PD-MaOEFO does not attain optimal results for all objectives, its performance gap remains marginal when compared to the algorithms that achieve optimal outcomes. This indicates that PD-MaOEFO exhibits certain deficiencies in handling linear optimization problems and spherical Pareto front optimization problems when benchmarked against the MaOEA-IGD and PeEA algorithms. Nevertheless, PD-MaOEFO maintains competitive advantages over other comparative algorithms, highlighting its robustness in many-objective optimization.

Table 3. Comparison of IGD^{+} mean and standard deviation of algorithms for DTLZ

Problem	M	D	hpaEA	MPSOD	MaOEAIT	PeEA	RVEA	NMPSO	MaOEAIGD	PD-MaOEFO
	3	7	1.4106e-2 (2.87e-4) -	3.6432e-1 (1.81e-1) -	1.4081e-1 (1.13e-1) -	1.6698e-2 (7.66e-4) -	1.4829e-2 (4.26e-4) -	1.7770e-2 (1.93e-2) -	3.1093e-1 (3.59e-1) -	**1.1554e-2 (1.43e-3)**
	5	9	4.3724e-2 (4.93e-4) -	2.8626e-1 (1.27e-1) -	3.6335e+0 (6.46e+0) -	4.4296e-2 (1.01e-3) -	4.6068e-2 (1.67e-4) -	4.0095e-2 (1.51e-3) -	2.4441e-1 (2.18e-1) -	**3.7179e-2 (1.23e-4)**
DTLZ1	8	12	9.2241e-2 (5.82e-2) -	9.2589e-2 (3.91e-2) -	1.0625e+1 (8.86e+0) -	7.0077e-2 (1.09e-3) -	7.9253e-2 (4.35e-4) -	9.7165e-2 (1.04e-2) -	3.4476e-1 (3.32e-1) -	**6.8694e-2 (8.81e-3)**
	10	14	8.3006e-2 (1.97e-2) =	1.4365e-1 (3.21e-2) -	1.8169e+1 (1.30e+1) -	7.1225e-2 (1.27e-3) +	6.8102e-2 (4.22e-4) +	9.9132e-2 (1.16e-2) -	**6.4381e-2 (4.24e-3) +**	7.4116e-2 (3.63e-3)
	15	19	9.4576e-2 (7.08e-3) +	**8.4528e-2 (1.12e-2) +**	1.2083e+1 (1.21e+1) -	1.0230e-1 (5.48e-3) +	1.1798e-1 (5.18e-3) =	2.6963e+1 (1.70e+1) -	2.9837e-1 (1.93e-1) =	1.1881e-1 (2.73e-2)
	3	12	2.1883e-2 (1.76e-4) -	2.3001e-2 (7.89e-5) -	4.9776e-1 (1.00e-1) -	2.5479e-2 (7.66e-4) -	2.2477e-2 (1.90e-5) -	2.0644e-2 (6.30e-4) -	2.6934e-2 (5.51e-5) -	**1.8050e-2 (4.83e-4)**
	5	14	6.8774e-2 (5.42e-4) +	7.3184e-2 (2.35e-4) -	6.2586e-1 (7.04e-2) -	8.8139e-2 (1.91e-3) -	7.1311e-2 (1.98e-5) -	6.9491e-2 (1.05e-3) =	**6.2344e-2 (2.08e-4) +**	6.9313e-2 (6.25e-4)
DTLZ2	8	17	1.4046e-1 (1.72e-2) -	1.2866e-1 (2.64e-4) -	6.7492e-1 (8.74e-2) -	1.5395e-1 (5.42e-3) -	1.2550e-1 (7.53e-6) -	1.3012e-1 (1.87e-3) -	1.2439e-1 (1.97e-3) -	**1.0713e-1 (8.38e-4)**
	10	19	2.1177e-1 (1.69e-2) -	1.7717e-1 (2.46e-4) -	8.5664e-1 (6.53e-2) -	1.8118e-1 (1.83e-2) -	1.7519e-1 (6.64e-5) -	1.6892e-1 (1.75e-3) -	1.7714e-1 (1.53e-3) -	**1.5317e-1 (8.50e-3)**
	15	24	3.6407e-1 (1.87e-2) -	2.3882e-1 (4.04e-4) =	9.4187e-1 (8.50e-2) -	**2.3432e-1 (3.22e-2) +**	2.3796e-1 (2.42e-3) +	3.8990e-1 (7.46e-2) -	4.5289e-1 (9.34e-2) -	2.3901e-1 (7.42e-4)
	3	12	3.2925e-2 (5.14e-3) -	3.5432e+1 (9.43e+0) -	3.1416e+0 (3.00e+0) -	3.2142e-2 (4.87e-3) -	1.8467e-1 (4.69e-1) -	5.9504e+0 (4.30e+0) -	1.0074e+1 (5.66e+0) -	**2.3500e-2 (1.77e-2)**
	5	14	9.9570e-2 (1.31e-2) -	3.4584e+1 (5.24e+0) -	1.4659e+1 (1.72e+1) -	9.4581e-2 (3.71e-3) -	8.1542e-2 (7.00e-3) -	1.5543e-1 (2.23e-1) -	9.7637e+0 (4.05e+0) -	**6.7282e-2 (5.18e-4)**
DTLZ3	8	17	3.4182e+0 (2.14e+0) -	2.3351e+1 (5.08e+0) -	3.0296e+2 (1.11e+2) -	1.6238e-1 (8.11e-3) -	1.2760e-1 (1.78e-3) -	6.1569e-1 (1.92e+0) -	9.0094e+0 (5.77e+0) -	**1.0693e-1 (8.54e-4)**
	10	19	9.4736e+0 (3.15e+0) -	2.5367e+1 (5.71e+0) -	3.2292e+2 (9.05e+1) -	1.9067e-1 (1.27e-2) -	1.7569e-1 (4.01e-4) -	2.9392e-1 (1.42e-1) -	4.8546e+0 (1.70e+0) -	**1.5059e-1 (2.17e-3)**
	15	24	1.3122e+1 (4.02e+0) -	1.1651e+1 (3.95e+0) -	3.7701e+2 (1.11e+2) -	**2.3427e-1 (2.04e-2) +**	2.3833e-1 (3.80e-3) +	4.4534e+0 (4.29e+0) -	3.6117e+0 (1.47e+0) -	2.3916e-1 (2.14e-3)
	3	12	1.8118e-1 (1.56e-1) -	2.3037e-2 (9.80e-5) -	5.3032e-1 (7.45e-2) -	2.6128e-2 (5.76e-4) -	2.2475e-2 (1.82e-5) -	3.0228e-2 (4.61e-2) -	1.1655e-1 (1.20e-1) -	**1.7807e-2 (2.93e-4)**
	5	14	1.8160e-1 (1.10e-1) -	7.4224e-2 (3.57e-4) -	5.8148e-1 (9.67e-2) -	9.9686e-2 (6.27e-2) -	7.1311e-2 (2.20e-5) -	6.8424e-2 (9.43e-4) =	7.9773e-2 (3.63e-2) -	**6.8065e-2 (7.89e-4)**
DTLZ4	8	17	1.4900e-1 (4.23e-2) -	1.2590e-1 (2.70e-4) -	8.2082e-1 (1.94e-1) -	1.6251e-1 (9.71e-3) -	1.2550e-1 (4.03e-6) -	1.7305e-1 (6.28e-2) -	1.4212e-1 (3.25e-2) -	**1.0722e-1 (2.38e-3)**
	10	19	1.7716e-1 (3.37e-3) -	1.6773e-1 (6.34e-4) -	6.5310e-1 (7.85e-2) -	1.9279e-1 (1.36e-2) -	1.7535e-1 (8.40e-5) -	1.6763e-1 (1.24e-2) -	1.7204e-1 (1.41e-3) -	**1.5086e-1 (5.00e-3)**
	15	24	2.5823e-1 (9.62e-3) -	2.3991e-1 (6.56e-4) -	1.0166e+0 (1.25e-1) -	2.5587e-1 (2.46e-2) =	2.3968e-1 (4.09e-3) =	5.8827e-1 (1.83e-1) -	2.5010e-1 (1.34e-2) -	**2.3832e-1 (3.72e-4)**
+/-/=			2/17/1	1/18/1	0/20/0	4/15/1	3/15/2	0/18/2	2/17/1	

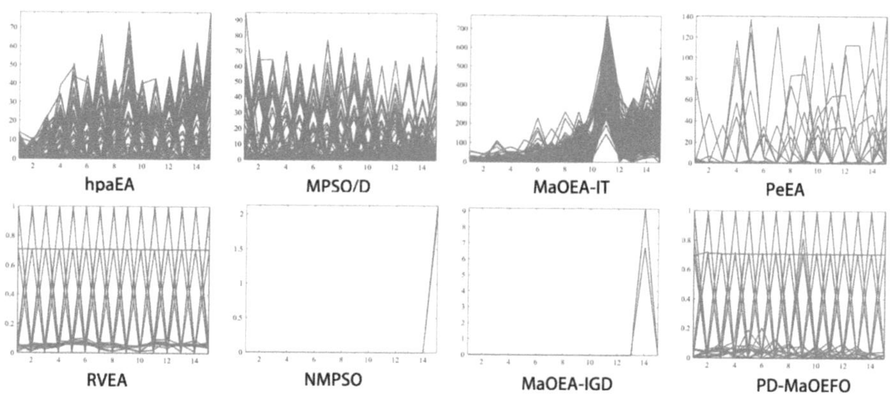

Fig. 4. The effect of 15-objective DTLZ3 optimization for each algorithm.

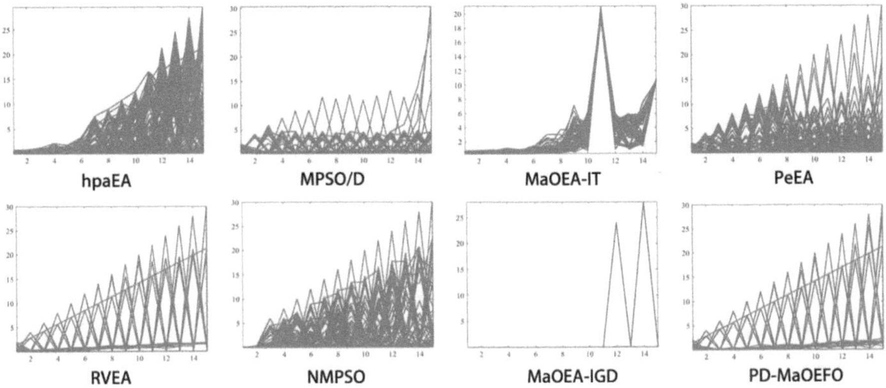

Fig. 5. The effect of 15-objective WFG5 optimization for each algorithm.

Table 4. Comparison of *HV* mean and standard deviation of algorithms for WFG

Problem	M	D	hpaEA	MPSOD	MaOEAIT	PeEA	RVEA	NMPSO	MaOEAIGD	PD-MaOEFO
WFG1	3	12	6.5200e-1 (1.36e-1) =	3.2992e-1 (3.39e-2) -	2.7235e-1 (8.06e-2) -	9.1393e-1 (8.00e-3) +	**9.2132e-1 (2.19e-2) +**	7.8533e-1 (4.78e-2) +	2.4533e-1 (1.42e-1) -	7.0012e-1 (5.90e-2)
	5	14	8.8809e-1 (7.85e-2) -	3.7291e-1 (2.56e-2) -	3.4278e-1 (1.02e-1) -	9.8877e-1 (1.74e-3) =	**9.9198e-1 (1.74e-2) +**	9.0783e-1 (6.13e-2) -	4.2177e-1 (1.92e-1) -	9.7390e-1 (3.44e-2)
	8	17	9.8114e-1 (2.89e-2) =	3.4158e-1 (2.43e-2) -	1.3023e-1 (6.69e-2) -	9.6836e-1 (1.07e-1) =	9.5249e-1 (7.84e-2) =	8.7567e-1 (5.35e-2) -	2.9145e-1 (7.60e-2) -	**9.9711e-1 (9.94e-4)**
	10	19	9.9761e-1 (3.46e-3) =	2.9552e-1 (1.93e-2) -	1.6626e-1 (9.37e-2) -	9.7459e-1 (1.03e-1) -	9.9744e-1 (4.60e-4) -	8.7159e-1 (8.72e-2) -	6.7818e-1 (2.83e-1) -	**9.9835e-1 (1.75e-3)**
	15	24	9.9782e-1 (3.46e-3) =	2.9572e-1 (2.80e-2) -	1.0870e-1 (3.23e-2) -	9.6245e-1 (1.16e-1) =	9.6850e-1 (6.64e-2) =	8.7125e-1 (5.65e-2) -	5.1535e-1 (2.50e-1) -	**9.9898e-1 (3.43e-4)**
WFG2	3	12	**9.2936e-1 (2.28e-3) +**	8.9343e-1 (7.22e-3) -	5.7841e-1 (7.63e-2) -	9.1018e-1 (7.18e-3) -	9.2484e-1 (1.25e-3) =	8.6684e-1 (1.36e-2) -	6.2557e-1 (7.87e-2) -	9.2585e-1 (2.80e-3)
	5	14	**9.9057e-1 (1.40e-3) +**	8.7818e-1 (8.88e-3) -	5.8465e-1 (4.19e-2) -	9.6544e-1 (6.31e-3) -	9.8978e-1 (2.53e-3) +	9.6175e-1 (1.24e-2) -	9.1675e-1 (2.91e-2) -	9.8138e-1 (2.57e-3)
	8	17	**9.9366e-1 (1.44e-3) +**	9.2528e-1 (8.61e-3) -	4.5098e-1 (5.54e-2) -	9.8623e-1 (4.51e-3) +	9.8795e-1 (3.55e-3) +	9.8480e-1 (8.19e-3) +	9.7136e-1 (3.56e-2) -	9.7621e-1 (5.33e-3)
	10	19	9.9342e-1 (1.24e-3) +	9.5194e-1 (7.04e-3) -	5.5635e-1 (5.04e-2) -	9.8984e-1 (5.00e-3) +	9.8821e-1 (4.03e-3) +	**9.9600e-1 (2.26e-3) +**	9.7426e-1 (5.44e-2) -	9.8495e-1 (4.44e-3)
	15	24	9.9088e-1 (2.69e-3) +	9.6422e-1 (6.59e-3) -	5.5064e-1 (6.30e-2) -	**9.9651e-1 (1.57e-3) +**	9.8069e-1 (5.55e-3) =	9.6914e-1 (3.52e-2) =	8.9439e-1 (1.19e-1) -	9.8009e-1 (1.69e-2)
WFG3	3	12	3.7454e-1 (1.23e-2) +	2.8598e-1 (9.49e-3) -	2.1421e-4 (9.50e-4) -	3.4945e-1 (9.53e-3) -	3.3789e-1 (8.50e-3) -	**4.0701e-1 (1.76e-3) +**	9.0804e-2 (6.32e-2) -	3.6047e-1 (7.97e-3)
	5	14	1.5219e-1 (1.52e-2) =	1.2968e-2 (1.85e-2) -	0.0000e+0 (0.00e+0) -	1.8291e-1 (1.60e-2) +	1.3371e-1 (2.89e-2) =	**1.9017e-1 (1.10e-2) +**	8.3467e-2 (2.54e-3) -	1.4623e-1 (2.93e-2)
	8	17	3.1186e-3 (5.73e-3) -	0.0000e+0 (0.00e+0) -	0.0000e+0 (0.00e+0) -	**8.6114e-2 (1.34e-2) +**	0.0000e+0 (0.00e+0) -	5.3564e-3 (8.23e-3) -	1.5799e-2 (1.59e-2) -	4.6846e-2 (3.57e-2)
	10	19	0.0000e+0 (0.00e+0) -	0.0000e+0 (0.00e+0) -	0.0000e+0 (0.00e+0) -	3.2053e-2 (1.36e-2) -	0.0000e+0 (0.00e+0) -	0.0000e+0 (0.00e+0) -	6.7202e-3 (1.18e-2) -	**4.2744e-2 (2.43e-2)**
	15	24	**0.0000e+0 (0.00e+0) =**	0.0000e+0 (0.00e+0) =	0.0000e+0 (0.00e+0) =	0.0000e+0 (0.00e+0) =	0.0000e+0 (0.00e+0) =	0.0000e+0 (0.00e+0) =	0.0000e+0 (0.00e+0) =	0.0000e+0 (0.00e+0)
WFG4	3	12	5.5264e-1 (6.94e-3) +	5.0608e-1 (3.98e-3) -	1.8712e-1 (4.22e-2) -	5.3887e-1 (2.15e-3) -	5.5007e-1 (1.52e-3) =	**5.6546e-1 (1.01e-3) +**	1.2819e-1 (6.42e-2) -	5.4882e-1 (3.57e-3)
	5	14	7.3986e-1 (2.67e-2) -	5.7782e-1 (7.47e-3) -	1.7293e-1 (3.40e-2) -	7.3465e-1 (6.53e-3) -	7.8564e-1 (1.51e-3) +	**7.9713e-1 (1.73e-3) +**	1.0722e-1 (3.37e-2) -	7.7933e-1 (4.32e-3)
	8	17	7.3246e-1 (2.84e-2) -	5.0130e-1 (4.57e-3) -	3.3229e-1 (2.63e-2) -	8.5638e-1 (8.46e-3) -	**9.1402e-1 (1.22e-3) +**	9.1032e-1 (3.86e-3) +	1.1560e-1 (3.87e-2) -	8.9087e-1 (6.13e-3)
	10	19	7.6662e-1 (2.72e-2) -	5.0326e-1 (1.13e-2) -	2.2837e-1 (3.19e-2) -	9.1827e-1 (7.03e-3) -	9.3837e-1 (6.92e-3) -	9.3223e-1 (4.75e-3) -	1.0707e-1 (3.32e-2) -	**9.6042e-1 (1.72e-3)**
	15	24	6.9479e-1 (4.23e-2) -	4.2223e-1 (1.40e-2) -	2.7579e-1 (3.51e-2) -	9.5362e-1 (7.13e-3) -	9.7055e-1 (7.94e-3) -	9.3520e-1 (1.66e-2) -	1.3050e-1 (5.44e-2) -	**9.8976e-1 (4.96e-4)**
WFG5	3	12	5.1832e-1 (5.83e-4) -	4.8332e-1 (1.44e-3) -	7.4318e-2 (3.00e-2) -	5.0245e-1 (2.61e-3) -	5.1609e-1 (5.38e-4) -	5.2239e-1 (3.93e-3) -	4.1565e-1 (8.79e-2) -	**5.2889e-1 (8.08e-4)**
	5	14	7.4067e-1 (5.31e-3) -	5.5170e-1 (5.65e-3) -	1.2544e-1 (2.92e-2) -	6.9405e-1 (5.04e-3) -	7.4292e-1 (4.96e-4) -	7.4602e-1 (3.42e-3) -	4.8589e-1 (2.36e-1) -	**7.4964e-1 (4.24e-3)**
	8	17	8.3585e-1 (1.58e-2) -	4.9612e-1 (1.57e-2) -	2.8088e-1 (1.92e-2) -	8.0759e-1 (6.63e-3) -	8.5161e-1 (8.50e-3) -	8.4731e-1 (2.97e-3) -	1.5271e-1 (1.81e-1) -	**8.5857e-1 (3.95e-4)**
	10	19	8.5311e-1 (1.72e-2) -	5.7410e-1 (2.11e-2) -	1.4125e-1 (2.48e-2) -	8.6643e-1 (3.79e-3) -	**9.0323e-1 (3.52e-4) +**	8.7926e-1 (2.98e-3) -	1.3361e-1 (1.42e-1) -	8.9512e-1 (1.94e-3)
	15	24	7.9005e-1 (1.89e-2) -	5.4931e-1 (2.99e-2) -	1.0967e-1 (1.90e-2) -	8.8758e-1 (9.61e-3) -	**9.1761e-1 (1.06e-4) +**	8.7619e-1 (1.68e-2) -	9.1801e-2 (3.01e-2) -	9.0892e-1 (5.77e-3)
WFG6	3	12	5.0363e-1 (9.99e-3) -	5.0933e-1 (9.60e-3) -	5.6627e-2 (1.53e-2) -	4.9593e-1 (1.79e-2) -	4.9797e-1 (1.13e-2) -	4.3888e-1 (3.97e-2) -	2.3759e-1 (1.41e-1) -	**5.2904e-1 (1.54e-2)**
	5	14	7.1264e-1 (1.35e-2) -	5.6217e-1 (1.57e-2) -	1.1454e-1 (2.26e-2) -	6.6533e-1 (1.91e-2) -	7.2569e-1 (1.49e-2) -	6.3553e-1 (1.71e-3) -	2.1488e-1 (7.46e-2) -	**7.4747e-1 (2.45e-2)**
	8	17	8.0336e-1 (2.15e-2) =	5.0563e-1 (4.33e-2) -	2.6213e-1 (3.88e-2) -	7.5270e-1 (2.75e-2) -	8.1950e-1 (3.18e-2) =	7.2237e-1 (1.42e-3) -	2.7972e-1 (1.51e-1) -	**8.3305e-1 (2.14e-2)**
	10	19	8.2163e-1 (1.61e-2) -	5.0158e-1 (4.93e-2) -	1.4932e-1 (7.94e-3) -	8.2180e-1 (1.17e-2) -	8.6369e-1 (3.50e-2) =	7.4910e-1 (1.50e-3) -	3.8125e-1 (2.10e-1) -	**8.7575e-1 (1.77e-2)**
	15	24	7.6537e-1 (2.69e-2) -	4.3331e-1 (5.60e-2) -	1.3475e-1 (1.81e-2) -	8.4701e-1 (3.64e-2) =	8.4595e-1 (4.87e-2) =	7.4045e-1 (3.48e-3) -	2.8539e-1 (2.30e-1) -	**8.6764e-1 (3.17e-2)**
WFG7	3	12	5.5717e-1 (8.79e-4) +	5.2856e-1 (2.46e-3) -	1.3742e-1 (2.49e-2) -	5.3875e-1 (4.34e-3) -	5.5087e-1 (1.17e-3) -	**5.6732e-1 (9.73e-4) +**	3.4073e-1 (1.56e-1) -	5.5614e-1 (1.21e-3)
	5	14	7.8539e-1 (5.35e-3) -	5.8545e-1 (7.09e-3) -	1.8139e-1 (2.15e-2) -	7.1583e-1 (2.19e-2) -	7.8741e-1 (1.04e-3) -	7.9056e-1 (4.99e-3) -	2.5151e-1 (1.17e-1) -	**8.0222e-1 (1.31e-3)**
	8	17	8.4666e-1 (2.28e-2) -	5.1647e-1 (1.78e-2) -	3.5045e-1 (2.24e-2) -	8.4323e-1 (1.30e-2) -	9.0918e-1 (2.98e-3) +	**9.1501e-1 (2.13e-3) +**	2.1219e-1 (7.23e-2) -	8.9410e-1 (4.34e-3)
	10	19	8.6935e-1 (2.51e-2) -	5.2152e-1 (1.88e-2) -	2.0652e-1 (2.19e-2) -	9.2435e-1 (5.50e-3) -	**9.5846e-1 (1.47e-3) +**	9.5019e-1 (2.05e-3) +	2.0488e-1 (7.10e-2) -	9.4504e-1 (4.17e-3)
	15	24	7.9957e-1 (2.54e-2) -	5.1317e-1 (3.68e-2) -	1.9384e-1 (1.91e-2) -	9.5923e-1 (6.10e-3) -	9.7937e-1 (2.27e-3) -	9.6449e-1 (1.26e-2) -	1.7841e-1 (6.18e-2) -	**9.8960e-1 (4.88e-4)**
WFG8	3	12	4.7100e-1 (3.18e-3) +	4.4157e-1 (3.55e-3) -	3.4140e-2 (1.88e-2) -	4.4501e-1 (3.39e-3) -	4.6582e-1 (1.98e-3) =	**4.8664e-1 (1.10e-3) +**	3.9746e-2 (5.97e-2) -	4.6442e-1 (3.41e-3)
	5	14	6.7025e-1 (2.77e-3) +	5.9928e-1 (5.56e-3) -	9.8334e-2 (1.75e-2) -	6.2686e-1 (8.21e-3) -	6.6793e-1 (5.53e-3) =	**6.9983e-1 (3.18e-3) +**	5.6243e-2 (5.87e-2) -	6.6573e-1 (6.18e-3)
	8	17	7.8111e-1 (1.89e-2) =	5.8091e-1 (3.20e-2) -	2.7543e-1 (4.35e-2) -	7.0940e-1 (4.51e-2) -	6.7938e-1 (8.24e-2) -	**8.0993e-1 (8.16e-3) +**	1.6501e-1 (7.20e-2) -	7.8011e-1 (1.97e-2)
	10	19	8.4220e-1 (2.21e-2) =	6.2373e-1 (2.39e-2) -	1.5547e-1 (1.20e-2) -	8.2708e-1 (6.97e-2) =	8.4119e-1 (7.40e-2) =	**9.0200e-1 (2.36e-2) +**	2.2130e-1 (7.24e-2) -	8.5386e-1 (2.08e-2)
	15	24	8.3466e-1 (2.02e-2) -	5.3244e-1 (8.07e-2) -	4.2437e-2 (1.43e-2) -	8.9903e-1 (8.37e-3) -	6.9278e-1 (1.70e-1) -	9.0725e-1 (6.16e-3) -	1.7691e-1 (6.09e-2) -	**9.1587e-1 (1.93e-2)**
WFG9	3	12	5.3179e-1 (3.59e-3) -	5.0808e-1 (2.69e-3) -	1.4469e-1 (2.60e-2) -	5.2583e-1 (2.97e-3) -	5.3307e-1 (3.22e-3) =	5.1686e-1 (5.60e-2) -	2.0770e-1 (4.43e-2) -	**5.3472e-1 (3.48e-3)**
	5	14	7.3357e-1 (8.59e-3) -	5.5910e-1 (1.10e-2) -	2.4191e-1 (2.31e-2) -	7.0943e-1 (8.33e-3) -	7.3907e-1 (1.11e-2) =	**7.4938e-1 (4.50e-2) +**	3.1349e-1 (1.82e-1) -	7.4117e-1 (6.90e-3)
	8	17	8.2282e-1 (1.55e-2) +	4.9457e-1 (2.08e-2) -	1.1311e-1 (2.07e-2) -	8.1594e-1 (4.08e-2) +	**8.4067e-1 (3.91e-2) +**	7.7797e-1 (8.85e-2) =	1.8359e-1 (7.04e-2) -	7.9206e-1 (3.58e-2)
	10	19	8.5407e-1 (1.66e-2) +	5.6399e-1 (2.91e-2) -	2.0182e-1 (2.52e-2) -	**8.9294e-1 (4.94e-3) +**	8.8498e-1 (3.84e-2) +	8.5124e-1 (6.21e-2) +	1.9203e-1 (6.32e-2) -	8.1357e-1 (5.41e-2)
	15	24	7.7189e-1 (2.08e-2) =	5.5448e-1 (6.00e-2) -	1.1586e-1 (2.94e-2) -	**9.0775e-1 (5.02e-2) +**	8.7566e-1 (5.50e-2) +	7.6480e-1 (1.16e-1) =	1.9496e-1 (1.32e-1) -	7.6307e-1 (5.68e-2)
+/-/=			12/23/10	0/44/1	0/44/1	9/30/6	14/16/15	17/24/4	0/44/1	

(2) WFG experimental results and analysis.

Table 4 summarizes the mean and standard deviation of the HV obtained from 20 independent runs of each algorithm on WFG1-WFG9. Out of the 45 test problems, PD-MaOEFO obtains the 18 best results and therefore the best overall performance in the WFG test suite. This was followed by NMPSO and RVEA, which obtained 12 and 7 best results, respectively. From the data in Table 4, it can be seen that PD-MaOEFO still has some shortcomings in dealing with test problems (WFG1-3) with irregular Pareto fronts, mainly because the reference vectors generated in the PD-MaOEFO algorithm do not completely and uniformly cover the irregular Pareto surfaces. Besides, PD-MaOEFO achieves a large advantage over other many-objective optimization algorithms, for example, the PD-MaOEFO algorithm achieves the best results for many times in the WFG4-WFG6 test problems, which shows that the PD-MaOEFO algorithm has the ability of jumping out of the local optimum, which is advantageous for solving the test problems of asymmetric mapping and complex objective function mapping.

Table 5 summarizes the mean and standard deviation of the IGD^+ obtained from 20 independent runs of each algorithm on WFG1-WFG9. Out of the 45 test problems, PD-MaOEFO obtains 17 best results and therefore the best overall performance in the WFG test suite. This was followed by RVEA and NMPSO, which obtained 15 and 9 best results, respectively. From the data in Table 5, it can be seen that PD-MaOEFO obtains three best results in both WFG4-7 and WFG9 test problems with different objectives. It shows that the PD-

Table 5. Comparison of IGD^+ mean and standard deviation of algorithms for WFG

Problem	M	D	hpaEA	MPSOD	MaOEAIT	PeEA	RVEA	NMPSO	MaOEAIGD	PD-MaOEFO
WFG1	3	12	4.4495e-1 (1.60e-1) =	1.2700e+0 (8.00e-2) -	1.3688e+0 (1.28e-1) -	9.2328e-2 (1.19e-2) +	1.3057e-1 (4.17e-2) +	3.2458e-1 (8.71e-2) +	1.3912e+0 (2.58e-1) -	4.7075e-1 (1.08e-1)
	5	14	2.7844e-1 (5.25e-2) =	1.6517e+0 (8.96e-2) -	1.8159e+0 (1.01e+0) -	1.4029e-1 (3.52e-3) +	1.6957e-1 (3.92e-2) +	7.5346e-1 (4.36e-1) -	3.2001e+0 (1.23e+0) -	2.7394e-1 (5.25e-2)
	8	17	2.6725e-1 (3.09e-2) =	2.0795e+0 (9.92e-2) -	2.3645e+0 (1.27e+0) -	2.5202e-1 (2.45e-1) +	1.4648e-1 (2.09e-2) +	2.7598e+0 (1.39e+0) -	8.0479e+0 (1.51e+0) -	3.3614e-1 (1.46e-1)
	10	19	2.5908e-1 (2.69e-2) -	2.4759e+0 (9.80e-2) -	2.6211e+0 (4.26e-1) -	2.4580e-1 (2.65e-1) =	1.1659e-1 (1.98e-3) +	3.3641e+0 (2.16e+0) -	6.7053e+0 (3.84e+0) -	2.1992e-1 (9.02e-2)
	15	24	3.3225e-1 (4.18e-2) -	2.7795e+0 (1.40e-1) -	4.2663e+0 (3.02e+0) -	3.1206e-1 (3.05e-1) -	1.2985e-1 (7.65e-3) +	5.3364e+0 (3.33e+0) -	1.3600e+1 (7.14e+0) -	2.5344e-1 (1.50e-1)
WFG2	3	12	5.0081e-2 (3.49e-3) +	7.1043e-2 (6.09e-3) -	7.6376e-1 (1.33e-1) -	5.7303e-2 (3.94e-3) -	7.0311e-2 (3.04e-3) -	1.2310e-1 (2.21e-2) -	8.0961e-1 (1.52e-1) -	5.4233e-2 (4.04e-3)
	5	14	1.7081e-1 (7.29e-3) -	1.6714e-1 (5.08e-3) -	1.5923e+0 (2.92e-1) -	1.2940e-1 (4.02e-3) -	1.4828e-1 (5.08e-3) -	2.6301e-1 (8.09e-2) -	1.3685e+0 (2.96e-1) -	1.2335e-1 (6.10e-3)
	8	17	2.6684e-1 (1.18e-2) -	1.9981e-1 (3.01e-3) -	2.9297e+0 (7.16e-1) -	2.0362e-1 (3.88e-2) -	1.6650e-1 (1.16e-2) -	4.5013e-1 (7.73e-2) -	1.1810e+0 (4.96e-1) -	1.5110e-1 (7.50e-3)
	10	19	3.0349e-1 (1.46e-2) -	2.0302e-1 (5.14e-3) -	3.6002e+0 (8.37e-1) -	1.8372e-1 (2.93e-2) -	1.3453e-1 (1.25e-2) =	4.4648e-1 (9.10e-2) -	1.4180e+0 (7.71e-1) -	1.3697e-1 (6.25e-3)
	15	24	7.6719e-1 (1.39e-1) -	1.8857e-1 (5.69e-3) =	5.3255e+0 (1.31e+0) -	3.2723e-1 (1.78e-1) -	1.7949e-1 (2.31e-2) =	2.7186e+0 (1.44e+0) -	3.9695e+0 (2.96e+0) -	2.0195e-1 (6.91e-2)
WFG3	3	12	7.9042e-2 (1.11e-2) +	2.0289e-1 (1.82e-2) -	1.0794e+0 (8.49e-2) -	1.0733e-1 (1.35e-2) +	1.6963e-1 (1.75e-2) -	2.7526e-2 (3.07e-3) +	2.3612e+0 (1.01e+0) -	1.2994e-1 (1.48e-2)
	5	14	2.9322e-1 (2.88e-2) +	5.3905e-1 (5.75e-2) -	1.7984e+0 (1.46e-1) -	3.0822e-1 (9.11e-2) +	4.2249e-1 (4.77e-2) -	1.9257e-1 (3.34e-2) +	4.9927e+0 (1.59e-2) -	3.5287e-1 (4.38e-2)
	8	17	5.7512e-1 (7.34e-2) +	9.8987e-1 (7.12e-2) +	1.2393e+0 (3.42e-1) =	7.2357e-1 (1.51e-1) +	1.9868e+0 (2.52e-1) -	4.4486e-1 (5.91e-2) +	8.0078e+0 (5.71e-2) -	1.2322e+0 (2.06e-1)
	10	19	5.9222e-1 (1.73e-1) +	1.1430e+0 (4.62e-2) +	3.7157e+0 (2.26e-1) -	8.6643e-1 (1.75e-1) +	2.4705e+0 (1.97e-1) -	5.8836e-1 (7.11e-2) +	1.0007e+1 (2.43e-2) -	1.7405e+0 (1.34e-1)
	15	24	1.0085e+0 (2.79e-1) +	2.6846e+0 (1.83e-1) +	4.6313e+0 (4.78e-1) -	1.4875e+0 (1.98e-1) +	3.7069e+0 (8.40e-2) =	6.7464e-1 (1.50e-1) +	1.3383e+1 (4.17e+0) -	3.9097e+0 (2.61e-1)
WFG4	3	12	1.0099e-1 (4.06e-2) =	1.4707e-1 (3.09e-3) -	9.2547e-1 (1.13e-1) -	1.1418e-1 (3.38e-3) -	9.0783e-2 (2.49e-3) =	7.3498e-2 (1.83e-3) +	1.5987e+0 (1.07e+0) -	8.9250e-2 (5.01e-3)
	5	14	6.6119e-1 (1.64e-1) -	5.4789e-1 (5.60e-3) -	3.9794e+0 (8.01e-1) -	4.7811e-1 (1.19e-2) -	3.3974e-1 (1.71e-3) -	3.2924e-1 (4.91e-3) =	2.0901e+0 (1.95e+0) -	3.2825e-1 (7.58e-3)
	8	17	3.6209e+0 (3.69e-1) -	1.0730e+0 (5.96e-3) -	6.4873e+0 (6.63e-1) -	9.8376e-1 (1.21e-2) -	7.5918e-1 (1.23e-2) -	8.0273e-1 (1.60e-2) -	1.9920e+0 (2.07e+0) -	7.1914e-1 (1.66e-3)
	10	19	5.4982e+0 (5.35e-1) -	1.2158e+0 (6.75e-3) -	1.0769e+1 (7.45e-1) -	1.1676e+0 (7.42e-3) -	8.9708e-1 (3.88e-3) +	1.0903e+0 (1.99e-2) -	1.5913e+0 (6.26e-3) -	9.5124e-1 (3.14e-2)
	15	24	1.4067e+1 (1.04e+0) -	1.4745e+0 (6.83e-3) -	1.5334e+1 (8.23e-1) -	4.0315e+0 (1.60e+0) -	1.3020e+0 (2.43e-2) -	1.5874e+0 (4.00e-2) -	1.4109e+1 (7.49e+0) -	1.2653e+0 (1.31e-2)
WFG5	3	12	1.3383e-1 (1.58e-3) -	1.6347e-1 (1.50e-3) -	1.1629e+0 (1.09e-1) -	1.6109e-1 (3.77e-3) -	1.3666e-1 (8.78e-4) -	1.2274e-1 (1.50e-3) -	4.0918e-1 (6.47e-1) -	1.2138e-1 (3.89e-3)
	5	14	4.1376e-1 (1.97e-2) -	5.6624e-1 (4.00e-3) -	2.8514e+0 (5.22e-1) -	5.1896e-1 (1.02e-2) -	3.9079e-1 (3.54e-4) -	3.8503e-1 (6.27e-3) -	1.1776e+0 (1.69e+0) -	3.5635e-1 (4.16e-3)
	8	17	1.1798e+0 (2.90e-1) -	1.1220e+0 (1.20e-2) -	4.9691e+0 (3.28e-1) -	1.0410e+0 (1.43e-2) -	7.8139e-1 (4.44e-4) +	9.0853e-1 (1.89e-2) -	9.9939e+0 (4.06e+0) -	8.0413e-1 (3.95e-2)
	10	19	2.1858e+0 (5.77e-1) -	1.2870e+0 (2.08e-2) -	1.0409e+1 (1.12e+0) -	1.2246e+0 (9.25e-3) -	9.4774e-1 (3.53e-3) +	1.1568e+0 (2.27e-2) -	1.3256e+1 (3.61e+0) -	9.8126e-1 (8.82e-3)
	15	24	9.3579e+0 (7.39e-1) -	1.6163e+0 (4.39e-2) -	1.6598e+1 (9.79e-1) -	3.0112e+0 (1.46e+0) -	1.5148e+0 (3.49e-1) -	4.5098e+0 (1.72e+0) -	2.1462e+1 (7.05e+0) -	1.3355e+0 (6.31e-3)
WFG6	3	12	1.5543e-1 (1.50e-2) -	1.4820e-1 (1.40e-2) -	1.2647e+0 (6.57e-2) -	1.7273e-1 (2.52e-2) -	1.6475e-1 (1.67e-2) -	2.7069e-1 (6.23e-2) -	7.2804e-1 (7.14e-1) -	1.1843e-1 (2.31e-2)
	5	14	4.5373e-1 (2.60e-2) -	6.0461e-1 (1.31e-2) -	3.3470e+0 (6.19e-1) -	5.6178e-1 (2.61e-2) -	4.1561e-1 (2.01e-2) -	5.5674e-1 (5.22e-3) -	1.3611e+0 (6.53e-1) -	3.7236e-1 (3.60e-2)
	8	17	1.3490e+0 (3.05e-1) -	1.1836e+0 (3.98e-2) -	5.5270e+0 (8.65e-1) -	1.0799e+0 (3.32e-2) -	8.6335e-1 (6.94e-2) -	1.0445e+0 (1.17e-2) -	5.4925e+0 (4.05e+0) -	8.1409e-1 (2.88e-2)
	10	19	2.2961e+0 (5.80e-1) -	1.3638e+0 (2.27e-2) -	1.0896e+1 (3.99e-1) -	1.2520e+0 (1.59e-2) -	9.9520e-1 (2.32e-2) +	1.2334e+0 (1.77e-2) -	5.8419e+0 (5.59e+0) -	1.0418e+0 (4.80e-2)
	15	24	9.8694e+0 (8.03e-1) -	1.6607e+0 (2.16e-2) =	1.6261e+1 (6.33e-1) -	3.0128e+0 (1.52e+0) =	1.3804e+0 (3.07e-2) +	1.7023e+0 (7.65e-2) =	1.4710e+1 (9.63e+0) -	2.3174e+0 (1.07e+0)
WFG7	3	12	7.9759e-2 (1.24e-3) =	1.2170e-1 (2.05e-3) -	9.8876e-1 (8.56e-2) -	1.1455e-1 (6.68e-3) -	8.9630e-2 (1.81e-3) -	7.2097e-2 (2.17e-3) +	8.5765e-1 (1.16e+0) -	8.0112e-2 (1.77e-3)
	5	14	3.9098e-1 (5.05e-2) -	5.4730e-1 (7.13e-3) -	2.9299e+0 (3.65e-1) -	4.9506e-1 (2.86e-2) -	3.4059e-1 (1.43e-3) -	3.2584e-1 (4.30e-3) -	1.4035e+0 (1.29e+0) -	3.1765e-1 (1.48e-2)
	8	17	1.7905e+0 (3.93e-1) -	1.0973e+0 (1.44e-2) -	5.1842e+0 (4.43e-1) -	9.8545e-1 (1.30e-2) -	7.4825e-1 (8.98e-3) -	8.0904e-1 (1.34e-2) -	6.7410e+0 (2.61e+0) -	7.2775e-1 (2.84e-3)
	10	19	3.3884e+0 (6.30e-1) -	1.2487e+0 (1.03e-2) -	1.0432e+1 (5.49e-1) -	1.1564e+0 (1.32e-2) -	9.2672e-1 (5.51e-3) -	1.0431e+0 (2.17e-2) -	9.4003e+0 (2.47e+0) -	9.0410e-1 (5.08e-3)
	15	24	1.1177e+1 (8.00e-1) -	1.4794e+0 (1.00e-2) +	1.5753e+1 (6.54e-1) -	2.7460e+0 (1.20e+0) =	1.2898e+0 (1.46e-2) +	4.0177e+0 (1.37e+0) -	1.8845e+1 (3.44e+0) -	1.7743e+0 (4.13e-1)
WFG8	3	12	2.1231e-1 (2.32e-3) +	2.4966e-1 (4.83e-3) -	1.4682e+0 (8.82e-2) -	2.5886e-1 (5.53e-3) -	2.2393e-1 (3.14e-3) =	1.9615e-1 (1.42e-3) +	1.2864e+0 (3.00e-1) -	2.2521e-1 (5.19e-3)
	5	14	6.5227e-1 (2.82e-2) -	7.3480e-1 (6.80e-3) -	3.5743e+0 (5.39e-1) -	7.1766e-1 (1.17e-2) -	6.2192e-1 (4.26e-3) =	5.8470e-1 (4.97e-3) +	1.9879e+0 (1.24e+0) -	6.2577e-1 (1.26e-2)
	8	17	2.2005e+0 (4.67e-1) -	1.4970e+0 (4.56e-2) =	5.9287e+0 (9.55e-1) -	1.8252e+0 (2.44e-1) -	1.1235e+0 (1.44e-1) +	1.4511e+0 (1.24e-1) =	7.4459e+0 (1.62e+0) -	1.4670e+0 (1.53e-1)
	10	19	2.9992e+0 (8.24e-1) -	1.7937e+0 (6.08e-2) =	1.1130e+1 (5.01e-1) -	2.7497e+0 (2.81e-1) -	1.1561e+0 (2.01e-1) +	1.2880e+0 (2.69e-1) +	8.9651e+0 (1.75e+0) -	1.8227e+0 (3.14e-1)
	15	24	1.0329e+1 (7.98e-1) -	1.7483e+0 (5.64e-2) +	1.5158e+1 (1.38e+0) -	6.7880e+0 (5.33e-1) -	1.8812e+0 (8.79e-1) +	5.9778e+0 (1.78e+0) -	1.9463e+1 (3.72e+0) -	5.3064e+0 (7.83e-1)
WFG9	3	12	1.0929e-1 (5.02e-3) -	1.2976e-1 (2.74e-3) -	9.9345e-1 (9.34e-2) -	1.2875e-1 (4.52e-3) -	1.0864e-1 (3.36e-3) -	1.3989e-1 (9.12e-2) -	6.4233e-1 (2.25e-1) -	9.9290e-2 (3.52e-3)
	5	14	4.2064e-1 (2.00e-2) -	5.6879e-1 (8.60e-3) -	2.0061e+0 (1.91e-1) -	4.9861e-1 (1.52e-2) -	3.8942e-1 (1.45e-2) -	3.8853e-1 (6.89e-2) =	1.3214e+0 (7.43e-1) -	3.6736e-1 (1.04e-2)
	8	17	1.3542e+0 (2.72e-1) -	1.1943e+0 (1.71e-2) -	5.8593e+0 (8.66e-1) -	1.0477e+0 (5.92e-2) -	9.0128e-1 (6.67e-2) -	1.0872e+0 (1.53e-1) -	8.1752e+0 (2.31e+0) -	8.0088e-1 (6.14e-2)
	10	19	2.3419e+0 (5.24e-1) -	1.4208e+0 (3.46e-2) -	9.9018e+0 (8.53e-1) -	1.2102e+0 (1.54e-2) -	9.8673e-1 (6.65e-2) +	1.3253e+0 (1.10e-1) -	1.0841e+1 (3.23e+0) -	1.1312e+0 (1.10e-1)
	15	24	9.6903e+0 (8.32e-1) -	2.2218e+0 (3.45e-1) +	1.4328e+1 (1.47e+0) -	4.7082e+0 (1.15e+0) -	1.3891e+0 (7.58e-2) +	3.6825e+0 (1.27e+0) =	1.9452e+1 (5.99e+0) -	3.0587e+0 (9.62e-1)
+/-/=			7/33/5	6/35/4	0/44/1	8/34/3	16/23/6	11/29/5	0/45/0	

MaOEFO algorithm is dominant in dealing with test problems with asymmetric, multi-peaked target space and highly complex test problems with high nonlinear dependencies among decision variables. Figure 5 shows the optimization of each algorithm on the WFG5 (M=15) test problem.

5 Summary and Future Work

In this paper, we propose a many-objective electric field optimization algorithm (PD-MaOEFO) guided by angular penalty distance and polarity division. To address the problem that it is difficult to balance convergence and diversity in many-objective space, this paper designs the charge function of particles based on APD to quantify the optimization potential of the particles; in order to cope with the problem of a single search strategy, the particles are divided into protons, electrons, and neutrons, and simulate their different rules of motion: protons focus on guiding, electrons are responsible for global searching, and neutrons are responsible for local optimization. The search efficiency of the algorithm many-objectives space is improved by the different motion strategies; a proton library updating strategy based on reference vectors and APD is proposed. By assigning a proton to each reference vector and combining with the APD screening mechanism, it ensures that the protons are balanced in terms of convergence and diversity. This algorithm is compared with seven many-objective optimization algorithms on 13 optimization test problems, and the experimental results show that PD-MaOEFO demonstrates superior performance in dealing with the DTLZ series of test problems, and its obtained HV and IGD^+ values obtain better results compared with other algorithms. Dealing with many-objective optimization problems with irregular PFs is a challenge for decomposition-based many-objective optimization algorithms. Therefore PD-MaOEFO performs poorly on WFG1-3. Therefore, in future work, it is necessary to add an adaptive reference vector adjustment strategy or combine methods such as machine learning for PD-MaOEFO so that PD-MaOEFO has the ability to handle many-objective optimization problems with irregular PFs.

References

1. Liu, B.-J., Bi, X.-J.: A constrained many-objective evolutionary algorithm based on angle information. J. Electron. **49**(11), 2208–2216 (2021)
2. Jiao, S.-M., Xie, L.-P., Zeng, J.-C.: Research on collective behavior based on artificial physics. Complex Syst. Complex. Sci. **9**(2), 72–78 (2012)
3. Xie, L.-P., Zeng, J.-C.: Global optimization algorithm based on artificial physics. J. Comput. Res. Dev. **48**(5), 848–854 (2011)
4. Xie, L.-P., Zeng, J.-C.: Global optimization algorithm inspired by artificial physics. Syst. Eng. Theory Pract. **30**(12), 2276–2282 (2010)
5. Xie, L.-P., Zeng, J.-C.: Artificial physics approach for swarm robot target search. Pattern Recogn. Artif. Intell. **22**(4), 647–652 (2009)
6. Suman, B., Kumar, P.: A survey of simulated annealing as a tool for single and multi-objective optimization. J. Oper. Res. Soc. **57**(10), 1143–1160 (2006)

7. Birbil, S.İ, Fang, S.C.: An electromagnetism-like mechanism for global optimization. J. Global Optim. **25**, 263–282 (2003)
8. Formato, R.A.: Central force optimization: a new nature-inspired computational framework for multi-dimensional search and optimization. In: Nature Inspired Cooperative Strategies for Optimization (NICSO 2007). Springer, Heidelberg (2008)
9. Petwal, H., Rani, R.: An improved artificial electric field algorithm for multi-objective optimization. Processes **8**(5), 584 (2020)
10. Abedinpourshotorban, H., Shamsuddin, S.M., Beheshti, Z., et al.: Electromagnetic field optimization: a physics-inspired metaheuristic optimization algorithm. Swarm Evol. Comput. **26**, 8–22 (2016)
11. Zhu, C., Zhang, Y., Wang, M., et al.: Optimization, validation and analyses of a hybrid PV-battery-diesel power system using enhanced electromagnetic field optimization algorithm and ϵ-constraint. Energy Rep. **11**, 5335–5349 (2024)
12. Chauhan, D.: Shivani: self-adaptive and locally-guided artificial electric field algorithm for global optimization with aggregative learning. Knowl.-Based Syst. **325**, 113835 (2025)
13. Zhou, B.-H., Wen, M.-D.: A multi-objective artificial electric field algorithm with reinforcement learning for milk-run assembly line feeding and scheduling problem. Comput. Ind. Eng. **190**, 110080 (2024)
14. Zhang, Q., Li, H.: MOEA/D: a multi-objective evolutionary algorithm based on decomposition. IEEE Trans. Evol. Comput. **11**(6), 712–731 (2007)
15. Liu, H.L., Gu, F., Zhang, Q.: Decomposition of a multi-objective optimization problem into a number of simple multi-objective subproblems. IEEE Trans. Evol. Comput. **18**(3), 450–455 (2013)
16. Cheng, R., Jin, Y., Olhofer, M., et al.: A reference vector guided evolutionary algorithm for many-objective optimization. IEEE Trans. Evol. Comput. **20**(5), 773–791 (2016)
17. Qiu, W., Zhu, J., Yu, H., et al.: An adaptive reference vector adjustment strategy and improved angle-penalized value method for RVEA. Complexity **2021**(1), 8870356 (2021)
18. Yang, L., Zhang, Y., Cao, J., et al.: A many-objective evolutionary algorithm based on reference vector guided selection and two diversity and convergence enhancement strategies. Appl. Soft Comput. **154**(1), 111369 (2024)
19. Dai, C., Wang, Y., Ye, M.: A new multi-objective particle swarm optimization algorithm based on decomposition. Inf. Sci. **325**, 541–557 (2015)
20. Lin, Q., Liu, S., Zhu, Q., et al.: Particle swarm optimization with a balanceable fitness estimation for many-objective optimization problems. IEEE Trans. Evol. Comput. **22**(1), 32–46 (2016)
21. Chen, H., Tian, Y., Pedrycz, W., et al.: Hyperplane assisted evolutionary algorithm for many-objective optimization problems. IEEE Trans. Cybern. **50**(7), 3367–3380 (2019)
22. Li, L., Yen, G.G., Sahoo, A., et al.: On the estimation of pareto front and dimensional similarity in many-objective evolutionary algorithm. Inf. Sci. **563**, 375–400 (2021)
23. Sun, Y., Yen, G.G., Yi, Z.: IGD indicator-based evolutionary algorithm for many-objective optimization problems. IEEE Trans. Evol. Comput. **23**(2), 173–187 (2018)
24. Sun, Y., Xue, B., Zhang, M., et al.: A new two-stage evolutionary algorithm for many-objective optimization. IEEE Trans. Evol. Comput. **23**(5), 748–761 (2018)

Advances in Multi-Agent Systems and Visual Perception for Complex Environments

AE-BiseNet: Anomaly-Enhanced Bilateral Segmentation Network for Industrial Defect Detection

Xinyang Wang[1,2], Xinwei Wu[1,2], Yongjie Hou[1,2], Hongbin Ma[1,2,3,4(✉)], and Ying Jin[1,2]

[1] School of Automation, Beijing Institute of Technology, Beijing 100081, China
mathmhb@bit.edu.cn
[2] State key Lab of Autonomous Intelligent unmanned Systems, Beijing Institute of Technology, Beijing 100081, China
[3] Beijing Institute of Technology, Zhuhai, China
[4] National Key Lab of Autonomous Intelligent Unmanned Systems, 100081 Beijing, China

Abstract. Industrial defect detection is of great significance in enhancing the level of manufacturing industries and achieving the high-end development of manufacturing. Segmentation is an important technology to defect defects. While complex and variable backgrounds, subtle and small defects, class imbalance and sparseness make it difficult to accurately identify defects. The article proposes Anomaly-enhanced Bilateral Segmentation Network (AE-BiseNet) to improve segmentation performance by feature contrasting spatial path with spatial defect enhancement auxiliary loss and local positive and negative representation contrasting auxiliary loss. The former creates a dual-branch structure to capture rich low-level spatial features and trained with auxiliary loss to distinguish normal area and defects. The latter extract the representative high-level features of normal area and defects and mitigate class imbalance problem. Both the auxiliary loss function use ground truth to help confirm normal features and defect features. The stamped parts data set is built to validate the model. Experiments show the proposed method reach mean IoU of 83.8% and defect IoU of 68.0% on the stamped parts data set and mean IoU of 88.1% and defect IoU of 79.4%.

Keywords: ESemantic Segmentation · Industrial Defect Detection · AE-BiseNet

1 Introduction

Industrial defect detection, which serves as a core component of the quality assurance system of intelligent manufacturing, is of strategic significance for enhancing the core competitiveness of the manufacturing industry. In high value-added industries such as aerospace, automotive manufacturing, and precision instruments, the microscopic surface defects (including scratches, micro-cracks, pits, etc.) are directly related to the service performance and safety of the products [1, 2]. Besides, such defects may trigger a chain reaction of equipment failure, leading to catastrophic accidents and huge

H. Ma et al. (Eds.): IWACIII 2025, CCIS 2780, pp. 563–578, 2026.
https://doi.org/10.1007/978-981-95-6730-0_38

economic losses [3, 4]. With the development of industrial intelligence, the pursuit of "zero-defect" production in intelligent manufacturing has driven the adoption of automated inspection technologies, which has become a key driving force for industrial upgrading.

With the raise of the image processing, technologies such as automatic defect detection based on machine vision commence to replace manual inspection for such technologies are not influenced by personal subjective judgments or fatigue and offer superior speed, accuracy, cost-effectiveness, and scalability [5]. Since AlexNet [6] take the crown in ImageNet Large Scale Visual Recognition Challenge and demonstrates significant superiority compared to traditional methods, deep learning has become the mainstream approach for various visual tasks like classification, object detection, segmentation and so on. Segmentation which aims to divide an image into meaningful parts that are easier to analyze and process and isolate regions of interest (such as objects or defects) from the background, play a crucial role in industrial defect detection, because it allows the detection system to focus on the exact location of defects, which may be too small or subtle for simple classification tasks or objection detect task. In many industrial settings, defects can be heterogeneous and involve subtle texture changes, scratches, or cracks, which means they are not easily distinguishable from the rest of the product. Precise boundaries provided by segmentation allow for a more nuanced understanding of the defect characteristics.

The fully convolutional network (FCN) method [7] first uses a fully connected neural network to perform pixel-by-pixel image classification and achieve image semantic segmentation. The SegNet [8] introduced the encoder-decoder structure of the segmentation network to capture high-level semantic information and obtain precise and smooth boundary. The DeepLabV3+ method [9] proposed atrous spatial pyramid pooling (ASPP) by several parallel atrous convolution with different rates to catch multiscale feature of the object. The BiseNet [10] proposes an innovative dual-branch model structure using two down-sampling branches (spatial path and context path) for extracting spatial detail information and semantic information, which provide accurate segmentation result with fewer parameters and higher real-time performance. The spatial information plays an important role in the defect detection for lots of defects are very minor, and can't be easily distinguish from normal area, while context information improves the robustness of the network to deal with subtle and complex texture changes and heterogeneous background.

However, those segmentation networks are designed for natural image data, not industrial defect image data. Industrial defect images often feature highly complex and varied backgrounds, which are usually composed of intricate textures, patterns, or machinery components without uniform or contain well-defined object boundaries. Sometimes it is even difficult to provide absolutely correct boundary for humans. Defects in industrial images such as tiny scratches, or slight surface deformations are often very subtle and small in size, which means defects may occur sporadically across large surfaces, with only a small fraction of the image being affected. Some defects might be irregular in shape, making it difficult for segmentation models that rely on geometric priors to accurately capture their boundaries. In industrial defect detection tasks, defects are often rare occurrences in a given image. This imbalance creates challenges for segmentation algorithms, which may be biased toward predicting the background (non-defective) class.

To overcome those difficulties, FDSNeT [11] uses ground truth information to guide the model to focus on the boundaries and handle the local similarity problem of different surface defects. However, it still has shortcomings in handling minor defects such as scratches. DFC method [12] proposes deep feature contrasting enhancing segmentation accuracy in scenarios with limited supervision and pixel-level labeled anomalous images. While it uses two backbone convolutional neural networks with identical structures, so consumes more time to train the model and do prediction.

In order to overcomes the sample unbalance problem, improve the ability of detecting fine defects and balance the real-time performance and accuracy of the model, this paper proposes an AE-BiseNet for industrial defect segmentation tasks. The main contributions of this paper are as follows:

1. A feature contrasting spatial path module trained with an auxiliary loss function using ground truth label to improve model's ability in capture the decisive feature of defects different from normal area, especially when dealing with complex normal surface and fine defects.
2. A local positive and negative representation contrasting auxiliary loss function to obtain pairs of local positive and negative representative semantic feature and enhance their discrimination. This can help to reduce imbalance problems and distinguish defects from heterogeneous normal area.
3. A stamped parts data set is built from Real factory production workshop. Experiments show proposed method reach mean IoU of 83.8% and defect IoU of 68.0%, increased by 1.2% and 0.6% comparing to BiseNet. Besides experiment on NEU-Seg data set shows the mean IoU of the proposed model is 88.1% and the IoU of the defects is 79.4%. Comparing to the BiseNet, mean IoU is increased by 1.2% and IoU of the defects is increased by 2.2%.

2 Method

As is shown in Fig. 1, the AE-BiseNet is based on BiseNet [17] and it consists of two branches: the spatial path and the context path.

In spatial path, a pair of CBR blocks consisting of convolutional layers with batch normalization and ReLU activation function are used to encode rich spatial information. Then the spatial information is confused and constrained by spatial defect enhancement auxiliary loss function. The context path adopts Resnet18 [13] to obtain high-level contextual semantic information. The last 2 layers (16x down and 32x down) of the Resnet output are processed by attention and fusion mechanism and then used as the output of context path. Next, the features from both path is confused and the prediction result is produced through segmentation head. The features from spatial path and FFM are put into the semantic local positive and negative representation contrasting auxiliary loss function function for auxiliary training.

The CBR blocks, ARM, Feature Fusion Module (FFM) and BiseNet SegHead are from BiseNet. The ARM is used to enhance feature representation by capture contextual information with global average pooling and adaptively recalibrate the channel

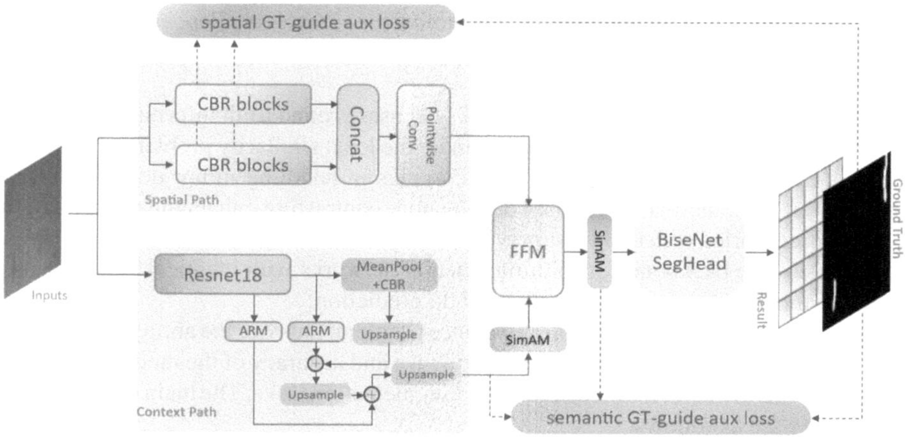

Fig. 1. Framework of the proposed AE-BiseNet

characteristic response. The FFM connect the outputs from two paths and performs weighted fusion on each channel of the concatenated feature. The segmentation head transformed the features into segmentation result which calculates the probabilities of various classifications for each pixel.

2.1 Feature Contrasting Spatial Path with Spatial Defect Enhancement Auxiliary Loss

Due to the low contrast and fine size of industrial defects (as is shown in Fig. 2), rich spatial detail information and high-contrast anomaly features representation are indispensable for anomaly segmentation.

As is shown in Fig. 3, the spatial path uses convolutional layers with Batch Normalization and ReLu activation function to catch spatial information. Simple network structure makes it easy to train and catch rich low-level spatial information to calculate smooth and precise segmentation boundaries. Compared with BiseNet, the number of channels generated by the convolutional layers in our spatial path is only half of that, making sure that dual CBR blocks will not introduce more parameters and reduce the prediction speed. The spatial path input the RGB images and output a feature map with 128 channels and a size of the 1/8 of the input image.

After the last convolutional layer with Batch Normalization and ReLu, A Simple, Parameter-Free Attention Module (SimAM) [14] is introduced to enhance the ability to identify defect features in the network. SimAM directly infers 3D weights, considering both spatial and channel dimensions and introduce no parameter into the existing model, which means it only consumes only a little computing resource and easily applied in any network structure. In SimAM, energy function for each neuron is defined as:

$$e_t(w_t, b_t, \mathbf{y}, x_i) = \frac{1}{M-1}\sum_{i=1}^{M-1}(-1-(w_t x_i + b_t))^2 + (1-(w_t t + b_t))^2 + \lambda w_t^2 \quad (1)$$

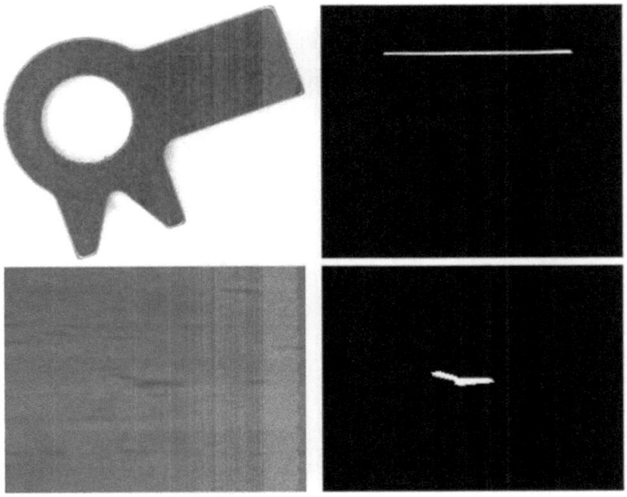

Fig. 2. Low-contrast minor defects

in which M is the number of the neurons, t and x_i are the target neuron and other neurons in a single channel of the input feature, w_t and b_t are weight and bias the transform. This energy measures the linear separability between one target neuron and other neurons. Lower energy reveals lower linear separability and more importance. The minimal energy can be computed with the following:

$$e_t^* = \frac{4(\hat{\sigma}^2 + \lambda)}{(t - \hat{\mu})^2 + 2\hat{\sigma}^2 + 2\lambda} \tag{2}$$

$$\hat{\mu} = \frac{1}{M}\sum_{i=1}^{M} x_i \ , \ \hat{\sigma}^2 = \frac{1}{M}\sum_{i=1}^{M} (x_i - \hat{\mu})^2 \tag{3}$$

The whole refinement phase of our module is:

$$\tilde{\mathbf{X}} = \mathrm{sigmoid}\left(\frac{1}{\mathbf{E}}\right) \odot \mathbf{X} \tag{4}$$

in which $\tilde{\mathbf{X}}$ is the output, $\mathbf{X}$ is the input, and $\mathbf{E}$ groups all e_t^* across channel and spatial dimensions.

SimAM is neuron-level attention, which assigns weights to each neuron (3D attention: channel × height × width), giving much finer resolution of the crucial areas. This help SimAM perform better to identify tiny, subtle, and irregular patterns of the industrial defects than other channel-wise or spatial-wise attention modules like SE [15] and CBAM [16]. Besides SimAM is truly parameter-free, relying on a closed-form energy function. This ensures no model size increase and stable training, which is important for lightweight defect detection systems and avoids overfitting to limited defect data, which is common in industrial inspection.

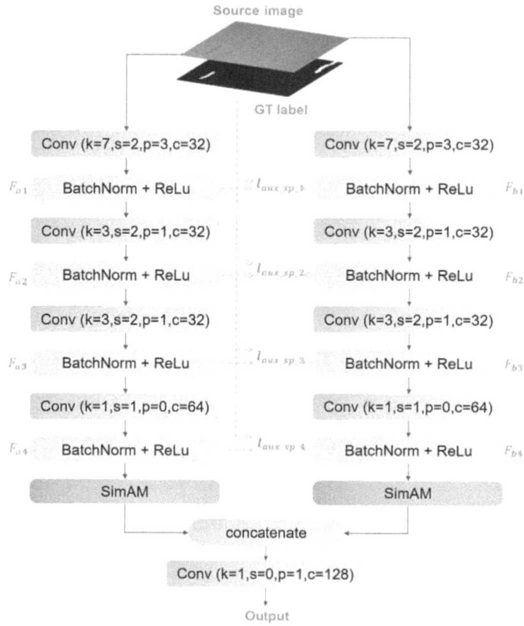

Fig. 3. Structure of the spatial path

The i th features from dual convolution layers with batch normalization and ReLu are constrained by auxiliary loss function $l_{aux\_sp\_i}$, to minimize the distance of normal features while maximizing the distance of abnormal features. In the article, the features from dual blocks are $[F_{a1}, F_{a2}, F_{a3}, F_{a4}]$ and $[F_{b1}, F_{b2}, F_{b3}, F_{b4}]$, $F_{ai}, F_{bi} \in R^{H_i, W_i, C_i}$. Cosine distance is calculated to measure the distance between F_{ai}, F_{bi}:

$$D_{\cos}(F_{ai}, F_{bi}) = \frac{1}{2} - \frac{1}{2}\frac{\sum_{d=1}^{C_i} F_{ai}^{(d)} F_{bi}^{(d)}}{\sqrt{\sum_{d=1}^{C_i}\left(F_{ai}^{(d)}\right)^2}\sqrt{\sum_{d=1}^{C_i}\left(F_{bi}^{(d)}\right)^2}} \tag{5}$$

The range of $D_{\cos}$ is [0, 1]. The small values of $D_{\cos}$ prove the similarity between features from dual blocks. The label is M, $M \in R^{H, W, 1}$. With bilinear interpolation, the label can be resized with the same size of F_{ai} as $\widetilde{M}_i \in R^{H_i, W_i, 1}$. So the elements in the mask to indicate the defect features M_{pi} is

$$M_{pi}(x, y) = \begin{cases} \tilde{M}_i(x, y), & \text{if } \tilde{M}_i(x, y) > 0.5 \\ 0 & else \end{cases} \tag{6}$$

for if $\tilde{m}$ is too small, the feature may be confused with too much normal feature. And corresponding, the elements in the mask to indicate the normal features M_{ni} is

$$\tilde{M}_{ni}(x, y) = \begin{cases} 1, & \text{if } \tilde{M}_i(x, y) = 0 \\ 0, & else \end{cases} \tag{7}$$

To minimize the distance of normal features, the auxiliary loss function is designed as:

$$l_{ni} = \frac{D_{\cos}(F_{ai}, F_{bi}) \odot M_{ni}}{\sum\sum M_{ni}(x, y)} \tag{8}$$

To maximize the distance of abnormal features, the auxiliary loss function is designed as:

$$l_{pi} = \frac{\mathrm{ReLU}((\beta_1 - D_{\cos}(F_{ai}, F_{bi})) \odot M_{ni})}{\sum\sum M_{ni}(x, y)} \tag{9}$$

So, the spatial defect enhancement auxiliary loss is:

$$l_{sp} = \frac{1}{L}\sum_{i=1}^{L}(l_{pi} + l_{ni}) \tag{10}$$

The auxiliary loss only works in training, and increase no calculation in prediction.

2.2 Local Positive and Negative Representation Contrasting Auxiliary Loss

The features and ground label are divided into $K \times K$ (as is shown in Fig. 4) local regions.

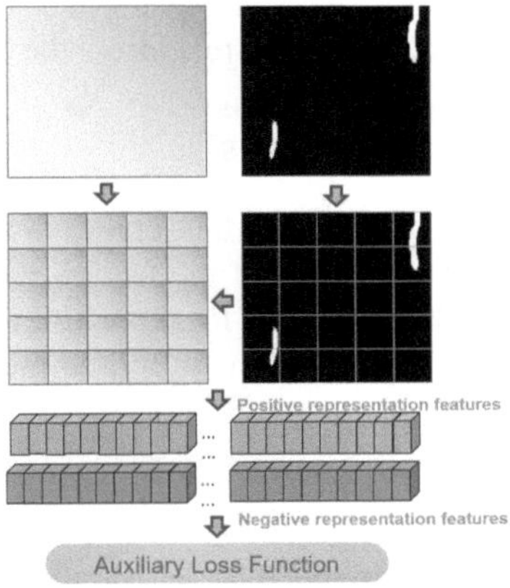

Fig. 4. Subdivide features to construct the loss function

In each region, a positive representation feature and a negative representation feature can be calculated with M_p and M_n calculated according with formula (6) and (7). For a feature map F, the positive representation feature in region k is:

$$f_p^{(k)} = \sigma_p^{(k)} \frac{F^{(k)} \odot M_p^{(k)}}{\sum_k \sum_{x,y} M_p^{(k)}(x, y)} \tag{11}$$

in which, $\sigma_p^{(k)}$ is used to indicate whether the local region is considered as a region with any defects. It is defined as:

$$\sigma_p^{(k)} = \begin{cases} 1, & \sum_{x,y} M_p^{(k)}(x, y) > 10 \\ 0, & \text{else} \end{cases} \tag{12}$$

For industrial defects have not clear and definite boundaries, some local regions may only include the edges of the defect labels. Those regions lack rich defect features and should not be considered as regions with defects.

And the negative representation feature in region k is:

$$f_n^{(k)} = \frac{F^{(k)} \odot M_n^{(k)}}{\sum_k \sum_{x,y} M_n^{(k)}(x, y)} \tag{13}$$

in which $k \in \{1, 2, \ldots, K^2\}$, $F^{(k)}, M_n^{(k)}, M_p^{(k)}$ are the sub-matrix of feature map F, positive label masks M_p and negative label mask M_n in the k th region. So, the auxiliary loss is as formular (14), in which K_n is the number of the negative representation features and K_p is the number of the positive representation features.

$$l_{cp} = \frac{\sum_{i=1}^{K_n} \sum_{j=1}^{K_n} \mathrm{D_{cos}}\left(f_n^{(i)}, f_n^{(j)}\right)}{K_n{}^2} + \frac{\sum_{i=1}^{K_n} \sum_{j=1}^{K_p} \mathrm{ReLu}\left(\beta_2 - \mathrm{D_{cos}}\left(f_n^{(i)}, f_p^{(i)}\right)\right)}{K_n K_p} \tag{14}$$

Through creating negative-positive feature pairs, the local positive and negative representation contrasting auxiliary loss help to address the class imbalance problem.

3 Experiment Results and Analysis

This section first introduces the experimental dataset settings, parameters of the training and evaluation indicators. Then the effects of the feature contrasting spatial path with spatial defect enhancement auxiliary loss and local positive and negative representation contrasting auxiliary loss is validated. Next the proposed method is compared with the state-of-the-art methods.

3.1 Experimental Dataset

Stamped Parts Data Set: Images in this data set is collected from the stamped parts with scratch. All the defective product images are captured from the actual production scene of the factory with camera MV-CU120-10GM, and they can represent the actual inspection requirements of the factory. The resolution of the camera is 4024 × 3036, while for the workpiece does not occupy the entire picture, all pictures are cropped to remain only the parts with workpiece. Average resolution is about 1530 × 1120. A total of 160 images is captured. After data augmentation, 1342 images are split as training set, and 49 images without data augmentation are split as the test set. The statistical analysis is shown in Fig. 5. Although every image in the data set contains at least a scratch, defect pixels are far less than normal pixels, showing enormous unbalance in class.

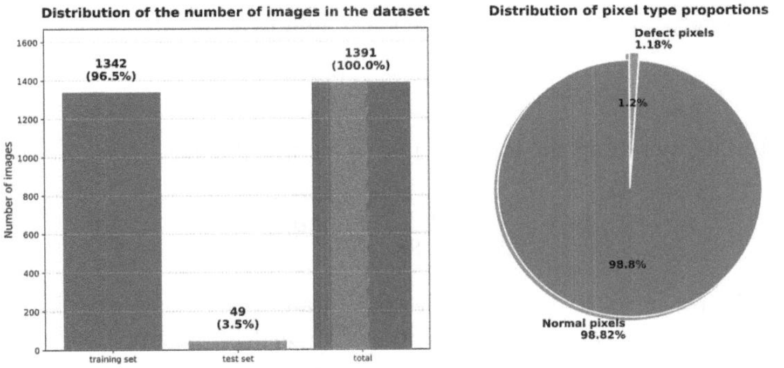

Fig. 5. Statistical analysis of stamped parts data set

NEU-Seg: This data set [17] contains three kinds of typical surface defects of the hot-rolled steel strip, including inclusion, patches and scratches.

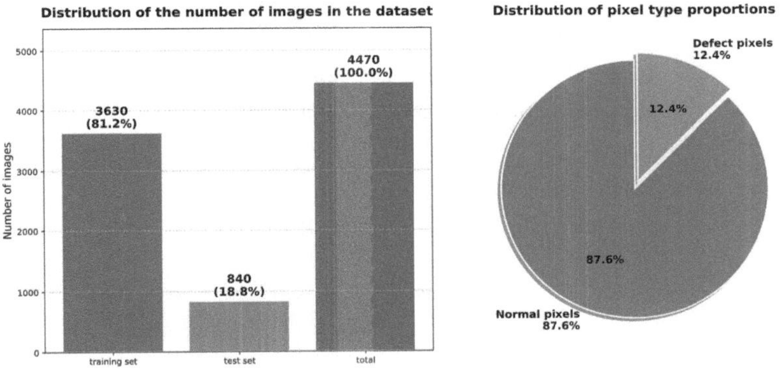

Fig. 6. Statistical analysis of NEU data set

The resolution of each image is 200 × 200. In the dataset, 3630 images are split as the training set, and 840 images are split as the test set. For the defect data is rare, original images have undergone data augmentation. The statistical analysis of NEU data set is shown as Fig. 6. Obviously normal pixels are farther more than normal pixels.

3.2 Parameters of the Training and Evaluation Indicators

The network experiments are conducted on a Ubuntu 20.04.6 LTS operating terminal system, using Python 3.10 language, Intel(R) Core(TM) i5–4590 CPU @ 3.30GHz(4 core and 1 Thread per core) with NVIDIA GeForce RTX 4060 (8GB) graphics configuration, trained and predicted under the Pytorch(v.2.0.0) framework environment, and accelerated using CUDA 11.1.

The SGD optimizer with momentum of 0.9 and weight decay of 0.0001 is employed. Besides, a hybrid learning rate schedule combining polynomial decay with warmup phases is employed. Before 5th epoch, the learning rate in the ith epoch is

$$LR(i) = 0.01 \times 0.1^{1-\frac{i}{5}}, \quad i < 5 \tag{15}$$

And after 500 iterations the learning rate in the ith epoch is:

$$LR(i) = \left(1 - \frac{i-4}{88}\right)^{0.7} LR(4), \; i \geq 5 \tag{16}$$

The batch size is set to 16, and network is trained for 92 epochs. In the training of NEU-seg, the images are resized to 256 × 256, and for stamped parts data set, the images are resized to 640 × 480. The training parameters in the network are shown in Table 1. Besides cross-entropy loss is adopted with OHEM [18], which automatically select difficult cases with high loss to improve training efficiency and segmentation performance.

Table 1. Training related parameters

Parameters	Parameters Values
Image Size (NEU-Seg)	256 × 256
Image Size (Stamped parts)	640 × 480
Epochs	92
Batch size	16
Strat learn rate	0.01

3.3 Evaluation Indicators

Intersection-Over-Union (IoU), also known as the Jaccard Index, is one of the most common and useful indicators to evaluate the performance of the segmentation network. It measures the degree of overlap between the predicted segmentation mask and the ground-truth mask, and is defined for a single class (or object) as

$$\text{IoU} = \frac{|P \cap G|}{|\, P \cup G \,|} = \frac{\sum [P_i \wedge G_i]}{\sum [P_i \vee G_i]} \tag{17}$$

where P denotes the set of pixels predicted as belonging to the target class, G denotes the set of ground-truth pixels of that class, The summations run over all pixel indices .

An IoU of 0 indicates no overlap (complete failure), whereas an IoU of 1 indicates perfect agreement.

Mean IoU (mIoU) is then taken over all classes:

$$\text{mIoU} = \frac{1}{C} \sum_{c=1}^{C} \text{IoU}_c \tag{18}$$

The F1 score is a statistical measure used to evaluate the accuracy of a model, especially in classification tasks with imbalanced data. In the context of segmentation, it quantifies the balance between precision and recall of the predicted mask relative to the ground truth. It is calculated as

$$\begin{aligned} \mathrm{F1} &= \tfrac{2PR}{P+R} \\ P = \tfrac{\mathrm{TP}}{\mathrm{TP+FP}}, R &= \tfrac{\mathrm{TP}}{\mathrm{TP+FN}} \end{aligned} \tag{19}$$

where P is precision, R is recall, TP denotes pixels correctly predicted as foreground, FP denotes background pixels incorrectly predicted as foreground and FN denotes foreground pixels missed by the prediction. F1 score is robust to class imbalance.

Macro F1 calculates the F1 score independently for each class, and then takes the unweighted average:

$$\text{Macro F1} = \frac{1}{C}\sum_{c=1}^{C}\mathrm{F1}_c \tag{20}$$

Micro F1 aggregates the true positives (TP), false positives (FP), and false negatives (FN) across all classes, then computes global precision and recall, and finally the F1 score:

$$\text{Micro F1} = \frac{\sum \mathrm{TP}_c}{2\sum \mathrm{TP}_c + \sum \mathrm{FP}_c + \sum \mathrm{FN}_c} \tag{21}$$

3.4 Experiment Result and Analysis

3.4.1 Ablation Experiments

To validate the efficacy of feature contrasting spatial path with spatial defect enhancement auxiliary loss and local positive and negative representation contrasting auxiliary loss, ablation experiments are performed on two data sets. BiseNet is used as baseline. On stamped parts data set, the IoU of normal area is 99.5%, the IoU of the defect area is 66.7%, mIoU is 83.1%, f1 score of the normal area is 99.8%, f1 score of the defects area is 80.0%, the macro f1 score is 89.9% and the micro f1 score is 99.5%. On NEU-Seg, the IoU of normal area is 96.5%, the IoU of the defect area is 77.2%, mIoU is 86.9%, f1 score of the normal area is 98.2%, f1 score of the defects area is 87.2%, the macro f1 score is 92.7% and the micro f1 score is 96.9%. The contrast models in the experiments are:

(1). The BiseNet with the efficacy of feature contrasting spatial path training with spatial defect enhancement auxiliary loss.
(2). The BiseNet training with local positive and negative representation contrasting auxiliary loss.
(3). AE-BiseNet proposed with both feature contrasting spatial path with spatial defect enhancement auxiliary loss and local positive and negative representation contrasting auxiliary loss.

The result on stamped parts data set is shown in Table 2. The result on NEU-Seg is shown in Table 3, in which (n) refer to the index of normal area and (d) refer to the index of defect area.

Table 2. Result of ablation experiments on stamped parts data set

Model	IoU(n)	IoU(d)	mIoU	F1(n)	F1(d)	Macro F1	Micro F1
Baseline	99.5%	66.7%	83.1%	**99.8%**	80.0%	89.9%	99.5%
(1)	99.5%	67.5%	83.5%	99.8%	80.6%	90.2%	99.5%
(2)	99.5%	67.8%	83.6%	99.8%	80.8%	90.3%	99.5%
(3)	**99.5%**	**68.0%**	**83.7%**	99.7%	**81.0%**	**90.4%**	**99.5%**

Table 3. Result of ablation experiments on NEU-Seg

Model	IoU(n)	IoU(d)	mIoU	F1(n)	F1(d)	Macro F1	Micro F1
Baseline	96.5%	77.2%	86.9%	98.2%	87.2%	92.7%	96.9%
(1)	96.7%	78.5%	87.6%	98.3%	87.9%	93.1%	97.0%
(2)	96.8%	79.0%	87.9%	98.4%	88.3%	93.3%	97.1%
(3)	**96.8%**	**79.4%**	**88.1%**	**98.4%**	**88.5%**	**93.4%**	**97.2%**

By comparing the result of Table 2, the AE-BiseNet performs best in IoU of defect, mIoU, f1 score of defect and Maro F1. For the proportion of normal pixels is 98.8%, the index of normal pixel is similar, and Micro F1 almost equal to the f1 scores of the normal area. The result illustrates that the introduction of the feature contrasting spatial path with spatial defect enhancement auxiliary loss and local positive and negative representation contrasting auxiliary loss can improve the ability of segmentation. Comparing the result of (1) and (2), it is obvious that the improvement caused by feature contrasting spatial path with spatial defect enhancement auxiliary loss is higher. So the low-level feature that can distinguish defects from normal area is more crucial for the segmentation of industrial defects than high-level feature. Comparing the result of (1) and (3), AE-BiseNet's index of normal area is the same with that of model (1), while AE-BiseNet performs better in the segmentation of defects area, indicating that local positive and negative representation contrasting auxiliary loss helps to recognize the defect that are infrequent in the data set. The result shown in Table 3 also support similar conclusion, while for proportion of the defect pixel is larger, Micro F1 also shows improvement.

3.4.2 Experiment of Auxiliary Loss Function

To research the influence of the feature maps input into spatial defect enhancement auxiliary loss on performance of segmentation, contrast experiments on the stamped parts data set are conduct. The models include:

1. The model in which $\{l_{aux\_sp\_i}, i \in \{1, 2, 3, 4\}\}$($\{l_{aux\_sp\_i}$ see in Fig. 3) are included in the auxiliary loss.
2. The model in which $\{l_{aux\_sp\_i}, i \in \{1, 2, 3\}\}$ are included in the auxiliary loss.
3. The model in which $\{l_{aux\_sp\_i}, i \in \{1, 2\}\}$ are included in the auxiliary loss.
4. The model in which $\{l_{aux\_sp\_i}, i \in \{2, 3, 4\}\}$ are included in the auxiliary loss.
5. The model in which $\{l_{aux\_sp\_i}, i \in \{3, 4\}\}$ are included in the auxiliary loss.

The result is shown in Table 4.

Table 4. Result of contrast experiments of features maps put into spatial defect enhancement auxiliary loss

Model	IoU(n)	IoU(d)	mIoU	F1(n)	F1(d)	Macro F1	Micro F1
(1)	99.5%	67.4%	83.5%	99.7%	80.6%	90.2%	99.5%
(2)	99.5%	66.9%	83.2%	99.7%	80.2%	90.0%	99.5%
(3)	**99.5%**	**68.0%**	**83.7%**	99.7%	**81.0%**	**90.4%**	**99.5%**
(4)	99.5%	67.1%	83.3%	**99.8%**	80.4%	90.1%	99.5%
(5)	99.5%	67.4%	83.5%	99.7%	80.5%	90.2%	99.5%

Result shows the layers put into the spatial defect enhancement auxiliary loss does not vary much more. When the lowest 2 layers are included in the auxiliary loss, the network perform best. So, the low-level and high resolution with less channels feature needs play important part in correct segmentation boundary, and are more suitable to distinguish defects according to the cosine distance.

For researching the influence of the feature maps input into local positive and negative representation contrasting auxiliary loss on performance of segmentation, contrast experiments are conduct. The models include:

(1). The model in which output of the spatial path, and output of the context path are put into the auxiliary loss.
(2). The model in which output of the spatial path, output of the context path, and output of the FFM are put into the auxiliary loss.
(3). The model in which output of the spatial path and output of the FFM are put into the auxiliary loss.
(4). The model in which output of the context path and output of the FFM are put into the auxiliary loss.

The result is shown in Table 5. It is showed that the layers put into the local positive and negative representation contrasting auxiliary loss impact nothing on the segmentation on the normal area. The feature map from the context path is most needed to get the intervention of the auxiliary. For the receptive field of the feature map from the context path is the biggest, and more susceptible to the influence of class imbalance. Putting both output of the context path and output of the FFM into the auxiliary loss obtains the highest improvement on the performance of defect segmentation.

Table 5. Result of contrast experiments of features maps put into local positive and negative representation contrasting auxiliary loss

Model	IoU(n)	IoU(d)	mIoU	F1(n)	F1(d)	Macro F1	Micro F1
(1)	99.5%	67.4%	83.5%	99.8%	80.5%	90.2%	99.5%
(2)	99.5%	66.9%	83.2%	99.8%	80.1%	90.0%	99.5%
(3)	99.5%	67.1%	83.3%	99.8%	80.3%	90.0%	99.5%
(4)	**99.5%**	**67.5%**	**83.5%**	**99.8%**	**80.6%**	**90.2%**	**99.5%**

3.4.3 Comparisons with State-of-the-arts.

The proposed method is compared with real-time state-of-the-art semantic segmentation methods on the stamped parts data set. Result of IoU of defects, mIoU, f1 score of defects, macro f1 score and FPS (frame per second) is shown in Table 6.

Table 6. Comparisons with state-of-the-art semantic segmentation methods on stamped parts data set

Model	IoU(d)	mIoU	F1(d)	Macro F1	FPS
FastSCNN [19]	63.1%	81.4%	77.4%	88.6%	170.0
Fdsnet [11]	61.1%	80.3%	75.9%	87.8%	163.8
Deeplabv3 [20]	67.9%	83.7%	80.9%	90.3%	17.6
DFCnet [12]	63.2%	81.3%	77.4%	88.6%	64.6
BiseNet [10]	66.7%	83.1%	80.0%	89.9%	**210.3**
AE-BiseNet	**68.0%**	**83.8%**	**81.0%**	**90.4%**	159.4

With input size of 640 × 480, the proposed method achieve the highest IoU of defects, mIoU, f1 score of defects and macro f1 score, with a fast and acceptable prediction speed. So it successfully achieves a good trade-off between speed and accuracy.

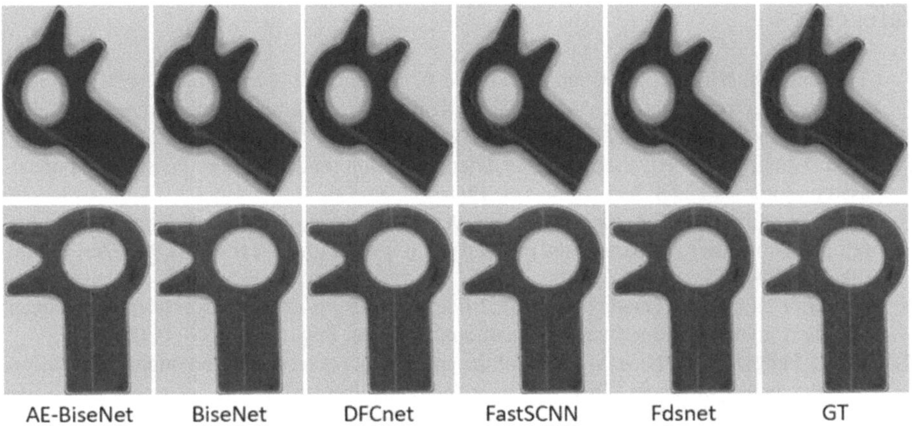

Fig. 7. The result of each network

The result of each network is shown in Fig. 7. AE-BiseNet is able to recognize complicated scratch with various features and shallow scratch. While BiseNet can't capture the shallow scratch and scratch with complicated shapes, DFCnet fails in detecting black scratch, FastSCNN only recognizes about half of the shallow scratch and fail in black scratch and complicated scratch, and Fdsnet preforms the worst with only a bit of preponderance in segmentation speed.

4 Conclusion

The article proposes an AE-BiseNet, an enhanced model based on BiseNet to improve the ability of segmentation to detect industrial defects. The AE-BiseNet introduced the feature contrasting spatial path with spatial defect enhancement auxiliary loss and local positive and negative representation contrasting auxiliary loss. The AE-BiseNet introduced the feature contrasting spatial path with spatial defect enhancement auxiliary loss to enhance the ability of model to extract the feature maps distinguishing normal areas and defects by creating a dual-branch layers and training with auxiliary loss to increase similarity between normal features from branches and decrease the similarity between defect features. And local positive and negative representation contrasting auxiliary loss alleviate the problem of class imbalance by calculating the representation features of normal area and defect area and enforce similarity constraints. The experiments show the proposed method effectively enhanced the performance of the model in identifying industrial defects, increase the mIoU by 0.6% and defect's IoU by 1.2% in stamped part data set, and increase the mIoU by 1.2% and defect's IoU by 2.4% in NEU-Seg data set.

Acknowledgments. Supported by the Science and Technology on Metrology and Calibration Laboratory (Grant No. JLKG2024001B004). Supported by the National Natural Science Foundation of China under Grant No.62473052 and National Key Laboratory under Grant 241-HF-D09-01.

References

1. Chenyang, D., Xiaowei, L., Chang, L., et al.: Combining ultrasonic guided wave and low-frequency electromagnetic technology for defect detection in high-temperature Cr-Ni alloy furnace tubes. Sci. Rep. **13**(1), 18592 (2023)
2. Li, X., Wang, C., Ju, H., et al.: Surface defect detection model for aero-engine components based on improved YOLOv5. Appl. Sci. **12**(14), 7235 (2022)
3. Wei, L., Solihin, M.I., Saruchi, S.A., Astuti, W., et al.: Surface defects detection of cylindrical high-precision industrial parts based on deep learning algorithms: a review. Oper. Res. Forum. **5**(3), 58 (2024)
4. Wenbin, H., Jinglin, L., Guangyi, L., et al.: Effect of defects in adhesive layer on the mechanical property of hybrid bond-riveted lap shear joint. J. Mech. Eng. **55**(24), 37–44 (2019)
5. Ren, Z., Fang, F., Yan, N., et al.: State of the art in defect detection based on machine vision. Int. J. Precis. Eng. Manuf. - Green Technol. **9**(2), 661–691 (2022)
6. Krizhevsky, A., Sutskever, et al.: ImageNet classification with deep convolutional neural networks. Commun. ACM. **60**(6), 84–90 (2017)
7. Shelhamer, E., Long, et al.: Fully convolutional networks for semantic segmentation. IEEE Trans. Pattern Anal. Mach. Intell. **39**(4), 640–651 (2017)
8. Vijay, B., Alex, K., Roberto, C.: SegNet: a deep convolutional encoder-decoder architecture for image segmentation. IEEE Trans. Pattern Anal. Mach. Intell. **39**(12), 2481–2495 (2017)
9. Chen, L.-C., Papandreou, G., et al.: DeepLab: semantic image segmentation with deep convolutional nets, atrous convolution, and fully connected CRFs. IEEE Trans. Pattern Anal. Mach. Intell. **40**(4), 834–848 (2018)
10. Yu, C., Wang, J., Peng, C., Gao, C., et al.: BiseNet: bilateral segmentation network for real-time semantic segmentation. Lect. Notes Comput. Sci. **11217**, 334–349 (2018)
11. Zhang, J., Ding, R., Ban, M., Guo, T.: FDSNeT: an accurate real-time surface defect segmentation network. In: 2022 IEEE International Conference on Acoustics, Speech and Signal Processing, pp. 3803–3807 (2022)
12. Wan, Q., Cao, Y., Gao, L., Li, X., Gao, Y.: Deep Feature Contrasting for Industrial Image Anomaly Segmentation. IEEE Trans. Instrum. Meas. **73**, 1–11 (2024)
13. He, K., Zhang, X., Ren, S., et al.: Deep residual learning for image recognition. In: 2016 IEEE Conference on Computer Vision and Pattern Recognition, pp. 770–778. CVPR (2016)
14. Yang, L., Zhang, R.Y., Li, L., et al.: SimAM: a simple, parameter-free attention module for convolutional neural networks. In: Tong Proceedings of the 38th International Conference on Machine Learning, pp. 11862–11874 (2021)
15. Hu, L., Shen, S., Albanie, G., Sun, Wu, E.: Squeeze-and-excitation networks. IEEE Trans. Pattern Anal. Mach. Intell. **42**(8), 2011–2023 (2020)
16. Woo, S., Park, J., Lee, J.Y., et al.: CBAM: convolutional block attention module. In: European Conference on Computer Vision. European Conference on Computer Vision, pp. 3–19 (2018)
17. Dong, H., Song, K., He, Y., et al.: PGA-net: pyramid feature fusion and global context attention network for automated surface defect detection. IEEE Trans. Industr. Inform. **16**(12), 7448–7458 (2020)
18. Shrivastava, A., Gupta, A., Girshick, R.: Training region-based object detectors with online hard example mining. In: 29th IEEE Conference on Computer Vision and Pattern Recognition, pp. 761–769. CVPR (2016)
19. Poudel, R.P., Liwicki, S., Cipolla, R.: Fast-scnn: Fast semantic segmentation network. arXiv preprint arXiv, 1902.04502 (2019)
20. Chen, L.C., Papandreou, G., Schroff, F., Adam, H.: Rethinking atrous convolution for semantic image segmentation. arXiv preprint arXiv, 1706.05587 (2017)

Adaptive Flight Control of Unmanned Aerial Vehicle Under Sensor Failure Mode

Xiaoyong Guo[1(✉)], Jianhong Wang[2], and Lin Hua[1]

[1] School of software, Henan University of Engineering, Zhengzhou 451191, China
xyguomath@126.com

[2] School of Electrical Engineering and Automation, Jiangxi University of Science and Technology, , Ganzhou 341000, China

Abstract. Drones are widely used in multiple industries due to their flexibility and mobility. However, sensor failure, as a key issue affecting its flight safety and control accuracy, requires efficient adaptive control methods to ensure system stability. This paper aims at the flight control problem of unmanned aerial vehicles (UAVs) in sensor failure scenarios and proposes a control strategy that integrates fault estimation and adaptive adjustment. By constructing a sliding mode observer and an adaptive controller, the flight stability and trajectory tracking accuracy under the fault state are achieved. This paper system model for unmanned aerial vehicle dynamics and kinematics analysis and fault diagnosis for narrative. Then, the sensor faults were estimated by using the sliding mode observer method and the effect was demonstrated on Matlab. Finally, the adaptive control system is designed based on sensor fault estimation, and the stability and effectiveness of adaptive flight control are verified on Matlab.

Keywords: Unmanned aerial vehicle · Sensor failure · Adaptive control · Mode control

1 Introduction

Unmanned technology is widely applied in both military and civilian fields. With multiple advantages, unmanned aerial vehicles (UAVs) have become an important part of modern aviation technology [1]. In the military field, it is used for reconnaissance and surveillance as well as tactical strikes. In the civilian field, it is used for aerial photography and mapping [2], material transportation, agricultural and forestry operations, emergency rescue and environmental monitoring [3], etc. However, drones are confronted with challenges such as sensor failures, which affect flight safety and control accuracy. Sensor faults come in various types and are caused by multiple factors. The study of adaptive flight control methods under sensor failure modes is of great significance for improving the safety and reliability of unmanned aerial vehicles. This article will conduct research around this issue and provide solutions for the reliable operation of unmanned aerial vehicles.

H. Ma et al. (Eds.): IWACIII 2025, CCIS 2780, pp. 579–590, 2026.
https://doi.org/10.1007/978-981-95-6730-0_39

2 Modeling of Unmanned Aerial Vehicle Systems

To reduce the difficulty of modeling and highlight the core issues, this study makes the following assumptions:

First, the curvature of the Earth is disregarded, and the surface of the Earth is approximately regarded as a plane, while ignoring the Coriolis force and centrifugal force generated by the Earth's rotation. The second is to determine that the acceleration due to gravity is constant, always vertically downward, and does not change with height. Thirdly, the unmanned aerial vehicle is regarded as a rigid body, ignoring its structural deformation and vibration. It is assumed that the mass is uniformly distributed and the center of mass is consistent with the geometric center. Fourth, the apparent inertial coordinate system is equivalent to the ground coordinate system. The fifth is to assume that there is no friction inside the unmanned aerial vehicle.

2.1 Kinetic Analysis

The dynamic model of unmanned aerial vehicles is based on Newton's second law, including the linear motion equation and the angular motion equation. The linear motion equation describes the motion of the UAV under the action of a combined external force, while the angular motion equation describes the rotational motion of the UAV under the action of a combined external moment. Through coordinate system transformation, the motion equations in the ground coordinate system can be converted into those in the body coordinate system.

According to Newton's second law, the linear motion equation of the unmanned aerial vehicle under the action of the resultant external force $\sum \boldsymbol{F}$is:

$$\sum \boldsymbol{F} = m\frac{d\boldsymbol{V}}{dt} \tag{1}$$

The angular motion equation of the unmanned aerial vehicle under the action of the resultant external moment $\sum \boldsymbol{M}$is:

$$\sum \boldsymbol{M} = \frac{d\boldsymbol{L}}{dt} \tag{2}$$

The angular motion equations of the unmanned aerial vehicle under the action of the combined external moment can be obtained as follows:

$$\begin{cases} G = \dot{p}I_x - \dot{r}I_{xz} + qr\left(I_z - I_y\right) - pqI_{xz} \\ M = \dot{q}I_y + pr(I_x - I_z) + \left(p^2 - r^2\right)I_{xz} \\ N = \dot{r}I_z - \dot{p}I_{xz} + pq\left(I_y - I_x\right) + qrI_{xz} \end{cases} \tag{3}$$

The following equations of the unmanned aerial vehicle dynamics model can be obtained through organization:

$$\begin{cases} \dot{p} = \left(\frac{(I_y - I_z)I_z - I^2{}_{xz}}{I_xI_z - I^2{}_{xz}}r + \frac{(I_x - I_y + I_z)I_{xz}}{I_xI_z - I^2{}_{xz}}p\right)q + \frac{I_z}{I_xI_z - I^2{}_{xz}}G + \frac{I_{xz}}{I_xI_z - I^2{}_{xz}}N \\ \dot{q} = \frac{I_z - I_x}{I_y}pr - \frac{I_{xz}}{I_y}\left(p^2 - r^2\right) + \frac{1}{I_y}M \\ \dot{r} = \left(\frac{(I_x - I_y)I_x + I^2{}_{xz}}{I_xI_z - I^2{}_{xz}}p - \frac{(I_x - I_y + I_z)I_{xz}}{I_xI_z - I^2{}_{xz}}r\right)q + \frac{I_{xz}}{I_xI_z - I^2{}_{xz}}G + \frac{I_x}{I_xI_z - I^2{}_{xz}}N \end{cases} \tag{4}$$

2.2 Kinematics Analysis

The motion model of UAVs is a very important part of the UAV motion equation. It can be used to describe the position and attitude changes of UAVs in three-dimensional space.

According to Euler's equation, the relationship between the angular velocity of unmanned aerial vehicles and their attitude angular velocity is:

$$\begin{bmatrix} \dot{\phi} \\ \dot{\theta} \\ \dot{\psi} \end{bmatrix} = \begin{bmatrix} 1 & \sin\phi\tan\theta & \cos\phi\tan\theta \\ 0 & \cos\phi & -\sin\phi \\ 0 & \sin\phi\sec\theta & \cos\phi\sec\theta \end{bmatrix} \begin{bmatrix} p \\ q \\ r \end{bmatrix} \tag{5}$$

Obtain the angular motion equations of the unmanned aerial vehicle:

$$\begin{cases} \dot{\phi} = p + q\sin\phi\tan\theta + r\cos\phi\tan\theta \\ \dot{\theta} = q\cos\phi - r\sin\phi \\ \dot{\psi} = q\sin\phi\sec\theta + r\cos\phi\sec\theta \end{cases} \tag{6}$$

2.3 Simplified Model

The motion model of UAVs is a very important part of the UAV motion equation. It can be used to describe the position and attitude changes of UAVs in three-dimensional space.

To simplify the control design, it is assumed that the yaw Angle of the unmanned aerial vehicle is zero when it is in a hovering state, and the changes in the three angles are very small. Based on these assumptions, the dynamic model of the unmanned aerial vehicle can be linearized to obtain a simplified model. Simplifying the model facilitates analysis and control design, and can better reflect the main dynamic characteristics of unmanned aerial vehicles. Linearizing the model results in:

$$\begin{cases} \ddot{x} = g\theta \\ \ddot{y} = g\phi \\ \ddot{z} = \frac{U_1}{m} - g \\ \ddot{\phi} = \frac{U_2}{I_x} \\ \ddot{\theta} = \frac{U_3}{I_y} \\ \ddot{\varphi} = \frac{U_4}{I_z} \end{cases} \tag{7}$$

The unmanned aerial vehicle system can be represented in the form of the following state matrix:

$$\begin{cases} \dot{x}(t) = Ax(t) + Bu(t) \\ y(t) = Cx(t) \end{cases} \tag{8}$$

Among them, the matrices A, B, C in the formula respectively represent the system state matrix, the system input matrix, and the system output matrix.

3 Fault Diagnosis of Sensors

3.1 Analysis of the Sources of Sensor Faults

The complexity of the UAV flight control system makes it difficult to fully ensure safety and effectiveness in actual operation [4], especially the reliability issue of feedback parameters from various high-precision sensors. Sensor failures can be classified into five major categories, as shown in Fig. 1.

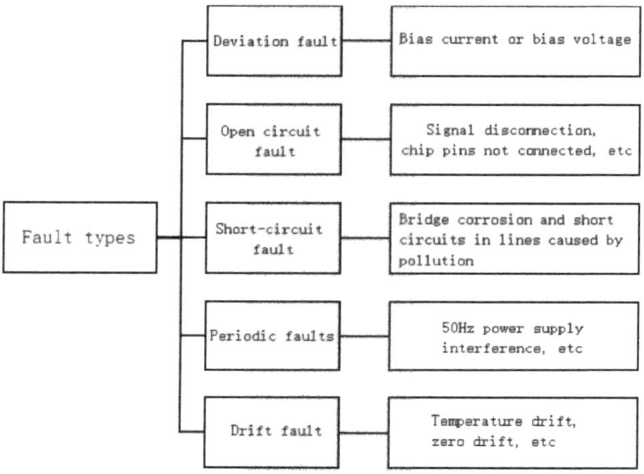

Fig. 1. Fault type.

3.2 Fault Diagnosis Method Based on Sliding Mode Observer

The sliding mode observer is an effective fault diagnosis tool [5]. By designing the sliding mode surface and observer gain, rapid detection and accurate estimation of sensor faults can be achieved [6]. The improved sliding mode observer introduces adaptive gain, which can dynamically adjust the gain according to the residuals, enhancing robustness and reducing chattering. The asymptotic stability of the observer was proved through Lyapunov stability analysis.

3.3 Improve the Observer Design

Sliding mode surface analysis, defining residual signals:

$$e(t) = y(t) - \hat{y}(t) \tag{9}$$

To estimate the system state, the following equivalent input sliding mode observer is proposed:

$$\begin{cases} \dot{\hat{x}}(t) = A\hat{x}(t) + Bu(t) + L(y(t) - \hat{y}(t)) + v(t) \\ \hat{y}(t) = C\hat{x}(t) \end{cases} \tag{10}$$

The sliding mode control term adopts an improved saturation function form:

$$\nu(t) = K(t)sat(e/\Phi) \tag{11}$$

Here, $\hat{x}(t)$ is the state estimate, $\hat{y}(t)$ is the output estimate, and L is the linear observer gain matrix, which is used to adjust the performance of the observer. $K(t)$ is the sliding mode gain matrix, where Φ is the boundary layer thickness parameter and $sat(\cdot)$ is the improved saturation function. This design effectively reduces the chattering phenomenon in the traditional sliding mode control.

$$sat(z) = z/(|z| + \varepsilon(z)) \tag{12}$$

On the sliding mode surface $s = \dot{e}(t) = 0$, the dynamic equation of the error is

$$\dot{e}(t) = \dot{y}(t) - \dot{\hat{y}}(t) \tag{13}$$

By designing an appropriate system, the state estimation error of the system can reach the sliding mode surface within a limited time and remain in motion on it. The sliding mode gain matrix adopts a diagonal array form, and its adaptive law is designed as:

$$K(t) = \text{diag}([k_1(t)\ k_2(t)\ k_3(t)]) \tag{14}$$

Add an adaptive link in the selection stage of the K value and introduce an adaptive sliding mode gain. When the residual increases, the gain is automatically increased to enhance robustness. When the residual is reduced, the gain is decreased and the chattering is reduced. Compared with fixed gain, adaptive gain reduces high-frequency jitter while ensuring detection sensitivity.

$$K_i = 0.01 \cdot \text{sign}(|e_i(t - \Delta t)| - |e_i(t)|) \tag{15}$$

Through the sliding mode equivalent control principle, the fault estimation value can be expressed as

$$\hat{f}(t) = \nu_{eq}(t) \tag{16}$$

Among them, ν_{eq} is the equivalent control signal:

$$\nu_{eq} = LPF(\nu) \tag{17}$$

In fault estimation, a first-order low-pass filter is added to extract the equivalent control quantity, and the adjustable time constant τ balances the response speed and smoothness.

$$\hat{f}_i(k) = \hat{f}_i(k-1) + \frac{Ts}{\tau}\left(\nu_i(k) - \hat{f}_i(k-1)\right) \tag{18}$$

Defining an evaluation function $J(t)$ and an adaptive threshold in fault diagnosis can determine whether a fault has occurred.

$$J(t) = \sqrt{\frac{1}{T}\int_{t-T}^{t} e^T(\tau)e(\tau)d\tau} \tag{19}$$

And adaptive threshold:

$$J_{th}(t) = \mu + k\sigma(t) \tag{20}$$

Here, μ represents the mean residual when there is no fault, σ is the standard deviation, and k is the safety factor.

When $J(t) > J_{th}(t)$ occurs, it is determined that a fault has occurred.

3.4 Simulated Analysis

The sensor for selecting the roll Angle φ experienced a deviation failure in the fifth second. Through system simulation, the estimations of the attitude Angle sensors were obtained respectively as shown in Fig. 2, and the comparisons between the estimated values and the true values of each attitude Angle are shown in Fig. 3.

The sensor with the selected roll Angle φ experienced a drift failure at the fifth second. Through system simulation, the estimates of the attitude Angle sensors were obtained respectively as shown in Fig. 4. The comparison between the estimated values and the true values of each attitude Angle is shown in Fig. 5.

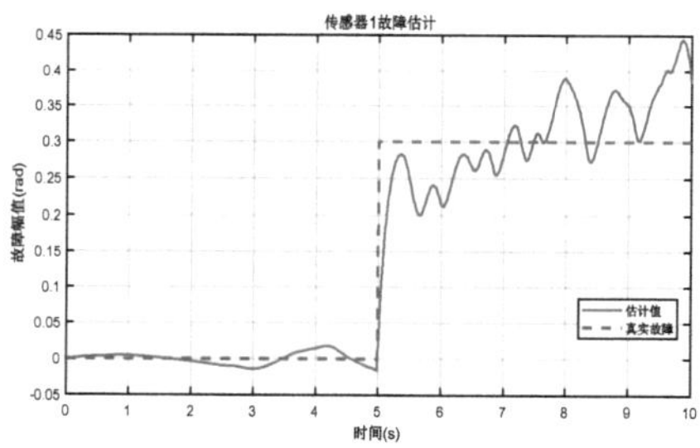

Fig. 2. Bias fault estimation.

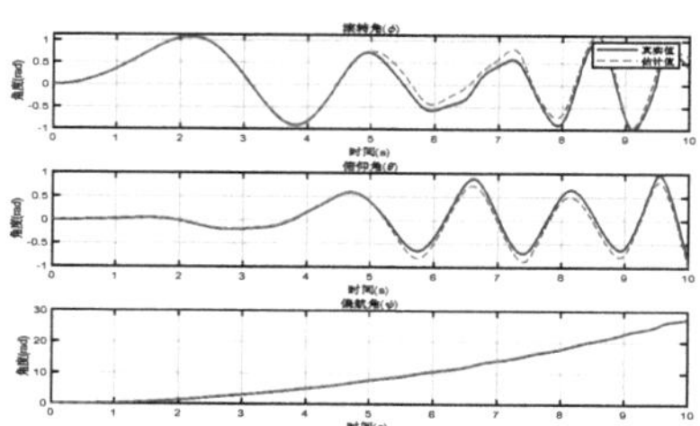

Fig. 3. Attitude Angle tracking.

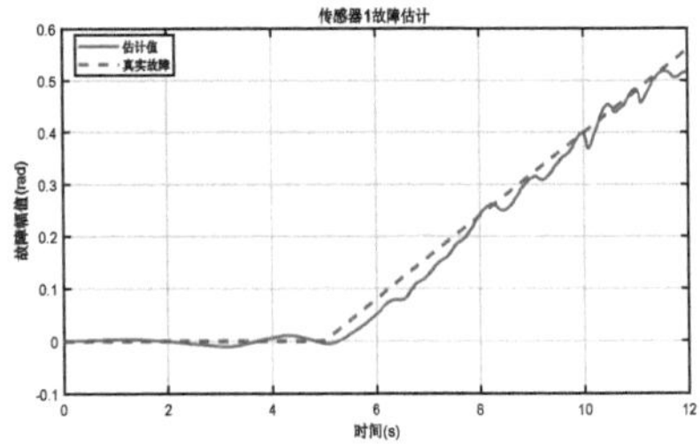

Fig. 4. Drift fault estimation.

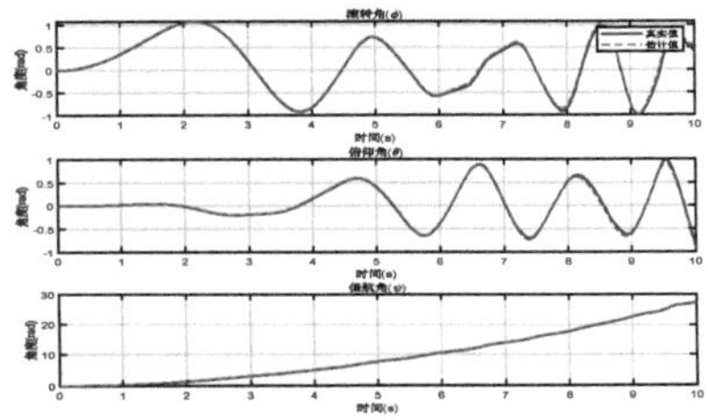

Fig. 5. Attitude Angle tracking.

4 Design of Adaptive Control Strategy

4.1 Control Strategy

This paper proposes an adaptive control strategy based on fault compensation. This strategy combines PD position control and sliding mode attitude control, compensating for sensor faults through fault estimation results to enhance the robustness of the system [7]. PD control is used to achieve precise position adjustment, while sliding mode control is used to achieve precise attitude control.

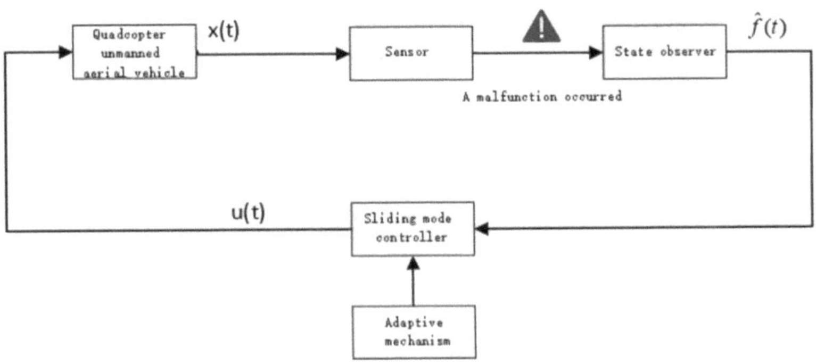

Fig. 6. Adaptive sliding mode controller control strategy.

4.2 Adaptive Controller Design

The design of the adaptive controller is based on the sliding mode control theory. Dynamic compensation for the attitude and trajectory of unmanned aerial vehicles is achieved by designing the sliding mode surface and controlling the gain. The adaptive gain is dynamically adjusted based on the error integral to balance robustness and buffeting.

For the UAV control system, it is divided into position control and attitude control. Position is controlled by PD, and attitude is controlled by sliding mode, as shown in Fig. 7.

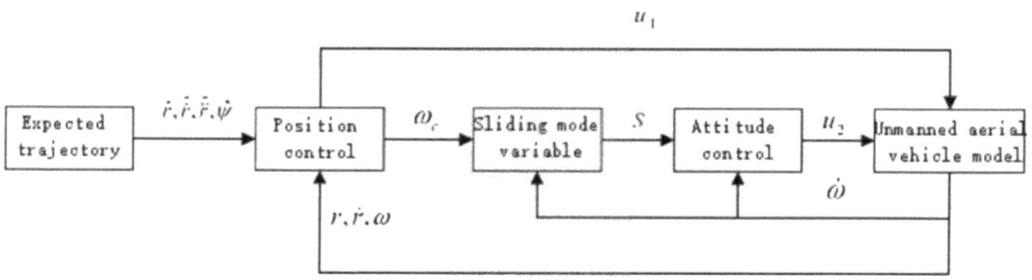

Fig. 7. Control block diagram of unmanned aerial vehicle.

1. The PD control

The location selection using PD control, is suitable for the control in UAV hover state. Its control law can be expressed as:

$$u_{pd} = K_p(p_d - p) + K_d(v_d - v) \tag{21}$$

The linear motion equation in the ground coordinates obtained from Chapter Two is

$$m\ddot{r} = \begin{bmatrix} 0 \\ 0 \\ -mg \end{bmatrix} + R_B \begin{bmatrix} 0 \\ 0 \\ F_1 + F_2 + F_3 + F_4 \end{bmatrix} \tag{22}$$

Substitute the approximate values to obtain the simplified linear motion model:

$$m\begin{bmatrix} \ddot{x} \\ \ddot{y} \\ \ddot{z} \end{bmatrix} = \begin{bmatrix} (c\psi s\theta + c\theta s\phi s\psi)u_1 \\ (s\psi s\theta - c\theta s\phi c\psi)u_1 \\ -mg + c\phi c\theta u_1 \end{bmatrix} \tag{23}$$

Find the expression of u_1:

$$u_1 = \begin{bmatrix} \ddot{x}_c \\ \ddot{y}_c \\ \ddot{z}_c + mg \end{bmatrix} .* R^* \begin{bmatrix} 0 \\ 0 \\ 1 \end{bmatrix} \tag{24}$$

The input reference trajectory is: $r_{ref} = \left[\hat{x}\ \hat{y}\ \hat{z}\ \hat{\psi}\right]^T$, and the required linear acceleration is calculated as follows:

$$\begin{cases} \ddot{x}_c = \hat{\ddot{x}} + k_{dx}\left(\hat{\dot{x}} - \dot{x}\right) + k_{px}(\hat{x} - x) \\ \ddot{y}_c = \hat{\ddot{y}} + k_{dy}\left(\hat{\dot{y}} - \dot{y}\right) + k_{py}(\hat{y} - y) \\ \ddot{z}_c = \hat{\ddot{z}} + k_{dz}\left(\hat{\dot{z}} - z\right) + k_{pz}(\hat{z} - z) \end{cases} \tag{25}$$

The solution formula can be used to obtain the attitude Angle required for the flight motion of the unmanned aerial vehicle as:

$$\begin{cases} \phi_c = \frac{1}{g}\left(\ddot{x}_c \sin\hat{\psi} - \ddot{y}_c \cos\hat{\psi}\right) \\ \theta_c = \frac{1}{g}\left(\ddot{x}_c \cos\hat{\psi} + \ddot{y}_c \sin\hat{\psi}\right) \\ \psi_c = \hat{\psi} \end{cases} \tag{26}$$

2. Adaptive sliding mode attitude control

Sliding mode controller for attitude control design. The part only known Angle motion equation is:

$$I\dot{\omega} = u - \omega \times I\omega \tag{27}$$

$$I\begin{bmatrix} \dot{p} \\ \dot{q} \\ \dot{r} \end{bmatrix} = \begin{bmatrix} L(F_2 - F_4) \\ L(F_3 - F_1) \\ M_2 + M_4 - M_1 - M_3 \end{bmatrix} - \begin{bmatrix} p \\ q \\ r \end{bmatrix} \times I\begin{bmatrix} p \\ q \\ r \end{bmatrix} \tag{28}$$

Here, I represents the moment of inertia and $\omega = \left[p\ q\ r\right]^T$ represents the angular velocity of the machine body.

When approximating $p \approx \dot{\varphi}, q \approx \dot{\theta}, r \approx \dot{\psi}$ under near-hovering conditions, the attitude dynamics equation can be expressed as:

$$\ddot{\omega} = u - \dot{\omega} \times I\dot{\omega} \tag{29}$$

$$\ddot{\omega} = -I^{-1}\dot{\omega} \times I\dot{\omega} + I^{-1}u \tag{30}$$

Define the error as:

$$e(t) = \omega - \omega_c \tag{31}$$

Since the flight control system of the unmanned aerial vehicle is a second-order system, the sliding mode surface is defined as:

$$s = \dot{e} + \lambda e \tag{32}$$

$$\dot{s} = \ddot{\omega} - \ddot{\omega}_c + \lambda(\dot{\omega} - \dot{\omega}_c) \tag{33}$$

$$\begin{aligned} \dot{s} = &-I^{-1}\dot{\omega} \times I\dot{\omega} + I^{-1}u_2 \\ &- \ddot{\omega}_c + \lambda(\dot{\omega} - \dot{\omega}_c) \end{aligned} \tag{34}$$

$$\dot{s} = -I^{-1}\dot{\omega} \times I\dot{\omega} - \ddot{\omega}c + \lambda(\dot{\omega} - \dot{\omega}_c) + I^{-1}u_2 \tag{35}$$

$$\dot{s} = \alpha + \beta u_{eq} \tag{36}$$

where

$$\alpha = -I^{-1}\dot{\omega} \times I\dot{\omega} - \ddot{\omega}_c + \lambda(\dot{\omega} - \dot{\omega}_c) \tag{37}$$

$$\beta = I^{-1} \tag{38}$$

Obtain the equivalent control of $\dot{s} = 0$:

$$u_{eq} = \beta^{-1}(-\alpha) \tag{39}$$

Add switching control in the control law to enable the system to maintain movement on a specific sliding mode surface:

$$u_{sm} = \beta^{-1} \times v \tag{40}$$

$$v = -K \cdot \text{sign}(s) \tag{41}$$

Adaptive switching gain, where K_0 is the initial gain and η is the adaptive coefficient. The gain is dynamically adjusted through error integration to balance robustness and chattering.

$$K(t) = K_0 + \int_0^t \eta \|s(\tau)\| d\tau \tag{42}$$

Using the fault estimation $\hat{f}(t)$ output by the observer, add the compensation term:

$$u_{comp} = -E\hat{f}(t) \tag{43}$$

Its E is the pseudo-inverse of the fault distribution matrix. The control law is selected as:

$$u_2 = \beta^{-1}(-\alpha + v) - E\hat{f}(t) \tag{44}$$

4.3 Simulated Analysis

In order to verify the effectiveness of the proposed sliding mode control compensation strategies in this chapter, for UAV dynamics, in this chapter, the design of the adaptive control system simulation, fault compensation and compensation for two kinds of circumstances, to determine the effectiveness of the control system.

The known offset happened in 5 s roll Angle sensor fault, fault compensation under the condition of without, The trajectories of UAV as shown in Fig. 8, trajectory began to shift from 5 s, affected the UAV flight.

When fault compensation in the amount of control, the trajectories of UAV as shown in Fig. 9, the actual trajectory are essentially coincident with the desired trajectory.

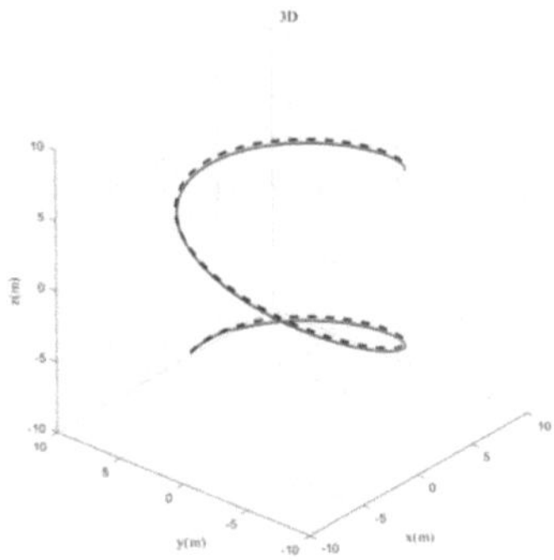

Fig. 8. Unfault-compensated motion trajectory.

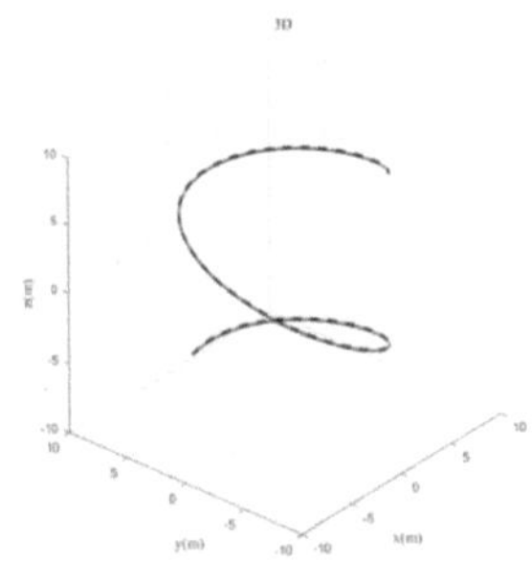

Fig. 9. The motion trajectory after fault compensation.

5 Summary and Prospect

This article around the UAV in sensor failure about the problem of how to realize the stable flight research, through the theoretical derivation, algorithm design and computer simulation, formed a complete research.

This study for the reliable operation of the UAV in complex environment provides effective solution, However, there is still room for improvement in aspects such as the efficiency of multi-sensor fusion diagnosis and the real-time computing performance of the controller. The research on adaptive flight control of UAVs in sensor failure modes is of great significance.

References

1. Dong, Z.: A brief discussion on the challenges in the development of UAV technology and future market trends [J]. Eng. Technol. **2016**(6), 233–233 (2016)
2. Liu, L., Wei, H., Zu, W., et al.: Application of drone oblique photogrammetry technology in the investigation of photovoltaic resources on the roofs of constructible cities [J]. Surv. Mapp. Bull. (12), 142–146 (2023)
3. Yin, Y., Li, L.Z.: Pre processing method for aerial triangulation of POS data in DJI drone images [J]. Northwest Hydroelectr. (02), 36–39 (2022)

4. Qi, J.: Research on Fault Diagnosis and Fault Tolerant Control of Quadcopter Drones [D]. Guangxi University of Science and Technology (2023)
5. Tan, J.: Research on Fault Diagnosis Technology of Flight Control System Based on Analytical Model [D]. Nanjing University of Aeronautics and Astronautics (2020)
6. Ruiqing, H.: Research on Fault Detection Method for UAV Flight Attitude [D]. Xi'an University of Technology (2021)
7. Chen, G., Jiang, L., Li, T.: Fault tolerant control for trajectory tracking of quadcopter unmanned aerial vehicles [J]. Comput. Eng. Appl. **55**(24), 259–264+270 (2019)

Dynamic Diffusion-Based Stochastic Trajectory Prediction with Time-Frequency Analysis

Jiaqi Ma[1], Fuji Fu[1], Dechen Hao[1], and Jinfu Yang[1,2](✉)

[1] Beijing University of Technology, Beijing, China
{majiaqi,fufj,dechenhao}@emails.bjut.edu.cn, jfyang@bjut.edu.cn
[2] Beijing Key Laboratory of Computational Intelligence and Intelligent System, Beijing, China

Abstract. Stochastic trajectory prediction plays a pivotal role in various applications, such as autonomous driving and robotic navigation. As one of the most mainstream approaches, diffusion-based stochastic trajectory prediction algorithms have achieved remarkable progress. However, they overlook the fine-grained temporal analysis of the trajectories, and the large number of diffusion steps makes it challenging to ensure real-time performance. To address these limitations, this paper proposes a Dynamic Diffusion-based stochastic trajectory prediction with Time-frequency Analysis (DDTA), which focuses on optimizing both the trajectory encoder and diffusion model. DDTA consists of two core components: the Multi-scale Time-frequency Analysis Module (MTAM) and the Dynamic Diffusion Module (DDM). (1) The MTAM, comprising a wavelet-based decomposer and a pyramid-based aggregator, is designed to analyze the instantaneous characteristics and behavioral patterns of temporal features and generate trajectory features through multi-scale fusion. (2) The DDM leverages trajectory features as a diffusion initialization prior and dynamically adjusts the number of diffusion steps to significantly reduce computational overhead. By combining the strengths of MTAM and DDM, DDTA enhances algorithm performance while effectively improving inference efficiency. Extensive comparative, ablation, visualization, and efficiency experiments conducted on three datasets (ETH-UCY, NBA, and SDD) demonstrate the effectiveness and superiority of the proposed method.

Keywords: Stochastic trajectory prediction · Multi-scale time-frequency analysis · Dynamic diffusion model

1 Introduction

Trajectory prediction aims to forecast the future trajectories of one or more agents based on their past movements. It is essential in applications such as autonomous driving [1], unmanned aerial vehicles and robotic navigation [2],

H. Ma et al. (Eds.): IWACIII 2025, CCIS 2780, pp. 591–605, 2026.
https://doi.org/10.1007/978-981-95-6730-0_40

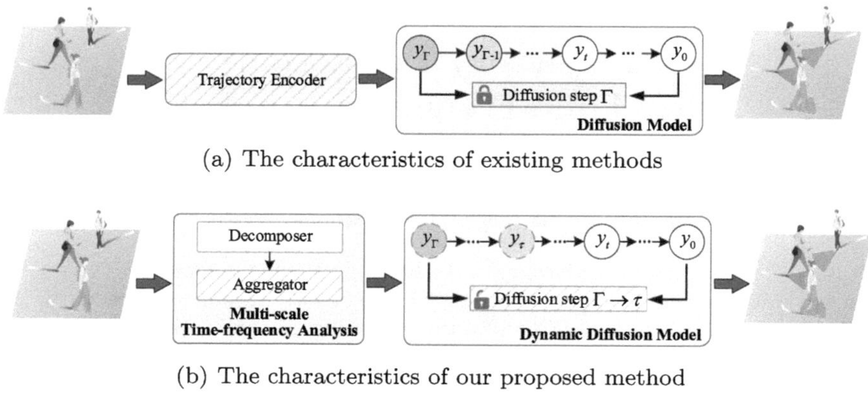

Fig. 1. Comparison of existing methods with our approach. (a) The characteristics of existing methods. Current approaches typically employ a trajectory encoder to extract high-dimensional trajectory embeddings, subsequently utilizing diffusion models to generate multimodal future trajectories. (b) The characteristics of our proposed method. Our proposed framework introduces a decomposer and an aggregator for multi-scale time-frequency analysis of historical trajectories. Furthermore, we use trajectory features as prior and dynamically adjust the number of diffusion steps to significantly enhance algorithm efficiency while maintaining prediction accuracy.

and intelligent surveillance [3]. Recently, stochastic trajectory prediction has garnered increasing attention due to its ability to model the probabilistic distribution of future trajectories, generating multiple plausible trajectory hypotheses.

However, accurately modeling the future trajectory distributions is a key challenge in stochastic trajectory prediction. Recent researches have achieved notable progress through the development of various deep generative models. Specifically, Variational Autoencoders (VAEs) [4,5] have been used to learn latent variables and generate multiple trajectory samples, while Generative Adversarial Networks (GANs) [6,7] employ generator-discriminator frameworks to produce and evaluate trajectory candidates. However, VAE-based methods often require complex loss designs for multimodal outputs, and GAN-based methods are prone to mode collapse. To overcome these limitations, recent advancements [8,9] have integrated Denoising Diffusion Probabilistic Models (DDPMs) [10,11] from image generation into trajectory prediction, achieving groundbreaking progress.

The classical framework of diffusion-based stochastic trajectory prediction methods [12,13] is illustrated in Fig. 1(a). The trajectory encoder extract high-dimensional trajectory embeddings from historical trajectory, which generally encapsulate the social context information. These embeddings are then used as conditional inputs to a diffusion model, which progressively generates the distribution of future trajectories. However, this framework exhibits two major limitations: (1) the trajectory encoder struggles to simultaneously capture both the instantaneous characteristics and behavioral patterns inherent in temporal trajectories, which affects the accurate feature representation; (2) the diffusion model utilizes a fixed and excessively large number of diffusion steps, compromising the real-time performance. In practical, target trajectory often exhibits both instantaneous characteristics and behavioral patterns, which are closely correlated with local and global features. Furthermore, as powerful generative models, diffusion models typically require a substantial number of denoising steps (diffusion steps) to ensure robust representational capacity, inevitably introducing significant computational overhead.

To address the aforementioned challenges, we propose Dynamic Diffusion-based stochastic trajectory prediction with Time-frequency Analysis, namely DDTA. As illustrated in Fig. 1(b), DDTA comprises two core components: (1) the Multi-scale Time-frequency Analysis Module (MTAM) and (2) the Dynamic Diffusion Module (DDM). The MTAM includes a Wavelet-based Decomposer (WDer) and a Pyramid-based Aggregator (PAor). The WDer applies Discrete Wavelet Transform (DWT) to decompose historical trajectory into multiple frequency bands, capturing both instantaneous characteristics (high-frequency bands) and behavioral patterns (low-frequency bands). The PAor then performs multi-scale analysis and fusion of these features to generate comprehensive trajectory representations for diffusion model. The DDM introduces a significant advancement by overcoming the limitation of fixed diffusion steps in traditional diffusion models through a diffusion initialization prior and a novel diffusion step reduction strategy. While conventional diffusion models typically employ a large, fixed number of diffusion steps to enhance generative capability, the DDM implements an initialization prior and a sequence of decreasing diffusion steps, maintaining the generative ability of the model while significantly reducing computational overhead.

In summary, the main contributions of this paper are:

- Propose a novel stochastic trajectory prediction framework based on dynamic diffusion with time-frequency analysis, which thoroughly analyzes the characteristics of temporal trajectories and the limitations of conventional diffusion models.
- Design a MTAM to integrate a WDer and a PAor to achieve comprehensive encoding of historical trajectories, enhancing the diffusion denoising process.
- Design a DDM to optimize the diffusion process through a diffusion initialization prior and a diffusion step reduction, reducing computational cost while maintaining model performance.

2 Related Work

2.1 Stochastic Trajectory Prediction

Early research can be broadly categorized into two main types based on generative models: VAE-based methods [5,14] and GAN-based methods [15,16]. (1) VAE-based methods focus on learning latent distributions to generate smooth and probabilistic future trajectories. For instance, Lee et al. [14] propose MUSE-VAE, a probabilistic modeling framework that leverages a cascaded conditional VAE architecture with a coarse-to-fine multi-factor prediction mechanism. Similarly, Qiang et al. [5] introduce a VAE-based framework that aligns the predicted distribution with the observed distribution. (2) GAN-based methods aim to improve trajectory diversity through adversarial training. For example, Pang et al. [15] develop BR-GAN, which incorporates multiple modules within a GAN framework to process pedestrian information. Additionally, Hou et al. [16] propose SSTAttGAN, which integrates a spatiotemporal attention mechanism into the GAN generator to enhance prediction accuracy.

2.2 Diffusion-Based Stochastic Trajectory Prediction

The remarkable success of DDPMs in image generation has garnered widespread attention. Rasul et al. [17] pioneer the application of diffusion models to multivariate probabilistic time-series forecasting, establishing foundational technical insights for trajectory prediction. Building on this, Gu et al. [8] propose MID, formulating trajectory prediction as an inverse process of motion uncertainty diffusion, and representing this process as a parameterized Markov chain. Subsequently, Mao et al. [18] introduce LED, whose core innovation lies in employing a trainable leapfrog initializer to learn the multimodal distribution of future trajectories. Lv et al. [13] develop LADM, integrating a pedestrian-group interaction module with an autoencoder-based diffusion architecture to enhance prediction precision. Wang et al. [19] propose TrajFine, which replaces traditional social relation modeling between agents with diffusion-based interaction learning, while synthesizing agent behaviors across heterogeneous scenarios.

Although existing diffusion-based methods have achieved significant progress, several critical limitations remain. Specifically, current approaches inadequately address multi-scale temporal analysis of sequential trajectories and suffer from compromised real-time performance due to the extensive diffusion steps.

3 Preliminaries

3.1 Problem Definition

Trajectory prediction aims to forecast the future motion trajectories of agents based on the observed historical trajectories. At each timestamp t, there are N agents in the scene $\mathbb{X}_N = \{\mathrm{X}_i | i = 1,2,...,N\}$. The historical trajectory of the i-th agent over T_{his} timestamps is $\mathrm{X} = \left[\mathrm{x}^{-T_{his}+1}, \mathrm{x}^{-T_{his}+2}, ..., \mathrm{x}^0\right] \in \mathbb{R}^{T_{his}\times 2}$ (omitting subscripts for clarity), where $\mathrm{x}^t \in \mathbb{R}^2$ presents the 2D spatial coordinates. The ground-truth (GT) future trajectory of the i-th agent over T_{pre} timestamps is $\mathrm{Y} = \left[\mathrm{y}^1, \mathrm{y}^2, ..., \mathrm{y}^{T_{pre}}\right] \in \mathbb{R}^{T_{pre}\times 2}$. To predict multimodal future trajectories, the stochastic trajectory prediction algorithm trains a parameterized model, which takes $\mathbb{X}_N$ as input, to generate future trajectory distribution P_θ. Based on P_θ, K future trajectory samples $\hat{\mathbb{Y}}_K = \left\{\hat{Y}_i | i = 1,2,..., K\right\}$ can be drawn, where the k-th trajectory over T_{pre} timestamps is denoted as $\hat{Y} = \left[y^1, y^2, ..., y^{T_{pre}}\right] \in \mathbb{R}^{T_{pre}\times 2}$.

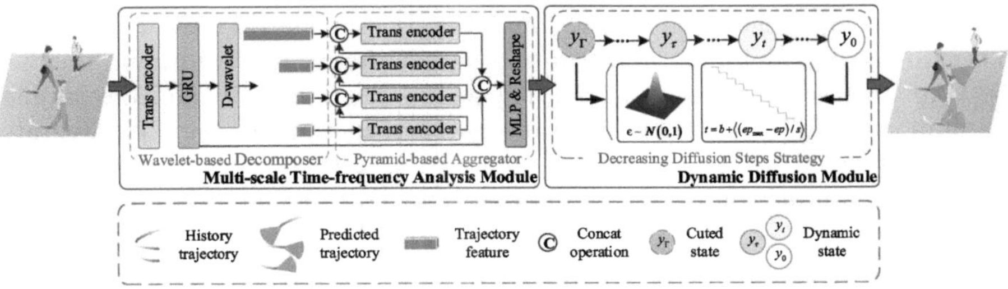

Fig. 2. The overall framework of DDTA. DDTA contains a MTAM and a DDM. The MTAM comprises WDer and PAor for multi-scale analysis of temporal features. The DDM integrates a diffusion initialization prior and a diffusion step reduction to reduce computational time.

3.2 Diffusion Model

As a classical generative algorithm, diffusion models aim to generate target data samples from noise (sampled from a simple distribution). Diffusion models typically consist of two processes: the forward process and the reverse process. Both processes are described by a parameterized Markov chain. (1)Forward process follows the Markov chain, gradually adding known noise from Y_0 to Y_Γ, where the variable at the current step depends only on the variable at the previous step. This process can be formulated as:

$$q\left(Y_{1:\Gamma}|Y_0\right) := \prod_{\nu=1}^{\Gamma} q\left(Y_\nu|Y_{\nu-1}\right), \tag{1}$$

$$q\left(Y_\nu|Y_{\nu-1}\right) := \mathcal{N}\left(Y_\nu; \sqrt{1-\beta_\nu}Y_{\nu-1}, \beta_\nu \mathrm{I}\right), \tag{2}$$

where Y_0 represents the input sample at the initial step, $q\left(Y_\nu|Y_{\nu-1}\right)$ denotes the conditional distribution at step ν, and β_ν is a positive constant that increases with ν. (2) Reverse process is a denoising inference step of the diffusion model, aiming to gradually recover the original data by removing the noise:

$$p_\theta\left(Y_{0:\Gamma}\right) := p\left(Y_\Gamma\right) \prod_{\nu=1}^{\Gamma} p_\theta\left(Y_{\nu-1}|Y_\nu\right), \tag{3}$$

$$p_\theta\left(Y_{\nu-1}|Y_\nu\right) := \mathcal{N}\left(Y_{\nu-1}; \epsilon_\theta\left(Y_\nu, t\right), \beta_\nu \mathrm{I}\right), \tag{4}$$

where $Y_\Gamma \sim \mathcal{N}\left(0,\mathrm{I}\right)$ is the initial noise sampled from a Gaussian distribution $p\left(Y_\Gamma\right)$, and $\epsilon_\theta\left(\cdot\right)$ represents a parameterized noise prediction model. The diffusion model predicts the noise at each step using a neural network and generates the sample distribution through a step-by-step denoising process, ultimately achieving the generation of data samples from noise.

4 Methodology

4.1 Framework Overview

As illustrated in Fig. 2, DDTA enables multi-scale analysis of temporal trajectories and dynamic adjustment of diffusion steps. This framework comprises two main components: the MTAM and the DDM. The MTAM employs the WDer to decompose temporal features into multi-frequency bands, and then use the PAor to fuse these bands at varying scales. The DDM dynamically adjusts the number of diffusion steps based on a step reduction strategy. It integrates MTAM-extracted trajectory features to assist in the denoising process, thereby reducing computational overhead and ensuring the real-time performance.

4.2 Multi-scale Time-Frequency Analysis Module

The MTAM, which comprises a WDer and a PAor, primarily accurately extracts temporal features, aiding the diffusion module in the denoising process.

1) Wavelet-based Decomposer: The trajectory of an agent exhibits temporal complexity, typically manifested as local instantaneous characteristics and global behavioral patterns. Instantaneous characteristics capture abrupt changes and fluctuations, while behavioral patterns represent the smooth or large-scale trends. The WDer sequentially employs a transformer-based encoder [20] and a GRU [21] to extract high-dimensional temporal features from X:

$$y' = f_{\text{GRU}}\left(f_{\text{trans},0}\left(\text{X}\right)\right), \tag{5}$$

where $y' \in \mathbb{R}^{B \times T_{his} \times C}$ and $C = 256$. To perform time-frequency analysis on y', the DWT [22] is introduced to decompose this features into J levels: $D_J = f_{\text{DWT}}\left(y'\right)$. When $J = 3$, $D = \{D_0, D_1, D_2, D_3\}$, where $D_0 \in \mathbb{R}^{B \times T_{his} \times \frac{C}{8}}$ represents the low-frequency band, $D_1 \in \mathbb{R}^{B \times T_{his} \times \frac{C}{8}}$, $D_2 \in \mathbb{R}^{B \times T_{his} \times \frac{C}{4}}$, and $D_3 \in \mathbb{R}^{B \times T_{his} \times \frac{C}{2}}$ denote the high-frequency bands.

2) Pyramid-based Aggregator: To better capture both the instantaneous characteristics and behavioral patterns of temporal features, the PAor employs transformer-based encoders to perform multi-scale processing of the wavelet bands. Specifically, the PAor assigns a transformer encoder to each wavelet component and fuses them through a concatenation operation:

$$D_i' = \begin{cases} f_{\text{trans},i+1}\left(\text{Concat}\left(D_i, D'_{i-1}\right)\right), & i = 1,2,3 \\ f_{\text{trans},i+1}\left(D_i\right), & i = 0. \end{cases} \tag{6}$$

Then, D_3' and y' are concatenated and fed into a MultiLayer Perceptron (MLP) for effective fusion: $\tilde{y}' = f_{\text{MLP}}\left(\text{Concat}\left(y', D'_3\right)\right)$. Finally, the trajectory features are obtained through: $y = \text{Reshape}\left(\tilde{y}'\right)$. The MTAM, based on WDer and PAor, performs both local and global analysis on the temporal features, thereby comprehensively capturing the trajectory temporal characteristics, which are subsequently used as input to the diffusion model.

4.3 Dynamic Diffusion Module

In trajectory prediction, diffusion models predict future trajectories from Gaussian noise based on historical trajectories. The diffusion model consists of two trainable components: (1) The scene interaction block is a transformer-based encoder designed to learn social embeddings between the target agent and other agents: $e = f_{\text{social}}(\mathrm{X}, \mathbb{X}_N)$. (2) The noise estimation block takes three inputs: the k-th noisy trajectory $Y_k^{\nu+1}$, the ν-th denoising step, and social embeddings, and predicts the noise in $Y_k^{\nu+1}$ to generate the denoised trajectory ϵ_θ^ν:

$$\epsilon_\theta^\nu = f_\epsilon\left(Y_k^{\nu+1}, \nu+1, e\right), \tag{7}$$

$$Y_k^\nu = \frac{1}{\sqrt{\alpha_\nu}}\left(Y_k^{\nu+1} - \frac{1-\alpha_\nu}{\sqrt{1-\bar{\alpha}_\nu}}\epsilon_\theta^\nu\right) + \sqrt{1-\alpha_\nu}\,\mathrm{z}, \tag{8}$$

where α_ν and $\bar{\alpha}_\nu$ denote parameters, $\bar{\alpha}_\nu = \prod_{i=1}^{\nu}\alpha_i$, and z represents the added noise, $\mathrm{z} \sim \mathcal{N}(\mathrm{z}; 0, \mathrm{I})$. In DDTA, DDM is designed to generate more accurate future trajectories while maintaining real-time performance.

The DDM incorporates trajectory features y as prior knowledge to assist the denoising process. Besides, the model requires more denoising steps due to limited fitting capability initially. As training progresses, fewer steps suffice for accurate predictions. Consequently, a diffusion step reduction strategy is designed, where the number of diffusion steps gradually decreases as the training epochs increase:

$$\nu = b + \langle (ep_{\max} - ep)/s \rangle, \tag{9}$$

where b and s represent the base and scaling factor, indicating the minimum number of diffusion steps and the range of variation, respectively. $ep_{\max}$ and ep denote the maximum and the current training epoch, respectively, while $\langle \cdot \rangle$ signifies the floor operation. The range of denoising steps ν is $[b, \tau]$, and when $\tau \leq \mathit{\Gamma}$, the number of denoising steps is significantly reduced. During model inference, each trajectory requires b denoising steps to generate the future trajectory distribution, which ensures high accuracy while significantly reducing computational overhead during both training and inference.

4.4 Training and Inference

During the training phase, we adopt a conventional training approach for the denoising module [10] to train the entire model. Specifically, the denoising module is responsible for estimating the noise added at each step of the diffusion model. Therefore, the total loss is expressed as:

$$\mathcal{L} = ||\epsilon - f_\epsilon\left(Y_k^{\nu+1}, \nu+1, e\right)||_2. \tag{10}$$

Note that the DDM utilizes the MTAM-extracted trajectory features as prior knowledge, allowing the parameters in the MTAM to be trained simultaneously. Additionally, the number of diffusion steps ν in the DDM is dynamically adjusted based on the training epoch, $\nu \in [b, \tau]$.

During inference, unlike conventional diffusion models [10], DDTM generates accurate future trajectories using only b diffusion steps, maintaining model performance while achieving significant inference speedup.

5 Experiments

5.1 Dataset and Metrics

1) Datasets: We evaluate DDTA on three trajectory prediction datasets, including ETH-UCY dataset [23,24], NBA SportVU (NBA) Dataset, and Stanford Drone Dataset (SDD) [25].

ETH-UCY dataset: The ETH-UCY dataset comprises pedestrian movement data from various locations and scenarios, including ETH, HOTEL, UNIV, ZARA1, and ZARA2. The dataset provides detailed annotations for each pedestrian, such as position, movement direction, and velocity. For the trajectory prediction task, we use data from four scenes for training and data from one scene for testing.

NBA dataset: The NBA dataset is derived from the SportVU tracking system used in NBA games, which precisely records the positions and movements of each player and the basketball. In the trajectory prediction task, we utilize 10 frames (2.0 s) of observed trajectories to predict the future trajectories over 20 frames (4.0 s).

SDD dataset: The SDD dataset is a comprehensive dataset collected from an aerial view of Stanford University campus, containing over 11,000 pedestrians. In order to align with previous work [25], we predict the future trajectories over 12 frames (4.8 s) based on an observation trajectory spanning 8 frames (3.2 s).

2) Metrics: To maintain consistency with most prediction algorithms, we adopt Average Displacement Error (ADE) and Final Displacement Error (FDE) as evaluation metrics. ADE calculates the mean distance errors between the predicted and GT trajectories across all timestamps. FDE computes the displacement error between the endpoints of the predicted and GT trajectories.

5.2 Implementation Details

DDTA comprises of two main components: the MTAM and the DDM. The MTAM includes a transformer-based encoder (256 dimensions, 2 attention heads) and a GRU with a hidden size of 256. In the DDM, the base and scaling factors are set to 5 and 50, respectively, corresponding to a diffusion step range of [5,10]. For training, we use the Adam optimizer with an initial learning rate of 10^{-2}, trained for 256 epochs with a batch size of 128. The learning rate decays by half every 32 epochs. All experiments are conducted in PyTorch on an NVIDIA RTX 4090 GPU and Ubuntu 20.04.

5.3 Performance Comparison

We compare the DDTA with existing state-of-the-art (SOTA) methods on three datasets. The compared methods include NMMP [7], MID [8], LED [18], MemoNet [26], PECNet [27], Trajectron++ [28], NPSN [29], GroupNet [30], SingularTrajectory [31], Social-GAN [32], SOPHIE [33], and DiffWT [34].

Table 1. Comparison of DDTA with SOTA methods on ETH-UCY dataset. The best and second-best results are in **bold** and underline, respectively.

Method	ETH		Hotel		Univ		Zara01		Zara02		AVG	
	ADE	FDE	ADE	FDE	ADE	FDE	ADE	FDE	ADE	FDE	ADE	FDE
NMMP	0.61	1.08	0.33	0.63	0.52	1.11	0.32	0.66	0.43	0.85	0.41	0.82
PECNet	0.54	0.87	0.18	0.24	0.35	0.60	0.22	0.39	0.17	0.30	0.29	0.48
Trajectron++	0.61	1.02	0.19	0.28	0.30	0.54	0.24	0.42	0.18	0.32	0.30	0.51
MemoNet	0.40	0.61	**0.11**	0.17	0.24	0.43	0.18	0.32	0.14	0.24	0.21	0.35
NPSN	0.40	0.76	0.12	0.18	0.22	0.41	**0.17**	**0.31**	**0.12**	0.24	0.21	0.38
MID	0.39	0.66	0.13	0.22	0.22	0.45	**0.17**	**0.30**	0.13	0.27	0.21	0.38
LED	0.39	0.58	**0.11**	0.17	0.26	0.43	0.18	0.26	0.13	**0.22**	0.21	0.32
SingularTrajectory	**0.35**	**0.42**	0.13	0.19	0.25	0.44	0.19	0.32	0.15	0.25	0.21	**0.32**
DDTA	0.37	0.50	**0.11**	**0.16**	**0.21**	**0.40**	**0.17**	**0.30**	**0.11**	0.23	**0.19**	**0.32**

Table 2. Comparison of DDTA with SOTA methods on NBA dataset, where the minADE20/minFDE20 (meters) are reported. The best and second-best results are in **bold** and underline, respectively.

Time(s)	PECNet	MemoNet	NPSN	GroupNet	MID	LED	DDTA
1.0	0.40/0.71	0.38/0.56	0.35/0.58	0.26/0.34	0.28/0.37	0.18/0.27	**0.16/0.24**
2.0	0.83/1.61	0.71/1.14	0.68/1.23	0.49/0.70	0.51/0.72	0.37/0.56	**0.35/0.52**
3.0	1.27/2.44	1.00/1.57	1.01/1.76	0.73/1.02	0.71/0.98	0.58/0.84	**0.55/0.77**
Total(4.0)	1.69/2.95	1.25/1.47	1.31/1.79	0.96/1.30	0.96/1.27	0.81/1.10	**0.77/1.00**

1) ETH-UCY dataset: The comparative experimental results on ETH-UCY dataset are shown in Table 1. Compared to existing SOTA methods, DDTA achieves the best performance in the Hotel and Univ, with ADE/FDE of 0.11/0.16 and 0.21/0.40, respectively, and demonstrates competitive results in the other three scenarios. Besides, DDTA exhibits superior generalization with the average 0.19/0.32 across all scenarios, further validating its effectiveness.

2) NBA dataset: The comparative experimental results on NBA dataset are presented in Table 2. Compared to existing SOTA methods, DDTA significantly outperforms others in both ADE and FDE across all timestamps. Specifically, DDTA achieves ADE/FDE of 0.16/0.24, 0.35/0.52, 0.55/0.77, and 0.77/1.00 at the 1 s, 2 s, 3 s, and 4 s benchmarks, respectively. Compared to the second-best performing LED [18], DDTA demonstrates improvements ranging from 0.02 to 0.1 across these timestamps, showcasing its superior performance.

3) SDD dataset: The comparative experimental results on SDD dataset are shown in Table 3. Compared to existing SOTA methods, DDTA achieves the

Table 3. Comparison of DDTA with SOTA methods on SDD, where the minADE20/minFDE20 (meters) are reported. The best and second-best results are in **bold** and underline, respectively.

Time(s)	Social-GAN	SOPHIE	MID	LED	DiffWT	DDTA
4.8	27.23/41.44	16.27/29.38	14.67/26.72	8.48/11.66	7.19/11.21	**6.47/10.45**

Table 4. The ablation experiments of MTAM and DDM on ETH-UCY dataset. The best results are in **bold**.

	AVG (ADE/FDE)	Training Time/batch size=128(ms)				
		ETH	Hotel	Univ	Zara01	Zara02
DDTA(w/o MTAM)	0.23/0.36	**~42**	**~44**	**~41**	**~42**	**~42**
DDTA(w/o DDM)	0.21/0.33	~395	~400	~389	~401	~399
DDTA	**0.19/0.32**	~43	~45	~42	~44	~43

Table 5. The sensitivity experiments of J levels in MTAM on ETH-UCY dataset. The best results are in **bold**.

J	ETH	Hotel	Univ	Zara01	Zara02	AVG
1	0.36/0.53	0.17/0.22	0.23/0.42	0.20/0.32	0.15/0.27	0.22/0.35
2	0.37/0.53	0.13/0.20	0.21/0.42	0.19/0.31	0.12/0.26	0.20/0.34
3	0.37/0.50	**0.11/0.16**	0.21/**0.40**	**0.17/0.30**	**0.11/0.23**	**0.19/0.32**
4	**0.36/0.48**	**0.11**/0.19	**0.20**/0.41	**0.17**/0.32	0.12/0.24	**0.19**/0.33
5	0.38/0.49	0.14/0.21	0.23/0.41	0.19/0.33	0.14/0.26	0.22/0.34

Table 6. The sensitivity experiments of base factor b in DDM on ETH-UCY dataset. The best results are in **bold**.

b	ETH	Hotel	Univ	Zara01	Zara02	AVG
3	0.36/0.49	0.17/0.24	0.23/0.43	0.19/0.31	0.13/0.25	0.22/0.34
5	**0.37**/0.50	**0.11/0.16**	**0.21/0.40**	**0.17/0.30**	**0.11/0.23**	**0.19/0.32**
10	**0.37/0.49**	**0.12**/0.18	0.21/0.42	0.18/0.33	0.12/**0.23**	0.20/0.33

best performance. Specifically, DDTA reduces ADE and FDE to 6.87 and 10.45, respectively, outperforming the state-of-the-art method DiffWT by 0.72 and 0.75, fully demonstrating its superiority.

Table 7. The sensitivity experiments of scaling factor s in DDM on ETH-UCY dataset. The best results are in **bold**.

s	$[b, \tau]$	ETH	Hotel	Univ	Zara01	Zara02	AVG
75	[5,8]	0.36/0.49	0.17/0.24	0.23/0.43	0.19/0.31	0.13/0.25	0.22/0.34
50	[5,10]	**0.37**/0.50	**0.11/0.16**	**0.21/0.40**	**0.17/0.30**	**0.11/0.23**	**0.19/0.32**
25	[5,15]	0.38/**0.49**	**0.11**/0.18	0.23/0.43	**0.17**/0.32	0.12/0.24	**0.20**/0.33

5.4 Ablation Studies

1) Effect of Each Component: We conduct ablation experiments on MTAM and DDM using the ETH-UCY dataset, as shown in Table 4. The DDTA (w/o MTAM) denotes the model without MTAM, which instead employs a conventional trajectory encoder to extract temporal features, and DDTA (w/o DDM) represents the model without DDM, which uses the fixed diffusion steps for denoising. As shown in Table 4, DDTA (w/o MTAM) requires the least training time per batch, as the diffusion initialization and step reduction strategy based on DDM significantly reduce computational time. Furthermore, DDTA (w/o DDM) achieves lower average ADE/FDE, indicating that the MTAM better captures the trajectories temporal characteristics. Overall, DDTA combines the MTAM and DDM, achieving the lowest error while significantly reducing computational overhead.

2) Sensitivity of MTAM: We analyze the sensitivity of the MTAM and conduct experiments on hyperparameter selection using the ETH-UCY dataset. In the MTAM, the number of wavelet components is determined by the level of wavelet decomposition, denoted as J, which reflects the granularity of temporal feature analysis. As shown in Table 5, we observe that as J increases from 1 to 3, the overall performance improves. This indicates that multi-scale analysis of temporal features by the MTAM enhances future trajectory prediction. However, when J increases from 3 to 5, the performance of model declines, suggesting that excessive decomposition of temporal features can undermine the effectiveness of feature representation. Therefore, we select $J = 3$ as the default value.

3) Sensitivity of DDM: In the DDM, the number of diffusion steps significantly impacts the generative capability of the diffusion model. The dynamic diffusion range is jointly determined by the base factor b and the scaling factor s. To this end, we design hyperparameter selection experiments on the ETH-UCY dataset, as shown in Tables 6 and 7, respectively. From Table 6, it can be observed that DDTA performs best when $b = 5$. When $b = 5$ is fixed, the dynamic diffusion range is determined by the scaling factor s, which is set to 75, 50, and 25, respectively. As shown in Table 7, when $s = 50$, the dynamic diffusion range is [5, 10], and DDTA exhibits the best performance. Therefore, we select $b = 5$ and $s = 50$ as the default values.

Fig. 3. Visualization of predicted trajectories on ETH-UCY dataset. (a)-(e) Predicted trajectories of ETH, Hotel, Univ, Zara01, Zara02, respectively. The future trajectories predicted by our proposed DDTA are closest to the GT trajectories.

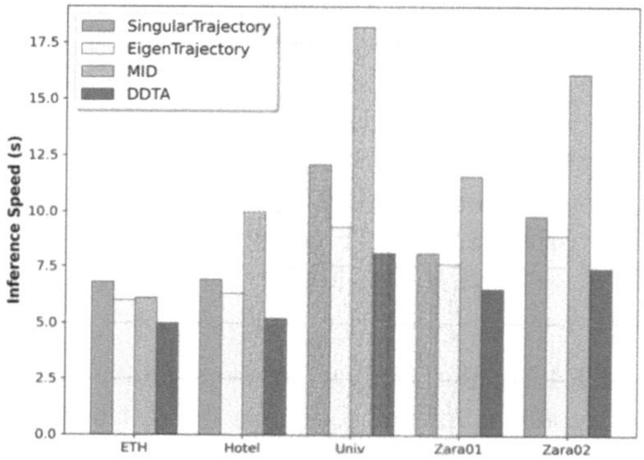

Fig. 4. Visualization of inference time of each scenario on ETH-UCY dataset. DDTA achieves the shortest inference time, demonstrating its superior efficiency.

5.5 Visualization Analysis

To comprehensively validate the effectiveness and superiority of the proposed method, we conduct a visual comparison experiment, as shown in Fig. 3. Specifically, we perform a qualitative analysis with GPGraph [35] and EigenTrajectory [36] across five scenarios in ETH-UCY dataset. The experimental results demonstrate that, in simple scenarios (e.g. Figs. 3(a), (b), and (d)), the predicted trajectories of DDTA closely align with the GT, while in more complex scenarios (e.g. Figs. 3(c) and (e)), DDTA exhibits strong robustness due to its inherent advantages. This is primarily attributed to DDTA's ability to perform multi-scale time-frequency analysis on historical trajectories, achieving precise trajectory prediction.

5.6 Efficiency Analysis

To further evaluate the computational efficiency of DDTA, this paper compares its inference speed with SOTA methods on ETH-UCY dataset. The selected comparative methods include SingularTrajectory [32], EigenTrajectory [36], and MID [8], as shown in Fig. 4. Specifically, the inference times of DDTA on the ETH, Hotel, Univ, Zara01, and Zara02 scenarios are 5.0 s, 5.1 s, 7.3 s, 6.2 s, and 7.4 s, respectively. In particular, within the 'Univ' and 'Zara02' scenarios, DDTA demonstrats approximately twice the inference speed of MID [8]. Moreover, while achieving efficient inference, DDTA also exhibits significantly lower prediction errors compared to the aforementioned comparative methods, further validating its superior performance in balancing inference efficiency and prediction accuracy.

6 Conclusion

This paper proposes DDTA, a stochastic trajectory prediction framework based on dynamic diffusion with time-frequency analysis, focusing on optimizing the trajectory encoder and the diffusion model design. DDTA comprises of two core modules: the Multi-scale Time-frequency Analysis Module (MTAM) and the Dynamic Diffusion Module (DDM). Specifically, the MTAM performs multi-scale feature analysis using a wavelet-based decomposer and a pyramid-based aggregator, generating comprehensive trajectory representations. The DDM employs the trajectory features as diffusion initialization prior and dynamically optimizes the denoising process through a diffusion step reduction strategy, significantly reducing computational overhead. Extensive experiments demonstrate that DDTA achieves state-of-the-art performance on three benchmark datasets.

Although the DDTA provides a novel paradigm for the design of trajectory encoders and diffusion models, it still relies on manually-tuned hyperparameters, such as the wavelet decomposition level J in MTAM, and the base factor b and scale factor s in DDM. In future work, we will explore adaptive optimization methods to reduce this reliance on manual tuning.

References

1. Kitchat, K., et al.: Pedcross: pedestrian crossing prediction for auto-driving bus. IEEE Trans. Intell. Transp. Syst. (2024)
2. Zhang, X., Liu, Y., Hu, D., Liu, L.: A maze robot autonomous navigation method based on curiosity and reinforcement learning. In: The 7th International Workshop on Advanced Computational Intelligence and Intelligent Informatics (IWACIII 2021), Article, no. M1-6, p. 1 (2021)
3. Jeon, Y., Tran, D.Q., Park, M., Park, S.: Leveraging future trajectory prediction for multi-camera people tracking. In: Proceedings of the IEEE/CVF Conference on Computer Vision and Pattern Recognition, pp. 5399–5408 (2023)
4. Yu, Q., Xu, Z., Zhou, Y., Liu, Z., Silamu, W.: Pedestrian trajectory prediction using spatio-temporal VAE. In: Chinese Conference on Pattern Recognition and Computer Vision (PRCV), pp. 297–311. Springer, Cham (2024)

5. Qiang, H., Xie, S., Guo, Z., Peng, X.: Socialvessel: VAE-inspired multi-vessel gated recurrent units for vessel trajectory prediction in port areas. In: 2023 IEEE 26th International Conference on Intelligent Transportation Systems (ITSC), pp. 4504–4509. IEEE (2023)
6. Wang, Z., Chen, X., Wen, S., Wang, Y.: Hei-GAN: a human-environment interaction based GAN for multimodal human trajectory prediction. In: Chinese Conference on Pattern Recognition and Computer Vision (PRCV), pp. 359–370. Springer, Cham (2023)
7. Hu, Y., Chen, S., Zhang, Y., Gu, X.: Collaborative motion prediction via neural motion message passing. In: Proceedings of the IEEE/CVF Conference on Computer Vision and Pattern Recognition, pp. 6319–6328 (2020)
8. Gu, T., et al.: Stochastic trajectory prediction via motion indeterminacy diffusion. In: Proceedings of the IEEE/CVF Conference on Computer Vision and Pattern Recognition, pp. 17 113–17 122 (2022)
9. Jiang, C., et al.: Motiondiffuser: controllable multi-agent motion prediction using diffusion. In: Proceedings of the IEEE/CVF Conference on Computer Vision and Pattern Recognition, pp. 9644–9653 (2023)
10. Ho, J., Jain, A., Abbeel, P.: Denoising diffusion probabilistic models. In: Advances in Neural Information Processing Systems, vol. 33, pp. 6840–6851 (2020)
11. Nichol, A.Q., Dhariwal, P.: Improved denoising diffusion probabilistic models. In: International Conference on Machine Learning, pp. 8162–8171. PMLR (2021)
12. Li, R., Li, C., Ren, D., Chen, G., Yuan, Y., Wang, G.: Bcdiff: bidirectional consistent diffusion for instantaneous trajectory prediction. In: Advances in Neural Information Processing Systems, vol. 36, pp. 14 400–14 413 (2023)
13. Lv, K., Yuan, L., Ni, X.: Learning autoencoder diffusion models of pedestrian group relationships for multimodal trajectory prediction. IEEE Trans. Instrum. Measur. (2024)
14. Lee, M., Sohn, S.S., Moon, S., Yoon, S., Kapadia, M., Pavlovic, V.: Muse-VAE: multi-scale VAE for environment-aware long term trajectory prediction. In: Proceedings of the IEEE/CVF Conference on Computer Vision and Pattern Recognition, pp. 2221–2230 (2022)
15. Pang, S.M., Cao, J.X., Jian, M.Y., Lai, J., Yan, Z.Y.: BR-GAN: a pedestrian trajectory prediction model combined with behavior recognition. IEEE Trans. Intell. Transp. Syst. **23**(12), 24 609–24 620 (2022)
16. Huo, J., et al.: Safety-aware vehicle trajectory prediction with spatio-temporal attentional GAN in hybrid transportation system. IEEE Trans. Instrum. Measur. (2023)
17. Rasul, K., Seward, C., Schuster, I., Vollgraf, R.: Autoregressive denoising diffusion models for multivariate probabilistic time series forecasting. In: International Conference on Machine Learning, pp. 8857–8868. PMLR (2021)
18. Mao, W., Xu, C., Zhu, Q., Chen, S., Wang, Y.: Leapfrog diffusion model for stochastic trajectory prediction. In: Proceedings of the IEEE/CVF Conference on Computer Vision and Pattern Recognition, pp. 5517–5526 (2023)
19. Wang, K.-L., Tsao, L.-W., Wu, J.-C., Shuai, H.-H., Cheng, W.-H.: Trajfine: predicted trajectory refinement for pedestrian trajectory forecasting. In: Proceedings of the IEEE/CVF Conference on Computer Vision and Pattern Recognition, pp. 4483–4492 (2024)
20. Vaswani, A.: Attention is all you need. In: Advances in Neural Information Processing Systems (2017)
21. Chung, J., Gulcehre, C., Cho, K., Bengio, Y.: Empirical evaluation of gated recurrent neural networks on sequence modeling. arXiv preprint arXiv:1412.3555 (2014)

22. Mallat, S.G.: A theory for multiresolution signal decomposition: the wavelet representation. IEEE Trans. Pattern Anal. Mach. Intell. **11**(7), 674–693 (1989)
23. Pellegrini, S., Ess, A., Schindler, K., Van Gool, L.: You'll never walk alone: Modeling social behavior for multi-target tracking. In: IEEE 12th International Conference on Computer Vision, pp. 261–268. IEEE (2009)
24. Lerner, A., Chrysanthou, Y., Lischinski, D.: Crowds by example. In: Computer Graphics Forum, vol. 26, no. 3. pp. 655–664. Wiley Online Library (2007)
25. Robicquet, A., Sadeghian, A., Alahi, A., Savarese, S.: Learning social etiquette: human trajectory understanding in crowded scenes. In: Leibe, B., Matas, J., Sebe, N., Welling, M. (eds.) ECCV 2016. LNCS, vol. 9912, pp. 549–565. Springer, Cham (2016). https://doi.org/10.1007/978-3-319-46484-8_33
26. Xu, C., Mao, W., Zhang, W., Chen, S.: Remember intentions: retrospective-memory-based trajectory prediction. In: Proceedings of the IEEE/CVF Conference on Computer Vision and Pattern Recognition, pp. 6488–6497 (2022)
27. Mangalam, K., et al.: It is not the journey but the destination: endpoint conditioned trajectory prediction. In: Vedaldi, A., Bischof, H., Brox, T., Frahm, J.-M. (eds.) ECCV 2020. LNCS, vol. 12347, pp. 759–776. Springer, Cham (2020). https://doi.org/10.1007/978-3-030-58536-5_45
28. Salzmann, T., Ivanovic, B., Chakravarty, P., Pavone, M.: Trajectron++: dynamically-feasible trajectory forecasting with heterogeneous data. In: Vedaldi, A., Bischof, H., Brox, T., Frahm, J.-M. (eds.) ECCV 2020. LNCS, vol. 12363, pp. 683–700. Springer, Cham (2020). https://doi.org/10.1007/978-3-030-58523-5_40
29. Bae, I., Park, J.-H., Jeon, H.-G.: Non-probability sampling network for stochastic human trajectory prediction. In: Proceedings of the IEEE/CVF Conference on Computer Vision and Pattern Recognition, pp. 6477–6487 (2022)
30. Xu, C., Li, M., Ni, Z., Zhang, Y., Chen, S.: Groupnet: multiscale hypergraph neural networks for trajectory prediction with relational reasoning. In: Proceedings of the IEEE/CVF Conference on Computer Vision and Pattern Recognition, pp. 6498–6507 (2022)
31. Bae, I., Park, Y.-J., Jeon, H.-G.: Singulartrajectory: universal trajectory predictor using diffusion model. In: Proceedings of the IEEE/CVF Conference on Computer Vision and Pattern Recognition, pp. 17 890–17 901 (2024)
32. Yu, C., Ma, X., Ren, J., Zhao, H., Yi, S.: Spatio-temporal graph transformer networks for pedestrian trajectory prediction. In: Vedaldi, A., Bischof, H., Brox, T., Frahm, J.-M. (eds.) ECCV 2020. LNCS, vol. 12357, pp. 507–523. Springer, Cham (2020). https://doi.org/10.1007/978-3-030-58610-2_30
33. Sadeghian, A., Kosaraju, V., Sadeghian, A., Hirose, N., Rezatofighi, H., Savarese, S.: Sophie: an attentive GAN for predicting paths compliant to social and physical constraints. In: Proceedings of the IEEE/CVF Conference on Computer Vision and Pattern Recognition, pp. 1349–1358 (2019)
34. Chen, X., Zeng, L., Gao, M., Ding, C., Bian, Y.: Diffwt: diffusion-based pedestrian trajectory prediction with time-frequency wavelet transform. IEEE Internet Things J. (2024)
35. Bae, I., Park, J.-H., Jeon, H.-G.: Learning pedestrian group representations for multi-modal trajectory prediction. In: European Conference on Computer Vision, pp. 270–289. Springer, Cham (2022)
36. Bae, I., Oh, J., Jeon, H.-G.: Eigentrajectory: Low-rank descriptors for multi-modal trajectory forecasting. In: Proceedings of the IEEE/CVF International Conference on Computer Vision, pp. 10 017–10 029 (2023)

Torque Distribution Strategy Based on Multi-mode Clutch Transmission

Yang Gao(✉), Xueliang Shan, Huanhuan Gong, Zhidong Gao, Ruiguang Wang, Zhengxing Dai, Jiangfeng Liu, and Yiqiang Liu

Aurobay(Ningbo) Intelligent Technology Co., Ltd, Ningbo 315336, China
auto_gaoyang@163.com

Abstract. Based on the research on multi-mode clutch transmission, the stability challenges under parallel driving conditions were explored. The study found that designing multi-mode clutches and coordinating control of multiple motors is the core method to solve this problem. The core lies in achieving efficient energy transfer through multi-mode clutches and coordinated control of multiple motors, to meet the stability and energy efficiency optimization goals of hybrid electric vehicles under different operating conditions. However, under rapid acceleration conditions, unbalanced torque distribution may lead to differences in the speeds of the inner and outer loops, causing the "de-parallel" phenomenon, which affects the acceleration performance and energy efficiency of the vehicle. To solve this problem, a dynamic power distribution control strategy for parallel driving was proposed. By calculating the minimum distribution ratio of residual power and dynamically adjusting the power distribution according to actual needs, the balance of angular acceleration between the inner and outer loops was achieved, avoiding the exit of the parallel working state. Experimental results show that during the simulation period, the optimized control strategy not only improved the stability and robustness of the system, but also enhanced energy efficiency, providing important reference basis for the design of parallel hybrid wide-area power grid systems. This research provides theoretical support and technical guidance for the stability of parallel multi-mode clutch transmission in practical applications.

Keywords: DHT · Multi-mode Clutch · Control Strategy · Torque Distribution · Vehicle Engineering

1 Introduction

With the increasingly serious global energy crisis and environmental pollution problems, hybrid vehicles as an efficient and clean alternative have received wide attention. In hybrid vehicles, multi-mode clutch transmission has become a research hotspot because of its high efficiency, multi-mode operation and wide applicability. However, in practical applications, there are still some challenges in power transmission and torque distribution of multi-mode clutch transmission under parallel driving conditions, especially when the driver presses the accelerator pedal, the system may appear "deparallel" phenomenon, which not only affects the driving performance of the vehicle, but also may lead to the

H. Ma et al. (Eds.): IWACIII 2025, CCIS 2780, pp. 606–618, 2026.
https://doi.org/10.1007/978-981-95-6730-0_41

reduction of energy utilization efficiency. The core of multi-mode clutch transmission lies in its unique mechanical structure and working principle. The transmission achieves flexible coupling between the engine, motor and wheels by separating the inner and outer rings. Under pure electric driving conditions, the engine and motor P1 are separated from torque transmission, and motor P2 drives the vehicle separately. Under the driving condition of series power generation, the engine generates electricity by driving motor P1 to provide energy support for motor P2. Under parallel driving conditions, the torque of the engine is transferred to the outer ring through the inner ring, and the motor P2 also participates in the power output. This multi-mode switching design enables the hybrid vehicle to operate efficiently under different working conditions. However, the stability and reliability of parallel driving conditions directly affect the overall performance of the vehicle, especially in the case of rapid acceleration, the dynamic response and torque distribution of the system become the key. In parallel driving conditions, the remaining torque of the engine (that is, the value of the engine torque minus the motor P1 generation torque) needs to be transferred to the inner and outer rings to achieve effective power output. However, when the inner ring speed is lower than the outer ring speed, the remaining torque of the engine can not be effectively transferred to the outer ring, and the system will give priority to the inner ring acceleration until the inner ring speed is consistent with the outer ring speed, in order to truly enter the parallel driving state. The existence of this speed difference not only affects the response speed of the system, but also may lead to energy loss and efficiency reduction. In the actual driving process, when the driver presses the accelerator pedal, the engine and the motor P2 increase the torque output to meet the driving demand. However, due to the unbalanced torque distribution, the P2 motor torque distribution may be too large, the engine torque distribution is too small, so that the outer ring speed is higher than the inner ring speed. The appearance of this speed difference will trigger the protection mechanism of the system and cause the system to exit the parallel mode, that is, the so-called "deparallel" phenomenon. This phenomenon will not only affect the acceleration performance of the vehicle, but also may lead to energy waste and a reduction in system efficiency [1, 2].

At present, the research on multi-mode clutch transmission is mainly concentrated in the following aspects: first, the transmission mechanical structure design and optimization; Second, energy management strategies under different working modes; The third is the simulation and analysis of dynamic performance. However, the research on the problem of "de-parallel" under the condition of rapid acceleration is still limited. Most of the existing researches focus on the stability analysis of the system and the improvement of the control strategy, but the dynamic relationship between the torque distribution and the speed difference is not deeply discussed. In addition, the existing research on the cause analysis and prevention methods of "shunt withdrawal" phenomenon also have some shortcomings, especially in practical applications, how to optimize the torque distribution strategy and speed control algorithm to avoid "shunt withdrawal" problem still need to be further studied.

The purpose of this thesis is to deeply study the problem of "deparallel" of multi-mode clutch transmission under the condition of rapid acceleration. By analyzing the power transmission mechanism and torque distribution characteristics of the system, the fundamental cause of "de-parallel" phenomenon is revealed, and a control strategy based on speed difference estimation and torque distribution optimization is proposed to improve the stability and efficiency of the system. Specifically, this paper will study from the following aspects: First, establish the dynamic model of multi-mode clutch transmission, analyze the relationship between the inner and outer ring speed and torque; Secondly, the dynamic response characteristics of torque distribution and speed difference under rapid acceleration condition are studied. Finally, a torque distribution optimization algorithm based on model prediction is designed to avoid the phenomenon of "deparalleling". Through this research, we hope to provide theoretical support and practical guidance for the optimization design of multi-mode clutch transmission and the improvement of control strategy, and further promote the development of hybrid electric vehicle technology.

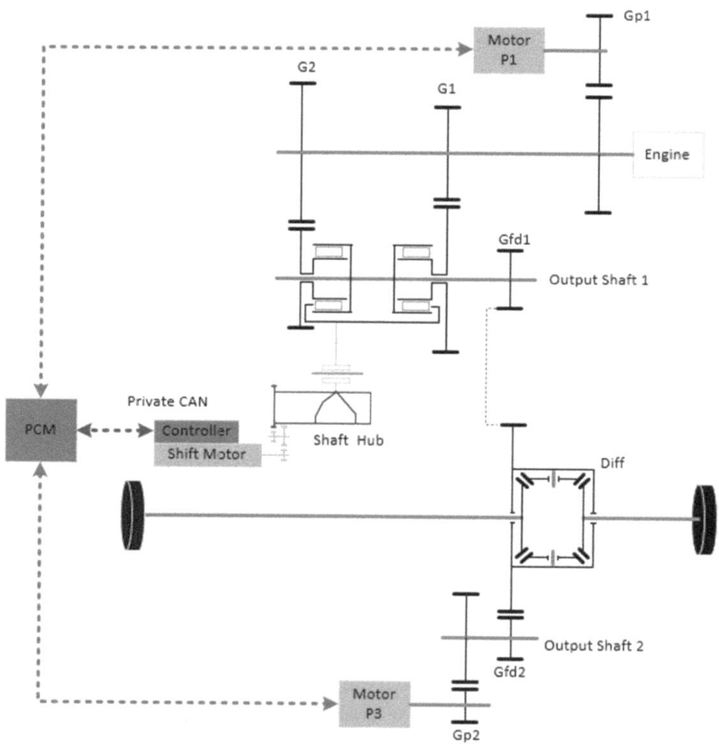

Fig. 1. Multi-mode clutch transmission transmission principle

2 Multi-mode Clutch Transmission Principle

The transmission principle of multi-mode clutch transmission is shown in Fig. 1. Multi-mode clutch transmission is a complex system which combines mechanical transmission and electric transmission. Its core lies in the design of multi-mode clutch and multi-motor coordination control. Through the separation of the inner ring and the outer ring, the transmission realizes the flexible coupling between the engine, the motor and the wheel, so as to meet the power demand and energy optimization objectives of the hybrid electric vehicle under different working conditions. The following will elaborate its transmission principle from three aspects: structure, working mode and power transmission mechanism. The mechanical structure of multi-mode clutch transmission is composed of two parts: inner ring and outer ring. The inner ring is directly and rigidly connected with the engine, and at the same time is connected with the generator motor P1; The outer ring is directly and rigidly connected to the wheel, and at the same time is connected to the drive motor P2. This design allows the inner ring and the outer ring to work independently or together under different working conditions, so as to achieve a variety of operating modes. For example, in pure electric driving conditions, motor P2 directly drives the outer ring and wheels, while the engine and motor P1 are completely out of torque transmission; Under the series power generation condition, the engine generates electricity by driving motor P1 to provide energy support for motor P2; In parallel driving conditions, the engine and motor P2 work together to drive the vehicle. The transmission principle of multi-mode clutch is the most complicated under parallel driving condition. The torque of the engine is transferred from the inner ring to the outer ring, while the motor P2 also applies additional driving torque to the outer ring. However, whether the speed of the inner ring and the outer ring is consistent directly affects the working state of the system. When the inner ring speed is lower than the outer ring speed, the remaining torque of the engine (that is, the value of the engine torque minus the motor P1 generation torque) cannot be transferred to the outer ring, but is mainly used to accelerate the inner ring until the inner ring speed is consistent with the outer ring speed, the system can truly enter the parallel running state. This process shows that the power transmission of multi-mode clutch not only depends on the distribution of torque, but also closely related to the matching of speed.

From the perspective of power transmission, the speed difference between the inner ring and the outer ring is determined by the respective moment of inertia and the net torque received. The angular acceleration of the inner ring is obtained by dividing the residual torque of the engine by the moment of inertia at the inner ring. By integrating the diagonal acceleration, the rotational speed of the inner ring can be obtained. Similarly, after subtracting the vehicle resistance moment from the driving torque of motor P2, divide the moment of inertia at the outer ring to obtain the angular acceleration of the outer ring; By integrating the angular acceleration of the outer ring, the rotational speed of the outer ring can be obtained. This dynamic relationship shows that the speed of the inner and outer rings is not only affected by their respective drive sources, but also by the inertia of the system and the external load. In the actual driving process, when the driver presses the accelerator pedal, the engine and the motor P2 increase the torque output to meet the driving demand. However, due to the unbalanced torque distribution, the P2 motor torque distribution may be too large, the engine torque distribution is too small, so that the outer ring speed is higher than the inner ring speed. The appearance of this speed difference will trigger the protection mechanism of the system and cause the system to exit the parallel mode, that is, the so-called "deparallel" phenomenon. This phenomenon will not only affect the acceleration performance of the vehicle, but also may lead to energy waste and a reduction in system efficiency. The transmission principle of multi-mode clutch transmission is based on the separation of inner ring and outer ring and the coordinated control of multiple motors. Through flexible torque distribution and speed matching, the system can achieve efficient power transfer and energy management under different working conditions. However, under the condition of rapid acceleration, the unbalanced torque distribution and the appearance of speed difference may lead to the deterioration of system performance and even the phenomenon of "deparallel". To solve this problem, the optimization of torque distribution, speed control algorithm and dynamic response of the system need to be deeply studied to ensure the stability and high efficiency of multi-mode clutch transmission in practical applications. The structure principle of multi-mode clutch is shown in Fig. 2 [3–7].

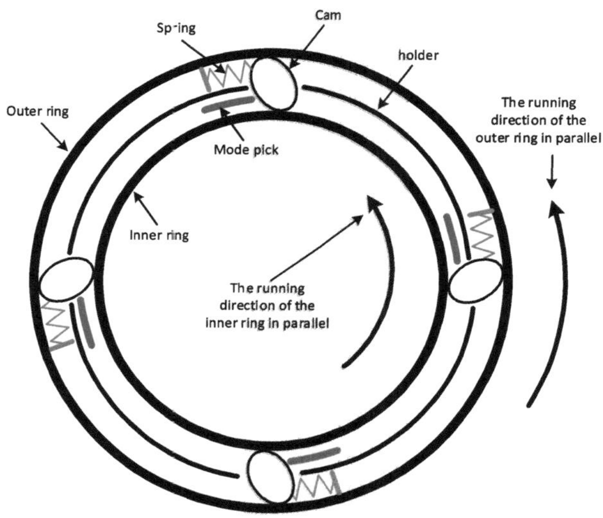

Fig. 2. Structure principle of multi-mode clutch

3 Simulation Platform of Multi-mode Clutch Transmission

In this paper, a complete control system simulation closed-loop system is established based on the vehicle. The system adopts multi-mode clutch structure and mainly includes four core components: hybrid transmission controller (PCM), multi-mode clutch transmission, engine and tire. Among them, the design of multi-mode clutch transmission is particularly complex, integrating P1 motor, P3 motor, shift motor and multi-mode clutch four key components, forming an efficient power transmission and distribution system. In terms of sensor configuration, the shift motor is equipped with a high-speed responsive speed sensor for accurate closed-loop speed control. After the speed signal of P1 motor is collected and processed by the control algorithm, it is used to calculate the real-time speed parameter of the inner ring of the multi-mode clutch. Similarly, the speed information of the P3 motor is also monitored in real time and used to calculate the speed data of the outer ring of the multi-mode clutch. By comparing the speed difference between the driving part and the driven part of the multi-mode clutch, the system can accurately calculate the slip value of the current gear, which provides an important basis for the subsequent control decision. The multi-dimensional signal acquisition system of the shift motor is particularly outstanding, which can not only obtain the speed signal, but also monitor its position signal and current signal in real time. These multi-dimensional information are integrated to accurately judge the current gear state and achieve higher precision closed-loop control. This multi-signal fusion control method significantly improves the response speed and control accuracy of the system. In terms of power distribution and energy management, the clever combination of P1 motor and engine shows a powerful function. The system can flexibly switch the working mode according to the actual demand: when the engine needs to start, the P1 motor can be used as the starting motor; When the vehicle is running, the P1 motor can be switched to a power generation mode to recover kinetic energy into electricity, achieving more efficient energy utilization; When the power demand increases, the P1 motor can work with the engine to provide additional driving force. In addition, the P3 motor takes on the main driving task during normal driving and is responsible for energy recovery during braking, further improving the overall efficiency of the system. This multi-mode clutch control system realizes the high efficiency and controllability of power transmission through exquisite mechanical design and advanced control algorithm. The multi-mode clutch vehicle system is shown in Fig. 3 [8–11].

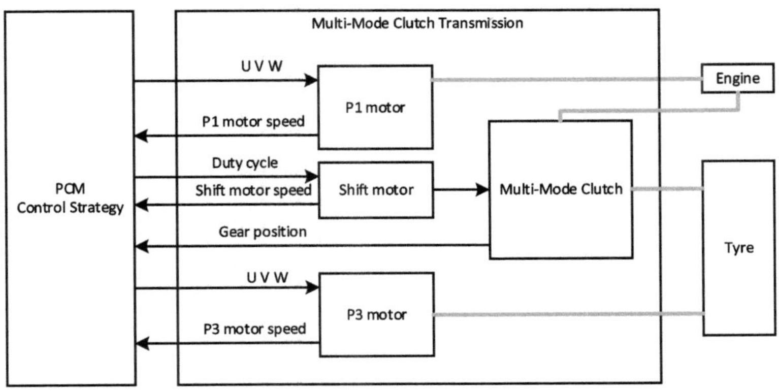

Fig. 3. Multi-mode clutch vehicle system

4 Torque Distribution Strategy of Multi-mode Clutch Transmission

In this paper, a parallel driving control strategy is developed, which aims to ensure the angular acceleration balance between the inner ring and the outer ring by rationally distributing the driving torque of the engine and the motor, and avoid the exit of the parallel operation. The core of the strategy is to calculate the minimum distribution ratio of the residual torque of the engine, and dynamically adjust the torque distribution according to the actual demand to ensure the stable operation of the system.

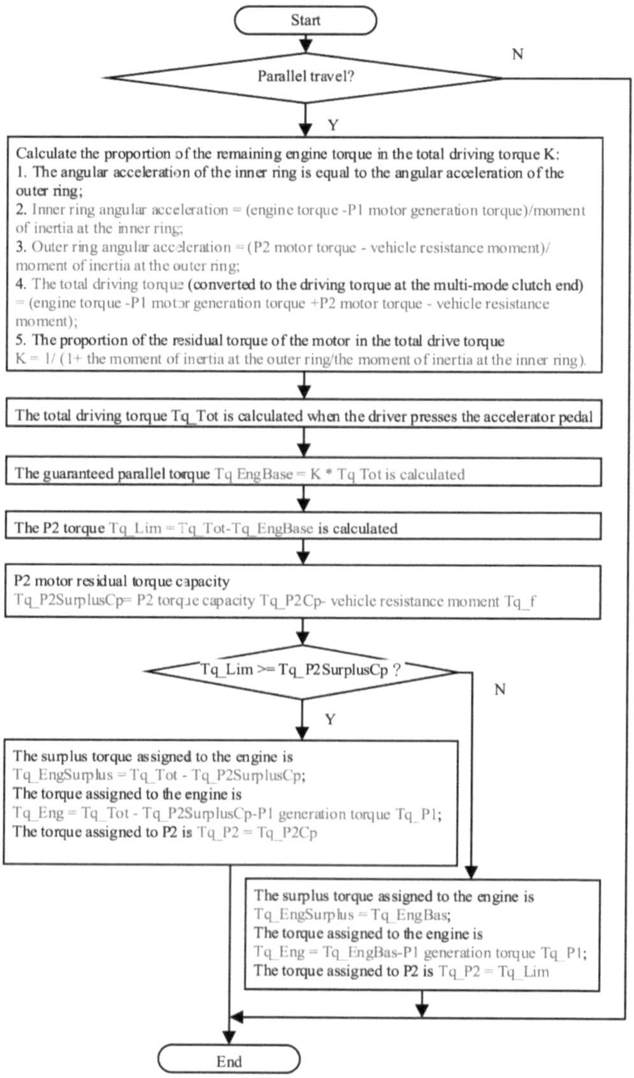

Fig. 4. Torque distribution strategy of multi-mode clutch

First, the system calculates the minimum distribution ratio of the residual torque of the engine in the total driving torque according to the condition that the angular acceleration of the inner ring and the outer ring is equal. The total driving torque is the result of converting the driving torque of the multi-mode clutch end. This ratio is the lowest distribution ratio of the remaining torque of the engine. When the actual distribution ratio of the residual torque of the engine is higher than this minimum ratio, the continuous stability of the parallel operation can be ensured. When the driver presses the accelerator pedal, the system first calculates the total drive torque. Then, according to the minimum distribution ratio of the residual torque of the engine, the minimum torque value assigned to the residual torque of the engine is calculated. Through this value, the torque limit value of P2 motor can be further calculated to guarantee the parallel working condition. The remaining torque capacity of a P2 motor is its maximum torque capacity minus the vehicle resistance moment. When the torque limit value of P2 motor is greater than or equal to the residual torque capacity of P2 motor, the system will adjust the distribution of residual torque of the engine to ensure that the maximum torque output of P2 motor is not exceeded. At this point, the total torque distribution of the engine will be adjusted according to the actual output of the P2 motor to ensure system balance. When the torque limit value of P2 motor is less than its residual torque capacity, the system will be strictly distributed in accordance with the minimum distribution ratio of the residual torque of the engine to ensure that the driving torque of the engine is not lower than the minimum requirement. At the same time, the torque distribution of the P2 motor will be adjusted according to the actual limit value to maintain the balance of the angular acceleration of the inner ring and the outer ring. Through this dynamic adjustment strategy, the system can effectively ensure the coordinated operation of the engine and P2 motor, and ensure that the angular acceleration of the inner ring is not lower than the angular acceleration of the outer ring during parallel driving, so as to avoid the exit of the parallel working condition. This control strategy not only improves the stability and reliability of the system, but also optimizes the energy utilization efficiency, which provides an important reference for the design and optimization of the parallel hybrid power system. The torque distribution strategy of multi-mode clutch transmission is shown in Fig. 4 [12–15].

5 Simulation Analysis

In order to study the effect of torque distribution strategy of multi-mode clutch on vehicle performance, a closed-loop simulation platform was established in Simulink environment. The sampling period is 10 ms. The platform can verify the power transmission characteristics of different control strategies under actual working conditions, focusing on the response characteristics of vehicle acceleration. The core of the simulation system is the simulation of the working principle of the multi-mode clutch, including the engagement/separation state of the clutch, the switching logic of the power source and the torque distribution algorithm. The simulation parameters are set as follows: the mass of the SUV vehicle is 2200 kg, the throttle opening is 80%, the initial state is the clutch is not engaged, the speed ratio of the motor P1 to the engine is 2.1, the main reduction ratio from the engine end to the differential is 2.3, the tire radius is 0.36 m, the main reduction ratio of the motor P3 to the differential is 1.2, the Gear1 speed ratio is 3.2, and the Gear2 speed ratio is 1.2. The moment of inertia of both motor P1 and engine is 0.16 kg·m^2, and the moment of inertia of motor P3 is 3.5 kg·m^2. The system design includes power source model, transmission system model and vehicle dynamics model. The power source model includes the characteristic curves of the internal combustion engine and the dual motor, and the driveline model covers the power transfer relationships of the multi-mode clutch, transmission and differential. The vehicle dynamics model considers the vehicle body mass, rolling resistance and driving force calculation. In the simulation process, the system will use the unoptimized torque distribution strategy and the optimized control strategy for comparative testing. The unoptimized strategy is based on a fixed torque distribution ratio, while the optimized strategy introduces state feedback and prediction algorithms. By comparing the vehicle acceleration response under the two strategies, the effectiveness and improvement space of the optimal control strategy can be evaluated. The simulation results will focus on the time response characteristics of acceleration, including the smoothness during initial acceleration, the ability to maintain acceleration under steady state, and the response time and overshoot of the system. These analyses will provide theoretical basis and experimental support for further optimization of multi-mode clutch control strategy.

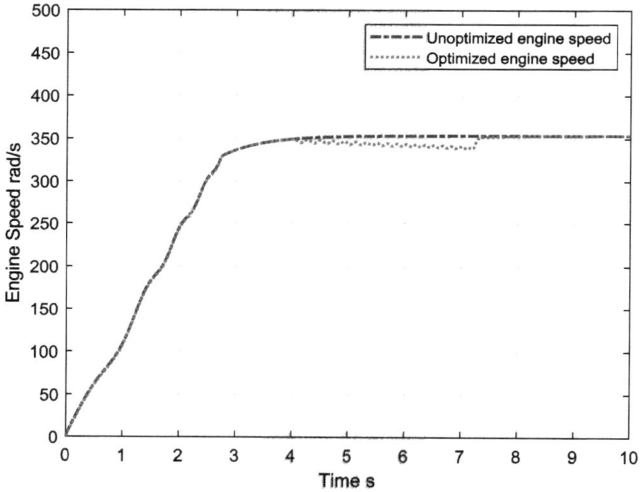

Fig. 5 Engine speed

Figure 5 shows the engine speed curve. In the period of 4 ~ 7.5 s, through comparative analysis, it is found that the engine speed is significantly lower when the control strategy is not optimized than when the control strategy is optimized. Specifically, the maximum speed difference reached 12 rad/s, which indicates that at this stage, the inner ring speed of the multi-mode clutch is relatively low, while the outer ring speed is relatively high. This speed difference is due to the different treatment of clutch engagement and torque distribution by the control strategy. The unoptimized strategy results in an inadequate torque distribution, resulting in a less compact clutch engagement, which makes the engine speed unable to effectively follow the dynamic change of the optimized strategy. This speed difference not only affects the operating state of the engine, but can also have a significant impact on the acceleration and overall performance of the vehicle.

Figure 6 shows the mark line activated in parallel mode. In the period of 4 to 7.5 s, the vehicle with the unoptimized control strategy appeared to exit the parallel mode, while the vehicle with the optimized control strategy did not appear this problem. This indicates that under the non-optimized control strategy, the engine output torque may be too small, resulting in the clutch unable to maintain the closed state, and thus open in advance. This phenomenon is related to the control strategy of torque estimation and clutch engagement logic processing is not accurate enough, resulting in the system misjudge the clutch engagement conditions. In contrast, the optimized control strategy can better adapt to the dynamic conditions and ensure that the torque output is consistent with the clutch state, thus avoiding the exit of the parallel mode. This difference not only reflects the advantages of the optimized control strategy in terms of stability and robustness, but also provides important support for vehicle ride comfort and energy efficiency optimization.

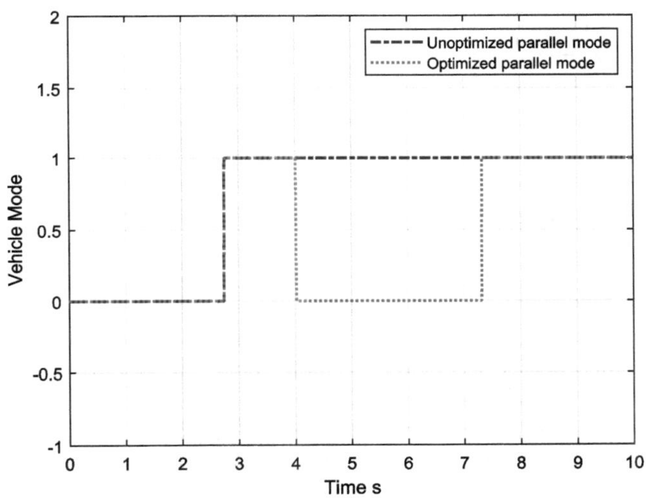

Fig. 6. The parallel mode activation flag

Figure 7 shows the acceleration curve of the vehicle. At 7.5 s, the vehicle with the unoptimized control strategy experienced a noticeable acceleration change due to the unexpected re-engagement of the clutch during this period, resulting in a shock response of the system. This sudden acceleration fluctuation will be directly transmitted to the vehicle driver and passengers, significantly affecting the vehicle's driving comfort. In contrast, vehicles with optimized control strategies exhibit more stable acceleration characteristics at the same point in time. This shows that the optimized control strategy can better coordinate the process of clutch engagement and disengagement, and reduce the clutch engagement frequency and impact degree. This improvement not only improves the ride comfort of the vehicle, but also effectively reduces the problem of declining ride comfort caused by unstable clutch operation. By optimizing the control strategy, the vehicle exhibits higher stability and better user experience during dynamic operation.

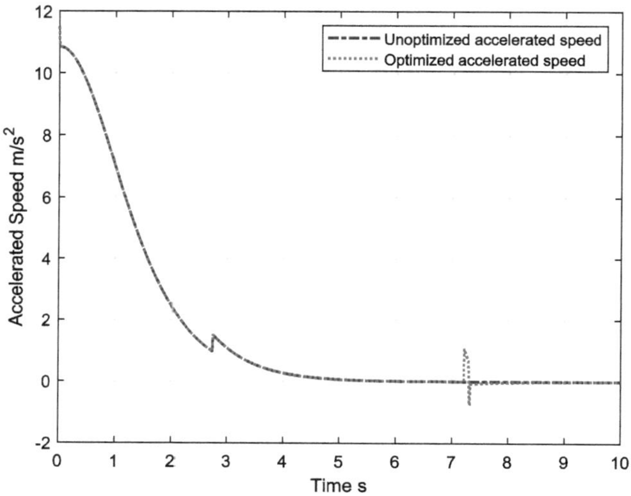

Fig. 7. Vehicle acceleration

6 Closing Remarks

In this paper, the transmission principle, control strategy and optimization method of multi-mode clutch transmission are discussed in depth, and the system instability caused by uneven torque distribution and speed difference under parallel driving conditions is mainly solved. The research shows that the multi-mode clutch transmission can achieve efficient power transmission and energy management under different working conditions through flexible coupling of inner ring and outer ring, combined with multi-motor coordination control. However, under rapid acceleration conditions, the imbalance of torque distribution may lead to the speed difference between the inner ring and the outer ring, resulting in the phenomenon of "deparallel", which affects the acceleration performance and energy efficiency of the vehicle. To solve this problem, a parallel driving control strategy based on dynamic torque distribution is proposed in this paper. By calculating the minimum distribution ratio of the remaining torque of the engine and dynamically adjusting the torque distribution according to the actual demand, the system can ensure that the angular acceleration of the inner and outer rings is balanced, thus avoiding the exit of the parallel operation. The experimental results show that the vehicle with optimized control strategy has significant advantages in engine speed, parallel mode stability and acceleration ride comfort. By better coordinating the clutch engagement and disengagement process, the optimization strategy significantly reduces the clutch engagement frequency and impact degree, ensures the smoothness of acceleration, and improves the ride comfort and ride comfort of the vehicle. In this study, a complete control system simulation closed-loop system is established, including hybrid transmission controller, multi-mode clutch transmission, engine and other components, and the precise control of the system is realized by multi-signal fusion. This control method significantly improves the response speed and control accuracy of the system, and provides an important reference for the design and optimization of the parallel hybrid power system. There are still some limitations in this study. First of all, the research is mainly based on simulation experiments, and the dynamic characteristics of actual vehicles may be more complex and need to be further verified. Secondly, the applicability of the optimal control strategy in more complex operating conditions still needs to be further discussed. In addition, the optimization of torque distribution algorithm and the feasibility of hardware implementation are also important directions of future research. In general, through the in-depth analysis and optimization of multi-mode clutch transmission, this study puts forward an effective solution, which provides theoretical support and technical guidance for the design and optimization of parallel hybrid power system. Future research could further integrate real-world vehicle testing, optimize control algorithms, and explore more efficient energy management strategies to drive the widespread use of multi-mode clutch transmissions in hybrid vehicles.

References

1. Yang, G., Mingqi, G., Ruiguang, W., et al.: Study on synchronization point self-learning strategy of 3DHT synchronizer. In: The 35th Chinese Control and Decision Conference, CCDC2023, pp. 2630–2635
2. Jie, Y., Xiangdong, H., Kegang, Z., et al.: Optimal control for gear shift of a pure electric vehicle's 2-speed transmission with single friction clutch. J. Mech. Eng. **52**, 92–100 (2016)
3. Yang, G., Mingqi, G., Ruiguang, W., et al.: Study on gear inhibition strategy of single hub multi-gear 3DHT. In: The 2023 Conference on Fully Actuated System Theory and Applications, CFASTA, pp. 486–490 (2023)
4. Hongkui, L.: Modeling of Synchronization Process and Research of Control Strategy in Wet Dual Clutch Transmission. Shanghai Jiao Tong University (2015)
5. Y. Yiquan, S. Jian, L. Liang. Multi-section optimization shift control method of dry clutch transmission. Trans. Chin. Soc. Agric. Mach. vol. 45. 2014, pp. 30–36.
6. Yang, G., Mingqi, G., Ruiguang, W., et al.: Study on shift strategy of single hub multi-gear actuator. In: 2023 International Conference on Service Robotics, ICoSR, pp. 86–89 (2023)
7. Qiong, L., Lina, R., Haiyan, Z., et al.: Gear shifting control of battery vehicle with 2-speed I-AMT. Automot. Eng. **35**, 1000–1003 (2013)
8. Yongdan, C., Junqiang, X., Huiyan, C.: AMT gear shift control strategy for parallel hybrid electric urban buses. J. Harbin Inst. Technol. **45**, 56–63 (2013)
9. Zhi, H.: A driving strategy for automated manual transmissions of heavy rucks. SAE-China Congress, 829–836 (2018)
10. Yang, G., Zhengxing, D., Yiqiang, L., et al.: Post-processing strategy of hub-type shift mechanism 3DHT after hitting teeth. In: 2024 International Annual Conference on Complex Systems and Intelligent Science, pp. 463–468. CSIS-IAC (2024)
11. Yang, G., Mingqi, G., Ruiyue, Z., et al.: Study on logic strategy for reference point recognition of 3DHT synchronizer. In: The 2023 Jiangsu Annual Conference on Automation, JACA, pp. 42–47 (2023)
12. Jianfei, L., et al.: Research on the verification method of shift selection for electronic mechanical automatic transmission. In: Internal Combustion Engine & Parts 16, pp. 1–3 (2022)
13. Chang, L., Haiyang, L., Yaogang, L., et al.: Simulation of Gearbox Shifting Process Based on ADAMS. Mech. Eng. Autom. **2**, 47–49 (2020)

14. Yan, Z., Han, Z., Yinfei, Z., et al.: Novel robust speed control method of agricultural electric vehicle Preshift process. Trans. Chin. Soc. Agric. Mach. **51**, 364–371 (2020)
15. B. Shanshan, Y. Zixin: Shift control of AMT transmission of two-speed pure electric vehicles based on fuzzy PI. In: Forestry Machinery & Woodworking Equipment, vol. 50, pp. 17–22 (2022)

A Multi-ship Intelligent Cooperative Control Framework Based on Hierarchical Reward

Yang Bai[1], Hao Li[1], Chi Zhang[1], and Sheng Rong Chia[2](✉)

[1] Beijing Jinghang Computation and Communication Research Institute, Beijing, China
[2] Beijing Institute of Technology, Beijing, China

Abstract. To address the challenge of balancing local decision-making and global coordination in a multi-ship-cooperative-control framework tasks, this paper proposes a hierarchical reward mechanism that divides rewards into three levels: basic, cooperative, and adaptive. The basic level guides navigation tracking and obstacle avoidance; the cooperative level maintains formation and relative positions; the adaptive level dynamically adjusts weights according to environmental complexity, achieving multi-scale refined incentives. Experimental results demonstrate that the proposed approach significantly improves formation stability and task efficiency in complex sea conditions.

Keywords: Reinforcement Learning · Hierarchical Reward · Multi-ship Cooperative Control

1 Introduction

In multi-agent systems (MAS), optimizing cooperative control strategies has become a core issue in marine engineering, intelligent transportation, and unmanned swarm applications [1,10,12,13]. In particular, for multi-ship cooperative tasks, coordinating individual vessels to achieve formation keeping, target tracking, and dynamic obstacle avoidance represents a key challenge in building intelligent surface fleet control systems. Although reinforcement learning (RL) has achieved remarkable success in single-agent control, its extension to multi-agent scenarios faces bottlenecks such as policy convergence difficulties due to environmental non-stationarity and the lack of stable global incentive mechanisms [6,11]. In terms of reward design, traditional methods typically rely on unified or static functions, which struggle to capture the complex interactions between individual behaviors and collective cooperation and cannot adapt flexibly in dynamic sea conditions [4]. Consequently, there is an urgent need for a multi-scale reward structure that can simultaneously drive local decision-making and ensure global cooperation, thereby enhancing learning stability and task execution efficiency.

H. Ma et al. (Eds.): IWACIII 2025, CCIS 2780, pp. 619–630, 2026.
https://doi.org/10.1007/978-981-95-6730-0_42

2 Related Work

To address the aforementioned challenges, hierarchical reward mechanisms have been widely explored in multi-ship intelligent cooperative control. A seminal result in early deep reinforcement learning (DRL) comes from Mnih et al.'s work on Atari games, where they employed a Deep Q-Network (DQN) to achieve end-to-end learning from pixel inputs to action outputs and surpassed human-level performance in several games, providing both theoretical and practical foundation for applying DRL in complex maritime scenarios [8]. However, this approach suffers from poor sample efficiency in long-horizon tasks and is difficult to transfer directly.

Subsequently, Gao et al. proposed a hierarchical reinforcement learning framework HRLfD-RbRS that integrates Learning from Demonstration (LfD). By employing reachability-based reward shaping to decompose the original task into high-level subgoal generation and low-level subgoal execution, they effectively accelerated convergence and enhanced policy performance [5]. To address multi-faceted reward requirements, Ding et al. introduced magnetic fieldâĂŞbased reward shaping, treating target positions as "magnetic poles" and guiding learning by simulating field intensity, which improved path quality and convergence speed in goal-conditioned RL [3]; Ranjan et al. drew on barrier functions, incorporating soft penalties for safety constraints into the reward to achieve joint optimization of obstacle avoidance and task objectives [9]. Furthermore, De Lellis et al. combined reward shaping with Control Lyapunov Functions (CLF) and used real-time Quadratic Programming (QP) to guarantee stability and safety [2]; Miranda et al. approached the problem from the perspective of policy generalization, employing multi-task training and adversarial example generation to significantly enhance navigation robustness in unseen environments [7].

Although these studies have proposed various approaches for hierarchical control and reward shaping, there remains a lack of a unified and extensible reward structure tailored to the complex requirements of multi-ship formation tasks, capable of coordinating individual and collective behavior across multiple scales. In addition, a systematic mechanism for balancing demonstration learning and exploration has not yet been established, and issues of transferability and robustness in dynamic marine environments remain insufficiently addressed. To this end, this paper constructs a hierarchical reward framework for multi-agent systems based on a layered design philosophy and proposes a corresponding control algorithm to enhance cooperative efficiency and environmental adaptability.

Building upon these observations, this work introduces a comprehensive multi-scale reward framework that integrates base rewards, cooperative rewards, and an adaptive reward layer, providing a unified and extensible structure suitable for multi-ship formation tasks. By dynamically adjusting reward weights in response to evolving environmental conditions, the proposed mechanism enhances robustness and learning efficiency, ensuring stable policy performance under uncertainty. Furthermore, close integration with Proximal Policy Optimization (PPO) accelerates convergence and encourages richer exploration behaviors, addressing limitations in sample efficiency and adaptability observed in prior studies. Overall, the main contributions of this study include: (i) a multi-scale reward design that captures both individual and collective objectives across multiple layers, (ii) an adaptive reward weighting mechanism that improves robustness in dynamic and uncertain marine environments, and (iii) seamless integration with PPO, enabling faster learning cycles while maintaining high-quality cooperative control. Through these innovations, the proposed framework provides a practical and theoretically grounded solution to the challenges of multi-agent maritime coordination.

3 Methodology

To address the coordination of multiple objectives in multi-ship cooperative control tasks, we propose a Hierarchical Reward (HR) mechanism. As shown in Fig. 1, the reward is divided into three levels:

- **Base Layer**: focuses on individual target tracking and heading control;
- **Cooperative Layer**: encourages formation keeping and optimization of inter-vessel relative positions;
- **Adaptive Layer**: dynamically adjusts incentive weights based on environmental complexity.

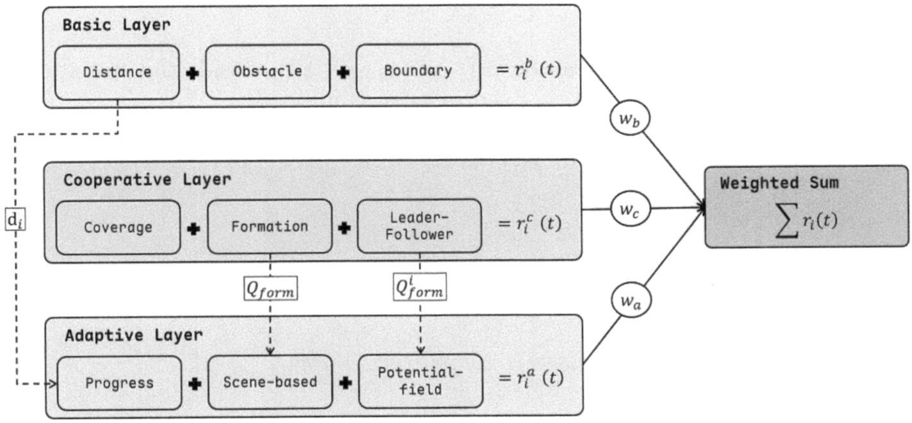

Fig. 1. Hierarchical Reward.

With this layered design, our approach not only provides fine-grained guidance of individual behavior at the local scale, but also maintains formation-wide coordination at the global scale and enhances robustness to sea-state variations through the adaptive mechanism. In the following subsections, we detail the definition and computation of each reward level and describe their integration into a multi-agent reinforcement learning algorithm.

3.1 Base Reward Component

As illustrated in Fig. 1, the base reward component comprises three parts: the goal proximity reward, the obstacle avoidance penalty, and the boundary penalty. These serve to encourage each vessel to approach the target, avoid collisions with obstacles, and remain within the operational area, respectively.

Goal Proximity Reward. Let the distance between vessel i and the target at time t be

$$d_i(t) = \left\|\mathbf{p}_i(t) - \mathbf{q}(t)\right\|.$$

We define the goal proximity reward with an exponential decay:

$$r_i^{\text{tar}}(t) = R_{\text{tar}} \exp\Big(-\frac{d_i(t)}{D_{\text{tar}}}\Big), \tag{1}$$

where R_{tar} is the base reward coefficient and D_{tar} is the decay distance scale.

Obstacle Avoidance Penalty. For each obstacle j, let $d_{ij}(t)$ denote the distance from vessel i. A decaying penalty is applied only when $d_{ij}(t) < R_{\text{obs}}$:

$$r_i^{\text{obs}}(t) = \sum_{j=1}^{O} P_{\text{obs}} \exp\Big(-\frac{d_{ij}(t)}{D_{\text{obs}}}\Big), \tag{2}$$

where P_{obs} is the obstacle penalty coefficient, D_{obs} is the decay scale, and O is the total number of obstacles.

Boundary Penalty. If vessel position $\mathbf{p}_i(t) = (x_i, y_i)$ falls outside the operational region $[m, L-m] \times [m, L-m]$, a fixed penalty is imposed:

$$r_i^{\text{bnd}}(t) = \begin{cases} P_{\text{bnd}}, & x_i(t) < m \text{ or } x_i(t) > L-m \text{ or } y_i(t) < m \text{ or } y_i(t) > L-m, \\ 0, & \text{otherwise.} \end{cases} \tag{3}$$

Combined Base Reward. Combining the above three components, the base reward for vessel i is

$$r_i^{(b)}(t) = r_i^{\text{tar}}(t) \;-\; r_i^{\text{obs}}(t) \;-\; r_i^{\text{bnd}}(t). \tag{4}$$

3.2 Cooperative Reward Component

As shown in Fig. 1, to promote fleet-wide cooperation, the cooperative reward consists of formation quality reward, encirclement quality reward, and leader-follower reward. Let the fleet size be N, the leader index be L, and the follower set be $F = \{1, \ldots, N\} \backslash \{L\}$. At time t, the cooperative reward for vessel i is defined as

$$r_i^{\text{coop}}(t) = w_{\text{form}}\, Q_{\text{form}}(t) + w_{\text{enc}}\, Q_{\text{enc}}(t) + w_{\text{lead}}\, r_i^{\text{lead}}(t), \tag{5}$$

where w_{form}, w_{enc}, and w_{lead} are weight coefficients for formation, encirclement, and leader-follower rewards, respectively.

Formation Quality Reward. Based on the leader's position and a predefined formation radius R_f, compute the ideal position $\mathbf{P}_i(t)$ for each follower $i \in F$ and define the deviation

$$\Delta_i(t) = \big\|\mathbf{p}_i(t) - \mathbf{P}_i(t)\big\|. \tag{6}$$

The total deviation is

$$D_{\text{form}}(t) = \sum_{i \in F} \Delta_i(t),$$

and the normalized formation quality is

$$Q_{\text{form}}(t) = \exp\Big(-\frac{D_{\text{form}}(t)}{|F|\, R_f}\Big). \tag{7}$$

Encirclement Quality Reward. For the target centroid $\mathbf{q}(t)$, let each vessel's radial distance be $r_i(t) = \|\mathbf{p}_i(t) - \mathbf{q}(t)\|$ and angle $\theta_i(t)$. Define the average radius deviation and maximum angular gap

$$\bar{r}(t) = \frac{1}{N}\sum_{i=1}^{N} r_i(t), \quad \Delta\theta_{\max}(t) = \max_i\big(\theta_{i+1}(t) - \theta_i(t)\big) \pmod{2\pi}.$$

Then the radial consistency and angular coverage qualities are

$$Q_{\text{rad}}(t) = \exp\Big(-\frac{\big|\bar{r}(t) - R_e\big|}{R_e}\Big), \quad Q_{\text{dist}}(t) = \exp\Big(-\frac{\Delta\theta_{\max}(t)}{2\pi/N}\Big), \tag{8}$$

combined into the encirclement quality

$$Q_{\text{enc}}(t) = Q_{\text{rad}}(t)\, Q_{\text{dist}}(t). \tag{9}$$

Leader-Follower Reward. The leader-follower reward incentivizes the leader to drive the mission and followers to maintain coordination. Let $\Pi(t)$ be the overall mission progress metric, then

$$r_i^{\text{lead}}(t) = \begin{cases} \Pi(t), & i = L, \\ Q_{\text{form}}^{(i)}(t), & i \in F, \end{cases} \tag{10}$$

where $Q_{\text{form}}^{(i)}$ is the pairwise formation quality between the leader and follower i.

Thus, the cooperative reward component provides multi-level incentives for formation stability, encirclement completeness, and leader-follower synergy in multi-ship cooperative tasks.

3.3 Adaptive Reward Mechanism

As depicted in Fig. 1, within the hierarchical reward framework, the adaptive reward dynamically adjusts each agent's incentive signals based on mission progress, current context, and potential field distribution. The adaptive reward for vessel i at time t is defined as

$$r_i^{(a)}(t) = w_p\, r_i^{\text{prog}}(t) + w_s\, r_i^{\text{sit}}(t) + w_u\, r_i^{\text{pot}}(t), \tag{11}$$

where w_p, w_s, and w_u are weight coefficients for the progress reward, situational reward, and potential-field reward, respectively.

Progress Reward. The progress reward measures the fleet's overall approach to the goal. With $d_i(t)$ denoting vessel i's distance to the target, $\exp(-d_i(t)/D_p)$ is its normalized progress. The average progress across the fleet defines each vessel's progress reward:

$$r_i^{\text{prog}}(t) = \frac{1}{N}\sum_{k=1}^{N} \exp\Big(-\frac{d_k(t)}{D_p}\Big). \tag{12}$$

Situational Reward. Let $S_i(t)$ be a situational indicator taking values in $\{$*approaching target, avoiding obstacle, maintaining formation, default*$\}$. Then

$$r_i^{\text{sit}}(t) = \begin{cases} \exp\left(-\dfrac{d_i(t)}{D_a}\right), & S_i = \text{approaching target}, \\ \exp\left(-\dfrac{\min\limits_j d_{ij}(t)}{D_e}\right), & S_i = \text{avoiding obstacle}, \\ Q_{\text{form}}^{(i)}(t), & S_i = \text{maintaining formation}, \\ 0, & S_i = \text{default}. \end{cases} \tag{13}$$

Here $d_{ij}(t)$ is the distance to obstacle j, and $Q_{\text{form}}^{(i)}$ is the leaderâĂŞfollower formation quality.

Potential-Field Reward. The potential-field reward consists of attractive, repulsive, and formation potential fields. Define

$$U_i^{\text{tar}}(t) = \exp\Big(-\frac{d_i(t)}{D_{pt}}\Big), \quad U_i^{\text{obs}}(t) = -\sum_{j=1}^{O} \exp\Big(-\frac{d_{ij}(t)}{D_{po}}\Big),$$

$$U_i^{\text{form}}(t) = \begin{cases} \exp\big(-\frac{|d_{iL}(t)-R_f|}{D_{pf}}\big), & i \in F, \\ 0, & \text{otherwise}, \end{cases}$$

then the potential-field reward for vessel i is

$$r_i^{\text{pot}}(t) = U_i^{\text{tar}}(t) + U_i^{\text{obs}}(t) + U_i^{\text{form}}(t). \tag{14}$$

4 Experiment

4.1 Experimental Setup

This study employs the Proximal Policy Optimization (PPO) algorithm as the baseline reinforcement learning method due to its proven stability and efficiency in multi-agent continuous control tasks. The core hyperparameters are set as follows: learning rate $\alpha = 3\times10^{-4}$, discount factor $\gamma = 0.99$, GAE coefficient $\lambda = 0.95$, PPO clipping threshold $\epsilon = 0.2$, batch size $B = 64$, entropy regularization coefficient $\beta = 0.01$, value loss coefficient $c_1 = 0.5$, and maximum gradient norm clipping $|\nabla|_{\max} = 0.5$.

We fix the number of ships at $ns = 5$ for all experiments to maintain consistency in multi-ship interactions.

The simulation environment consists of a bounded 2D maritime space of size $500\,\mathrm{m} \times 500\,\mathrm{m}$, containing randomly distributed static obstacles and dynamic currents. The initial positions of the five ships are uniformly sampled within a designated start zone, and the target locations are fixed at the opposite corners of the environment. This setup ensures diverse navigation challenges while keeping the evaluation consistent across different parameter settings.

A hierarchical reward structure is adopted to guide learning. The base layer reward R_{base} combines goal-reaching progress and collision avoidance penalties to ensure safe navigation. The cooperative layer reward R_{coop} evaluates formation maintenance and adherence to leader-follower strategies, promoting coordinated behaviors among ships. The adaptive layer dynamically adjusts the relative weights of R_{base} and R_{coop} based on the environmental complexity, balancing convergence speed and overall policy performance. Specifically, the total reward is computed as:

$$R_{\text{total}} = w_b R_{\text{base}} + w_c R_{\text{coop}}, \quad w_b, w_c \text{ adaptively adjusted.}$$

This comprehensive setup allows for rigorous evaluation of the multi-ship cooperative control policy under varied conditions and provides a strong foundation for subsequent experiments analyzing reward parameter sensitivity and the effect of adaptive feedback mechanisms.

4.2 Experimental Results and Analysis

In this section, we present the key comparative results to evaluate the proposed hierarchical reward framework, focusing on two aspects: Reward Parameter Sensitivity and the Effect of the Adaptive Feedback Mechanism. The experiments aim to identify optimal reward parameter configurations and demonstrate the efficacy of adaptive weighting in multi-ship cooperative control tasks.

Experiment: Reward Parameter Sensitivity. The first set of experiments investigates how variations in the formation maintenance weight w_f and the cooperative reward ratio s_c influence overall system performance. Specifically, we explore

$$w_f \in \{0.25, 0.5, 0.75\}, \quad s_c \in \{0.4, 0.8, 1.2\},$$

resulting in a total of 3×3 parameter combinations, while keeping all other hyperparameters at their default values.

Figure 2 illustrates the average formation quality under different parameter settings. The heatmap clearly shows that an intermediate formation maintenance weight ($w_f = 0.5$) combined with a moderate cooperative reward ratio ($s_c = 0.8$) yields the highest formation stability. This indicates a delicate balance between individual vessel control and coordinated multi-agent behavior: overly low w_f values reduce formation adherence, whereas excessively high w_f values hinder flexibility and responsiveness.

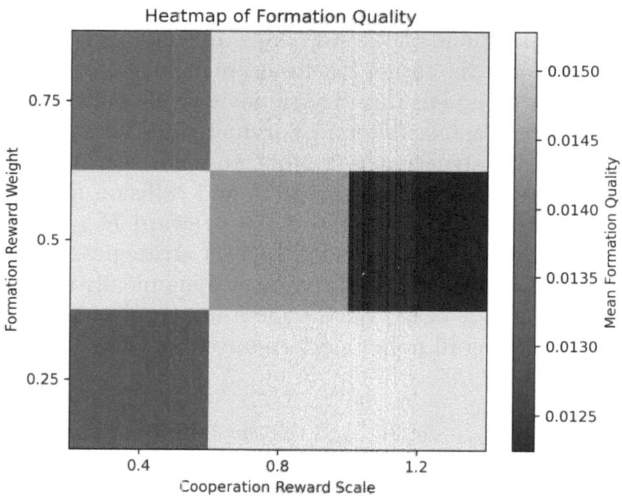

Fig. 2. Average formation quality heatmap under different combinations of formation maintenance weight w_f and cooperative reward ratio s_c. Darker colors indicate better formation stability. The results suggest that an intermediate weight $w_f = 0.5$ and moderate cooperation factor $s_c = 0.8$ achieve an optimal trade-off between individual vessel autonomy and cooperative behavior.

In addition, Fig. 3 depicts the number of replay corrections across different parameter settings, providing insight into the learning dynamics and sample efficiency. Moderate replay correction counts are indicative of stable learning while avoiding excessive re-sampling of past experiences. The results show that $w_f = 0.5$ and $s_c = 0.8$ maintain replay corrections at an optimal level, confirming that this configuration supports both policy convergence and learning efficiency.

Overall, these experiments demonstrate that the proposed hierarchical reward structure is sensitive to the formation maintenance weight and cooperative reward ratio. The optimal combination of $w_f = 0.5$ and $s_c = 0.8$ not

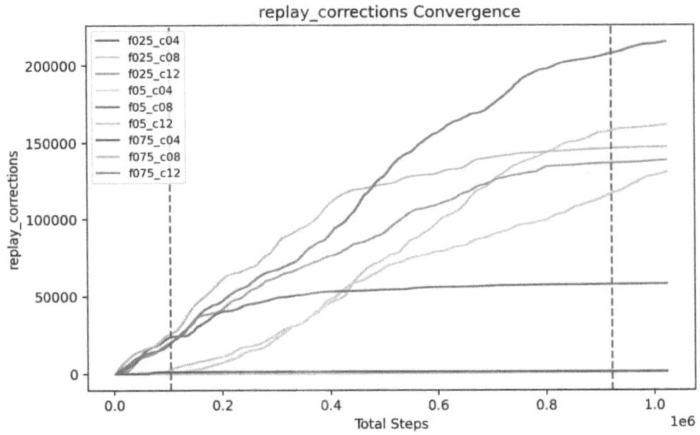

Fig. 3. Replay correction count across different reward parameter settings. Optimal settings ($w_f = 0.5$, $s_c = 0.8$) achieve a balance between learning stability and sample efficiency, preventing over-sampling while ensuring sufficient policy updates.

only achieves superior formation quality but also ensures efficient policy updates, highlighting the importance of carefully tuning reward parameters in multi-agent reinforcement learning tasks. Splitting the figures allows readers to interpret each metric independently, providing clearer insight into the effects of different reward parameter configurations on both task performance and learning stability.

Ablation Study: Effect of Adaptive Feedback Mechanism. In this ablation study, we investigate the role of the adaptive feedback mechanism in enhancing learning efficiency and policy robustness. Specifically, we compare the system performance with and without the adaptive layer, using the average clip fraction and policy entropy as primary metrics. These metrics provide complementary insights: the clip fraction reflects the stability of policy updates under PPO optimization, while policy entropy indicates the diversity of agent exploration behavior.

Figure 4 presents the convergence curves of the average clip fraction. A lower clip fraction corresponds to smoother and more stable policy updates. It can be observed that the inclusion of the adaptive feedback mechanism leads to a more rapid and consistent reduction in clip fraction. This suggests that dynamically adjusting reward weights allows the policy to converge faster while maintaining stable learning dynamics, preventing excessive oscillations during training.

Figure 5 depicts the evolution of policy entropy over training iterations. Higher entropy values signify richer exploratory behavior, which is essential for effective reinforcement learning in complex and dynamic environments. The results show that the adaptive feedback mechanism facilitates a faster increase in policy entropy, indicating that agents are exploring the action space more effectively. This enhanced exploration contributes to improved policy generalization

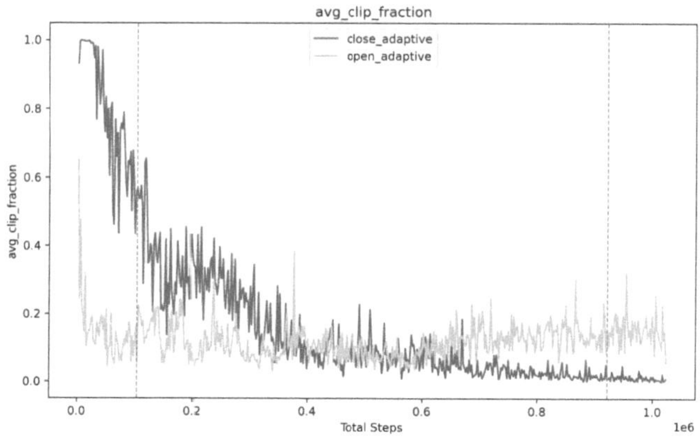

Fig. 4. Convergence of the average clip fraction with and without the adaptive feedback mechanism. Lower clip fractions indicate more stable updates during PPO optimization. The adaptive feedback mechanism consistently reduces the clip fraction more rapidly, demonstrating accelerated policy convergence and improved learning stability.

and robustness, particularly in uncertain maritime scenarios where environmental conditions and ship interactions are highly variable.

Taken together, these ablation experiments demonstrate that the adaptive feedback mechanism substantially improves learning efficiency and policy robustness. By dynamically modulating reward weights according to environmental complexity and task progress, the adaptive layer not only accelerates convergence but also encourages broader exploration, leading to more resilient multi-ship cooperative behaviors. These results underscore the critical role of adaptive reward mechanisms in multi-agent reinforcement learning, particularly in scenarios with high uncertainty and complex interactions.

Summary. Through the comprehensive experiments and ablation studies presented above, we have systematically validated the effectiveness of our hierarchical reward structure in multi-ship cooperative control. The reward parameter sensitivity analysis identifies the optimal balance between formation maintenance and cooperative behavior, showing that moderate values of w_f and s_c achieve both high formation stability and efficient policy updates. Furthermore, the ablation study highlights the critical role of the adaptive feedback mechanism: by dynamically adjusting reward weights in response to environmental complexity, it accelerates policy convergence, stabilizes learning, and encourages richer exploration. Collectively, these results demonstrate that the combination of hierarchical rewards and adaptive feedback is not only effective but also essential for achieving robust, high-performance multi-ship coordination under uncertain and dynamic maritime conditions. This summary underscores the practical significance of our approach, providing clear guidelines for parameter selection and confirming the necessity of adaptive mechanisms in reinforcement learning for complex, real-world scenarios.

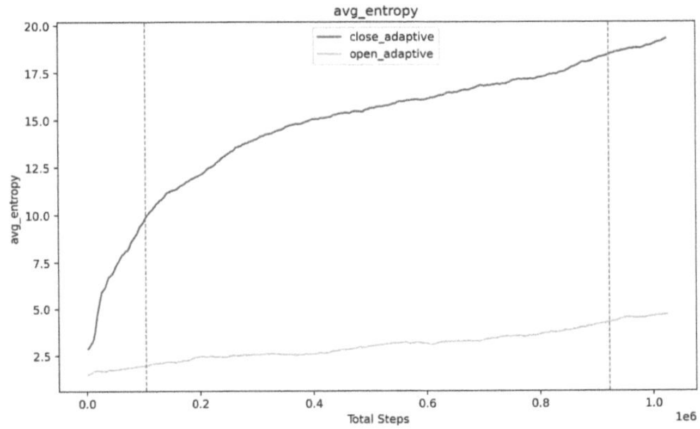

Fig. 5. Policy entropy evolution under adaptive feedback. Higher entropy reflects richer exploration behavior in reinforcement learning. The adaptive feedback mechanism accelerates the growth of entropy, suggesting enhanced exploration, better policy generalization, and increased robustness in dynamic maritime environments.

5 Conclusion

In this paper, we tackled the critical challenge of balancing local decision-making with global coordination in multi-ship cooperative control by proposing a hierarchical reward framework composed of three distinct layers. The base layer ensures reliable goal tracking and collision avoidance, the cooperative layer promotes formation integrity and leader-follower behaviors, and the adaptive layer dynamically adjusts the contribution of each layer according to environmental complexity. Integrated with the Proximal Policy Optimization (PPO) algorithm, our extensive experiments demonstrate that this hierarchical reward structure significantly enhances both formation stability and learning efficiency across a variety of scenarios, including different vessel counts, environmental disturbances, and varying ship speeds. The ablation study highlights the indispensable role of the adaptive feedback mechanism, showing that dynamically modulating reward weights accelerates convergence, stabilizes policy updates, and encourages broader exploration. Overall, the proposed approach provides a robust, efficient, and extensible framework for complex maritime multi-ship coordination, offering practical guidance for real-world deployment and establishing a foundation for future research in multi-agent reinforcement learning for dynamic, uncertain environments.

References

1. Alqudsi, Y., Makaraci, M.: Uav swarms: research, challenges, and future directions. J. Eng. Appl. Sci. **72**(1), 12 (2025)
2. De Lellis, F., Coraggio, M., Russo, G., Musolesi, M., di Bernardo, M.: Guaranteeing control requirements via reward shaping in reinforcement learning. IEEE Trans. Control Syst. Technol. (2024)
3. Ding, H., Tang, Y., Wu, Q., Wang, B., Chen, C., Wang, Z.: Magnetic field-based reward shaping for goal-conditioned reinforcement learning. IEEE/CAA J. Automatica Sinica **10**(12), 2233–2247 (2023)
4. Du, H., Gou, F., Cai, Y.: Scalable safe multi-agent reinforcement learning for multi-agent system. arXiv preprint arXiv:2501.13727 (2025)
5. Gao, X., Liu, J., Wan, B., An, L.: Hierarchical reinforcement learning from demonstration via reachability-based reward shaping. Neural Process. Lett. **56**(3), 184 (2024)
6. Liang, J., et al.: A review of multi-agent reinforcement learning algorithms. Electronics **14**(4), 820 (2025)
7. Miranda, V.R., Neto, A.A., Freitas, G.M., Mozelli, L.A.: Generalization in deep reinforcement learning for robotic navigation by reward shaping. IEEE Trans. Industr. Electron. **71**(6), 6013–6020 (2023)
8. Mnih, V., et al.: Human-level control through deep reinforcement learning. Nature **518**(7540), 529–533 (2015)
9. Ranjan, A., Agrawal, S., Jain, A., Jagtap, P., Kolathaya, S., et al.: Barrier functions inspired reward shaping for reinforcement learning. In: 2024 IEEE International Conference on Robotics and Automation (ICRA), pp. 10807–10813. IEEE (2024)
10. Wu, Z., Pan, L., Yu, M., Liu, J., Mei, D.: A game-based approach for designing a collaborative evolution mechanism for unmanned swarms on community networks. Sci. Rep. **12**(1), 18892 (2022)
11. Xiao, W., Yuan, L., He, L., Ran, T., Zhang, J., Cui, J.: Multigoal visual navigation with collision avoidance via deep reinforcement learning. IEEE Trans. Instrum. Meas. **71**, 1–9 (2022)
12. Xie, S., Hu, J., Ding, Z., Arvin, F.: Collaborative overtaking of multi-vehicle systems in dynamic environments: a distributed artificial potential field approach. In: 2021 20th International Conference on Advanced Robotics (ICAR), pp. 873–878. IEEE (2021)
13. Zakartchouk, A., Jr., Morishita, H.M.: Cooperative control of marine vehicles using the receding horizon concept and consensus theory. J. Ocean Eng. Mar. Energy **10**(3), 611–625 (2024)

Attention-Enhanced Hybrid Graph Neural Network for 3D Localization of Rock Fracture Events

Bingchen Li[1], Jianing Sun[1], Heyao Li[1,2], Chengzhi Zhang[1], Weixian Teng[1], Haoyuan Song[1], Xianrui Ji[1,2], and Zhibin Yao[2(✉)]

[1] The School of Future Technologies, Northeastern University, Shenyang 110819, China

[2] The State Key Laboratory of Intelligent Deep Metal Mining and Equipment, Northeastern University, Shenyang 110819, China

yaozhibin@mail.neu.edu.cn

Abstract. Accurate localization of rock fracture events is critical for timely warning the rockburst events in microseismic monitoring systems. Conventional physics-based inversion techniques and data-driven neural networks rely heavily on P- and S-wave arrival-time, and their performance often degrades substantially when suffer from abnormal arrival-time. To address this limitation, we propose an Attention-Enhanced Hybrid Graph Neural Network (AE-HGNN) for three-dimensional (3d) event localization. In this framework, a graph representation is constructed by embedding both sensor coordinates and arrival-time information, thereby enabling the effective modeling of spatial—temporal dependencies. Subsequently, the architecture incorporates a Graph Attention Network (GAT) module for temporal feature extraction and an attention-enhanced Graph Isomorphism Network (GIN) module for spatial feature learning. Finally, three independent output heads are employed to predict the 3d coordinates of fracture events along each axis, achieving precise localization. Experimental evaluations on localization datasets demonstrate that AE-HGNN consistently outperforms both traditional inversion-based approaches and purely learning-based models in terms of accuracy. Moreover, case studies from Tunnel Boring Machine (TBM) excavation projects highlight the practical applicability and engineering significance of the proposed method in real-world scenarios.

Keywords: Microseismic monitoring · rockburst warning · microseismic source location · graph neural network

1 Introduction

Microseismic event localization serves as a critical technique for monitoring the initiation and evolution of geological hazards (e.g., rockbursts) during construction processes. It furnishes essential data to mitigate casualties and property losses, thereby ensuring the safe and steady progression of deep underground

H. Ma et al. (Eds.): IWACIII 2025, CCIS 2780, pp. 631–642, 2026.
https://doi.org/10.1007/978-981-95-6730-0_43

engineering projects, including mines and tunnels [1,2]. As mining depths continue to increase, the geological conditions become progressively intricate. Microseismic signals are notably susceptible to perturbations from rock anisotropy, sensor deployment inaccuracies, and ambient noise interference. Conventional localization methodologies have gradually manifested inherent limitations in positioning precision, necessitating the development of more robust technical frameworks to overcome such constraints [3,4].

The core early algorithm for microseismic event localization is the Geiger method [5]. This method constructs a system of equations using the wave velocity difference and arrival time difference between P-waves and S-waves, and solves for hypocenter coordinates and origin time via the least squares method. Subsequent researchers have conducted extensive optimization work on the Geiger method [6–12]. Specifically, Inglada [11] improved this method and proposed the efficient explicit Inglada method, which only requires P-wave arrival time data from 4 sensors and their coordinates to construct a system of hyperbolic equations for inverting hypocenter coordinates. Christy [12] further implemented iterative optimization on the Inglada method, and the resulting localization accuracy was significantly superior to that of non-iterative schemes.

However, the aforementioned algorithms have issues of no solution or low computational efficiency in solving hypocenter coordinates. With the development of optimization algorithms such as the simplex method and genetic algorithms, researchers have gradually applied them to hypocenter coordinate inversion [13–16]. Among these, Prugger and Gendzwill [13] introduced the simplex method into the field of seismic localization. This method eliminates the need for derivative calculations and matrix transposition operations, which can significantly reduce computational complexity. Additionally, they incorporated the L1 norm to minimize the interference of residuals on arrival time data. Shen et al. [16] proposed a localization method combining a time difference extraction algorithm with a genetic algorithm; this method can localize hypocenters outside the sensor network, with accuracy comparable to that of hypocenters inside the localization network. Nevertheless, both types of optimization algorithms still have limitations: the simplex method excessively relies on the selection of initial hypocenter positions and is prone to falling into local optimal solutions; while genetic algorithms, although not dependent on initial hypocenter positions, suffer from low computational efficiency.

The successful application of deep learning techniques in computer vision domains—such as object detection and image segmentation—has provided a novel technical pathway for addressing the traditional geophysical problem of microseismic event localization [17,18]. A subset of researchers has thus initiated explorations into integrating deep learning into microseismic localization studies. The core network architectures of deep learning employed for microseismic localization include Multilayer Perceptrons (MLPs), Convolutional Neural Networks (CNNs), Recurrent Neural Networks (RNNs), and Graph Neural Networks (GNNs) [19,20]. A defining characteristic of these methods lies in their data-driven paradigm: models are trained on large datasets of labeled microseismic data, leveraging neural networks to excavate latent mapping relationships between the data and hypocenter parameters (e.g., coordinates, origin time). This ultimately enhances localization accuracy and enables adaptation to the complexity of diverse geological conditions.

In specific research endeavors, Zhang et al. [21] designed a microseismic event localization network using a convolutional neural network; however, this network suffers from overfitting. Furthermore, Zhang et al. [22] constructed a Fully Convolutional Network (FCN) model, which directly learns localization features associated with hypocenter positions from raw microseismic waveform data, thereby significantly improving signal processing efficiency.

The emergence of Graph Neural Networks (GNNs) has opened new possibilities for joint spatiotemporal modeling in microseismic localization. Zhan et al. [23] proposed an LGF microseismic event localization model—a hybrid framework integrating Long Short-Term Memory (LSTM), GNN, and FCN models. Through the effective integration of these three models, the accuracy of microseismic event localization is enhanced.

Nevertheless, despite the phased progress achieved by existing deep learning methods in the field of microseismic localization, a systematic review of the current research status reveals three core issues in state-of-the-art models that demand urgent resolution: First, the design of feature architectures remains incomplete. Most models fail to establish a *"temporal-spatial-triaxial"* collaborative feature framework, which easily leads to information loss or weight imbalance in one dimension. For example, the Z-axis feature often exhibits low localization accuracy due to its excessively low weight proportion. Second, noise suppression mechanisms are overly simplistic. Traditional fixed-threshold filtering or basic attention mechanisms struggle to adaptively identify anomalous data in complex scenarios (e.g., geological structure noise, abnormal arrival times caused by random sensor failures), thereby compromising the reliability of localization results. Third, spatiotemporal correlation modelling is inadequate. Some models only focus on the single-dimensional modeling of either temporal dynamic features or spatial structural features, failing to achieve deep collaborative fusion of the two. This makes it difficult to accurately match the spatiotemporal coupling characteristics inherent in the propagation process of microseismic waves.

2 Related Work

2.1 Graph Attention Network

The Graph Attention Network (GAT), proposed by Veličković et al. [24], is a graph-structured data-oriented neural network that assigns differential weights to neighboring nodes via masked self-attentional layers (no complex computations required). It provides critical technical support for the AE-BGNN model to capture local temporal correlations in sensor networks.

In microseismic localization, sensor networks can be abstracted as "node (sensor)—edge (spatial/signal propagation correlation)" graphs. GAT dynamically adjusts attention weights based on sensor data indicators (e.g., SNR, waveform integrity)—assigning higher weights to sensors with clear P/S-wave arrival times and lower weights to noise-disturbed sensors—to improve local feature reliability. However, traditional GAT relies on node degree for attention calculation: uneven sensor layouts (e.g., dense vs. sparse distribution due to geological/engineering constraints) cause imbalanced attention, masking effective features of sparse-area sensors and limiting localization precision.

GATv2 (optimized GAT, Brody et al. [25]) addresses this limitation by abandoning node degree-dependent weight normalization. Instead, it adapts attention to nodes of different degrees via dynamic query-key matching. For microseismic localization, this means GATv2 can accurately capture temporal correlations (e.g., microseismic wave arrival time differences, waveform similarity) between sparsely distributed sensors and their neighbors—even in complex scenarios like deep mines (where faults/karst caves hinder dense deployment—providing optimized support for AE-BGNN's "local temporal dynamic correlation modeling" module.

2.2 Graph Isomorphism Network

Graph Isomorphism Networks (GIN) capture global graph structural features via learnable aggregation functions, laying a technical foundation for AE-BGNN to integrate sensor network global spatial constraints. Proposed by Brody et al. [25], GIN fuses local node features with global graph structure through a "node feature update—graph-level feature aggregation" two-layer architecture. Its learnable aggregation parameters generate differentiated global features for different graphs, avoiding the insufficient global perception of traditional GCNs.

In microseismic localization, sensor network layout (total count, regional density, relative orientation) directly affects wave propagation time accuracy. GIN converts these spatial constraints into feature information: enhancing spatial correlation weights in dense areas (improving noise robustness via data redundancy) and supplementing missing information in sparse areas (reducing spatial gaps from node scarcity).

GIN's aggregation design (summation + learnable parameters) is highly compatible with microseismic spatial characteristics—unlike traditional GCNs' fixed average aggregation, it adjusts weights based on a sensor's network role (e.g., potential hypocenter region, critical propagation path). This enables AE-BGNN to combine GATv2-based local temporal correlation capture with GIN-based global spatial constraints, achieving "local-global" collaborative modeling and overcoming standalone GATv2's insufficient global perception.

3 Method

In this work, an Attention-enhanced Hybrid Graph Neural Network (AE-HGNN) is proposed to predict the location of the rock fracture microseismic events. Specifically, AE-HGNN firstly constructs a graph representation to capture the spatial—temporal dependencies of sensors. Subsequently, an attention mechanism is further utilized to suppress the impact of abnormal arrival-time. Finally, this information is fed into a mixed-GNN to extract features and achieve a high-precision 3D coordinate prediction of rock fracture microseismic events. The whole framework of AE-HGNN is shown in Fig. 1.

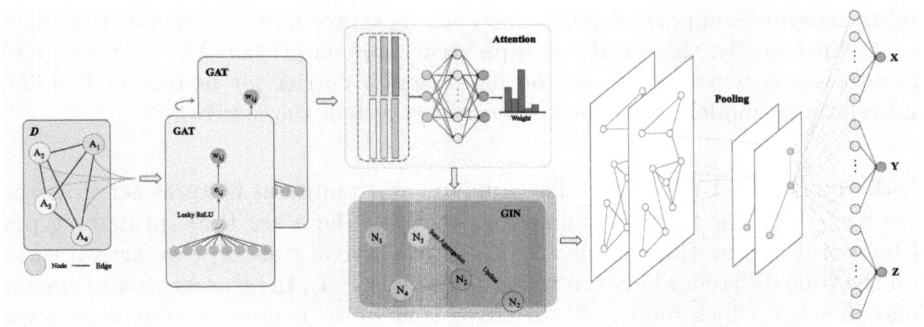

Fig. 1. The overview of AE-HGNN, which comprises three key components: (1) a GAT for temporal feature extraction, (2) an attention mechanism for smoothing abnormal arrival-time, and (3) a GIN for spatial feature learning.

3.1 Problem Formulation

Let define the dataset as $\mathcal{D} = \{(\mathbf{s}_i, \mathbf{t}_i, \mathbf{v}) \in \mathbb{R}^{3\times2\times2}\}_{i=1}^{8}$ where $\mathbf{s}_i = (s_{x,i}, s_{y,i}, s_{z,i})$ denotes the 3D spatial coordinates of each sensor $A_i, /; i = 1, 2, ..., 8$, $\mathbf{t}_i = (t_{p,i}, t_{s,i})$ represents the recorded the arrival-time of P-wave and S-wave, and $\mathbf{v} = (v_p, v_s)$ corresponds to their respective velocities. The primary objective of our method is to predict the 3D location of a rock fracture source, denoted by $\mathbf{c} = (c_x, c_y, c_z)$. This is formalized as a mapping function:

$$\mathbf{c} = f(\mathbf{s}, \mathbf{t}, \mathbf{v}|\theta), \tag{1}$$

where $f(\cdot)$ denotes the localization function parameterized by θ.

3.2 Graph Representation Constructing

For data-driven approaches, an appropriate representation of sensor observations (serving as the model input) plays a critical role in determining the prediction accuracy. Traditional methods usually directly use the discrete information captured by sensors as the inputs of neural networks. This design always results in low-accuracy predictions because of the issues of imbalance in three-axis features and interference from abnormal arrival times. To address these issues,

AE-HGNN explores the structured spatio—temporal information of the original discrete information and processes it to construct a graph representation. Specifically, the originally discrete sensor measurements are reformulated into structured graph representations that explicitly capture the relationship among sensors in terms of coordinate, inter-sensor distances and their arrival-time. In this work, the undirected complete graph $\mathcal{G} = (\mathcal{V}, \mathcal{E})$ is constructed based on four fixedly deployed sensors, where each sensor is represented as a node $\mathbf{v} \in \mathcal{V}$, and the spatio—temporal dependencies among sensor pairs are encoded as edges $\mathbf{e} \in \mathcal{E}$. Specifically, this yields a graph structure comprising $|\mathcal{V}| = 4$ nodes and $|\mathcal{E}| = 6$ edges, which serves as the fundamental carrier for feature embedding and relational modelling in the subsequent stages of the network.

Node Features Encoding. The fundamental temporal features serve as the core basis for rock fracture source localization. There are three primary types of temporal information in this graph: P-wave arrival time, S-wave arrival time, and the time difference between them. For a sensor A_i, the P-wave arrival time is denoted as$t_{p,i}$, which represents the travel time of the primary compression wave from the source to sensor A_i. This value constitutes the primary temporal indicator for estimating the source-to-sensor distance. Similarly, the S-wave arrival time is given by$t_{s,i}$, which reflects the travel time of the S-wave to sensor A_i. Furthermore, the relative difference between the two arrivals encodes spatial distance information. Specifically, the time difference is defined as $\Delta t_{ps,i} = t_{s,i} - t_{p,i}$, and it serves as the core input feature in our method.

In addition, the coordinates of each sensor are incorporated into the node feature encoding to explicitly represent their spatial characteristics. Thus, the feature subset for sensor A_i can be expressed as $\mathbf{v}_i = \{t_{p,i}, t_{s,i}, \Delta t_{ps,i}, \mathbf{s}_i\}$. The same construction is applied to the other sensors.

Edge Features Encoding. For the features of edge, they also incorporate two categories of information: (i) Relative time differences and (ii) Relative coordinates. This design ensures both the temporal and spatial features across each sensor.

The temporal features for each sensor A_i, $i = 0, 1, \ldots, 8$, are computed with respect to a reference sensor A_j. Specifically, the relative P-wave arrival-time difference is defined as $\Delta t_{p,ij} = t_{p,i} - t_{p,j}$, quantifying the differential travel time of the P-wave between sensor A_i and the reference sensor A_j. To further enhance temporal robustness, the relative S-wave arrival-time difference is analogously defined as $\Delta t_{s,ij} = t_{s,i} - t_{s,j}$, capturing the differential propagation of the S-wave between sensor A_i and A_j.

For the spatial features, the single linear distance usually impacts the spatial diffraction distance, followed an error prediction. To eliminate this problem, we enhance the representation of distance, by including three categories of information: linear distance, cylindrical diffraction distance, and distance ratio. These features provide essential constraints for modelling microseismic wave propagation paths.

First, the linear distance between two sensors A_i and A_j is defined as:

$$\mathrm{dist}_{\mathrm{lin}}(A_i, A_j) = \sqrt{(s_{x,j} - s_{x,i})^2 + (s_{y,j} - s_{y,i})^2 + (s_{z,j} - s_{z,i})^2}, \tag{2}$$

which directly reflects their Euclidean separation and serves as a basic indicator of spatial correlation.

Second, to account for cylindrical monitoring scenarios such as tunnels and mines, the cylindrical diffraction distance modifies the traditional linear propagation assumption to better approximate the actual engineering environment. Its derivation is as follows: Compute the axial displacement along the cylindrical axis: $\Delta s_{x,ij} = |s_{x,j} - s_{x,i}|$; Compute the minimal central angle between the two sensors in the circular cross-section:

$$\Delta\phi_{i,j} = \min\left(|\phi_i - \phi_j|,\ 2\pi - |\phi_i - \phi_j|\right) \tag{3}$$

where ϕ_i and ϕ_j denote the polar angles of A_i and A_j in the Y–Z plane; Combining these terms with the cylindrical radius $r = 3.8\,\mathrm{m}$, the cylindrical diffraction distance is approximated as:

$$\mathrm{dist}_{\mathrm{cyl}}(A_i, A_j) = \sqrt{(\Delta s_{x,ij})^2 + (r \cdot \Delta\phi_{i,j})^2}. \tag{4}$$

Finally, the distance ratio quantifies the deviation of the actual propagation path from the straight-line distance and is defined as:

$$\gamma_{A_i,A_j} = \frac{\mathrm{dist}_{\mathrm{lin}}(A_i, A_j)}{\mathrm{dist}_{\mathrm{cyl}}(A_i, A_j) + \epsilon}, \tag{5}$$

where $\epsilon = 10^{-6}$ is a small constant introduced to avoid division by zero. This ratio provides a corrective factor for subsequent localization tasks by explicitly characterizing the deviation between idealized linear propagation and the cylindrical propagation path.

Thus, the edge feature subset for sensor A_i and A_j can be expressed as $\mathbf{e}_{ij} = \{\Delta t_{p,ij}, \Delta t_{s,ij}, \mathrm{dist}_{\mathrm{lin}}(A_i, A_j), \mathrm{dist}_{\mathrm{cyl}}(A_i, A_j), \gamma_{A_i,A_j}\}$.

3.3 Attention-Enhanced Hybrid Graph Neural Network

In microseismic monitoring systems, the accuracy of P-wave and S-wave arrival-times is highly susceptible to variations in the type of rock surrounding the sensor and environmental noise. To mitigate these limitations, we propose an attention-based adaptive smoothing mechanism to suppress abnormal arrival-time values. The smoothed features are then processed by a hybrid GNN, enabling more robust spatio—temporal representation learning.

Totally, the framework of our method consists of three major components:

Arrival-Time Smoothing Module: By assigning adaptive weights to arrival-time features, this module suppresses noise-induced uncertainty and strengthens the reliability of temporal information. Mathematically, given temporal feature $\mathbf{T} = \{\mathbf{t}_i\}$, the attention output is

$$\tilde{\mathbf{T}}_i = \alpha_i \cdot \mathbf{T}_i, \tag{6}$$

$$\alpha_i = \frac{\exp\left(\text{GAT}(\mathbf{v}_i, \mathbf{e}_i)\right)}{\sum_k \exp\left(\text{GAT}(\mathbf{v}_i, \mathbf{e}_i)\right)}, \tag{7}$$

where $\alpha_{i,j}$ are normalized attention coefficients and derived from two GAT modules.

Hybrid Graph Convolutional Layer ($\mathcal{H}$): This component integrates spatial correlation features (linear distance, cylindrical distance, and the distance ratio) and temporal correlation features (the arrival-time of P- and S-waves and relative differences).

$$\mathbf{H} = \sigma\big(\text{GIN}(\tilde{\mathbf{T}})\big). \tag{8}$$

Three-Axis Independent Output Heads ($\mathcal{O}_X, \mathcal{O}_Y, \mathcal{O}_Z$): Instead of a single regression head, three parallel MLPs are employed to predict source coordinates along each axis separately:

$$\hat{c}_x = \mathcal{O}_X(\mathbf{h}), \quad \hat{c}_y = \mathcal{O}_Y(\mathbf{h}), \quad \hat{c}_z = \mathcal{O}_Z(\mathbf{h}), \tag{9}$$

where $\mathbf{h}$ is the representation derived from the final layer in GIN. This design enforces axis-wise balancing, ensuring that accuracy is not dominated by any single spatial dimension.

The outputs of the three branches are then concatenated to form the final predicted seismic source position: $\hat{\mathbf{c}} = \begin{bmatrix} c_x, & c_y, & c_z \end{bmatrix}^\top \in \mathbb{R}^3$, and the final loss function is shown below:

$$\ell = MSE(\mathbf{c}, \hat{\mathbf{c}}), \tag{10}$$

where $MSE(\cdot)$ indicates the Mean Squared Error.

4 Experiments

In this section, we present a series of experiments designed to validate the proposed method, encompassing both controlled laboratory studies and two real-world case applications

4.1 Experiment Settings

Datasets. The dataset employed in this study consists of three primary components: 1) signals generated through manual hammering on the tunnel wall, 2) signals obtained from internal hammering after drilling, and 3) signals identified in proximity to actual rockburst occurrences using recent state-of-the-art localization techniques. In total, the dataset comprises 50,000 samples dedicated to source localization tasks, of which 80% (40,000 samples) are allocated for training and 20% (10,000 samples) for testing. Each sample contains sensor coordinates, P-wave and S-wave arrival-time, corresponding wave velocities, and the source location.

Baselines. To comprehensively evaluate the performance of the proposed microseismic event location method, we selected three representative baseline methods for comparative experiments. These methods span a range of technical approaches, from traditional physical models to modern deep learning. These included the combined P-wave and S-wave time difference localization method (CDT), the LSTM-GNN-FCNN (LGF), and the LSTM-FCNN. We replicated the models from published papers and applied them to our data.

Implementation Detail. This study employed an end-to-end training strategy, using the Adam optimizer (weight decay set to $1e^{-4}$ to prevent overfitting) for parameter updates. The initial learning rate was set to 0.001 (selected via grid search in the range $[0.0001, 0.0003, 0.001, 0.003, 0.01],$), combined with the CosineAnnealingLR scheduler (T_{max} = epochs) for cosine annealing learning rate adjustment. Training was limited to a maximum of 300 epochs, with an early stopping strategy (patience = 20, stopping after 20 consecutive epochs with no improvement in validation set loss) to prevent overfitting. The batch size was set to 128 samples (chosen to balance GPU memory limitations and training stability).

4.2 Evaluate Metrics

This study established a multi-level evaluation metric system to assess the positioning performance of the AE-HGNN model comprehensively. Key metrics include the mean absolute error (MAE), which calculates the L_1 distance in 3d space; and positioning accuracy statistics, which measure the percentage of correctly located events within specified error thresholds (5 m, 10 m, and 15 m). Robustness analysis includes statistical distribution analysis of 3d positioning errors (including minimum, maximum) to assess model stability.

4.3 Results

Laboratory Results. We evaluated the model using the basic error metrics: mean absolute error (MAE). Our method achieved **12.803 m**. While lower than LSTM (**16.312 m**), it still holds a significant advantage over other methods (CDT: **23.111 m**, LGF: **13.396 m**).

Regarding error metrics across various spatial dimensions, our model exhibits excellent MAE performance in the X, Y, and Z axes. In particular, our method demonstrates strong accuracy in both the Y and Z axes, with a particularly strong MSE of **16.327 m** on the Y axis, significantly outperforming other methods. This demonstrates that our method surpasses existing models in localization accuracy across all spatial dimensions.

In robustness analysis, we conducted a detailed evaluation of the 3d localization error distribution. Our method performed well for both the minimum localization error (**0.464 m**) and the maximum localization error (**60.134 m**), particularly in extreme cases. Our model was able to better control localization error than CDT (**206.327 m**) and LSTM (**8.492 m**). The results of these experiments are shown in Table 1

Table 1. Performance Comparison of Microseismic Event Localization Methods

Evaluation Metrics	CDT	LGF	LSTM-FCNN	Proposed Method
Basic Error Metrics				
Mean Absolute Error (MAE) [m]	23.110	13.396	16.312	**12.803**
Dimensional Error Metrics				
X-axis MAE [m]	22.978	11.084	14.201	**11.993**
Y-axis MAE [m]	23.870	18.991	20.669	**16.327**
Z-axis MAE [m]	22.486	10.114	14.066	**10.091**
Localization Accuracy Statistics				
Error $\leq$ 5 m Percentage [%]	6.45	9.30	0.00	**30.23**
Error $\leq$ 10 m Percentage [%]	14.07	18.60	6.98	**44.19**
Error $\leq$ 15 m Percentage [%]	37.37	37.20	18.60	**53.49**
Robustness Analysis (3D Localization Error Distribution)				
Minimum [m]	1.010	1.825	7.973	**0.464**
Maximum [m]	206.327	97.923	87.492	**60.134**

Real-World Cases. To further evaluate the effectiveness of the proposed AE-HGNN, we conducted case studies on two representative real-world events: a rockburst incident and a strip-shaped-rock controlled fracture. In the rockburst case, the proposed model successfully localized the event with high accuracy, closely matching the location of rockburst.

In the strip-shaped-rock, AE-HGNN maintained robust performance (70.3% of the predicted locations of fracture events are within a 5-m radius of the strip), as the attention-based refinement module effectively suppressed the influence of abnormal arrivals and preserved global spatial—temporal consistency. The results of these experiments are shown in Fig. 2

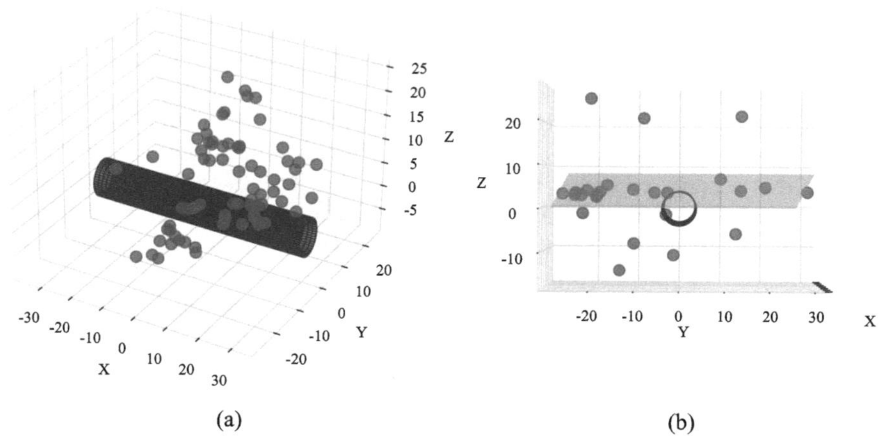

Fig. 2. The results of two true cases, which (a) indicates a rockburst event and (b) is a rock strip.

5 Conclusion

In this study, we proposed an Attention-Enhanced Balanced Hybrid Neural Network (AE-HGNN) for accurate 3d localization of rock fracture events. By integrating sensor coordinates and arrival-time information into a graph representation, the proposed framework effectively captures spatial—temporal dependencies. Furthermore, the attention-based refinement module adaptively reweights arrival-time features to preserve global distribution consistency, thereby enhancing localization robustness. Extensive experiments on datasets demonstrate that AE-HGNN consistently outperforms conventional physics-based inversion approaches and purely data-driven neural networks in terms of both accuracy and generalization. Case studies from TBM excavation projects further confirm the practical utility of the method in real-world microseismic monitoring and early rockburst warning.

Acknowledgments. This study was funded by Northeastern University 19th (2025) Undergraduate Innovation Training Program "Intelligent Localization Algorithm for Deeply Tunnel Microseismic Events" (grant number 250263).

Disclosure of Interests. The authors have no competing interests to declare that are relevant to the content of this article.

References

1. Li, S.C., Zhang, Q.S., Liu, B.: Key technologies for advanced geological prediction and control of water and mud inrush disasters in deep-buried tunnels. Chinese J. Rock Mech. Eng. **38**(01), 1–20 (2019)
2. He, M.C., Xie, H.P., Peng, S.P.: Research progress and challenges of rock mechanics in deep mining. Chin. J. Rock Mech. Eng. **40**(02), 225–242 (2021)
3. Wang, S.R., Li, N., Zhang, X.D.: Experimental study on the influence of rock anisotropy on microseismic wave propagation characteristics. Rock Soil Mech. **41**(05), 1678–1686 (2020)
4. Liu, X.X., Huang, G., Yong, R.: Study on optimal layout method of microseismic monitoring sensors in mines. J. Min. Saf. Eng. **35**(03), 562–568 (2018)
5. Geiger, L.: Seismicity and the structure of the Earth's interior. Bull. Seismol. Soc. Am. **2**(4), 185–251 (1912)
6. Lee, W.H.K.: A computer program for determining hypocenter, magnitude, and first motion pattern of local earthquakes. U.S. Geol. Surv. Open File Rep. **75**(311) 1–116 (1975)
7. Aki, K., Lee, W.H.K.: Determination of three-dimensional velocity anomalies under a seismic array using first P arrival times from local earthquakes: 1. A homogeneous initial model. J. Geophys. Res. **81**(23), 4381–4399 (1976)
8. Klein, F.W.: Hypocenter location program HYPOINVERSE: Part I. Users guide to versions 1, 2, 3 and 4. Part II. Source listings and notes. U.S. Geological Survey Open-File Report 78–694 (1978)
9. Lienert, B.R., Berg, F.L.: N: HYPOCENTER: an earthquake location method using centered, scaled, and adaptively damped least squares. Bull. Seismol. Soc. Am. **76**(3), 771–783 (1986)

10. Nelson, G.D., Vidale, J.E.: Earthquake locations by 3-D finite-difference travel times. Bull. Seismol. Soc. Am. **80**(2), 395–410 (1990)
11. Inglada, V.: The calculation of the stove coordinates of a Nahbebens [Note: Referring to hypocenter coordinates calculation]. Gerlands Beiträge zur Geophysik **19**, 73–98 (1928)
12. Christy, J.J.: A comparative study of the 'miner' and least squares location techniques as used for the seismic location of trapped coal miners. Unpublished master's thesis, Pennsylvania State University, University Park, PA, USA (1982)
13. Prugger, A.F., Gendzwill, D.J.: Microearthquake location: a nonlinear approach that makes use of a simplex stepping procedure. Bull. Seismol. Soc. Am. **78**(2), 799–815 (1988). https://doi.org/10.1785/BSSA0780020799
14. Thurber, C.H.: Nonlinear earthquake location: theory and examples. Bull. Seismol. Soc. Am. **75**(3), 779–790 (1985)
15. Li, G., Chen, J., Han, M., et al.: Accurate microseismic event location inversion using a gradient-based method. In SPE Annual Technical Conference and Exhibition (SPE-159187-MS) (2012). https://doi.org/10.2118/159187-MS
16. Shen, J.-S., Zeng, X.-D., Li, W., Jiang, M.-S.: An FBG acoustic emission source locating system based on PHAT and GA. Optoelectron. Lett. **13**(5), 330–334 (2017). https://doi.org/10.1007/s11801-017-7129-z
17. Krizhevsky, A., Sutskever, I., Hinton, G.E.: ImageNet classification with deep convolutional neural networks. Commun. ACM **60**(6), 84–90 (2017)
18. Long, J., Shelhamer, E., arrell, T.: Fully convolutional networks for semantic segmentation. In: Proceedings of the IEEE Conference on Computer Vision and Pattern Recognition, pp. 3431–3440 (2015)
19. Hinton, G.E., Osindero, S., Teh, Y.W.: A fast learning algorithm for deep belief nets. Neural Comput. **18**(7), 1527–1554 (2006)
20. LeCun, Y., Bengio, Y., Hinton, G.: Deep learning. Nature **521**(7553), 436–444 (2015). https://doi.org/10.1038/nature14539
21. Zhang, X., Yuan, C., Zhang, J., et al.: Automatic microseismic detection and location via the deep-convolutional neural network. In: SEG Technical Program Expanded Abstracts (2018). https://doi.org/10.1190/segam2018-2997548.1
22. Zhang, X., Zhang, J., Yuan, C., et al.: Locating induced earthquakes with a network of seismic stations in Oklahoma via a deep learning method. Sci. Rep. **10**(1), 1941 (2020). https://doi.org/10.1038/s41598-020-58908-5
23. Zhan, K., Wen, X., Xu, R., et al.: LGF SeismoLocator: a deep learning model for precision microseismic event localization in coal mines. Rock Mech. Rock Eng. **57**(12), 10717–10730 (2024)
24. Veličković, P., Cucurull, G., Casanova, A., Romero, A., Lio, P., Bengio, Y.: Graph attention networks. *arXiv preprint*arXiv:1710.10903 (2017)
25. Brody, S., Alon, U., Yahav, E.: How attentive are graph attention networks? In: Advances in Neural Information Processing Systems, vol. 34, pp. 21090–21101 (2021)

Author Index

H. Ma et al. (Eds.): IWACIII 2025, CCIS 2780, pp. 643–645, 2026.
https://doi.org/10.1007/978-981-95-6730-0

MIX
Papier aus verantwortungsvollen Quellen
Paper from responsible sources
FSC® C105338
FSC
www.fsc.org

If you have any concerns about our products,
you can contact us on
ProductSafety@springernature.com

In case Publisher is established outside the EU,
the EU authorized representative is:
Springer Nature Customer Service Center GmbH
Europaplatz 3, 69115 Heidelberg, Germany

Printed by Libri Plureos GmbH
in Hamburg, Germany